Ecological Methods

Ecological Methods

Fourth Edition

P. A. Henderson
T. R. E. Southwood

Revised By

P. A. Henderson

WILEY Blackwell

Library of Congress Cataloging-in-Publication Data

Names: Henderson, P. A. | Southwood, Richard, Sir.
Title: Ecological methods / P.A. Henderson and T.R.E. Southwood.
Description: Fourth edition / revised by P.A. Henderson. | Chichester, West Sussex :
    John Wiley & Sons, Inc., 2016. | Includes bibliographical references and index.
Identifiers: LCCN 2015033630 (print) | LCCN 2015043753 (ebook) | ISBN 9781118895276
    (cloth : alk. paper) | ISBN 9781118895283 (pbk. : alk. paper) | ISBN 9781118895252
    (Adobe PDF) | ISBN 9781118895269 (ePub)
Subjects: LCSH: Ecology–Technique. | Animal populations.
Classification: LCC QH541.28 .S68 2016 (print) | LCC QH541.28 (ebook) | DDC 577–dc23
LC record available at http://lccn.loc.gov/2015033630

A catalogue record for this book is available from the British Library.

Wiley also publishes its books in a variety of electronic formats. Some content that appears in print
may not be available in electronic books.

Cover images: courtesy of P. A. Henderson

Set in 9.5/12.5pt Palatino by Aptara Inc., New Delhi, India

Printed in Singapore by C.O.S. Printers Pte Ltd

1   2016

Contents

A colour plate section falls between pages 300 and 301

Prefaces

Preface to fourth edition

My coauthor for the Third Edition, and the original author of *Ecological Methods*, Sir Richard Southwood FRS, known to his many colleagues and friends as Dick, died on 26th October 2005. For those interested in reading about his long and highly distinguished career his Wikipedia page http://en.wikipedia.org/wiki/Richard_Southwood will direct you to obituaries and his Royal Society biographical memoir. Dick was a wonderful man to work with and a fine head of department. While achieving senior academic positions at an unusually young age, he retained an open and pleasant manner, a love of natural history, and remained accessible to the most junior members of staff, all of whose names he would invariably know. When he was head of the Department of Zoology at Imperial College and I was a first-year undergraduate, I was astonished that he knew the name of every undergraduate in his department.

I was keen to revise *Ecological Methods* in part to honour Dick's memory, but also because, I feel, the book still serves a useful purpose acting as an ecologist's handbook of methods and sources of information. While The Web now gives ecologists, even in isolated spots, access to a huge amount of information it can be difficult to glean the full range of possibilities for experimental approaches, sources of information and sampling gears. The old problem of how to design a successful sampling scheme and build samplers remains with us.

The trends in ecological research that we noted in the Preface to the Third Edition have continued at an accelerating pace. Computation and data handling has advanced greatly, and the present edition includes many references to R, the computing language and environment for statistical analysis and graphics. The dramatic growth in R packages for ecologists, all of which are offered free of charge, is one of the most important developments since the publication of the Third Edition. I have included many examples of R code in the present edition. Electronic developments in radar, sonar, remote sensing satellites, miniature tags, geographical positioning, movement detectors, lights, digital cameras, mobile phones and batteries have all greatly increased the opportunities for data acquisition. These advances, combined with novel biochemical techniques such as species detection from amplified DNA fragments, are creating tremendous opportunities for ecological research. We now have tools and resources that would have seemed incredible to a 1950s or 1960s ecologist. Yet, many of the techniques we use are still based on the ideas developed and refined between 1930 and 1980. Indeed, some of our sampling methods would have been familiar to our hunter–gatherer ancestors. One of my aims has been to maintain continuity with this great body of earlier knowledge. In part, this is because earlier papers are able to describe techniques and equipment in far more detail than is normal today. But, it is also the case that our predecessors often had great insight, and in many cases we can re-apply their ideas using our superior electronics and data handling to good effect. It is heartening to note that as journals have fully digitised their back numbers, many earlier papers are being regularly cited.

Early ecologists suffered from a lack of long-term time series. With each passing decade datasets are becoming larger and the opportunities for more detailed analysis of temporal dynamics increases. In addition, remote sensing and large-scale observation, as undertaken in particular by bird and butterfly watchers, has greatly extended the opportunity for spatial analysis. Recent concerns about species loss, habitat destruction and fragmentation, and the effects of climate change are dependent on the collection and analysis of temporally and spatially extensive data sets. The collection and handling of these data and the computation of indices of change, species richness and diversity are important fields which continue to develop.

Dick Southwood is still included as an author of *Ecological Methods* Fourth Edition because there are still many parts of the book which were originally authored by him and have been little changed. However, I answer for the inadequacies of the present edition. Finally, I would like to express my gratitude to all my colleagues, including Clive Hambler and Anne Magurran in particular, who have directed me to interesting work and techniques for inclusion. This edition has been much improved by the careful and accurate work of the copyeditor, Mr William Down. The book would have many irritating errors without his thoughtful attention to detail. A considerable task for a book of this size and complexity.

Peter Henderson
Lymington, February 2015

Preface to third edition

We have been encouraged to prepare this third edition by the continuing use of the earlier editions. In doing so we have been struck by contrast between the advances in some areas, especially data handling, and the enduring value of various other techniques. Ecology has continued its advance into the popular and political domain, though far from everything that has gained the 'eco-' prefix falls within our purview. The underlying paradigm of ecology has however shifted. In particular the concept of the metapopulation is now recognised as central to the understanding of the distribution and abundance of animals and its exploration aided by the accessibility of numerous data sets, often with large temporal or spatial scales. The availability of molecular techniques has encouraged the consideration of genetic and phylogenetic aspects and has permitted the growth of quantitative comparative analyses described by P.H. Harvey and M.D. Pagel (1991) *The Comparative Method in Evolutionary Biology* (Oxford).

The extent to which we have felt that revision was necessary has varied greatly in different parts of the book. Where there has been little change in the method, we have retained the early references. We have done this on the bases that journal editors formerly permitted more detailed description of methods, that these papers will not be located by computer searches and that these pioneers continue to deserve credit. Other portions have required considerable modification; we have deleted one chapter (12) and added a new one (15), as well as reorganising the structure in some places. Although the primary focus remains on insects, which are in terms of species the dominant animals, we have taken the opportunity to explicitly expand the coverage to all major macroscopic groups.

The widespread availability of high capacity PCs, with software packages and access to the internet, has totally changed the speed and ease of handling (and sometimes accessing) data. We have therefore given references to some relevant software packages and web sites, whilst eliminating many descriptions of time consuming graphical methods. However, we believe that the advice given in the preface to the second edition is even more applicable today. The researcher, who relies entirely on the output of a computer, is in danger of drawing false conclusions and overlooking possible insights. It is essential to understand the features of the data (are there any outliers?), the assumptions of the methods, the biological basis of the analysis and to acquire a feel for the capabilities and responses of the species under study.

The interpretation of 'ecological methods' remains as described in the Preface to the First Edition, namely those methods peculiar to ecologists, either in their origin or in modification. Just as the measurement of physical factors, using the methods of the physical sciences, has always been outside the book's scope, so are the methods of molecular biology. These are described in a number of works such as *Molecular markers for population ecology* (1998), *Ecology* **79**, 359–425, edited by A.A. Snow and P.G. Parker.

We are most grateful to many ecologists who sent comments on the second edition. In particular generous help has been given by Drs C. Henderson, D.J. Rogers, A.E. Magurran, W.D. Hamilton, G.R.W. Wint and Mr C. Hambler.

T. R. E. Southwood
P. A. Henderson
Oxford, October 1998

Preface to second edition

In the twelve years since the First Edition was prepared there have been remarkable developments in ecology. The subject has changed its lay image, from a rather recondite branch of biology, to something that is widely considered 'good', but only vaguely understood. The public's focus on environmental problems and the insights into these that ecology can provide are a great challenge to ecologists to develop their subject: they need to be able to provide reliable quantitative inputs for the management of the biosphere. The enormous volume of work that it has been necessary to review for this edition is evidence of the extent to which ecologists are seeking to meet this challenge.

I believe that the theme of the first edition, the need for precise measurement and critical analysis, is equally valid today; although many recent studies show levels of sophistication that were beyond my wildest hopes when I embarked on the preparation of the first edition. In his review of the first edition Dr R.R. Sokal was kind enough to say it was an 'unusual book' for it covered both traps and mathematical formulae, topics that were usually of interest to different people. This, I am glad to say, is now no longer generally true. The computer and the electronic calculator have revolutionised the handling of ecological data, but neither can make a 'silk purse' of sound insight, out of a 'sow's ear' of unreliable raw data or confused analytical procedure. More than ever the ecologist needs to keep his biological assumptions in mind and remember the value of preliminary simple graphical analysis as a means of recognising new patterns and gaining fresh insights.

It has been a gratifying, though exhausting, experience preparing the new edition! Progress has been so rapid in several areas that some chapters have been completely or largely rewritten (e.g. 11, 12, 13), whilst most have large new sections. It has been necessary to be highly selective in the additions to the bibliographies, even so there are nearly a thousand new entries and only a few older references could be deleted. I hope that, with the advent of *Ecological Abstracts*, the selective nature of the bibliographies will not handicap workers. As the mathematical side of ecology has grown, so have the problems of notation and it is now quite impossible (without extending far beyond the roman and greek alphabets!) to retain a unique notation throughout. Apart from widespread and generally accepted symbols, I have merely aimed to be consistent within a section.

I am most grateful to many ecologists who have helped me in this revision by sending me reprints of their papers or notes on difficulties and errors in the first edition. Detailed criticism, advice and help, including access to unpublished work has been generously given by M.H. Birley, P.F.L. Boreham, M.J.W. Cock, G.R. Conway, M.P. Hassell, R.M. May, A. Milne, S. McNeill, G. Murdie, G.A. Norton, S. Parry, P.M. Reader, G. Seber, N.E.A. Scopes, L.R. Taylor, R.A.J. Taylor, T.W. Tinsley, J.M. Webster, R.G. Wiegert and I.J. Wyatt. I am deeply indebted to them, and to others, especially Margaret Clements and my wife, who have assisted with patience and forbearance in the essential tasks associated with the preparation of the manuscript of the new edition.

T. R. E. Southwood
Imperial College
Silwood Park, Ascot.
October 1977

Preface to first edition

This volume aims to provide a handbook of ecological methods pertinent for the study of animals. Emphasis is placed on those most relevant to work on insects and other non-microscopic invertebrates of terrestrial and aquatic environments, but it is believed that the principles and general techniques will be found of value in studies on vertebrates and marine animals.

The term ecology is now widely used in the field of social, as well as biological, science; whilst the subject of ecology, covering as it does the relationship of the organism to its environment, has many facets. It is, in fact, true to say that the ecologist may have need of recourse to almost all of the methods of the biologist and many of those of the physical scientist: the measurement of the physical factors of the environment may be a particularly important part of an ecologist's work and he will refer to books such as R.B. Platt & J.E. Griffiths' (1964) *Environmental Measurement and Interpretation*.

There are, however, certain methods that are peculiar to the ecologist, those concerning the central themes of his subject, the measurement, description and analysis of both the population and the community. These are *ecological* methods (as opposed to 'methods for ecologists' which would need to cover everything from laboratory workshop practice to information theory); they are the topic of this book. During the 10 years that I have been giving advanced and elementary courses on ecological methods at Imperial College, London, and at various Field Centres, the number and range of techniques available to the ecologist have increased enormously. It has been the comments of past students on the utility of these courses in helping to overcome the difficulties of coping with the scattered and growing literature that have encouraged me to attempt the present compilation. I am grateful to many former students for their criticisms and comments, as I am to the members of classes I was privileged to teach at the University of California, Berkeley, and at the Escuela Nacional de Agricultura, Mexico, whilst writing this book.

Although the general principles of most methods are of wide application, the study of a particular animal in a particular habitat may require certain special modifications. It is clearly impossible to cover all variants and therefore the reader is urged to consult the original papers that appear relevant to his problem. I am grateful to my publishers for agreeing to the publication of the extensive bibliographies, it is hoped that these will provide many leads on specific problems; they are, however, by no means exhaustive.

The present book is designed to be of use to those who teach the practical aspects of animal ecology in schools, training colleges and universities; insects, being numerically the dominant component of the macrofauna of terrestrial and many aquatic habitats, almost invariably come to the forefront of ecological field work. This volume is intended as an aid to all who need to measure and compare populations and communities of animals, not only for the research ecologist, but also for the conservationist and the economic entomologist. Population measurement is as necessary in the assessment of the effects of a pesticide and in the determination of the need for control measures, as it is in intensive ecological studies. It is frequently pointed out that ecological theories have outstripped facts about animal populations and I trust that it is not too presumptuous of me to hope that this collection of methods may encourage more precise studies and more critical analysis of the assembled data so that, in the words of O.W. Richards, we may have 'more light and less heat', in our discussions.

The topics have been arranged on a functional basis, that is, according to the type of information given by a particular method. As a result some techniques are discussed in several places, e.g. radiotracers will be found under marking methods for absolute population estimates (chapter 3), the measurement of predation and dispersal (chapter 9) and the construction of energy budgets (chapter 14). By its very nature ecology cannot be divided into rigid compartments, but frequent crossreferences in the text, together with the detailed contents list and index, should enable the reader to find the information he needs. The sequence of chapters parallels, to a large extent, the succession of operations in a piece of intensive research.

It is a pleasure to express my great indebtedness to colleagues who have read and criticised various chapters in draft: Dr N.H. Anderson (ch. 6), Dr R.E. Blackith (ch. 2 & 13), Dr J.P. Dempster (ch. 1, 2 & 3), Mr G.R. Gradwell (ch. 10), Dr C.S. Holling (ch. 12), Mr S. Hubbell (ch. 14), Dr C.B. Huffaker (ch. 2), Dr G.M. Jolly, (section II of ch. 3), Dr C.T. Lewis, (section I of ch. 3), Dr R.F. Morris (ch. 10), Dr O.H. Paris (ch. 9 & 14), Mr L.R. Taylor (ch. 2 & 4), Professor G.C. Varley (ch. 10) and Dr N. Waloff (ch. 1, 2 & 10); frequently these colleagues have also made available unpublished material; they are of course in no way responsible for the views I have expressed or any errors. For access to 'in press' manuscripts, for unpublished data and for advice on specific points I am grateful to: Drs J.R. Anderson, R. Craig and D.J. Cross, Mr R.J. Dalleske, Drs W. Danthanarayana, H.V. Daly, E.A.G. Duffey, P.J.M. Greenslade, M.P. Hassell, P.H. Leslie, J. MacLeod, C.O. Mohr, W.W. Murdoch and F. Sonleitner, Mr W.O. Steel, Drs A.J. Thorsteinson, R.L. Usinger, H.F. van Emden, E.G. White, D.L. Wood and E.C. Young. Ecologists in all parts of the world have greatly helped by sending me reprints of their papers. I have been extremely fortunate too in the assistance I have received in translating; Mrs M. Van Emden has generously made extensive translations of works in German, and with other languages I have been helped by Dr F. Baranyovits (Hungarian), Dr T. Bilewicz-Pawinska (Polish), Mr Guro Kuno (Japanese), Dr P. Stys (Czechoslovakian) and Dr N. Waloff (Russian).

Much of the manuscript was prepared whilst I held a visiting professorship in the Department of Entomology and Parasitology of the University of California, Berkeley; I am indebted to the Chairman of that Department, Dr Ray F. Smith, for his interest and the many kindnesses and facilities extended to me and to the Head of my own Department, Professor O.W. Richards, F.R.S., for his support and advice. I wish to thank Mrs M.P. Candey and Mrs C.A. Lunn for assisting me greatly in the tedious tasks of preparing the bibliographies and checking the manuscript. My wife has encouraged me throughout and helped in many ways, including typing the manuscript.

T.R.E. Southwood
London, October 1965

About the Companion Website

This book is accompanied by a companion website:

www.wiley.com/go/henderson/ecologicalmethods

The website includes:

- PowerPoint slides of all the figures from the book, for downloading
- PDF files of all the tables from the book, for downloading
- Additional resources

1 Introduction to the Study of Animal Populations

Information about animal populations is sought for a variety of purposes; but *the object* of a study will largely determine the methods used and thus this must be clearly defined at the outset. Very broadly, studies may be divided into *extensive* and *intensive* (Morris, 1960). Extensive studies are carried out over larger areas or longer time periods than intensive studies, and are frequently used to provide information on distribution and abundance for conservation or management programmes (e.g. monitoring the status of mammal populations; Macdonald *et al.*, 1998). Recent developments in remote sensing capability and geographical information system software have given great impetus to extensive studies in recent years (see Chapter 15). For pests or parasites they provide assessments of incidence or damage, and may also guide the application of control measures. Aphids, for example, are regularly monitored by the Rothamsted suction trap survey (http://www.rothamsted.ac.uk/insect-survey/STAphidBulletin.html; Harrington & Woiwod, 2007). In extensive surveys, an area will often be sampled once or at the most a few times per study period. The timing of sampling in relation to the life cycle of the animal is obviously of critical importance, and for many species, only limited stages in the life cycle can be sampled. Extensive studies produce information about the spatial pattern of populations, and it is often possible to relate the level of the population to edaphic, oceanographic or climatic factors. Recently, long-term data have proved invaluable to quantify the effects of climate change on distribution and phenology.

Intensive studies involve the *repeated* observation of the population of an animal. Usually, information is acquired on the sizes of the populations of successive developmental stages so that a life-table or budget may be constructed. Then, using this table an attempt is made at determining the factors that influence population size and those that govern or regulate it (Varley & Gradwell, 1963; Sibley & Smith, 1998). It is important to consider at the start the type of analysis (see Chapter 11) that will be applied and so ensure that the necessary data are collected in the best manner. Intensive studies may have even more limited objectives, such as the determination of the level of parasitism, the amount of dispersal, or the overall rate of population change.

The census of populations and the stages at which mortality factors operate are necessary first stages in the estimation of the productivity (see Chapter 14) of ecosystems.

In survey and conservation work, the species make-up of the community and changes in its diversity associated with human activities are most frequently the features it is desired to measure. The estimation of species richness invariably requires

Ecological Methods, Fourth Edition. P. A. Henderson and T. R. E. Southwood.
© 2016 John Wiley & Sons, Ltd. Published 2016 by John Wiley & Sons, Ltd.
Companion Website: www.wiley.com/go/henderson/ecologicalmethods

repeated sampling of the habitat, with special methods of analysis needed to estimate the total species number (see Chapter 13). The relative abundance of species is a key attribute which may give insight into the functioning and health of a community, though difficulties usually arise because of the impossibility of recording the abundance of all the species living within a habitat with equal efficiency.

1.1 Population estimates

Population estimates can be classified into a number of different types; the most convenient classification is that adopted by Morris (1955), although he used the terms somewhat differently in a later study (Morris, 1960).

1.1.1 Absolute and related estimates

For large animals that are easily observed and have small, countable, populations such as rhinos, elephants, tigers, whales or some birds, it may be possible to express the global or metapopulation size as a total number of individuals. However, for most animals, numbers will be expressed as a density per unit area or volume or per unit of the habitat. Such estimates are given by distance sampling and related techniques (Chapter 9), marking and recapture (Chapter 3), by sampling a known fraction of the habitat (Chapters 4–6) and by removal sampling and random walk techniques (Chapter 7).

1.1.1.1 Absolute population

This is defined as the number of animals per unit area. For planktonic animals, the number per unit volume can be more appropriate. It is almost impossible to construct a budget or to study mortality factors without the conversion of population estimates to absolute figures. This is because other measures of habitat are variable. For example, the amount of plant available to an insect is always changing; further, insects often move from the plant to the soil at different developmental stages. The importance of obtaining absolute estimates cannot be overemphasised.

1.1.1.2 Population intensity

This is the number of animals per unit of habitat, for example per leaf, per shoot, per plant, per host. Such a measure is often, from the nature of the sampling, the type first obtained (see also p. 146). When the level of the animal population is being related to habitat availability or plant or host damage, it is more meaningful than an estimate in absolute terms. It is also valuable when comparing the densities of natural enemies and their prey. However, the number of habitat units per area should be assessed, for differences in plant density can easily lead to the most intense population being the least dense in absolute terms (Pimentel, 1961). When dealing with different varieties of plants, differences in leaf area may account for apparently denser populations, in absolute terms, on certain varieties (Bradley, 1952). Thus, the choice of the leaf or of the

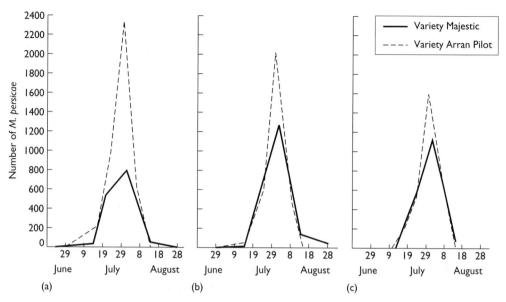

Figure 1.1 **The influence of habitat unit on relative population levels when these are measured in terms of population intensity; the populations of *Myzus persicae* on different varieties of potato. (a) Per 100 leaves; (b) per plant; (c) per plant, corrected for proportions of upper, middle, and lower leaves. (Adapted from Broadbent, 1948.)**

plant as the unit for expressing population intensity can affect the relative population estimate obtained (Broadbent, 1948) (Fig. 1.1). Similarly, with litter fauna – owing to the effects of seasonal leaf fall – the intensity measure (on animals/weight of litter) will give a different seasonal picture from an absolute estimate per square metre (Gabbutt, 1958). These examples also underline the importance of absolute estimates when interest lies primarily in the animal population.

1.1.1.3 Basic population

In some habitats, especially forests and orchards, it is often convenient to have an intermediate unit between that used for measuring intensity and absolute measures of ground area, for example 1 m^2 of branch surface (Morris, 1955) or branches of apple trees (Lord, 1968).

1.1.2 *Relative estimates*

These estimates, in which the number caught cannot be expressed as a density or intensity per area or habitat unit, allow only comparisons in space or time. They are especially useful in extensive work on species distributions, monitoring changes in species richness, environmental assessments, recording patterns of animal activity, and investigating the constitution of a polymorphic population. The methods employed are either the catch per unit effort type or various forms of trapping, in which the number of individuals caught depends on a number of factors besides population density (Chapter 7). There is no hard and fast line between relative and absolute methods, for absolute methods of sampling are seldom 100% efficient and relative methods can sometimes be corrected in various ways to give density estimates.

Relative methods are important in applied areas such as fisheries or game management, where most of the information available may be derived from fishing or hunting returns. In fisheries research, catch per unit effort is often difficult to calculate from landing statistics because of changes in catch efficiency as fishing technology and the economy changes. The behaviour of fishermen frequently changes with the abundance of the target species.

1.1.3 *Population indices*

These are generated when the animals themselves are not counted, but their products (e.g. frass, webs, exuviae, tubes, nests) or effects (especially plant damage) are recorded.

Both, population indices and relative estimates of population can sometimes be related to absolute population (if this is measured at the same time) by regression analysis. If such a study has been based on sufficient data, subsequent estimates from relative methods or indices can be converted to absolute terms using various correction factors; such an approach is common in fisheries research (e.g. Beverton and Holt, 1957).

1.2 Errors and confidence

The statistical errors of various estimates can usually be calculated and the upper and lower boundaries of an estimate are referred to as the fiducial limits (the estimate (x) being expressed as $x \pm y$ where y = fiducial limits). These are sometimes incorrectly referred to as 'confidence' limits, but the distinction between the two terms is in practice unimportant. The fiducial limits are calculated for a given probability level, normally 0.05, which means that there are only five chances in 100 that the range given by the fiducial limits does not include the true value (hence the expressions 5% probability level and 95% fiducial limits). If more samples are taken the limits will be narrower, but the estimate may not move closer to the actual value. Biologists are often worried that assumptions about sampling efficiency (e.g. does the 'knockdown' method really collect all the weevils on a tree?) may be incorrect. Intuitively, most believe – quite correctly – that this estimate should be compared with another method that has *different* assumptions. If the estimates are of the same order of magnitude, then the investigator can have much greater confidence that the result of the study is not misleading. It is therefore sound practice for the ecologist to *contemporaneously estimate the population or other variables by more than one method*.

When two estimates have been obtained from different sampling procedures, and provided they are internally consistent (e.g. a t-test shows that the means are not significantly different), they may be combined to give a weighted mean, weighting each estimate inversely as its variance (Cochran, 1954). Under some circumstances Bayes' theorem could be used (see p. 88, equation 3.29) to give the combined estimate. Laughlin (1976) has suggested that the ecologist may be satisfied with a higher probability level (say 0.2) and thus narrower fiducial limits for estimates based on more than one method, because such estimates have a qualitative, biological assurance, additional to that from the consistency of the data, that the true mean lies close to the estimate.

Most population studies are based on sampling, and the values obtained are considered to have a generality that scales with the area from which the samples were drawn. All estimates have fiducial limits and the level of accuracy that should be aimed at is difficult to determine. Morris (1960) has aptly said that '...we are not likely to learn what precision is required by pessimistic contemplation of individual fiducial limits'. Excessive concerns about accuracy can always be used by those who favour the warmth of the hearth to being in the field. As the amount of time and labour that can be put into a problem is limited, it should always be borne in mind that the law of diminishing returns applies to the reduction of the statistical errors of sampling. In the long run, more knowledge of the ecology of the animal may be gained by studying other areas, by making other estimates, or by taking further samples than by straining for a very high level of accuracy in each operation. Against this must be set the fact that when animals are being extracted from samples, the errors all lie below the true value as animals will occasionally be missed. A number of very carefully conducted control samples may allow a correction factor to be applied, but the percentage of animals missed may vary with density; sometimes, more are overlooked at the lowest densities (Morris, 1955).

An alternative to sampling is the continuous, or regularly repeated, study of a restricted cohort, such as the population of an aphid on a particular leaf or leaf-miners on a bough. These studies have a very high level of accuracy but they sacrifice generality. A combination of some cohort studies with larger-scale sampling often provides valuable insights.

References

Beverton, R.J.H. & Holt, S.J. (1957) *On the dynamics of exploited fish populations*. Ministry of Agriculture, Fisheries and Food Great Britain, HMSO, London.

Bradley, R.H.E. (1952) Methods of recording aphid (Homoptera: Aphidae) populations on potatoes and the distribution of species on the plant. *Can. Entomol.* **84**, 93–102.

Broadbent, L. (1948) Methods of recording aphid populations for use in research on potato virus diseases. *Ann. Appl. Biol.* **35**, 551–66.

Cochran, W.G. (1954) The combination of estimates from different experiments. *Biometrics* **10**, 101–29.

Gabbutt, P.D. (1958) The seasonal abundance of some arthropods collected from oak leaf litter in S. E. Devon. *Proc. X Int. Congr. Entomol.* **2**, 717.

Harrington, R. & Woiwod, I. (2007) Foresight from hindsight: the Rothamsted Insect Survey. *Outlook Pest Manag.* **18**, 9–14.

Laughlin, R. (1976) Counting the flowers in the forest: combining two population estimates. *Aust. J. Ecol.* **1**, 97–101.

Lord, F.T. (1968) An appraisal of methods of sampling apple trees and results of some tests using a sampling unit common to insect predators and their prey. *Can. Entomol.* **100**, 23–33.

MacDonald, D.W., Mace, G., & Rushton, S. (1998) Proposals for future monitoring of British mammals. Department of the Environment, London, 374 pp.

Morris, R.F. (1955). The development of sampling techniques for forest insect defoliators, with particular reference to the spruce budworm. *Can. J. Zool.* **33**, 225–294.

Morris, R.F. (1960) Sampling insect populations. *Annu. Rev. Entomol.* **5**, 243–64.

Pimentel, D. (1961) The influence of plant spatial patterns on insect populations. *Ann. Entomol. Soc. Am.* **54**, 61–9.

Sibley, R.M. & Smith, R.H. (1998) Identifying key factors using lambda contribution analysis. *J. Anim. Ecol.* **67**, 17–24.

Varley, G.C. & Gradwell, G.R. (1963) The interpretation of insect population changes. *Proc. Ceylon Assoc. Adv. Sci.* **18**(D), 142–56.

2 The Sampling Programme and the Measurement and Description of Dispersion

2.1 Preliminary sampling

2.1.1 Planning and fieldwork

At the outset you must be quite clear as to the problem you are proposing to investigate. As it is normally impossible to count and identify all the animals in a habitat, it is necessary to estimate the population by sampling. Naturally, these estimates should have the highest accuracy commensurate with the effort expended. This requires a plan that includes a sampling programme laying down the number of samples, their distribution and size. While the statistical principles for sampling are given in texts (e.g. Hansen *et al.*, 1953; Stuart, 1962; Yates, 1953; Elliott, 1977; Cochran, 1977; Seber, 1982; Green, 1979; Mead, 1990; Scheiner & Gurevitch, 1993; Underwood, 1997; Lohr, 2009), there is no universal sampling method. *The importance of careful formulation of the hypothesis for test cannot be overstressed.*

In community studies, preliminary work should consider species richness and potential problems with species identification. The appropriate degree of taxonomic discrimination must be decided, as it is important to maintain a consistent taxonomy. Sample sorting and species identification are often the most labour-intensive parts of a study, and it may be useful to process a trial sample to assess the effort required. Saila *et al.* (1976) found that a single marine grab sample cost US$25 to collect and US$300 to sort. By 2015, it is not uncommon for the cost of picking and identifying to species the contents of marine benthic grab samples to cost £450 (US$700) each. The planning of the timing of sampling requires knowledge of life cycles. Preliminary work will be necessary to gain some knowledge of the distribution of the animal(s). From these observations, an estimated cost per unit sample in terms of time, resources and money can be made.

The first decision concerns the scale of the universe to be sampled. Whether this is to be a single habitat (e.g. field, woodland, or pond) or representatives of the habitat type from a wide geographical area will depend on whether an intensive or an extensive study is planned. The correct definition of the target population or community is essential; if too small, it may not produce results representative of the structure as a whole; if too large, it will waste resources. Sale (1998) argued that the appropriate scale when sampling reef fish is determined by the extent of movement of the life stage under study, and the same approach can be usefully applied to terrestrial organisms. A good description of this decision process for plankton is given by Omori & Ikeda (1984).

The second decision must be to define the *accuracy* or *precision* of the population estimates required. *Accuracy* measures how close an estimate is to the real value, while *precision* measures the reproducibility of the estimate. If a systematic error is intrinsic to a sampling method then increased sampling may improve the precision, but it will not increase the accuracy of the estimate. The decision on accuracy is taken by considering both the objectives of the study and the variability of the system under study. For example, many species of insect pest exhibit ten- or even hundred-fold population change within a single season, and recruitment in fish stocks can vary tenfold between years (Cushing, 1992). For such variable insect species an estimate of population density with a standard error of about 25% of the mean, enabling the detection of a doubling or halving of the population, is sufficiently accurate for damage control (Church & Strickland, 1954). For life-table studies, more especially on natural populations, a higher level of accuracy – frequently set at 10% – will be necessary.

Conservationists often seek to build species inventories. Here, completeness will replace accuracy as a measure of quality. Whereas, for large mammals or birds the aim might be to record all resident species, for high-diversity groups such as beetles the objective may be set at only 5–10%. Community studies often aim to generate summarising statistics such as measures of diversity or species richness (Chapter 13), which can be used to compare localities or changes through time. The accuracy and precision of these estimates must be carefully considered if changes are to be detected at the desired resolution.

In extensive work, the amount of sampling in a particular locality will be limited and therefore a further decision concerning both the age-group to be studied and the timing of sampling will need to be made. For insects, the best stage for sampling may be that most closely correlated with the amount of damage (Burrage & Gyrisco, 1954). Alternatively, if the purpose of the survey is to assess the necessity for control, the timing should be such that it will give advanced information of an outbreak (Gonzalez, 1970). Direct sampling methods such as sweep-netting or counts of individuals per leaf are often most effective for less-mobile immature stages (Martini *et al.*, 2012). For marine organisms, timing is often determined by the reproductive cycle. For example, in temperate waters, marine benthic surveys carried out in autumn will show a population dominated by recent recruits. The same survey carried out in the spring will show the resident community which has survived both competition for space and the rigours of the winter (Bamber, 1993). In freshwater habitats, samples collected in spring and autumn can differ markedly. Carlson *et al.* (2013), in a study of Swedish streams, found chironomids to be most abundant in autumn, which they related to the timing of emergence. Other seasonal changes were linked to the harsh conditions during the spring ice melt when water discharge is high and temperatures low. Similarly, large seasonal changes occur in tropical waters such as the Amazon linked to the flood cycle. Henderson *et al.* (1998) reported order-of-magnitude changes in floodplain fish density between high- and low-water seasons as typical.

While the life history stage to be sampled will depend on the objectives, the stage must not be one whose numbers change greatly with time and it must be present for a period sufficient to allow the survey to be completed. Further, the easier the stage is to sample and count the better – birds, butterflies and some mammals may be most easily counted during courtship and mating. Extensive surveys can result in different areas holding populations that differ in their position in the reproductive cycle. The reliability of samples in an extensive survey may be particularly sensitive to current weather conditions (Harris *et al.*, 1972). The influence of latitudinal gradients on the timing of sampling can be important. In extensive surveys of marine plankton, the timing of blooms or larval production can vary by one month over 1 degree of latitude.

Although the preliminary sampling and analysis of the assembled data will provide a measure of many of the variables, the actual decisions must still, in many cases, be a matter of judgement. Furthermore, as the density changes so too will many of the statistical parameters, and a method that is suitable at a higher density may be found inadequate if the population level drops. Shaw (1955) found that Thomas and Jacob's (1949) recommendation for sampling potato aphids – one upper, middle and lower leaf from each of 50 plants – was unsatisfactory in Scotland, in certain years, because of the lower densities.

Detailed and now classical accounts of the development of sampling programmes for various insects are given by Morris (1955, 1960), LeRoux & Reimer (1959), Harcourt (1961, 1964), Lyons (1964) and Coulson *et al.* (1975, 1976), amongst others. Terrestrial sampling programmes for vertebrates, small mammals and birds are described by Caughley (1977), Smith *et al.* (1975) and Blondel & Frochet (1987), respectively. For aquatic systems, estuarine and coastal sampling programmes are described for many groups by Baker & Wolff (1987), for sampling marine benthos by Eleftheriou (2013), and for zooplankton by Omori & Ikeda (1984).

The next step is to subdivide for sampling purposes. These subdivisions may be orientated with respect to an environmental gradient. If none is apparent, it is often convenient to divide the area into regions of regular shape, but this may be neither possible nor desirable. Ideally, the within-subdivision variability should be much less than the between-subdivision variability in habitat. The habitat must be considered from the biological angle and a decision made as to whether it might need further division. If it is woodland, for example, the various levels of the tree – upper, middle and lower canopy and probably the tips and bases of the branches – would on a priori grounds be considered as potentially different divisions. The aspect of the tree might also be important. In herbage or grassland, if leaves or another small sampling unit is being taken, the upper and lower parts of the plants should be treated separately. Similarly, aquatic samples may be taken from different depths, position with respect to flow, and from different plants or substrates.

It is often of value to take at least two different sized sampling units (Waters & Henson, 1959). In benthic surveys a 0.1 m^2 grab sample suitable for sampling macroinvertebrates can itself be core sampled for the meiofauna prior to processing (McIntyre and Warwick, 1984). In an insect study, the aim should be to sample towards the smallest possible limit, for example a leaf blade or half-leaf. As a general principle, a higher level of reproducibility is obtained (for the same cost) by taking more smaller units

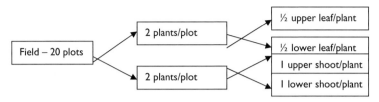

Figure 2.1   Example of a simple sampling plan.

than by taking fewer large ones.[1] Small sampling units may also enable precision to be increased by distinguishing between favourable and unfavourable microhabitats; Condrashoff (1964) found with a leaf miner that the upper and lower leaf surfaces should be considered as separate units. Two examples of any one size-sampling unit should be taken within each sampling plot or subsection. For example, for a field crop a preliminary plan could be as in Fig. 2.1, which gives a total of 160 samples.

In community studies where it is essential to gain some idea of species diversity and the species–area relationship, a number of sampling methods may be needed. When benthic sampling for example, a pilot study might use dredges to obtain a general idea of animal presence and distribution and a limited number of grab samples to estimate densities. During a pilot study a record should be kept of the cost of each part of the sampling routine, normally expressed in man-hours.

2.1.2 Statistical aspects

Before the data gathered in the preliminary samples can be analysed, some aspects of data organisation and statistics need to be considered.

2.1.2.1 Data storage, software and statistical analysis considerations

A sampling plan needs to consider the processes involved in: (1) data acquisition; (2) organisation and storage; (3) analysis; and (4) presentation. Smooth and rapid progress along this chain almost always uses computers, but problems arise if data cannot be easily transferred. If different software products and operating systems are used, then compatibility must be considered. Computational power is now so great that even small, portable, machines can handle 1000×1000 arrays of data, which is quite sufficient for most researchers' needs as few of us will ever count the abundance of 1000 taxa in 1000 samples. However, computational demands soon start to rise to demanding levels when automated forms of data acquisition are used. For example,

---

[1]The only disadvantage of sampling by small units is the number of zeros that may result at low densities; this truncation may make analysis difficult and has led to the suggestion that larger sized samples should be taken (Pradham & Menon, 1945; Spiller, 1952). The decision must be related to the density of the animal, although moderate truncation can be overcome by suitable transformation; in other cases it may be necessary to increase the size of the unit.

the analysis of sonar data for fish counting or recordings of bat echolocation calls can rapidly generate huge amounts of digital information that needs to be analysed. Particular care is needed when automated recording is to be undertaken to ensure the data can be analysed within the required time scale. Both software and hardware capability need to be considered for each stage of the study. During data acquisition, it is important to assess the data storage and processing requirements. Data has only been acquired when it has been processed into a usable form. Data acquisition and processing rates are often different. A common problem is the analysis of video recordings, for example, the study of salmon movement through a fish pass. In one such study over a 10-week period about 1680 h of video was recorded of the water in the fish pass. It was originally planned to use an image analysis program to identify and to count the fish on the video. However, at the end of the fieldwork it was found that the image analysis software worked at less than one-tenth of real time, and required over 16 800 h of computer time to analyse the images. In practice, this meant that image analysis would only just be complete by the beginning of the next salmon migratory season. The only way to acquire the fish count data within the required time was to employ three people for 6 weeks to view the video and manually input observations.

A wide range of powerful statistical and data manipulation packages, such as R, SAS, SPSS, Systat or Genstat, are available. Many ecologists now use R, the open source statistical analysis environment available from the Comprehensive R Archive Network (CRAN) at http://cran.r-project.org, because it is not only free but also offers a huge range of statistical tests and procedures, including more advanced methods such as general and generalised linear models, time series analysis, multivariate analysis, cluster methods and a wide range of graphical displays such as heat maps (see Plate 7). It is, however, a far from intuitive environment within which to work, and those with less programming experience can find it difficult to use and will usually prefer commercial statistical packages.

Many software products are available for initial data organisation. Those most frequently used by ecologists are spreadsheets such as Excel, which has become the industry standard. When choosing the product to use, it is important to consider carefully its capability to hold, manipulate and convert the data into other formats. A general data file format that can be imported and used by many programs is a csv file. Much historical data from the 1970s and 1980s has become difficult to access because it was recorded on tape or disc formats that are no longer used. This should be less of a problem in the future as data sets can be uploaded to web-based data storage. Consideration should be given to long-term data archiving and maintenance.

Ecological data sets of animal abundance can often be organised within a three-dimensional array of species (or other taxa) by station (locality) by time. This is the data structure that spreadsheets are designed to handle. If a spreadsheet has the capability to hold the data sets then they are often the preferred software for data organisation and elementary analysis for they allow the rapid production of summary statistics, tables and graphs. Specialist database applications for desktop computers such as Microsoft Access, Filemaker Pro, Alpha Five, Paradox and Lotus Approach can also be used to hold and organise data. These programs are better than spreadsheets if there is a need to store and relate together different types of data, but they are usually inferior to spreadsheets in their ability to undertake calculations and plot graphs. The ease and ability to move data into other programs should be considered before committing to a particular database or spreadsheet. The three most common requirements are to export data to statistical, graphical or word-processing packages. Large statistical software packages often offer their own spreadsheet capability, as do graphics packages such as SigmaPlot (http://www.sigmaplot.co.uk/products/sigmaplot/sigmaplot-details.php).

A common problem is backward compatibility. While most new programs will be able to handle older file formats, it is often impossible to import new formats into older software packages. Care must be taken if an old, but well tested and valued program is to be used, to ensure that it can run with current data formats. It is often good practice to store numerical data in a highly general file format such as a 'csv' (comma separated values) file. This is an ideal format for data that will be analysed using R or many multivariate packages; in R, the function read.csv is used to load csv files. Spreadsheets such as Excel offer the option to export or save data as a csv file. Care must also be taken if data are to be moved between machines using differing operating systems, for example Windows and Apple PCs.

Particular care is needed to define date and number formats, as Europe and the USA differ in day and month order. The real number represented as 3,14 in continental Europe is 3.14 in Britain or the USA. There are also many formats used to represent missing values.

2.1.2.2   The normal distribution and transformations

The most important and commonly used of the theoretical distributions is the normal or Gaussian distribution. This has a probability curve that is a symmetrical bell-shape.

The probability density of a normal variable $P(x)$ is:

$$P(x) = \frac{1}{\sqrt{2}\sigma}e^{\left[-\frac{1}{2}\left(\frac{x-\mu}{\sigma}\right)^2\right]}$$ (2.1)

where μ is the mean and σ the standard deviation.

This distribution is symmetrical about the mean and the shape is determined by the standard deviation (Fig. 2.2).

The normal distribution has a key role in statistics because the mean of many random variables independently drawn from the same distribution is distributed approximately normal, irrespective of the form of the original distribution. This is termed the 'central limit theorem'. Thus, if an ecologist calculates a series of means from samples obtained from a population which conforms to a non-normal distribution, such as a negative binomial, the distribution of the means will be normal. While many distributions obtained from observation (e.g. the heights of humans) are approximately normal, the spatial distribution of individuals is seldom, if ever, normal. The importance of the normal distribution to ecologists arises solely from the fact that many statistical methods, such as linear regression, assume that the errors in measurement are normally distributed. General linear models, which include analysis of variance, assume homogeneity of variance and normality of error. These assumptions need to be tested

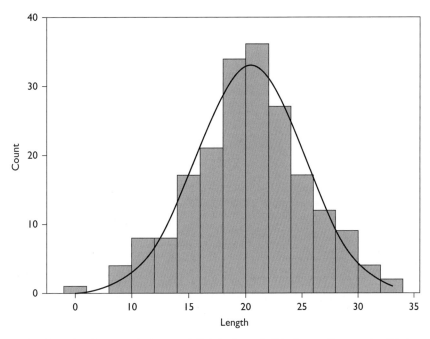

Figure 2.2   A randomly generated normal distribution of 200 observations and a fitted normal distribution. The values were generated in Excel using the formula = NORMINV(RAND(),20,5).

to ensure that statistical significance is correctly determined. The *glm()* and related functions in R offer a range of diagnostics to ensure the assumptions are met. While it is viewed that analysis of variance is quite robust to deviations from normality – and in some respects more so than the χ^2 test (Reimer, 1959; Abrahamsen & Strand, 1970) – data whose frequency distribution is considerably skewed and with the variance closely related to the mean cannot be analysed without the risk of errors.

Ecological data is usually skewed. The skew can be negative or positive and the terminology is confusing as a negative skew is biased towards the right (Fig. 2.3).

Such distributions are usually transformed by taking logarithms or square-roots. For example, if the square-root transformation were applied to 9, 16 and 64 they would become 3, 4 and 8, and it will be observed that this reduces the spread of the larger values. The interval between the second and third observations (16 and 64) is on the first scale nearly 7 times that between the first and second observations; when transformed, the interval between the second and third observations is only 4 times that between the first and second. A transformation of this type would tend to 'push' the long tail of a skew distribution in, so that the curve becomes more symmetrically bell-shaped. If all the observations (x) are positive and no a priori reason exists to choose between the many types of transformation possible, the Box–Cox empirical method can be used (Box and Cox, 1964). The general transformation function from x to x' is:

$$x' = \frac{(x^\lambda - 1)}{\lambda}, \lambda \neq 0 \quad \text{and} \quad x' = lnx, \lambda = 0 \tag{2.2}$$

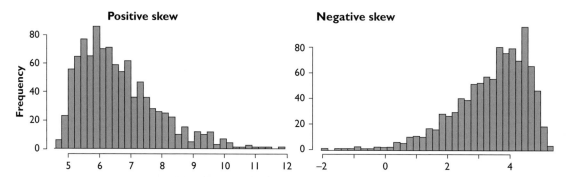

Figure 2.3   Examples of positive and negative skew. These distributions were generated in R using the sn package. The negative skew was generated with the following code:

```
library(sn)
r <- rsn(1000, 5, 2, -10)
hist(r, breaks=30, density=20)
```

The best transformation to normality is the value of exponent, λ, which maximises the log likelihood (L). When fitting a linear model using R, a suitable value of λ can be identified using the *boxcox* function in the MASS package.

Examination of the general transformation function (2.2) will show that certain values of λ correspond to commonly used transformations; these are listed in Table 2.1. If the calculated value for λ falls close to that corresponding to a common transformation, then this transformation should be used. Thus, if $\lambda = 0.1$ then a value of a value of $\lambda = 0$ would be appropriate and a logarithmic transformation used.

Taylor's power law (see p. 37), which relates the variance to the mean, has been used to identify suitable transformations. If the goal is to minimise the squared errors in predicting the mean, then better methods are sometimes available. For example, if replicate counts are taken from a Poisson-distributed population, but sampling error allows the mean to vary from count to count, it can be shown that the variance–mean relationship is:

$$\sigma^2 = \mu + c^2\mu^2 \tag{2.3}$$

where c is the coefficient of variation of the Poisson means.

Following the development of Generalised Linear Models (GLMs) (McCullugh & Nelder, 1989) in which all additive effects and the relationship between mean and

Table 2.1   Values of γ obtained by fitting the Box–Cox transformation function and their corresponding transformations.

λ	Transformation
1	linear $y = ax + b$
0.5	Square-root $y = $ SQRT(x)
0	logarithmic
−0.5	reciprocal square-root
−1	reciprocal $y = 1/x$

variance are specified separately, the need for variance-stabilising transformations prior to statistical analysis has been made unnecessary. When dealing with count data, when there cannot be negative counts, the basic GLM is a Poisson model with a log link (e.g. Roslin *et al.*, 2006, in a study of the leaf miner, *Tischeria ekebladella*, on oak, see p. 17). The *glm()* function in R will meet most ecologists' needs. It must be stressed that transformation does, however, lead to difficulties, particularly in the consideration of the mean and other estimates (see below). It should not be undertaken routinely, but only when the conditions for statistical tests are grossly violated (LeRoux & Reimer, 1959; Finney, 1973). In many applications, GLMs should be used rather than seeking a suitable transformation.

In order to overcome difficulties with zero counts in log transformations a constant (normally 1) is generally added to the original count (x); this is expressed as 'log $(x + 1)$'. Anscombe (1948) has suggested that a better transformation would be obtained by taking log $(x + k/2)$, where k is the dispersion parameter of the negative binomial (see below). As k is frequently in the region of 2, in many cases this refinement would have little effect. Andersen (1965) has shown that if the mean and k are very small (less than 3, and approaching zero, respectively) then the variance will not be stabilised by $k^{1/2}$ or any of the common transformations. However, if independent samples are pooled, or the size of the sampling unit is increased, the data may be satisfactorily transformed.

It has been customary to transform percentage or proportion data using an arc-sine transformation (calculated in R using *(sin⁻¹sqrt(.01∗%cover)))*. Generally, this can be avoided by the choice of a GLM binomial model with a logit link function. One field where Crawley (2005) still recommends the use of the arc-sine transformation is for percentage cover data as collected by botanists. After transformation, such data can be analysed using traditional linear methods.

The use of transformations can lead to problems and, as stressed above, transformation should not be routinely undertaken. If the fiducial limits are calculated from the transformed mean this may be erroneous (Abrahamsen & Strand, 1970). The biological interpretation of estimates based on transformed data is often difficult. There is indeed much to commend the use of the arithmetic mean (that based on the untransformed data) in population studies (van Emden *et al.*, 1961; Lyons, 1964), and if the distribution of the animal is random (see below) the fiducial limits are available in tables (Pearson & Hartley, 1958). If data have been transformed, the means of the untransformed and transformed values should be provided; back transforms of, say, geometric means from logarithmic transformations are more difficult to interpret and contain biases unless the variances are small (Finney, 1973). Beauchamp & Olson (1973) suggest that the bias will be corrected by using the log-normal distribution for de-transforming.

2.1.2.3 Checking the adequacy of the transformation

An adequate transformation should eliminate or considerably reduce two attributes of the data that are easily tested for. These are:

1. Excess skewness of the frequency curve causing large deviations from normality, which can be detected either from a probit plot or by statistical test (Snedecor & Cochran, 1989). In R, a Q-Q plot using the function *qqnorm()* produces a plot in

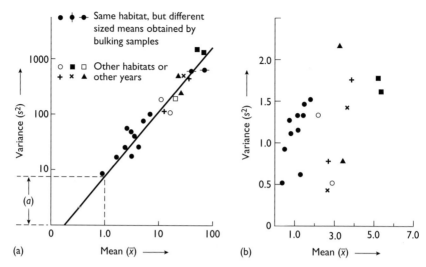

Figure 2.4 (a) The plot of variance against the mean on a log/log scale to obtain the constant *a* of Taylor's power law, data from samples of olive scales, *Parlatoria oleae*, per twig; (b) The same data transformed to $x^{0.4}$, showing the relative independence of the variance from the mean.

which departures from a straight line suggest violations of normality; alternatively, the *shapiro.test(x)* function performs a Shapiro–Wilk test The *nortest* package in R offers five omnibus tests for normality:

2. The dependence of the variance on the mean (instability or non-homogeneity of the variance) which may be shown graphically (Fig. 2.4). In R, the *bartlett.test( )* function provides parametric and the *fligner.test( )* function non-parametric tests for homogeneity of variances.

Generally, as Hayman and Lowe (1961) have pointed out, '…as non-normality must be extreme to invalidate the analysis of variance it is better to concentrate on stabilising the variance of the samples'. A correct transformation for this property will also ensure the third property necessary for the analysis of variance, the additivity of the variance (Bliss & Owen, 1958). Indeed, all three properties are related and for practical purposes the distinction between transformation for normality and that for stabilising the variance need not be emphasised. The adequacy of a transformation in stabilising the variance may be tested for graphically (Fig. 2.4b) or by using the *hovPlot()* function in the HH package in R.

2.2 The sampling programme

2.2.1 *The number of samples per habitat unit (e.g. plant, host or puddle)*

There are two aspects: first, whether different regions of the unit need to be sampled separately; and second, the number of samples within each unit or subunit (if these are necessary) that should be taken for maximum efficiency. Although the habitat unit could, for example, be the fleece of a sheep, a bag of grain, a rock in a stream, or an area of seabed, for convenience the word plant will, in general, be used in its place in the discussion below.

2.2.1.1 Subdivision of the habitat

If the distribution of the population throughout the habitat is biased towards certain subdivisions, but the samples are taken randomly, then what LeRoux & Reimer (1959) aptly term *systematic errors* will arise. This can be overcome either by sampling so that the differential number of samples from each subdivision reproduces in the samples the gradient in the habitat, or by regarding each part separately and correcting at the end. The question of the estimation of the area or volume of the plant is discussed in Chapter 4.

2.2.1.1.1 Subdivision of plants

The amount of subdivision of the plant that various workers have found necessary varies greatly. On apple, the eggs, larvae and pupae of the tortricid moth, *Archips argyrospilus*, were found for most of the year to be randomly distributed over the tree so that only one level (the lower for ease) needed to be sampled (Paradis & LeRoux, 1962). In contrast, on the same trees and in the same years, the immature stages of two other moths showed marked differences between levels at all seasons (LeRoux & Reimer, 1959). With the spruce budworm, *Choristoneura fumiferana*, Morris (1955) found that there were '…substantial and significant differences from one crown level to another', and that there was a tendency for eggs and larvae to be more abundant at the top levels, but there was no significant difference associated with different sides of the same tree. A similar variation with height was found with the eggs of the larch sawfly, *Pristiphora erichsonii*, although here it was concluded that in view of the cost and mechanical difficulties of stratified sampling at different heights a reasonable index of the population would be obtained by sampling the mid-crown only (Ives, 1955). When studying all of the organisms on aspen, *Populus tremuloides*, Henson (1954) found it is necessary to sample at three different levels of the crown. Variability within a single tree is an important factor for herbivores (Suomela & Ayres, 1994). Roslin *et al.* (2006) report on the performance of the lepidopteron leaf miner, *Tischeria ekebladella*, on oak. Oak, *Quercus robur*, leaves and moth larvae were sampled at five hierarchical scales: (1) across individual oak stands; (2) among trees within stands; (3) among branches within trees; (4) among shoots within branches; and (5) among leaves or samples of several leaves within branches. When modelling leaf mine count data, a Generalised Linear Mixed Model with a log link function and Poisson-distributed errors was used, and it was concluded that the greatest variation in leaf mine abundance was between shoots within branches. This was greater than the variation between branches, within a tree, or between trees within a single stand. Thus, it was concluded that a single oak is '…a mosaic of heterogeneous resources'.

Even with field crops height often needs to be considered: Broadbent (1948) recommended that potato aphids be estimated by picking three leaves – lower, middle and upper – from each plant, and when estimating the population of the European corn borer, *Ostrinia nubilalis*, Hudson & LeRoux (1961) showed that the lower and upper halves of the maize stem needed to be considered separately as the former contained the majority of the larvae. The age of the plant may also affect dispersion (Hirata, 1962; Peng & Brewer, 1994). In contrast, Teulon & Cameron (1995) found few significant differences in the distribution of thrips' eggs and adults at different heights in sugar maple trees.

2.2.1.1.2 *Aspect*

Aspect is sometimes important; for example, in Nova Scotia in the early part of the season the codling moth lays mostly on the south-east of the apple trees, but later this bias disappears (MacLellan, 1962). Aspect has also been found to influence the distribution on citrus of the long-tailed mealy bug, *Pseudococcus adonidum* (Browning, 1959), the eggs of the oak leafroller moth, *Archips* (Ellenberger & Cameron, 1977), and of three species of mite, each of which was most prevalent on a different side (Dean, 1959), but not that of the pine beetle, *Dendroctonus* (Dudley, 1971). Variations in the spatial distribution of similar species in the same habitat, which complicates a sampling programme designed to record both, has also been recorded for two potato aphids by Helson (1958). Some insects are distributed without bias on either side of the mid vein of leaves, so that they may be conveniently subsampled. Similarly, Nelson *et al.* (1957) recorded that, when estimating populations of sheep keds, *Melophagus ovinus*, the fleece of only one side need be sampled.

Occasionally, sampling may be restricted to one part of a plant upon which a large and constant proportion of the population of the animal lives. Wilson (1959) showed that in Minnesota, 84% of the eggs of the spruce budworm, *Choristoneura fumiferana*, are laid on the tips of the branches and if sampling is confined to these, rather than entire branches, the sampling time may be reduced by up to 40%.

Aspect can also be important in aquatic systems. For example, in the northern hemisphere north-facing rocks in the inter-tidal zone offer the coolest, most shady habitat available, and thus tend to favour red algae and its associated fauna (Baker & Crothers, 1987). Conversely, in Caithness, at the northern limit of its range, the inter-tidal barnacle, *Chthamalus* is more abundant on south-facing slopes. In aquatic habitats, orientation with respect to flow and the degree of exposure are important (Baker & Crothers, 1987). Intertidal communities were classified in terms of their exposure to wave action by Ballentine (1961). Plankton sampling may be undertaken at set depths. However, when a general measure of abundance throughout the water column is required, a common practice is to collect what are termed double-oblique plankton net hauls by lowering and raising the net while fishing (Omori & Ikeda, 1984). For example, Henderson *et al.* (1984) used samples collected in this way to estimate growth and mortality of larval herring, *Clupea harengus*, which were unevenly distributed with respect to depth (Henderson, 1987).

2.2.1.2 Hierarchical design and nested analysis of variance

A hierarchical design is one in which, for example, a number of plants are sampled from each of a number of plots from within a number of fields. If a certain number of samples are collected randomly within each strata this is often termed *nested random sampling*, and is analysed by a *nested analysis of variance*. There may be two or three hierarchical levels, rarely more. Examples of such sampling programs are Bancroft & Brindley (1958), Harcourt (1961), Buntin & Pedigo (1981) and Steffey *et al.* (1982). The R function *aov* can handle nested and hierarchical models.

2.2.1.3   The number of samples per subdivision

To determine the optimum number of samples per plant (or part of it) (n), the variance of within-plant samples (s_s^2) must be compared with the variance of the between-plant samples (s_p^2) and set against the cost of sampling within the same plant (c_s) or of moving to another plant and sampling within it (c_p):

$$n_s = \sqrt{\frac{s_s^2}{s_p^2} \cdot \frac{c_p}{c_s}} \tag{2.4}$$

If the interplant variance, s_p^2 is the major source of variance, and unless the cost of moving from plant to plant is very high, n will be of the order of one or less (which means one in practice). Interplant variance has been found to be much greater than within-plant variance in many insect populations, such as the spruce sawfly, *Diprion hercyniae* (Prebble, 1943), the lodgepole needle miner, *Recurvaria starki* (Stark, 1952), the cabbage aphid, *Brevicoryne brassicae* (Church & Strickland, 1954), the spruce budworm, *Choristoneura fumiferana* (Morris, 1955), the diamondback moth, *Plutella maculipennis* (Harcourt, 1961a), the cabbage butterfly, *Pieris rapae* (Harcourt, 1962), the Western pine beetle, *Dendroctonus brevicomis* (Dudley, 1971), the pine chermid, *Pineus pinifoliae* (Ford & Dimond, 1973), and the spider mite, *Panonychus ulmi* (Herbert & Butler, 1973). In most of these examples the within-plant variance was small so that only one sample was taken per plant or per stratum of that plant, although of course when this is done, the within- and between-tree variances cannot be separated. However, with *Eoreuma loftini* on sugar cane the within-plant variance, from stalk to stalk, was found to be high, as was field-to-field variance, the lowest value being within a field (Meagher *et al.*, 1996). On apples the within-tree variance of insect numbers may also be high, particularly at certain seasons, when as many as seven samples may be taken from a single tree (LeRoux & Reimer, 1959; LeRoux, 1961; Paradis & LeRoux, 1962).

Often, a considerable saving in cost without loss of accuracy in the estimation of the population, but with loss of information on the sampling error, may be obtained by taking at random a number of subsamples which are bulked before sorting and counting. This is especially true where the extraction process is complex, as with soil samples; Jepson & Southwood (1958) bulked four random, 76.25 mm (3 inch) row samples of young oat plants and soil to make a single 305 mm (1 ft) row sample that was then washed and the eggs of the frit fly, *Oscinella frit*, extracted. Such a process gave a mean as accurate as that obtained by washing all the 76.25 mm samples separately at greater cost. Paradis & LeRoux (1962) sampled the eggs of a tortricid moth, *Archips argyrospilus*, on apple by bulking 25 cluster samples.

2.2.2   *The sampling unit, its selection, size and shape*

The criteria for the sample unit are broadly those of Morris (1955):

1. It must be such that all units of the universe have an equal chance of selection.
2. It must have stability (or if not, its changes should be easily and continuously measured – as with the number of shoots in a cereal crop).
3. The *proportion* of the population using the sample unit as a habitat must remain constant.
4. The sampling unit must lend itself to conversion to unit areas.
5. The sampling unit must be easily delineated in the field.
6. The sampling unit should be of such a size as to provide a reasonable balance between the variance and the cost.
7. The sampling unit must not be too small in relation to the animal's size as this will increase edge effect errors.
8. The sampling unit for mobile animals should approximate to the average ambit of an individual. This 'condition' is particularly significant in studies on dispersion involving contiguous sampling units, Lloyd (1967) has suggested that a test of the appropriate size would be provided by several series of counts of animals in contiguous quadrats conforming to a Poisson series.

A sampling unit defined in relation to the animal's ambit or territory (e.g. the gallery of a bark-beetle) will give different information from one defined in terms of the habitat (e.g. Cole, 1970). This re-emphasises the need to be very clear as to objectives and hypothesis before commencing a sampling programme.

To compare various sampling units in respect to variance and cost it is generally convenient to keep one or other constant. The same method of sampling must, of course, be used throughout. From preliminary sampling the variances of each of the different units (s_u^2) can be calculated; these should then be computed to a common basis, which is often conveniently the size of the smallest unit. For example, if the smallest unit is 1 m of row, then the variance of 2 m row unit will be divided by 2 and those of 4 m by 4. The costs will similarly be reduced to a common basis (C_u). The relative net cost for the same precision for each unit will then be proportional to:

$$C_u s_u^2 \qquad (2.5)$$

where C_u = cost per unit on a common basis and s_u^2 variance per unit on a common basis. Alternatively, the relative net precision of each will be proportional to the reciprocal of Equation (2.5). The higher this reciprocal, the greater the precision for the same cost.

A full treatment of the methods of selecting the optimum size sampling unit is given in Cochran (1977) and other textbooks, but as population density – and hence variance – is always fluctuating, too much stress should not be placed on a precise determination of optimum size of the sampling unit. An example of the determination of optimal sample size is Zehnder's (1990) study of larval and adult Colorado potato beetle *Leptinotarsa decemlineata*. Zehnder found that while a five-stem sampling unit was most efficient, a three-stem unit would reduce costs with only a small reduction in sampling precision. Pieters (1977) compared the estimates of mean density and precision of arthropod samples collected from cotton using a D-Vac® suction sampler, and found that mean estimates increased with reduced sample size while precision remained unchanged; this led to the conclusion that small sample sizes are better. Similarly, when sampling for benthic macrofauna no advantage has been generally shown for 0.2 m² over 0.1 m² grab samplers (McIntryre *et al.*, 1984). With insects on plants, the nature of the plant usually restricts the possible sizes to, for example, half-leaf, single leaf, or shoot (see p. 147). Because the sample size range initially chosen for examination is often arbitrary, it may not include the optimal sample size.

The shape of the sampling unit when this is of the quadrat type, rather than a biological unit, is theoretically of importance because of the bias introduced by edge effects. These are minimal with circles, maximal with squares and rectangles, and intermediate with hexagons (Seber, 1982), because they are proportional to the ratio of sample unit boundary length to sampling unit area. If the total habitat is to be divided into numbered sampling units (for random number selection), then circular units are impractical because of the gaps, and it is doubtful if the reduction of error from the use of hexagons normally justifies the difficulties of lay-out. Clearly, the larger the sampling unit, proportionally smaller is the boundary edge effect. The size of the organism will also influence this effect: the larger it is in relation to the sample size, the greater the chance of an individual lying across a boundary. This problem has been investigated for subcortical insects where the damage to the edge individuals by the punch, and the curved nature of the sampled substrate pose special problems (Safranyik & Graham, 1971). In general, edge effects can be minimised by a convention (e.g. of the animals crossing the boundaries only those on the top and left-hand boundaries are counted).

2.2.3 *The number of samples*

Precision is measured by the scatter about the mean of the results obtained and is sometimes expressed as the coefficient of variation of the samples. This is different from the accuracy of a method, which measures how close the estimate is to the actual population. The total number of samples depends on the degree of precision required. This may be expressed either in terms of achieving a standard error of the mean within a predetermined magnitude, or as a probability that the estimated mean is within a selected value of the mean (Karandinos, 1976; Ruesink, 1980). For many purposes, a standard error of 5% of the mean is satisfactory. Within a homogeneous habitat the number of samples (n) required is given by:

$$n = \left(\frac{s}{E\bar{x}} \right)^2 \tag{2.6}$$

where s = standard deviation, $\bar{x}$ = mean and E = the predetermined standard error as a decimal of the mean (i.e. for a standard error of ±5%, then $E = 0.05$). This expression compares the standard deviation (s) of the observations with the standard error (E)

acceptable for the contrasts we need to make; it will be noted from this equation that in any given situation the value of the standard error will change with the square root of the number of samples: thus, a large increase in n is necessary to bring about a small improvement in s.

Where sampling is necessary at two levels, for example a number of clusters per tree, the number of units (n) that need to be sampled at the higher level, such as trees (LeRoux & Reimer, 1959; Harcourt, 1961b), is given by:

$$n_t = \frac{\left(\frac{s_s^2}{n_s}\right) + S_p^2}{(\bar{x} + E)^2} \tag{2.7}$$

where n_t = the number of samples within the habitat unit (calculated as above), s^2 = variance within the habitat unit, S_p^2 = variance between the habitat unit (= interplant variance), $\bar{x}$ = mean per sample (calculated from the transformed data and given in this form and E as above.

If the dispersion of the population is well described by either the Poisson or negative binomial distribution, the desired number of samples is given by:

$$N = \frac{1}{(E^2 \bar{x})}$$

and

$$n = \frac{\frac{1}{\bar{x}} + \frac{1}{k}}{E^2} \tag{2.8}$$

where k is the dispersion parameter of the negative binomial (see p. 30).

The second approach uses confidence limits so that the estimate is within a selected distance from the mean with a given probability. The general formula then becomes:

$$n = \left(\frac{ts}{D\bar{x}}\right)^2 \tag{2.9}$$

where t = 'Student's t' of standard statistical tables, and depends on the number of samples and approximates to 2 for more than 10 samples at the 5%, level, and D = the predetermined half-width of the confidence limits for the estimation of the mean expressed as a decimal (e.g. $\pm$ 10% = 0.1). It will be seen that normally this gives a similar estimate to Equation (2.6), provided the values of E and D are adjusted to accord with their meanings. The procedure is, perforce, somewhat approximate, depending on the preliminary estimates of the mean and standard deviation, and the inclusion of t does perhaps give it a slightly bogus air of precision! Additionally, there is normally a biological aspect to consider, as population characteristics change with time, so will the optimal number of samples (e.g. Bryant, 1976; Kapatos et al., 1977).

If the distribution of the animal can be described by a Poisson or negative binomial distribution, or if Taylor's power law applies, then s^2 can be substituted by other expressions (Buntin, 1994). For example, the formula when Taylor's power law applies is:

$$n = t^2 a \bar{x}^{b-2} D^{-2} \tag{2.10}$$

where parameters a and b are those in Equation (2.34). Ward *et al.* (1986) in a study of the cereal aphid, *Sitobion avenae*, on winter wheat demonstrated that the mean and variance conformed to Taylor's power law with the same equation throughout the season and were able to use this relationship to apply Equation (2.10) throughout the season.

Equation (2.10) should be used with caution. Because, as discussed in Section 2.1.2, Taylor's power law may give a poor description of the variance–mean relationship at low densities; Riddle (1989) showed Equation (2.10) to substantially underestimate the required sample size for low-density benthic animals. Shelton & Trumble (1991) concluded that Equation (2.10) should only be used for $b < 2.0$. For convenience, Table 2.2 gives the number of samples for different levels of precision, mean density and clumping as calculated by Elliot & Drake (1981) using Equation (2.10) with $a = 1$. Because the estimated number of samples required is the product of the tabulated value and the

Table 2.2   Number (n) of sampling units per sample required for different levels of precision (95% CL expressed as a percentage of the sample mean) for different values of the mean number caught per sample (x), and for different values of b (Equation (2.4)). The value of a in Equation (2.10) is assumed to be 1. To find n, multiply the value of n found in the table by the actual value of a. (Data from Elliot & Drake, 1981.)

Precision D%	Sample mean (x)								Index of aggregation (b)
	0.5	1	5	10	20	50	100	1000	1
10	800	400	80	40	22	10	6	1	1
20	200	100	22	12	7	3	1	1	1
40	50	26	7	4	1	1	1	1	1
60	24	13	3	1	1	1	1	1	1
80	14	8	1	1	1	1	1	1	1
100	10	6	1	1	1	1	1	1	1
10	566	400	178	126	90	56	40	15	1.5
20	140	100	45	32	24	16	12	5	1.5
40	35	26	13	10	8	6	5	1	1.5
60	17	13	7	5	4	2	1	1	1.5
80	11	8	5	3	2	1	1	1	1.5
100	8	6	2	2	1	1	1	1	1.5
10	400	400	400	400	400	400	400	400	2
20	100	100	100	100	100	100	100	100	2
40	26	26	26	26	26	26	26	26	2
60	13	13	13	13	13	13	13	13	2
80	8	8	8	8	8	8	8	8	2
100	6	6	6	6	6	6	6	6	2
10	283	400	894	1265	1789	2828	4000	12649	2.5
20	70	100	224	316	447	707	1000	3162	2.5
40	19	26	56	79	112	177	250	791	2.5
60	10	13	26	35	50	79	111	351	2.5
80	6	8	14	20	28	44	63	198	2.5
100	4	6	9	13	18	28	40	126	2.5

power law parameter, a, the value of which can itself change with sampling conditions and sample size (Taylor, 1984), the estimated sample number is sensitive to sample size which must be chosen appropriately.

Another type of sampling programme concerns the measurement of the frequency of occurrence of a particular organism or event; examples are the frequency of occurrence of galls on a leaf or of a certain genotype in the population (Oakland, 1950; Cornfield, 1951; Henson, 1954; Cochran, 1977). Before an estimate can be made of the total number of samples required, an approximate value of the probability of occurrence must be obtained. For example, if it is found in a preliminary survey that 25% of the leaves of oak trees bear galls, the probability is 0.25. The number of samples (N) is given by:

$$N = \frac{t^2 p(1 - p)}{D^2} \tag{2.11}$$

where p = the probability of occurrence (i.e. 0.25 in the above example), and D is as defined for Equation (2.10).

If it is found that the leaves (or other units) are distributed differently in the different parts of the habitat, they should be sampled in proportion to the variances. For example, Henson (1954) found from an analysis of variance of the distribution of the leaf-bunches of aspen that the level of the crown from which the leaves had been drawn caused a significant variation and, when this variance was portioned into levels, the values were: lower 112 993, middle 68 012, upper 39 436. Therefore, leaf-bunches were sampled in the ratio of 3:2:1 from these three levels of the crown. Note that if the Taylor's power law constants a and b, are substituted in Equation (2.19) using the equalities $p = a$ and $1 - p = \bar{x}^{-b-2}$ we obtain Equation (2.12).

Worked examples of the calculation of sample number for a variety of sampling programs are given by Greenwood (1996).

2.2.4   The pattern of sampling

Again, it is important to consider the object of the programme carefully. If the aim is to obtain estimates of the mean density for use in, for example, life-tables, then it is desirable to minimise variance. But, if the dispersion (= distribution = pattern) of the animal is of prime interest then there is no virtue in a small variance.

In order to obtain an unbiased estimate of the population, the sampling data should be collected at *random* – that is, so that every sampling unit in the universe has an equal chance of selection. In the simplest form – the *unrestricted random sample* – the samples are selected by the use of random numbers from the whole area (universe) being studied (random number tables are available in many statistical works, or may be generated using a computer; failing these sources the last two figures in the columns of numbers in most telephone books provide a substitute). The position of the sample site is selected on the basis of two random numbers giving the distances along two coordinates, the point of intersection is taken as the centre or a specified corner of the sample. If the size of the sample is large compared with the total area, then the area should be divided in plots which will be numbered and selected using a single random number (e.g. Lloyd, 1967). Such a method eliminates any personal choice by the worker whose bias in selecting sampling sites may lead to large errors (Handford, 1956).

However, a random choice method is not an efficient way to minimise the variance, since the majority of the samples may turn out to come from one area of the field. The method of *stratified random sampling* is therefore to be preferred for most ecological work (Yates & Finney, 1942; Abrahamsen, 1969), where the area is divided into a number of equal-sized subdivisions or strata and one sample is randomly selected from each strata. Alternatively, if the strata are unequal in size, the number of units taken in each part is proportional to the size of the part; this is referred to as self-weighting (Wadley, 1952). Such an approach maximises the accuracy of the estimate of the population, but an exact estimate of sampling error can only be obtained if additional samples are taken from one (or two) strata (Yates & Finney, 1942). The taking of one sample randomly and the other a fixed distance from it has been recommended by Hughes (1962) as a method of mapping aggregations. The fixed distance must be less than the diameter of the aggregations that are assumed circular; the standard error cannot be calculated. However, the method has been found useful for soil and benthic faunas (Gardefors & Orrage, 1968).

When it is apparent that there are systematic variations in the density of the study animal across the study area, stratified sampling should be used. In general, the strata are chosen to minimise within-stratum variance. Individual strata need not form continuous patches within the study area. When the habitat is stratified, biological knowledge can often be used to eliminate strata in which few animals would be found. Such a restricted universe will give a greater level of precision for the calculation of a mean than an unrestricted and completely random sample with a wide variance. Prebble (1943), with a pine sawfly, found that satisfactory estimates of the pupae were only obtained if sampling was limited to the areas around the bases of the trees. The variance of completely random sampling throughout the whole forest was too great, as many areas were included that were unsuitable pupation sites (see also Stark & Dahlsten, 1961).

The other approach is the *systematic sample*, taken at a fixed interval in space (or time). In general, such spatial data cannot be analysed statistically, but Milne (1959) has shown that if the *centric systematic area-sample* is analysed as if it were a random sample, the resulting statistics are '…at least as good, if not rather better', than those obtained from random sampling. The centric systematic sample is the one drawn from the exact centre of each area or stratum, and its theoretical weakness is that it might coincide with some unsuspected systematic distribution pattern. As Milne points out, the biologist should, and probably would, always watch for any systematic pattern, either disclosing itself as the samples are recorded on the sampling plan or apparent from other knowledge. Such a sampling programme may be carried out more quickly than the random method, and so has a distinct advantage from the aspect of cost (see also p. 20). Systematic sampling is often used in marine studies where the primary aim is to map distribution with respect to environmental gradients or suspected sources of pollution. In the absence of environmental gradients, marine benthic or plankton sampling is often undertaken at regularly distributed stations (McIntryre et al., 1984; Omori & Ikeda, 1984). This allows easier contour mapping of animal density, particularly if computers are used.

An example of an unbiased systematic method is given by Anscombe (1948). All the units (e.g. leaves) are counted systematically (e.g. from top to bottom and each stem in turn). Subsequently, every time a certain number (say 50) is reached, that unit is sampled and the numbering is commenced again from 1; only one allocation of a random number is needed and that is the number (say somewhere between 1 and 20) allotted to the first unit.

Biologists often use methods for random sampling that are less precise than the use of random numbers, such as throwing a stick or quadrant or the haphazard selection of sites. Such methods are not strictly random; their most serious objection is that they allow the intrusion of a personal bias, quite frequently marginal areas tend to be under-sampled (nobody wants their quadrat to disappear into a bed of nettles or a small pond!).

It may be worthwhile doing an extensive trial comparing a simple haphazard method with a fully randomised or systematic one, especially if the cost of the latter is high when compared with the former. Spiller (1952) found that scale insects on citrus leaves could be satisfactorily sampled by walking round the tree, clockwise and then anti-clockwise, with the eyes shut and picking leaves haphazardly. For assessing the level of red bollworm eggs, *Diparopsis castanea*, to determine the application of control measures, Tunstall & Matthews (1961) recommended two diagonal traverses across the field, counting the eggs at regular intervals.

Bias may intrude due to causes other than personal selection by the worker. Grains of wheat that contain the older larvae or pupae of the grain weevil, *Sitophilus granarius*, are lighter than uninfested grains. Hence, the most widespread method of sampling is to spread the grains over the bottom of a glass dish and then scoop up samples of a certain volume; as the lighter infected grains tend to be at the top this method can easily overestimate the population of these stages (Howe, 1963). In contrast, for the earlier larval instars, before they have appreciably altered the weight of the grain, such a simple method gives reliable results (Howe, 1963). It was undoubtedly this difference that led to Krause & Pedersen (1960) stressing the need for samples to contain a relatively high proportion of the same stage if good replication was to be obtained. The distribution of parasitised animals may be such that they are more vulnerable to capture. For example, roach infested with *Ligula intestinalis* swim close to the surface (Loot *et al.*, 2002).

When choosing a sampling method it is important to choose a method with a low and stable level of bias, as this cannot be reduced by increasing sample number (Dahlsten *et al.*, 1990).

2.2.5 *The timing of sampling*

In the absence of sampling constraints, the seasonal timing of sampling will be determined by the life cycle (e.g. Morris, 1955). In extensive work when only a single stage is being sampled, it is obviously most important that this operation should coincide with peak numbers (e.g. Edwards, 1962). This can sometimes be determined by phenological considerations (Unterstenhofer, 1957), but the possibility of a control population in an outdoor cage (Harcourt, 1961a) to act as an indicator should be borne in mind. The faster the development rate, the more critical is the timing. With intensive studies

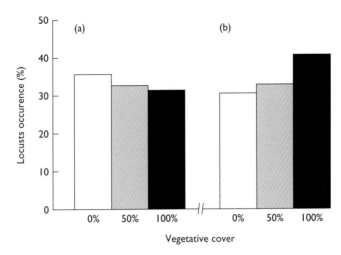

Figure 2.5 The variation in the distribution of adults of the Moroccan locust, *Dociostaurus maroccanus*, at different times of day. (a) Early morning before 09:30 h; (b) morning, after 09:00 h and before noon. The histograms show the relative numbers on bare ground and areas with moderate vegetation. (Data from Dempster, 1957.)

that are designed to provide a life-table, regular sampling will be needed throughout the season. It is not always realised that the time of day at which the samples are taken may also have a considerable effect. The diurnal rhythms of the insects may cause them to move from one part of the habitat to another, as Dempster (1957) found with the Moroccan locust, *Dociostaurus maroccanus* (Fig. 2.5).

Many grassland insects move up and down the vegetation not only in response to weather changes but also at certain times of the day or night (p. 279); during the day a quite large proportion of active insects may be airborne [cf. the observations of Southwood *et al.* (1961) on the numbers of adults of the frit fly, *Oscinella frit*, on oats]. There is a marked periodicity of host-seeking behaviour in many blood-sucking invertebrates (e.g. Camin *et al.*, 1971; Corbet & Smith, 1974). Diurnal and tidal changes in activity and distribution are common in aquatic organisms. Plankton may become concentrated in surface waters at night, and fish that remain hidden by day become active and vulnerable to trapping at night. Sampling methods that depend on the activity of the organism, such as gill nets for fish or pitfall traps for spiders, insects and small mammals, will vary in efficiency as activity levels change. Aquatic insects may emerge at a particular time of day or phase of the lunar cycle (Corbet, 1964). The ecologist may find that some sampling problems can be overcome, or at least additional information gained, if they work at night, dusk or dawn, rather than during conventional working hours.

2.3 Dispersion

The dispersion of a population, the description of the pattern of the distribution or disposition of the animals in space, is of considerable ecological significance. Not only does it affect the sampling programme and the method of analysis of the data, but it may also be used to give a measure of population size (nearest-neighbour and related techniques) and, in its own right, is a description of the condition of the population. Changes in the dispersion pattern should be considered alongside changes in size when interpreting population dynamics. For example, if a mortality factor reduces the clumping of a sessile organism this is an indication that it acts most severely on the highest densities, or if the dispersion of a population becomes more regular then intensification of competition should be suspected (Iwao, 1970a). An understanding of dispersion is vital in the analysis of predator–prey and host–parasite relationships (Crofton, 1971; Murdie & Hassel, 1973; Hassel & May, 1974; Anderson, 1974).

There is a vast literature on the spatial analysis of plants and animals; Perry *et al.* (2002) reviewed the statistical methods for quantifying spatial pattern, and Fortin & Dale (2005) provided a good ecological introduction to the huge field of spatial analysis. Two other important recent works on species distribution modelling are those of Franklin (2010) and Peterson *et al.* (2011). In R, species distribution modelling methods are available in the *dismo* and *raster* packages (http://cran.r-project.org/web/packages/raster/vignettes/Raster.pdf).

2.3.1 *Mathematical distributions that serve as models*

It is necessary to outline some of the mathematical models that have been proposed to describe the distribution of organisms in space; for a fuller treatment, reference can be made to early works such as Anscombe (1950), Wadley (1950), Cassie (1962) and Katti (1966) and to textbooks, for example Bliss & Calhoun (1954) and Patil & Joshi's (1968) dictionary of distributions. In R, the *fitdistr()* function in the MASS package and the *fitdistrplus* function in the fitdistrplus package (see p. 30 for an example application) undertake maximum likelihood fitting of the following univariate distributions; beta, Cauchy, chi-squared, exponential, *f*, gamma, geometric, log-normal, lognormal, logistic, negative binomial, normal, Poisson, *t*, and Weibull.

2.3.1.1 Binomial family

In the binomial distribution the variance is less than the mean, for the Poisson it is equal to the mean and for the negative binomial it is greater than the mean. A full discussion of variance and its calculation will be found in most statistical textbooks (e.g. Bailey, 1959; Zar, 1984), but the estimated variance (s^2) of a distribution may be calculated using:

$$s^2 = \frac{\sum x^2 - \left[\frac{(\sum x)^2}{N}\right]}{N-1} \tag{2.12}$$

where $\sum$ = the sum of x = various values of the number of animal/sample and N = number of samples.

The central place in this family is occupied by the *Poisson series* which describes a *random distribution* (see Fig. 2.6). It is important to realise that this does not mean an even

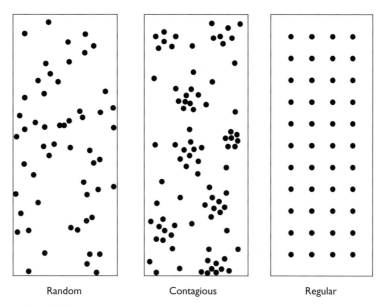

Random   Contagious   Regular

Figure 2.6   Three different types of spatial distribution.

or uniform distribution (Fig. 2.6), but that there is an equal probability of an organism occupying any point in space and that the presence of one individual does not influence the distribution of another. Because, for the Poisson distribution, the variance (S^2) is equal to the mean (x) the probability of finding a certain number (x) of animals is described by one parameter as follows:

$$p_x = e^{-\bar{x}} \frac{\bar{x}^x}{x!} \tag{2.13}$$

where e = base of natural (Napierian) logarithms.

The goodness of fit of a set of data to the Poisson distribution may be tested by a χ^2 on the observed and expected values.

Occasionally, it may be found that the variance is less than the mean; this implies a more regular (or uniform or even) distribution than is described by a Poisson series (Fig. 2.6).

Most commonly in ecological studies the variance will be found to be larger than the mean, that is, the distribution is contagious[2] (Fig. 2.6)-the population is clumped

---

[2]The term 'contagious' is a mathematical one coined in connection with work on epidemiology, and has certain implications that to some extent make its use in ecology inappropriate (Waters & Henson, 1959). An alternative is the term 'over-dispersion', first introduced into ecology by Romell (1930), with its opposite – for more uniform spacing – 'under-dispersion'; unfortunately, however, the use of these terms has been reversed by some ecologists, therefore the terms used here are contagious and regular, which are also those commonly used in plant ecology (Greig-Smith, 1978).

or aggregated. Many contagious populations that have been studied can adequately be expressed by the negative binomial distribution. Examples of its use in insect and marine benthic populations have been provided by Bliss & Owen (1958), Lyons (1964), and Harcourt (1965). This distribution is described by two parameters, the mean and the exponent k, which is a measure of the amount of clumping and is often referred to as the dispersion parameter. The mean and variance of the negative binomial are given in terms of c and k by:

$$\mu = \frac{k}{c},$$ (2.14)

and

$$\sigma^2 = \frac{k}{c} + \frac{k}{c^2}.$$ (2.15)

Defining for convenience $p = c/(c + 1)$ and $q = 1 - p$, the probabilities of 0,1,2,3. ...... events (animals per sample) is given by the terms of the expansion of $p^k \left(1 - q\right)^{-k}$

   For which the first few terms are

$$p^k \left\{ 1, kp, \frac{k(k + 1)}{2!} q^2, k(k + 1) \frac{(k + 2)}{3!} q^3, \ldots\ldots\ldots \right\}.$$ (2.16)

Generally, for natural populations, values of k are in the region of 2; as $k \to \infty$ the distribution tends to the Poisson, whilst as $k \to 0$ the distribution tends to the logarithmic series. The value of k is not a constant for a population, but often increases with the mean (Anscombe, 1949; Bliss & Owen, 1958; Waters & Henson, 1959) (see p. 41). The truncated Poisson may be of value where the distribution is non-random, but the data are too limited to allow the fitting of the negative binomial (Finney & Varley, 1955).

2.3.1.1.1   *Calculating k of the negative binomial*
This is easily accomplished using standard statistical software. In R the *fitdistr()* function in the MASS package can be used. A package with enhanced features to handle both censored and non-censored data is *fitdistrplus* (Delignette-Muller *et al.*, 2013). A simple example application modified from the *fitdistrplus* help file and using data on the *Toxocara cati* abundance in feral cats living on Kerguelen island (Fromont *et al.*, 2001) is given below.

```
#load add-on package
library(fitdistrplus)
#Select a data set of parasite incidence in cats
data(toxocara)
number <-toxocara$number
number
#Fit of a negative binomial distribution to data
fitnb <- fitdist(number,"nbinom")
summary(fitnb)
#Plot the fit
plot(fitnb
```

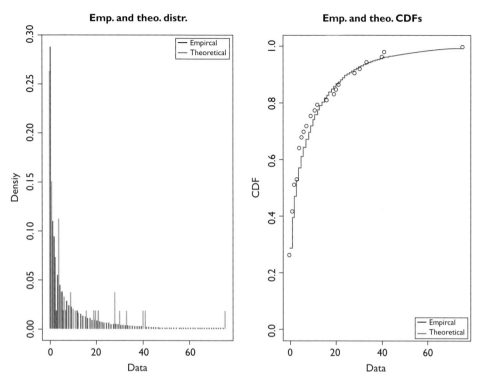

Figure 2.7   Plots of the observed and fitted negative binomial distribution generated by the fit-distrplus package in _R_.

The parameter estimates are termed mu, which is the mean and size, _l_ which is the dispersal parameter (_k_), of the negative binomial distribution respectively. The summary function gives the goodness-of-fit statistics Loglikelihood, AIC and BIC. The default plot of the observed and predicted distributions are shown in Fig. 2.7.

2.3.1.1.2   _Calculating a common_ k

Samples may be taken from various fields or other units and each will have a separate _k_. The comparison of these and the calculation of a common _k_ (if there is one) will be of value in transforming the data for the analysis of variance and for sequential sampling (p. 51). The simplest method is the moment or regression method (Bliss & Owen, 1958; Bliss, 1958). Two statistics are calculated for each unit:

$$x^1 = \bar{x}^2 - \left(\frac{s^2}{N}\right) \qquad (2.17)$$

$$y^1 = s^2 - \bar{x} \qquad (2.18)$$

where $\bar{x}$ = the mean, S^2 = variance and N = number of individual counts on which $\bar{x}$ is based. When y^1 is plotted against x^1 (Fig. 2.8) (including occasional negative or zero values of y^1) the regression line of y^1 on x^1 passes through the origin and has the slope 1.

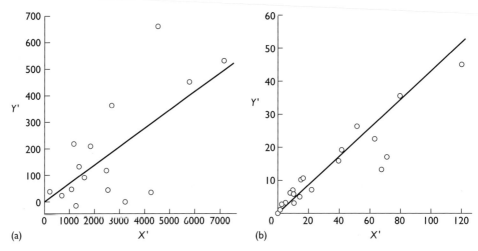

Figure 2.8   Regression estimate of a common *k* for: (a) Colorado beetle, *Leptinotarsa*, in eight plots within each of 16 blocks; (b) wireworms (Coleptera, Elateridae) in 175 sampling units in each of 24 irrigated fields. (Data from Bliss & Owen, 1958.)

An approximate estimate of the common k (k_c) is given by:

$$\frac{1}{k_c} = \frac{\sum y^1}{\sum x^1} \tag{2.19}$$

It may be apparent from the plotting of y^1 on x^1 that a few points lie completely outside the main trend, and therefore although their exclusion will mean that the resultant k is not common to the whole series of samples, it is doubtful if the k, derived by including them would really be meaningful.

A further graphical test of the homogeneity of the samples is obtained by plotting $(1/k) = (y^1/x^1)$ against the mean, $\bar{x}$, for each sub-area or group of samples. If there is neither trend nor clustering (Fig. 2.9), we may regard the fitting of a common k as justified.

Fairly rough estimates of k_c are usually adequate, but in critical cases a weighted estimate of k_c should be obtained and whether or not this lies within the sampling error can be tested by the computation of χ^2. The method is given by Bliss (1958) and Bliss & Owen (1958). The latter authors describe another method of calculating a common k that is especially suitable for field experiments arranged in restricted designs (e.g. randomised blocks).

2.3.1.2   Logarithmic and other contagious models

A number of other mathematical models have been developed to describe various non-random distributions, and several of these may have more than one mode. Anscombe (1950) and Evans (1953) have reviewed these and they include the Thomas, Neyman's types A, B and C and the Polya-Aeppli. The Thomas (1949) type is based on the

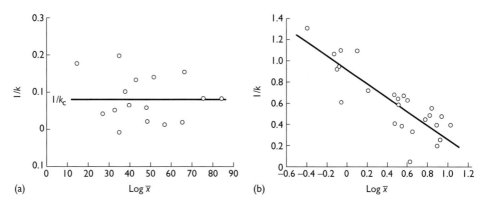

Figure 2.9   The relationship of 1/k to the mean for: (a) Colorado beetle; (b) wireworms, based on the same data as in Fig. 2.8. (Data from Bliss & Owen, 1958.)

assumption of randomly distributed colonies whose individual populations are values plus one, from a Poisson series. Neyman (1939) distributions are similar and are intended to describe conditions found soon after insect larvae hatched from egg batches; the modes are equally spaced. Skellam (1958) has shown that Neyman's type A is particularly applicable where the organisms occur in compact clusters, but that it can be used as an approximation in certain conditions when the clustering is less compact. The Polya-Aeppli distribution describes the situation when an initial wave of simultaneous invaders have settled and produced clusters of offspring; it may have one or two modes.

As mentioned above, the logarithmic model describes situations for which the negative binomial would give a very small value of k. The logarithmic series (Fisher et al., 1943), which is derived from the negative binomial with k tending to zero and with the zero readings neglected, and the *discrete and truncated* (and censored) *log-normal* distributions (Preston, 1948; Grundy, 1952) have been found of most value in the description of the relationship between numbers of species and numbers of individuals and are discussed later (p. 506). Also they have been found to give a reasonable description of the distribution of the individuals of some insects, including the citrus scale insect, *Aonidiella ornatum* (Spiller, 1952) and the eggs of the larch sawfly, *Pristiphora erichsonii* (Ives, 1955).

Working with plankton, with large mean values, Cassie (1962) suggested that the action of a series of environmental factors led to the population being distributed in a succession of Poisson series, the means of the series being themselves distributed according to the log-normal model. He called this the *Poisson log-normal*, and it differs from the negative binomial mainly in the left-hand flank, as ordinarily plotted, where it allows for fewer zero values.

2.3.1.3   Implications of the distribution models and of changes in the type of distribution

Neyman's distributions are based on precise models, and as Upholt & Craig (1940) found, if the biological assumptions underlying the distribution are not fulfilled it will not adequately describe the dispersion of the population. It is perhaps for this reason that Neyman's distributions have been used relatively little by ecologists, most preferring the other distributions that can be derived from a number of different hypotheses. The negative binomial can arise in at least five different ways (Anscombe, 1950; Waters & Henson, 1959):

1. **Inverse binomial sampling**: If a proportion of individuals in a population possess a certain character, the number of samples, in excess of k, that have to be taken to obtain k individuals with this character will have a negative binomial distribution with exponent k.
2. **Heterogeneous Poisson sampling**: If the mean of a Poisson distribution varies randomly between samples, under certain conditions, a negative binomial results (Pielou, 1969). A biological example of this is the observation of Ito et al. (1962) that a series of counts of a gall-wasp on chestnut trees were distributed as a Poisson for each single tree, but when the counts from all trees were combined they were described by a negative binomial. The distribution of oribatid mites in the soil (Berthet & Gerard, 1965) and of a tapeworm in fish (Anderson, 1974) also appear to arise according to this model.
3. **Compounding of Poisson and logarithmic distributions**: If a number of colonies are distributed as a Poisson, but the number of individuals per colony follows a logarithmic distribution, the resulting distribution per unit area (i.e. independent of colonies) will be a negative binomial. Counts of bacteria (Quenouille, 1949) and the dispersion of eggs of the cabbage butterfly, *Pieris rapae* (Kobayashi, 1966), have been shown to satisfy this model.
4. **Constant birth–death–immigration rates**: The former two expressed per individual and the immigration rate per unit of time, will lead to a population whose size will form a negative binomial series.
5. **True contagion**: Where the presence of one individual in a unit increases the chance that another will occur there also.

The logarithmic series (Fisher et al., 1943) can also be derived from several modes of population growth (Kendall, 1948; Shinozaki & Urata, 1953).

It is clear, therefore, that from mathematical considerations alone, it is unsound to attempt to analyse the details of the biological processes involved in generating a distribution from the mathematical model it can be shown to fit or, more often, approximately fit (Waters & Henson, 1959). There is value in expressing the various possible mechanisms in biological terms: this has been done for parasite–host interactions by Crofton (1971) and Anderson (1974): these then provide alternative hypotheses that may be tested by other observations.

The extent of clumping and the changes in it provide important evidence about the population. The uses and interpretation of the parameters will be discussed later (see p. 40), but here changes in the actual type of distribution will be discussed. The main distinction lies between regular, random and contagious distributions (see Fig. 2.6), with the respective implications that the animals compete (or at least tend to keep apart), have no effect on each other, or are aggregated or clumped. One must be careful to define the spatial scale over which the dispersion is described. The behavioural significance of the spacing pattern within the colony area (see below) is different from that between colonies.

The first difficulty is that the sampling method chosen by the experimenter may affect the apparent distribution (Waters & Henson, 1959), as is shown by the example from the work of Ito *et al.* (1962) as quoted above and that of Shibuya & Ouchi (1955), who found that the distribution of the gall-midge, *Asphondylia*, on soya bean was contagious if the plant was taken as the sampling unit, but random if the numbers per pod were considered. The contagion appeared to be due to there being more eggs on those plants with more pods. The effect of the size of the sampling unit on the apparent distribution has also been demonstrated for chafer beetle larvae (Burrage & Gyrisco, 1954) and grasshopper egg pods (Putnam & Shklov, 1956). Careful testing of the size and pattern of sampling, as suggested above, will help to discover whether the contagion is an artefact or a reflection of the mixture of dispersion within 'colony areas' and the pattern of 'colony' distribution. Kennedy & Crawley (1967) aptly described the dispersion of the sycamore aphid, *Drephanosiphon platanoides*, as 'spaced out gregarious'; they pointed out that at close proximity, how close depending on the size of their 'reactive envelopes', animals may repel each other (e.g. caddis larvae; Glass & Boubjerg, 1969), but on a larger scale will often show grouping or contagion. This may arise from patchiness of the habitat (Ferguson *et al.*, 1992), including differential predation (Waters & Henson, 1959) or from the behaviour of the animals themselves, or a combination of both. The behaviour leading to aggregation in the absence of special attractive areas in the habitat may be of two types: inter-individual attraction; or the laying of eggs (or young) in groups (Cole, 1946).

The dispersion of the initial insect invaders of a crop is often random; this randomness may be real or an artefact due to the low density relative to the sample size (see below). The distribution of aphids in a field during the initial phase of infestation is random, becoming contagious as each aphid reproduces (Sylvester & Cox, 1961; Shiyomi & Nakamura, 1964), although there may be differences between species (Kieckhefer, 1975; Badenhauser, 1994). The egg masses of many insects are randomly distributed (e.g. Chiang & Hodson, 1959) and the individual eggs and young larvae are clumped; however, the dispersion of the larvae may become random, or approach it, in later instars. Such changes in distribution with the age of the population have been observed in wireworms (Salt & Hollick, 1946), the rice stem borer, *Chilo simplex* (Kono, 1953), a chafer beetle, *Amphimallon majalis* (Burrage & Gyrisco, 1954), the cabbage butterfly, *Pieris rapae* (Ito *et al.*, 1960), the diamond-back moth, *Plutella maculipennis* (Harcourt, 1961a), white grubs (*Phyllophaga*) (Guppy & Harcourt, 1970), leafhoppers (Nestel & Klein, 1995), and others. It is tempting to infer from such changes that either mortality or emigration or both are density-dependent; however, as Morisita (1962) has shown, this conclusion may not be justified with non-sedentary animals. Such changes could result from an alteration of the size of the area occupied by the colony relative to that of the sample, or from the decrease in population density (see below). In contrast, the tendency to aggregate may be such that density-dependent mortality is masked, as in the pear lacebug, *Stephanitis nashi*, where even after the bugs have been artificially removed so as to produce a random distribution, it returns to non-randomness in a few days (Takeda &

Hukusima, 1961), while in the cabbage root fly, *Erioschia brassicae*, k is similar for all immature stages (Mukerji & Harcourt, 1970).

Changes in the density of an animal often lead to changes, or at least apparent changes, in the distribution. When the population is very sparse the chances of individuals occurring in any sampling unit is so low that their distribution is effectively random. The random distribution of low populations and the contagious distributions of higher ones have often been observed; for example, with wireworms (Finney, 1941), grasshopper egg pods (Davis & Wadley, 1949), a ladybird beetle, *Epilachna 28-maculata* (Iwao, 1956), the cabbage butterfly, *Pieris rapae* (Harcourt, 1961b), the pea aphid, *Acyrthosiphon pisum* (Forsythe & Gyrisco, 1963) and cotton insects (Kuehl & Fye, 1972). However, with the Nantucket pine tip moth, *Rhyacionia frustrana*, as the population density increases still further the distribution tends towards the Poisson again – that is, the k-value becomes high (Waters & Henson, 1959). Populations of several other forest insects (Waters & Henson, 1959) and of the periodical cicadas (Dybas & Davis, 1962) show a tendency to become more random at higher densities (p. 28). Indeed, it can be envisaged that at even higher densities the distribution would pass beyond the Poisson and become regular. Dispersions approaching a regular distribution have been rarely observed: examples include the bivalve, *Tellina* (Holme, 1950), ants (Waloff & Blackith, 1962), aphids (Kennedy & Crawley, 1967) and a coral (Stimson, 1974).

In different habitats the type of distribution may change, as Yoshihara (1953) found with populations of the winkle, *Tectarius granularis* (Mollusca) on rough and smooth rocks, although it is difficult to separate this habitat effect from associated changes in density.

Within the same habitat different species will usually show different dispersion patterns, as has been found with leaf hoppers (Homoptera: Auchenorrhyncha) on rice plants (Kuno, 1963). These differences can arise from several biological causes: one species may aggregate more than another in the same habitat because it disperses less, or because it reproduces more, or because only certain parts of the habitat are suitable for it. However, Abrahamsen & Strand (1970) suggest that, for many soil organisms there may be a relationship between k and $\bar{x}$ that can be expressed by the regression:

$$k = 0.27 + 0.334\sqrt{\bar{x}}. \tag{2.20}$$

2.3.1.4 Comparison of aggregation indices

When a population is sampled four basic bits of information are available:

1. the estimate ($\bar{x}$) of the true mean, m:
2. the estimate (s^2) of the true variance σ^2:
3. the locality of the sample: and
4. the size (unit).

The indices used for the description of animal populations are derived from various arrangements of this information. The simplest approach is the variance/mean ratio:

$s^2/\bar{x}$ which as described above is unity for a Poisson (random) distribution. The Index of Clumping of David & Moore (1954) is simply:

$$I_{DM} = \frac{s^2}{\bar{x}} - 1 \qquad (2.21)$$

which gives a value of zero for a random (non-clumped) population.

A more general approach to the variance–mean relationship is given by Taylor's Power Law:

$$s^2 = a\bar{x}^b. \qquad (2.22)$$

This empirical relationship holds for many species (Taylor *et al.*, 1978; Taylor, 1984; Perry, 1981), and has been found to be of considerable practical value in designing sampling programmes. Such a relationship can be shown to arise as a result of a variety of mechanisms, both stochastic demographic events and density-dependent-related behavioural processes (Anderson *et al.*, 1982; Perry, 1988); and also from its underlying mathematical basis (Tokeshi, 1995).

Pattern analysis in plant communities was developed by Greig-Smith (1978). The method is based on increasing quadrat size; where a peak arises in the plot of a measure of variance against quadrat size this is considered as the 'clump area'. This method can only be applied to contiguous quadrats; Morisita (1954, 1959, 1962, 1964) developed an index I_δ, that provided similar information but could be used on scattered quadrats. Zahl (1974) has suggested that Greig-Smith's approach can be made more robust by the use of the S-method of Scheffé.

With mobile animals, the degree of crowding experienced by an individual interested Lloyd (1967), who devised an Index of Mean Crowding ($\dot{m}$):

$$\dot{m} = \mu + \left(\frac{\sigma^2}{m} - 1 \right), \qquad (2.23)$$

where μ and σ^2 are the true mean and variance, respectively. Thus, $\dot{m}$ is the amount by which the variance mean ratio exceeds unity added to the mean. If the distribution conforms to a Poisson (random), then $\frac{\sigma^2}{\bar{x}} = 1$ and $\dot{m} = \mu$. With estimates based on samples the Index of Mean Crowding, $\dot{x}$ is given by:

$$\dot{x} \approx \bar{x} + \left(\frac{s^2}{\bar{x}} - 1 \right). \qquad (2.24)$$

For contagious dispersions described by the negative binomial [Equation (2.14)]:

$$s^2 = \bar{x} + \frac{\bar{x}^2}{k} \qquad (2.25)$$

so Equation (2.24) may be written as:

$$\dot{x} = \bar{x} + \frac{\bar{x}}{k} \qquad (2.26)$$

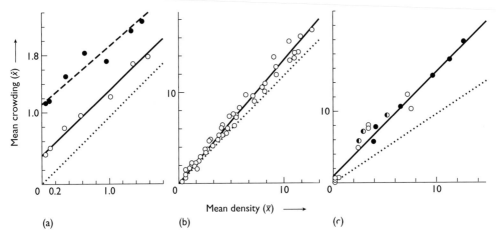

Figure 2.10   Iwao's patchiness regression. (a) Potato beetles, *Epilachna*, number of egg masses per plant (o) and individual eggs (•). (b) Spruce budworm, *Choristoneura*, numbers of larvae/twig. (c) Pea aphid, *Macrosiphum*, total aphids (excluding 1ˢᵗ instar) per shoot. (Data from Iwao, 1968.)

On rearranging Equation (2.26)

$$\frac{\dot{x}}{\bar{x}} = 1 + \frac{1}{k} \qquad (2.27)$$

It can be seen that the reciprocal of k is the proportion by which mean crowding exceeds mean density. Lloyd (1967) termed $\frac{\dot{x}}{\bar{x}}$ the 'patchiness'.

A clarification and unification of these various approaches was achieved by Iwao (1968, 1970a, 1970b, 1970c, 1972), who initially demonstrated that the relationship of mean crowding $\dot{x}$ to mean density $\bar{x}$ for a species could be expressed over a range of densities by a linear regression:

$$\dot{x} = \alpha + \beta\bar{x} \qquad (2.28)$$

(Iwao used the expression $\dot{m}$-m method, but in reality one is normally working with estimates ($\dot{x}$ and $\bar{x}$), rather than the true mean and its derivatives. α and β are characteristic for the species and a particular habitat (Fig. 2.10), and their interpretation will be discussed later. Subsequently, Iowa developed the ρ-index (Iwao, 1972):

$$\rho_1 = \frac{\dot{x}_i - \dot{x}_{i-1}}{\bar{x}_i - \bar{x}_{i-1}} \qquad (2.29)$$

where $\dot{x}_i$, and $\bar{x}_i$ are the mean crowding index and mean density for the ith-sized quadrat sample. Thus, with a series of different-sized quadrats, values of ρ may be plotted for the second and subsequent quadrat sizes (Fig. 2.11).

Changes in the value of the ρ-index will indicate clump or territory size in a manner analogous, but theoretically preferable, to Greig-Smith's method.

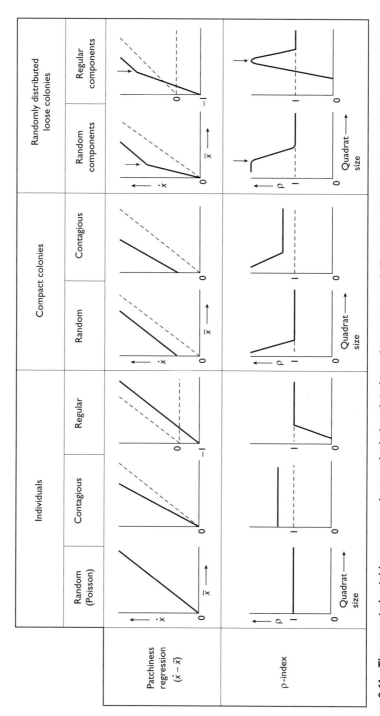

Figure 2.11 The expected patchiness regressions and ρ-index plots for various patterns of dispersion. (Data from Iwao, 1972.)

$\dot{x}$ expresses 'crowding', Lloyd derives this by estimating the mean number minus one present in the area. Working with mites on leaves, Tanigoshi *et al.* (1975) restored the one to give an index they have called 'mean colony size' ($\dot{C}$)

$$\dot{C} = \dot{x} + 1. \tag{2.30}$$

Lloyd (1967) termed this 'mean demand' – the mean number of individuals per quadrat per individual – and used it as an expression of trophic demand. The concept of mean colony size is useful, but only likely to be meaningful as formulated when the sampling unit, as with spider mites on a leaf, contains an entire colony.

Thus, all the above-described indices are related to each other and to the simple concept of the comparison of the estimate of the variance (s^2) with that for the mean $\bar{x}$:

$$1 + \frac{s^2 - \bar{x}}{\bar{x}^2} = 1 + \frac{1}{k} = \frac{\dot{x}}{\bar{x}} = \frac{\dot{C} - 1}{\bar{x}} \approx I_\delta.$$

Bringing in the third piece of information, quadrat size, the ρ-index is approximately the same as plotting the changes in I_δ with quadrat size; it is the proportion by which mean crowding changes against mean density with increased sample area. Essentially, in Greig-Smith's method variance is plotted against sample area.

Perry & Hewitt (1991) criticised the above-described methods because they lacked a direct relationship between their components and the movements of the individual animals. Further, they do not use the available spatial information. A methodology called SADIE – Spatial Analysis by Distance Indices – was developed (Perry & Hewitt, 1991; Aliston, 1994; Perry, 1994, 1995) which involves the use of computer simulation. The approach is to compare the observed spatial arrangement with arrangements derived from it, which are as crowded, regular or random as possible. Perry (1994) gives examples where the method has been applied to a variety of animals and plants over a range of spatial scales; the software, which can be obtained at nominal cost (Perry, 1995), was recently applied to the analysis of earthworm distribution (Richard *et al.*, 2012).

2.3.2 *Biological interpretation of dispersion parameters*

2.3.2.1 Index of dispersion: the departure of the distribution from randomness

If the dispersion follows the Poisson distribution (p. 29) the mean will equal the variance and therefore departures of the coefficient of dispersion (or variation) $\frac{s^2}{\bar{x}}$ from unity will be a measure of the departure from the Poisson (from randomness). This is tested by calculating the index:

$$I_D = \frac{s^2 (n - 1)}{\bar{x}} \tag{2.31}$$

where n = the number of samples and I_D is approximately distributed as χ^2 with $(n - 1)$ degrees of freedom, so that if the distribution is in fact Poisson, the value of I_D will *not* lie outside the limits (taken as 0.95 and 0.05) of χ^2 for $(n - 1)$ as given in standard tables[3]. The index is therefore used as a test criterion for the null hypothesis that the pattern is random. The coefficient of variation or mean/variance ratio is a descriptive sample statistic that will approach zero for regularly distributed organisms, whilst a large value implies aggregation.

As Naylor (1959) has pointed out, the indices may be added if they are from the same-sized samples. Early examples of the use of this index are in Salt & Hollick's (1946) studies on wireworms, Naylor's (1959) on *Tribolium*, Nielsen's (1963) on *Culicoides*, and Milne's (1964) on the chafer beetle, *Phyllopertha*. In contagious populations the value of this index is influenced by the density and size of the sampling unit (i.e. by the size of the mean).

2.3.2.2 'K' of the negative binomial: an index of aggregation in the population

If the negative binomial (p. 30) can be fitted to the data the value of k gives a measure of dispersion; the smaller the value of k the greater the extent of aggregation, whereas a large value (over about 8) indicates that the distribution is approaching a Poisson, that is, virtually random. This may be appreciated from the relationship of k to the coefficient of variation, cv:

$$cv = \frac{s}{\bar{x}} = \left[\frac{1}{k} + \frac{1}{\bar{x}}\right]^{\frac{1}{2}} \tag{2.32}$$

Clearly, the smaller k, the larger the cv. The value of k is often influenced by the size of the sampling unit (Cole, 1946; Morris & Reeks, 1954; Waters & Henson, 1959; Harcourt, 1961a; Ho, 1993), and therefore comparisons can only be made using the same-sized unit. But, within this restriction it does provide a useful measure of the degree of aggregation of the particular population, varying with the habitat and the developmental stage (Hairston, 1959; Waters & Henson, 1959; Harcourt, 1961a; Dybas & Davis, 1962). Examples of these variations are given in Table 2.3. As Waters and Henson (1959) have pointed out, the degree of aggregation of a population, which k expresses, could well affect the influence of predators and parasites.

The aggregation recognised by the negative binomial may be due either to active aggregation by the insects or to some heterogeneity of the environment at large (microclimate, soil, plant, natural enemies; see p. 390). Dr R. E. Blackith has suggested that if the mean size of a clump is calculated using Arbous & Kerrich's (1951) formula, and is found to be *less than 2*, then the 'aggregation' would seem to be due to some environmental effect and not to an active process. Sedentary animals, a large number of which are killed between settling and sampling, could be an exception in having a behavioural cause for an aggregation size of less than 2. Aggregations of 2 or more insects could be caused by either factor. The mean number of individuals in the aggregation is

---

[3]For values of larger than those given in tables, ξ^2 may be calculated (see p. 86).

Table 2.3 The variation *k* of the negative binomial with some factors. a, b and c with sampling unit; d with developmental stage and change in density. Block *a*: Eriophyes leaf galls on *Populus tremuloides*. Block *b*: Nantucket pine tip moth, *Rhyacionia frustrana* data from Waters & Henson, 1959. Blocks *c* and *d*: cabbage white butterfly, *Pieris rapae* data from Harcourt, 1961.

	Sampling unit	Mean number per unit	K
a.	Single leaf	0.21 + 0.01	0.0614 + 0.0013
	1 leaf-bunch	0.88 + 0.04	0.8830 + 0.0024
	2 leaf-bunches	1.65 + 0.08	0.1150 + 0.0035
	5 leaf-bunches	3.78 + 0.26	0.1740 + 0.0064
	Branch	14.23 + 1.68	0.2000 + 0.0119
b.	Tip of shoot	0.49 + 0.03	0.449 + 0.021
	Branch whorl	2.36 + 0.36	0.214 + 0.035
	Tree	13.13 + 4.21	0.253 + 0.074
	Sampling unit	*k* for 1st instar larvae	*k* for 4th instar larvae
c.	Quadrant of cabbage	1.38	1.96
	Half cabbage	2.28	4.24
	Whole cabbage	2.32	4.28
	Stage	Mean density	*k*
d.	Egg	9.5	3.1
	1st instar	5.6	2.8
	2nd instar	4.4	2.8
	3rd instar	4.0	4.6
	4th instar	3.6	5.1
	5th instar	2.6	7.8
	Pupa	1.7	2.3

calculated by:

$$\lambda = \frac{\bar{x}}{2k}v \qquad (2.33)$$

where v is a function with a χ^2 distribution with $2k$ degrees of freedom and λ = the number of individuals in the aggregation for the probability level allocated to v. To find the mean size of the 'aggregate', the value at the 0.5 probability level is used. $2k$ degrees of freedom will usually be fractional, but an adequate χ^2 can be calculated by reference to graphs or by proportionality.

To take two examples from Table 2.3d, the eggs and pupae of *Pieris rapae* -

Eggs: $\lambda = (9.5/3.1 \times 2) \times 5.55 = 8.5$
Pupae $\lambda = (1.7/2.3 \times 2) \times 3.54 = 1.3$

From this it can be concluded that the 'clumping' of eggs could be due to a behavioural cause – the females tending to lay a number of eggs in close proximity or because of environmental heterogeneity – with only certain areas being suitable for oviposition. 'Clumping' of the pupae is due to environmental causes, however.

The values of k and the mean for $\lambda = 2$ are plotted in Fig. 2.12, from which it can be ascertained for a particular population whose k and mean are known whether the mean

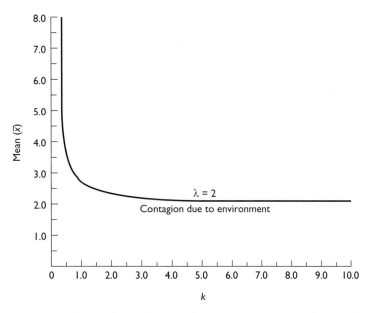

Figure 2.12   The cause of contagion in the data: the plot of mean aggregation (λ) of two individuals for various values of the mean and k of the negative binomial. Populations whose value for the mean plotted against k lies below the line may be considered to exhibit contagion because of some environmental factor; in those populations whose value lies above the line, contagion could be due either to an active behavioural process or to the environment.

'aggregation' size is above or below 2. The negative binomial has been found useful in describing the pattern of predator attacks on patches of prey (Hassell, 1978). May (1978) has shown that the k of the attack pattern is the reciprocal of the coefficient of dispersion (or variation) ($\frac{s^2}{\bar{x}}$) of the numbers of predators per patch. It has been argued that it is a useful model for the dynamics of animal aggregation and density dependence (Taylor *et al.*, 1979).

2.3.2.3   'b' of Taylor's power law: a coefficient of spatial variability for the species

Taylor's power law, which is widely used in the study of spatial distributions, describes the relationship between the sample variance (s^2) and sample mean ($\bar{x}$) of the number of individuals per quadrat. If a series of locations are each divided into a number of equal quadrats, and the number of organisms in each quadrat counted, it is observed then the variance is not independent of the mean. A plot of log s^2 against log $\bar{x}$ shows an approximately linear increase (see Fig. 2.13) which was shown by Taylor (1961, 1965) and by Taylor *et al.* (1976, 1978) to fit a power law:

$$s^2 = a\bar{x}^b \tag{2.34}$$

where a and b are constants; a is largely a sampling factor, while b is sometimes used as an index of aggregation characteristic of the species.

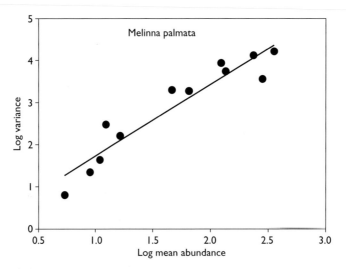

Figure 2.13 **The relationship between log variance and log mean for the polycheate *Melinna palmata* sampled at 12 stations within Milford Haven. At each station, five replicate grab samples were taken. A straight line was fitted by linear regression, giving the equation $s^2 = 0.045\bar{x}^{1.69}$.**

This relationship holds for a variety of distributions from regular through random to highly aggregated. The same relationship between the mean $\bar{x}$ and the variance s^2 was discovered independently by Fracker & Brischle (1944) for quadrat counts of the currant, *Ribes*, and by Hayman & Lowe (1961) for counts of the cabbage aphid, *Brevicoryne brassicae*. The applicability of Taylor's power law is discussed by Routledge & Swartz (1991). They held that it had no unique advantage over a quadratic of the form:

$$s^2 = a_1\bar{x} + a_2\bar{x}^2 \tag{2.35}$$

and noted that Taylor's Power Law was particularly poor in predicting the variance for means close to zero. Clark & Perry (1994) concluded that Taylor's power law should probably not be applied in situations where, for the majority of the sites, fewer than 15 samples had been collected because of bias in parameter estimation.

The extent to which the parameter b of Taylor's power law (p. 43) is a constant for a species has been a matter of controversy. It may be concluded that it provides a description of the spatial structure of a species in a particular environment. There are many examples of it being relatively constant for a species (Taylor, 1961; Taylor *et al.*, 1978; Banerjee, 1976; Perry, 1988; Sawyer, 1989). However, as Riley *et al.* (1992) found when working with a weevil on peppers, although b did not vary significantly over time or with changes in population density, it was sensitive to the variety of the host plant and the type of sample. Bliss (1971) has pointed out that special techniques should be applied for the calculation of this regression as the independent variate is not error-free. The same criticisms apply even more forcefully to Iwao's patchiness regression (see below), but in practice the more accurate approach is seldom required. As shown by Iwao (1970c), Taylor's power law will not fit certain theoretical distributions, although the biological frequency (and significance) of these is uncertain. It remains a simple and useful description of species distribution (Taylor, 1970; Usher, 1971; Kieckhefer, 1975; Egwuatu & Taylor, 1975; Croft *et al.*, 1976; Ng *et al.*, 1983; Pena & Baranowski, 1990; Steiner, 1990; Pringle & Giliomee, 1992; Peng & Brewer, 1994). High values of *b* show strong contagion; however, if the regression line crosses the Poisson line (slope $b = 1$) this shows a change to random distribution at lower densities. Hence, unless $a = 1$, the value of *b* in itself cannot be taken as a test of randomness, as shown by George (1974) for zooplankton. The constant *a* is largely a scaling factor related to sample size, but, if the sample size is larger than the colony size but refers to a biologically meaningful entity (e.g. a leaf for spider mites), then *a* is also related to colony size. It would seem that the amount of scatter of the points around the line (see Fig. 2.4) which can be approximately estimated as the variance around the appropriate regression line is a measure of the effect of habitat variation on the extent of aggregation.

2.3.2.4 Lloyd's mean crowding and Iwao's patchiness regression indices for the population and species

The index of mean crowding $\dot{x}$ (Lloyd, 1967) is given approximately in Equation (2.24). If the underlying distribution is known to fit the negative binomial, then Equation (2.26) is applicable. This is an efficient estimate provided that the number of samples (N) is large. When using Equation (2.26), the most accurate method for estimating k should be used. A general formula is:

$$\dot{x} = \bar{x} + \left(\frac{s^2}{\bar{x}} - 1 \right) \left(1 + \frac{s^2}{N\bar{x}^2} \right) \tag{2.36}$$

the bracket term correcting for sampling bias and k being calculated using the R function fitdistrplus (p. 30). The appropriate large-sample variance of the estimate is:

$$\mathrm{var}\,(\dot{x}) \approx \left[\frac{1}{N} \left(\frac{\dot{x}}{\bar{x}} \right) \left(\frac{s^2}{\bar{x}} \right) \left(\dot{x} + \frac{2s^2}{\bar{x}} \right) \right] \tag{2.37}$$

Lloyd (1967) discusses the problems of applying these to rare species and the statistical requirement, which is difficult for the biologist to meet, that one is a priori assured of encountering the species. The index $\dot{x}$ is the mean number of other individuals per sample unit per individual and is thus an expression of the intensity of interaction between individuals. In an extensive comparison of all the indices described hitherto applied to samples of *Heliothis* (cornborer) eggs, Terry *et al.* (1989) found that $\dot{x}$ agreed least with the other indices, but it has been found useful with adult pear thrips, *Psylla pyricola*, which do not form colonies but have a contagious distribution (Burts & Brunner, 1981).

As will be seen from the first formula for the calculation of $\dot{x}$, when the population is randomly distributed the mean crowding index equals the mean density. Many examples have been given in the previous section of increased contagion with increased density. Iwao (1968) showed that $\dot{x}$ is related to the mean ($\bar{x}$) over a series of densities:

$$\dot{x} = \alpha + \beta\bar{x} \tag{2.38}$$

This is based on the changing relationship of Lloyd's Index of Patchiness, $\frac{\dot{x}}{\bar{x}}$, and may be termed Iwao's Patchiness Regression. The constant α indicates the tendency to crowding (+ve) or repulsion (-ve) and Iwao (1970b) termed it the 'Index of Basic Contagion'. It is a property of the species, and the mathematical basis suggests that it may warrant a biological interpretation (Tokeshi, 1995).

The coefficient β is related to the pattern in which the organism utilises its habitat, and Iwao named it the 'Density contagiousness coefficient'. The theoretical effects of various underlying mathematical distributions on α and β and their application to many sets of field data are described by Iwao (1968, 1970a, 1970b, 1972) and by Iwao & Kuno (1971), and provides a measure of the numbers in the minimum clump in the sample unit; for example, 38 for the individual eggs of *Epilachna* (Fig. 2.10a) or 21.5 for *Macrosiphon* (excluding 1st instar) (Fig. 2.10c). A negative value of *a* shows there is a tendency for the animals to repel each other although, of course, there will not be actual points in this region (see Fig. 3 of Iwao & Kuno, 1971). The coefficient β expresses the extent to which the colonies (as defined by β) are contagious at higher densities, for example in *Choristoneura* (Fig. 2.10b) or *Macrosiphum* (Fig. 2.10c). Although a single 'patchiness regression' can often be fitted for data from different areas, significantly different values of β could theoretically arise where the pattern of spatial heterogeneity differed between the two areas. Many comparative applications of Taylor's power law and Iwao's patchiness regression to insect sampling data have found the former to provide a more consistent fit (e.g. Pena & Duncan, 1992). Strong density-dependent mortality would influence the value of β, whereas density-independent mortality should alter the value of α but leave β unaffected (Iwao, 1970).

2.3.2.5 Iwao's ρ index: a measure of colony area

The area occupied by a colony may be determined by the use of Iwao's, (1972) index given in Equation (2.29).

Clearly, this approach is only possible with animals living in habitats where sample size can be regularly increased. The index shows a sharp change in magnitude at the quadrat size corresponding to colony area (Fig. 2.11).

2.3.2.6 Breder's equations: a measure of the cohesion of aggregations

Working on fish schools, Breder (1954) found that the distance between individuals varied only slightly, and such uniform or regular distribution is indeed characteristic of the distribution of 'crowded' individuals (p. 29). With such schools or social swarms, when the individuals are their normal distance (d) apart the attraction (a) between individuals can be said to equal the repulsion (r), so that c, a measure of the cohesion in the equation $c = (a - r)/d^2$, is zero. Negative values of c will show 'overcrowding', and there

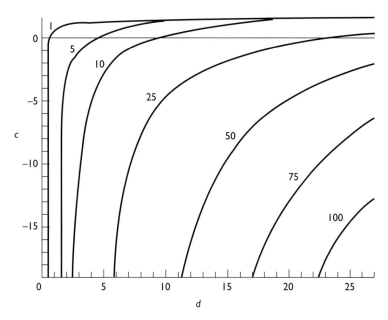

Figure 2.14 Breder's curves for the cohesion of aggregations. The normal distance between individuals as a decimal fraction of the animal total length gives the value of d for $c = 0$, the appropriate curve for r ($= \sqrt{r_1 \times r_2}$) may be read off. (Adapted from Breder, 1954.)

will be a tendency for the individuals to move apart; positive values that normal aggregation has not been obtained – these will approach +1 asymptotically if the attractive force is taken as 1. If r_1 and r_2 are the repulsive actions of two animals on each other, then the equation becomes:

$$c = 1 - r_1 r_2 / d^2 \tag{2.39}$$

or

$$r_1 r_2 = d_s^2 \tag{2.40}$$

where d_s = the mean normal distance between individuals in the aggregation because, by definition, $c = 0$ under these conditions.

The mean distance between individuals in a moving school may be measured from photographs, and so the value for r is found. It is then possible to plot c against d for a series of values (see Fig. 2.12). The angle at which the curves cross the 0 value is of interest: the steeper the line the tighter the aggregation, angles of intersection of 40–50° being characteristic of fish that keep apart by a distance considerably less than their own length; smaller angles indicate looser aggregations, both spatially and probably, Breder suggests, in their tendency to split up. The values of r given in Fig. 2.14 represents $\sqrt{r_1 r_2}$.

Some insects, such as locusts and army worms, form swarms many of the characters of which would seem to parallel fish schools, and therefore the equations developed by Breder might well be used for comparisons with these insects. From Breder's equations it would seem that the smaller the mean distance, in terms of their total lengths, between individuals the more stable the aggregation. Extending this approach, various models have been developed for simulating the dynamics of fish schools and these may be used to analyse behaviour (Gotceitas & Colgan, 1991; Niwa, 1994) and the movement of locusts and other swarming organisms which have both attraction and repulsion interactions (e.g. Romanczuk & Schimansky-Geier, 2012).

2.3.2.7 Deevey's coefficient of crowding

This measure was used by Deevey (1947) to express the extent of crowding in barnacle populations, where the individuals actually impinge on one another as they grow and is given by:

$$C_c = 2\pi r^2 N^2 \tag{2.41}$$

where r = the radius of the fully grown animal and N = the density of the animal per unit area. The units of r and N must of course be comparable (e.g. cm and cm^2) and then C_c will define the number of contacts per unit area. This coefficient could be used with any sedentary, approximately circular animal or with plant galls on leaves; when a realistic value can be given to r, for example the radius of a circular home range (p. 410), it could also be used with mobile forms.

2.3.3 Nearest-neighbour and related techniques: measures of population size or of the departure from randomness of the distribution

There are basically two separate approaches: one may either select an individual at random and measure the distance between it and its nearest neighbour (true nearest-neighbour techniques), or one may select a point and measure the distance between this point and the nearest or n^{th} nearest individuals (sometimes referred to as closest individual techniques). From each of these, conclusions may either be drawn about the departure of the distribution from random, if the population density is known from some other method or, alternatively, if one can assume the distribution to be random, its density can be estimated (Morisita, 1954). Clearly, tests of randomness must be carried out. If the population has been sampled with quadrats and the mean is greater than one, the Poisson Index of Dispersion (see above) is a satisfactory test. Other test procedures are given by Seber (1982).

These techniques are most easily used with stationary, discrete, easily mapped organisms such as trees, and have been used extensively by botanists (Greig-Smith, 1978). With animals, their mobility and the risk that one will fail to find the nearest neighbour limit their application (Turner, 1960). However, such methods have been applied to studies on grasshoppers (Blackith, 1958) and frogs (Turner, 1960) and, with more success, to studies of fairly conspicuous and relatively stationary animals (e.g. snails; Keuls et al., 1963) or of well-marked colonies (e.g. ant mounds; Waloff & Blackith, 1962; Blackith et al., 1963). Nearest-neighbour and related techniques could also be used with other relatively sedentary animals such as barnacles, limpets, scale insects, tube-building animals and gall formers (especially on leaves). The use of photography in mapping the natural distributions of snails has been described by Heywood & Edwards (1961). This can be a valuable aid to the application of these techniques, to not only fairly sedentary organisms, but also where the substratum is suitable, to far more mobile forms. The various dolichopodid and other flies on mud are an example of a community that might be studied in this way. There are, furthermore, a few other techniques, such as the 'squashing' or 'imprinting' method for estimating mite numbers on leaves (p. 153), or the deadheart method for estimating stem borers (p. 343) that do, or could easily, provide maps of the distribution of the individuals to which nearest-neighbour or closest individual methods could be applied. Henson (1961) has used a combination of photography and nearest-neighbour methods to study the aggregations of a scolytid beetle in the laboratory (see also p. 148–149).

2.3.3.1   Nearest-neighbour methods

In this method a point is selected at random, and then one searches around in tight concentric rings until an animal is found. The searching is then continued until its nearest neighbour is found and the distance between these two animals is measured. Strictly speaking, this is not a random method (Pielou, 1969); a truly random method would involve numbering all the animals in advance. The extent to which the method is robust in respect of this violation is not clear, and thus in general the closest individual method is to be preferred (Seber, 1982). There are two expressions; the simplest is due to Clark & Evans (1954):

$$m = \frac{1}{4\bar{r}^2} \tag{2.42}$$

where m = density per unit area, and $\bar{r}$ = mean distance between nearest neighbours. As has been pointed out by Blackith (1958), this formula is based on the mean r squared, whereas Craig's (1953) formula is based more correctly on the sum of r squared; the latter formula is, however, less easy to use and describes the density in a quarter segment. Turner (1960) has concluded that for randomly distributed populations Clark & Evans's formula is fairly satisfactory.

The importance of a number of population estimates by different methods is emphasised elsewhere (p. 4). In fairly homogeneous habitats (e.g. field crops), unless the species obviously aggregates, these methods would provide a useful 'order of magnitude check' on estimates obtained by other methods such as marking and recapture. Like mark and recapture' they are independent of sample size. If, however, the density has been measured by one of these other methods, the departure of the distribution from randomness will be given by the value of the numerator in the expressions (Clark & Evans, 1954; Waloff & Blackith, 1962):

Uniform, hexagonal spacing competition:

$$\bar{x}^2 = \frac{1.154}{m} \tag{2.43}$$

Random:

$$\bar{x}^2 = \frac{0.25}{m} \tag{2.44}$$

where $\bar{x}$ = mean distance between nearest neighbours, and m = number of individuals per unit area.

Smaller values of the numerator will indicate aggregation or clumping due to environmental or behavioural factors. These estimates should be based on large samples and, of course, the density and distance units of measurement must be the same.

The use of the second, third,..... n^{th} nearest neighbours, as discussed by Morisita (1954) and Thompson (1956), not only enables more accurate density determinations to be made but also the dispersion pattern to be detected over a larger area. However, as Waloff & Blackith (1962) pointed out, competition effects detectable between nearest neighbours may be masked at greater distances because of heterogeneity of the habitat. Thompson discusses the statistic:

$$2\pi D \sum r_n^2 \tag{2.45}$$

where D = density and r = distance, is distributed as a χ^2 with $2N_n$ degrees of freedom (N = number of observations). This value may be calculated and compared with the expected value under the hypothesis of randomness (the χ^2 value for $2N$, is found from tables). A probability of χ^2 greater than 0.95 indicates significant aggregation, whilst a probability of less than 0.05 indicates significant regularity.

A simpler but – according to Thompson – less reliable comparison is given by the proportionality constants in the equation:

$$\bar{r}_n^2 = \frac{p}{D} \tag{2.46}$$

where p = proportionality constant and other symbols are as above. For a random distribution for $n = 1$ (i.e. the nearest neighbour) as indicated above $p = 0.25$; for $n = 2$, $p = 0.87$; for $n = 3$, $p = 0.97$; for $n = 4$, $p = 1.05$.

2.3.3.2 Closest individual or distance method

In this method a point is selected at random and the distance between this and the nth nearest individual is measured. The main theoretical difficulty is the possibility of the random point being closer to the boundary than to an individual. Practically, this method is more time-consuming than the nearest neighbour method, because of the need to find the third or more distant individual. However, in spite of these problems, because of the difficulties of true random selection of an animal, as referred to above (difficulties that apply less to botanical studies), this method is to be preferred.

The method has also been derived, apparently independently, by Keuls *et al.* (1963) and used for population estimations of the snail *Limnaea* (= *Galba*) *truncatula*. It depends on the population being randomly distributed, although if the discontinuities are small, the errors that are made will tend to cancel each other out. The basic equation is:

$$m = \frac{r-1}{\pi} \cdot \frac{1}{a_n^2} \tag{2.47}$$

where r = the rank of the individual in distance from the randomly selected point, for example for the nearest neighbour $n = 1$, for the second nearest 2; a = the distance between the randomly selected point and the individual and other symbols are as above. This basic equation is used in the ordered distance estimators in Chapter 9 to estimate density.

Various indices of non-randomness have been derived from Equation (2.47) (for a fuller discussion see Pielou, 1969 and Seber, 1982). Two such indices are:

1. Pielou's index:

$$I_a = \bar{a}_i^2 D\pi \tag{2.48}$$

   which requires a knowledge of the density per unit area (D) as well as the mean distance from a random point to the closest individual (a_i), $I_a = 1$ for random distributions, > 1 for contagious, and < 1 for regular. This index has been used by Underwood (1976) with intertidal molluscs.
2. Eberhardt (1967) index:

$$I_E = \frac{\sum \left(\bar{a}_i^2 \right)}{\left(\sum \left(\bar{a}_i \right) \right)^2} \tag{2.49}$$

   This has the advantage that density need not be known, but the departure from randomness cannot easily be determined. $I_E = 1.27$ for a random distribution, being greater or less for contagious and regular, respectively.

2.4 Sequential sampling

In this type of sampling, the total number of samples taken is variable and depends on whether or not the results obtained so far give a definite answer to the question posed about the frequency of occurrence of an event (i.e. the abundance of an insect). It is of particular value for the assessment of pest density in relation to control measures, when these are applied only if the pest density has reached a certain level.

2.4.1 Sampling numbers

Extensive preliminary work is necessary to establish the type of distribution (the relationship of the mean and the variance) and the density levels that are permissible and those that are associated with extensive damage. From such data it would be possible to plan to take a fixed number of samples that would enable one to be sure, in all instances, what was the population level. However, this would frequently lead to unnecessary sampling and a sequential plan usually allows one to stop sampling as soon as enough data have been gathered. For example, to take extremes, if 80 insects per shoot, or more, caused severe damage and no insects were found in the first five samples, common sense would tempt one not to continue for the additional, say 15, samples. Sequential sampling gives an exact measure (based on the known variance), so that with extremely high and extremely low populations very few samples need be taken and the expenditure of time and effort (cost) is minimal. As described later, this approach may also be used to obtain population estimates with a fixed level of precision (Kuno, 1963; Green, 1970).

As the distribution of most species can be fitted to the negative binomial, Iwao's patchiness regression or Taylor's power law, these are three types of distribution most commonly used to describe the population and formulae are given only for them; the principles are the same as for other distributions. Although the negative binomial, being a probability model, is in principle to be preferred (Nyrop & Binns, 1991), variation in the value of k may indicate that the use of this distribution alone is inappropriate (Nyrop & Binns, 1991; Peng & Brewer, 1994; Nyrop et al., 1995). Many workers have found that Taylor's power law was the most robust (e.g. Pena & Baranowski, 1990; Pena & Duncan, 1992; Steiner, 1990; Riley et al., 1992; Ho, 1993; Walker & Allsopp, 1993; Meagher et al., 1996).

The method is described by Guenther (1977), Cho (1980), Nyrop & Binns (1991) and in many textbooks. An early application was in Stark (1952) to normalised data for a needle miner (Lepidoptera). Whilst mostly applied to the monitoring of invertebrate (particularly insect) pests, it has also been applied to benthic community monitoring for environmental impact assessment (Resh & Price, 1984), the state of amphipod populations in acid-sensitive lakes (France, 1992), and stock assessments of fish (Weitman et al., 1979).

The first decision must be to fix the population level or levels related to the infestation classes. The most widely used plans have two levels, the higher-level plan where action is required and a lower-level plan where it definitely unnecessary. For example, we wish to distinguish between hypothesis H_1 that there are 200 or more egg masses per branch, sufficient to cause heavy damage, and hypothesis H_0 that there are 100 or fewer egg masses per branch, insufficient to cause damage. Although plans having only a single level have been developed (Nyrop et al., 1989; Legg et al., 1994 this account will be based on the more traditional two-level plans.

The second decision concerns the level of probability of incorrect assessment one is prepared to tolerate; there are two types of error:

ψ = the probability of accepting H_1, when H_0 is the true situation
ω = the probability of accepting H_0 when H_1, is the true situation

Let us say that the same level is accepted for both, 1 false assessment in 20, that is a probability level of 0.05 for ψ and ω. Different levels may be necessary because the costs of failing to apply control measures when an outbreak occurs may be much higher

Table 2.4   Values needed for the assessment of which hypothesis to accept (see text for details).

	Infestation level	
	H_0	H_1
Mean $= kp$	kp_0 (e.g. 100)	kp_1 (e.g. 200)
$p = kp/k$	p_0	p_1
$q = 1 + p$	q_0	q_1
Variance $= kpq$	kp_0q_0	kp_1q_1

than those arising from the application of the treatment that is proved not to have been required.

For an animal whose distribution corresponds to the negative binomial, the following values in Table 2.4 need to be calculated; the common k having been found already (p. 31) and the means being fixed as above (Oakland, 1950; Morris & Reeks, 1954).

The next aim is to plot the two lines (Fig. 2.15) that mark the 'acceptance' and 'rejection' areas (these terms have come from quality control work where sequential sampling was originally developed).

The formulae for the lines are:

$$d_0 = \theta n + h_o \qquad (2.50)$$

$$d_1 = \theta n + h_1 \qquad (2.51)$$

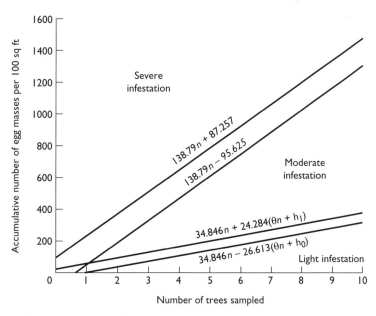

Figure 2.15   A sequential sampling chart with two sets of acceptance and rejection lines. (Adapted from Morris & Reeks, 1954.)

where d = cumulative number of insects, n = number of samples taken, θ = the slope of the lines given by

$$\theta = k\frac{\log \frac{q_1}{q_o}}{\log \frac{p_1 q_o}{p_o q_1}}$$ (2.52)

and h_o and h_1 are the lower and upper intercepts given by

$$h_o = \frac{\log\left[\frac{\omega}{(1-\varphi)}\right]}{\log\left(\frac{p_1 q_o}{p_o q_1}\right)}$$ (2.53)

$$h_1 = \frac{\log\left[\frac{1-\omega}{(\varphi)}\right]}{\log\left(\frac{p_1 q_o}{p_o q_1}\right)}$$ (2.54)

Once these calculations are made, a graph similar to Fig. 2.15 can be drawn up, and it will be clear that occasionally (especially when the true population lies between the population levels chosen, i.e. 100 and 200 in this example) the cumulative total continues to lie in the uncertain zone. In these circumstances an arbitrary upper limit must be set to the number of samples taken. This limit (x), often known as the truncation mark, will be decided on considerations of accuracy and cost (Nyrop *et al.*, 1989). Normally it will be laid down that if after x samples the result still lies in the uncertain zone treatment will be applied.

If Iwao's Patchiness Regression (p. 45) is used, the lines can be best calculated for successive values of n (number of samples taken) (Iwao, 1975):

$$d_1 = nx_c + t\sqrt{n\left[(\alpha+1)x_c + (\beta-1)x_c^2\right]}$$ (2.55)

$$d_0 = nx_c - t\sqrt{n\left[(\alpha-1)x_c + (\beta-1)x_c^2\right]}$$ (2.56)

where x_c = the critical density, α and β are the parameters from the patchiness regression, and t is taken from statistical tables for the acceptable error levels for φ and ω (for the d_1 equation).

Using the parameters of Iwao's regression and Taylor's power law respectively, Kuno (1969) and Green (1970) devised methods that allowed population estimates to be obtained with a fixed level of precision. In Green's method, the cumulative number of animals to be sampled (T_n) is given by:

$$T_n^{b-2} = \frac{D_0^2}{\alpha}n^{b-1}$$ (2.57)

where α and b are the intercept and slope, respectively, in Taylor's power law (see p. 43), n is sample size, and D_0 is the chosen level of precision. This will be linear in its log form:

Table 2.5   A sequential sampling table, giving the uncertainty band for various numbers of samples. (Data from Morris, 1954).

Sample tree	Moderate versus severe infestation uncertainty band (cumulative total of insects)
1	42–225
2	183–366
3	323–506
4	460–643
5	599–782
6	738–921

$$\log T_n = \frac{\log\left(D^2_0/\alpha\right)}{b-2} + \frac{b-1}{b-2}\log n \tag{2.58}$$

The decision lines can be generated by computer simulations, and a number of algorithms have been described (Fowler & Lynch, 1987; Nyrop & Binns, 1991; Brewer *et al.*, 1994). Although Taylor's power law has been found to be a less-biased predictor of variance than Iwao's regression, the parameters of these relationships are not as constant as theory suggests; rather, they vary from location to location, as well as in the same location with time. No method can be used to give plans that may be uncritically used at different times or in different places to give estimates with the required precision (Trumble *et al.*, 1989; Brewer *et al.*, 1994; Naranjo & Flint, 1995). Simulations may be undertaken, and it is possible to determine operating characteristic curves, sample number curves and the influence of the truncation mark for the different plans on the basis of information on the populations of the organism to be sampled. This will permit a choice of plan based on a consideration of the cost (number of samples) and benefit (accuracy). The actual field work may be carried out based on the graph, a sequential table (see Table 2.5) or the data fed into a computer.

General accounts of the utility of sequential sampling plans for insects have been given in reviews ranging over time from Ives (1954) to Pedigo & Buntin (1994). In addition to the papers already referred to, plans have been developed for around two hundred species and /or situations; Fowler & Lynch (1987) and Pedigo & Buntin (1994) provide bibliographies.

2.5 Presence or absence sampling

It is sometimes very costly to estimate the actual numbers in a sample, but presence or absence can be easily assessed; this is true for example of aphids and spider mites where the populations may be strongly clumped. The formulae used depend on assumptions concerning the distribution. This method provides an alternative approach for the development of sequential sampling plans as a basis for pest control decisions.

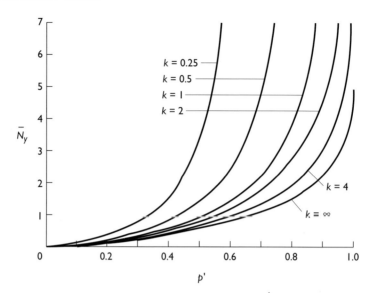

Figure 2.16   Relationship between mean population density ($\bar{N}_y$), and the probability of cell occupancy (p') for selected values of k. (Data from Gerrard & Cook, 1972.)

If the dispersion can be described by the negative binomial distribution with a known k, the probability of a particular mean population per unit (P_y) being reached can be estimated (Gerrard & Chang, 1970; Wilson & Gerrard, 1971; Gerrard & Cook, 1972; Maillloux & Bostanian, 1989):

$$(P_y) = k\left[\left(\frac{1}{1-p'}\right)^{\frac{1}{k}} - 1\right]$$

(2.59)

where p' = probability of the animal being present in the sampling unit determined on the basis of presence or absence sampling.

The form of the relationship is shown in Fig. 2.16. It is reliable only if the critical density levels are related to values of p' less than about 0.8; above this, the uncertainty associated with the predictions is too great. Obviously, the sampling unit can be varied to achieve this relationship but for it to retain its value for rapid assessment, the new unit must be easily recognised. Furthermore, any major changes in the value of k with density would invalidate this approach.

Other approaches have been developed that do not depend on assuming a particular model for the dispersion. In a situation where the total number of potential sampling units (Φ) is meaningful and may be known – for example, young trees in a plantation – the decision can be made (with a certain level of probability of it being untrue, α) that there is a zero or very low population (p_0) if a certain number of samples from which the organism is absent are drawn in succession. The number of such zero samples is given by (Kuno, 1991) as:

$$n_o = \Phi(1 - \alpha^{1/\Phi p_o}).$$

(2.60)

However, for satisfactory levels of confidence the number of successive samples must be large in relation to Φ.

Of wider application is an approach that depends on a knowledge of the relationship between the number of samples where the animal is absent (or present in very low numbers) and the mean of the population /unit, the population intensity (P_y). In a general form this is (Gerrard & Chang, 1970):

$$\ln(P_y) = a + b \ln(-\ln p_0) \qquad (2.61)$$

where a and b are constants for the animal in habitats of the type studied and p_0 is the probability of zero (or very low – if such a level has been chosen) numbers in a particular sampling unit. The values of a and b can be calculated from the regression of P_y on p_0 from preliminary field data. However, the estimates of the variance is less straightforward (Nachman, 1984; Kuno, 1986; Nyrop *et al.*, 1989; Binns & Bostanian, 1990; Tonhasca *et al.*, 1994) but is important for determining the relationship of the numbers of samples to be taken to the level of confidence in the decision. There is not only sampling error, but error in the estimation of a and b and 'error' which is a reflection of real biological variation in the dispersion, which would be reflected in changes in k of the negative binomial. Schaalje *et al.* (1991) provide a detailed review and recommendations based on simulations and field data.

2.6 Sampling a fauna

With faunal surveys, the problem of ensuring the detection of an adequate proportion of the species present must be considered. When sampling vegetation, Gleason (1922) pointed out that, as the area sampled increased the numbers of hitherto unrecorded species added decreased, a concept referred to as the species–area curve, and one that has been often used and disputed in plant ecology (Goodall, 1952; Evans *et al.*, 1955; Greig-Smith, 1978; Watt, 1964). The logarithmic series and log-normal distributions (p. 506) express the same basic assumptions in another way.

One approach to this problem is therefore to take a number of preliminary samples and calculate the index of diversity based on the log series (p. 506). If one can make an estimate of the total number of individuals in the fauna, then the theoretical total number of species present can be arrived at from the index of diversity, and the total number of individuals that should be collected to find y species present can be determined. A number of other estimates for the total number of species are presented in Chapter 13. The value of y will, of course, depend on the aim of the study. Unfortunately, as will be pointed out later (Chapter 13), this approach has a number of practical and theoretical limitations. The practical ones stem from the mosaic nature of most habitats (Hairston, 1964). Unless sampling is by a light trap or some other device that is independent of microhabitats, more new species are added by further samples than would be expected. The theoretical problems arise from the discovery that the log-normal distribution provides a better description of the phenomenon than the simpler logarithmic series (see p. 507 for a further discussion).

A number of nonparametric methods for estimating species richness have been developed. A particularly simple and occasionally effective estimator for the total number of species S_T in a locality is (Chao, 1984):

$$S_T = S_{obs} + \left(\frac{a^2}{2b} \right) \tag{2.62}$$

where S_{obs} is the observed number of species in a sample, a is the number of species represented by only a single individual, and b is the number of species represented by two individuals. Subsequently, Chao (1987) gave the variance of this estimate as:

$$\text{var}(S_T) = b \left[\left(\frac{\frac{a}{b}}{4} \right)^4 + \left(\frac{a}{b} \right)^3 + \left(\frac{\frac{a}{b}}{2} \right)^2 \right]. \tag{2.63}$$

If only presence–absence data are available, a similar approach can be taken (Chao, 1984):

$$S_T = S_{obs} + \left(\frac{L^2}{2M} \right) \tag{2.64}$$

where L is the number of species that occur in only one sample and M is the number of species that occur in only two samples. The variance is calculated using Equation (2.63), with L and M substituting a and b, respectively. These estimates are probably best considered as the lower bound of total species number (see Colwell & Coddington, 1994). In Chapter 13, other methods for estimating species number are presented, but these require more data and thus are less useful during the planning stage.

Sometimes, the mosaic nature of the habitat can be utilised to facilitate the planning of an optimal sampling programme. This is especially true where the effects of pollution are being studied in streams, but might be extended to other situations where a factor is acting simultaneously over a wide range of special habitats within a major habitat or community type. An example is the effect of aerial spraying on the fauna of a mixed woodland. If a fixed number of samples of the fauna are taken – say from three different zones of a stream: riffles, pools and marginal areas – the first few samples will be found to provide a more complete picture of the total observed fauna in one zone than in the other (Gaufin *et al.*, 1956). Because of the different diversities and microdistributions of the fauna in the different zones, and also possibly because of the differing efficiencies of the collecting methods under these conditions, the rates of accumulation of new species vary from zone to zone. A zone in which a high percentage of the total species taken in a large number of samples were recorded in the first few samples can have a smaller number of samples drawn from it than a zone in which the rate of accumulation is less rapid.

A method of calculating the average number of new species contributed by the *K*th successive sample (when *K* can equal any number from 1 upwards) has been devised by Gaufin *et al.* (1956), who also provide a table of the coefficients necessary for up to a total of ten samples. Their formula is:

$$S_k = \frac{1}{S} \sum_{i=1}^{n-k+1} a_i K S_i \tag{2.65}$$

where S_K = the number of new species added by the Kth sample, n = the total number of samples taken in the preliminary survey, S = the total number of species taken in the n samples, S_i = the number of different species appearing in i out of n samples and a_i K are coefficients by which successive values of S_i are multiplied in calculating the summation item. By summing a series of values of S_K (for K = 1 to a predetermined number K'), it is possible to compute the average number of species found in K' samples. A method of calculating the standard errors of such estimates has been developed by Harris (1957).

2.7   Biological and other qualitative aspects of sampling

Hitherto, stress has been laid on the statistical aspects of sampling; however, there are certain biological problems that are of cardinal importance. The ecologist should always remember that the computation of fiducial limits of estimates only tells the consistency of the samples collected, and the statistical techniques cannot be blamed if these have been consistently excluding a major part of the population. This is indeed the most serious biological problem in all sampling work, namely to ensure that there is not a part of the population with a behavioural pattern or habitat preference such that it is never sampled (Macleod, 1958).

Less dangerous because the phenomenon will be recorded, although it may well be misinterpreted, is the tendency for the behaviour of the animal to change and affect its sampling properties. The reactions of the fly, *Meromyza, to* weather conditions (Hughes, 1955), and of the larvae of different species of corixid water-bug to light (Teyrovsky, 1956), lead to errors when net sampling is carried out under different conditions (p. 259). Many – perhaps almost all – animals alter their behaviour with age, and this may lead to a confusion between a change in behaviour and death. For example, older females of the mirid grass bug, *Leptopterna dolobrata*, spend more time on the base of the grass shoots and less on the tops than do males. If a marking and recapture experiment is carried out and the bugs collected by sweeping, the females will appear to be much shorter-lived than the males. However, this is because as they age, the females enter the sampling zone less frequently; they are actually longer-lived than the males.

Variations in the distribution of an animal may well be linked with some character of the environment or its host plant. It is important to distinguish this variation from sampling variance; otherwise, the fiducial limits of the estimates may become so wide that no conclusions can be drawn (cf. Prebble, 1943). In other words, systematic errors will arise (p. 17). When population trend is eliminated, approximately 50% of the variance of the counts of populations of the sugar cane froghopper, *Aeneolamia varia*, on sugar cane stools could be accounted for in terms of the size of the stools (Fewkes, 1961). Hence, more 'accurate' estimates would be obtained by taking stool size into consideration and adjusting population estimates, or by allowing for it in the analysis. This is particularly important when comparisons of froghopper numbers are being made between insecticide-treated fields; the fields are of different ages and hence with stools of different sizes. Many other examples could be given of environmental factors influencing distribution. They are commonly microclimatic in nature; for example, soil moisture affects the distribution of the cocoons of a sawfly (Ives, 1955). Even when the environment appears uniform the distribution may be systematically nonrandom; for example, the densest populations of the beetle, *Tribolium*, in flour are always adjacent to the walls of the container (Cox & Smith, 1957). Covariance analyses are often helpful in such situations.

2.8 Jack knife and Bootstrap techniques

It is frequently difficult to estimate the accuracy of ecological indices because no equation exists for the variance of the statistic and the sampling distribution is unknown. Jack knife and bootstrap methods, which offer a nonparametric means of estimating parameters, are now widely used by ecologists to estimate the accuracy of population parameter estimates and indices for diversity or similarity. Both techniques are based on the repeated estimation of the parameters of interest on a subsample of the data, for which a computer is essential. Perhaps their most important application is for the estimation of error limits and confidence intervals for estimated parameters. There are computer programs for their calculation, and an example using R is given below.

The jack knife method uses less computational effort than a bootstrap analysis and is thus often quicker and easier to undertake. However, it cannot be used to estimate confidence intervals. The general scheme is as follows:

1. Consider a situation where a parameter is to be estimated from n samples; for example, a species diversity index from 10 kicknet samples from a stream.
2. Use all n samples to calculate the parameter of interest, E, the diversity index.
3. Now remove one of the samples at a time and recalculate the parameter of interest, E_i.
4. Calculate the pseudovalues for the parameter of interest:

$$\phi_i = nE - (n-1)E_i \qquad (2.66)$$

where ϕ = pseudovalue for jack knife estimate i, n = the total number of replicates, E = the estimate for all n samples, E_i = the estimate with sample i removed
5. Estimate the mean and standard error of the parameter of interest from the n pseudovalues. The jack knife estimate of bias is E – the mean of the pseudovalues and the estimate of the standard error of the sample is simply the standard error of the pseudovalues:

$$se = \sqrt{\frac{\sum (\phi_i - \bar{\phi})^2}{n(n-1)}} \qquad (2.67)$$

where ϕ is the mean of the pseudovalues.

The bootstrap method differs from the jack knife in the means of sampling the original data set. The subsets of observations used to repeatedly calculate the parameter of interest are each selected at random with replacement from the original data set of n observations with each observation only included once. This procedure is repeated, often hundreds of times, so that many combinations of the original observations are generated. For example, if a sample comprised 10 species and 100 individuals, then each subset of data would be a randomly selected 100 individuals taken from a population that represented all the species according to their original frequency of observation. Each of these samples is then used to generate estimates of the parameter of interest (e.g. a diversity index) as for the jack knife method. The bootstrap estimate of the bias is simply found by subtracting the observed parameter value from the mean parameter obtained from the bootstrap replicates. The estimate of the standard error of the sample is simply the standard error of the replicate parameter estimates:

$$se = \sqrt{\frac{\sum (\phi_i - \bar{\phi})^2}{b - 1}} \tag{2.68}$$

where ϕ is the mean of the replicates and b is the number of bootstrap samples taken. A number of methods can be used to estimate confidence intervals. The percentile bootstrap is the simplest and most commonly used. The 2.5 and 97.5 percentiles of the bootstrap distribution of the parameter are used as the 95% confidence intervals. For example, if 1000 bootstrap replicates have been taken, then these are arranged in order of magnitude and the 2.5[th] percentile calculated as the average of the 25th and 26th smallest values and the 97.5 percentile as a the average of the 975th and 976th smallest values.

Bootstrap and jack knife estimates are used to estimate species richness (see Chapter 13), population size (see Chapter 3) and population growth rates (see Chapter 12). Introductions to bootstrap and jack knife methods are given by Efron & Tibshirani (1986), Scheiner & Gurevitch (1993) and Potvin & Roff (1993). Jack knife and bootstrap calculations are easily undertaken in R. As an example, the following code calculates bootstrap confidence intervals for the coefficient of variation of a series of observation held in the vectors obs.

```
#An example to bootstrap confidence intervals for the
#Coefficient of variation
# Enter your data into a vector x
obs <-c(2, 3, 5, 2, 4, 2, 1, 3, 3,5,3,5,6,3,2)
#generate a vector to hold 1000 bootstrap values
bootest <-numeric(1000)
#Define function to calculate coefficient of variation
CV <- function(obs) sqrt(var(obs))/mean(obs)
#Calculate CV
CV(obs)
#Bootstrap 1000 CV estimates for (i in 1:1000) bootest[i] <- CV(sample(obs,replace=T))
#calculate the mean of the bootstrap estimates
mean(bootest)
#Calculate the variance
```

```
var(bootest)
#calculate value at upper97.5%
quantile(bootest,0.975)
#calculate value at lower 2.5%
quantile(bootest,0.025)
#Calculate bias as the difference between original and mean bootstrap CV
bias <- mean(bootest) - CV(obs)
# bootstrap-corrected estimate of the CV is just the original estimate minus the bias,
CV(obs) - bias
# Assuming normality, the approximate 95% confidence interval is given by
CV(obs) - bias - 1.96*sqrt(var(bootest))
CV(obs) - bias + 1.96*sqrt(var(bootest))
 # Efron's confident limit
quantile(bootest,0.975)
quantile(bootest,0.025)
#Hall's confidence limits
2*CV(obs) - quantile(bootest,0.025)
2*CV(obs) - quantile(bootest,0.975)
```

References

Abrahamsen. G. (1969) Sampling design in studies of population densities in Enchytracidae (Oligochaeta). *Oikos* **20**, 54–66.

Abrahamsen, G. & Strand, L. (1970) Statistical analysis of population density data of soil animals, with particular reference to Enchytraeidae (Oligochaeta). *Oikos* **21**, 276–84.

Aliston, R. (1994) *Statistical analysis of animal populations.* PhD thesis, University of Kent, England.

Andersen, F.S. (1965) *The negative binomial distribution and the sampling of insect populations. Proc. XII Int. Congr. Entomol.*, 395.

Anderson, R.M. (1974) Population dynamics of the cestode *Caryophyllaeus laticep, v* (Pallas, 1781) in the bream (*Abramis brama L.*). *J. Anim. Ecol.* **43**, 305–21.

Anderson, R.M., Gordon, D.M., Crawley, M.J., & Hassell, M.P. (1982) Variability in abundance of animal and plant species. *Nature (Lond)* **296**, 245–8.

Anon. (1955) British Instrument Industries' exhibition. *Engineering* **180**, 116.

Anscombe, F.J. (1948) On estimating the population of aphids in a potato field. *Ann. Appl. Biol.* **35**, 567–71.

Anscombe, F.J. (1949) The statistical analysis of insect counts based on the negative binomial distribution. *Biometrics* **5**, 165–73.

Anscombe, F.J. (1950) Sampling theory of the negative binomial and logarithmic series distributions. *Biometrika*, 358–82.

Arbous, A.G. & Kerrich, J.E. (1951) Accident statistics and the concept of accident-proneness. *Biometrics* **7**, 340–432.

Badenhauser, I. (1994) Spatial patterns of alate and apterous morphs of the *Brachycaudus helichrysi* (Homoptera: Aphididae) in sunflower fields. *Environ. Entomol.* **23**, 1381–90.

Bailey, N.T.J. (1959) *Statistical methods in biology.* English Universities Press, London.

Baker, J.M. & Crothers, J.H. (1987) Intertidal rock. In: Baker, J.M.W. & Wolff, W.J. (eds), *Biological surveys of estuaries and coasts*. Cambridge University Press, Cambridge, New York, New Rochelle.

Baker, J.M. & Wolff, W.J. (eds) (1987) *Biological Surveys of Estuaries and Coasts. Estuarine and brackish water sciences association handbook*. Cambridge, Cambridge University Press.

Ballentine, W.J. (1961) A biologically-defined exposure scale for the comparative description of rocky shores. *Field Studies* **1**, 1–19.

Bamber, R.N. (1993) Changes in the infauna of a sandy beach. *J. Exp. Mar. Biol. Ecol.* **172**, 93–107.

Bancroft, T.A. & Brindley, T.A. (1958) Methods for estimation of size of corn borer populations. *Proc. Xth Int. Congr. Entomol.* **2**, 1003–1014.

Banerjee, B. (1976) Variance to mean ratio and the spatial distribution of animals. *Experientia* **32**, 993–4.

Beauchamp, J.I. & Olson, I.S. (1973) Corrections for bias in regression estimates after logarithmic transformation. *Ecology* **54**, 1403–7.

Berthet P. & Gerard, G. (1965) A statistical study of microdistribution of Oribatei (Acari). Part 1. The distribution pattern. *Oikos* **16**, 214–27.

Binns, M. R. & Bostanian, N.J. (1990) Robustness in empirically based binomial decision rules for integrated pest management. *J. Econ. Entomol.* **83**, 420–7.

Blackith, R.E. (1958) Nearest-neighbour distance measurements for the estimation of animal populations. *Ecology* **39**, 147–50.

Blackith, R.E., Siddorn, J.W., Waloff, N., & Van Enden, H.F. (1963) Mound nests of the yellow ant, *Lasius flavus* L., on water-logged pasture in Devonshire. *Entomol. Monthly Mag.* **99**, 48–9.

Bliss, C.I. (1958) The analysis of insect counts as negative binomial distributions. *Proc. Xth Int. Congr. Entomol.* **2**, 1015–32.

Bliss, C.I. (1971) The aggregation of species within spatial units. In: Patil, G.P., Pielou, E.C., & Waters, W.E (eds), *Statistical Ecology I : Spatial Patterns and Statistical distributions*. Pennsylvania State University Press, Philadelphia.

Bliss, C.I. & Calhoun, D.W. (1954) *An outline of biometry*. Yale Co-operative Corp., New Haven.

Bliss, C.I. & Fisher, R.A. (1953) Fitting the negative binomial distribution to biological data and note on the efficient fitting of the negative binomial. *Biometrics* **9**, 176–200.

Bliss, C.I. & Owen, A.R.G. (1958) Negative binomial distributions with a common k. *Biometrika* **45**, 37–58.

Blondel, J. & Frochet, B. (eds) (1987) Bird census and atlas studies. *Proceedings, IX International Conference on Bird Census and Atlas Work*, University de Dijon, Dijon.

Box, G.E.P. & Cox, D.R. (1964) An analysis of transformations. *J. Roy. Statist. Soc.* **26**, 211–52.

Breder, C.M. (1954) Equations descriptive of fish schools and other animal aggregations. *Ecology* **35**, 361–70.

Brewer, M.J., Legg, D.E., & Kaltenbach, J.E. (1994) Comparison of three sequential sampling plans using binomial counts to classify insect infestation with respect to decision thresholds. *Environ Entomol.* **23**, 812–26.

Broadbent, L. (1948) Methods of recording aphid populations for use in research on potato virus diseases. *Ann. Appl. Biol.* **35**, 551–66.

Browning, T.O. (1959) The long-tailed mealybug, *Pseudococcus adonidum* L. in South Australia. *Aust. J. Agric. Res.* **10**, 322–37.

Bryant, D.G. (1976) Sampling populations of *Adelges piceae* (Homoptera: Phylloxeridae) on Balsam fir, *Abies balsamea*. *Can. Entomol.* **108**, 1113–24.

Buntin, G.D. (1994) Developing a primary sampling program. In: Pedigol, P. & Buntin, G.D. (eds), *Handbook of Sampling Methods for Arthropods in Agriculture*. CRC Press, Boca Raton.

Buntin, G.D. & Pedigo, L.P. (1981) Dispersion and sequential sampling of green cloverworm eggs in soybeans. *Environ. Entomol.* **10**, 980–5.

Burrage, R.H. & Gyrisco, G.G. (1954) Estimates of populations and sampling variance of European chafer larvae from samples taken during the first, second and third instar. *J. Econ. Entomol.* **47**, 811–17.

Burts, E.C. & Brunner, J.F. (1981) Dispersion statistics and sequential sampling plan for adult pear psylla. *J. Econ Entomol.* **74**, 291–4.

Camin, J.H., George, J.F., & Nelson, V.E. (1971) An automatic tick collector for studying the rhythmicity of 'drop off' Ixodidae. *J. Med. Entomol.* **8**, 394–8.

Carlson, P.E., Johnson, R.K., & McKie, B.G. (2013) Optimizing stream bioassessment: habitat, season, and the impacts of land use on benthic macroinvertebrates. *Hydrobiologia* **704**, 363–73.

Cassie, R.M. (1962) Frequency distribution models in the ecology of plankton and other organisms. *J. Anim. Ecol.* **31**, 65–92.

Caughley, G. (1977) *Analysis of vertebrate populations*. John Wiley & Sons, New York.

Chao, A. (1984) Non-parametric estimation of the number of classes in a population. *Scand. J. Statist.* **11**, 265–70.

Chao, A. (1987) Estimating the population size for capture-recapture data with unequal matchability. *Biometrics* **43**, 783–91.

Chiang, H.C. & Hodson, A.C. (1959) Distribution of the first-generation egg masses of the European corn borer in corn fields. *J. Econ. Entomol.* **52**, 295–9.

Cho, C.-K. (1980) *An introduction to software quality control*. John Wiley & Sons, New York.

Church, B.B. & Strickland, A.H. (1954) Sampling cabbage aphid populations on Brussels sprouts. *Plant Pathol.* **3**, 76–80.

Clark, P.I. & Evans, F.C. (1954) Distance to nearest neighbor as a measure of spatial relationships in populations. *Ecology* **35**, 445–53.

Clark, S.J. & Perry, J.N. (1994) Small sample estimation for Taylor's power law. *Environ. Ecol. Statist.* **1**, 287–302.

Cochran, W.G. (1977) *Sampling techniques*. 3rd edn. John Wiley & Sons, New York.

Cole, L.C. (1946) A theory for analysing contagiously distributed populations. *Ecology* **279**, 329–41.

Cole, W.E. (1970) The statistical and biological implications of sampling units for mountain pine beetle populations in lodgepole pine. *Res. Popul. Ecol.* **12**, 243–8.

Colwell, R.K. & Coddington, J.A. (1994) Estimating terrestrial biodiversity through extrapolation. *Philos. Trans. Roy. Soc. (Series B)* **345**, 101–18.

Condrashoff, S.F. (1964) Bionomics of the aspen leaf miner, *Phyllocnistis populiella* Cham. (Lepidoptera: Gracillariidae). *Can. Entomol.* **96**, 857–74.

Corbet, P.S. (1964) Temporal patterns of emergence in aquatic insects. *Can. Entomol.* **96**, 264–79.

Corbet, P.S. & Smith, S.M. (1974) Diel periodicities of landing of multiparous and parous *Aedes aegypti* (L.) at Dar es Salaam, Tanzania (Diptera, Culicidae). *Bull. Entomol. Res.* **64**, 111–21.

Cornfield, J. (1951) The determination of sample size. *Am. J. Pub. Health* **41**, 654–61.

Coulson, R.N., Hain, F.P., Foltz, J.L., & Mayyasi, A.M. (1975) Techniques for sampling the dynamics of Southern Pine beetle populations. *Texas Agric. Exp. Stn. Misc. Pub.* **1185**, 3–18.

Coulson, R.N., Pulley, P.E., Foltz, J.L., & Martin, W.C. (1976) Procedural guide for quantitatively sampling within-tree populations of *Dendroctonus frontalis. Texas Agric. Exp. Stn. Misc. Pub.* **1267**, 3–26.

Cox, D.R. & Smith, W. L. (1957) On the distribution of *Tribolium confusum* in a container. *Biometrika* **44**, 328–35.

Craig, C.C. (1953) On a method of estimating biological populations in the field. *Biometrika* **40**, 216–18.

Crawley, M.J. (2005) Statistics: An Introduction using R. John Wiley. ISBN 9780470022979.

Croft, B.A., Welsh, S.M., & Dover, M.J. (1976) Dispersion statistics and sample size estimates for populations of the mite species *Panonychus ulmi* and *Amblyseius fallacis* on apple. *Environ. Entomol.* **5**, 227–34.

Crofton, H.D. (1971) A quantitative approach to parasitism. *Parasitology* **62**, 179–93.

Cushing, D.H. (1992) *Climate and Fisheries.* Academic Press, London.

Dahlsten, D.L., Rowney, D.L., Cooper, W.A., Tait, S.M., & Wenz, J.M. (1990) Long-term population studies of the Douglas-fir moth in California. In: Watt, A.D., Leather, S.R., Hunter, M.D., & Kidd, N.A.C. (eds), *Population Dynamics of Forest Insects.* Intercept Press, Andover, UK, pp. 45–58.

David, F.N. & Moore, G.P (1954) Notes on contagious distributions in plant populations. *Ann. Bot. Lond. New Ser.* **18**, 47–53.

Davis, E.G. & Wadley, F. (1949) Grasshopper egg-pod distribution in the northern Great Plains and its relation to egg-survey methods. *USDA Circ.* **816**, 16 pp.

Dean, H.A. (1959) Quadrant distribution of mites on leaves of Texas grapefruit. *J. Econ. Entomol.* **52**, 725–7.

Deevey, E.S. (1947) Life tables for natural populations of animals. *Quart. Rev. Biol.* **22**, 283–314.

Delignette-Muller, M., Pouillot, R., Denis, J., & Dutang, C. (2013) fitdistrplus: Help to Fit of a Parametric Distribution to Non-Censored or Censored Data. R package version 1.0-1, URL http://cran.r-project.org/web/packages/fitdistrplus/.

Dempster, J.P. (1957) The population dynamics of the Moroccan locust (*Dociostaurus maroccanus* Thunberg) in Cyprus. *Anti-Locust Bull.* **27**, 1–60.

Dudley, C. (1971) A sampling design for the egg and first instar larval populations of the western pine beetle, *Dendroctonus brevicomis* (Coleoptera: Scolytidae). *Can. Entomol.* **103**, 1291–313.

Dybas, H.S. & Davis, D.D. (1962) A population census of seventeen-year periodical cicadas (Homoptera: Cicadidae: Magicicada). *Ecology* **43**, 432–44.

Eberhardt, L.L. (1967) Some developments in 'distance sampling'. *Biometrics* **23**, 207–16.

Edwards, R.L. (1962) The importance of timing in adult grasshopper surveys. *J. Econ. Entomol.* **55**, 263–4.

Efron, B. & Tibshirani, R. (1986) Bootstrap methods for standard errors, confidence intervals, and other measures of statistical accuracy. *Statist. Sci.* **1**, 54–77.

Egwuatu, R.I. & Taylor, T.A. (1975) Aspects of the spatial distribution of *Acanthomia tomentosicollis* Stal. (Heteroptera, Coreidae) in *Cajanus cajan* (Pigeon Pea). *J. Econ. Entomol.* **69**, 591–4.

Ellenberger, J.S. & Cameron, E.A. (1977) The spatial distribution of oak leaf roller egg masses on primary host trees. *Environ. Entomol.* **6**, 101–6.

Elliot, J.M. & Drake, C.M. (1981) A comparative study of seven grabs used for sampling macroinvertebrates in rivers. *Freshwater Biol.* **11**, 99–120.

Elliott, J.M. (1977) *Some Methods for the Statistical Analysis of Samples of Benthic Invertebrates.* Freshwater Biology Association.

Eleftheriou, A. (ed.) (2013) *Methods for the Study of Marine Benthos.* 4th edn. John Wiley & Sons.

Emden, H.F.V., Jepson, W.F., & Southwood, T.R.E. (1961) The occurrence of a partial fourth generation of *Oscinella frit* L. (Diptera: Chioropidae) in southern England. *Entomol. Exp. Appl.* **4**, 220–5.

Evans, D.A. (1953) Experimental evidence concerning contagious distributions in ecology. *Biometrika*, **40**, 186–211.

Evans, F.C., Clark, P.I., & Brand, R.H. (1955) Estimation of the number of species present in a given area. *Ecology* **36**, 342–3.

Ferguson, C.S., Linit, M.J., & Krause, G. (1992) Dispersion and density of Asiatic oak weevil (Coleoptera: Curculionidae) relative to oak density. *Environ. Entomol.* **21**, 247–52.

Fewkes, D.W. (1961) Stool size as a factor in the sampling of sugarcane froghopper nymph populations. *J. Econ. Entomol.* **54**, 771–2.

Finney, D.I. & Varley, G.C. (1955) An example of the truncated Poisson distribution. *Biometrics* **11**, 387–94.

Finney, D.J. (1941) Wireworm populations and their effect on crops. *Ann. Appl. Biol.* **28**, 282–95.

Finney, D.J. (1973) Transformation of observations for statistical analysis. *Cotton Growers' Rev.* **50**, 1–14.

Fisher, R.A., Corbet, A.S., & Williams, C.B. (1943) The relation between the number of species and the number of individuals in a random sample of an animal population. *J. Anim. Ecol.* **12**, 42–58.

Ford, R.P. & Dimond, J.B. (1973) Sampling populations of pine leaf chermid *Pirieus pinifoliae* (Homoptera: Chermidae) II. Adult Gailicolae on the secondary host. *Can. Entomol.* **105**, 1265–74.

Forsythe, H.Y. & Gyrisco, G.G. (1963) The spatial pattern of the pea aphid in alfalfa fields. *J. Econ. Entomol.* **56**, 104–7.

Fortin, M.-J. & Dale, M.R.T. (2005) *Spatial Analysis: A Guide for Ecologists.* Cambridge University Press.

Fowler, G.W. & Lynch, A.M. (1987) Sampling plans in insect pest management based on Wald's sequential probability ratio test. *Environ. Entomol.* **16**, 345–54.

Fracker, S.B. & Brischle, H.A. (1944) Measuring the local distribution of Ribes. *Ecology* **25**, 283–303.

France, R. (1992) Use of sequential sampling of amphipod abundance to classify the biotic integrity of acid-sensitive lakes. *Environ. Manag.* **16**(2), 157–166.

Franklin, J. (2010) *Mapping Species Distributions: Spatial Inference and Prediction.* Cambridge University Press.

Fromont, E., Morvilliers, L., Artois, M., & Pontier, D. (2001) Parasite richness and abundance in insular and mainland feral cats. *Parasitology* **123**, 143–51.

Gardefors, D. & Orrage, L. (1968) Patchiness of some marine bottom animals. A methodological study. *Oikos* **19**, 311–21.

Gaufin, A.R., Harris, E.K., & Walter, H.J. (1956) A statistical evaluation of stream bottom sampling data obtained from three standard samples. *Ecology* **51**, 643–8.

George, D.C. (1974) Dispersion patterns in the zooplankton populations of a eutrophic reservoir. *J. Anim. Ecol.* **43**, 537–51.

Gerrard, D.I. & Chang, H.C. (1970) Density estimation of corn rootworm egg populations based upon frequency of occurrence. *Ecology* **51**, 237–45.

Gerrard, D.I. & Cook, R.D. (1972) Inverse binomial sampling as a basis for estimating negative binomial population densities. *Biometrics* **28**, 971–80.

Glass, L.W. & Boubjerg, R.V. (1969) Density and dispersion in laboratory populations of caddis fly larvae (Cheumatopsyche Hydropsychidae). *Ecology* **50**, 1082–4.

Gleason, H.A. (1922) On the relation between species and area. *Ecology* **3**, 158–62.

Gonzalez, D. (1970) Sampling as a basis for pest management strategies. *Tall Timbers Conf. Ecol. Animal Control by Habitat Management* **2**, 83–101.

Goodall, D.W. (1952) Quantitative aspects of plant distribution. *Biol. Rev.* **27**, 194–245.

Gotceitas, V. & Colgan, P. (1991) Assessment of patch profitability and ideal free distribution – the significance of sampling. *Behaviour* **119**, 65–76.

Green, R.H. (1970) On fixed precision level sequential sampling. *Res. Popul. Ecol.* **12**, 249–51.

Green, R.H. (1979) *Sampling Design and Statistical Methods for Environmental Biologists.* John Wiley & Sons, New York.

Greenwood, J.J.D. (1996) Basic techniques. In: Sutherland, W.J. (ed.) *Ecological Census Techniques.* Cambridge University Press, Cambridge.

Greig-Smith, P. (1978) *Quantitative Plant Ecology.* 3rd edn. Butterworths, London.

Grundy, P.M. (1952) The fitting of grouped truncated and grouped censored normal distributions. *Biometrika* **39**, 252–9.

Guenther, W.C. (1977) *Sampling Inspection in Statistical Quality Control.* Oxford University Press, New York.

Guppy, I.C. & Harcourt, D.G. (1970) Spatial pattern of the immature stages and teneral adults of *Phyllophaga* spp. (Coleoptera: Scarabaeidae) in a permanent meadow. *Can. Entomol.* **102**, 1354–9.

Hairston, N.G. (1959) Species abundance and community organisation. *Ecology* **40**, 404–16.

Hairston, N.G. (1964) Studies on the organization of animal communities. *J. Anim. Ecol.* **33** (suppl.), 227–239.

Handford, R.H. (1956) Grasshopper population sampling. *Proc. Entomol. Soc. B. C.* **52**, 3–7.

Hansen, M.H., Hurwitz, W.N., & Madow, W.G. (1953) *Sample Survey Methods and Theory.* Vol. 1, John Wiley & Sons, New York.

Harcourt, D.G. (1961a) Design of a sampling plan for studies on the population dynamics of the diamondback moth, *Plutella maculipennis* (Curt.) (Lepidoptera: Plutellidae). *Can. Entomol.* **93**, 820–31.

Harcourt, D.G. (1961b) Spatial pattern of the imported cabbageworm, *Pieris rapae* (L.) (Lepidoptera: Pieridae), on cultivated Cruciferae. *Can. Entomol.* **94**, 945–52.

Harcourt, D.G. (1962) Design of a sampling plan for studies on the population dynamics of the imported cabbageworm, *Pieris rapae* (L.) (Lepidoptera: Pieridae). *Can. Entomol.* **94**, 849–59.

Harcourt, D.G. (1964) Population dynamics of *Leptinotarsa decemlineata* (Say) in Eastern Ontario. II Population and mortality estimation during six age intervals. *Can. Entomol.* **96**, 1190–8.

Harcourt, D.G. (1965) Spatial pattern of the cabbage looper, *Trichoplusia ni*, on Crucifers. *Ann. Entomol. Soc. Am.* **58**, 89–94.

Harris, E.K. (1957) Further results in the statistical analysis of stream sampling. *Ecology* **38**, 463–8.

Harris, J.W.E., Collis, D.G., & Magar, K.M. (1972) Evaluation of tree-beating method for sampling defoliating forest insects. *Can. Entomol.* **104**, 723–9.

Hassel, M.P. & May, R.M. (1974) Aggregation of predators and insect parasites and its effect on stability. *J. Anim. Ecol.* **43**, 567–94.

Hassell, M.P. (1978) *The Dynamics of Arthropod Predator–Prey Systems.* Princeton University Press, New Jersey.

Hayman, B.I. & Lowe, A.D. (1961) The transformation of counts of the cabbage aphid (*Brevicoryne brassicae* (L.). *N. Z. J. Sci.* **4**, 271–8.

Helson, G.A.H. (1958) Aphid populations: Ecology and methods of sampling aphids *Myzus persicae* (Sulz.) and *Aulacorthum solani* (Kltb.). *N. Z. Entomol.* **2**, 20–3.

Henderson, P.A. (1987) The vertical and transverse distribution of larval herring in the River Blackwater estuary, Essex. *J. Fish Biol.* **31**, 281–90.

Henderson, P.A., Hamilton, W.D., & Crampton, W.G.R. (1998) Evolution and diversity in the Amazonian floodplain communities. In: Newbury, D.M., Prins, H.H.T., & Brown, N.D. (eds), *Dynamics of Tropical Communities*. The 37th British Ecological Society, pp. 385–419.

Henderson, P.A., Whitehouse, J.W., & Cartwright, G. (1984) The growth and mortality of larval herring, *Clupea harengus* in the River Blackwater estuary, 1978–1980. *J. Fish Biol.* **24**, 613–22.

Henson, W.R. (1954) A sampling program for Poplar insects. *Can. J. Zool.* **32**, 421–33.

Henson, W.R. (1961) Laboratory studies on the adult behaviour of *Conopthorus coniperda* (Schwarz) (Coleoptera: Scolytidae). II. Thigmotropic aggregation. *Ann. Entomol. Soc. Am.* **54**, 810–19.

Herbert, H.J. & Butler, K.P. (1973) Sampling systems for European red mite, *Panonychus ulmi* (Acarina: Tetranychidae) eggs on apple in Nova Scotia. *Can. Entomol.* **105**, 1519–23.

Heywood, I. & Edwards, R.W. (1961) Some aspects of the ecology of *Potamopyrgus jenkinsi* Smith. *J. Anim. Ecol.* **31**, 239–50.

Hirata, S. (1962) Comparative studies on the population dynamics of important Lepidopterous pests on cabbage. 2. On the habits of oviposition of *Pieris rapae crucivora*, *Plusia nigrisigna* and *Manestra (Barathra) brassicae* on cabbage plants. *Jap. J. Appl. Entomol. Zool.* **6**, 200–7.

Ho, C.-C. (1993) Dispersion statistics and sample size estimates for *Tetranychus kanzawai* (Acari: Tetranychidae) on Mulberry. *Environ. Entomol.* **22**, 21–5.

Holme, N.A. (1950) Population dispersion in *Tellina tenuis* Da Costa. *J. Mar. Biol. Assoc. UK* **299**, 267–80.

Howe, R.W. (1963) The random sampling of cultures of grain weevils. *Bull. Entomol. Res.* **54**, 135–46.

Hudson, M. & LeRoux, E.J. (1961) Variation between samples of immature stages, and of mortalities from some factors, of the European corn borer, *Ostrinia nubilalis* (Hubner) (Lepidoptera: Pyralidae) on sweet corn in Quebec. *Can. Entomol.* **93**, 867–88.

Hughes, R.D. (1955) The influence of the prevailing weather on the numbers of MeroMyza variegata Meigen (Diptera, Chloropidae) caught with a sweepnet. *J. Anim. Ecol.* **24**, 324–5.

Hughes, R.D. (1962) The study of aggregated populations. In Murphy, P.W. (ed.) *Progress in Soil Zoology*. Butterworths, London.

Ito, Y., Gotoh, A., & Miyashita, K. (1960) On the spatial distribution of *Pieris rapae crucivora* population. *Jap. J. Appl. Entomol. Zool.* **4**, 141–5.

Ito, Y., Nakamura, M., Kondo, M., Miyashita, K., & Nakamura, K. (1962) Population dynamics of the chestnut gall-wasp, *Dryocosmus kuriphilus* Yasumatsu (Hymenoptera: Cynipidae). 11. Distribution of individuals in bud of chestnut tree. *Res. Popul. Ecol.* **4**, 35–46.

Ives, W.G.H. (1954) Sequential sampling of insect populations. *Forestry Chron.* **30**, 287–91.

Ives, W.G.H. (1955) Effect of moisture on the selection of cocooning sites by larch sawfly, *Pristiphora erichsonii* (Hartwig). *Can. Entomol.* **87**, 301–11.

Iwao, S. (1956) The relation between the distribution pattern and the population density of the large twenty-eight-spotted lady beetle, *Epilachna 28-maculata* Motschulsky, in eggplant field. Pattern of the spatial distribution of insect 6. *Jap. J. Ecol.* **5**, 130–5.

Iwao, S. (1968) A new regression method for analysing the aggregation pattern of animal populations. *Res. Pop. Ecol.* **10**, 1–20.

Iwao, S. (1970a) Analysis of contagiousness in the action of mortality factors on the western tent caterpillar population by using the m–m relationship. *Res. Popul. Ecol.* **12**, 100–10.

Iwao, S. (1970b) Analysis of spatial patterns in animal populations: progress of research in Japan. *Rev. P. I. Protec. Res.* **3**, 41–54.

Iwao, S. (1970c) Problems of spatial distribution in animal population ecology. In: Patil, G.P. (ed.), *Random Counts in Biomedical and Social Sciences*. Penn State University Press, Philadelphia.

Iwao, S. (1972) Application of the *m-m method to the analysis of spatial patterns by changing the quadrat size. *Res. Popul. Ecol.* **14**, 97–128.

Iwao, S. (1975) A new method of sequential sampling to classify populations relative to a critical density. *Res. Popul. Ecol.* **16**, 281–8.

Iwao, S. & Kuno, E. (1971) An approach to the analysis of aggregation pattern in biological populations. In: Patil, G.P., Pielou, E.C., & Waters, W.E. (eds), *Statistical Ecology 1: Spatial Patterns & Statistical Distributions*. Pennsylvania State University Press, Philadelphia.

Jepson, W.F. & Southwood, T.R.E. (1958) Population studies on *Oscinella frit* L. *Ann. Appl. Biol.* **46**, 465–74.

Kapatos, E., McFadden, M.W., & Pappas, S. (1977) Sampling techniques and preparation of partial life tables for the olive fly, *Dacus oleae* (Diptera: Trypetidae) in Corfu. *Ecol. Entomol.* **2**, 193–6.

Karandinos, M.G. (1976) Optimum sample size and comments on some published formulae. *Bull. Entomol.* **22**, 417–21.

Katti, S.K. (1966) Interrelations among generalized distributions and their components. *Biometrics*, **22**, 44–52.

Kendall, D.G. (1948) On some modes of population growth leading to R. A. Fisher's logarithmic series distribution. *Biometrika* **35**, 6–15.

Kennedy, J.S. & Crawley, L. (1967) Spaced-out gregariousness in sycamore aphids *Drepanosiphum platanoides* (Schrank) (Hemiptera Callaphididae). *J. Anim. Ecol.* **369**, 147–70.

Keuls, M., Over H. I., & De Wit, C.T. (1963) The distance method for estimating densities. *Stat. Neerland.* **17**, 71–91.

Kieckhefer, R.W. (1975) Field populations of cereal aphids in South Dakota spring grains. *J. Econ. Entomol.* **68**, 161–4.

Kobayashi, S. (1966) Process generating the distribution pattern of eggs of the common cabbage butterfly, *Pieris rapae crucitora*. *Res. Popul. Ecol.* **8**, 51–60.

Kono, T. (1953) Basic unit of population observed in the distribution of the rice-stem borer, *Chilo simpler*, in a paddy field. *Res. Popul. Ecol.* **2**, 95–105.

Krause, G.F. & Pedersen, J.R. (1960) Estimating immature populations of rice weevils in wheat by using subsamples. *J. Econ. Entomol.* **53**, 215–16.

Kuehl, R.O. & Fye, R.E. (1972) An analysis of the sampling distributions of cotton insects in Arizona. *J. Econ. Entomol.* **65**, 855–60.

Kuno, E. (1963) A comparative analysis on the distribution of nymphal populations of some leaf- and planthoppers on rice plant. *Res. Popul. Ecol.* **5**, 31–43.

Kuno, E. (1969) A new method of sequential sampling to obtain the population estimates with a fixed level of precision. *Res. Popul. Ecol.* **11**, 127–36.

Kuno, E. (1986) Evaluation of statistical precision and design of efficient sampling for the population estimation based on frequency of occurrence. *Res. Popul. Ecol.* **28**, 305–19.

Kuno, E. (1991) Verifying zero-infestation in pest control: a simple sequential test based on the succession of zero-samples. *Res. Popul. Ecol.* **33**, 29–32.

Legg, D.E., Nowierski, R.M., Feng, M.G., Peairs, F.B., Hein, G.L., Elberson, L.R., & Johnson, J.B. (1994) Binomial sequential sampling plans and decision support algorithms for managing the Russian wheat aphid (Homoptera: Aphididae) in small grains. *J. Econ. Entomol.* **87**, 1513–33.

LeRoux, E.J. (1961) Variations between samples of fruit, and of fruit damage mainly from insect pests, on apple in Quebec. *Can. Entomol.* **93**, 680–94.

LeRoux, E.J. & Reimer, C. (1959) Variation between samples of immature stages and of mortalities from some factors of the Eye-spotted Bud Moth, *Spilonota ocellana* (D & S) (Lepidoptera: Olethreutidae), and the Pistol Casebearer, *Coleophora serratella* (L.) (Lepidoptera: Coleophoridae), on apple in Quebec. *Can. Entomol.* **91**, 428–49.

Lloyd, M. (1967) Mean crowding. *J. Anim. Ecol.* **36**, 1–30.

Lohr, S. (2009) *Sampling: design and analysis.* Cengage Learning.

Loot, G., Aulagnier, S., Lek, S., Thomas, F., & Guégan, J.F. (2002) Experimental demonstration of a behavioural modification in a cyprinid fish, *Rutilus rutilus* (L.), induced by a parasite, *Ligula intestinalis* (L.). *Can. J. Zool.* **80**, 738–44.

Lyons, L.A. (1964) The spatial distribution of two pine sawflies and methods of sampling for the study of population dynamics. *Can. Entomol.* **96**, 1373–407.

MacLellan, C.R. (1962) Mortality of codling moth eggs and young larvae in an integrated control orchard. *Can. Entomol.* **94**, 655–66.

Macleod, J. (1958) The estimation of numbers of mobile insects from low incidence recapture data. *Trans. R. Entomol. Soc. Lond.* **110**, 363–92.

Maillloux, G. & Bostanian, N.J. (1989) Presence-absence sequential decision plans for management of *Lygus lineolaris* (Hemiptera: Miridae) on strawberry. *Environ. Entomol.* **18**, 827–34.

Martini, X., Seibert, S., Prager, S.M., & Nansen, C. (2012) Sampling and interpretation of psyllid nymph counts in potatoes. *Entomol. Exp. Appl.* **143**, 103–110.

May, R.M. (1978) Host–parasitoid systems in a patchy environment: a phenomenological study. *J. Anim. Ecol.* **47**, 833–43.

McCullagh, P. & Nelder, J.A. (1989) *Generalized Linear Models.* Vol. 37. CRC Press.

McIntyre, A.D., Elliott, J.M., & Ellis, D.V. (1984) Meiofauna techniques. In: Holme, N.A. & McIntyre, A.D. (eds), *Methods for the Study of Marine Benthos.* Blackwell Scientific Publications, Oxford.

Mead, R. (1990) *The Design of Experiments: Statistical Principles for Practical Applications.* Cambridge University Press.

Meagher, R.L., Wilson, L.T., & Pfannenstiel, R.S. (1996) Sampling *Eoreuma loftini* (Lepidoptera:Pyralidae) on Texas sugarcane. *Environ. Entomol.* **25**, 7–16.

Milne, A. (1959) The centric systematic area-sample treated as a random sample. *Biometrics* **15**, 270–97.

Milne, A. (1964) Biology and ecology of the garden chafer, *Phyllopertha horticola*(L.), IX. Spatial distribution. *Bull. Entomol. Res.* **54**, 761–95.

Morisita, M. (1954) Estimation of population density by spacing method. *Mem. Fac. Sci. Kyushu Univ. E* **1**, 187–97.

Morisita, M. (1959) Measuring of the dispersion of individuals and analysis of the distributional patterns. *Mem. Fac. Sci. Kyushu Univ. E (Biol.)* **2**, 215–35.

Morisita, M. (1962) I_d-index, a measure of dispersion of individuals. *Res. Popul. Ecol.* **49**, 1–7.

Morisita, M. (1964) Application of I_d-index to sampling techniques. *Res. Popul. Ecol.* **69**, 43–53.

Morris, R.F. (1955) The development of sampling techniques for forest insect defoliators, with particular reference to the spruce budworm. *Can. J. Zool* **33**, 225–94.

Morris, R.F. (1960) Sampling insect populations. *Annu. Rev. Entomol.* **5**, 243–64.

Morris, R.F. & Reeks, W.A. (1954) A larval population technique for winter moth, *Operophtera brumata* (Linn.) (Lepidoptera: Geometridae). *Can. Entomol.* **86**, 433–8.

Mukerji, M.K. & Harcourt, D.G. (1970) Spatial pattern of the immature stages of *Hylemya brassicae* on cabbage. *Can. Entomol.* **102**, 1216–22.

Murdie, G. & Hassel, M.P. (1973) Food distribution, searching success and predator–prey models. In: Bartlett, M.S. & Hiorns, R.W. (eds), *The Mathematical Theory of the Dynamics of Biological Populations*. Academic Press, London.

Nachman, G. (1984) Estimates of mean population density and spatial distribution of *Tetranychus urticae* (Acarina: Tetranychidae) and *Phytoseiulus persimilis* (Acarina: Phytoseiidae) based upon the proportion of empty sampling units. *J. Appl. Ecol.* **21**, 903–13.

Naranjo, S.E. & Flint, H.M. (1995) Spatial distribution of adult *Bemisia tabaci* (Homoptera:Aleyrodidae) in cotton and development and validation of fixed-precision sampling plans for estimating population density. *Environ. Entomol.* **24**, 261–70.

Naylor, A.F. (1959) An experimental analysis of dispersal in the flour beetle, *Tribolium confusum*. *Ecology* **40**, 453–65.

Nelson, W.A., Slen, S.B., & Banky, E.C. (1957) Evaluation of methods of estimating populations of sheep ked, *Melopgagus ovinus* (L)(Diptera: Hippoboscidae), on mature ewes and young lambs. *Can. J. Anim. Sci.* **37**, 8–13.

Nestel, D. & Klein, M. (1995) Geostatistical analysis of leafhopper (Homoptera: Cicadellidae) colonisation and spread in deciduous orchards. *Environ. Entomol.* **24**, 1032–9.

Neyman, J. (1939) On a new class of 'contagious' distributions, applicable in entomology and bacteriology. *Ann. Math. Stat.* **10**, 35–57.

Ng, Y.-S., Trout, J.R., & Ahmad, S. (1983) Spatial distribution of the larval populations of the Japanese beetle (Coleoptera: Scarabaeidae) in turf grass. *J. Econ. Entomol.* **76**, 26–30.

Nielsen, B. (1963) The biting midges of Lyngby Aamose (Culicoides: Ceratopogonidae). *Natura Jutlandica* **10**, 46 pp.

Niwa, H.-S. (1994) Self-organizing model of fish schooling. *J. Theoret. Biol.* **171**, 123–36.

Nyrop, J.P., Agnello, A.M., Kovach, J., & Reissig, W.H. (1989) Binomial sequential classification sampling plans for European red mite (Acari: Tetranychidae) with special reference to performance criteria. *J. Econ. Entomol.* **82**, 482–90.

Nyrop, J.P. & Binns, M. (1991) Quantitative methods for designing and analyzing sampling programs for use in pest management. In: Pimental, D. (ed.) *CRC Handbook of Pest Management*. Boca Raton, Florida.

Nyrop, J.P., Villani, M.G., & Grant, J.A. (1995) Control decision rule for European Chafer (Coleoptera: Scarabaeidae) larvae infesting turfgrass. *Environ. Entomol.* **24**, 521–8.

Oakland, G.B. (1950) An application of sequential analysis to whitefish sampling. *Biometrics* **6**, 59–67.

Omori, M. & Ikeda, T. (1984) *Methods in Marine Zooplankton Ecology*. John Wiley & Sons, New York.

Paradis, R.O. & LeRoux, E.J. (1962) A sampling technique for population and mortality factors of the fruit-tree leaf roller, *Archips argyrospilus* (WLK) (Lepidoptera: Tortricidae), on apple in Quebec. *Can. Entomol.* **94**, 561–73.

Patil, G.P. & Joshi, S.W. (1968) *A Dictionary and Bibliography of Discrete Distributions*. Oliver & Boyd, Edinburgh.

Pearson, E.S. & Hartley, H.O. (1958) *Biometrika Tables for Statisticians*. Cambridge University Press, Cambridge.

Pedigo, L.P. & Buntin, G.D. (eds) (1994) *Handbook of Sampling Methods for Arthropods in Agriculture*. CRC Press, Boca Raton, Ann Arbor.

Pena, J.E. & Baranowski, R.M. (1990) Dispersion indices and sampling plans for the broad mite (Acari: Tarsonemidae) and the citrus rust mite (Acari: Eriophyidae) on limes. *Environ. Entomol.* **19**, 378–82.

Pena, J.E. & Duncan, R. (1992) Sampling methods for *Prodiplosis longifila* (Diptera: Cecidomyiidae) in limes. *Environ. Entomol.* **21**, 996–1001.

Peng, C. & Brewer, G.J. (1994) Spatial distribution of the red sunflower seed weevil (Coleoptera: Curculionidae) on sun flower. *Environ. Entomol.* **25**, 1101–5.

Perry, J.N. (1981) Taylor's power law for dependence of variance on mean in animal populations. *Appl. Statist.* **30**, 254–63.

Perry, J.N. (1988) Some models for spatial variability of animal species. *Oikos* **51**, 124–30.

Perry, J.N. (1994) Chaotic dynamics can generate Taylor's power law. *Proc. Roy. Soc. Lond. B.* **257**, 221–6.

Perry, J.N. (1995) Spatial analysis by distance sampling. *J. Anim. Ecol.* **64**, 303–14.

Perry, J.N., Bell, E.D., Smith, R.H., & Woiwod, I.P. (1996) SADIE: software to measure and model spatial pattern. *Aspects Appl. Biol.* **46**, 95–102.

Perry, J.N. & Hewitt, M. (1991) A new index of aggregation for animal counts. *Biometrics* **47**, 1505–18.

Perry, J.N., Liebhold, A.M., Rosenberg, M.S., Dungan, J., Miriti, M., Jakomulska, A., & Citron-Pousty, S. (2002) Illustrations and guidelines for selecting statistical methods for quantifying spatial pattern in ecological data. *Ecography* **25**(5), 578–600.

Peterson, A.T. (2011) Ecological niches and geographic distributions (MPB-49) (No. 49). Princeton University Press.

Pielou, E.C. (1969) *An Introduction to Mathematical Ecology*. Wiley-Interscience, New York, London.

Pieters, E.P. (1977) Comparison of sample-unit sizes for D-Vac sampling of cotton arthropods in Mississippi. *J. Econ. Entomol.* **71**, 107–8.

Potvin, C. & Roff, D.A. (1993) Distribution free and robust statistical methods: Viable alternative to parametric statistics? *Ecology* **74**, 1617–28.

Pradhan, S. & Menon, R. (1945) Insect population studies. I. Distribution and sampling of spotted bollworm of cotton. *Proc. Natl Inst. Sci. India* **6**(2), 61–73.

Prebble, M.L. (1943) Sampling methods in population studies of the European spruce sawfly, *Gilpinia hercyniae* (Hartig.), in Eastern Canada. *Trans. R. Soc. Canada* **III**(V), 93–126.

Preston, F.W. (1948) The commonness and rarity of species. *Ecology* **29**, 254–83.

Pringle, K.L. & Giliomee, J.H. (1992) Dispersion statistics and sample size estimates for monitoring mite populations in commercial apple orchards. *J. Appl. Ecol.* **29**, 143–9.

Putnam, L.G. & Shklov, N. (1956) Observations on the distribution of grasshopper egg-pods in Western Canadian stubble fields. *Can. Entomol.* **88**, 110–17.

Quenouille, M.H. (1949) A relation between the logarithmic, Poisson, and negative binomial series. *Biometrics* **5**, 162–4.

Reimer, C. (1959) Statistical analysis of percentages based on unequal numbers, with examples from entomological research. *Can. Entomol.* **91**, 88–92.

Resh, V.R. & Price, D.G. (1984) Sequential sampling: a cost-effective approach for monitoring benthic invertebrates in environmental impact assessments. *Environ. Manag.* **8**, 75–80.

Richard, B., Legras, M., Margerie, P., Mathieu, J., Barot, S., Caro, G., Desjardins, T., Dubs, F., Dupont, L., & Decaëns, T. (2012) Spatial organization of earthworm assemblages in pastures of northwestern France. *Eur. J. Soil Biol.* **53**, 62–9.

Riddle, M.J. (1989) Precision of the mean and the design of benthos sampling. *Mar. Biol. (Berlin)* **103**(2), 225–30.

Riley, D.G., Schuster, D.J., & Barfield, C.S. (1992) Sampling and dispersion of pepper weevil (Coleoptera: Curculionidae) adults. *Environ. Entomol.* **21**, 1013–21.

Romell, L.G. (1930) Comments on Raunkiær's and similar methods of vegetation analysis and the 'Law of Frequency'. *Ecology*, **11**, 589–96.

Romanczuk, P. & Schimansky-Geier, L. (2012) Swarming and pattern formation due to selective attraction and repulsion. *Interface Focus* **2**(6), 746–56.

Roslin, T., Gripenberg, S., Salminen, J.P., Karonen, M., B O'Hara, R., Pihlaja, K., & Pulkkinen, P. (2006) Seeing the trees for the leaves – oaks as mosaics for a host-specific moth. *Oikos* **113**(1), 106–20.

Routledge, R. D., & Swartz, T. B. (1991). Taylor's power law re-examined. *Oikos*, 107–112.

Ruesink, W.G. (1980) Introduction to sampling theory. In: Kogan, M. & Herzog, D.C. (eds), *Sampling Methods in Soybean Entomology*. Springer-Verlag, New York.

Safranyik, L. & Graham, K. (1971) Edge-effect bias in the sampling of sub-cortical insects. *Can. Entomol.* **103**, 240–55.

Saila, S.B., Pikanowski, R.A., & Vaughan, D.S. (1976) Optimum allocation strategies for sampling benthos in New York Bight. *Estuarine Coastal Mar. Sci.* **4**, 119–28.

Sale, P.F. (1998) Appropriate spatial scales for studies of reef-fish ecology. *Aust. J. Ecol.* **23**, 202–8.

Salt, C. & Hollick, F.S. (1946) Studies of wireworm populations. 11. Spatial distribution. *J. Exp. Biol.* **23**, 1–46.

Sawyer, A.J. (1989) Inconstancy of Taylor's *b*: simulated sampling with different quadrat sizes and spatial distributions. *Res. Popul. Ecol.* **31**, 11–24.

Schaalje, G.B., Butts, R.A., & Lysyk, T.J. (1991) Simulation studies of binomial sampling: a new variance estimator and density predictor, with special reference to the Russian Wheat Aphid (Homoptera: Aphididae). *J. Econ. Entomol.* **84**, 140–7.

Scheiner, S.M. & Gurevitch, J. (eds) (1993) *Design and Analysis of Ecological Experiments*. Chapman & Hall, New York.

Seber, G.A.F. (1982) *The Estimation of Animal Abundance and Related Parameters*. Griffin, London.

Shaw, M.W. (1955) Preliminary studies on potato aphids in north and north-east Scotland. *Ann. Appl. Biol.* **43**, 37–50.

Shelton, A.M. & Trumble, J.T. (1991) Monitoring insect populations. In: Pimentel, D. (ed.) *Handbook of Pest Management in Agriculture*, 2nd edn. CRC Press, Boca Raton.

Shibuya, M. & Ouchi, Y. (1955) Pattern of spatial distribution of soy bean pod gall midge in a soy bean field. *Oyo-Kontya* **11**, 91–7.

Shinozaki, K. & Urata, N. (1953) Apparent abundance of different species and heterogeneity. *Res. Popul. Ecol.* **2**, 8–21.

Shiyomi, M. & Nakamura, K. (1964) Experimental studies on the distribution of the aphid counts. *Res. Popul. Ecol.* **6**, 79–87.

Skellam, J.G. (1958) On the derivation and applicability of Neyman's type A distribution. *Biometrika* **45**, 32–6.

Smith, M.H., Gardner, R.H., Gentry, J.B., Kaufman, D.W., & O'Farrell, M.H. (eds) (1975) *Density Estimations of Small Mammal Populations. Small Mammals, Their Productivity and Population Dynamics*. Cambridge, Cambridge University Press.

Snedecor, G.W. & Cochran, W.G. (1989) *Statistical Methods*. University of Iowa Press, Ames, IA.

Southwood. T.R.E., Jepson, W.F., & Van Emden, H.F. (1961) Studies on the behaviour of *Oscinella frit* L. (Diptera) adults of the panicle generation. *Entomologia Exp. Appl.* **4**, 196–210.

Spiller, D. (1952) Truncated log-normal distribution of red scale (*Aonidella aurantiim Mask.*) on citrus leaves. *N. Z. J. Sci. Tech. (B)* **33**, 483–7.

Stark, R.W. (1952) Sequential sampling of lodgepole needle miner. *Forestry Chron.* **28**, 57–60.

Stark, R.W. & Dahlsten, D.L. (1961) Distribution of cocoons of a Neodiprion sawfly under open-grown conditions. *Can. Entomol.* **93**, 443–50.

Steffey, K.L., Tollefson, J.J., & Hinz, P.N. (1982) Sampling plan for population estimation of northern and western corn rootworm adults in Iowa cornfields. *Environ. Entomol.* **11**, 287–291.

Steiner, M. (1990) Determining population characteristics and sampling procedures for the Western Flower Thrips (Thansanoptera: Thripidae) and the predatory mite *Amblyseius cucumeris* (Acari: Phytoseiidae) on greenhouse cucumber. *Environ. Entomol.* **19**, 1605–13.

Stimson, J. (1974) An analysis of the pattern of dispersion of the hermatypic coral, *Pocillopora meandrina* var. *nobilis* Verrill. *Ecology* **55**, 445–9.

Stuart, A. (1962) *Basic Ideas of Scientific Sampling*, Griffin, London.

Suomela, J. & Ayres, M.P. (1994) Within-tree and among-tree variation in leaf characteristics of mountain birch and its implications for herbivory. *Oikos* **70**, 212–22.

Sylvester, E. & Cox, E.L. (1961) Sequential plans for sampling aphids on sugar beets in Kern County, California. *J. Econ. Entomol.* **54**, 1080–5.

Takeda, S. & Hukusima, S. (1961) Spatial distribution of the pear lace bugs, *Stephanitis naski* Esaki et Takeya (Hemiptera: Tingitidae) in an apple tree and an attempt for estimating their populations. *Res. Bull. Fac. Agric., Gifu Univ.* **14**, 68–77.

Tanigoshi, L.K., Browne, R.W., & Hoyt, S.C. (1975) A study on the dispersion pattern and foliage injury by *Tetranychus medanieli* (Acarina: Tetranychidae) in simple apple ecosystems. *Can. Entomol.* **107**, 439–46.

Taylor, L.R. (1961) Aggregation, variance and the mean. *Nature* **189**, 732–5.

Taylor, L.R. (1965) A natural law for the spatial disposition of insects. *Proc. XII Int. Congr. Entomol.* 396–7.

Taylor, L.R. (1970) Aggregation and the transformation of counts of Aphisfabae Scop. on beans. *Ann. Appl. Biol.* **65**, 181–9.

Taylor, L.R., Kempton, R.A., & Woiwood. I.P. (1976) Diversity statistics and the log-series model. *J. Anim. Ecol.* **45**, 255–72.

Taylor, L.R. (1984) Assessing and interpreting the spatial distributions of insect populations. *Annu. Rev. Entomol.* **29**, 321–57.

Taylor, L.R., Woiwod, I. P., & Perry, J.N. (1979) The negative binomial as a dynamic ecological model for aggregation, and the density dependence of (k). *J. Anim. Ecol.* **48**, 289–304.

Taylor, L.R., Woiwood, I. P., & Perry, J.N. (1978) The density-dependence of spatial behaviour and rarity of randomness. *J. Anim. Ecol.* **47**, 383–406.

Terry, I., Bradley, J.R., & Van Duyn, J.W. (1989) *Heliothis zea* (Lepidoptera: Noctuidae) eggs in soybeans: within-field distribution and precision level sequential count plans. *Environ. Entomol.* **18**, 908–16.

Teulon, D.A.J. & Cameron, E.A. (1995) Within-tree distribution of pear thrips (Thysanoptera: Thrioidae) in sugar maple. *Environ. Entomol.* **24**, 233–8.

Teyrovsky, V. (1956) Fotopathie larey kiestanek (Corixinae). *Acta Unir. Agric. Silv. Brunn.* **2**, 147–77.

Thomas, M. (1949). A generalization of Poisson's binomial limit for use in ecology. *Biometrika*, 18–25.

Thomas, M. & Jacob, F.H. (1949) Ecology of potato aphids in north Wales. *Ann. Appl. Biol.* **30**, 97–101.

Thompson, H.R. (1956) Distribution of distance to nth neighbour in a population of randomly distributed individuals. *Ecology* **37**, 391–4.

Tokeshi, M. (1995) On the mathematical basis of the variance-mean power relationship. *Res. Popul. Ecol.* **37**, 43–8.

Tonhasca, A., Palumbo, J.C., & Byrne, D.N. (1994) Binomial sampling plans for estimating *Bemisia tabacci* populations in Cantaloupes. *Res. Popul. Ecol.* **36**, 181–6.

Trumble, J.T., Brewer, M.J., Shelton, A.M., & Nyrop, J.P. (1989) Transportability of fixed-precision sampling plans. *Res. Popul. Ecol.* **31**, 325–42.

Tunstall, J.P. & Matthews, G.A. (1961) Cotton insect control recommendations for 1961–2 in the Federation of Rhodesia and Nyasaland. *Rhodesia Agric. J.* **58**(5), 289–99.

Turner, F.B. (1960) Size and dispersion of a Louisiana population of the cricket frog, *Acris gryllus. Ecology* **41**, 258–68.

Underwood, A.J. (1976) Nearest neighbour analysis of spatial dispersion of intertidal Prosobrunel Gastropods within two substrata. *Oecologia* **26**, 257–66.

Underwood, A.J. (1997) *Experiments in Ecology: Their Logical Design and Interpretation Using Analysis of Variance.* Cambridge University Press, Cambridge.

Unterstenhofer, G. (1957) The basic principles of plant protection field tests. *Hofchen Briefe* **10**, 173–236.

Upholt, W.M. & Craig, R. (1940) A note on the frequency distribution of black scale insects. *J. Econ. Entomol.* **33**, 113–14.

Usher, M.B. (1971) Properties of the aggregations of soil arthropods, particularly Mesostigmata (Acarina). *Oikos* **22**, 43–9.

Wadley, F.M. (1950) Notes on the distribution of insect and plant populations. *Ann. Entomol. Soc. Am.* **43**, 581–6.

Wadley, F.M. (1952) Elementary sampling principles in entomology. United States Department of Agriculture, Agricultural Research Administration, Bureau of Entomology and Plant Quarantine.

Walker, P.W. & Allsopp, P.G. (1993) Sampling distributions and sequential sampling plans for *Eumargarodes laingi* and *Promargarodes* spp. (Hemiptera: Margarodidae) in Australian Sugarcane. *Environ. Entomol.* **22**, 10–15.

Waloff, N. & Blackith, R.E. (1962) The growth and distribution of the mounds of *Lasius flavus* (Fabricius) (Hym.: Formicidae) in Silwood Park, Berkshire. *J. Anim. Ecol.* **31**, 421–37.

Ward, S.A., Chambers, R.J., Sunderland, K., & Dixon, A.F.G. (1986) Cereal aphid populations and the relation between mean density and spatial variance. *Neth. J. Plant Pathol.* **92**, 127–32.

Waters, W.E. & Henson, W. R. (1959) Some sampling attributes of the negative binomial distribution with special reference to forest insects. *Forest Sci.* **5**, 397–412.

Watt, A.S. (1964) The community and the individual. *J. Econ. Entomol.* **52**, 618–21.

Weithman, A.S., Reynolds, J.B., & Simpson, D.E. (1979) Assessment of structure of large-mouth bass [*Micropterus salmoides*] stocks by sequential sampling. In: *Proceedings Annual Conference - Southeastern Association of Fish and Wildlife Agencies (USA)*, vol. 33, pp. 415–24

Wilson, L.F. (1959) Branch tip sampling for determining abundance of spruce budworm egg masses. *J. Econ. Entomol.* **52**, 618–21.

Wilson, L.F. & Gerrard, D.J. (1971) A new procedure for rapidly estimating European pine sawfly (Hymenoptera: Diprionidae) population levels in young pine plantations. *Can. Entomol.* **103**, 1315–22.

Yates, F. (1953) *Sampling Methods for Censuses and Surveys.* C. Griffin, London.

Yates, F. & Finney, D.J. (1942) Statistical problems in field sampling for wireworms. *Ann. Appl. Biol.* **29**, 156–67.

Yoshihara, T. (1953) On the distribution of *Tectarius granularis*. *Res. Popul. Ecol.* **2**, 112–22.

Zahl, S. (1974) Application of the S-method to the analysis of spatial pattern. *Biometrics* **30**, 513–24.

Zar, J.H. (1984) *Biostatistical Analysis*. Prentice-Hall, Englewood Cliffs, NJ.

Zehnder, G.W. (1990) Evaluation of various potato plant sample units for cost-effective sampling on Colorado potato beetle (Coleoptera: Crysomelidae). *J. Econ. Entomol.* **83**, 428–33.

3 Absolute Population Estimates Using Capture–Recapture Experiments

When an absolute estimate of population size is required, these methods are the main alternative approach to those based on the count of animals within a fixed unit of the habitat. They offer the advantage that their accuracy does not depend on an assessment of the amount of habitat, but their disadvantage is that accuracy does depend on capturing a large proportion of the population. For many invertebrates they will only produce order of magnitude estimates of population size as available resources will not allow even 1% of the population to be marked. However, even an order of magnitude estimate may be satisfactory as a second, independent, check on population size; as stressed in Chapter 1 it is wise, whenever possible, to use more than one independent method. Seber (1982, 1986) reviews the statistical theory that underlies this approach. Originally developed to estimate population size, capture–recapture methods are now frequently used for the estimation of birth, death and emigration rates. Lebreton *et al.* (1992) review methodologies for the estimation of survival rates.

Populations can be classified as either *open* or *closed*. A closed population remains unchanged over the period of study, whereas an open population changes because of some combination of birth, death, and emigration. Reviews of methods applicable to closed and open populations are those of White *et al.* (1982) and Chao (2001) for closed, and Pollock *et al.* (1990) for open populations, respectively. Descriptions of the application of the methodology to particular animal groups are given by Nichols *et al.* (1981) for birds, by Burnham *et al.* (1987) for fish, and by Eberhardt *et al.* (1979) for marine mammals. Recent advances in capture–recapture modelling are reviewed in Amstrup *et al.* (2010).

A mark–recapture method was first used for ecological study in 1896 by Petersen to estimate plaice, *Platichthys platessa*, populations; later, Lincoln (1930) independently developed the method to estimate waterfowl populations. The principle used to calculate population size in these early studies, in which the population was assumed closed, still applies. If a sample from a population is marked, returned to the original population and then, after complete mixing, re-sampled, the number of marked individuals in the second sample would have the same ratio to the total numbers in the second sample as the total of marked individuals originally released would have to the total population. Historically, a basic prerequisite to the use of these methods was a technique for marking the animals so that they can be released unharmed and unaffected into the wild and recognised again on recapture. Such techniques may also be used in other

---

types of study, such as dispersal, longevity and growth; however, for convenience all marking methodologies for animals will be discussed in the second half of this chapter.

It is now possible to identify individual animals using unique DNA signatures as genetic tags and therefore to undertake mark–recapture noninvasively by the collection of hair, feathers, and faeces. Lukacs & Burnham's (2005) review of capture–recapture methods applicable to genetic sampling and the R package capwire has been developed for this type of sampling.

3.1 Capture–recapture methods

Various assumptions underlie all methods of capture–recapture analysis. If the particular animal does not fulfil one or more conditions it might be possible to allow for this to some extent, but a method of analysis should not be applied without ensuring, as far as is practicable, that its inherent assumptions are satisfied. Parsimony, the use of the smallest possible number of parameters to model a situation, is essential, so though assumptions can be avoided by adding further parameters to a model, this inevitably leads to an increase in the variability of parameter estimates (Cormack, 1979; Begon, 1983). In the following account, models are presented in order of increasing complexity and decreasing number of underlying assumptions. The search for parsimony has led to the development of computer programs to identify and fit the most appropriate model from the wide range available (Crosbie & Manly, 1985; Lebreton & Clobert, 1986, Lebreton et al., 1992). A list of computer software for the analysis of capture–recapture data is available at http://www.phidot.org/software/, and from the Patuxent Software Archive – the USGS at http://www.mbr-pwrc.usgs.gov/software.html. Many resources are now available using R; for example, the RMark library is a collection of R functions that interface to MARK for analysis of capture–recapture data.

At the onset of study neither the size of the population nor the correct assumptions to make about the capture probabilities are known. The ecologist therefore faces the task of both estimating the population parameters and identifying the most appropriate model for the purpose. Fieldwork should not be undertaken until the applicability of the method has been assessed. The effort that will be expended on a capture–recapture study will be determined by resource availability and the percentage accuracy required (see Chapter 2). The latter is defined as:

$$A = \pm 100 \left(\frac{\text{Estimated} - \text{True}}{\text{True}} \right) \tag{3.1}$$

Robson & Reiger (1964) suggest that A should range from ±50% for preliminary surveys, ±25% for management work, to ±10% for research work. Examination of published studies suggests that these ranges have rarely been achieved. Capture–recapture methods are labour-intensive, and quite high proportions of the population must be marked if even 50% accuracy is to be gained. Having defined the required accuracy and made a guess at population size, it is helpful to first estimate the sampling effort that would be required for a simple Peterson–Lincoln index estimation (see p. 87). All other methods will require greater effort. If the required effort is feasible, simulated data can then be explored to gain insight into the performance of the proposed models and their sensitivity to violations in the underlying assumptions.

3.1.1   Assumptions common to most methods

There are four assumptions:

1. The marked animals are not affected (either in behaviour or life expectancy) by being marked, and the marks will not be lost.
2. The marked animals become completely mixed in the population.
3. The probability of capturing a marked animal is the same as that of capturing any member of the population; that is, the population is sampled randomly with respect to its mark status, age and sex. Termed 'equal catchability', this assumption has two aspects: first, that all individuals of the different age groups and of both sexes are sampled in the proportion in which they occur; second, that all the individuals are equally available for capture irrespective of their position in the habitat.
4. Sampling must be at discrete time intervals and the actual time involved in taking the samples must be small in relation to the total time. Chao (2001) briefly reviews continuous-time models.

As the violation of these assumptions may invalidate the capture–recapture method, care must be taken to test them, as will be discussed below.

3.1.1.1   Marking has no effect

Manly (1971a) has described a test to determine if the survival of individuals immediately after marking and release is different from that of animals that have been marked for some while (mortalities associated with marking are termed Type I losses in fisheries science). It is thus a test of the effect of the marking process and not of the influence of bearing a mark. It depends on each animal bearing a unique mark and may be used in conjunction with the actual data derived from a series of, at least four, marking occasions. The difference (y) between the survival of animals newly exposed to marking and others is given by:

$$y = \ln\left(\frac{r_i'}{r_i} \times \frac{u_i}{R_i}\right) \tag{3.2}$$

where r_i' = the number of marked animals in the ith sample that are recaptured again later, r_i = the total marked animals in the ith sample, u_i = the total unmarked animals in the ith sample, and R_i = the total animals marked on the ith occasion and subsequently recaptured. Now, if the process of marking has no effect on the animal, y will have an expected value of zero. The significance of any value of y can be determined by the statistic g, derived by dividing y by its variance:

$$g = \frac{y}{\left(\frac{1}{r_i'} + \frac{1}{R_i} - \frac{1}{r_i} - \frac{1}{u_i}\right)^{\frac{1}{2}}} \tag{3.3}$$

g will behave as a random normal variate with zero mean and variance. The one-tail test is normally justified because an increase in survival due to the marking process can be ruled out; thus, at the 5% level $g < 1.64$ or at 10% level $g < 1.29$ may be taken as showing that the marking process has no effect on survival. Robson (1969) has developed a similar test.

3.1.1.2   Marks will not be lost

Mark loss may be detected by double marking (Seber & Felton, 1981), where each marked individual carries two different marks. The total number of recaptures is given by:

$$r = r_a + r_b + r_{ab} \tag{3.4}$$

where r_a is the number of recaptures with only type a tag, r_b the number with only type b tag, and r_{ab} those with both a and b tags.

Seber (1982) shows that if

$$k = \frac{r_a r_b}{(r_a + r_{ab})(r_b + r_{ab})} \tag{3.5}$$

and

$$c = \frac{1}{1-k} \tag{3.6}$$

then the estimated number of recaptures in the absence of mark loss is given by:

$$\hat{r} = c(r_a + r_b + r_{ab}). \tag{3.7}$$

This estimate should be compared with r to determine the extent of the problem.

The probability of losing mark b or a is

$$r_b / r_b + r_{ab} \tag{3.8}$$

and

$$r_a / r_a + r_{ab} \tag{3.9}$$

respectively.

3.1.1.3 Equal catchability

The violation of this assumption resulting in population underestimates is common (Cormack, 1972; Bohlen & Sundstrom, 1977; Caughly, 1977). Eberhardt (1969) gives the three general causes:

1. Differences in behaviour near traps.
2. Learning, leading to trap addiction or shyness.
3. Unequal probability of capture because of trap positions.

Roff (1973) reviewed the statistical tests then available for the purpose of testing for equal catchability and, from simulations, concluded that under practical conditions they may be unreliable. An extreme example of such unreliability results if certain individuals cannot be caught by the chosen sampling method (Boonstra & Krebs, 1978). This warning is useful; the ecologist must use intuition and any biological indications for confirmation that the population is homogeneous (White, 1975). Reasons for and examples of unequal catchability are listed below but, as shown by Krebs *et al.* (1994) in a study of feral house mice, there can be more than one reason for variable and low catchability and this can change over the course of a study.

1. Subcategories of the population are differently sampled because, by the nature of the habitat, only part of the population is available. This problem is particularly acute with subcortical, wood-boring and subterranean insects. It is well exemplified by ants: the number and composition of the foragers of an ant colony are so variable that unless one has a considerable knowledge of these aspects, or can sample from within the nest (Stradling, 1970), Petersen–Lincoln Index and related methods cannot be used for population estimations (Ayre, 1962; Erickson, 1972). The marking and recapture of foraging ants may however be used to estimate the foraging area (Kloft *et al.*, 1965) and, over a short period of time, the number of foragers (F) can be estimated using the following expression (Holt, 1955):

$$F = R_f \times (T_0 + T_i) \tag{3.10}$$

where F = the number of foragers; R_f = rate of flow of foragers per unit time (i.e., the average number of ants passing fixed points on all routes from a nest per minute); T_0 = average time (in the same units as R_f) spent outside the nest (found by marking individuals on the outward journey and timing their return); and T_i = average time spent inside the nest (found by marking ants as they enter and timing their reappearance). In *Pogonomyrmex badius* only about 10% of the total population forage above ground in any two-week period (Golley & Gentry, 1964), whilst in the wood ant, *Formica*, a constant group of individuals forage such that if their number (*F*) is estimated, colony size (*C*) is given by (Kruk-de-bruin *et al.*, 1977):

$$\log \widehat{C} = (\log \widehat{F} + 0.75)/1.01 \tag{3.11}$$

The estimation of the number of uncatchable individuals using double sampling is discussed by Aebischer (1986). The approach is to apply two different sampling methods, such as *trapping* and *sighting*. Some animals, which may be impossible to trap, may still be observed. For example, in a recapture experiment ducks were caught on the nest; however, some ducks could not be caught because they ran off when the recorder approached. By using binoculars, the nesting ducks could be counted and the two types of information combined to obtain a population estimate.

2. Subcategories of the population are differentially sampled because differences due to sex and/or age or other causes. Lomnicki (1969), for example, found that land snails are very nonrandom in their movements, some individuals being more inclined to recapture than others. In many animals, especially when trapping techniques are used, differences will be associated with differences in sex or age. Methods to adjust for changes linked to breeding behaviour are given in Nichols *et al.* (1994). It is indeed a good principle to enumerate and estimate males and females separately, at least in the first instance; many workers have found striking differences in their behaviour (e.g. Jackson, 1933a; Cragg & Hobart, 1955; Macleod, 1958). Initially, different stages or age classes should also be considered separately and their survival rates determined using one of the formulae given below; if their survival rates differ significantly, one must continue to consider them separately. Indeed, every attempt should be made to ensure that the population estimated is as homogeneous as possible. Provided there is no mortality, a χ^2 goodness-of-fit test may be applied to a table of the releases, recaptures and nonrecaptures of the different subclasses (Seber, 1982; White, 1975).

3. There is a periodicity in the availability of subcategories. This problem has been particularly encountered with mosquitoes, where an initial period of emigration (Sheppard *et al.*, 1969) and the gonotrophic and linked feeding cycles invalidate the equal catchability assumption and may lead to fivefold overestimates (Conway *et al.*, 1974). Other animals show periodicities in their behaviour and therefore, particularly when sampling is by a behavioural method (e.g. trapping) and the cycle is long compared with the sampling interval, workers should consider this problem. With the Fisher & Ford method (see below) this may be allowed for by introducing additional terms into the model (Sheppard *et al.*, 1969; Conway *et al.*, 1974). This modification requires prior, independently gained, knowledge of the periodicities. The extent to which the incorporation of these additional terms improves the relationship of the expected and observed recaptures gives a measure of their value, but a precise statistical test is not available.

4. The processes of capturing and marking affect catchability. This can have two origins: the initial experience may alter catchability or it may be a cumulative effect. In the former case (more relevant to vertebrates), a regression method analogous to removal trapping (p. 268) may be used to estimate total population (Marten, 1970). The effect of repeated capture may be tested for by Leslie's test for random recaptures demonstrated below (Orians & Leslie, 1958).

3.1.1.3.1 *Statistical tests to detect variation in catchability*

A number of tests are available.

Cormack's test

This method is appropriate if there are three sampling periods. Cormack (1966) gives two methods, the first for experiments where the capture technique is changed at each sampling occasion, and the second where the same sampling methodology is used on all three occasions. Let

n_1 = the number caught and marked in period 1,
r_{10} = the number only caught in period 1,
r_{12} = the number only caught in periods 1 and 2
r_{13} = the number only caught in periods 1 and 3
r_{123} = the number caught in periods 1, 2 and 3.

For both methods 1 and 2 equal catchability is tested using the hypothesis that the coefficient of variation in catchability is zero.

For method 1, the test statistic is:

$$z = \frac{(r_{123}r_{10} - r_{12}r_{13})\sqrt{n_1}}{\sqrt{(r_{12} + r_{123})(r_{13} + r_{123})(n_1 - r_{12} - r_{123})(n_1 - r_{13} - r_{123})}} \tag{3.12}$$

where z is the standard normal deviate. If z is negative, equal catchability can be assumed.

For method 2, first calculate estimates of the coefficient of variation (CV) and variance (s^2) of the catchability as follows:

$$\hat{C}V = \frac{(r_{123}\ r_{10}) - (r_{12}\ r_{13})}{(r_{12} + r_{123})(r_{13} + r_{123})} \tag{3.13}$$

and

$$\hat{s}^2 = \frac{r_{123}^2 n_{12}}{(r_{12} + r_{123})^2(r_{13} + r_{123})^2}\left[\frac{1}{r_{123}} - \frac{1}{n_1} - \frac{r_{12} + r_{13}}{(r_{12} + r_{123})(r_{13} + r_{123})}\right]. \tag{3.14}$$

Then use these values to derive the final estimates

$$CV = \frac{1 - 2\hat{C}V - \sqrt{1 - 8\hat{C}V}}{2(1 + \hat{C}V)} \tag{3.15}$$

$$s^2 = \frac{(3 - \sqrt{1 - 8\hat{C}V})^4(\hat{s}^2)}{16(1 - 8\hat{C}V)(1 + \hat{C}V)^4} \tag{3.16}$$

The test statistic is

$$z = \frac{CV}{s^2} \tag{3.17}$$

where z is the standard normal deviate. Unequal catchability is significant at the 5% level for $z > 1.64$. If d is greater than 0.125, the equations cannot be solved because the test assumptions are violated. This indicates unequal catchability.

Zero-truncated Poisson test

This test is more powerful than Cormack's but requires more than three samples (Caughly, 1977). It is zero-truncated because there is no information on the animals which avoided capture. The expected frequency distribution given equal catchability is determined by the mean

$$\bar{x} = \frac{\sum f_x x}{\sum f_x} \qquad (3.18)$$

where x is the number of times captured and f_x the number of animals caught exactly x times.

Given p sampling periods, if $x/s < 0.25$ then a Poisson distribution can be fitted to the data. When $x/s > 0.25$ a binomial should be fitted (Seber, 1982; see Chapter 2, pp. 28 for R functions to fit Poisson and binomial distributions). The observed and expected is compared using a χ^2 test.

Chapman's test

Chapman (1954) devised a nonparametic test for closed populations. Let

r_{ij} = the number of animals marked in period i and first recaptured in period j,
n_i = the total number of animals caught in period i,
p = the number of sampling periods, u_i = the number of unmarked animals caught in sample i. Construct the array:

$$
\begin{bmatrix}
\dfrac{r_{12}}{u_1 n_2} & \dfrac{r_{13}}{u_1 n_3} & \dfrac{r_{14}}{u_1 n_4} \cdots \cdots & \dfrac{r_{1p}}{u_1 n_p} \\[2ex]
\vdots & \dfrac{r_{23}}{u_2 n_3} & \dfrac{r_{24}}{u_2 n_4} \cdots \cdots & \dfrac{r_{2p}}{u_2 n_p} \\[2ex]
\vdots & \vdots & \vdots & \vdots \\[1ex]
\vdots & \vdots & \vdots & \dfrac{r_{p-1,p}}{u_{p-1} n_p}
\end{bmatrix}
$$

The test statistic, X, is defined using the sum of the differences between successive values in each row as follows:

$$X = D_1 + D_2 + D_3 \cdots \cdots \cdots + D_{p-2}$$

where D_i is the number of negative differences in the ith row. Table 3.1 gives critical values for the X statistic.

Table 3.1   Critical values for the X statistic.

Sampling periods,p	Observed number of negative differences							
	0	1	2	3	4	5	6	7
5	0.0035	0.0590	0.3056	0.6944	0.9410	0.9965	1.0000	1.0000
6	—	0.0012	0.0172	0.1052	0.3392	0.6608	0.8948	0.9828
7	—	—	0.0001	0.0020	0.0166	0.0627	0.2010	—
8	—	—	—	—	0.0001	0.0009	0.0323	0.1103

Leslie's test

A comparison is made of the actual and expected variances of a series of recaptures of individuals known to be alive throughout the sampling period; individual marks must have been used. This test is best illustrated by a worked example taken from Leslie's appendix, based on the recaptures of shearwaters; with insects, of course, the recapture periods would be days or weeks rather than years.

Thirty-two individuals marked in 1946 were recovered for the last time in 1952; therefore, they were available for recapture in the years 1947–51 inclusive and the following two tables can be prepared (Tables 3.2 and 3.3).

The actual sum of squares:

$$\sum (x - \bar{x})^2 = \sum x^2 f(x) - \frac{\left[\sum xf(x)\right]^2}{N}$$

$$= 69 - 30.03 = 38.97.$$

The expected variance:

$$\sigma^2 = \frac{\sum n_1}{N} - \frac{\sum (n_i^2)}{N^2}$$

$$= 31/32 - 199/32^2 = 0.9688 - 0.1943 = 0.7745$$

Then, $X^2 = 38.97/0.7745 = 50.32$.

Table 3.2   Leslie's test: year-by-year analysis.

Year	No. of recaptures in each year (n_i)
1947	7
1948	7
1949	6
1950	4
1951	7
n_i	=31

Table 3.3 Leslie's test: all years analysis.

No. of recaptures for each individual X	Frequency of X f(x)
0	15
1	7
2	7
3	2
4	1
5	0
	$N = \Sigma f(x) = 32$

X^2 may be treated as equivalent to a X^2 and for degrees of freedom (N − 1) of between 20 and 30, the probability of a value as great as or greater than this can be assessed from X^2 tables; probabilities of less than 0.50 imply that capture is not random. With values over 30, use can be made of the fact that

$$\sqrt{2\chi^2} - \sqrt{2df - 1}$$

is approximately normally distributed about a mean of zero without standard deviation. In other words, one calculates the value for this expression and looks up its probability in the table of the normal deviate. In the present case:

$$\sqrt{2 \times 50.32} - \sqrt{(2 \times 32)} - 1 = 10.03 - 7.81 = 2.22$$

The probability of a deviate as great as this is somewhere between 0.025 and 0.020, and therefore the shearwaters were not collected randomly. Leslie has suggested that the test should only be used when the number of individuals is 20 or more and the number of occasions on which recapture was possible is at least 3. This test will not distinguish whether the higher catchability of some individuals is due to catching effects or, as Lomnicki (1969) presumed with land snails, inherent individual differences. Carothers (1973) developed an extension of Leslie's test more appropriate when the number of occasions on which recapturing occurs is small.

3.1.2 Estimating closed populations

These methods assume that the population does not change over the period of study. It is therefore essential that this period is short compared to life expectancy. Further, individuals must not be able to enter or leave the study area. Methods in which the probability of capture is not assumed constant require a series of occasions (at least two) on which animals previously marked are recaptured marked again and released. We shall first describe the simple Petersen–Lincoln index which, as it uses only one period of recapture, assumes a constant probability of capture.

3.1.2.1 The Petersen–Lincoln estimator

Provided that the four common assumptions (see Section 3.1.1) are satisfied, together with those of closure and constant probability of capture, it is legitimate to estimate the total population from the simple index used by Lincoln (1930):

$$\hat{N} = \frac{an}{r} \tag{3.19}$$

where $\hat{N}$ = the estimate of the number of individuals in the population, a = total number marked in the first sample, n = total number of individuals in the second sample, and r = total recaptures (individuals marked at time 1 and re-caught at time 2).

When n is predetermined and approximately equal to a, the variance of this estimate is (Bailey, 1952):

$$\text{var}\hat{N} = \frac{a^2 n (n - r)}{r^3} \tag{3.20}$$

If the second sample consists of a series of subsamples and a large proportion of the population has been marked, then it is possible to utilise the recovery ratios ($= r/n$) in each to calculate the standard error of the estimated population (Welch, 1960):

$$\hat{N} = \frac{a}{R_T} \tag{3.21}$$

where R_T = the recovery ratio (r/n) based on the total animals in all of the samples. The variance is approximately:

$$\text{var}(\hat{N}) = \left(\frac{a}{R_T^2} \right)^2 \times R_T \frac{(1 - R_T)}{y} \tag{3.22}$$

where y = total animals in subsamples. This approach is only valid if the marked individuals are distributed randomly in the subsamples, which may be tested by comparing the observed and theoretical variance (see p. 85).

Chapman (1951) shows that Equation (3.19) is the best estimate of $\hat{N}$ as $\hat{N} \to \infty$ but gives a large bias for small samples. When $n_1 + n_2 \geq \hat{N}$ a better estimator is:

$$N^* = \frac{(a + 1)(n + 1)}{(r + 1)} - 1. \tag{3.23}$$

Using a Poisson approximation to the hypergeometric distribution, Seber (1970) gives an approximate estimate of the variance:

$$\text{var}(N^*) = \frac{(a + 1)(n_2 + 1)(a - r)(n_2 - r)}{(r + 1)^2 (r + 2)}. \tag{3.24}$$

The above calculations are applicable to large samples where the value of r is fairly large; (Bailey, 1951, 1952) has suggested that with small samples, a less-biased estimate based on a binomial approximation to the hypergeometric distribution is:

$$\hat{N} = \frac{n(a+1)}{r+1} \qquad (3.25)$$

and an approximate estimate of the variance is:

$$\text{var}(\hat{N}) = \frac{a^2(n_2+1)(n_2-r)}{(r+1)^2(r+2)}. \qquad (3.26)$$

Bonett (1988) compares the bias of the different estimates. The above methods are based on what is referred to as *direct sampling* in which the size of n_2 is predetermined. If possible, *inverse sampling*, where the number of marked animals to be captured (r) is predetermined has the advantage of giving unbiased estimates of population size ($\hat{N}$) and variance as follows (Bailey, 1952):

$$\text{var}(\hat{N}) = \frac{(a-r+1)(a+1)n_2(n_2-r)}{r^2(r+1)} \qquad (3.27)$$

The sample size required to given a 95% probability of achieving an accuracy (A) (p. 78) can be obtained using computer programs. Examination of the results often shows that it is often necessary to mark >50% of the population to achieve modest accuracy (Roff, 1973). It may well show that the proposed project is impossible.

Seber (1982) argues that, given two population estimates N_a and N_b with variances v_a and v_b, the null hypothesis $H_0 : N_a = N_b$ can be tested using:

$$z = \frac{N_a - N_b}{\sqrt{(v_a^* + v_b^*)}}, \qquad (3.28)$$

which is approximately unit normal when H_o is true. Inference relating to the comparison of Petersen–Lincoln estimates is also discussed by Skalski *et al.* (1983). A potentially useful model for small r is a Bayesian model (Gaskell & George, 1972), which takes account of prior knowledge about the distribution of N. Gaskell & George's formula is essentially:

$$\hat{N} = \frac{an_2 + 2N'}{r+2} \qquad (3.29)$$

where N' is a prior estimate obtained, for example, from quadrat sampling or nearest-neighbour-type techniques. A Bayesian approach is also adopted by Zucchini & Channing (1986).

While simple, the Petersen–Lincoln index is still much used, and some example applications are available in Bozeman *et al.* (1985), Gaillard *et al.* (1986), Baker &

Herman (1987), Menkens & Anderson (1988), Gatz & Loar (1988), Poljak *et al.* (1989), Law (1994), and Mazzotti (1995), Baber *et al.* (2010), and Williams (2013).

3.1.2.2   Other single mark methods

A variant on the Peterson–Lincoln estimation is where a single marking period is followed by a sequence of samples, each comprising animals which are permanently removed from the population. Such a procedure is common in fisheries studies where commercial landings yield the recaptures. Termed a mark–removal model, it has been studied in detail by Skalski & Robson (1982). Population size can be estimated using a least-squares method (Paloheimo, 1963). For fish populations, the Petersen–Lincoln method has also been combined with age–length data to estimate production (Newman & Martin, 1983).

Another variant discussed by Minta & Mangel (1989) is a capture–resight experiment where initially marked animals are noted but not captured during subsequent recording periods. The technique successfully estimated populations of badgers, bison, and crested porcupines.

3.1.2.3   The Schnabel census

These are methods for closed populations using more than two periods of capture (Schnabel, 1938). Darroch (1958) showed that the maximum likelihood estimator of population size is the largest root of the polynomial:

$$\left(1 - \frac{r}{N}\right) = \prod_{i=1}^{s}\left(1 - \frac{n_i}{N}\right), \tag{3.30}$$

where $r = \Sigma a_w$ and a_w is the number of animals with a particular capture history; for example, a_{14} represents the animals caught on the 1st and 4th sampling occasions.

When $s = 2$, Equation (3.30) reduces to the simple Peterson–Lincoln index and for $s = 3$ it is:

$$N^2(m_2 + m_3) - N(n_1n_2 + n_1n_3 + n_2n_3) + n_1n_2n_3 = 0.$$

where n_i and m_i are the total number of captures and recaptures at time i, respectively.

For $s > 3$, Equation (3.30) needs to be solved iteratively.

A simpler approach is just an extension of the Petersen–Lincoln method to p sampling periods. First estimate N for each sampling step i using:

$$\hat{N}_i = \frac{(M_i + 1)(n_i + 1)}{m_i + 1} - 1 (i = 2, 3. \dots p), \tag{3.31}$$

where $M_i = \Sigma\, n_i - m_i.$

The variance v^*_i of this estimate is:

$$v^*_i = \frac{(M_i + 1)(n_i + 1)(M_i - m_i)(n_i - m_i)}{(m_i + 1)^2 (m_i + 2)} \tag{3.32}$$

Then calculate the average population number,

$$\bar{N} = \frac{\sum \hat{N}_i}{(s - 1)} \tag{3.33}$$

If animals are sampled one at a time we have a particular type of Schnabel census which can be analyzed using Craig's method (p. 92).

Examples of the application of the Schnabel census are Poljak *et al.* (1989), Bozeman *et al.* (1985) and Belmar-Lucero *et al.* (2012).

3.1.2.4 Selection from a family of models with differing assumptions for a closed population

Given a series of samples, some of the stringent assumptions underlying the Petersen–Lincoln index can be relaxed. A general approach to the selection and analysis of suitable models has been described by Otis *et al.* (1978) and White *et al.* (1982), who consider a family of models which are systematically tested for applicability. Unlike the Petersen–Lincoln model, the assumption that the probability of capture (p) is constant can now be relaxed. The models they consider are as follows:

M_0 – Constant p;
M_t – Time varying p;
M_b – a behavioural response to capture changes in p;
M_h – there is heterogeneity in p for unknown reasons.
Plus, the four mixed models, M_{bh}, M_{th}, M_{tb}, M_{tbh}.

Chao (1987) describes a point estimator and its associated confidence interval for the M_h model, and subsequently in Chao (1989) considered the application of M_h and M_t models to sparse populations.

By using a computer program each of these models can be fitted to the data and the 'best' estimator identified (Pollock, 1974, 1975, 1981; White *et al.*, 1982). The program MARK can fit 12 different models which differ in the assumptions used to estimate the size of a closed population. MARK can be downloaded from http://warnercnr.colostate.edu/~gwhite/mark/mark.htm, and an introductory text MARK is available from http://www.phidot.org/software/mark/docs/book/. The RMark package is an R interface for MARK.

Loglinear models (Cormack, 1989; these are discussed further on p. 102) can also be applied to closed populations.

As an example of the ease of application of closed models, the following R listing fits *M0, Mt, Mh Chao (LB), Mh Poisson2, Mh Darroch, Mh Gamma3.5, Mth, Chao (LB), Mth*

Poisson2, Mth Darroch, Mth Gamma3.5, Mb and *Mbh* models using the Rcapture package in R. The data set here comprises the capture histories of 68 snow shoe hares sampled over five periods.

```
require(Rcapture)
#Get hare data set
data(hare)
#Print data
hare
#Make calculations
hare`closed`estimates <-closedp(hare)
#Print results
hare`closed`estimates
```

The output is as follows:

Abundance estimations and model fits:

	abundance	stderr	deviance	df	AIC
M0	75.4	3.5	68.516	61	154.707
Mt	75.1	3.4	58.314	56	154.505
Mh Chao (LB)	79.8	6.4	58.023	58	150.214
Mh Poisson2	81.5	5.7	59.107	60	147.298
Mh Darroch	90.4	11.6	61.600	60	149.791
Mh Gamma3.5	100.6	21.7	62.771	60	150.961
Mth Chao (LB)	79.6	6.3	47.115	52	151.305
Mth Poisson2	81.1	5.6	48.137	55	146.327
Mth Darroch	90.5	11.7	50.706	55	148.896
Mth Gamma3.5	101.6	22.4	51.956	55	150.147
Mb	81.1	8.3	67.027	60	155.217
Mbh	74.2	14.6	63.257	59	153.447

The AIC value indicates that the *Mh* model estimate of 81.5 is the best.

3.1.2.5 Frequency of capture methods

The information gained from a series of samples may be considered another way. If the frequency of capture follows a particular distribution (f_1 … animals captured once, f_2 … animals captured twice, f_3 … … f_x), then the term f_0 represents those animals that have not been marked or captured at all, and the sum of all the terms will represent the total population. Theoretically, problems do not arise from unequal catchability, a non-Poisson distribution may be fitted that describes the deviations from random in the probability of capture. Tanton (1965) fitted a zero-truncated negative binomial to the frequency of recapture of wood mice, *Apodemus*. Further details of such models, which are particularly useful in small mammal work (e.g. Marten, 1970) are given by Seber (1982).

3.1.2.6 Craig's method

This method, devised for use with butterflies, assumes that the population is both closed and each individual has a constant probability of capture. In practice the butterflies, or other highly mobile but colonial animals, are collected randomly, marked and immediately released, after recording the number of times, if any, the animal had previously been captured. The effective sample size is thus one. Craig (1953) assumed that the frequency of recapture could be described by two mathematical models: the truncated Poisson (p. 30) and the Stevens distribution function. On the basis of these two models, six different methods can be used to estimate the size of the population: three based on moments, and three on maximum likelihood: Craig found, however, that they gave similar results. The first[1] of the two simplest models, based on the Poisson, is:

$$\hat{N} = \frac{\sum x f_x^2}{\left(\sum x^2 f_x - \sum x f_x\right)} \tag{3.34}$$

where N = the estimate of population, x = the number of times an individual had been marked, f_x = the frequency with which individuals marked x times had been caught, and thus $\Sigma x f_x$ = the total number of different times animals were captured (namely, the number caught once + 2 × the number caught twice + 3 × the number caught thrice, etc.). This method is simple, allowing a direct solution, but is subject to greater sampling error than Equation (3.35). It is useful for obtaining a trial value for use in solving Equation (3.35) (and also in the other methods given by Craig).

The second model is:

$$\log \hat{N} - \log\left(\hat{N} - \sum f_x\right) = \frac{\sum x f_x}{N} \tag{3.35}$$

where the symbols are as above, Σf_x being the total different individual animals caught (namely the number caught once + the number caught twice + the number caught thrice, etc.). Craig (1953) gives formulae for the variances of these estimates:

For Equation (3.34):

$$\mathrm{var}(\hat{N}) = \frac{2N}{\left(\sum x f_x\right)^2} \tag{3.36}$$

For Equation (3.35):

$$\mathrm{var}(\hat{N}) = \frac{N}{\left(e^{\Sigma x f_x} - 1 - \sum x f_x\right)} \tag{3.37}$$

It is clear that this method is based on different assumptions from the Peterson-Lincoln index. It demands that the animals be very mobile so that their chances of recapture are effectively random immediately after release, and yet they must not leave the habitat. It has been used with butterflies and might be applied to other large conspicuous flying or

---

[1]The actual numbering of these methods is the reverse of that given by Craig, but it is most logical here to give the simplest method first.

very mobile animals under certain circumstances. Phillips & Campbell (1970) showed that the Schnabel census method is more robust when portions of the population are inaccessible.

3.1.2.7   Change in ratio methods (Kelker's selective removal)

This method uses the natural marks of a population, normally the difference between the sexes, but theoretically any other recognisable distinction could be used (e.g. the different morphs of a polymorphic species). The proportion of the different forms or components are determined, a known number of one form is removed ('selective removal') and the new ratio is found; from the change in the ratio the total population can be estimated:

$$\hat{N} = \frac{K_\alpha}{\left[D_{\alpha 1} - \frac{(D_{\beta 1} D_{\alpha 2})}{D_{\beta 2}}\right]} \tag{3.38}$$

where α and β are the components (e.g. sexes) in a population, α is the component, part of which is removed during the interval between times 1 and 2, and β is the component (or components), the member which are not removed, then K_α = number of component α that are killed; D = the proportion of the population represented by a component at time 1 or 2; thus, $D_{\alpha 1}$ = the proportion of α in the population as a decimal before any were removed; $D_{\beta 1}$ = the proportion of the population as a decimal represented β after K individuals of α were removed. Such an approach is clearly most applicable with game, where individuals of certain sex or age class are culled (e.g. Kelker, 1940; Hanson, 1963); it is sometimes referred to as the dichotomy method (Chapman, 1954).

The mathematical background has been developed by Chapman (1955), who compared the results from this method with those given by the Petersen–Lincoln-type capture–recapture analysis. Chapman showed that a capture–recapture estimation procedure will yield more information for equal effort. A further objection to the use of this method with insect populations is that the selective removal and the resulting atypical unbalance could seriously prejudice any further study of the population; in populations of game animals such removal may be one of the 'normal' mortality factors.

This method does not depend on the assumption that initial capture does not alter the probability of subsequent recapture, and this is an advantage over those methods that utilise artificial marks and might indicate conditions for its use, perhaps only as a test of estimates derived by Lincoln-type methods. Chapman (1955) gives a technique for combining the two approaches. Seber (1982) provides a comprehensive review of this approach.

3.1.3   *Estimations for open populations*

Open populations change under some combinations of birth, death, immigration and emigration during the course of the capture–recapture study. By necessity, such populations must be subject to at least three periods of sampling.

3.1.3.1 Review of method development

If the animals are marked on a series of two or more occasions, then an allowance may be made for the loss of marked individuals between the time of the initial release and the time when the population is estimated. These methods have been reviewed by Seber (1982, 1986). Their history and development are given below. A simple, non-algebraic development of the main formulations is provided by Cormack (1973), who shows how the estimates are based on the largest 'known groups'.

Jackson (1933a, 1937, 1953), Dowdeswell *et al.* (1940) and Fisher & Ford (1947) were the first to devise methods for estimating the population using the data from a series of marking and recapture occasions. Besides assumptions 1 to 4 listed above (see p. 79), they also assumed a constant probability of capture and survival rate; even though this was known to be an approximation it was considered necessary for an algebraic solution.

Jackson developed two methods for his work on the tsetse fly, where he calculated a theoretical recapture immediately after release by either the 'positive method', where the loss ratio was calculated for that day over all release groups, and the 'negative method', where the loss ratio was calculated from the subsequent daily percentage losses for the two release groups. It is possible that the negative method might still be found useful in a situation, as provided by the use of radioactive isotopes, where marking is carried out on a limited number of occasions followed by a long series of recaptures.

Fisher's method is often referred to as the 'trellis method', as the data are initially set out in a trellis diagram; details of its working are given by Dowdeswell (1959). The survival rate has to be determined by trial and error, and by using computers this comparatively robust method has enjoyed a revival. A useful discussion and summary of these early methods is given by MacLeod (1958). Although these methods are of historical interest and have been widely used, more modern methods are to be recommended.

Bailey (1951, 1952), who introduced maximum likelihood techniques into capture–recapture analysis and so was able to calculate the variances of his estimates, made an important advance. The equations in Bailey's triple-catch method are simple and may be solved directly to provide estimates of various population parameters.

Another significant step was made at about the same time by Leslie (1952) and Leslie & Chitty (1951), who compared three different methods of classifying the animals according to their marks. In Leslie's method A (as mentioned above), the animals were classified according to the occasion on which they were marked, and thus an animal bearing several marks would give several entries in the recapture table. Jackson, Fisher and Bailey all also used this method of classification in drawing up their recapture tables, but Leslie and Chitty (1951) showed that it could lead to a loss of information. Leslie's method B, in which the animals were classified according to the date on which they were last marked (ignoring all earlier marks), was completely efficient under the assumptions he made. Leslie (1952) gives a full account of his method and a worked example, but his formulae require solution by iterative methods when more than three sampling occasions are considered. Jolly (1963) simplified the calculations by providing formulae for explicit solutions (i.e., direct not iterative). This method is particularly applicable if a fairly large number of individuals have been recaptured several times (multiple recaptures) in a long series of samples, a situation that often arises with work on mammals.

Working on blowflies, where the incidence of recapture was very low, Macleod (1958) found that none of the then known methods were usable, and he derived two formulae (based on the Lincoln index) which gave estimates (albeit rather approximate) of the total population. For both methods, the mortality rate had to be ascertained independently (in Macleod's case in the laboratory), and for one method the percentage of flies not immigrating over a given time was measured by a separate experiment.

All of the above methods are based on deterministic models which assume that the survival rate over an interval is an exact value, whereas it would be more correct to state that, in Nature, an animal has a probability of surviving over the interval. This probability is well expressed by a stochastic model, but initially it was thought that the computations arising from a stochastic model would be too complex. Darroch (1958, 1959) showed that, under certain conditions, a fully stochastic model, giving explicit solutions for the estimation of population parameters, was possible, and Seber (1965) and Jolly (1965) have independently extended this method to cover situations in which there is both loss (death and emigration) and dilution (births and immigration). Their methods give similar solutions, except that Jolly's method makes an allowance for the common experience of animals dying after capture.

The Jolly–Seber method efficiently groups the data and is fully stochastic; however, its reliability strictly depends on the assumption that every marked animal has the same probability of survival. Effectively, the probability of any animal surviving through any period is not affected by its age at the start of the period. Manly & Parr (1968) pointed out that with short-lived adult insects (e.g. many Lepidoptera) studied over their emergence period, the animals marked on the ith occasion are likely, at that time, to be younger than those marked on the first occasion. If all deaths were independent of age, this in itself would not invalidate the Jolly–Seber method, but this cannot be assumed. They therefore devised a method free of this assumption, based on intensity of sampling, but depending on a relatively high frequency of multiple recapture. The term 'multiple recapture' has been used extensively and is appropriate with mark-grouping methods (e.g. Bailey's) where the number of marks an animal bears (or the number of occasions on which it has been captured) affects its classification. However, it will be noted that in the date-grouping methods no significance is attached to any mark other than the last and hence, as Jolly (1965) points out, the term multiple recapture should not be used in connection with these methods.

The Jolly–Seber method was developed further by Buckland (1980) for the analysis of tags returned from dead animals. He also modified the method to situations where cohorts differ in catchability, and provided methods for estimating the survival curve (Buckland, 1982; Buckland et al., 1983). Trap shyness was discussed by Buckland & Hereward (1982). Generalisations to cope with survival and capture probabilities which vary with capture history were developed by Robson (1969) and Pollock (1975), and age-dependent capture and survival were considered by Pollock (1981). Pollock & Mann (1983) considered this problem in fisheries research. Pollock et al. (1990) reviewed the Jolly–Seber model and gave details of the computer software used. A contemporaneous development has been the application of loglinear models (Cormack, 1989, 1993; Rivest & Baillargeon, 2007). These methodologies are available in R. The Rcapture package uses Poisson regressions fitted with the glm function to fit both open and closed models.

3.1.3.2   Choice of method

Comparative studies on actual or simulated data have been made by several groups, including Parr (1965), Sheppard et al. (1969), Manly (1970), Roff (1973), Bishop & Sheppard (1973), and Schwarz et al. (1993). Two approaches to the method choice can be taken.

The first (systematic) approach is to consider the 'goodness of fit' of each member of a family of models which represent all the different assumptions possible (like the approach of Otis et al. (1978) and White et al. (1982) for closed populations). In practice, this is not always possible because of problems fitting some models and measuring the goodness of fit. The log-linear formulation (Cormack, 1989) is useful for this approach as all the standard models can be represented using a consistent formulation. Each candidate model may then be solved using generalised linear models. For example, in R, the Rcapture package which uses Poisson regressions fitted with the glm function to fit open models.

The second approach is to identify the assumptions that can be made prior to the experiment, to select a parsimonious model, and then design the experiment to collect the data required to meet the desired accuracy. This approach will require preliminary experimentation or prior knowledge. It is prudent to apply more than one method and compare the results. Robust methods which use a mix of models should also be considered when undertaking longer-term studies (see p. 99). The following general notes on the merits of the various methods may be of use.

Bailey's Triple Catch and the Jolly–Seber method both provide estimates of the variance; however, these have been shown to be related to the population estimate and are particularly unreliable for small samples. The Jolly–Seber method population estimate is usually reliable when 9% or more of the population is sampled and the survival rate is not less than 0.5 (Bishop & Sheppard, 1973). It is less sensitive to age-dependent variations in the mortality rate than Fisher & Ford's method, but is not of course independent of them. Except when there is only a very limited number of sampling occasions, the Jolly–Seber method is superior to that of Bailey.

The Jolly–Seber method may seriously overestimate the survival rate (Bishop & Sheppard, 1973): nonetheless, it remains in its various parameterisations the most useful method and gives good estimates if a high proportion of the population can be sampled. Manly & Parr's method is not affected by age-dependent mortality, but does require the sampling of a relatively high proportion of the population (>25% for populations under 250, >10% for larger populations; Manly, 1970). When these conditions can be achieved it should be used if mortality is thought to be related to age, particularly if it is high early in the life span studied. Manly & Parr's method produces larger standard errors for the population size estimates than the Jolly–Seber method because the data is used less efficiently. The method of Manley et al. (2003) offers considerable advantages when working on animals that can be easily and accurately aged, as it requires only two sampling occasions and the calculations can be undertaken using any software offering logistic regression.

The Fisher–Ford trellis method is no longer popular and is not offered by the standard computer packages. It is robust for small samples and, provided that the survival rate remains relatively constant, gives population estimates not dissimilar to those of the Jolly–Seber method. It has the advantage that periodicity of availability for sampling, known from biological knowledge, may be incorporated into the trellis model (see Conway *et al.*, 1974).

The above methods utilise batch marks, but when it is desired to model individual behaviour then individual marks need to be used and a log-linear formulation is required (Cormack, 1981, 1989).

3.1.3.3   The Fisher–Ford Method

This method has been superseded by other methods and is now little used. Details of the method are given by Fisher & Ford (1947), and in an expository form by Dowdeswell (1959) and Parr (1965). Example applications of the Fisher–Ford method are Rosewell & Shorrocks (1987) and Wilkin & Scofield (1991). Cianci *et al.* (2013) used a logistic regression model based on the principle of the Fisher–Ford model.

3.1.3.4   Bailey's triple-catch method

As indicated above, this method is based on a deterministic model of survival and uses an inefficient method of grouping the data, although the latter is not a serious fault if very few animals are recaptured more than once (Sonleitner & Bateman, 1963). The method has the great advantage that the formulae for the calculation of the various parameters are relatively simple; Bailey (1951) gives the mathematical background and Bailey (1952) and Macleod (1958) describe the process. The method was found to be particularly appropriate by Coulson (1962) in a study on craneflies, *Tipula*, with very short life expectancies, but series of triple-catch estimates may be used to estimate a population over a longer period (e.g. Iwao *et al.*, 1963). It seems that the results are most reliable when large numbers are marked and recaptured (Parr, 1965). The method has been used with a variety of insects, other invertebrates including snails (Woolhouse & Chandiwana, 1990) and lobster (Evans & Lockwood, 1994), as well as vertebrates such as water snakes (King *et al.*, 2006).

For the purposes of illustration, the sampling is said to take place on days 1, 2 and 3. In practice, the intervals between these sampling occasions may be of any length, provided they are long enough to allow for ample mixing of the marked individuals with the remainder of the population, and not so long that a large proportion of the marked individuals have died. With large samples the population on the second day is estimated:

$$\widehat{N}_2 = \frac{a_2 n_2 r_{31}}{r_{21} r_{32}} \tag{3.39}$$

where a_2 = the number of newly marked animals released on the second day; n_2 = the total number of animals captured on the second day and r = recaptures with the first subscript representing the day of capture and the second the day of marking; thus r_{21} = the number of animals captured on the second day that had been marked on the first, r_{31} = the number of animals captured on the third day that had been marked on the first. It is clear that we are really concerned with the number of marks and hence the same animal could contribute to r_{31} and r_{32}. The logic of the above formula may be seen as follows:

$$\widehat{N}_2 = \frac{\widehat{a}_1 n_2}{r_{21}}$$ (3.40)

which is the simple Lincoln index with $\widehat{a}_1$ = the estimate of the number of individuals marked on day 1 that are available for recapture on day 2. Now, if the death-rate is constant:

$$\frac{\widehat{a}_1}{a_2} = \frac{r_{31}}{r_{32}}$$

Substituting for $\widehat{a}_1$ in the above equation we have the estimate of population size given in Equation (3.39). The large-sample variance of the estimate is:

$$\text{var}\widehat{N}_2 = \widehat{N}_2^2 \left(\frac{1}{r_{21}} + \frac{1}{r_{32}} + \frac{1}{r_{31}} - \frac{1}{n_2} \right)$$ (3.41)

Where the numbers recaptured are fairly small there is some advantage in using 'Bailey's correction factor'; that is, the addition of 1 so that:

$$\widehat{N}_2 = \frac{a_2(n_2 + 1)r_{31}}{r_{21}r_{32}}$$ (3.42)

with approximate variance:

$$\text{var}\widehat{N}_2 = \widehat{N}_2^2 - \frac{a_2^2 (n_2 + 1)(n_2 + 2)(r_{31} - 1)r_{31}}{(r_{21} + 1)(r_{21} + 2)(r_{32} + 1)(r_{32} + 2)}$$ (3.43)

The loss rate, which is compounded of the numbers actually dying and the numbers emigrating, is given by:

$$\gamma = \ln \left(\frac{a_2 r_{31}}{a_1 r_{32}} \right)^{\frac{1}{t_1}}$$ (3.44)

where t_1 = the time interval between the first and second sampling occasions. The dilution rate, which is the result of births and immigration, is given by:

$$\beta = \ln \left(\frac{r_2 n_3}{n_2 r_{31}} \right)^{\frac{1}{t_2}}$$ (3.45)

where t_2 = the time interval between the second and third sampling occasions. Both of these are measures of a rate per unit of time (the unit being the units of t). Bailey (1952) gives formulae for their variance.

3.1.3.5 Jolly–Seber stochastic method

The advantages of the stochastic model on which this method is based have already been discussed and its general utility has been indicated (p. 95). The program MARK (downloaded from: http://warnercnr.colostate.edu/~gwhite/mark/mark.htm) offers a number of different variants on the Jolly–Seber method for the estimation of an open population. These different formulations are the POPAN; the Link–Barker and Pradel-recruitment; and the Burnham JS and Pradel- formulations. All of these variants use the same data entry format, so more than one can be tried without cost. The one to choose will depend on the information required. POPAN can assume losses on capture, and provides estimates of population size and net births. Link–Barker can assume losses on capture, and provides estimates of recruitment and population growth, λ. Pradel-recruitment assumes no losses on capture and provides estimates of recruitment and λ. Burnham JS can assume losses on capture, and provides estimates of abundance, net births and λ. Finally, Pradel-λ can assume losses on capture and estimates λ. In practice, the Burnham JS frequently experiences difficulties in finding a solution.

Examples of using the Jolly–Seber method, often with comparisons to other methods, are:

- Insects: Gouteux & Buckland, 1984; Komazaki & Sakagami, 1989; Nelemans *et al.*, 1989; Ohgushi & Sawada, 1998.
- Other invertebrates: Wilkin & Scofield, 1991; Nakai *et al.*, 1995).
- Fish: Schwarz *et al.*, 1993; Law, 1994.
- Amphibians: Freeland, 1986.
- Reptiles: Le Gall *et al.*, 1986; Woodward *et al.*, 1987.
- Birds: Boano & Cucco, 1991; Loery & Nichols, 1985; Spendelow & Nichols, 1989; Loery *et al.*, 1987.
- Diving mammals: Baker and Herman, 1987.
- Terrestrial mammals: Baker and Herman, 1987; Paradis *et al.*, 1993.

3.1.3.6 Robust design

While the Jolly–Seber method has been the standard approach to the estimation of demographic parameters in long-term capture–recapture studies of wildlife and fish species, it can give biased estimates of population size and recruitment. Pollock (1982) proposed a sampling scheme in which a series of closely spaced samples were separated by longer intervals such as a year. In this nested sampling structure, closed population models can be applied to groups of samples collected close together in time while the population is assumed open to migration, recruitment and death for the longer time intervals. For this 'robust design' Pollock suggested a flexible *ad hoc* approach that combines the Jolly–Seber estimators with closed population estimators, to reduce bias caused by unequal catchability, and to provide estimates for parameters that are unidentifiable by the Jolly–Seber method alone. Kendall *et al.* (1995) describe the analysis of data obtained using the robust design which has been recommended for small mammals studies by Nichols *et al.* (1984) and Nichols & Pollock (1990). Rivest & Gaétan (2004) applied the log-linear approach in a robust design with an example application of the method to red-back vole in Canada. Their analysis is available in the R package, Rcapture.

3.1.3.7   Manly and Parr's and Manly *et al.*'s aging methods

The assumption that survival probability is equal is often quite unjustified in entomological studies (Iwao *et al.*, 1966). To overcome this, Manly and Parr (1968) devised a method based on the intensity of sampling, each animal being considered to have the same chance of capture on the ith occasion, so that

$$\hat{N}_i = \frac{n_i}{p_i}$$

where p_i = the sampling intensity. In practice:

$$\hat{p}_i = \frac{r_i}{\hat{M}_i}$$

where r is the number recaptured and M the total number marked in the population. Thus,

$$\hat{N}_i = \frac{an_i}{r_i}$$

the basic Lincoln–Petersen index. The animals must be individually marked or have date-specific marks. Manly & Parr prepared an individual animal table allocating the symbol x for the first or last occurrence of an individual mark, y for the intermediate occasion when individual mark was captured, and z for occasions when it was there, but not recaptured – that is, the blanks left between the two x values after the y values have been inserted. A greatly curtailed example is given in Table 3.4. The sampling intensity will known to be present before and after. For any day these are given by $\sum y_i$ and $\left(\sum y_i + \sum z_i\right)$ respectively. Then:

$$\hat{p}_i = \frac{\sum y_i}{\left(\sum y_i + \sum z_i\right)}$$

so

$$\hat{N}_i = \frac{n_i}{p_i} = \frac{n_i\left(\sum y_i + \sum z_i\right)}{\sum y_i} \qquad (3.46)$$

Table 3.4 Manly and Parr's method of recording data for their method of estimation (obviously one would normally mark far more than three animals on each occasion).

	Sampling days					
	1	2	3	4	5	6
1	x	y	z	x	—	—
2	x	z	y	z	x	—
3	x	—	—	—	—	—
4		x	z	x	—	—
5		x	y	z	y	x
6		x	z	y	x	—
7			x	y	z	x
8			x	z	x	—
9			x	y	z	x
$\Sigma y_i =$		1	2	3	1	
$\Sigma z_i =$		1	3	3	2	

and survival may be estimated as:

$$\Phi_{i-(i+1)} = \frac{r_{i,i+1}}{n_i b_{i+1}} \tag{3.47}$$

where $r_{i,i+1}$ = the animals caught in both i^{th} and $(i+1)^{\text{th}}$ samples.

The births, number of new animals entering the population, are estimated by:

$$\hat{B}_{i-(i+1)} = \hat{N}_{i+1} - \hat{\Phi}_i \hat{N}_i$$

Manly (1969) gives the variance of the population estimate as:

$$\text{var}(\hat{N}_i) = \frac{\hat{N}_i(\hat{N}_i - n_i)(\hat{N}_i - (\Sigma y_i + \Sigma z_i))}{n_i(\Sigma y_i + \Sigma z_i)} \tag{3.48}$$

As mentioned above, this method is not robust unless a fairly large sample is taken (Manly, 1970). Seber (1982) shows how the necessary data may be tabulated in a form similar to that of the Jolly–Seber method. This would undoubtedly be more convenient with large sets of data, but Manly & Parr's original arrangement is simpler for exposition.

Manly et al. (2003) extended this method to include data on the age of the animals, and demonstrated that the method could be analysed using logistic regression. The first consequence of incorporating age data into Manly and Parr's method is that the size of an open population can be estimated with data from just two years (sampling periods). This is because captured animals aged 2 or more in the second year are known to have been in the population in the first sampling year. Age information can also eliminate some possibilities. For example, if an animal first captured in year 5 is aged 2, then it was only previously available for capture in year 4, assuming that animals of age 0 cannot be caught. Models without age data would include the possibility it was alive in years 1, 2 and 3. A second advantage is that, by using age, the standard errors of the population estimates are reduced. These methods are particularly suited to long-lived mammals and birds that can be accurately aged and are sampled annually.

3.1.3.8   Cormack's log-linear method

The log-linear formulation (Cormack, 1981, 1989) can describe many of the models proposed for the estimation of population parameters by capture–recapture. All animals in the population are considered to be multinomially distributed between 2^s capture histories, where s is the number of samples. The expected number of individuals showing each capture history is given by the product of the probability of the event(s) and the size of the population. By taking logs, these equations are made linear and can then be solved using standard computer packages. Examples of its use are given in Cormack (1989) and Baker (1990). In R, the Rcapture package (Baillargeon and Rivest, 2007) uses Poisson regressions fitted with the glm function to fit open models.

The following listing gives a simple example of the use of Rcapture to obtain open population estimates using an example data set available within Rcapture.

```
#Load the Rcapture package
require(Rcapture)
#Open a data set – this example comprises a 6-year study of Eider Duck
# by Coulson and analyzed in Cormack (1989)
data(duck)
#print out the data
duck
#Run an open model using the openp() function
op.m1 <- openp(duck, dfreq=TRUE)
#print the results
op.m1
```

The data comprises the lists of capture histories over the 6 years (p1...p6) and the observed frequency of each capture history. So in the truncated section of the data shown below birds captured on all 6 years had a frequency of 13.

	p1	p2	p3	p4	p5	p6	freq
[1,]	1	1	1	1	1	1	13
[2,]	0	1	1	1	1	1	4
[3,]	1	0	1	1	1	1	5
[4,]	0	0	1	1	1	1	4
[5,]	1	1	0	1	1	1	7
[6,]	0	1	0	1	1	1	3
[7,]	1	0	0	1	1	1	7
[8,]	0	0	0	1	1	1	6

The output generated includes tests for a trap effect, capture and survival probabilities, estimated abundances and new arrivals.

3.2 Methods of marking animals

Marking techniques are often classified as either group (batch) or individual (tagging) techniques. While methods such as fin clipping, branding or staining are often used in a manner that precludes individual identification, this need not always be so. For example, a range of coloured dyes may be applied to different areas of the body. The number of unique marks available (C) is given by:

$$C = (x - 1)^n - 1$$

where, x is the number of colours or types of mark and n is the number of sites to hold the mark or tag. Batch methods enable a large number of animals to be marked in the same way and are perfectly adequate for most capture–recapture population estimations and dispersal studies. Almost all methods are capable of one or two variants so that two or three groups may be marked differently.

If each individual can be separately marked, additional information can be obtained on longevity and dispersal and, if they can be aged and sexed initially, survival can be related to these and other characters. Birth- and death rates may be more easily calculated and the excessive handling of animals, recaptured more than once, with its attendant problems, may be avoided (Dobson & Morris, 1961). Individual marks may also, for example, allow the assessment of the rate of oviposition in the field by weighing individual females (Richards & Waloff, 1954), and they provide a method for assessing the randomness of recapture. The extent to which the higher cost, in terms of effort, of marking animals individually is justified will depend on the percentage of the marked individuals that are recovered; high recovery rates justify elaborate marking programmes; with low recovery rates individual marking is seldom justified.

Below are reviewed the principal techniques available. Marking methods have also been reviewed in Stonehouse (1977) and a number of guides applicable to particular animal groups, including birds (Calvo & Furness, 1992) and fish (Laird & Stott, 1978) have been published. The chosen method must seek to meet the following list of requirements.

1. Marking should not affect growth, longevity or behaviour of the animals. An attempt should always be made to confirm that this is true in the particular case under investigation. For example, although the pigments used in most markers may be non-toxic, the solvents are often toxic. This may be checked in the laboratory or field cage by keeping samples of living marked and unmarked individuals and comparing longevity (e.g. Stobow et al., 1992; Kideys & Nash, 1993; Morgan & Walsh, 1993; Montgomery et al., 1995) or in the field by comparing the longevity of individuals bearing differing numbers of marks (e.g. Richards & Waloff, 1954; Dobson & Morris, 1961; Thorne et al., 1996). Newly emerged insects may be more sensitive to the toxic substances used in markers than older insects (Jackson, 1948; Dobson et al., 1958) and the attachment of labels to their wings, which has no effect on old insects, may cause distortion due to interference with the blood circulation

(Waloff, 1963). Marking can also cause disease often in unexpected ways. For example, ear tags on the white-footed mouse (*Peromyscus leucopus*) were found by Ostfeld *et al.* (1993) to result in increased tick infestations, probably because they interfered with grooming. External tagging and branding can cause wounds, leading to infections. Calvo and Furness (1992), in a review of the effects of marking on birds, points out that very few studies tested for harmful effects or data biases caused by marking, but where assessments have been made it is clear that all methods of marking can have adverse effects.

2. It should not affect vulnerability to predation. An animal's natural camouflage may be lost making it more liable or, as Hartley (1954) observed with marked snails, less liable to predation. This effect is difficult to assess; it can to some extent be avoided by marking in inconspicuous places or by the use of fluorescent powders (Pal, 1947), dyes in powder form (Quarterman *et al.*, 1954), phenolphthalein solution (Peffly & Labrecque, 1956), tetracycline (Wastle *et al.*, 1994), internal tags such as coded wire tags or radioactive tracers whose presence is only detectable by the use of a special technique after recapture. The effect on predation of a conspicuous, but convenient, marking method could be measured by marking further individuals with one of these invisible methods and comparing longevity. The effect of a conspicuous mark can be checked through choice experiments in the laboratory: Buckner (1968) confirmed that small mammals are not influenced if their prey was stained with vital dyes. Special care should be taken to avoid atypical predation when animals are released (see p. 107).

3. The mark must be durable. Some paints, particularly cellulose lacquers, may flake off; adhesives can lose strength (Treble *et al.*, 1993); student's oil paints and powdered dyes wash off, some fluorescent powders may lose this property on exposure to sunlight (Polivka, 1949) or be abraded during collection (Dow, 1971), tags may be shed or ejected (Wisniewolski and Nabialek, 1993; Hampton & Kirkwood, 1990), and a radioactive isotope could decay and/or be excreted by the animal. Animals will lose marks on their cuticle when they moult. Laboratory tests of durability are not always reliable; Blinn (1963) found that cellulose lacquers would remain on the shells of land snails for two years in the laboratory, but they only lasted about one year in the field. Wineriter & Walker (1984) reported on the durability of insect marking materials. The rate at which a radioactive isotope is lost may depend on the animal's diet and other factors.

4. The mark must be easily detected. However, a conspicuous mark on animals sampled by a method that relies on the sight of the collector, may result in the over-reporting of marked individuals (Edwards, 1958).

5. The objectives must justify the cost. The amount of effort that can be put into a marking programme and the choice of marking method need to be related to the percentage of recoveries that can be expected. A high cost per individual marked will be justified where the recovery rate is high or the animals have a special significance.

6. Generally, marks must be easily applied in the field and without the need for anaesthetic. Ethical aspects of the distress and harm caused to the animals need to be considered (Putman, 1995).

7. Consideration may also have to be given to the effect of preservatives on the mark and its toxicity, appearance and safety if the marked animals can be seen, caught or eaten by the public.

After selecting an appropriate marking technique, it is important to consider the handling and release of the animals. During a marking campaign animals must be captured, marked, held for recovery, and released. At each stage in this chain undue stress must be minimised and this requires planning. When badly undertaken it results in the death or injury of many animals, a result which is probably more common than many biologists would hope.

3.2.1 Handling techniques

Handling can be stressful, even fatal, and should be carefully considered. Most experimental designs assume that the marked animals behave normally, but if handling is poor, this will not be so. For example, in order to radio-tag freshwater dolphins a standard method is to trap the animals in nets and then to harpoon them – an unpleasant procedure which may change their behaviour, at least when in the presence of human beings. Netting can be highly stressful to fish, and it may only be possible to capture fish at set localities or times in the tidal cycle. This has resulted in the capture of salmon in freshwater pools followed by their release in estuarine water, and the behavioural consequences of such an abrupt change cannot be known. The trapping techniques that may be used are considered in Chapters 4, 5 and 6. Various methods are described below where the animal is marked without capture; these clearly should avoid handling problems.

If the animals are to be marked by spraying or dusting this can be done while they are still active, either in the field or in small cages. However, for marking with paints, especially for precise spotting and for most methods of labelling and mutilation, it is necessary for the animal to be still. Insects, crustaceans, molluscs, birds and small mammals can often be held in the hand or between the fingers; where this is impossible the animal must be held by some other method, anaesthetised or chilled. Fish can be placed between two boards fixed to give a V-shaped cross-section.

Invertebrates may be held still under a net or with a hair. One end of the hair is fixed and the other has a piece of Plasticine on it; the animal is placed underneath and the hair may be tightened under a stereoscopic microscope by applying pressure to the Plasticine (Banks, 1955). More robust insects may be held immobile on top of a cork by a piece of Terylene net which is kept taut by a ring (of about 5 cm diameter, but of course depending on the size of the animal). This ring is pushed down into the surrounding Plasticine (Fig. 3.1b) and the animal can be marked through the holes of the net (Murdoch, 1963). Conway and coworkers (1974) held mosquitoes for spot marking between diaphragms of nylon mesh and stockinette. These devices may then be placed under a microscope.

Insects can also be held by suction (Hewlett, 1954). Muir (1958) found that a mirid bug could be conveniently held by the weak suction of a pipette, formed from the ground-down point of a coarse hypodermic needle, connected to a water pump. The picking up

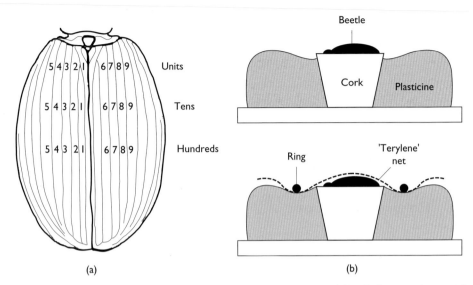

Figure 3.1   (a) A 'mutilation system' for marking carabid beetles individually by scraping certain positions on the elytra, applied to *Agonum fuliginosum*; (b) A device for holding a hard-bodied animal whilst it is marked. (Adapted from Murdoch, 1963.)

of the bugs by this weak suction was facilitated by coating the inside of the glass dish, in which they were held, by an unsintered dispersion of 'Fluon' GPI, polytetrafluoroethylene (PTFE), which presented an almost frictionless surface (Muir, 1958; Radinovsky & Krantz, 1962). Butterflies may be held in a clamp (Horn, 1976).

If none of these methods can be used, or if the insects are so active that they cannot even be handled and counted, they may have to be immobilised. Chilling at a temperature between 1 and 5 °C is probably the best method, and may be done in a lagged tank surrounded by an ice-water mixture (MacLoed & Donnelly, 1957). Alternatively an anaesthetic may be employed: carbon dioxide is often used and is easily produced from 'dry ice' (Caldwell, 1956) or bought in cartridges designed to carbonate drinks (Dobson, 1993; Killick-Kendrick, 1993). Caldwell (1956) gives a table of recovery times of houseflies from carbon dioxide anaesthesia at 21 °C (Table 3.5).

When working with the earwig, *Forficula*, Lamb & Wellington (1974) found that although activity may be quickly resumed after carbon dioxide anaesthesia of less

Table 3.5   Recovery times (minutes) of houseflies from carbon dioxide anaesthesia. (Data from Caldwell, 1956.)

Exposure time	Recovery time
up to 5 min	1–2 min
5–30 min	3–5 min
30–60 min	5–10 min

than 1 minute, normal behaviour could be affected for many hours; hence, they recommended a 24-h recovery period. Dalmat (1950) found that female blackflies (Simulidae) laid more eggs than normal after being subjected to carbon dioxide, but Edgar (1971) found this did not affect the development of the spider, *Lycosa lugubris*. Other anaesthetics are ether, chloroform, nitrogen and nitrous oxide. Honey bees are prematurely aged by these compounds (Ribbands, 1950; Simpson, 1954). For insects, the use of anaesthetics should, wherever possible, be avoided in ecological and behavioural studies.

The use of anaesthetic while marking fish is advised by Laird & Stott (1978); MS 222 and benzocaine are the two most commonly used. Wedermeyer (1970) suggested that benzocaine was better than MS 222 at reducing the stress experienced by fish. Drug immobilisation using dart guns is often used for the capture of large mammals and arboreal animals (Eltringham, 1977).

3.2.2 Release

The release of animals after marking is an operation that is too often casually undertaken. Some methods allow animals to be marked without capture, while others allow for them to be marked in the field and immediately released. However, the animal is often incarcerated, handled and much disturbed before being released. It should not be surprising if they show a high level of activity immediately after release; indeed, Greenslade (1964) recorded with individually marked ground beetles that there was far greater movement on the day after release than subsequently. Two approaches can be used to minimise this effect.

If the animal has a marked periodicity of movement (i.e., it is strictly diurnal or nocturnal) then it should be released during its inactive period. For example, radioactively tagged individuals of the frit fly (*Oscinella frit*) were released at dusk (its period of activity is from dawn to late afternoon) (Southwood, 1965). Animals that are active at most times of the day may be restrained from flying immediately after release by covering them with small cages (Evans & Gyrisco, 1960). The release sites should be chosen carefully. It is especially important to avoid the release of small flying insects in the middle of the day when their escape flights may carry them beyond the shelter of the habitat into winds or thermals that can transport them for miles. Of course, only apparently healthy, unharmed individuals should be released.

The release points should be scattered throughout the habitat, as it is essential that the marked animals mix freely with the remainder of the population; for example, Muir (1958) returned the arboreal mirids he had marked to all parts of the tree. Very sedentary animals may indeed invalidate the use of the capture–recapture method if they do not move sufficiently to re-mix after marking, as Edwards (1961) found, surprisingly, with a population of the grasshopper, *Melanoplus*. The extent of the re-mixing may be checked, to some degree, by a comparison of the ratio of marked to unmarked individuals in samples from various parts of the habitat; the significance of the difference may be tested by a χ^2 test (Iwao *et al.*, 1963).

The release of disorientated or unadapted animals can also result in high predation losses. Frequently, marked crustaceans and fish have adapted their colouration to the conditions within the holding tanks so that they have no natural camouflage when released. A general feature of predatory fish is that they recognise disorientation and quickly attack. Thus, a common feature of fish tagging studies is a high mortality rate shortly after release. Fish tagging in the Amazon has proved difficult because of piranha predation. Fish should be released as near as possible to their place of capture and shoaling species should be released in a group. It has frequently been found that tagged fish are particularly vulnerable to predatory birds such as cormorants.

3.2.3   Surface marks using paints and solutions of dyes

The hard cuticle or shell of arthropods and molluscs allows the use, in terrestrial environments, of a wide range of paints and surface treatments. In aquatic or moist habitats the range is much more limited. Hagler & Jackson (2001) reviewed the possible approaches, including ball-point pens and marker pens.

Artist's oil paint is an extensively used marking material; it can, of course, be obtained in a variety of colours and has been used successfully for marking moths (Collins & Potts, 1932), tsetse flies (Jackson, 1933a), bed bugs (Mellanby, 1939), locusts and grasshoppers (Richards & Waloff, 1954), mirids (Muir, 1958), flies (Cragg & Hobart, 1955; Dobson *et al.*, 1958), beetles (Mitchell, 1963) and others. However, Davey (1956) found such oil paints toxic to certain locusts, though this may have been the effect of the diluent and they are, of course, slow-drying. Artist's poster paints have been used to mark mosquitoes (Gillies, 1961; Slooff & Herath, 1980), as have 'Humbrol' enamel paints (Trpis & Hausermann, 1986). More recently, various paint markers have been used (e.g. Sugimoto *et al.*, 1994).

Nitrocellulose lacquers or paints (e.g. model aircraft dope) and alkyl vinyl resin paints are quick-drying and have been used by a number of workers; on snails, where they were applied to a small area on the underside of the shell, from which the periostracum had been scraped (Sheppard, 1951), and on grass hoppers (Richards & Walloff, 1954), where they were found to be less satisfactory than artist's oil paints. Other arthropods to which they have been applied include ants (Holt, 1955), lace bugs (Southwood & Scudder, 1956), dragonflies (Corbet, 1952), mites (Hunter, 1960), tipulid flies (Freeman, 1964), various beetles (Greenslade, 1964; Malter, 1996; Ohguski & Sawada, 1998) and mosquitoes (Sheppard *et al.*, 1969). Fluorescent lacquer enamels or fluorescent pigments with gum arabic glue plus a trace of detergent have been used to mark tsetse flies (Jewell, 1956, 1958; McDonald, 1960), chafer beetles (Evans & Gyrisco, 1960) and lepidopterous caterpillars (Wood, 1963). Animals marked in these ways may be spotted after dark in the field at distances of up to 8–10 m by the use of a beam of ultraviolet light produced by a battery-powered lamp (McDonald, 1960). Solutions of fluorescent dyes (principally rhodamine B) in alcohol or acetone have been used to mark mosquitoes (Chang, 1946) and *Drosophila* (Wave *et al.*, 1963) and the tick, *Argus* (Medley & Ahrens, 1968). Felt-tip pens provide a convenient way for marking some large insects (Iwao *et al.*, 1966), while Porter & Jorgensen (1980) marked ants with coloured fluorescent ink extracted from 'magic-marker' pens. Wada *et al.* (1975) used an aqueous fluorescent solution which comprised 1.0% yellow 8G, 1.0% Kaycal BZ, 0.1% rhodamine 6G and 0.5% crystal violet to stain mosquitoes. Reflecting paints may also be used to mark animals for detection at night, and have the advantage that they can be seen with a small hand-held torch for up to 10 m. Nail varnish has been found effective for gastropods (Fenwick & Amin, 1983; Woolhouse, 1988; Gosselin, 1993).

Aluminium paint adheres well to scraped areas of the elytra of carabid beetles (Murdoch, 1963), while solutions of various stains in alcohol, such as eosin, orange G and Congo red, have been used to mark adult Lepidoptera (Nielsen, 1961). Petroleum-based inks (e.g. Easterbrook Flowmaster) have been found particularly useful, staining the integument below the scales (Wolf & Stimmann, 1972). Working on house flies, Peffly & Labrecque (1956) used a 6% solution of phenolphthalein in acetone; the marked flies were identified on recapture by placing them in 1% sodium hydroxide solution, whereupon they became purple. Fales *et al.* (1964) used waterproof inks to mark face flies.

An interesting recent idea is marking the myrid, *Lygus hesperus*, adults externally by submersion in rabbit immunoglobulin G (IgG) solution which presumably will be undetectable to other animals while allowing detection using an immunological test, even in samples obtained from the guts of predators (Hagler *et al.*, 1992). The distribution of some insects has been studied by feeding the adults dyes such as Rhodamine B, which give the eggs a distinctive colour (Narayanaswamy *et al.*, 1994).

Henderson & Holmes (1985) marked the carapace of the common shrimp *Crangon crangon* using cyanoacrylate instant glue mixed with a coloured power or dye. This mixture hardened and bonded to the surface, even underwater. The mark was found to last at least one month or until the shrimp moulted.

Commercial hair dye and lighteners can be used to stain feathers (Ellis & Ellis, 1975). Bird feathers can be marked by dipping in dyes, but the application of paint or dyes to the skin, fur or feathers of a vertebrate rarely gives a reliable, lasting mark. Marking methods for birds are discussed by Calvo and Furness (1992). The feathers of incubating herring gulls, *Larus argentatus*, can be marked by applying a dye (Rhodamine B, malachite green or picric acid) to their eggs (Belant & Seamans, 1993). These authors discussed the merits of oil-based silica gel, petroleum jelly and vegetable shortening as a carrier for the dyes, and acetic acid, isopropyl alcohol, propylene glycol as fixatives. Turtles (Hain, 1965), Nile crocodiles, *Crocodylus niloticus*, (Pooley, 1962) and snakes (Parker, 1976) have all been given temporary marks using enamel paints, lacquers and other quick-drying materials.

3.2.3.1 Application methods

When the paints or solutions are in their most concentrated form they are most conveniently applied by the use of an entomological pin, a sharpened match-stick, a single bristle, or even a fine dry grass stem (Jackson, 1933a; Corbet, 1952; Muir, 1958; Hunter, 1960). With quick-drying cellulose lacquer it may be necessary to dilute them slightly with acetone or another solvent; if this is not done a fine skin may form over the droplet on the pin and it will not adhere firmly to the animal. With artist's oil paints, dyes in solution or diluted lacquers a brush may be used (Wood, 1963), although generally this method has no advantage over the use of a pin and frequently leads to the application of too large a mark. If the mark covers any of the sense organs or joints, the specimen will have to be discarded. Slooff and Herath (1980) applied poster paints to mosquitoes with a tiny loop of 0.0024 gauge wire. The loop was dipped first into the paint and then into 96% ethanol; this made the paint form a small blob prior to application on the scutum of the mosquito. Freeman (1964) found that a fine syringe was suitable for applying cellulose lacquers to tipulid flies.

The paints or solutions may be further diluted with acetone, dilute alcohol or other solvents and sprayed on. This may be done with a hand atomiser (e.g. a nasal spray) while the insects are contained in a small wire cage (Leeuwen, 1940; Evans & Gyrisco, 1960). Mortality during marking by this method can be reduced if, immediately after spraying, the insects are quickly dried in the draught from an electric fan (Leeuwen, 1940). Large numbers of moths may be rapidly marked, and this technique has been extended to field marking of locusts (Davey, 1956) and butterflies (Nielsen, 1961). By using a spray gun individuals can be marked at a distance of 5 m or more. Davey showed that for the same cost (labour and time), almost ten times as many locusts could be marked with this method than by that involving the capture and handling of each individual.

If small labels can be attached to the insects' wings (see below) there is no problem in marking each individual, and Nielsen (1961) records that in the laboratory he marked butterflies individually on the wing with a rubber stamp. With most small invertebrates, individual marks have to be obtained by a combination of spots in various positions, the numerical range of the coding often being increased by the use of various colours. It is generally wise to follow the policy of Michener *et al.* (1955) and ensure that all individuals bear the same number of marks, so that if one is lost number 121 does not become say 21, but is immediately recognised as an 'unreadable' mark. It is also desirable to use the minimum number of colours in marking any individual insect, as the change in colours increases the handling time. Obviously, a change in colour between say the ranges 1–99 and 100–199 is less of a practical problem, than the use of two colours in a single mark. The actual pattern will depend on the size and shape of the insect, the number of colours available, and the number of individual marks required. Patterns for bodies and wings have been devised for dragonflies (Borror, 1934; Parr *et al.*, 1968), bed bugs (Mellanby, 1939), tsetse flies (Jackson, 1953), bees (Frisch, 1950), grasshoppers (Richards & Waloff, 1954; Nakamura *et al.*, 1971; White, 1970), snails (Blinn, 1963), craneflies (Freeman, 1964), mosquitoes (Sheppard *et al.*, 1969; Conway *et al.*, 1974) and lepidoptera (Ehrlich & Davidson, 1960; Brussard, 1971; Dempster, 1971). The system of Richards & Waloff (1954) is logical and versatile, enabling up to 999 individuals to be marked with continuous numbering (Fig. 3.2). On the right of the thorax is a spot that represents the unit, on the left is one that represents the tens, the head spot the hundreds; 1–5 and 10–50 are white spots, 6–0 and 60–00 are red; ten different colours are used for the head mark including the zero class (for 199). With this system the addition of a single further mark would allow another 10 000 individuals to be numbered. With smaller insects the number of spots can be reduced by increasing the number of colours used, as Richards & Waloff did with the hundred mark; however, the practical problem of switching quickly between more than three colours is serious. The rapid reading of the mark is another advantage, and Richards

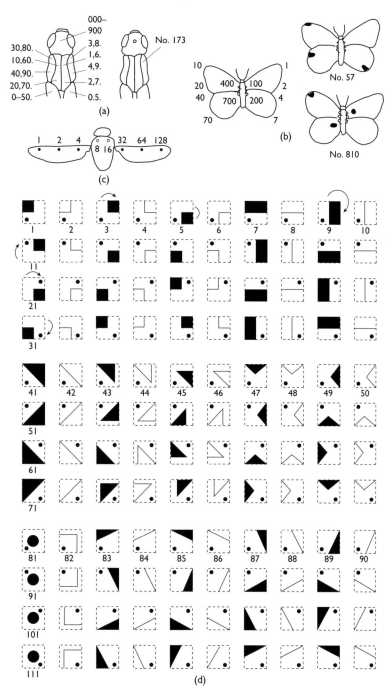

Figure 3.2 Systems for marking insects individually using colour and position codes. (a) Richards and Waloff's (1954) decimal system; (b) Brussard's (1971) modification of Ehrlich and Davidson's (1960) 1–2–4–7 system; (c) Sheppard et al.'s (1969) binomial system; (d) White's (1970) 'shape code' (see text for further explanations).

& Waloff's code is particularly clear in this respect because the same colours and corresponding positions represent the same figures in units and tens.

Richards & Waloff's code may be considered a decimal code, based on tens. Ehrlich & Davidson (1960) developed a 1–2–4–7 marking system, later modified by Brussard (1971), to mark up to 1000 individuals with one colour (Fig. 3.2). A binomial system for marking mosquitoes was developed by Sheppard *et al.* (1969), so that up to 255 individuals could be marked with a single colour (see Fig. 3.2); however, as with Brussard's method the number of marks is variable and the quick and accurate reading of the mark a matter of experience. A more elaborate code of shapes has been devised by White (1970) (Fig. 3.2). The system is self-checking (if marks are lost), and a single colour may be used up to 160, but as White points out rapid reading becomes more difficult beyond 120.

If the body and/or wings are unsuitable for spotting, the legs may be marked, although the removal of such marks by the animals during cleaning movements is a real risk. Kuenzler (1958) was able to number lycosid spiders by marking their legs with white enamel paint, and Corbet (1956) hippoboscid flies by marking both the body and the legs.

3.2.4   Dyes and fluorescent substances in powder form

Hairy insects or vertebrates may be marked by dusting them with various dyes in powder form; this is most easily done by applying the dusts from a powder dispenser (insufflator) or by producing a dust storm in a cage with a jet of air (e.g. produced by a bicycle pump). Frankie (1973) describes a suitable apparatus, and Williams *et al.* (1979) a unit for marking large numbers of stable flies, *Stomoxys calcitrons*. Only a very small quantity of powder is necessary. Such methods have been applied mostly to terrestrial insects.

Non-fluorescent dyes that have been found useful are the rotor and waxoline group. The marked insects are recognised by laying them on a piece of white filter paper and dropping acetone onto them, when a coloured spot or ring forms beneath those that have been marked. As the testing involves the killing of insects this method is not suitable for recapture work, and laboratory tests have shown that with blowflies the mark may only last for one week and seldom for more than two weeks (MacLeod & Donnelly, 1957). However, dyes of two different colours may be applied to the same insect and distinguished in the spotting. This method has been used for calypterate flies by Quarterman *et al.* (1954) and MacLeod & Donnelly (1957). It is possible that, under certain circumstances, it could be used for marking large aggregated populations in the field.

Fluorescent substances, detected by placing the animals under an UV lamp, have also been used extensively for marking (e.g. Foott, 1976; Wheye & Ehrlich, 1985; Aars *et al.*, 1995). 'Day-glo' fluorescent pigments, which are activated by blue visible light and UV, are used to mark insects. Service (1995) found that 'series A' pigments gave good adhesion. Examples of their use are Sheppard *et al.* (1973) with tabanids, and Isaacs & Byrne (1998) with whiteflies; the latter group achieved large-scale marking (for dispersal studies) by spraying the dust over a whole field with a tractor-mounted boom. Other fluorescent powders designed to mark insects are available from USR Optrix,

Hackettstown, NJ, USA. Multiple marking is possible, although combinations some-times produce distinct fluorescences (Moth & Barker, 1975), and because they may be detected without killing the insect, they are suitable for use in capture–recapture pop-ulation estimations (Crumpacker, 1974). Although with the stable fly, *Stomoxys*, field cage experiments suggested that they could adversely affect longevity (Labrecque *et al.*, 1975), this was not detected in trials with *Drosophila* (Moth & Barker, 1975). For some powders, a better adhesion can be obtained by mixing one part of the dye with six parts of gum arabic, adding water until a paste is formed, then drying the paste and pulveris-ing it in a mortar. This powder is applied to the insects that are then placed in a high humidity; the gum arabic particles absorb sufficient moisture to make them adhere to the insect. This method has been used for marking mosquitoes by Reeves *et al.* (1948) and Sinsko & Craig (1979). The movements of foraging bees have been studied by mark-ing them with a fluorescent powder as they leave the hive; this is conveniently done by forcing them to walk between two strips of velveteen that has been liberally dusted with the marker (Smith & Townsend, 1951); tabanids caught in a canopy trap (p. 300) have been self-marked in a similar way (Sheppard *et al.*, 1973). Bees have also been marked when visiting flowers by dusting the flowers with a mixture of the fluorescent powder and a carrier such as talc or lycopodium dust. The bees leave a trail of pow-der, which can be detected after dark with an UV lamp, on the other flowers they have visited (Smith, 1958; Johansson, 1959). Fluorescein and rhodamine B have been found useful in this work, as all of the bees leaving the hive are marked and the marks last for weeks (Smith & Townsend, 1951). The possibility that unmarked individuals may bear a few particles that will fluoresce under UV light should be remembered when using these markers. Wild-caught mosquitoes have been found with fluorescent blue, pur-ple, green, white, yellow and orange spots (Reeves *et al.*, 1948), and therefore the use of rhodamine B, which fluoresces red, was recommended. Fluorescent-dyed melamine copolymer resins were found effective for marking adult lesser grain borers *Rhyzop-ertha dominica* (Dowdy & McGaughey, 1992). Fluorescent labels can be used to detect night-foraging animals such as bats (Griffin, 1970). Nocturnal observation techniques in general are reviewed by Hill & Clayton (1985).

An ingenious self-marking method for newly emerged calypterate flies has been devised by Norris (1957). The principle here is that the soil or medium which con-tains the fly puparia is covered with a mixture of about 40 parts to one of fine sand and fluorescent powder, or the puparia are coated with dye; as the flies emerge a small quantity of the dust adheres to the ptilinum; and when after emergence the ptilinum is retracted the dust becomes lodged in the ptilinal suture. Sometimes in an exami-nation of the faces of such marked flies in UV light this suture will be seen to shine vividly, but a more reliable technique is to crush the whole head on a filter paper while at the same time adding a small amount of the appropriate solvent: the mark may be seen on the paper, using UV if necessary (Steiner, 1965). Under laboratory conditions, Cook & Hain (1992) showed that the same principle could be applied to the bark bee-tles *Dendroctonus frontalis* and *Ips grandicollis* which were self-marked with fluorescent powders upon emergence from treated logs. Niebylski & Meek (1989) describe a self-marking device for mosquitoes emerging from ditches. The apparatus holds folds of cheese cloth impregnated with fluorescent powder which attaches to emerging adults as they push past. A trap for marking newly emerged black flies *Simulium venustum* and *Stegopterna mutata* with fluorescent dust is described by Dosdall *et al.* (1992).

3.2.5 *Pollen*

One of the most effective powders for marking insects is pollen, as it has evolved to adhere to insect surfaces and the grains are distinctive in form (Hagler & Jackson, 2001). Pollen marking is rarely undertaken because pollen analysis is time-consuming and the great need to ensure that the experiment has not been contaminated with naturally produced pollen.

3.2.6 *Marking formed by feeding on or absorption of dyes*

As arthropods lose external marks when they moult, the marking of the tissues of invertebrates with a vital dye incorporated into the food, is a valuable tool. Moreover, its significance is enhanced if it can be detected without killing the animal and if the mark is retained from larval to adult life (see below). Many workers have screened a wide range of dyes, but few have been found satisfactory: the majority are either rapidly excreted or prove toxic (Reeves *et al.*, 1948; Zacharuk, 1963; Daum *et al.*, 1969; Hendricks, 1971). Calco oil red [N-1 700 American Cyanamid Co.], if fed to immature stages, has been found to mark adults and sometimes the resultant eggs (but not the larvae) of several cotton insects (*Heliothis, Platyedra, Anthonomus*) (Daum *et al.*, 1969; Hendricks & Graham, 1970; Graham & Mangum, 1971). The dye is fed in a natural oil (e.g. cotton seed) with the larval diet at a concentration of about 0.01% dye per unit diet: it is detected in the adult by crushing the abdomen. Similar results have been obtained with oil-soluble 'Deep Black BB' and 'Blue II' (BASF Corp., USA) (Hendricks, 1971). Sawfly larvae have been marked by feeding them, in the last instar, on foliage treated with solutions of rhodamine B (series 4; 3.7 g l^{-1}) and Nile Blue Sulphate (series 5; 0.4 g l^{-1}) (Heron, 1968). The cocoons, adults and eggs were all marked and the dyes were visible externally, with rhodamine B fluorescing bright yellow-orange under UV light. Mosquitoes (Reeves *et al.*, 1948), *Drosophila* (Wave *et al.*, 1963) and the eyes of a small fish (O'Grady & Hoy, 1972) have also been marked with rhodamine B. The eggs of the uzi fly, *Exorista bombycis* (Louis), were stained pink when the adults mere fed with 0.15% rhodamine-B, without any effect on adult longevity (Narayanaswamy *et al.*, 1994). Houseflies have been marked with thiorescin (Shura-Bura & Grageau, 1956) and fluorescein (Zaidenov, 1960) although in some cases only a small proportion of the insects could be induced to feed on the solution. South (1965) marked slugs by feeding them on agar jelly containing 0.2% neutral red; in this case, the digestive gland became deeply stained and the colour was easily visible through the foot. Bloodsucking Diptera have been marked by allowing them to feed on a cow to which 200 ml of an aqueous solution containing 4 g trypan blue had been administered intravenously over a 20-min period. The dyestuff can be detected using a paper chromatography technique, in which the gut contents of the fly is mixed with 0. 1 M sodium hydroxide solution and applied to a narrow strip of Whatman No. 1 chromatographic filter paper. When the paper is developed in 0. 1 M sodium hydroxide solution the trypan blue remains at the origin, while other marks due to the gut contents move away (Knight & Southon, 1963). Haematophagous animals may also be marked with specific agglutinins (Cunningham *et al.*, 1963). These dyes or stains may often be detected in the faeces of predators that have fed on marked animals (Hawkes, 1972).

Much less work has been undertaken on aquatic organisms. However, Dr Ilse Walker (pers. commun.) found that copepod guts could be stained so as to be visible while alive by feeding them cooked beetroot. Young fish have been marked by adding vital stains to their food (Loab, 1966) or to the food of the mother (Bagenal, 1967).

Various compounds can permanently stain bone, teeth, spines, scales, shell or otoliths as they are formed, creating a mark which can be used to measure the subsequent growth rate of the structure. These techniques are termed 'date banding' or 'time stamps'. Tetracycline antibiotics, because of their properties of localisation in hard tissues such as bone, low toxicity and fluorescence, have been used with success to mark many different animal groups. A bibliography of the marking of fish with tetracycline and its effects is has been produced by Wastle *et al.* (1994). Some recent examples of oxytetracycline application by injection are, Pacific sailfish, black marlin (Speare, 1992), larval southern brook lamprey *Icthyomyzon gagei* (Medland & Beamish, 1991) and subadult red drums *Sciaenops ocellatus* (Bumgaurdner, 1991). Examples of batch immersion marking for larval or young fish are; whitefish *Coregonus lavaretus* (Beltran *et al.*, 1995), shad *Alosa sapidissima* (Hendricks *et al.*, 1991), grayling *Thymallus thymallus* (Nagiec *et al.*, 1995), striped bass *Morone saxatilis* (Secor *et al.*, 1991) and walleye *Stizostedion vitreum* (Brooks *et al.*, 1994). The survival of walleye fry, *Stizostedion vitreum*, following immersion in tetracycline has been shown to be high (Peterson & Carline, 1996). Acute toxicities of oxytetracycline hydrochloride and calcein to juvenile striped bass *Morone saxatilis* following immersion marking are reported by Bumguardner & Kin (1996). Successful marking of otolith and bone within the egg is reported for Arctic charr *Salvelinus alpinus* (Rojas Beltran *et al.*, 1995) and for coregonids (Ruhle & Winecki Kuhn, 1992).

Oxytetracycline can also be administered in the food, for example in juvenile cod *Gadus morhua* (Pedersen & Carlsen, 1991; Nordeide *et al.*, 1992) and red fox, *Vulpes vulpes*, striped skunks, *Mephitis mephitis*, and raccoon, *Procyon lotor*. Oxytetracycline has been assessed as a method of marking individual that have taken bait containing the rabies vaccine (Nunan *et al.*, 1994). Tetracycline administered orally as a single 300 mg dose marked white-tailed deer for at least 150 days (Van Brackle *et al.*, 1994). Tetracycline treatment of bait also appeared to reduced its palatability.

Tetracycline staining can also be applied to the skeletal structures of lower animals such as molluscs (Ekaratne & Crisp, 1982). Examples include abalone, *Haliotis rubra* (Day *et al.*, 1995), whelk, *Buccinum undatum* (Kideys & Nash, 1993), bryozonans (Barnes, 1995) and the sponge, *Clathrina cerebrum* (Bavestrello *et al.*, 1993). Abelone, *Haliotis rubra*, has also been marked by injection, which Pirker & Schiel (1993) found caused more stress to the animal than immersion.

Other florescent compounds can also be used. The otoliths of cod, *Gadus morhua*, have been marked by immersing the eggs, larvae and juveniles in solutions of alizarin complex I (AC) (Blom *et al.*, 1994). Strontium, which is chemically similar to calcium and can substitute for this element in bony tissues of fishes, has been advocated as an inexpensive and efficient means of chemically marking juvenile salmonid scales (Snyder *et al.*, 1992).

3.2.7   *Marking by injection, panjet or tattooing*

For vertebrates such as fish, dyes can be applied by panjet inoculator, injection or tattooing machine. An individual marking scheme using a Panjet and Alcian blue applied to large numbers of brown trout is described by Bridcut (1993). Care must be taken to ensure the force applied to the skin is not sufficient to cause internal injuries. Ducklings have been dyed by injecting the egg prior to hatching (Evans, 1951).

Tattooing has been frequently used. For small mammals, individual marking for life can be achieved by tattooing the ear using tattooing tongs (Klimisch, 1986). Honma *et al.* (1986) considered ear tattooing superior to clipping, tagging and tattooing for marking small mammals. Joly & Miaud (1989) noted that although tattooing had been little used for urodeles it offered advantages over other methods. Tattoos are not always permanent; for example, on sturgeon, *Acipenser transmontanus*, they lasted less than one year (Rien *et al.*, 1994).

It has been found possible to mark crayfish by injecting a small amount of 'Bates numbering machine ink' into the centre of the abdomen; the black, blue and red inks were found to be non-poisonous (Slack, 1955; Black, 1963); similarly, Indian ink may be used in fish (O'Grady & Hoy, 1972). This method might be applied to other large arthropods that have an area of almost transparent cuticle. Latex injections have been used for crayfish (Slack, 1955) and fish (Davis, 1955). Coloured monofilament tags have been injected between the integument and the muscle on the ventral surface of prawns (Teboul, 1993). When applied to fish, such marks are often lost by ejection through the skin (Arnold, 1966).

Coded wire tags injected into crustaceans and fish are commonly used (e.g. Morrison *et al.*, 1990; Fitz & Wiegert, 1991; Nass & Bocking, 1992; Bergstedt *et al.*, 1993; Ingram, 1993; Frith & Nelson, 1994), and may be detected magnetically. An alternative is to use visible implant tags which are injected just under the skin where they can be seen and read. Unfortunately, these are susceptible to shedding (Niva, 1995), although coded wire tags injected into trout so as to be visible were found to have excellent retentions of 96% in postocular tissue and 99% in adipose fins after 238 days (Oven & Blankenship, 1993).

More recently, passive integrated transponder (PIT) tags only a few millimetres in length have been developed for the individual tagging of aquatic animals. These are now used extensively on fish, though other applications include flatback turtles, *Natator depressus* (Parmenter, 1993) and red king crab, *Paralithodes camtschaticus* (Donaldson *et al.*, 1992). The presence of these tags is detected electronically. The small size of PIT tags suggests they have potential application for studies on larger terrestrial invertebrates.

3.2.8   *External tags*

Bands and rings are used extensively in studies on birds, reptiles and mammals. Ear tags originally developed for domestic animals are used on large mammals, such as seals (Testa & Rotherly, 1992; Prestrud, 1992). A great variety of external tags are commercially available for fish (see Laird and Stott, 1978), and these can also be used for other animals groups, such as rattlesnakes (Smith, 1994). Rice & Taylor (1993) describe a waistband for Anurans, while a permanent method for tagging growing macrocrustaceans is described by Schmalbach *et al.* (1994); these animals are often problematical because they moult, even as adults. A nylon filament tag attached to the prawn, *Macrobrachium rosenbergii*, did not interfere with moulting or growth over a 17-month period. Molluscs can also be tagged; for example, abalone, *Haliotis rubra*, were individually marked using a nylon rivet to attach a numbered disc to the trema (Prince, 1991).

The small size of most insects often precludes these convenient methods. Butterflies and locusts have, however, been marked by attaching small labels with a word or a code written in waterproof black ink to part of their wings; in Lepidoptera, the area should first be denuded of scales. Earlier workers used paper or cellophane stuck on with an adhesive, such as 'Durofix', which was later replaced by 'Sellotape'. Freilich (1989) found that plastic tags attached with cyanoacrylate instant glue gave reliable results for benthic macroinvertebrates, and stoneflies retained tags for six months in a circular laboratory stream tank, and for up to 81 days under field conditions in a turbulent stream. Klock *et al.* (1953) devised a machine that glued lengths of coloured thread to anaesthetised flies. Punched ferrous labels have been used on bees, and these may be magnetically removed (Gary, 1971). It is possible that recent advances will soon allow large-bodied insects to be tagged internally using coded wire or PIT tags.

Small (1.0–1.5 mm diameter), lightweight plastic tags, produced using photographic film with a minimum of equipment, provide a means for individually marking insects without reliance on complex colour-coding or other schemes. Rubink (1988) describes their use in mark–recapture studies of native screwworm, *Cochliomyia hominivorax*, in Central America. Sullivan *et al.* (1990) attached numbered, coloured discs, measuring 2.5 mm in diameter to the shells of the planorbid snail, *Helisoma duryi*, with a waterproof epoxy which gave 100% tag retention for over 5 months, without significant mortality.

A recent development has been miniature radar tags for flying insects (Riley, 1996; Roland *et al.*, 1996). These tags, which reflect harmonic radar, were originally fitted to walking beetles (Mascanzoni & Wallin, 1986). The tags used by Roland *et al.* (1996) weighed 0.4 mg and comprised a Schottky low-barrier diode mounted in the centre of an 8-cm dipole of very fine aluminium wire. When mounted with rubber cement to the thorax or abdomen, with the dipole trailing behind the insect, they were successfully fitted to Apollo butterfly, *Parnassius sminthus*, common Alpine butterfly, *Erebia epipsodea*, forest tent moth, *Malacosoma disstria*, the tachinid fly, *Patelloa pachypyga*, and sarcophagid fly, *Arachnidomyia aldrichi*. The transmitter–receiver used to track the insects can be carried in a backpack.

3.2.9 *Branding*

Hot branding is commonly used for large vertebrates, while hot-wire branding can also be used for fish (Moav *et al.*, 1960). Freeze-branding has been applied to a variety of animals including fish (Peterson & Key, 1992), diving mammals (though not always with success; Tanaka & Kato, 1987) and small mammals (Ohwada, 1991). In mammals, freeze-branding produces a white mark by killing the pigment-producing cells. Peterson & Key (1992) compared survivorship, tag retention and growth of juvenile walleyes, *Stizostedion vitreum*, that had been hand-tagged with binary coded wire microtags (BCWMTs) or cold-branded with liquid nitrogen. It was concluded that BCWMTs were superior to cold branding. There are reports of tumours linked to freeze-branding in cattle (Yeruham *et al.*, 1993).

3.2.10 Mutilation

This method is also more widely used with vertebrates – especially fish, amphibians and reptiles (Laird and Stott, 1978) – than with the smaller insects. In this case, a mark – in order to be easily visible – may be proportionally so large as to affect the insect's behaviour. Lepidoptera have been marked by clipping their wings (Querci, 1936), beetles by damaging their elytra in various ways, such as incising the edges (Grilm, 1959), punching or burning small holes (Skuhravy, 1957; Schjotz-Christensen, 1961) or by scraping away the surface of the elytra between certain striae (Murdoch, 1963). Individual marks can be given to beetles by cutting small triangular and square notches in particular positions on the elytra, using a pair of scissors (Goldwasser *et al.*, 1993). Crabs have been marked by cutting some of the teeth on the carapace (Edwards, 1958), and orthopteroids by notching the pronotum and amputating tegmina (Gangwere *et al.*, 1964). Fin-clipping is commonly applied to fish (Laird and Stott, 1978), but this can be problematical particularly in tropical waters where regeneration is fast, natural fin damage common, and infection rife. Rien *et al.* (1994), when working with the white sturgeon, *Acipenser transmontanus*, compared a variety of tag and mutilation marks and concluded that lateral scute removal made a mark that lasted for at least 2 years.

An interesting variant on these procedures was the use of skin autografts by Rafinski (1977) to mark amphibians. The grafts were successful in more than 95% of cases and individuals could still be recognised 3 years later.

For snakes, Spellerberg & Prestt (1977) considered carefully executed scale clipping to be the most humane and reliable permanent marking method, while Hutton & Woolhouse (1989) marked crocodiles by coded toe clipping or tail scute notching.

3.2.11 Natural marks, parasites and genes

Higher animals especially may be individually recognised by their unique markings or scars; for example, characteristic fluke markings may be used to identify sperm whales (Dufault & Whitehead, 1995). Fish populations have been distinguished by the incidence of parasitic infections (Margolis, 1963; Templeman & Fleming, 1963; Bamber & Henderson, 1985) or by meristic characteristics such as gill raker form or number, otolith shape or scale circuli. Dispersal may be studied by the use of mutant genes (Peer, 1957; Levin, 1961; Hausermann *et al.*, 1971) or, when various genotypes are clearly distinct, by their different proportions in adjacent colonies (e.g. Sheppard, 1951; Richards and Waloff, 1954; Goodhart, 1962). However, it should be remembered that selection may operate differentially in the different colonies, perhaps on young stages before the genotype becomes identifiable. Different sexes and age classes also provide naturally marked groups.

3.2.12 Rare elements

Many elements, when exposed to a source of neutrons, become radioactive and emit a characteristic spectrum of gamma-rays; this process is termed 'neutron activation'. Various workers have marked animals by incorporating rare elements into them and subsequently recognising the mark by neutron activation and gamma spectroscopy. The great advantage of the method is that, at least in theory, the mark may be retained from larval to adult life and self-marking is possible. For example, if a small quantity of a rare element was mixed into a mosquito's breeding pool, any adults emerging should be detectable any time in their life; hence, the contributions of different breeding sites to a population could be determined. The disadvantages are that the equipment is extremely expensive (but is often available in nuclear physics or engineering centres), the procedure for gamma spectroscopy for rare elements is fatal for the animals (so it cannot be used in recapture studies), and there are often difficulties – undoubtedly of a basic physiological nature – in obtaining an adequate quantity of the rare element absorbed into the body and retained in the tissues (rather than just held in the gut).

Dysprosium, first used by Riebartsch (1963), has been found to be a life-long marker for some Lepidoptera (Jahn *et al.*, 1966) and *Drosophila* (Richardson *et al.*, 1969). Rubidium has been used to mark the larvae of cabbage looper, *Trichoplusia ni* (Berry *et al.*, 1972), bollworm *Helicoverpa zea* (Stadelbacher, 1991) tobacco budworm, *Heliothis virescens* (Stadelbacher, 1991), gypsy moth, *Lymantria dispar* (Fleischer *et al.*, 1990), and the dipterans, *Culicoides variipennis* (Holbrook *et al.*, 1991) and *Aedes aegypti* (Reiter *et al.*, 1995). Europium has also been used for Lepidoptera (Jahn *et al.*, 1966; Ito, 1970), manganese for fruit fly, *Ceratitis* (Monro, 1968), and lutetium, samarium and europium for the larvae of clams, barnacles, and polychaetes (Levin *et al.*, 1993).

Caesium, disprosium, rubidium and strontium, applied as chlorides to the foliage of host plants, have been found to mark phytophagous insects, including those feeding on flower heads (Fleischer *et al.*, 1990; Stadelbacher, 1991; Dempster *et al.*, 1995). Berry *et al.* (1972) found a straight-line relationship between the concentrations of rubidium in foliar spray and in the male moths that developed from the larvae that had been reared on the sprayed plants. However, with the pea aphid, *Acyrthosiphon pisum*, the biological half-life was only about one day and detectable quantities remained for only four days after the aphid had left the plant (Frazer & Raworth, 1974). Ito (1970) showed that the incorporation of europium (44 or 80 ppm) into the diets of larvae of the moths, *Hyphantria* and *Spodoptera*, did not affect growth and survival. The tick, *Dermacentor variabilis* was marked by injecting the host mice with RbCl (Burg, 1994).

Alternatively, the rare elements may be applied topically, essentially being used as labels. Cerium (50 μg $CeCl_3$ per insect in an alcoholic solution), which is absorbed into the cuticle of insects, has been used by Rahalkar *et al.* (1971) to mark weevils, *Rhynchophorus elytra*, and gold was used similarly by Bate *et al.* (1974) to mark the elm bark beetle. Emerging calypterate diptera may self-mark by contamination of the ptilinal suture. Various dyes are often used for this purpose (p. 112), and rare elements may be even more effective. Haisch *et al.* (1975) marked emerging *Rhagoletis* with dysprosium and samarium, mixed with fine sand and silica gel (1:1 by weight) at concentrations of 0.1% by weight. Provided that the pupa were 3 cm deep, the error in self marking was about 2%.

The trace elements within an animal can vary with the habitat, potentially allowing the origin of individuals to be identified by their 'chemoprint'. The identification of breeding site or place of origin has been applied to fish, birds and insects. Dempster *et al.* (1986) applied this approach to the identification of brimstone butterfly, *Gonepteryx rhamni*, populations, but found that differences related to the place of origin faded with age.

3.2.13 Protein marking

One of the more recent developments has been the ability to mark insects with vertebrate immunoglobulins and other proteins which are detected using an enzyme-linked immunosorbent assay (ELISA). Hagler *et al.* (2009), for example, reported on the feasibility of using rabbit immunoglobulin protein to mark termites for mark–recapture studies. When the termites were marked either externally by topical spraying or internally via their food, it was concluded that the termites retained the mark for at least 35 days.

3.2.14 Radioactive isotopes

Radioactive labelling, which was most popular soon after radioisotopes became readily available, is now rarely used to mark animals for the study of population size or dispersal. It has been largely replaced by safer techniques such as rare element tracers, protein marks and fluorescent powders. In many countries there are health and safety restrictions on the use of radioisotopes in the field, and these must be checked before use. Useful bibliographies to the early literature on the entomological use of radiotracers are given in Anon (1963) and Jenkins (1963); methodological details for the use of radioisotopes are given in Knoche (1991) and the earlier versions of *Ecological Methods*. Radioactive techniques for small mammal marking are reviewed by Linn (1977).

3.2.15 Radio and sonic tags

Radio tags transmit at radio frequencies and thus can be used in both terrestrial and aquatic environments. Sonic tags emit ultrasound and are only useful in water, where the signal is detected with a hydrophone. These tags are used to study movement, the animal's immediate environment, and its physiological activity. In addition to transmitting a signal which can be used to track the individual they can also transmit information on, for example, temperature, salinity, heart rate and depth of dive (pressure) (Amlaner & Macdonald, 1980). The key problem with these tags is their size and weight, which creates difficulties of attachment and interference with the animal's movement. The batteries that must be included within the tag limit miniaturisation. Alternative technologies for simply tracking animals are the much smaller passive integrated transponder (PIT) (see p. 116) or radar tags (p. 117), which do not need an *in situ* power supply.

References

Aars, J., Andreassen, H.P., & Ims, R.A. (1995) Root voles: Litter sex ratio variation in fragmented habitat. *J. Anim. Ecol.* **64**(4), 459–72.

Aebischer, N.J. (1986) Estimating the proportion of uncatchable animals in a population by double-sampling. *Biometrics* **42**(4), 973–80.

Amlaner, C.J. & Macdonald, D.W. (eds) (1980) *A Handbook on Biotelemetry and Radio tracking.* Pergamon Press, Oxford.

Amstrup, S.C., McDonald, T.L., & Manly, B.F. (eds) (2010) *Handbook of capture-recapture analysis.* Princeton University Press.

Anon. (1963) Radioisotopes and ionizing radiations in entomology. *Bibl. Ser. Int. Atomic Energy Ag.*, **9**, 414 pp.

Ayre, G.L. (1962) Problems in using the Lincoln Index for estimating the size of ant colonies (Hymenoptera: Formicidae). *J. N. Y. Entomol. Soc.* **70**, 159–66.

Baber, I., Keita, M., Sogoba, N., Konate, M., Doumbia, S., Traoré, S.F., Ribeiro, J.M.C., & Manoukis, N.C. (2010) Population size and migration of *Anopheles gambiae* in the Bancoumana Region of Mali and their significance for efficient vector control. *PloS One*, **5**(4), e10270.

Bagenal, T.B. (1967) A method of marking fish eggs and larvae. *Nature* **214**, 113.

Bailey, N.T.J. (1951) On estimating the size of mobile populations from recapture data. *Biometrika* **38**, 293–306.

Bailey, N.T.J. (1952) Improvements in the interpretation of recapture data. *J. Anim. Ecol.* **21**, 120–7.

Bailey, S.F., Eliason, D.A., & Iltis, W.C. (1962) Some marking and recovery techniques in *Culex tarsalis* coq. flight studies. *Mosquito News* **22**, 1–10.

Baillargeon, S., Rivest, L.P. (2007) Rcapture: Loglinear models for capture–recapture in R. *J. Statist. Software*, **19**(5); available at: http://www.jstatsoft.org/v19/i05

Baker, C.S. & Herman, L.M. (1987) Alternative population estimates of humpback whales (*Megaptera novaeangliae*) in Hawaiian waters (USA). *Can. J. Zool.* **65**(11), 2818–21.

Baker, S.G. (1990) A simple EM algorithm for capture–recapture data with categorical covariates. *Biometrics* **46**(4), 1193–200.

Bamber, R.N. & Henderson, P.A. (1985) Diplostomiasis in the sand smelt from the Fleet, Dorset and its use as a population indicator. *J. Fish Biol.* **26**, 223–9.

Banks, C.J. (1955) The use of radioactive tantalum in studies of the behaviour of small crawling insects on plants. *Br. J. Anim. Behav.* **3**, 158–9.

Barnes, D.K.A. (1995) Seasonal and annual growth in erect species of Antarctic bryozoans. *J. Exp. Mar. Biol. Ecol.* **188**(2), 181–98.

Bate, L.C., Lyon, W.S., & Wollerman, E.H. (1974) Gold tagging of elm bark beetles and identification by neutron activation analysis. *Radiochem. Radioanal. Lett.* **17**, 77–85.

Bavestrello, G., Cattaneo Vietti, R., Cerrano, C., & Sara, M. (1993) Rate of spiculogenesis in *Clathrina cerebrum* (Porifera: Calcispongiae) using tetracycline marking. *J. Mar. Biol. Ass. UK* **73**(2), 457–60.

Begon, M. (1983) Abuses of mathematical techniques in ecology: applications of Jolly's capture–recapture method. *Oikos* **40**, 155–158.

Belant, J.L. & Seamans, T.W. (1993) Evaluation of dyes and techniques to color-mark incubating herring gulls. *J. Field Ornithol.* **64**(4), 440–51.

Belmar-Lucero, S., Wood, J.L.A., Scott, S., Harbicht, A.B., Hutchings, J.A., & Fraser, D.J. (2012) Concurrent habitat and life history influences on effective/census population size ratios in stream-dwelling trout. *Ecol. Evol.* **2**, 562–73.

Beltran, R.R., Champigneulle, A., & Vincent, G. (1995) Mass-marking of bone tissue of *Coregonus lavaretus* L. and its potential application to monitoring the spatio-temporal distribution of larvae, fry and juveniles of lacustrine fishes. *Hydrobiologia* **301**, 399–407.

Bergstedt, R.A., Swink, W.D., & Seelye, J.G. (1993) Evaluation of two locations for coded wire tags in larval and small parasitic phase sea lampreys. *N. Am. J. Fish. Manag.* **13**(3), 609–12.

Berry, W.L., Stimmann, M.W., & Wolf, W.W. (1972) Marking of native phytophagous insects with rubidium: a proposed technique. *Ann. Entomol. Soc. Am.* **65**, 236–8.

Bishop, J.A. & Sheppard, P.M. (1973) An evaluation of two capture-recapture models using the technique of computer simulation. In: Bartlett, M.S. & Hiorns, R.W. (eds), *The Mathematical Theory of the Dynamics of Biological Populations*. Academic Press, London, pp. 235–52.

Black, J.B. (1963) Observations on the home range of stream-dwelling crawfishes. *Ecology* **44**, 592–5.

Blinn, W.C. (1963) Ecology of the land snails *Mesodon thyroidus* and *Allogona profunda*. *Ecology* **44**, 498–505.

Blom, G., Nordeide, J.T., Svasand, T., & Borge, A. (1994) Application of two fluorescent chemicals, alizarin complexone and alizarin red S, to mark otoliths of Atlantic cod, *Gadus morhua* L. *Aquacult. Fish. Manag.* **25**(Suppl. 1), 229–43.

Boano, G. & Cucco, M. (1991) Annual survival rates of a marsh warbler, *Acrocephalus palustris*, breeding population in Northern Italy. *Riv. Ital. Ornitol.* **61**(1–2), 10–18.

Bonett, D.G. (1988) Small sample bias and variance of Lincoln–Petersen and Bailey estimators. *Biometr. J.* **30**(6), 723–7.

Boonstra, R. & Krebs, C.J. (1978) Pitfall trapping of *Microtus townsendii*. *Ecology* **60**, 567–73.

Borror, D.J. (1934) Ecological studies of *Argia moesta* Hagen (Odonata: Coenag-Rionidae) by means of marking. *Ohio J. Sci.* **34**, 97–108.

Bohlin, T. & Sundström, B. (1977) Influence of unequal catchability on population estimates using the Lincoln index and the removal method applied to electro-fishing. *Oikos*, **28**, 123–29.

Bozeman, E.L., Helfman, G.S., & Richardson, T. (1985) Population size and home range of American eels (*Anguilla rostrata*) in a Georgia (USA) tidal creek. *Trans. Am. Fish. Soc.* **114**(6), 821–5.

Bridcut, E.E. (1993) A coded Alcian blue marking technique for the identification of individual brown trout, *Salmo trutta* L.: An evaluation of its use in fish biology. *Biol. Environ.* **93B**(2), 107–10.

Brooks, R.C., Heidinger, R.C., & Kohler, C.C. (1994) Mass-marking otoliths of larval and juvenile walleyes by immersion in oxytetracycline, calcein, or calcein blue. *N. Am. J. Fish. Manag.* **14**(1), 143–50.

Brownie, C., Hines, J.E., & Nichols, J.D. (1986) Constant-parameter capture–recapture models. *Biometrics* **42**(3), 561–74.

Brussard, P.F. (1971) Field techniques for investigations of population structure in a ubiquitous butterfly. *J. Lepidop. Soc.* **25**, 22–9.

Buckland, S.T. (1980) A modified analysis of the Jolly–Seber capture–recapture model. *Biometrics* **36**, 419–35.

Buckland, S.T. & Hereward, A.C. (1982) Trapshyness of yellow wagtails (*Motacilla flava flavissima*) at a premigratory roost. *Ringing Migration* **4**, 15–23.

Buckland, S.T., Rowley, I., & Williams, D.A. (1983) Estimation of survival from repeated sightings of tagged galahs. *J. Anim. Ecol.* **52**, 563–73.

Buckner, C.H. (1968) Reactions of small mammals to vital dyes. *Can. Entomol.* **100**, 476–7.

Bumgaurdner, B.W. (1991) Marking subadult red drums with oxytetracycline. *Trans. Am. Fish. Soc.* **120**(4), 537–40.

Bumguardner, B.W. & Kin, T.L. (1996) Toxicity of oxytetracycline and calcein to juvenile striped bass. *Trans. Am. Fish. Soc.* **125**(1), 143–5.

Burg, J.G. (1994) Marking *Dermacentor variabilis* (Acari: Ixodidae) with rubidium. *J. Med. Entomol.* **31**(5), 658–62.

Burnham, K.P., Anderson, D.R., White, G.C., Brownie, C., & Pollock, K.H. (1987) *Design and analysis methods for fish survival experiments based on release-recapture*. Am. Fish. Soc. Monograph, Bethesda, Maryland.

Caldwell, A.H. (1956) Dry ice as an insect anaesthetic. *J. Econ. Entomol.* **49**, 264–5.

Calvo, B. & Furness, R.W. (1992) A review of the use and the effects of marks and devices on birds. *Ringing and Migration* **13**, 129–36.

Carothers, A.D. (1973) The effects of unequal catchability on Jolly–Seber estimates. *Biometrics* **29**, 79–100.

Caughley, G. (1977) *Analysis of Vertebrate Populations*. John Wiley & Sons, New York.

Cianci, D., Van Den Broek, J., Caputo, B., Marini, F., Torre, A.D., Heesterbeek, H., & Hartemink, N. (2013) Estimating mosquito population size from mark–release–recapture data. *J. Med. Entomol.* **50**(3), 533–42.

Chang, H.T. (1946) Studies on the use of fluorescent dyes for marking *Anopheles quadrimaculatus* Say. *Mosquito News* **6**, 122–5.

Chao, A. (1987) Estimating the population size for capture–recapture data with unequal catchability. *Biometrics* **43**(4), 783–92.

Chao, A. (1989) Estimating population size for sparse data in capture–recapture experiments. *Biometrics* **45**(2), 427–38.

Chao, A. (2001) An overview of closed capture–recapture models. *J. Agric. Biol. Environ. Statist.* **6**(2), 158–75.

Chapman, D.G. (1951) Some properties of the hypergeometric distribution with applications to zoological sample censuses. *Univ. Calif. Publ. Stat.* **1**(7), 131–60.

Chapman, D.G. (1954) The estimation of biological populations. *Ann. Math. Statist.* **25**, 1–15.

Chapman, D.G. (1955) Population estimation based on change of composition caused by a selective removal. *Biometrika* **42**, 279–90.

Collins, C.W. & Potts, S.F. (1932) Attractants for the flying gypsy moths as an aid to locating new infestations. *USDA Tech. Bull.* **336**, 43 pp.

Conway, G.R., Tripis, M., & McClelland, G.A.H. (1974) Population parameters of the mosquito *Aedes aegypti* (L.) estimated by mark–release–recapture in a suburban habitat in Tanzania. *J. Anim. Ecol.* **43**, 289–304.

Cook, S.P. & Hain, F.P. (1992) The influence of self-marking with fluorescent powders on adult bark beetles (Coleoptera: Scolytidae). *J. Entomol. Sci.* **27**(3), 269–79.

Corbet, G.B. (1956) The life-history and host relations of a hippoboscid fly *Ornithomyia fringillina* Curtis. *J. Anim. Ecol.* **25**, 403–20.

Corbet, P.S. (1952) An adult population study of *Pyrrhosoma nymphula* (Sulzer); (Odonata: Coenagrionidae). *J. Anim. Ecol.* **21**, 206–22.

Cormack, R.M. (1966) A test for equal catchability. *Biometrika* **22**, 330–42.

Cormack, R.M. (1972) The logic of capture–recapture estimates. *Biometrics* **28**, 337–43.

Cormack, R.M. (1973) Commonsense estimates from capture–recapture studies. In: Bartlett, M.S. & Hiorns, R.W. (eds), *The Mathematical Theory of the Dynamics of Biological Populations*. Academic Press, London, pp. 225–34.

Cormack, R.M. (1979) Models for capture–recapture. In: Cormack, R.M. & Patil, G.P. (eds), *Sampling Biological Populations*. International Cooperative Publishing House, Fairfield, Maryland, pp. 217–55.

Cormack, R.M. (1981) Log-linear models for capture–recapture experiments on open populations. In: Hiorns, R.W. & Cooke, D. (eds), *The Mathematical Theory of the Dynamics of Biological Populations*. Academic Press, London, pp. 197–216.

Cormack, R.M. (1989) Log-linear models for capture–recapture. *Biometrics* **45**(2), 395–414.

Cormack, R.M. (1993) Variances of mark-recapture estimates. *Biometrics* **49**, 1188–93.

Coulson, J.C. (1962) The biology of *Tipula subnodicornis*, Zetterstedt, with comparative observations on *Tipula paludosa* Meigen. *J. Anim. Ecol.* **31**, 1–21.

Cragg, J.B. & Hobart, J. (1955) A study of a field population of the blowflies *Lucilia ceasar* (L.) and *L. serricata* (MG.). *Ann. Appl. Biol.* **43**, 645–63.

Craig, C.C. (1953) On the utilisation of marked specimens in estimating populations of flying insects. *Biometrika* **40**, 170–6.

Crosbie, S.F. & Manly, B.F.J. (1985) A new approach to parsimonius modelling of capture–mark–recapture experiments. *Biometrics* **41**, 385–98.

Crumpacker, D.W. (1974) The use of micronized fluorescent dusts to mark adult *Drosophila pseudoobscura*. *Am. Midl. Nat.* **91**, 118–29.

Cunningham, M.P., Harley, J.M., & Grainge, E.B. (1963) The labelling of animals with specific agglutinins and the detection of these agglutinins in the blood meals of *Glossina*. *Rep. E. Afr. Tryp. Res. Org.* **1961**, 23–4.

Dalmat, H.T. (1950) Studies on the flight range of certain Simuiiidae, with the use of aniline dye marker. *Ann. Entomol. Soc. Am.* **43**, 537–45.

Darroch, J.N. (1958) The multiple recapture census. I. Estimation of a closed population. *Biometrika* **45**, 343–51.

Darroch, J.N. (1959) The multiple-capture census. II. Estimation when there is immigration or death. *Biometrika* **46**, 336–51.

Daum, R.J., Gast, R.T., & Davich, T.B. (1969) Marking adult boll weevils with dyes fed in a cottonseed oil bait. *J. Econ. Entomol.* **62**, 943–4.

Davey, J.T. (1956) A method of marking isolated adult locusts in large numbers as an aid to the study of their seasonal migrations. *Bull. Entomol. Res.* **46**, 797–802.

Davis, C.S. (1955) The injection of latex solution as a fish marking technique. *Invest. Indiana Lakes Streams* **4**, 111–16.

Day, R.W., Williams, M.C., & Hawkes, G.P. (1995) A comparison of fluorochromes for marking abalone shells. *Mar. Freshwater Res.* **46**(3), 599–606.

Dempster, J.P. (1971) The population ecology of the cinnabar moth, *Tyria jacobaeae* L. (Lepidoptera, Arctiidae). *Oecologia* **7**, 26–67.

Dempster, J.P., Atkinson, D.A., & Cheesman, O.D. (1995) The spatial population dynamics of insects exploiting a patchy food resource I. Population extinctions and regulation. *Oecologia* **104**, 340–53.

Dempster, J.P., Lakhani, K.H., & Coward, P.A. (1986) The use of chemical composition as a population marker in insects: a study of the brimstone butterfly. *Ecol. Entomol.* **11**, 51–65.

Dobson, J.R. (1993) A portable carbon dioxide dispenser and its uses in field entomology. *Antenna* **17**, 10–13.

Dobson, R.M. & Morris, M.G. (1961) Observations on emergence and life-span of wheat bulb fly, *Leptohylemyia coarctata* (Fall.) under field-cage conditions. *Bull. Entomol. Res.* **51**, 803–21.

Dobson, R.M., Stephenson, J.W., & Lofty, J.R. (1958) A quantitative study of a population of wheat bulb fly, *Leptohylemyia coarctata* (Fall.) in the field. *Bull. Entomol. Res.* **49**, 95–111.

Donaldson, W.E., Schmidt, D., Watson, L., & Pengill, Y.D. (1992) Development of a technique to tag adult red king crab, *Paralithodes camtschatica* (Tilesius, 1815), with passive integrated transponder tags. *J. Shellfish Res.* **11**(1), 91–4.

Dosdall, L.M., Galloway, M.M., & Gadawski, R.M. (1992) New self-marking device for dispersal studies of black flies (Diptera: Simuliidae). *J. Am. Mosquito Control Assoc.* **8**(2), 187–90.

Dow, R.P. (1971) The dispersal of *Culex nigripalpus* marked with high concentrations of radiophosphorus. *J. Med. Entomol.* **8**, 353–63.

Dowdeswell, W.H. (1959) *Practical Animal Ecology.* Methuen, London.

Dowdeswell, W.H., Fisher, R.A., & Ford, E.B. (1940) The quantitative study of populations in the lepidoptera. 1. *Polyommatus icarus rott. Ann. Eugen.* **10**, 123–36.

Dowdy, A.K. & McGaughey, W.H. (1992) Fluorescent pigments for marking lesser grain borers (Coleoptera: Bostrichidae). *J. Econ. Entomol.* **85**(2), 567–9.

Dufault, S. & Whitehead, H. (1995) An assessment of changes with time in the marking patterns used for photoidentification of individual sperm whales, *Physeter macrocephalus. Mar. Mammal Sci.* **11**(3), 335–43.

Eberhardt, L.L. (1969) Population estimates from recapture frequencies. *J. Wildlife Manag.* **33**, 28–39.

Eberhardt, L.L., Chapman, D.G., & Gilbert, J.R. (1979) A review of marine mammal census methods. *Wildlife Monogr.* 3–46.

Edgar, W.D. (1971) The life-cycle, abundance and seasonal movement of the wolf spider, *Lycosa (Pardosa) lugubris*, in central Scotland. *J. Anim. Ecol.* **40**, 303–22.

Edwards, R.L. (1958) Movements of individual members in a population of the shore crab, *Carcinus maenas* L., in the littoral zone. *J. Anim. Ecol.* **27**, 37–45.

Edwards, R.L. (1961) Limited movement of individuals in a population of the migratory grasshopper, *Melanoplus bilituratus* (Walker) (Acrididae) at Kamloops, British Columbia. *Can. Entomol.* **93**, 628–31.

Ehrlich, P.R. & Davidson, S.E. (1960) Techniques for capture–recapture studies of lepidoptera populations. *J. Lepidop. Soc.* **14**, 227–9.

Ekaratne, S.U.K. & Crisp, D.J. (1982) Tidal micro-growth bands in tertidal gastropod shells, with an evaluation of band-dating techniques. *Proc. Roy. Soc. (B)* **214**, 305–23.

Ellis, D.H. & Ellis, C.H. (1975) Color marking Golden eagles with human hair dyes. *J. Wildlife Manag.* **39**, 446–7.

Eltringham, S.K. (1977) Methods of capturing wild animals for marking purposes. In: Stonehouse, B. (ed.), *Animal Marking.* The MacMillan Press Ltd, London, pp. 13–23.

Erickson, J.M. (1972) Marks.975) Color marking Golden eagles with human hair *Pogonomyrmex* and colonies: an evaluation of the ^{32}P technique. *Ann. Entomol. Soc. Am.* **65**, 57.

Evans, C.D. (1951) A method of color-marking young waterfowl. *J. Wildlife Manag.* **15**, 101–3.

Evans, C.R. & Lockwood, A.P.M. (1994) Population field studies of the Guinea chick lobster (*Panulirus guttatus* Latreille) at Bermuda: Abundance, catchability, and behavior. *J. Shellfish Res.***13**(2), 393–415.

Evans, W.G. & Gyrisco, G.G. (1960) The flight range of the European chafer. *J. Econ. Entomol.* **53**, 222–4.

Fales, J.H., Bodenstein, P.F., Mills, G.D., & Wessel, L.H. (1964) Preliminary studies on face fly dispersion. *Ann. Entomol. Soc. Am.* **57**, 135–7.

Fenwick, A. & Amin, M.A. (1983) Marking snails with nail varnish as a field experimental technique. *Ann. Trop. Med. Parasitol.* **77**, 387–90.

Fisher, R.A. & Ford, E.B. (1947) The spread of a gene in natural conditions in a colony of the moth *Panaxia dominula* L. *Heredity* **1**, 143–74.

Fitz, H.C.W. & Wiegert, R.G. (1991) Tagging juvenile blue crabs, *Callinectes sapidus*, with microwire tags. Retention, survival, and growth through multiple molts. *J. Crustac. Biol.* **11**(2), 229–35.

Fleischer, S.J., Ravlin, F.W., Delorme, D., Stipes, R.J., & McManus, M.L. (1990) Marking gypsy moth (Lepidoptera: Lymantriidae) life stages and products with low doses of rubidium injected or implanted into pin oak. *J. Econ. Entomol.* **83**(6), 2343–8.

Foott, W.H. (1976) Use of fluorescent powders to monitor flight activities of adult *Glischrochilus quadrisignatus* [Coleoptera: Nitidulidae]. *Can. Entomol.* **108**, 1041–4.

Frankie, G.W. (1973) A simple field technique for marking bees with fluorescent powders. *Ann. Entomol. Soc. Am.* **66**, 690–1.

Frazer, B.D. & Raworth, D.A. (1974) Marking aphids with rubidium. *Can. J. Zool.* **529**, 1135–6.

Freeland, W.J. (1986) Populations of cane toad, *Bufo marinus*, in relation to time since colonization. *Aust. Wildlife Res.* **13**(2), 321–30.

Freeman, B.E. (1964) A population study of *Tipula* species (Diptera, Tipulidae). *J. Anim. Ecol.* **33**, 129–40.

Freilich, J.E. (1989) A method for tagging individual benthic macroinvertebrates. *J. N. Am. Bentholog. Soc.* **8**(4), 351–4.

Frisch, K.V. (1950) *Bees, Their Vision, Chemical Senses, and Language*. Ithaca, New York.

Frith, H.R. & Nelson, T.C. (1994) Abundance, age, size, sex and coded wire tag recoveries for chinook salmon escapements of Campbell and Quinsam Rivers, 1993. *Can. Manuscript Rep. Fish Aquatic Sci.* **2251**, I–IX, 1–59.

Gaillard, J.M., Boisaubert, B., Boutin, J.M., & Clobert, J. (1986) Estimation of population numbers from capture–mark–recapture data: Application to roe deer. *Gibier Faune Sauvage* **3**, 143–58.

Gangwere, S.K., Chavin, W., & Evans, F.C. (1964) Methods of marking insects, with especial reference to Orthoptera (Sens. Lat.). *Ann. Entomol. Soc. Am.* **57**, 662–9.

Gary, N.E. (1971) Magnetic retrieval of ferrous labels in a capture–recapture system for honey bees and other insects. *J. Econ. Entomol.* **64**, 961–5.

Gaskell, T.L. & George, B.J. (1972) A Bayesian modification of the Lincoln index. *J. Appl. Ecol.* **9**, 377–84.

Gatz, A.J.J.R. & Loar, J.M. (1988) Petersen and removal population size estimates: Combining methods to adjust and interpret results when assumptions are violated. *Environ. Biol. Fish.* **21**(4), 293–308.

Gillies, M.T. (1961) Studies on the dispersion and survival of *Anopheles gambiae* Giles in East Africa, by means of marking and release experiments. *Bull. Entomol. Res.* **52**, 99–127.

Goldwasser, L., Schatz, G.E., & Young, H.J. (1993) A new method for marking Scarabaeidae and other Coleoptera. *Coleopt. Bull.* **47**, 21–6.

Golley, F.B. & Gentry, J.B. (1964) Bioenergetics of the southern harvester ant, *Pogonomyrmex badius. Ecology* **45**, 217–25.

Goodhart, C.B. (1962) Thrush predation on the snail *Cepaea hortensis. J. Anim. Ecol.* **279**, 47–57.

Gosselin, L.A. (1993) A method for marking small juvenile gastropods. *J. Mar. Biol. Assoc. UK* **73**(4), 963–6.

Gouteux, J.P. & Buckland, S.T. (1984) Tsetse fly ecology in the preforested area of Ivory Coast: 8. Population dynamics. *Cahiers OR. STOM Ser. Entomol. Med. Parasitol.* **22**, 19–34.

Graham, H.M. & Mangum, C.L. (1971) Larval diets containing dyes for tagging pink bollworm moths internally. *J. Econ. Entomol.* **64**, 376–9.

Greenslade, P.J.M. (1964) The distribution, dispersal and size of a population of *Nebria brevicollis* (F.), with comparative studies on three other Carabidae. *J. Anim. Ecol.* **33**, 311–33.

Griffin, D.R. (1970) Migrations and homing of bats. In: Wimsatt, W.A. (ed.) *Biology of Bats.* Academic Press, London.

Grilm, L. (1959) Seasonal changes of activity of the Carabidae. *Ekol. Poiska A* **7**, 25568.

Hagler, J.R., Baker, P.B., Marchosky, R., Machtley, S.A., & Bellamy, D.E. (2009) Methods to mark termites with protein for mark–release–recapture and mark–capture type studies. *Insectes Sociaux* **56**(2), 213–20.

Hagler, J.R., Cohen, A.C., Bradley Dunlop, D., & Enriquez, F.J. (1992) New approach to mark insects for feeding and dispersal studies. *Environ. Entomol.* **21**(1), 20–5.

Hagler, J.R. & Jackson, C.G. (2001) Methods for marking insects: current techniques and future prospects. *Annu. Rev. Entomol.* **46**(1), 511–43.

Hain, M.L. (1965) Ecology of the lizard *Uta mearnsi* in a desert canyon. *Copeia* **1965**, 78–81.

Haisch, A., Stark, H., & Forster, S. (1975) Markierung von Fruchtfliegen und ihre Erkennung durch indikatoraktivierung. *Entomologia Exp. Appl.* **18**, 31–43.

Hampton, J.K. & Kirkwood, G.P. (1990) Tag shedding by southern bluefin tuna *Thunnus maccoyii*. *US Nat. Mar. Fish. Serv. Fish. Bull.* **88**(2), 313–22.

Hanson, W.R. (1963) Calculation of productivity, survival and abundance of selected vertebrates from sex and age ratios. *Wildl. Monogr.* **9**, 1–60.

Hartley, P.H.T. (1954) Back garden ornithology. *Bird Study* **1**, 18–27.

Hausermann, W., Fay, R.W., & Hacker, C.S. (1971) Dispersal of genetically marked female *Aedes aegypti* in Mississippi. *Mosquito News* **31**, 37–51.

Hawkes, R.B. (1972) A fluorescent dye technique for marking insect eggs in predation studies. *J. Econ. Entomol.* **65**, 1477–8.

Henderson, P.A. & Holmes, R.H.A. (1987) On the population biology of the common shrimp *Crangon crangon* (L.) (Crustacea: Caridea) in the Severn Estuary and Bristol Channel. *J. Mar. Biol. Assoc.* UK **67**, 825–47.

Hendricks, D.E. (1971) Oil-soluble blue dye in larval diet marks adults, eggs, and first stage F, larvae of the pink bollworm. *J. Econ. Entomol.* **64**, 1404–6.

Hendricks, D.E. & Graham, H.M. (1970) Oil-soluble dye in larval diet for tagging moths, eggs, and spermatophores of tobacco budworms. *J. Econ. Entomol.* **63**, 1019–20.

Hendricks, M.L., Bender, T.R.J.R., & Mudrak, V.A. (1991) Multiple marking of American shad otoliths with tetracycline antibiotics. *N. Am. J. Fish. Manag.* **11**(2), 212–19.

Heron, R.J. (1968) Vital dyes as markers for behavioural and population studies of the larch sawfly, *Pristiphora erichsonii* (Hymenoptera: Tenthredinidae). *Can. Entomol.* **1009**, 470–5.

Hewlett, P.S. (1954) A micro-drop applicator and its use for the treatment of certain small insects with liquid insecticide. *Ann. Appl. Biol.* **41**, 45–64.

Hill, S.B. & Clayton, D.H. (1985) *Wildlife after dark: a review of nocturnal observation techniques.* Occasional paper, 17, James Ford Bell Museum of Natural History, University of Minnesota.

Holbrook, F.R., Belden, R.P., & Bobian, R.J. (1991) Rubidium for marking adults of *Culicoides variipennis* (Diptera: Ceratopogonidae). *J. Med. Entomol.* **28**(2), 246–9.

Holt, S.J. (1955) On the foraging activity of the wood ant. *J. Anim. Ecol.* **24**, 1–34.

Honma, M., Iwaki, S., Kast, A., & Kreuzer, H. (1986) Experiences with the identification of small rodents. *Exp. Anim.* **35**(3), 347–52.

Horn, H.S. (1976) A clamp for marking butterflies in capture–recapture studies. *J. Lepidop. Soc.* **30**(2), 145–6.

Hunter, P.E. (1960) Plastic paint as a marker for mites. *Ann. Entomol. Soc. Am.* **53**, 698.

Hutton, J.M. & Woolhouse, M.E.J. (1989) Mark–recapture to assess factors affecting the proportion of a Nile crocodile population seen during spotlight counts to monitor crocodile abundance. *J. Appl. Ecol.* **26**, 381–95.

Ingram, B.A. (1993) Evaluation of coded wire tags for marking fingerling golden perch, *Macquaria ambigua* (Percichthyidae), and silver perch, *Bidyanus bidyanus* (Teraponidae). *Aust. J. Mar. Freshwater Res.* **44**(6), 817–24.

Isaacs, R. & Byrne, D.N. (1998) Aerial distribution, flight behaviour and eggload: their interrelationship during dispersal by the sweet potato white fly. *J. Anim. Ecol.* **67**, 741–50.

Ito, Y. (1970) A stable isotope, europium-151 as a tracer for field studies of insects. *Appl. Entomol. Zool.* **5**, 175–81.

Iwao, S., Kiritani, K., & Hokyo, N. (1966) Application of a marking and recapture method for the analysis of larval-adult populations of an insect, *Nezara viridula* (Hemiptera: Pentatomidae). *Res. Pop. Ecol.* **8**, 147–60.

Iwao, S., Mizuta, K., Nakamura, H., Oda, T., & Sato, Y. (1963) Studies on a natural population of the large 28-spotted lady beetle, *Epilachna vigintioctomaculata* Motschulsky. 1. Preliminary analysis of the overwintered adult population by means of the marking and recapture method. *Jap. J. Ecol.* **13**, 109–17.

Jackson, C.H.N. (1933a) On a method of marking tsetse flies. *J. Anim. Ecol.* **2**, 289–90.

Jackson, C.H.N. (1933b) On the true density of tsetse flies. *J. Anim. Ecol.* **2**, 204–09.

Jackson, C.H.N. (1937) Some new methods in the study of *Glossina morsitans. Proc. Zool. Soc. Lond.* **1936**, 811–96.

Jackson, C.H.N. (1948) The analysis of a tsetse-fly population. *III. Ann. Eugen.* **14**, 91–108.

Jackson, C.H.N. (1953) A mixed population of *Glossina morsitans* and *G. swynnertoni. J. Anim. Ecol.* **22**, 78–86.

Jahn, E., Lippay, H., Weidinger, N., & Schwach, G. (1966) Untersuchungen uber die usbreitung von Nonnenfaltem durch Markierung mit Seltenen erden. *Anz. Schadlingskunst* **39**, 17–22.

Jenkins, D.W. (1963) Use of radionuclides in ecological studies of insects. In: Schultz, V. & Klement, A.W. (eds), *RadioEcology.* Rheinhold, New York, pp. 431–43.

Jewell, G.R. (1956) Marking of tsetse flies for their detection at night. *Nature* **178**, 750.

Jewell, G.R. (1958) Detection of tsetse fly at night. *Nature* **181**, 1354.

Johansson, T.S.K. (1959) Tracking honey bees in cotton fields with fluorescent pigments. *J. Econ. Entomol.* **52**, 572–7.

Jolly, G.M. (1963) Estimates of population parameters from multiple recapture data with both death and dilution – a deterministic model. *Biometrika* **50**, 113–28.

Jolly, G.M. (1965) Explicit estimates from capture–recapture data with both death and immigration – stochastic model. *Biometrika* **52**, 225–47.

Joly, P. & Miaud, C. (1989) Tattooing as an individual marking technique in urodeles. *Alytes* **8**(1), 11–16.

Kelker, G.H. (1940) Estimating deer populations by a differential hunting loss in the sexes. *Proc. Utah Acad. Sci. Arts Lett.* **17**, 65–9.

Kendall, W.L., Pollock, K.H., & Brownie, C. (1995) A likelihood-based approach to capture–recapture estimation of demographic parameters under the robust design. *Biometrics* **51**(1), 293–308.

Kideys, A.E. & Nash, R.D.M. (1993) A note on the mortality associated with fluorescent marking in a gastropod. *Turkish J. Zool.* **17**(2), 175–8.

Killick-Kendrick, R. (1993) 'Sodastream' CO_2 dispenser for killing insects. *Antenna* **17**, 115–16.

King, R.B., Queral-Regil, A., & Stanford, K.M. (2006) Population size and recovery criteria of the threatened Lake Erie watersnake: integrating multiple methods of population estimation. *Herpetological Monographs*, December 2006, Vol. **20**, No. 1, pp. 83–104.

Klimisch, J.J. (1986) Ear tattooing technique for individual identification of white small rodents. *Z. fur Versuchstierk.* **28**(6), 277–81.

Klock, J.W., Pimentel, D., & Stenburg, R.L. (1953) A mechanical fly-tagging device. *Science* **118**, 48–9.

Kloft, W., Holldobler, B., & Haishch, A. (1965) Traceruntersuchungen zur Abgrenzurg von Nestarealen Holzzerstorender Rossameisen (*Camponolus herculeanus* L. und *C. ligniperda* Satr.). *Entomologia Exp. Appl.* **8**, 20–6.

Knight, R.H. & Southon, H.A.W. (1963) A simple method for marking haematophagous insects during the act of feeding. *Bull. Entomol. Res.* **54**, 379–82.

Knoche, H.W. (1991) *Radioisotopic Methods for Biological and Medical Research*. Oxford University Press, Oxford.

Komazaki, S. & Sakagami, Y. (1989) Capture–recapture study on the adult population of the white spotted longicorn beetle, *Anoplophora malasiaca* (Thomson) (Coleoptera: Cerambycidae), in a citrus orchard. *Appl. Entomol. Zool.* **24**(1), 78–84.

Krebs, C.J., Singleton, G.R., & Kenney, A.J. (1994) Six reasons why feral house mouse populations might have low recapture rates. *Wildlife Res.* **21**(5), 559–67.

Kruk-de-bruin, M., Rost, L.C.M., & Draisama, F.G.A.M. (1977) Estimates of the number of foraging ants with the Lincoln-index method in relation to the colony size of *Formica polyctena*. *J. Anim. Ecol.* **46**, 457–70.

Kuenzler, E.J. (1958) Niche relations of three species of Lycosid spiders. *Ecology* **39**, 494–500.

Labrecque, G.C., Bailey, D.L., Meifert, D.W., & Weidhaas, D.E. (1975) Density estimates and daily mortality rate evaluations of stable fly (*Stomoxys calcitrans* (Diptera: Muscidae) populations in field cages. *Can. Entomol.* **107**, 597–600.

Laird, M.L. & Stott, B. (1978) Marking and tagging. In: Bagenal, T. (ed.) *Methods for Assessment of Fish Production in Fresh Waters*. Blackwell Scientific Publications, Oxford, pp. 84–100.

Lamb, R.I. & Wellington, W.G. (1974) Techniques for studying the behavior and ecology of the European earwig, *Forficula auricularia* (Dermaptera: Forficulidae). *Can. Entomol.* **106**, 881–8.

Law, P.M.W. (1994) Simulation study of salmon carcass survey capture–recapture methods. *Calif. Fish Game* **80**, 14–28.

Le Gall, J.Y., Bosc, P., Chateau, D., & Taquet, M. (1986) An estimation of the number of adult female green turtles *Chelonia mydas* per nesting season at Tromelin and Europa (Indian Ocean) (1973–1985). *Oceanogr. Trop.* **21**(1), 3–22.

Lebreton, J.D. & Clobert, J. (1986) User's manual for the program Surge, version 2.0. Montpellier, CEPE/CNRS.

Lebreton, J.D., Burnham, K.P., Clobert, J., & Anderson, D.R. (1992) Modeling survival and testing biological hypotheses using marked animals: A unified approach with case studies. *Ecol. Monogr.* **62**(1), 67–118.

Leeuwen, E.R.V. (1940) The activity of adult codling moths as indicated by captures of marked moths. *J. Econ. Entomol.* **33**, 162–6.

Leslie, P.H. (1952) The estimation of population parameters from data obtained by means of the capture–recapture method II. The estimation of total numbers. *Biometrika* **39**, 363–88.

Leslie, P.H. & Chitty, D. (1951) The estimation of population parameters from data obtained by means of the capture–recapture method. 1. The maximum likelihood equations for estimating the death-rate. *Biometrika* **38**, 269–92.

Levin, L.A., Huggett, D., Myers, P., Bridges, T., & Weaver, J. (1993) Rare earth tagging methods for the study of larval dispersal by marine invertebrates. *Limnol. Oceanogr.* **38**(2), 346–60.

Levin, M. (1961) Distribution of foragers from honey bee colonies placed in the middle of a large field of alfalfa. *J. Econ. Entomol.* **54**, 431–4.

Lincoln, F.C. (1930) Calculating waterfowl abundance on the basis of banding returns. *USDA Circ.* **118**, 1–4.

Linn, I.J. (1977) Radioactive techniques for small mammal marking. In: Stonehouse, B. (ed.) *Animal Marking*. The Macmillan Press Ltd., London.

Loab, H.A. (1966) Marking brown trout fry with the dye Sudan Black. *N. Y. Fish Game J.* **13**, 109–18.

Loery, G. & Nichols, J.D. (1985) Dynamics of a black-capped chickadee (*Parus atricapillus*) population, 1958–1983. *Ecology* **66**(4), 1195–203.

Loery, G., Pollock, K.H., Nichols, J.D., & Hines, J.E. (1987) Age-specificity of black-capped chickadee survival rates: Analysis of capture–recapture data. *Ecology* **68**(4), 1038–44.

Lomnicki, A. (1969) Individual differences among adult members of a snail population. *Nature* **223**, 1073–4.

Lukacs, P.M. & Burnham, K.P. (2005) Review of capture–recapture methods applicable to noninvasive genetic sampling. *Mol. Ecol.* **14**, 3909–19.

Macleod, J. (1958) The estimation of numbers of mobile insects from low-incidence recapture data. *Trans. R. Entomol. Soc. Lond.* **110**, 363–92.

MacLeod, I. & Donnelly, J. (1957) Individual and group marking methods for fly population studies. *Bull. Entomol. Res.* **48**, 585–92.

Malter, S.F. (1996) Interpatch movement of the red milkweed beetle, *Tetraopes tetraphthalmus*: individual responses to patch size and isolation. *Oecologica* **105**, 447–53.

Manly, B.F.J. (1969) Some properties of a method of estimating the size of mobile animal populations. *Biometrika,* **56**, 407–10.

Manly, B.F.J. (1970) A simulation study of animal population estimation using the capture–recapture method. *J. Appl. Ecol.* **7**, 13–39.

Manly, B.F.J. (1971a) Estimates of a marking effect with capture–recapture sampling. *J. Appl. Ecol.* **8**, 181–9.

Manly, B.F.J. (1971b) A simulation study of Jolly's method for analysing capture–recapture data. *Biometrics* **27**(2), 415–24.

Manly, B.F., McDonald, T.L., Amstrup, S.C., & Regehr, E.V. (2003) Improving size estimates of open animal populations by incorporating information on age. *BioScience* **53**(7), 666–9.

Manly, B.F. J. & Parr, M.J. (1968) A new method of estimating population size survivorship, and birth rate from capture–recapture data. *Trans. Soc. Br. Entomol.* **189**, 81–9.

Margolis, L. (1963) Parasites as indicators of the geographical origin of sockeye salmon *Oncorhynchus nerka* (Walbaum) occurring in the North Pacific Ocean and adjacent seas. *Bull. Int. N. Pacif. Fish. Commun.* **11**, 101–56.

Marten, G.G. (1970) A regression method for mark-recapture estimates with unequal catchability. *Ecology* **51**, 291–5.

Mascanzoni, D. & Wallin, H. (1986) The harmonic radar: a new method of tracing insects in the field. *Ecol. Entomol.* **11**, 387–390.

Mazzotti, S. (1995) Population structure of *Emys orbicularis* in the Bardello (Po delta, Northern Italy). *Amphib. Reptil.* **16**(1), 77–85.

McDonald, W.A. (1960) Nocturnal detection of tsetse flies in Nigeria with ultra-violet light. *Nature* **185**, 867–8.

Medland, T.E. & Beamish, F.W.H. (1991) Lamprey statolith banding patterns in response to temperature, photoperiod, and ontogeny. *Trans. Am. Fish. Soc.* **120**(2), 255–60.

Medley, J.G. & Ahrens, E.H. (1968) Fluorescent dyes for marking and recovering fowl ticks in poultry houses treated with insecticides. *J. Econ. Entomol.* **61**, 81–4.

Mellanby, K. (1939) The physiology and activity of the bed-bug (*Cimex lectularius* L.) in a natural infestation. *Parasitology* **31**, 200–11.

Menkens, G.E.J.R. & Anderson, S.H. (1988) Estimation of small-mammal population size. *Ecology* **69**(6), 1952–9.

Michener, C.D., Cross, E.A., Daly, H.V., Rettenmeyer, C.W., & Wille, A. (1955) Additional techniques for studying the behaviour of wild bees. *Insectes Sociaux* **2**, 237–46.

Minta, S. & Mangel, M. (1989) A simple population estimate based on simulation for capture–recapture and capture–resight data. *Ecology* **70**(6), 1738–51.

Mitchell, B. (1963) Ecology of two carabid beetles, *Bembidion lampros* (Herbst.) and *Trechus quadristriatus* (Schrank.). *J. Anim. Ecol.* **32**, 377–92.

Moav, R., Wohlfarth, G., & Lahman, M. (1960). An electric instrument for brandmarking fish. *Bamidgeh*, **12**, 92–95.

Monro, J. (1968) Marking insects with dietary manganese for detection by neutron activation. *Ecology* **49**, 774–6.

Montgomery, S.S., Brett, P.A., Blount, C., Stewart, J., Gordon, G.N.G., & Kennelly, S.J. (1995) Loss of tags, double tagging and release methods for eastern king prawns, *Penaeus plebejus* (Hess). Laboratory and field experiments. *J. Exp. Mar. Biol. Ecol.* **188**, 115–31.

Morgan, M.J. & Walsh S.J. (1993) Evaluation of the retention of external tags by juvenile American plaice (*Hippoglossoides platessoides*) using an aquarium experiment. *Fish. Res.* **16**(1), 1–7.

Morrison, J.K., Coyle, C.L., & Bertoni, S.E. (1990) Histological effect of tagging chum and coho salmon fry with coded wire tags. *Progr. Fish Cult.* **52**(2), 117–19.

Moth, I.I. & Barker, J.S.F. (1975) Micronized fluorescent dusts for marking drosophila adults. *J. Nat. Hist.* **91**, 393–6.

Muir, R.C. (1958) On the application of the capture–recapture method to an orchard population of *Blepharidopterus angulatus* (Fall.) (Hemiptera-Heteroptera, Miridae). *Rep. East Malling Res. Sta.* **1957**, 140–7.

Murdoch, W.W. (1963) A method for marking Carabidae (Col.). *Entomol. Monthly Mag.* **99**, 22–4.

Nakai, K., Musashi, T., Inoguchi, N., Saido, T., Kishida, T., & Matsuda, H. (1995) Estimation of survival rate in the early stage of released artificial young abalone by the capture–recapture data. *Nippon Suisan Gakkaishi* **59**(11), 1845–50.

Nakamura, K., Ito, Y., Nakamura, M., Matsumoto, T., & Hayakawa, K. (1971) Estimation of population productivity of *Parapleurus alliaceus* Germar (Orthoptera: Acridiidae) on a *Miscanthus sinensis* Anders. grassland, I Estimation of population parameters. *Oecologia (Berl.)* **7**, 1–15.

Nagiec, M., Czerkies, P., Goryczko, K., Witkowski, A., & Murawska, E. (1995) Mass-marking of grayling, *Thymallus thymallus* (L.), larvae by fluorochrome tagging of otoliths. *Fish. Manag. Ecol.* **2**(3), 185–95.

Narayanaswamy, K.C., Kumar, P., Manjunath, D., & Datta, R. K. (1994) Determination of flight range of the uzi fly, *Exorista bombycis* (Louis) (Diptera: Tachinidae) through marking technique by adding dye to the adult diet. *Ind. J. Sericult.* **33**(1), 40–3.

Nass, B.L. & Bocking, R.C. (1992) Abundance, age, size, sex and coded wire tag recoveries for chinook salmon escapements of Kitsumkalum River, 1989–1990. *Can. Manuscr. Rep. Fish. Aquat. Sci.* **2147**, 1–61.

Nelemans, M.N.E., Den Boer, P.J., & Spee, A. (1989) Recruitment and summer diapause in the dynamics of a population of *Nebria brevicollis* (Coleoptera: Carabidae). *Oikos* **56**, 157–69.

Newman, R.M. & Martin, F.B. (1983) Estimation of fish production rates and associated variances. *Can. J. Fish. Aquat. Sci.* **40**, 1729–36.

Nichols, J., Hines, J.E., Pollock, K.H., Hinz, R.L., & Link, W.A. (1994) Estimating breeding proportions and testing hypotheses about costs of reproduction with capture–recapture data. *Ecology* **75**(7), 2052–65.

Nichols, J.D., Noon, B.R., Stokes, S.L., & Hines, J.E. (1981) Remarks on the use of mark–recapture methodology in estimating avian population size. *Studies Avian Biol.*, **6**, 121–36.

Nichols, J.D. & Pollock, K.H. (1990) Estimation of recruitment from immigration versus in situ production using Pollock's robust design. *Ecology* **71**(1), 21–6.

Nichols, J.D., Pollock, K.H., & Hines, J.E. (1984) The use of a robust capture–recapture design in small mammal population studies: A field example with *Microtus pennsylvanicus*. *Acta Theriol.* **29**(26-36), 357–65.

Niebylski, M.L. & Meek, C.L. (1989) A self-marker device for emergent adult mosquitos. *J. Am. Mosquito Contr. Assoc.* **5**, 86–90.

Nielsen, E.T. (1961) On the habits of the migratory butterfly, *Ascia monuste* L. *Biol. Meddr. Dan. Vid. Selsk.* **23**, 1–81.

Niva, T. (1995) Retention of visible implant tags by juvenile brown trout. *J. Fish Biol.* **46**(6), 997–1002.

Nordeide, J.T., Holm, J.C., Ottera, H., Blom, G., & Borge, A. (1992) The use of oxytetracycline as a marker for juvenile cod (*Gadus morhua* L.). *J. Fish Biol.* **41**(1), 21–30.

Norris, K.R. (1957) A method of marking Calliphoridae (Diptera) during emergence from the puparium. *Nature* **180**, 1002.

Nunan, C.P., Macinnes, C.D., Bachmann, P., Johnston, D.H., & Watt, I.D. (1994) Background prevalence of tetracycline-like fluorescence in teeth of free ranging red foxes (*Vulpes vulpes*), striped skunks (*Mephitis mephitis*) and raccoons (*Procyon lotor*) in Ontario, Canada. *J. Wildlife Dis.* **30**(1), 112–14.

O'Grady, I.I. & Hoy, I.B. (1972) Rhodamine-B and other stains as markers for the mosquito fish, *Gambusia affinis*. *J. Med. Entomol.* **9**, 571–4.

Ohgushi, T. & Sawada, H. (1998) What changed the demography of an introduced population of an herbivorous lady beetle? *J. Anim. Ecol.* **67**, 679–88.

Ohwada, K. (1991) Permanent marking of colored mice using dry ice. *Jikken dobutsu. Experimental Animals*, **40**, 395–9.

Orians, G.H. & Leslie, P.H. (1958) A capture–recapture analysis of a shearwater population. *J. Anim. Ecol.* **27**, 71–86.

Ostfeld, R.S., Miller, M.C., & Schnurr, J. (1993) Ear tagging increases tick (*Ixodes dammini*) infestation rates of white footed mice (*Peromyscus leucopus*). *J. Mammal.* **74**(3), 651–5.

Otis, D.K., Burnham, K.P., White, G.C., & Anderson, D.R. (1978) Statistical inference from capture data on closed animal populations. *Wildl. Monogr.* **62**, 1–135.

Oven, J.H. & Blankenship, H.L. (1993) Benign recovery of coded wire tags from rainbow trout. *N. Am. J. Fish. Manag.* **13**(4), 852–5.

Pal, R. (1947) Marking mosquitoes with fluorescent compounds and watching them by ultraviolet light. *Nature* **160**, 298–9.

Paloheimo, J.E. (1963) Estimation of catchabilities and population sizes of lobsters. *J. Fish Res. Board Can.* **20**, 59–88.

Paradis, E., Guedon, G., & Pradel, R. (1993) Estimation of sex- and age-related survival rates in a microtine population. *J. Wildlife Manag.* **57**(1), 158–63.

Parker, W.S. (1976) Population estimates, age structure, and denning habits of whipsnakes, *Masticophis taeniatus taeniatus*. *Copeia* **1972**, 892–5.

Parmenter, C.J. (1993) A preliminary evaluation of the performance of passive integrated transponders and metal tags in a population study of the flatback sea turtle (*Natator depressus*). *Wildlife Res.* **20**(3), 375–81.

Parr, M.J. (1965) A population study of a colony of imaginal *Ischnura elegans* (van der Linden) (Odonata: Coenagriidae) at Dale, Pembrokeshire. *Field. Stud.* **2**(2), 237–82.

Parr, M.J., Gaskell, T.I., & George, B.J. (1968) Capture–recapture methods of estimating animal numbers. *J. Biol. Educ.* **2**, 95–117.

Pedersen, T. & Carlsen, B. (1991) Marking cod (*Gadus morhua* L.) juveniles with oxytetracycline incorporated into the feed. *Fish. Res.* **12**(1), 57–64.

Peer, D.F. (1957) Further studies on the mating range of the honey bee, *Apis mellifera* L. *Can. Entomol.* **89**, 108–10.

Peffly, R.L. & Labrecque, G.C. (1956) Marking and trapping studies on dispersal and abundance of Egyptian house flies. *J. Econ. Entomol.* **49**, 214–17.

Peterson, D.L. & Carline, R.F. (1996) Effects of tetracycline marking, transport density, and transport time on short-term survival of walleye fry. *Progr. Fish Cult.* **58**(1), 29–31.

Peterson, M.S. & Key, J.P. (1992) Evaluation of hand tagging juvenile walleyes with binary coded wire microtags. *N. Am. J. Fish. Manag.* **12**(4), 814–18.

Phillips, B.F. & Campbell, N.A. (1970) Comparison of methods of estimating population size using data on the whelk *Dicathais aegrota* (Reeve). *J. Anim. Ecol.* **399**, 753–9.

Pirker, J.G. & Schiel, D.R. (1993) Tetracycline as a fluorescent shell-marker in the abalone *Haliotis iris*. *Mar. Biol.* **116**(1), 81–6.

Polivka, J. (1949) The use of fluorescent pigments in a study of the flight of the Japanese beetle. *J. Econ. Entomol.* **42**, 818–21.

Poljak, A., Huber, D., & Greguric, J. (1989) Estimation of street pigeon (*Columba livia*) population size in Zagreb (Yugoslavia). *Larus* **42**, 141–50.

Pollock, K.H. (1974) *The assumption of equal catchability of animals in tag–recapture experiments.* Ithaca, New York, Cornell University, 82.

Pollock, K.H. (1975) A k-sample tag–recapture model allowing for unequal survival and catchability. *Biometrika* **62**, 577–83.

Pollock, K.H. (1981) Capture–recapture models allowing for age-dependent survival and capture rates. *Biometrics* **37**, 521–9.

Pollock, K.H. (1982) A capture–recapture design robust to unequal probability of capture. *J. Wildlife Manag.* **46**, 752–7.

Pollock, K.H. & Mann, R.H.K. (1983) Use of an age-dependent mark–recapture model in fisheries research. *Can. J. Fish. Aquat. Sci.* **40**, 1449–55.

Pollock, K.H., Nichols, J.D., Brownie, C., & Hines, J.E. (1990) Statistical inference for capture–recapture experiments. *Wildlife Monogr.* **107**, 1–98.

Pooley, A.C. (1962) The Nile crocodile (*Crocodylus niloticus*). *Lammergeyer* **2**, 1–55.

Porter, S.D. & Jorgensen, C.D. (1980) Recapture studies of the harvester ant *Pogonomytmex owyheei* Cole, using a fluorescent marking technique. *Ecol. Entomol.* **5**, 263–9.

Prestrud, P. (1992) Denning and home range characteristics of breeding arctic foxes in Svalbard. *Can. J. Zool.* **70**(7), 1276–83.

Prince, J.D. (1991) A new technique for tagging abalone. *Aust. J. Mar. Freshwater Res.* **42**(1), 101–6.

Putman, R.J. (1995) Ethical considerations and animal welfare in ecological field studies. *Biodivers. Conserv.* **4**(8), 903–15.

Quarterman, K.D., Mathis, W., & Kilpatrick, J.W.X. (1954) Urban fly dispersal in the area of Savannah, Georgia. *J. Econ. Entomol.* **47**, 405–12.

Querci, O. (1936) Aestivation of Lepidoptera. *Entomol. Rec.* **48**, 122.

Radinovsky, S. & Krantz, G.W. (1962) The use of Fluon to prevent the escape of stored-product insects from glass containers. *J. Econ. Entomol.* **55**, 815–16.

Rafinski, J. (1977) Autotransplantation as a method for permanent marking of urodele amphibians. *J. Herpetol.* **11**, 241–7.

Rahalkar, G.W., Mistry, K.B., Harnalkar, M.R., Bharathan, K.G., & Gopalayengar, A.R. (1971) Labelling adults of red palm weevil (*Rhynchophorus ferruginous*) with cerium for detection by neutron activation. *Ecology* **52**, 187–8.

Reeves, W.C., Brookman, B., & Hammon, W.M. (1948) Studies on the flight range of certain *Culex* mosquitoes, using a fluorescent-dye marker, with notes on Culiseta and Anopheles. *Mosquito News* **8**, 61–9.

Reiter, P., Amador, M.A., Anderson, R.A., & Clark, G.G. (1995) Short report: Dispersal of *Aedes aegypti* in an urban area after blood feeding as demonstrated by rubidium-marked eggs. *Am. J. Trop. Med. Hyg.* **52**(2), 177–9.

Ribbands, C.R. (1950) Changes in behaviour of honeybees, following their recovery from anaesthesia. *J. Exp. Biol.* **27**, 302–10.

Rice, T.M. & Taylor, D.H. (1993) A new method for making waistbands to mark anurans. *Herpetol. Rev.* **24**(4), 141–2.

Richards, O.W. & Waloff, N. (1954) Studies on the biology and population dynamics of British grasshoppers. *Anti-Locust Bull.* **17**, 1–182.

Richardson, R.H., Wallace, R.J., Gage, S.J., Bouchey, G.D., & Dennell, M. (1969) XII. Neutron activation techniques for labelling *Drosophila* in natural populations. In: Wheeler, M.R. (ed.) *Studies in Genetics V*. University of Texas Publication 6918, Houston, pp. 171–86.

Riebartsch, K. (1963) Inaktive markierung von insekten mit dysprosium. *Nachrbl. Dt. Pflschut.-dienst. Stuttg.* **15**, 154–7.

Rien, T.A., Beamesderter, R.C.P., & Foster, C.A. (1994) Retention, recognition, and effects on survival of several tags and marks for white sturgeon. *Calif. Fish Game* **80**(4), 161–70.

Riley, J.R. (1996) Tracking bees with harmonic radar. *Nature* **379**, 29–30.

Rivest, L.P., & Baillargeon, S. (2007) Applications and extensions of Chao's moment estimator for the size of a closed population. *Biometrics* **63**, 999–1006.

Rivest, L.P. & Gaétan, D. (2004) Loglinear models for the robust design in mark–recapture experiments. *Biometrics* **60**(1), 100–7.

Robson, D.S. (1969) Mark–recapture methods of population estimation. In: Johnson, N.L. & Smith, H. (eds), *New Developments in Survey Sampling*. John Wiley, New York.

Robson, D.S. & Regier, H.A. (1964) Sample size in Petersen mark–recapture experiments. *Trans. Am. Fish. Soc.* **93**, 215–26.

Roff, D.A. (1973) On the accuracy of some mark–recapture estimators. *Oecologia* **12**, 15–34.

Rojas Beltran, R., Gillet, C., & Champigneulle, A. (1995) Immersion mass-marking of otoliths and bone tissue of embryos, yolk-sac fry and fingerlings of Arctic charr *Salvelinus alpinus* (L.). *Nordic J. Freshwater Res.* **71**, 411–18.

Roland, J., McKinnon, G., Backhouse, C., & Taylor, P.D. (1996) Even smaller radar tags on insects. *Nature* **381**, 120.

Rosewell, J. & Shorrocks, B. (1987) The implications of survival rates in natural populations of *Drosophila*: Capture–recapture experiments on domestic species. *Biol. J. Linn. Soc.* **32**(4), 373–84.

Rubink, W.L. (1988) A photographic technique for producing high-quality insect tags for mark/release/recapture studies. *Entomol. News* **99**(3), 167–71.

Ruhle, C. & Winecki Kuhn, C. (1992) Tetracycline marking of coregonids at the time of egg fertilization. *Aquat. Sci.* **54**(2), 165–75.

Schjotz-Christensen, B. (1961) For plantnga biologien hos amara infirma dft. Og harpalus neglectus serv. *Flora og Fauna* **67**, 8–12.

Schmalbach, A.E., Quackenbush, L.S., & Melinek, R. (1994) A method for tagging the Malaysian prawn *Macrobrachium rosenbergii*. *Aquaculture* **122**(2-3), 147–59.

Schnabel, Z.E. (1938) The estimation of the total fish population of a lake. *Am. Math. Monogr.* **45**, 348–52.

Schwarz, C.J., Bailey, R.E., Irvine, J.R., & Dalziel, F.C. (1993) Estimating salmon spawning escapement using capture–recapture methods. *Can. J. Fish. Aquat. Sci.* **50**(6), 1181–97.

Seber, G.A.F. (1965) A note on the multiple-recapture census. *Biometrika* **52**, 249.

Seber, G.A.F. (1970) Estimating time -specific survival and reporting rates for adult birds from band returns. *Biometrika* **57**, 313–18.

Seber, G.A.F. (1982) *The Estimation of Animal Abundance and Related Parameters*. Griffin, London.

Seber, G.A.F. (1986) A review of estimating animal abundance. *Biometrics* **42**, 267–92.

Seber, G.A.F., & Felton, R. (1981) Tag loss and the Petersen mark–recapture experiment. *Biometrika*, **68**, 211–19.

Secor, D.H., White, M.G., & Dean, J.M. (1991) Immersion marking of larval and juvenile hatchery-produced striped bass with oxytetracycline. *Trans. Am. Fish. Soc.* **120**(2), 261–6.

Service, M.W. (1995) *Mosquito Ecology Field Sampling Methods*, 2nd edition. Chapman & Hall, 988 pp.

Sheppard, P.M. (1951) Fluctuations in the selective value of certain phenotypes in the polymorphic land snail *Cepaea nemoralis* (L.). *Heredity* **5**, 125–54.

Sheppard, P.M., Macdonald, W.W., Tonn, R.I., & Grab, B. (1969) The dynamics of an adult population of *Aedes aegypti* in relation to dengue haemorrhagic fever in Bangkok. *J. Anim. Ecol.* **38**, 661–702.

Sheppard, D.C., Wilson, G.H., & Hawkins, J.A. (1973) A device for self-marking of Tabanidae. *J. Environ. Entomol.* **2**, 960–1.

Shura-Bura, B.L. & Grageau, U.L. (1956) Fluorescent analysis studies of insect migration. [In Russian.] *Entomol. Obozr.* **35**, 760–3.

Simpson, J. (1954) Effects of some anaesthetics on honeybees: nitrous oxide, carbon dioxide, ammonium nitrate, smoker fumes. *Bee World* **35**, 149–55.

Sinsko, M.J. & Craig, G.B. (1979) Dynamics of an isolated population of *Aedes triseriatus* (Diptera: Culicidae). I Population size. *J. Med. Entomol.* **15**, 89–98.

Skalski, J.R. & Robson, D.S. (1982) A mark and removal field procedure for estimating population abundance. *J. Wildlife Manag.* **46**, 752.

Skalski, J.R., Robson, D.S., & Simmons, M.A. (1983) Comparative census procedures using single mark–recapture methods. *Ecology* **65**, 1006–015.

Skuhravy, V. (1957) Studium pohybu nekterych Stfevlikovitych znackovanim jedincu. *Acia Soc. Ent. Bohem.* **53**, 171–9.

Slack, K.V. (1955) An injection method for marking crayfish. *Prog. Fish. Cult.* **17**, 36–8.

Slooff, A. & Herath, P.R.J. (1980) Ovarian development and biting frequency in *Anopheles calicifacies* Giles in Sri Lanka. *Trop. Geog. Med.* **32**, 306–11.

Smith, C. (1994) Fish tags for observing free ranging rattlesnakes. *Herpetol. Rev.* **25**(2), 58.

Smith, M.V. (1958) The use of fluorescent markers as an aid in studying the forage behaviour of honeybees. *Proc. Xth Int. Congr. Entomol.* **4**, 1063.

Smith, M.V. & Townsend, G.F. (1951) A technique for mass-marking honeybees. *Can. Entomol.* **83**, 346–8.

Snyder, R.J., McKeown, B.A., Colbow, K., & Brown, R. (1992) Use of dissolved strontium in scale marking of juvenile salmonids: Effects of concentration and exposure time. *Can. J. Fish. Aquat. Sci.* **49**(4), 780–2.

Sonleitner, F.I. & Bateman, M.A. (1963) Mark–recapture analysis of a population of Queensland fruit-fly, *Dacus tryoni* (Frogg.) in an orchard. *J. Anim. Ecol.* **32**, 259–69.

South, A. (1965) Biology and ecology of *Agriolimax reticulatus* (Müll.) and other slugs: spatial distribution. *J. Anim. Ecol.* **34**, 403–17.

Southwood, T.R.E. (1965) Migration and population change in *Oscinella frit* L.(Diptera) on the oat crop. *Proc. XII Int. Congr. Entomol.* **4**, 420–1.

Southwood, T.R.E. & Scudder, G.G.E. (1956) The bionomics and immature stages of the thistle lace bugs (*Tings ampliata* H.-S. and *T. Cardui* L.; Hem., Tingidae). *Trans. Soc. Br. Entomol.* **12**, 93–112.

Speare, P. (1992) A technique for tetracycline injecting and tagging billfish. *Bull. Mar. Sci.* **51**(2), 197–203.

Spellerberg, I.F. & Prestt, I. (1977) Marking snakes. In: Stonehouse, B. (ed.) *Animal Marking*. The MacMillan Press Ltd, London.

Spendelow, J.A. & Nichols, J.D. (1989) Annual survival rates of breeding adult roseate terns. *Auk* **106**(3), 367–74.

Stadelbacher, E.A. (1991) Bollworm and tobacco budworm (Lepidoptera: Noctuidae): Labeling of adults produced in wild geranium, *Geranium dissectum*, treated with rubidium or strontium chloride. *J. Econ. Entomol.* **84**(2), 496–501.

Steiner, L.F. (1965) A rapid method for identifying dye-marked fruit flies. *J. Econ. Entomol.* **589**, 374–5.

Stobow, T., Fowler, G.M., & Sinclaire, A.F. (1992) Short term tagging mortality of laboratory held juvenile Atlantic herring (*Clupea harengus harengus*). *J. N. W. Atlantic Fish. Sci.* **12**, 27–33.

Stonehouse, B. (ed.) (1977) *Animal Marking: Recognition Marking of Animals in Research*. The MacMillan Press Ltd, London

Stradling, D.J. (1970) The estimation of worker ant populations by the mark–release–recapture method: an improved marking technique. *J. Anim. Ecol.* **39**, 575–91.

Sudia, T.W. & Linck, A.J. (1963) Methods for introducing radionuclides into plants. In: Schultz, V. & Klement, A.W.E. (eds), RadioEcology. Rheinhold, New York, pp. 417–23.

Sugimoto, T., Sakuratani, Y., Setokuchi, O., Kamikado, T., Kiritani, K., & Okada, T. (1994) Using the mark-and-release method in the estimation of adult population of sweet potato weevil, *Cylas formicarius* (Fabricius) in a sweet potato field. *Appl. Entomol. Zool.* **29**(1), 11–19.

Sullivan, J.J., Bishop, H.S., Schneider, K.R., & Rodrick, G.E. (1990) Marking snails with numbered, colored discs. A technique for identifying individual specimens. *Trop. Med. Parasitol.* **41**(3), 289–90.

Tanaka, S. Takao, K., & Kato, N. (1987) Tagging techniques for bottlenose dolphins *Tursiops truncatus. Nippon Suisan Gakkaishi,* **53**(8), 1317–1325.

Tanton, M.T. (1965) Problems of live-trapping and population estimation for the wood mouse. *J. Anim. Ecol.* **38**, 511–29.

Teboul, D. (1993) Internal monofilament tags used to identify juvenile *Penaeus monodon* tag retention and effects on growth and survival. *Aquaculture* **113**, 167–70.

Templeman, W. & Fleming, A.M. (1963) Distribution of *Lernaeocera branchialis* L. on the cod as an indicator of cod movements in the Newfoundland area. *Spec. Publ. Int. Commun. NW. Atlant. Fish.* **4**, 381–22.

Testa, J.W. & Rotherly, P. (1992) Effectiveness of various cattle ear tags as maker for Weddell seals. *Mar. Mammal Sci.* **8**(4), 344–53.

Thorne, B.L., Russek-Cohen, E., Forschler, B.T., Breisch, N.L., & Traniello, J.F. (1996) Evaluation of mark–release–recapture methods for estimating forager population size of subterranean termite (Isoptera: Rhinotermitidae) colonies. *Environ. Entomol.* **25**(5), 938–51.

Treble, R.J., Day, R.W., & Quinn, T.J. (1993) Detection and effects on mortality estimates of changes in tag loss. *Can. J. Fish. Aquat. Sci.* **50**(7), 1435–41.

Trpis, M. & Hausermann, W. (1986) Dispersal and other population parameters of *Aedes aegypti* in an African village and their possible significance in epidemiology of vector-borne diseases. *Am. J. Trop. Med. Hyg.* **35**, 1263–79.

Van Brackle, M.D., Linhart, S.B., Creekmore, T.E., Nettles, V.F., & Marchinton, R.L. (1994) Oral biomarking of white-tailed deer with tetracycline. *Wildlife Soc. Bull.* **22**(3), 483–8.

Wada, Y., Suenaga, O., & Miyagi, I. (1975) Dispersal experiment of *Aedes togoi. Trop. Med.* **16**, 137–46.

Waloff, Z. (1963) Field studies on solitary and transient desert locusts in the red sea area. *Anti-Locust Bull.* **40**, 1–93.

Wastle, R.J., Babaluk, J.A., & Decterow, G.M. (1994) A bibliography of marking fishes with tetracyclines including references to effects on fishes. *Can. Tech. Rep. Fish. Aquat. Sci.* **1951**, 1–26.

Wave, H.E., Henneberry, T.J., & Mason, H.C. (1963) Fluorescent biological stains as markers for *Drosophila. J. Econ. Entomol.* **56**, 890–91.

Wedermeyer, G. (1970) Stress of anaesthesia with MS-222 and benzocaine in rainbow trout (*Salmo gairdneri*). *J. Fish. Res. Bd Can.* **27**, 909–14.

Welch, H.E. (1960) Two applications of a method of determining the error of population estimates of mosquito larvae by the mark and recapture technique. *Ecology* **41**, 228–9.

Wheye, D. & Ehrlich, P.R. (1985) The use of fluorescent pigments to study insect behaviour: investigating mating patterns in a butterfly population. *Ecol. Entomol.* **10**, 231–4.

White, E.G. (1970) A self-checking coding technique for mark–recapture studies. *Bull. Entomol. Res.* **60**, 303–7.

White, E.G. (1975) Identifying population units that comply with capture–recapture assumptions in an open community of alpine grasshoppers. *Res. Popul. Ecol.* **16**, 153–87.

White, G.C., Anderson, D.R., Burnham, K.P., & Otis, D.L. (1982) *Capture–recapture and removal methods for sampling closed populations.* Los Alamos National Laboratory, Los Alamos, New Mexico.

Wilkin, P.J. & Scofield, A.M. (1991) The structure of a natural population of the medicinal leech, *Hirudo medicinalis,* at Dungeness, Kent (England, UK). *Freshwater Biol.* **25**(3), 539–46.

Williams, D.F., Patterson, R.S., & LaBrecque, G.C. (1979) Marking large numbers of stable flies *Stomoxys calcitrans* (L.) for a sterile release program. *Mosquito News* **39**, 146–8.

Williams, E.C., Jr (2013) A study of the box turtle, *Terrapene carolina carolina* (L), population in Allee Memorial Woods. *Proc. Indiana Acad. Sci.* **71**, 399–406.

Wineriter S.A. & Walker, T.J. (1984) Insect marking techniques: durability of materials. *Entomol. News* **95**, 117–23.

Wisniewolski, W. & Nabialek, J. (1993) Tag retention and survival of fish tagged in controlled pond experiments. *Aquat. Sci.* **55**(2), 143–52.

Wolf, W.W. & Stimmann, M.W. (1972) An automatic method of marking cabbage looper moths for release-recovery identification. *J. Econ. Entomol.* **65**, 719–22.

Wood, G.W. (1963) The capture–recapture technique as a means of estimating populations of climbing cutworms. *Can. J. Zool.* **41**, 47–50.

Woodward, A.R., Hines, T.C., Abercrombie, C.L., & Nichols, J.D. (1987) Survival of young American alligators on a Florida lake (USA). *J. Wildlife Manag.* **51**(4), 931–7.

Woolhouse, M.E.J. (1988) A mark–recapture method for ecological studies of schistosomiasis vector snail populations. *Ann. Trop. Med. Parasitol.* **82**, 485–97.

Woolhouse, M.E.J. & Chandiwana, S.K. (1990) Population biology of the freshwater snail *Bulinus globosus* in the Zimbabwe highveld. *J. Appl. Ecol.* **27**, 41–59.

Yeruham, I., Perl, S., Yakobson, B., Ben Yosef, H., & Lampert, M. (1993) Skin tumors in cattle following tattooing by liquid nitrogen. *Israel J. Vet. Med.* **48**(1), 38–40.

Zacharuk, R.Y. (1963) Vital dyes for marking living elaterid larvae. *Can. J. Zool.* **41**, 991–6.

Zaidenov, A.M. (1960) Study of housefly (Diptera, Muscidae) migrations in Chita by means of luminescent tagging. [In Russian.] *Entomol. Obozr.* **39**, 574–84. (Trans. *Entomol. Rev.* **39**, 406–14.)

Zucchini, W. & Channing, A. (1986) Bayesian estimation of animal abundance in small populations using capture–recapture information. *S. African J. Sci.* **82**(3), 137–40.

4 Absolute Population Estimates by Sampling a Unit of Habitat – Air, Plants, Plant Products and Vertebrate Hosts

This is one of four approaches to the absolute population estimate, the other three methods being: (i) distance or nearest-neighbour (p. 367); (ii) mark–recapture (p. 77); and (iii) removal trapping (p. 268). In this approach the habitat is sampled together with the animals it contains. Hence, two separate measurements have to be made: the total number of animals in the unit of the habitat sampled; and the total number of these units in the whole habitat of the population being studied. The second measurement may involve using techniques of the botanist, forester, surveyor or hydrologist and cannot be considered in detail here. The first measurement concerns the extraction of animals from the samples and sometimes the taking of samples. This chapter, and the next two chapters, will be concerned mainly with these problems in five habitats – two biotic (plants and vertebrate hosts) and three physical (air, soil and water).

It is important to remember that, if sampling is unbiased, the errors in the population estimates obtained by this method will normally lie below the true value. This is because extraction efficiency can never be more than 100% – one cannot normally find more animals than are present! As the basis of the approach is to multiply the mean count per sampling unit by the typically large number of such units in the habitat, the underestimation of total number may, in terms of individuals, be large.

4.1 Sampling from the air

In some ways air is the simplest of the five habitats as it is homogeneous in all environments and permits a universal solution which stems from the work of C. G. Johnson and L. R. Taylor. These investigators developed suction traps, standardised them, and measured their efficiency so precisely that aerial populations can be assessed with a greater level of accuracy than those in most other habitats. The only environmental variable for which allowance needs to be made is wind speed. There is no evidence that insects are in any way attracted or repelled by the traps (Taylor, 1962a), although precautions must be taken to avoid the development of a charge of static electricity on

Ecological Methods, Fourth Edition. P. A. Henderson and T. R. E. Southwood.
© 2016 John Wiley & Sons, Ltd. Published 2016 by John Wiley & Sons, Ltd.
Companion Website: www.wiley.com/go/henderson/ecologicalmethods

the traps (Maw, 1964). Only a very small proportion of the insects collected in suction traps are damaged by them.

The basic features of the suction trap are an electric fan that pulls or drives air through a fine gauze cone; this filters out the insects, which are collected in a jar or cylinder. The trap may be fitted with a segregating device which separates the catch according to a predetermined time interval, and this provides information not only about the numbers flying but also about the periodicity of flight. Few insects are active equally throughout 24 hours (Lewis & Taylor, 1965), and therefore the daily catch divided by the daily intake of air of the fan does not present as true a picture of aerial density as when each hour is considered separately.

Other traps for flying insects, sticky and water traps and suspended nets are greatly influenced by wind speed, their efficiencies being low at low wind speeds. Although Taylor (1962b) has shown how it is possible to convert sticky trap and suspended net catches into absolute densities if the wind speeds are known, these methods are used in general for the relative estimation of populations and are discussed in Chapter 7.

4.2   Sampling apparatus

4.2.1   *Exposed cone (Johnson–Taylor) suction trap*

Originally developed in the 1950s (Johnson, 1950; Taylor, 1951; Johnson & Taylor, 1955a), the air (and insects) is drawn through the fan first and subsequently via a cone mesh (c. 10.25 cm^{-1}) where the insects are impinged and then drawn down to a collecting tube at the apex of the cone (Fig. 4.1). The trap is often fitted with a disc dropping mechanism to temporally segregate the catch. Goodenough *et al.* (1983) describes a method for the segregation of the catch when it is preserved in liquid. Both 9-in and 12-in Johnson–Taylor insect suction traps are commercially produced by the Burkard Manufacturing Co., Rickmansworth, England; http://www.burkard.co.uk. In these models the sample is collected in a plastic bottle with side overflow ports, and in standard models the catch can be segregated at pre-set time intervals of 5 min to 1 h. Exposed cone samplers using Vent-Axia axial fans (Vent-Axia, Ltd, Fleming Way, Crawley, West Sussex, RH10 2NN, England; http://www.vent-axia.com/contact-us) can be easily built.

The efficiency of Johnson–Taylor suction traps changes with wind speed and they are most suitable for sheltered positions between plants. For example, using a 9-in trap, cross-winds of more than 22.4 km h^{-1} gave significant reductions in the air intake of the mouth of the fan (Johnson, 1950; Taylor, 1955). If the trap is standing amongst a crop, so that the cone is sheltered from the effects of the cross-winds which only impinge on the mouth of the fan, the delivery loss is only 0.5% per km h^{-1} cross-wind (Taylor, 1955). Johnson–Taylor traps have been used extensively for the study of aphids and other small, weak flying insects (e.g. Russian wheat aphid, *Diuraphis noxia*, Havelka *et al.*, 2014). Examples of their application to other groups are: aerial spiders (Dean & Sterling, 1985, 1990); pollen beetles in rape fields (Sedivy, 1993); Psocoptera (Locatelli & Limonta, 1993, 1994); and lesser grain borers, *Rhyzopertha dominica* (Leos Martinez *et al.*, 1986).

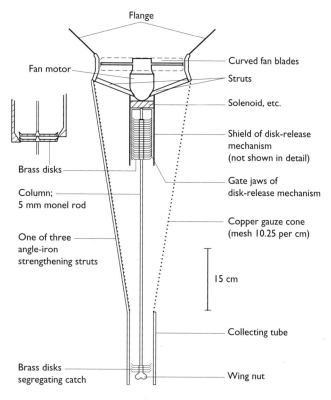

Figure 4.1 Cross-section of the exposed cone or Johnson–Taylor suction trap.

Numerous small suction traps have been designed. An inexpensive and easily constructed suction trap suitable for monitoring the flights of aphids is described by Allison & Pike (1988). For sites lacking mains electricity, Belding *et al.* (1991) describe a small suction trap powered by deep cycle battery, re-charged by a photovoltaic panel. This trap, suitable for sampling weakly flying insects such as aphids, has four 12-V, brushless, DC fans which are turned on and off by a timer under photoelectric control. Wainhouse (1980) describes a portable, battery-driven suction trap for sampling small insects.

Because suction traps can catch large numbers of insects, Hobbs & Hodges (1993) describe an optical method for automatic recording and classification of the insects catches. This is based on the principle that when the sample is illuminated against a dark background the amount of scattered light can be used to measure insect presence and size.

4.2.2 *Enclosed cone types of suction trap including the Rothamsted 12 m trap*

In these traps the fan unit is at the bottom of a metal cylinder and the metal gauze collecting cone opens to the mouth of the trap (Fig. 4.2), so that the insects are filtered out before the air passes through the fan. Because the cone and mouth of the fan are

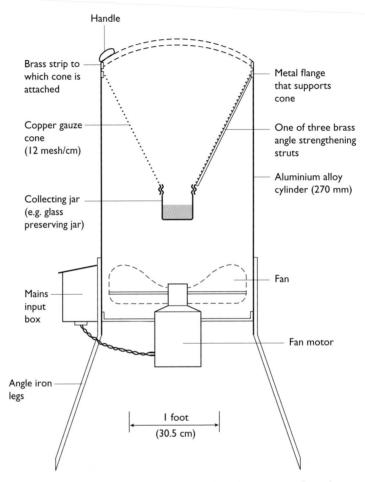

Figure 4.2   Cross-section of the 18-inch propeller enclosed cone type of suction trap.

protected from cross-winds they have a better performance than exposed-cone designs at higher wind speeds. Johnson & Taylor (1955a) originally developed various traps in this class; including a large but lightweight model. The 18-in propeller trap has proved a versatile and robust model that is relatively easily constructed from widely available components. It is powered by a longitudinally mounted, 400-W electric motor which drives a 18-in (46-cm) -diameter propeller fan mounted in a galvanised iron cylindrical duct. The air-filtering cone is of 32-mesh (= 12 mesh cm^{-1}) copper gauze and at its base the glass collecting jar screws into a socket; the jar contains 70% alcohol. To empty the trap the cone is removed and the jar unscrewed (Fig. 4.2).

The Rothamsted 12-m trap is a centrifugal flow trap which has proved valuable for continuous monitoring of insect abundance (Taylor & Palmer, 1972; Taylor, 1974) because it operates uniformly over a wide range of conditions. In the original design, the cylinder enclosing the cone could be extended vertically to sample at heights of up to 12.2 m with a narrow, high-velocity inlet and an expansion chamber to reduce air speed before filtering. The collecting cone is at the base of the trap so that it may be emptied at ground level. The traps were powered by a 1.5 B.H.P., 1440 rev min^{-1} motor with direct drive to a 30.5 cm-diameter horizontal discharge, forward-bladed centrifugal fan; the air-filtering cone was 15.74 mesh cm^{-1} gauze and the diameter of the cylinder, 25.4 cm. A network of 12.2-m suction traps has been operated by the Rothamsted Insect Survey since 1964 to monitor the aerial movement of aphids and other insects. Many variants on this design have evolved in other countries. A standard design for a '12-metre' suction trap is described by Macaulay *et al.* (1988). In the USA, networks of Allison–Pike traps (which only sample 20% of the volume of the Rothamsted trap) have been established for the same purpose (Allison & Pike, 1988).

4.2.3 *Rotary and other traps*

The rotary or whirligig trap consists of a gauze net that is rotated at a speed of about 16 km h^{-1} on the end of an arm (Fig. 4.3); this type of trap has been developed a number of times, sometimes with a single net and sometimes with a net at each end of the arm (Williams & Milne, 1935; Chamberlin, 1940; Stage *et al.*, 1952; Nicholls, 1960; Vité & Gara, 1961; Gara & Vité, 1962; Prescott & Newton, 1963). The greatest efficiency will be obtained if the net is built so as to sample isokinetically – that is, the airflow lines in it are straight; Taylor (1962b) has described such a net (Fig. 4.3). Rotary traps sample more or less independently of wind speed, until the wind speed exceeds that of the trap (Taylor, 1962b). Experiments indicate that they sample 85% of the flying population, but Juillet (1963) has suggested that some insects may be able to crawl out and escape; this can be prevented by placing adequate baffles inside the net. It is also probable that strong flyers with good vision may avoid the area of the moving trap. It is not possible to time-segregate the catches of rotary traps. A large rotary trap designed to measure aerial spider density was described by Topping *et al.* (1992). A comparison of the catch of spiders and other invertebrate groups between this trap's catch and that of a suction trap showed that the rotary trap was operating at a greater efficiency for most groups. However, the two traps caught equal numbers of spiders and non-staphylinid beetles.

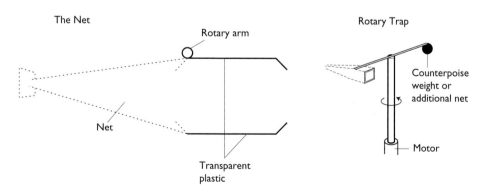

Figure 4.3 A rotary trap, with isokinetic net.

4.3   Comparison and efficiencies of the different types of suction traps

Three factors influence the choice of the type of suction trap: (i) the density of the insects being studied; (ii) the wind speeds in the situation where sampling is proposed; and (iii) the necessity or otherwise for information on periodicity. If information is required on periodicity, then a trap with a catch-segregating mechanism must be used, and this rules out the 18-in propeller trap. The other two factors generally give the same indications, for the sparser the insects the larger the desired air intake (so as to sample an adequate number), and the stronger the winds the stronger (larger) the trap should be. Now at ground level and up to 1–1.5 m amongst vegetation, insect populations are usually dense and wind speeds seldom exceed 10 km h^{-1}; under these conditions the 9-in exposed cone type is generally used. In more exposed situations (e.g. above short turf) or on towers or cables many feet above the ground, the enclosed cone traps should be used; the greater the height the more powerful the winds and the sparser the insects and, therefore, the stronger the fan required. Where information on periodicity can be foregone, the 18-in propeller trap provides a robust, simple and relatively inexpensive piece of equipment.

The average air deliveries of the traps given by Taylor (1955 and pers. commun.) are: 9-in Vent-Axia, 7.84 m^3 min^{-1}; 18-in propeller trap, 77.37 m^3 min^{-1}; centrifugal flow traps 97.6 m^3 min^{-1}. There is, however, a certain amount of individual variation, which over a long period could lead to significant differences in catches. Taylor (1962b) has suggested that unless the actual deliveries of the traps have been checked in the field, differences in catch should exceed 6% for 9-in traps or 10% for 18-in traps, before they be attributed to population differences.

The effects of wind speed on the absolute efficiencies of the traps vary according to the size of the insects as well as the performance of the fan. Taylor (1962a) has shown that the efficiency of a trap for a particular insect is given by the formula:

$$E = (W + 3)(0.0082CE - 0.123) + (0.104 - 0.159 \log i)$$

where E = the log efficiency of the trap, W = wind speed in m h^{-1}, CE = coefficient of efficiency of the trap and i = insect size in mm^2. The coefficient of efficiency of a trap = (35.3 × vol sampled in m^3 h^{-1})$^{0.3}$ / 0.39 inlet diameter of the fan in cm)$^{0.5}$; those for the above traps are: 9-in, 8.7; 12-in, 12. 1; 18-in, 12.1; and 30-in, 13.4. Sizes (i.e. length × wing span) of various insects (in mm^2) given by Taylor (1962a) are listed in Table 4.1.

Estimates of aphid density determined by radar have been compared with simultaneous catches from both a Rothamsted 12.2-m and an aerofoil trap (Schaefer et al., 1985). The 12.2-m trap effectiveness was found to decrease by a factor of two for each 2.4 m s^{-1}

Table 4.1   Sizes (length × wing-span) of some insects (in mm^2)

Minute Phoridae and Sphaeroceridae	1–3
Oscinella, Psychodidae and small Aphididae	3–10
Drosophila, Chlorops, large Aphididae	10–30
Fannia, Musca	30–100
Calliphora, Lucilia	100–300
Sarcophaga, Tabanus	300–1000

(about 5 mph) increase in average windspeed. The two trap sensitivities did not differ significantly in their catches, but the results were significantly different (P <0.001) from those calculated using the above formula, which is based upon a comparison of catches from suction traps with rotary (whirligig) and a tow net catches. Absolute calibration of the aerofoil trap was achieved using a remote-sensing IRADIT infrared system to measure the aerial density of aphid-sized insects near to the trap inlet in very light winds when the effectiveness was about unity, and the 12.2-m trap was predicted to perform similarly.

4.3.1 Conversion of catch to aerial density

The numbers of insects caught is divided by the volume of air sampled; the actual crude delivery figures, given above, may be used for suction traps close to the ground if the wind speed seldom exceeds 5 mph; in other situations appropriate corrections should be made (Taylor, 1955). This figure is then corrected for the efficiency of extraction, of the particular insect, by use of the formula for efficiency given immediately above, though this may need modification in the light of Schaefer's findings. This formula gives a negative value in logs and this is the proportion by which the actual catch is less than the real density.

Where catches have been segregated at hourly intervals the tables of conversion factors may be used (Table 4.2); Taylor (1962b) also gives conversion factors for 12-in and 30-in traps. The conversion factors are given in logs so that:

Log catch/hour + conversion factor = log density/10^6 ft^3 of air.

To convert to log density per m^3, divide by 28317.

Table 4.2   Conversion factors, log *F*, for suction trap catches. (Data from Taylor, 1962.)

9-INVENT-AXIATRAP							
	Wind speed (miles/h)						
Insect size (mm^2)	**0–2**	**2–4**	**4–6**	**6–8**	**8–10**		
1–3	1.89	2.00	2.12	2.23	2.34		
3–10	1.97	2.08	2.20	2.31	2.42		
10–30	2.05	2.16	2.28	2.39	2.50		
30–100	2.13	2.24	2.36	2.47	2.58		
100–300	2.21	2.32	2.44	2.55	2.66		
300–1000	2.29	2.40	2.52	2.63	2.74		
18-IN PROPELLERTRAP							
	Wind speed (miles/h)						
Insect size (mm^2)	**0–2**	**2–4**	**4–6**	**6–8**	**8–10**	**10–14**	**14–20**
1–3	0.85	0.89	0.95	1.01	1.07	1.15	1.29
3–10	0.93	0.97	1.03	1.09	1.15	1.23	1.37
10–30	1.01	1.05	1.11	1.17	1.23	1.31	1.45
30–100	1.09	1.13	1.19	1.25	1.31	1.39	1.53
199–300	1.17	1.21	1.27	1.33	1.39	1.47	1.61
300–1000	1.25	1.29	1.35	1.41	1.47	1.55	1.69

4.3.2   Conversion of density to total aerial population

Johnson (1957) has shown that the density of insects (*f*) at particular height (*z*) is given by:

$$f_z = C(z + z_e)^{-\lambda}$$

where f_z = density at height z, C = a scale factor dependent on the general size of the population in the air, λ = an index of the profile and of the aerial diffusion process, and z_e = a constant added to the actual height and possibly related to the height of the boundary layer: the height at which the insects' flight speed is exceeded by the wind speed (Taylor (1958). This may be solved by iteration using a computer or by the graphical method described by Johnson (1957).

The total number of insects between heights z_1 and z_2 (*n*) may then be integrated as follows:

$$N = \frac{C}{1 - \lambda}[(z_2 + z_e)^{1-\lambda} - (z_1 + z_e)^{1-\lambda}]$$

the symbols are as above.

With individual species of insect, the profiles may be found to be rather irregular and approximate integrations may then be made graphically (Johnson *et al.*, 1962).

4.4   Sampling from plants

In many ways this is the most difficult habitat from which to sample invertebrates. It differs from the soil and the air, both in being much more heterogeneous and in continually changing. It is frequently convenient to take a part of the plant as the sampling unit; the resulting population estimate is not an absolute one, but a measure of *population intensity*. The distinction between these terms has already been discussed (p. 2), but the point is so vital that some further elaboration here is justified. If the amount of damage to the plant is the primary concern it may seem reasonable to make all estimates in terms of numbers of, say, mites per leaf. This is the intensity of mites that the tree has to withstand, and if mite populations were being related to some index of the health of the tree such an estimate of 'population intensity' is probably the most relevant. But if the study is concerned with the changes in the numbers of mites in a season a series of estimates of population intensity could easily be misleading. If the number of mites per leaf fell throughout the summer this could be due to an actual reduction in the mite population or to an increase in the numbers of leaves. In order to determine which of these explanations is correct the number of leaves per branch would have to be counted on each sampling date, and the mite population could then be expressed in terms of numbers per branch. This is sometimes called a '*basic population estimate*', but it differs from the measure of population intensity only in degree and not fundamentally, for from year to year the number of branches or indeed the number of trees in the forest or orchard will alter. It is only when these units are also counted and the whole converted into numbers per unit surface area of soil (commonly per square metre or hectare) that it is possible to make a valid comparison between populations differing in time (and space), which is the essence of many population studies.

The labour of obtaining the additional data on the density of the plant unit is often comparatively slight, and it cannot be stressed too strongly that, whenever possible, this should always be done as it makes the resulting data more meaningful and useful for testing a wider range of hypotheses.

4.4.1   Assessing the plant

The simplest condition is found in annual crops where the regular spacing and uniform age of the plants makes the estimation of the number of shoots or plants per length of row a relatively easy matter: the variance of even a few samples is often small and the determination of the number of row length units per unit of area (e.g. per m^2), is also straightforward. With natural herbaceous vegetation there will usually be far more variability both in the number of individual host plants per unit area and in their age and form. Although the number of plants per unit area is sometimes an adequate measure, the variation in size and age often means that weight of plant or the number of some part of the plant frequented by the insect (e.g. the flower head) gives a more adequate measure. Occasionally, botanical measures such as frequency, as measured by a point quadrate (Greig-Smith, 1964; Mueller-Dombois & Ellenburg, 1974), are useful.

The most complex situation is found when one attempts to assess the habit of arboreal insects. The leaf is often, but not always, the sampling unit (e.g. Richards & Waloff, 1961); its age, aspect and height above the ground need to be considered as well as the age of the tree. Frequently, a stratified random sampling plan minimises the variance of the population estimate. For example, to sample the citrus mealy bug, Browning (1959) found the variance was minimal if each aspect of the tree was sampled in turn and individual leaves picked haphazardly rather than sampling the mealy bugs on all the leaves on a single branch.

The precise estimation of the number of leaves per branch is often a minor piece of research on its own. Actual numbers may be counted or dry weight used as the measure. In some trees the actual weight of the foliage may continue to increase after the leaves have ceased elongating (Ives, 1959). The diameter of the trunk may be taken as a cruder measure of foliage weight and of shoot number in young growing trees, especially conifers (Harris, 1960).

4.4.2   Determining the numbers of invertebrates

Irrespective of habitat, it is difficult to devise an 'all-species method', as the more mobile animals have a greater tendency to escape, small species to be missed, and large or firmly attached species more difficult detach or remove.

4.4.2.1 Direct counting

With large conspicuous animals, including insects, it is sometimes possible to count all the individuals; for example, Moore (1964) was able to make a census of the male dragonflies over a pond using binoculars. Smaller insects on distant foliage, such as the upper parts of trees, may be counted using a powerful tripod-mounted telescope, capable of short-range work and with a wide aperture (e.g. 'Questar'); a transect may be made through the canopy of a tree by systematically turning the elevation adjustment of the telescope (L.E. Gilbert, pers. commun.). This should prove a very useful method for the survey of forest insects. Direct counts may also be used to follow a particular cohort (see p. 5) from which we can obtain true estimates of population; the difficulties of determining the true mean and other parameters, which may be severe in highly aggregated populations, are avoided.

More often, a part of the habitat is delimited and the animals within the sample counted. This is easily accomplished for relatively immobile ground-living animals such as the snail, *Theba pisana* (Baker & Vogelzang, 1988; Baker & Hawke, 1990), where quadrats may be placed upon the ground and the animals immediately counted. With care, quadrats can also be used to estimate fairly mobile, but relatively large, insects such as grasshoppers (Richards & Waloff, 1954) or froghoppers (cercopids). These quadrats are made of wire and it is often useful to attach white marker-ribbons. They are placed in the field and left undisturbed for some while before a count is made. Each quadrat is approached carefully and the numbers of insects within it counted as they fly or hop out. Alternatively, a high-sided, box quadrat, possibly driven into the soil, can be used to retain less-mobile insects while collecting or counting is in progress (Balogh & Loksa, 1956). In such cases and with other relatively immobile large insects, e.g. the Wart-biter cricket, *Decticus verrucivorus* (Cherrill & Brown, 1990) the sunn pest, *Eurygaster integriceps* (Banks & Brown, 1962), or lepidopterous larvae and pupae (Arthur, 1962), a count is made as soon as the quadrat is in position.

Where the insects are restricted to particular plants then these or particular parts of them, such as fruits or flowers, may be collected or examined and the numbers on each sample counted. Care must be exercised, as many insects can be dislodged as the samples are taken (Satchell & Mountford, 1962) or small animals overlooked (Condrashoff, 1967). Forest and orchard pests are often assessed as numbers per shoot and crop insects as numbers per plant or per leaf (Day & Crute, 1990). Aphids on crops such as wheat may be sampled *in situ* by inspecting a set number of tillers (e.g. Cannon, 1986). It will commonly be found that the position of the shoot or leaf – whether it is in the upper, middle or lower part of the tree or plant – influences the numbers of insects upon it, and due allowance should be made for this stratification by a subdivision of the habitat (p. 9). When small abundant insects are actually being counted it is desirable that the sampling unit should be as small as possible. Shands *et al.* (1954) showed that various aphids on potatoes could be adequately estimated by counting, not every aphid on a selected leaf, but only those on the terminal and two basal leaflets; even a count based on a half of each of those leaflets could be satisfactory.

Photographs have long been used to count insects (e.g. recording aphids on rose shoots; Maelzer, 1976: density of the scale insect, *Trialeurodes vaporariorum*, on beans;

Ru-Mei, 1985). Digital photography and the ability to analyse images and automatically count objects has greatly increased the power of photographic methods. Video cameras can also be used to record insects. Systems exist for automatically counting flying insects such as mosquitoes using optical counters. A problem with photographic methods for small insects is ensuring accurate identification to species.

Insects in stored products, especially cacao, are often estimated by 'snaking'. The bag is opened and the contents tipped onto a floor in a long wavy line; the insects may be counted directly and a large number will often be found in the last piece tipped out, known as the 'tail'. This method, although technically providing an absolute type of sample, is really more properly regarded as a relative method and is often necessary to establish the basis of a presence or absence sampling programme (p. 55) for extensive work.

4.4.2.2   The separation of exposed small animals from the foliage on which they are living

4.4.2.2.1   *Knockdown: by chemicals, jarring and heat*

Knockdown by chemicals and jarring are discussed more fully below, as they are more generally used when the whole twig or tree is the sampling unit. Here, we are concerned with their use with animals on herbaceous plants and with animals on trees when the leaf is the sampling unit. Some insects can be made to drop off foliage by exposing them to the vapours of certain chemicals. With the wide range of organic substances available, this approach could undoubtedly be extended; some recognised repellents (Jacobson, 1966) might be screened. Aphids can be made to withdraw their stylets and will mostly drop off the plant if exposed to the vapour of methyl isobutyl ketone (Gray & Schuh, 1941; Helson, 1958) or 1-BHC (Way & Heathcote, 1966). Aphids that do not cover the host plant with a thick deposit of honey dew and are fairly active may be separated from the foliage by exposing them to this vapour and then shaking them in the type of sampling can described by Gray & Schuh (1941) (Fig. 4.4a). With the pea aphid, *Acyrthosiphon pisum*, these authors shook the can 50 times following a 5-min interval and found that after one such treatment an average of 89.3% of the aphids were extracted, after 10 min (and 100 shakes) 97%, and after 15 min, 99.1%. In the six samples tested the lowest extraction rate was 98.5% after 15 min; the numbers extracted from each sample in the can were between 3000 and 7000. Walker *et al.* (1972) found with *Schizaphis* that, because of entanglement with honeydew, extractions were in the region of 85% after 15 min exposure, shaking more than 50 times did not increase the recovery efficiency (and the effect from 25 to 50 times was only marginal). Laster & Furr (1962) found that the fauna of sorghum heads could be killed by insecticide and collected in a paper cone strapped round the stem and stabilised with a little sand in the base.

Thrips may be extracted from flowers and grassheads, as well as foliage, by exposing them to turpentine vapour (Lewis, 1960). A convenient apparatus has been described by Lewis, who found the method was over 80% efficient for adults, but confirmed Taylor & Smith's (1955) conclusion that it was unsatisfactory for larvae. Lewis & Navas (1962) obtained thrips and some other insects that were overwintering in bark crevices by breaking the bark sample into small pieces and sieving; this dislodges the majority,

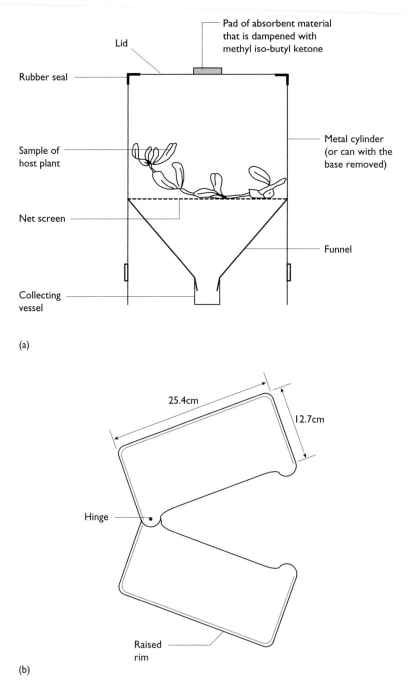

Figure 4.4   (a) Aphid sampling can. (Adapted from Gray & Schuh, 1941); (b) Sampling tray for flea-beetle, weevils and others on small crop plants. (Adapted from La Croix, 1961.)

but to obtain 97–100% extraction efficiencies, the fragments were exposed to turpentine vapour.

Heat was used by Hughes (1963) to stimulate, *Brevicoryne brassicae*, to leave cabbage leaves, and Hoerner's (1947) onion thrip extractor worked on the same principle; these techniques can be considered modifications of the Berlese funnel (p. 238). An ultrasonic vibrator (c. 20 kHz, 100 W output) was found to dislodge eriophyid mites from grass (Gibson, 1975). The foliage was placed in 50% or 75% alcohol (ethanol) and exposed for between 45 and 60 s (maximum) (see also p. 163).

Insects that fall from their host plant when disturbed (e.g. some beetles) can be sampled from young field crops by enclosing the base of the plant in a hinged metal tray (Fig. 4.4b). The plant is then tapped (jarred) and beetles from it may be counted as they fall onto the tray. This method was developed by La Croix (1961) for flea beetles, *Podagrica*, on cotton seedlings; however, they must be counted quickly before they hop off the tray (this is facilitated by painting the tray white).

4.4.2.2.2 *Brushing*

Mites, and apparently other small animals apart from insects, are removed from leaves if these are passed between two spiral brushes revolving at high speed in opposite directions (Henderson & McBurnie, 1943; Chant, 1962). Mite brushes are available from BioQuip (http://www.bioquip.com/AboutUs/default.htm). A YouTube video showing the application of a mite brush to remove and count spider mites from strawberry leaves is available at: https://www.youtube.com/watch?v=mkY-BNZ3Jf4.

4.4.2.2.3 *Washing*

Small animals, principally mites, aphids and thrips, can be washed off herbaceous plants or single leaves with various solutions. This approach can be more efficient, both in terms of accuracy and of cost, than direct counting. With aphids the extraction will be facilitated if they are first exposed to the vapour of methyl isobutyl ketone (= 4-methyl 2-pentanone) (Pielou, 1961). Dilute soap, detergent or alcohol solutions are often adequate for washing (e.g. Taylor & Smith, 1955), but with eggs a solvent must be used that will dissolve the cement. Benzene, heated almost to boiling point, removes many of the eggs of mites on fruit-tree twigs (Morgan *et al.*, 1955), as does hot sodium hydroxide solution (Kobayashi & Murai, 1965). Benzene is not only highly inflammable, but also carcinogenic; hence, such procedures should be undertaken in a fume cupboard with appropriate safety precautions. If conifer foliage with moth eggs is soaked for 30 h in a 1.5% solution of sodium hydroxide, a 96% recovery rate may be achieved (Condrashoff, 1967). Petrol has also been found satisfactory for washing eggs of the corn earworm, *Heliothis zea*, from the silks of maize (Connell, 1959). Thrips may be washed off foliage with ethanol (Le Pelley, 1942), the solution filtered, and the thrips separated from the other material by shaking in benzene and water; the thrips pass into the benzene where they may be counted (see also p. 231) (Bullock, 1963). Moth eggs may be freed from bark and moss by soaking for 45 min in 2% bleach solution (sodium hypochlorite solution; 5.25% free chlorine diluted with 98 parts water); the mixture is then agitated and passed through a sieve, which separates the larger debris (Otvos & Bryant, 1972). Wardlow

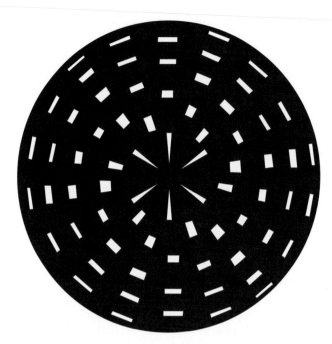

Figure 4.5   A counting grid. (Adapted from Strickland, 1954.) All animals that are wholly within the clear areas or overlap the left-hand or outer margins should be counted, and these will be one-sixth of the total.

& Jackson (1984) extracted apple rust mite, *Aculus schlechtendali*, from buds by using methanol.

The animals removed by washing are often so numerous that they require automatic counting methods. An alternative approach is to count a sample; this may be done by means of a counting grid or disc (Fig. 4.5) (Strickland, 1954; Morgan *et al.*, 1955). The animals are mixed with 70% alcohol or some other solution, agitated in a Petri dish, and allowed to settle. The dish is stood above a photographically produced counting disc and illuminated from below. Strickland (1954) found that the percentage error of the count for aphids using the disc in Fig. 4.5 ranged from 5 (with about 3000 aphids per dish) to 15 (with fewer than 500 aphids per dish).

If the number of animals is too great for the counting grid to be used directly, the 'suspension' of the animals in a solution may be aliquoted (Newell, 1947). The diluted samples may then be estimated with a counting disc or volumetrically, or the two methods combined. In the volumetric method the animals are allowed to settle or are filtered out of the solution; they are then estimated as, for example, so many small specimen-tubefuls, the numbers in a few of the tubes are counted partially by the use of a counting disc or *in toto* (Banks, 1954; Walker *et al.*, 1972). If the population consists of very different-sized individuals the volumetric method may lead to inaccuracies, unless special care is taken. This variation can, however, be used to aid estimation: Pielou (1961) showed that if apple aphids, *Aphis pomi*, were shaken with alcohol and the solution placed in Imhoff sedimentation cones (wide-mouthed graduated cones), the sediment stratifies into layers, first the larger living aphids, then the smaller living aphids, and finally dead aphids and cast skins. The volumes of these can easily be read off and calibrated for the various categories. Another method of separating the instars based on the same principle has been developed by Kershaw (1964). A routine for sub-sampling when collections vary enormously in size is described by Corbet (1966).

4.4.2.2.4 *Imprinting*

This technique has been used only with mites and their eggs; the infested leaves are placed between sheets of glossy absorbent paper and passed between a pair of rubber rollers (e.g. a household wringer); where each mite or its egg has been squashed a stain is left (Venables & Dennys, 1941; Austin & Massee, 1947). Summers & Baker (1952) used the same method to record the mites beaten from almond twigs. The advantage of this method is its speed and the provision of a permanent record; furthermore, with certain precautions, nearest-neighbour-type techniques (p. 367) might be applied to the resulting marks. However, Chant & Muir (1955) found with the fruit-tree red spider mite, *Panonychus ulmi*, that the number recorded by the imprinting method was significantly less than the number recorded by the brushing method (see above), which is therefore to be preferred in general. Trumble (1994) describes the imprinting procedure for two-spotted spider mite, *Tetranychus urticae*, on strawberry. The leaves are crushed on filter disks impregnated with bromophenol blue, which turns green in the presence of protein from the mites.

4.4.2.3 The expulsion of animals from tall vegetation

4.4.2.3.1 *Jarring or beating*

This is a collector's method in which, originally, the tree was hit sharply with a stick and the insects collected in an umbrella held upside down under the tree! The umbrella is now replaced by a beating tray which is basically a cloth-covered frame, flat or slightly sloping towards the centre, that is large enough to collect all the insects that drop off the tree. The colour of the cloth should produce the maximum contrast with that of the insect being studied, and the insects are rapidly collected from the tray by an aspirator or pooter (see Fig. 7.5). In general, this is only a relative method but with some insects such as leaf beetles (Chrysomelids), many weevils and lepidopterous larvae that fall, rather than fly, from the host plant when disturbed a sufficiently high proportion may be collected for it to be regarded as an absolute method (e.g. Richards & Waloff, 1961; White, 1975). Weather, strength of the blow and time of day can all change efficiency (Harris *et al.*, 1972; Le Blanc *et al.*, 1984). As very small animals (e.g. mites) may be overlooked on a beating tray and active ones may escape, more accurate counts can sometimes be obtained by fastening a screen over a large funnel. The twigs are tapped on the screen and mites and insects funnelled into a container (Steiner, 1962). The beating tray and funnel methods may be combined: Wilson (1962) attached a removal jar below a hole in the centre of the tray, while Coineau (1962) placed a grid above a funnel in the centre of a net tray; this grid helps to separate twigs and other debris from the insects. Arthropods on row crops (e.g. soybeans) may be collected by quietly unrolling a length of oil-cloth ground sheet (or polythene) between two adjacent rows and briskly shaking the plants over it; the method is time-consuming but seems to provide absolute estimates of large easily dislodged species (e.g. pentatomids) (Rudd & Jensen, 1977).

Although the tree is generally hit sharply with a stick, Legner & Oatman (1962) suggest the use of a rubber mallet, though Richards & Waloff (1961) found that some host plants (e.g. *Sarothamnus*) could be damaged by such violence (so leading to a change in the habitat); rather, they recommended vigorous shaking by hand.

Basset *et al.* (1997) reviewed the non-fogging methods for sampling arthropods in tree canopies. Following sampling in Papua New Guinea for leaf-feeding adult beetles, they concluded that the most species were collected by beating and hand-collecting.

4.4.2.3.2   *Chemical knockdown*

The most widespread approach to the sampling of arboreal insects is now based on this use of insecticides (Stork *et al.*, 1997). The pyrethrin group are particularly useful as they affect most arthropods and initially cause enhanced activity that can lead to the insect falling from the tree. The pioneering work was undertaken in plantations and orchards (Collyer, 1951; Gibbs *et al.*, 1968) when the tree was shrouded and the total fauna collected from large sheets spread on the ground below. These early studies were usually concerned with one or two pest species which could be rapidly picked out from the mass of material.

Around 1980, the method was used to obtain information on the total fauna of trees, but if all the specimens are saved, sampling a whole tree produces a very large number of individuals (and a substantial task), but gives no measure of intra-tree variance (see p. 16). Sampling efficiency was therefore increased by taking a limited sample from under any individual tree or any small area of forest; a small (normally 5–10) number of sampling sheets or trays were either suspended or placed on the ground under each tree (Moran & Southwood, 1982; Southwood *et al.*, 1982; Erwin, 1983; Stork, 1987; Chey *et al.*, 1998). Where the ground cover is low and sparse, or when it can be removed, the required samples may be collected on cloth sheets that are either spread directly on the ground or stretched on a metal frame with short legs. This design is similar to the 'frass collector' (see Fig. 8.1); it keeps the cloth above the vegetation and provides a precise area. One square metre is a convenient size, and insects may be removed (with a pooter) virtually as soon as they have fallen. With tall trees, particularly in rain forest, it is necessary to suspend the collecting devices below the canopy (Erwin, 1989; Stork & Hammond, 1997); these should retain the specimens that fall as they cannot be continuously tended, and funnel trap designs (see Fig. 10.4b) have been found effective. Washing the sides down with ethanol from a wash bottle may clear them. It is always important to seek to avoid disturbing the ground-cover fauna before sampling; otherwise, its members may fly into the trees and provide an atypical element in the subsequent sample.

If the trees to be sampled are less than about 15 m in height, the spray may be applied from a standard back-pack sprayer, using an extension lance if necessary. In rain forests it may be necessary to use an even finer spray from a 'mist-blower' or 'fogger' (e.g.

Swingfog SN50; see http://www.swingtec.de/thermal-fogging-machines/swingfog-sn-50.htm) and allow this to drift and penetrate the canopy, this process may be repeated at short intervals.

Paarmann & Kerck (1997) found that if natural pyrethrum (without synergist, 1% diluted in diesel oil) was used as the knockdown agent, many insects would recover and often survive to develop and breed normally. This enables biological investigations to be made on species that are not otherwise easily collected. It also has the advantage that sampling will have less effect on the population because the considerable part of the population that falls outside the area of the sampling devices may recover. Recovery rates differ between groups, so there is a risk of another type of bias if further studies are to be made in the same locality. Most synthetic pyrethrins are more toxic than the natural compound and few insects recover; however, knockdown may be more complete, especially if – as with neo-pybuthrin – this is quick.

The proportion of the fauna sampled is influenced by many factors, including: the method; the type of insect (the highly mobile and the highly cryptic, e.g. borers, are underrepresented); the type of tree; and the weather conditions. However, most studies have found that around 75% of the total population fell within the first hour; this figure may be increased if the tree can be repeatedly shaken or jarred towards the end of the period or shortly afterwards (Southwood *et al.*, 1982; Adis *et al.*, 1997). Yanoviak *et al.* (2003), in a study of arthropods in Costa Rican cloud forest, found that arthropods in epiphytes were poorly sampled by pyrethrin insecticide fogging because the killed animals were retained within the epiphyte mat.

In Britain, where the fauna is relatively well known, the entomofauna of a tree species as revealed by knockdown sampling was compared by Southwood *et al.* (1982) with that obtained by collation from faunal lists; they concluded that the two approaches gave comparable data.

4.4.2.3.3   *Collection of naturally descending animals*

Many arboreal larvae pupate in the soil, and these may be collected as they descend to provide a population estimate of the total numbers at that particular developmental stage (see p. 375). Cone (1963) found that the vine weevil, *Otiorrhynchus sulcatus*, dropped off its host plant at daybreak and the population could therefore be determined by placing funnel-type traps under the grape vines.

4.4.3   *The extraction of animals from herbage and debris*

Here, one is concerned with animals living in the herbage layer above the soil and the problem is not, as in the above section, to make them fall down, but to get them to come up either artificially or naturally.

4.4.3.0.4   *Suction apparatus*

Based on the principles of domestic vacuum cleaners, a number of types of suction sampler for animals living on or amongst the herbage down to the soil surface have been developed. Today, the three most widely used purpose-built commercial units are the

D-Vac, Univac and Vortis (Plate 4). Cheap machines can be made by converting those designed for other material such as garden litter (Wright & Stuart, 1992), and designs have been published which offer advantages of cost, weight and sampling efficiency over the established commercial models (e.g. Stewart & Wright, 1995; Harper & Guynn, 1998). Doğramaci *et al.* (2011) describe a design for a low-cost backpack-mounted petrol power suction sampler based on a Troy-Bilt 31 cc 2-Cycle Blower/Vacuum (Model # TB320BV; http://www.troybilt.com). Suction sampling is only really effective in dry, upright vegetation less than about 15 cm high.

The D-Vac

The D-Vac (available from Rincon-Vitova Insectaries, P.O. Box 1555, Ventura, CA 93002-1555, USA; see http://www.rinconvitova.com/corporat.htm) was developed by Dietrick *et al.* (1959), the Model 24 uses a 20-cm (8-inch) hose (Fig. 4.6). The efficiency of this sampler is greatly affected by the mode of operation; the nozzle must be moved vertically into the foliage and down to the ground surface, although even when this is done, if the ground is subsequently searched carefully, a number of beetles, isopods and other 'heavy' arthropods will usually be found to have been missed. If the nozzle is swept horizontally across the foliage very small catches are obtained (Richmond & Graham, 1969). Several comparisons of the D-Vac with other methods have shown efficiencies of

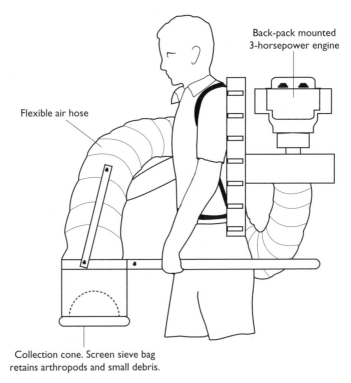

Back-pack mounted 3-horsepower engine

Flexible air hose

Collection cone. Screen sieve bag retains arthropods and small debris.

Figure 4.6   A model 24 D-vac suction sampler.

around 50% to 70%. Some of these studies have been concerned with coccinellid beetles and other predators on cotton (Turnbull & Nicholls, 1966; Leigh *et al.*, 1970; Shepard *et al.*, 1972; Smith *et al.*, 1976). Dewar & Dean (1982) compared D-vac, direct counting and washing sampling methods for sampling aphids on wheat, and concluded that the efficiency of the D-vac declined with aphid density and varied with species and life history stage. The D-vac cannot be recommended for grassland Colleoptera sampling, for which reported sampling efficiencies range between only 7% and 27% (Duffey, 1980) and 1% and 17% (Morris & Rispin, 1987) of that obtained using heat extraction. However, Tormala (1982) found suction sampling to be the most effective method for grassland *Auchenorhyncha* (Hemiptera), but efficiency can vary seasonally (Duffy, 1980) because larvae are not caught as efficiently as adults (Andrezejewska, 1965). Perfect *et al.* (1983) discuss the precision of population estimates obtained for planthoppers, and leafhoppers on rice, while Hand (1986) determined the capture efficiency for aphids on grasses and cereals. While effective, the D-Vac is heavy and difficult to handle. In order to increase efficiency, Turnbull & Nicholls (1966) reduced the nozzle and hose diameters to obtain a nozzle speed approaching 90 m h^{-1}. The model 24 can be fitted with a 4-inch adapter hose for greater suction.

The Univac

The Univac was developed by Arnold *et al.* (1973) from an original design by Johnson *et al.* (1957), and is manufactured by the Burkard Manufacturing Co., Rickmansworth, England; http://www.burkardscientific.co.uk/agronomics/pdf/UnivacSuctionSampler.pdf. This is a light backpack unit (total weight 11kg) which uses a 6-cm intake nozzle. The vacuum unit is powered by a two-stroke engine. The area to be sampled is enclosed by forcing a small metal cylinder (e.g. a bottomless bucket or a special cylinder; Hower & Ferguson, 1972) into the soil; the area delimited is then systematically worked over with the nozzle and a mixture of animals and plant debris is trapped in the collecting bag. This may be sorted by hand, but many workers have found a Berlese-type apparatus and/or flotation increased the ease and often the efficiency of recovery (Dietrick *et al.*, 1959; Turnbull & Nicholls, 1966) (pp. 238, 229; see also Chapter 5). Henderson & Whitaker (1976) found extraction rates of 70–98% for most groups on short to medium (15 cm) length grass. In taller grass the rates were lower and always below 50% for mites. The Burkard Manufacturing Co. also produces the Thornhill portable suction sampler, which is a petrol-driven portable suction sampler similar to the Univac, but with increased suction tube area up to 18 inches. This is designed for effective sampling in crops, grasslands and scrublands.

The Vortis

The Vortis sampler (also manufactured by the Burkard Manufacturing Co.; http://www.burkard.co.uk/vortis.htm), unlike the Univac and D-vac samplers, does not have an animal-collecting bag through which the main air stream must pass (Fig. 4.7). Because the collected material does not impede the air flow, the Vortis has the advantage that the flow rate will not decline as material is sucked-up. The air is sucked in above ground level and is turned through 180° as it lifts loose particles up the 10 cm-diameter main tube (collecting area 0.2 m^2). A series of vanes create a

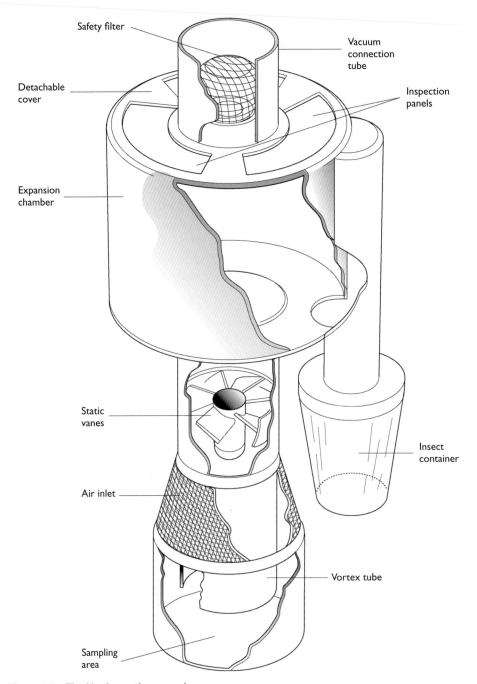

Safety filter

Vacuum
connection
tube

Detachable
cover

Inspection
panels

Expansion
chamber

Static
vanes

Insect
container

Air inlet

Vortex tube

Sampling
area

Figure 4.7   The Vortis suction sampler.

vortex so that when the insects reach the expansion chamber they fall freely into an externally attached collected pot or bag. The Vortis has the advantage that the air flow is constant and cannot be impeded, so that a less-powerful, lighter motor can be used. Brook *et al.* (2008) reported the efficiency of the Vortis as a sampler for the collection of beetles (Coleoptera), true bugs (Hemiptera: Heteroptera), planthoppers (Hemiptera: Auchenorrhyncha) and spiders (Araneae) in grassland. They found that sampling durations of 16 s collected 90% of individuals within the tube diameter of 15.7 cm. Generally, about 55 sub-samples were needed to collect 90% of the species of these groups present. The efficiency declines with the height of the sward, however. Brook *et al.* recommended that, to measure invertebrate diversity and assemblage structure, a minimum of 110 samples, each of 16-s duration, be taken.

Owing to the formation of a water film, in which animals may become trapped, inside the suction hose, it is difficult to sample damp herbage with the narrow suction-hose types of apparatus, but the use of a clear hose enables a check to be kept on this potential source of error. Arnold *et al.* (1973) used 'Vacuflex' hose that, although being made of clear plastic, is reinforced with a spiral wire. When sampling arable habitats with the Univac it may be an advantage to use two bags – one of nylon net which retains the insects, outside of which is a larger cloth bag to prevent the soil passing through the fan. The narrow suction-hose type of sampler, with its high nozzle wind speed and fan, the efficiency of which does not fall off when impeded, is particularly suited for work with the fauna of the lower parts of vegetation and the soil surface. The D-vac sampler overcomes the problem of adhesion to the hose by having the collecting bag incorporated in the widened nozzle (Fig. 4.6), while a squirrel-cage type of fan produces a strong air current, as long as it is not impeded.

In conclusion, the efficiency of suction samplers is affected by several factors, but of particular importance are nozzle wind speed (for heavy and/or tenacious species) and speed of enclosure (for active flying species). The Univac and Vortis are excellent on the latter count, but the nozzle wind speed of the D-vac is with several types of collecting head on the borderline of efficient extraction. Leigh *et al.* (1970) consider that the D-vac needs to be standardised for the particular conditions before it may be considered as an extraction method for absolute estimates. Low efficiencies for the D-vac occur if the animals are very active (escaping as the sampler approaches) or if they can cling close to the soil surface, when the nozzle speeds of the normal D-vac (c. 60 km h^{-1}) are inadequate to dislodge them. Other samplers seem more efficient under a wider range of conditions (Johnson *et al.*, 1957; Heikinheimo & Raatikainen, 1962). The method is usually more satisfactory than sweeping (p. 276), which can give biased results. Doxon *et al.* (2011) compared sweep-netting and D-vac samples in an area of vegetation comprising sage brush and grasses. Diptera, Homoptera, and Hymenoptera dominated the D-vac samples, while the orders Homoptera, Orthoptera, and Araneae dominated the sweep-net samples. The mean size of invertebrates collected and overall invertebrate biomass were greater for sweep-netting than for D-vac sampling. The D-vac was more effective at collecting small (e.g. <5 cm) invertebrates, whereas sweep-netting captured large (>5 cm) Orthopteran and Lepidopteran larvae, at higher rates.

Larger models seem efficient; Kirk & Bottrell (1969) devised a tractor-mounted model and the 'McCoy Insect Collector', that incorporated blowing – a principle first used by

Santa (1961) – as well as sucking, in a 'lorry-sized' motor vehicle. Its efficiency is high, and McCoy & Lloyd (1975) show it can detect two boll weevils in 10 acres of cotton. These vehicular machines that sample from such a large proportion of the habitat, overcome one of the main difficulties with smaller samplers, namely the high variance of the samples due to the patchy nature of many, especially natural, habitats. Theoretically, the answer lies in taking more samples, but the worker easily becomes over-burdened with material to sort. For samples from large vehicular machines, mechanised and computerised methods of sorting have been developed.

Small suction samplers have been used to obtain invertebrates from rodents' nests (Levi et al., 1959) and flowers (Kennard & Spencer, 1955).

4.4.3.0.5   *Cylinder or covering method*

The basis of this method is the enclosure of an area of herbage within a covered cylinder; the animals are collected by some method often after being knocked down by an insecticide. Kretzschmar (1948) used a modification of the Romney (1945) technique for sampling soya-bean insects by placing plastic transparent base plates beneath the plants on the day before sampling. These greatly facilitated the subsequent collection of the insects, but tests should be made to ensure that their presence does not influence the density of insects on the plants above. In order to prevent the escape of active insects, Balogh & Loksa (1956) attached the sampling cylinder to the end of a long pole and rapidly brought this down on to the crop. These workers killed the insects with powerful fumigants such as carbon disulphide and hydrogen cyanide, and then removed them by hand. As it is difficult to reach and search the base of a tall cylinder, Skuhravy et al. (1959) and Skuhravy & Novák (1961) used two cylinders and removed the tall outer one after the insects had been knocked down. Cherry et al. (1977) combined this method with the 'tent' technique (see below). Smaller plants have been sampled by enclosing them in an impermeable plastic tent, which is then flooded with carbon dioxide.

The combination of the 'covering method' with a suction apparatus for collection probably provides the most efficient method for sampling the fauna of herbage. The most appropriate cylinder or tent (see below) and suction apparatus will depend on the animal, the vegetation, and the availability of services (electricity, easy carriage of heavy equipment). Turnbull & Nicholls (1966) devised a 'quick trap' (Fig. 4.8) which they used in conjunction with a modified D-vac sampler. The quick trap is put into position several hours prior to sampling, with the net portion folded (the dotted outline in Fig. 4.8). When the sample is to be taken, the suspension cord – which can (and should) extend some distance from the trap – is pulled, which jerks it out of the slot at the top of tripod leg A. The tension springs expand the trap into position, but the operator should immediately check that the base ring is in close contact with the ground. The suction sampler hose is inserted through the top of the trap and the area systematically worked over. The mixture of small arthropods and debris can be sorted by hand or by a combination of funnel extractors and flotation (Dondale et al., 1971) (see Chapter 5 and pp. 238 and 229). Turnbull & Nicholls (1966) made most useful comparisons of this method with several others (Fig. 4.9). The extent of the advantage compared with the sweep net or normal D-vac sampler varied between taxa.

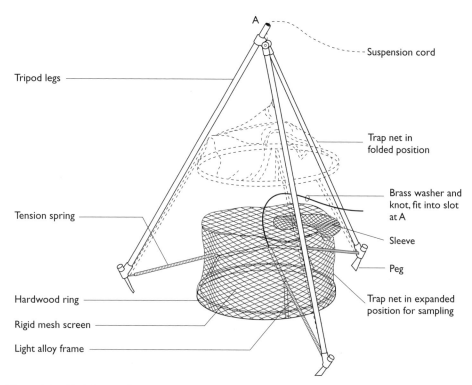

Figure 4.8 'Quick-trap'. (Adapted from Turnbull & Nicholls, 1966.)

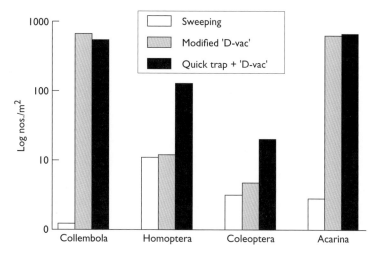

Figure 4.9 Comparison of population estimates for different taxa using three methods of sampling: sweep net; modified D-vac; and quick-trap covering method plus D-vac. (Adapted from Turnbull & Nicholls, 1967.)

4.4.3.0.6 *Tents for sampling strongly phototactic animals*

A large muslin covered cage or 'tent' (Fig. 4.10) is quickly put into position when the animals are least active, usually at dusk. When they are active again (on the following morning) the animals (if large) may be directly collected from the sides of the tent; alternatively, it may be covered with a black shroud leaving only the muslin-topped celluloid collecting cylinder exposed at the apex. After some time (about 15 min in bright sunshine) the majority of the animals will be in the cylinder, which can be quickly removed and is conveniently carried back to the laboratory by pushing its base into appropriately sized rings of Plasticine on a metal tray. If possible, someone should then get inside the 'tent' and remove the remaining animals from the insides of the walls; on a dull day, quite a large proportion of the total population may fail to enter the

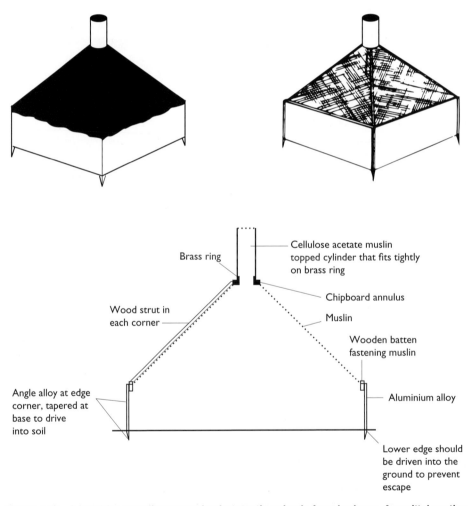

Figure 4.10 A 'tent' for sampling strongly phototactic animals from herbage. A sagittal section through one corner and a side, with sketches of its appearance shrouded and unshrouded.

collecting cylinder. This method has been used successfully with blowflies, *Lucilia* (MacLeod & Donnelly, 1957), mosquitoes (Chinaev, 1959) and frit flies, *Oscinella frit* (Southwood *et al.*, 1961). A miniature tent for use with mites is described by Jones (1950), while McGovern *et al.* (1973) have devised a model that they used to separate phototactic (and resilient) animals (e.g. boll weevils) from the debris in suction apparatus samples.

An apparatus that is intermediate between a cylinder and a tent has been devised by Wiegert (1961) for sampling grasshoppers, cercopids and other insects that readily leave the vegetation if disturbed. It consists of a truncated nylon-covered strap-iron cone, topped with a metal funnel and a collecting jar. As with the tent or cylinder, this is placed rapidly over the vegetation that is then agitated. The sampling cone is then gradually moved on to a flat sheet, of wood or other substance, that is placed adjacent to it. Eventually, most of the vegetation can be slid out of the cone whilst the active insects remain; it is then inverted and the insects jarred down into the collecting container. Wiegert considered the efficiency for the froghopper, *Philaenus*, greater than 90%. The leafhopper, *Empoasca*, was efficiently collected by Cherry *et al.* (1977) in a rather similar device: a black plastic dust (garbage) bin, with a collecting cylinder (as used in tents, Fig. 4.10) in its base, was quickly lowered, upside down, suspended from the end of a long pole.

4.4.3.0.7 *Extraction by heat-drying and/or flotation*

A further extension of the use of the animal's reactions to certain physical conditions leads logically to the removal of samples of herbage or debris to the laboratory, where they are subjected to drying by heat. This is the basis of the Berlese–Tullgren funnel. This principally is a method for soil and litter, and hence its many modifications are discussed in Chapter 6. It is frequently utilised to extract animals from herbage samples, especially the mixtures that result from suction sampling (see above) (e.g. Dietrick *et al.*, 1959; Dondale *et al.*, 1971). Also, adult beetles and late instar larvae from grain samples may be extracted using this approach (Smith, 1977). It is probable that it would be appropriate for many foliage insects that are moderately mobile: special attention would need to be given to the design of the funnel if they were strongly phototactic. Harris (1971) found the method fully efficient compared with hand-sorting, for the larvae of the pear psyllid, *Psylla pyricola*, provided that the heat source was not too powerful. Flotation methods (see Chapter 5) may also be useful for the material from a suction sampler (Dondale *et al.*, 1971). With animals that rest in litter, such as weevils and leaf-beetles, it may be possible (given a plentiful supply of water), to float the animals off the sample in the field. Wood & Small (1970) did this by enclosing the area in a solid metal frame and floating off the beetles and litter, through an overflow, on to a fine sieve.

4.4.4 *Methods for animals in plant tissues*

Although leaf-miners may be directly counted (p. 343), it is seldom possible to estimate eggs imbedded in plant tissue or larval or adult insects boring into the stems of herbaceous plants or trees so simply. The main impetus for the development of techniques for such animals has been provided by work on insect pests of stored grain and timber. Doubtless, considerable advances will be made when the approaches developed in this field are applied and extended with insects in the tissues of herbaceous plants.

4.4.4.0.8 Dissection

This method is widely applicable – if tedious. Sometimes it is possible to limit the dissection of stems, fruits or other parts to those that are obviously damaged (see Chapter 8).

4.4.4.0.9 Bleaching and/or selective staining

The plant tissue may be rendered transparent by treatment in 'Eau de Javelle', lactophenol (Koura, 1958; Carlson & Hibbs, 1962), or 10% sodium hydroxide solution (Apt, 1950) so that the insects become visible; this method has been used with grains and lentils. Plant parasitic nematodes are assessed by simultaneously bleaching the plant tissues and staining the worms in a mixture of lactophenol and cotton blue (Goodey, 1957). The eggs of various Hemiptera in potato leaves can be detected by bleaching the leaves (boiled in water until limp and then in 95% alcohol over a water bath) and then staining in a saturated solution of methyl red and differentiating in a slightly alkaline solution (containing sodium hydroxide) of the same dye; the leaf tissues become orange or yellow, but the eggs remain bright red and may be counted using transmitted light (Curtis, 1942). However, Carlson & Hibbs (1962) found that the same eggs could be counted after merely boiling for 1 min in lactophenol, when the leaf tissues become bleached and egg proteins coagulated. Insect egg plugs in grains have been found to be stained selectively by gentian violet (when moist) (Goosens, 1949) or by various alkaloids, derberine sulphate, chelidonium and primuline that fluoresce yellow, orange and light blue under ultraviolet light (Milner *et al.*, 1950). If the grain is crushed the insect protein may be recognised by the ninhydrin test (Dennis & Decker, 1962).

4.4.4.0.10 X-rays and CT scanners

In 1924, Yaghi suggested that X-rays could be used in the detection of wood-boring insects, but it is only since the 1950s that the principle has been extensively applied, more particularly to insects in homogeneously textured materials: in grain (Milner *et al.*, 1950, 1953; Pedersen & Brown, 1960) and other seeds (Simak, 1955). It is also of wide applicability to the assessment of insects in moist plant tissues: for wood-borers (Berryman & Stark, 1962; Bletchly & Baldwin, 1962; Amman & Rasmussen, 1969), stem-borers (Goodhue & van Emden, 1960) and bollworms (Graham *et al.*, 1964). The cavity made by the larva or adult is much more easily detected in the radiographs than the difference between animal and plant tissues. Where one has to rely on the latter distinction, as with the blackcurrant gall mite, *Cecidophytopsis ribis*, then the value of the method may be limited; long exposures to maximise the distinction will lead to blurred images (Smith, 1960). It has generally been found that 'soft' X-rays (5-35 kV) are best for this work; the thicker the plant material the greater the voltage required for the same exposure time or the greater the exposure time, which needs to be increased as the water content of the plant tissue rises (Amman & Rasmussen, 1969). It has been shown by De Mars (1963) that X-ray detection of a bark beetle, *Dendroctonus*, is virtually as efficient as dissection, eight times faster and less than one-quarter the cost, even when allowance is made for materials. Wickman (1964) reached similar conclusions for its use with Siricidae. The interpretation of X-rays of bark insects is discussed by Berryman (1964). Appropriate safety procedures should be followed when using X-rays. Medical scanners, ultrasound and nuclear magnetic resonance could be used in certain circumstances, but their costs generally rule them out. Recently, Tarver *et al.* (2006) used a micro-computed tomography scanner to observe cowpea seeds infested with cowpea weevils, *Callosobruchus maculates*; similarly, Jenning & Austin (2011) studied non-destructively the tunnels of larval xiphydriid woodwasps.

4.4.4.0.11   *Methods based on the different mass of the whole and the infested material*

Where the plant material is of fairly uniform weight (e.g. grain), advantage may be taken of the fact that material in which insects have, or are, feeding will be lighter. The lighter grain can be made to float (e.g. in 2% ferric nitrate solution; Apt, 1952) or to fall more rapidly when projected (Katz, Farrell & Milner, 1954). Such an approach naturally leads on to flotation techniques that are discussed in Chapter 6.

4.4.4.0.12   *Acoustic detection*

Insects living in plant material often make sounds, normally from their feeding. The aural detection of this sound, particularly in grain-infesting species was first proposed by Adams *et al.* (1953), while Street (1971) used a modified gramophone cartridge transducer. Recent advances have allowed the construction of electronic acoustic detection devices capable of identifying the acoustic signature of feeding or some other activity (Mankin *et al.*, 2011). For example, Mankin *et al.* (2002) used a portable low-frequency acoustic system to detect termite infestations in urban trees. Acoustic detection is particularly useful for the detection of insect infestations in stored products (e.g. Hagstrum *et al.*, 1996 for insects in wheat).

4.4.5   *Special sampling problems with animals in plant material*

4.4.5.1   The marking of turf samples

Van Emden (1963) has described a method whereby small areas of turf may be marked and subsequently found and recognised, without impairing mowing or grazing or attracting the attention of vandals. Wire rings, marked with a colour code, were sunk into the soil around the samples and here they could remain for weeks or even years. Their approximate positions were recorded by reference to a grid and a diagonal tape, and their condition or other relevant ecological information noted. After the desired period, the rings may be rediscovered and the ecological changes in the precise area observed, by plotting the approximate position with reference to the grid and finding the actual ring with a metal detector.

4.4.5.2 The sampling of bulk grain

When studying insect populations in stored grain, it is often desired to draw samples from known depths ensuring that they are not contaminated by insects or grain from the outer layers. Burges (1960) has developed a method for doing this: a hollow spear with a lateral aperture is forced into the required position, after which the desired sample is sucked out through its shaft by a domestic vacuum cleaner. Other types of aspirator for sampling grain have been described by Chao & Peterson (1952) and Ristich & Lockard (1953).

4.4.5.3 The sampling of bark

A circular punch that cuts 0.1 ft^2 (92.9 cm^2) samples from bark was devised by Furniss (1962). The punch is made from a segment of a 4 in (10 cm)-diameter steel pipe fitted with a central handle so that it can be rapidly and symmetrically hammered into the bark. Very thick bark may need to have the outer layer shaved off first if Scolytidae are being sampled and the phloem layer is required; but not, of course, if the crevice fauna is being studied (Lewis & Navas, 1962). There is also a problem of estimating the actual area sampled that will vary with the curvature of the trunk or bough (Dudley, 1971, p. 21).

4.5 Sampling from vertebrate hosts

There are two important variables to be considered when attempting to obtain complete samples of vertebrate ectoparasites from their hosts. First, the readiness or otherwise with which they will leave their hosts; second, whether the host can be killed or not. Furthermore, if the host is killed, can the skin be destroyed whilst obtaining the parasites or must it be retained in good condition? Sampling from living animals will generally be more likely to provide absolute population counts of those parasites that are readily dislodged from the host (e.g. fleas, hippoboscids) than for those that are more firmly attached (e.g. lice, mites); to estimate these the host will often need to be killed and even the skin destroyed. After collection each host should be kept isolated by being placed into individually sealed bags or containers to avoid cross-infection which can rapidly occur. Clayton & Walter (1997) provide a detailed review of the application of methods to birds.

4.5.1 Sampling from living hosts

Care must be taken in handling the host, for Stark & Kinney (1962) found that many fleas would rapidly leave a struggling or agitated host.

4.5.1.1 Searching

This method is usually satisfactory only with relatively large parasites (e.g. ticks), active ones (e.g. hippoboscids; Ash, 1952) or those that sit on exposed and hairless parts of the body. In the rare cases where the host is almost completely hairless, minute parasites can be counted using this method (e.g. on human, Johnson & Mellanby, 1942). Nelson *et al.* (1957) compared various relative sampling techniques for the sheep ked, *Melophagus ovinus*, with the 'total live counts' (MacLeod, 1948) and with the 'picked-off' count; in the latter method every ked that can be found on the sheep is picked off. It was shown that the 'total live count', in which the living keds are counted as the fleece is parted, underestimated the total population; the actual mean estimates for different breeds of sheep ranged from 61% to 81% of the 'picked-off' count. This experiment emphasises the importance of confirming that apparently absolute methods are really giving estimates of the total population. Ash (1960) sampled lice from living birds by removing a proportion of the feathers from the area frequented; the lice on these were determined and the total feathers in the area counted. Clayton and Walther (1997) considered visual examination of birds to be suitable for some ticks and feather mites, but fumigation (p. 167) yielded a higher proportion of the parasite load (Walther & Clayton, 1997). Buxton (1947) determined the head louse populations of men by shaving the scalp and dissolving the hair. Clayton and Drown (2001) found that only about 9% of Rock Pigeon lice were observed by visual examination. However, the total was a good predictor of the total parasite load.

Koop & Clayton (2013) compared the efficiency of visual searching and dust-ruffling with Z3 Flea and Tick powder to quantify the abundance of the louse, *Brueelia nebulosa*, on starling, *Sturnus vulgaris*. The birds were then killed and the total louse count found by washing. Visual examination and dust-ruffling each accounted for a relatively small proportion of total lice present (14% and 16%, respectively), but the authors concluded that both methods were significant predictors of abundance and can be used to reliably quantify the relative abundance of lice on European Starlings and other similar-sized passerines.

4.5.1.2 Combing

A fine-toothed comb may be used to remove the ectoparasites on living mammals, though with lice only a small part of the population is removed (see below). It is often an advantage if the host is anaesthetised with ether (Mosolov, 1959), but this also affects the parasites and is strictly speaking a combination of combing and fumigation (see below).

4.5.1.3 Fumigation

This technique is particularly successful for Hippoboscidae and fleas (Siphonaptera) and is used for obtaining these parasites from birds at ringing stations and elsewhere during ecological studies on vertebrates. The hosts are collected from a trap and immediately placed in a white cloth (linen) bag for transportation to the laboratory. Here, they may be exposed to an anaesthetic such as ether (Mosolov, 1959; Janion, 1960), chloroform (Williamson, 1954), ethyl acetate or carbon dioxide that will dislodge the parasites. Janion merely sprinkled the bags containing the hosts (mice in his work) with ether and after a few seconds of gentle squeezing the mouse was released, the fleas remaining in the bag; Mosolov removed the rodents from the bag and shook and

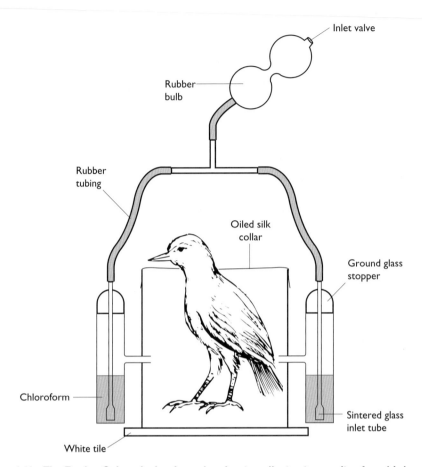

Figure 4.11   The Fowler–Cohen design for a chamber to collect ectoparasites from birds.

combed them with forceps and obtained a wide range of parasites. He points out that as larval ectoparasites are often pale and the adults dark they are most readily seen if they are shaken on to coloured paper. The 'Fair Isle apparatus' described by Williamson (1954) was designed to limit exposure of the host to the anaesthetic. Basically, the bird is placed in a jar from which its head protrudes and into which the anaesthetic can be introduced. Fowler & Cohen (1983) give an improved design (Fig. 4.11) in which a rubber diaphragm placed in the lid of the chamber has a slit cut into it through which the head of the bird can be pushed. The body of the bird within the chamber is then exposed to the anaesthetic. Bear (1995) described further improvements to the apparatus, and Walther & Clayton (1996) noted that a higher recovery rate was achieved by ruffling the feathers when the bird was removed. This apparatus removed up to 80% of ectoparasites (Fowler & Cohen, 1983; Poiani, 1992), but it does not sample the head!

A simple and effective method for birds is to dust with insecticidal power and then ruffle the feathers over a coloured surface (Clayton & Walther, 1997; Koop & Clayton, 2013). Malcomson (1960) dusted the plumage with pyrethrum powder and allowed the bird to flutter for about 5 min under an inverted paper cone. Pyrethrum produces a rapid knockdown and has no side effects on birds or mammals (Casida, 1973; Jackson, 1985). However, it is not 100% effective and superior results may be obtained using commercial formulations sold for treating pets that combine pyrethrin with a pyrethrin derivative and piperonyl butoxide. These are sold as dust and in aerosol form, and Clayton & Walther (1997) found the latter to work better on birds. The silicon aerogel Dri-Die 67, which kills arthropods by desiccation, was considered by Clayton & Walther (1997) to be 100% effective within 3 h but is long-lasting and can remove oil from feathers or fur, resulting in the host suffering from exposure during wet weather. Chlordimeform causes ticks to detach from their hosts (Gladney *et al.*, 1974). Kalamarz (1963) found that lice were efficiently removed from the host after fumigation if it was vacuum-cleaned with a special brush nozzle.

4.5.1.4 Trapping

Ectoparasites which leave the host after feeding may be trapped. Bird ticks have been sampled by placing the host in a cage over a pan of water into which the parasites fall after feeding (Krantz, 1978; Sonenshine, 1993). Castro (1973) used a funnel trap to collect mites emerging from the feathers of house sparrows.

4.5.2 *Sampling from dead hosts*

4.5.2.1 Searching and combing

These methods, when combined with brushing in all directions with a stiff brush, will provide many specimens from mammal pelts (Spencer, 1956), but often this is really only a relative method. With birds, however, Ash (1960) considered that a feather-by-feather search was the only efficient method for lice. Janzen (1963) used a novel principle that may be of wider application, to obtain almost absolute samples of the beaver beetle, *Platypsyllus castoris*. The host's pelts were first frozen and then brought into a warm environment, the beetles would move away from the chilled pelts and the majority were found by searching; most of the remainder were extracted by combing.

4.5.2.2 Fumigation

If the mammals are trapped live, but can be sacrificed, fumigation with hydrogen cyanide (produced from calcium cyanide) for about 15 min, followed by combing, will yield most parasites (Ellis, 1955; Murray, 1957). Bird ectoparasites been sampled by fumigation with methyl bromide (Harshbarger & Raffensperger, 1959) and ethyl actetate (Clayton *et al.*, 1992), followed by ruffling the feathers.

4.5.2.3 Dissolving

The pelt of the host is dissolved, most conveniently by incubating with the proteolytic enzyme trypsin for two days, followed by boiling in 10% caustic potash for some minutes (Cook, 1954; Clayton & Walther, 1997). After such treatment, virtually only the ectoparasites remain and even minute Listrophorid mites can be recovered. This method is generally associated with Hopkins (1949), who records the following comparison with searching and beating; the latter method had yielded 31 lice (larvae and adults) from three pelts; when these were dissolved, a further 1208 specimens were recovered. Although this method is very useful for mammal ectoparasites, differing results have been reported for bird ectoparasites, Ash (1960) found it unsatisfactory, but Clayton *et al.* (1992) obtained recovery rates for adult lice of between 91% and 100%.

4.5.2.4   Clearing

The pelt is shaved and the hair placed in lysol or some other cleaning medium when the lice and their eggs may be counted under a microscope (Murray, 1957). If the exact distribution of the parasites is to be determined, Murray recommends stunning the animals, soaking their coats with ether or chloroform to kill all lice *in situ*, and then placing the animal in a closed jar until dead. The hair is then shaved from each area (these may be delimited by a grid) and mounted in Berlese's mounting medium.

4.5.2.5   Washing

Large numbers of ectoparasites, especially mites, can be removed by washing the pelt or the animal in a solution (<5%) of detergent (Lipovsky, 1951). Optimal results are obtained by placing the host body in a container with 1–2% detergent solution and using a mechanical shaker such as a paint mixer. Henry & McKeewer (1971) found this method when applied to rats to be >90% efficient for mites, fleas and lice, but only 66% efficient for ticks, while McGroarty & Dobson (1974) found efficiencies of >95% for lice and >85% for feather mites from the bodies of house sparrows.

4.5.3   *Sampling from vertebrate 'homes'*

Birds' and rodents' nests, bat roosts and other vertebrate 'homes' may usually be treated as litter and their fauna extracted using modified Berlese funnels (Sealander & Hoffman, 1956; Levings & Windsor, 1985; Gwiazdowicz et al., 2012; Wolfs et al., 2012; see also other methods described in Chapter 6). However, Woodroffe (1953) concluded that there was no absolute method, and found a combination of warming and sieving most efficient. Drummond (1957) and Wasylik (1963) each collected continuous samples of the mites in mammal and avian nests by placing funnels or gauze-covered tubes below the nests; such traps basically resemble pitfall traps (p. 287) and provide relative estimates. Fumigation has also been used. Brown & Brown (2015) sampled swallow, *Petrochelidon pyrrhonota*, nests for ectoparastic swallow bugs, *Oeciacus vicarious*, by lightly misting the outside of cliff swallow nests and adjacent nesting substrate with a dilute solution of the insecticide Dibrom.

References

Adams, R.E., Wolfe, J.E., Milner, M., & Shellenberger, J.A. (1953) Aural detection of grain infested internally with insects. *Science* **118**, 163–4.

Adis, J., Paarmann, W., De Fonseca, C.R.V., & Rafael, J.A. (1997) Knockdown efficiency of natural pyrethrum and survival rate of living arthropods obtained by canopy fogging in central Amazonia. In: Stork, N.E., Adis, J., & Didham, R.K. (eds), *Canopy Arthropods*. Chapman & Hall, London, pp. 67–84.

Allison, D. & Pike, K.S. (1988) An inexpensive suction trap and its use in an aphid monitoring network. *J. Agric. Entomol.* **5**(2), 103–8.

Amman, G.D. & Rasmussen, L.A. (1969) Techniques for radiographing and the accuracy of the X-ray method for identifying and estimating numbers of the mountain pine beetle. *J. Econ. Entomol.* **62**, 631–4.

Andrzejewska, L. (1965) Stratification and its dynamics in meadow communities of Auchenorhyncha (Homoptera). *Ekologia Polska, Seria A* **13**, 685–715.

Apt, A.C. (1950) A method for detecting hidden infestation in wheat. *Milling Prod.* **15**(5), 1.

Apt, A.C. (1952) A rapid method of examining wheat samples for infestation. *Milling Prod.* **17**(5), 4.

Arnold, A.J., Needham, P.H., & Stevenson, J.H. (1973) A self-powered portable insect suction sampler and its use to assess the effects of azinphos methyl and endosulfan on blossom beetle populations on oil seed rape. *Ann. Appl. Biol.* **75**, 229–33.

Arthur, A.P. (1962) A skipper, *Thymelicus lineola* (Ochs.) (Lepidoptera: Hesperiidae) and its parasites in Ontario. *Can. Entomol.* **94**, 1082–9.

Ash, J.S. (1952) Records of Hippoboscidae (Dipt.) From Berkshire and Co. Durham in 1950, with notes on their bionomics. *Entomol. Monthly Mag.* **88**, 25–30.

Ash, J.S. (1960) A study of the Mallophaga of birds with particular reference to their Ecology. *Ibis* **102**, 93–110.

Austin, M.D. & Massee, A. M. (1947) Investigations on the control of the fruit tree red spider mite (*Metatetranychus ulmi* Koch) during the dormant season. *J. Hort. Sci.* **239**, 227–53.

Baker, G.H. & Hawke, B.G. (1990) Life-history and population-dynamics of *Theba pisana* (Mollusca, Helicidae) in a cereal pasture rotation. *J. Appl. Ecol.* **27**, 16–29.

Baker, G.H. & Vogelzang, B.K. (1988) Life history, population-dynamics and polymorphism of *Theba pisana* (Mollusca, Helicidae) in Australia. *J. Appl. Ecol.* **25**, 867–87.

Balogh, I. & Loksa, I. (1956) Untersuchungen ilber die Zoozonose des Luzernenfeldes. Strukturz6nologische Abhandlung. *Acta Zool. Hung.* **2**, 17–114.

Banks, C.I. & Brown, F.S. (1962) A comparison of methods of estimating population density of adult Sunn Pest, *Eurygaster integriceps* Put. (Hemiptera, Scutelleridae) in wheat fields. *Entomologia Exp. Appl.* **5**, 255–60.

Banks, C.J. (1954) A method for estimating populations and counting large numbers of *Aphis fabae* Scop. *Bull. Entomol. Res.* **45**, 751–6.

Bear, A. (1995) An improved method for collecting bird ectoparasites. *J. Field Ornithol.* **66**, 212–14.

Basset, Y., Springate, N.D., Aberlenc, H.P., & Delvare, G. (1997) A review of methods for sampling arthropods in tree canopies. *Canopy Arthropods* **35**, 27–52.

Belding, M.J., Isard, S.A., Hewings, A.D., & Irwin, M.E. (1991) Photovoltaic-powered suction trap for weakly flying insects. *J. Econ. Entomol.* **84**(1), 306–10.

Berryman, A.A. (1964) Identification of insect inclusions in X-rays of Ponderosa pine bark infested by western pine beetle, *Dentroctonus brevicomis* Le Conte. *Can. Entomol.* **96**, 883–8.

Berryman, A.A. & Stark, R.W. (1962) Radiography in forest entomology. *Ann Entomol. Soc. Am.* **55**, 456–66.

Bletchley, J.D. & Baldwin, W.J. (1962) Use of X rays in studies of wood boring insects. *Wood* **27**, 485–8.

Brook, A.J., Woodcock, B.A., Sinka, M., & Vanbergen, A.J. (2008) Experimental verification of suction sampler capture efficiency in grasslands of differing vegetation height and structure. *J. Appl. Ecol.* **45**(5), 1357–63.

Brown, C.R. & Brown, M.B. (2015) Ectoparasitism shortens the breeding season in a colonial bird. *Roy. Soc. Open Sci.* **2**, 140508.

Browning, T.O. (1959) The long-tailed mealybug *Pseudococcus adonidium* L. in South Australia. *Aust. J. Agric. Res.* **10**, 322–39.

Bullock, J.A. (1963) Extraction of Thysanoptera from samples of foliage. *J. Econ. Entomol.* **569**, 612–14.

Burges, H.D. (1960) A spear for sampling bulk grain by suction. *Bull. Entomol. Res.* **51**, 1–5.

Buxton, P.A. (1947) *The Louse.* 2nd edn. Arnold, London, pp. 164.

Cannon, R.J.C. (1986) Summer populations of the cereal aphid *Metopolophium dirhodum* (Walker) on winter-wheat. 3. Contrasting years. *J. Appl. Ecol.* **23**, 101–14.

Carlson, O.V. & Hibbs, E.T. (1962) Direct counts of potato leafhopper, *Empoasca. fabae*, eggs in *Solanum* leaves. *Ann. Entomol. Soc. Am.* **55**, 512–15.

Chamberlin, I.C. (1940) A mechanical trap for the sampling of aerial insect Populations. *USDA Bur. Entomol. Pl. Quar. ET* **163**, 12.

Chant, D.A. (1962) A brushing method for collecting mites and small insects from leaves. *Prog. Soil Zool.* **1**, 222–5.

Chant, D.A. & Muir, R.C. (1955) A comparison of the imprint and brushing machine methods for estimating the numbers of the fruit tree red spider mite, *Metaletranychus ulmi* (Koch), on apple leaves. *Rep. E. Malling Res. Sta. (A)* **19549**, 141–5.

Chao, Y. & Peterson. A. (1952) A new type of aspirator. *J. Econ. Entomol.* **45**, 751.

Cherrill, A.J. & Brown, V.K. (1990) The habitat requirements of adults of the wart-biter *Decticus verrucivorus* (L) (Orthoptera, Tettigoniidae) in southern England. *Biol. Conserv.* **53**, 145–57.

Cherry, R.H., Wood, K.A., & Ruesink, W.G. (1977) Emergence trap and sweep net sampling for adults of the potato leafhopper from alfalfa. *J. Econ. Entomol.* **70**, 279.

Chey, V.K., Holloway, J.D., Hambler, C., & Speight, M.R. (1998) Canopy knockdown of arthropods in exotic plantations and natural forest in Sabah, north-east Borneo, using insecticidal mist-blowing. *Bull. Entomol. Res.* **88**, 15–24.

Chinaev, P.P. (1959) Methods in quantitative sampling of bloodsucking mosquitoes (Diptera, Culicidae). [In Russian.] *Entomol. Obozr.* **38**, 757–65. (Transl. *Entomol. Rev.* **38**, 679–86.)

Casida, J.E. (1973) *Pyrethrum, the Natural Insecticide.* Academic Press, New York.

Castro, S.D. (1973) A method for collection of the quill mites, *Syringophiloides minor* (Berlese) (Prostigmata: Syringophilidae), from living birds. *J. Med. Entomol.* **10**, 524.

Clayton, D.H. & Drown, D.M. (2001) Critical evaluation of five methods for quantifying chewing lice (Insecta: Phthiraptera). *J. Parasitol.* **87**, 1291–300.

Clayton, D.H., Gregory, R.D., & Price, R.D. (1992) Comparative ecology of neotropical bird lice (Insecta, Phthiraptera). *J. Animal Ecol.* **61**, 781–95.

Clayton, D.H. & Walther, B.A. (1997) Collection and quantification of arthropod parasites of birds. In: Clayton, D.H. & Moore, A.J. (eds), *Host–Parasite Evolution: General Principles and Avian Models.* Oxford University Press, Oxford, UK, pp. 419–40.

Coineau, Y. (1962) Nouvelles methodes de prospection de la faune entomologique des plantes herbacées et ligneuses. *Bull. Soc. Entomol. Fr.* **67**, 115–19.

Collyer, E. (1951) A method for the estimation of insect populations of fruit trees. *Rep. E. Malling Res. Sta.* **1949–50**, 148–51.

Condrashoff, S.F. (1967) An extraction method for rapid counts of insect eggs and small organisms. *Can. Entomol.* **99**, 300–3.

Cone, W.W. (1963) The black vine weevil, *Brachyrhinus sulcatus*, as a pest of grapes in South Central Washington. *J. Econ. Entomol.* **56**, 677–80.

Connell, W.A. (1959) Estimating the abundance of corn ear worm eggs. *J. Econ. Entomol.* **52**, 747–9.

Cook, E.F. (1954) A modification of Hopkins' technique for collecting ectoparasites from mammal skins. *Entomol. News* **15**, 35–7.

Corbet, P.S. (1966) A method for sub-sampling insect collections that vary widely in size. *Mosquito News* **26**, 420–4.

Curtis, W.E. (1942) A method of locating insect eggs in plant tissues. *J. Econ. Entomol.* **35**, 286.

Day, K.R. & Crute, S. (1990) The abundance of spruce aphid under the influence of an oceanic climate. *Pop. Dyn. Forest Insects*, 25–33.

Dean, D.A. & Sterling, W.L. (1985) Size and phenology of ballooning spiders at 2 locations in eastern Texas (USA). *J. Arachnol.* **13**(1), 111–20.

Dean, D.A. & Sterling, W.L. (1990) Seasonal patterns of spiders captured in suction traps in eastern Texas (USA). *Southwest. Entomol.* **15**(4), 399–412.

De Mars, C.J. (1963) A comparison of radiograph analysis and bark dissection in estimating numbers of western pine beetle. *Can. Entomol.* **95**, 1112–16.

Dennis. N.M. & Decker, R.W. (1962) A method and machine for detecting living internal insect infestation in wheat. *J. Econ. Entomol.* **55**, 199–203.

Dewar, A.M. & Dean, G.J. (1982) Assessment of methods for estimating the numbers of aphids (Hemiptera: Aphididae) in cereals. *Bull. Entomol. Res.* **72**, 672.

Dietrick, E.J., Schlinger, E.I., & Bosch, R.V.D. (1959) A new method for sampling arthropods using a suction collecting machine and modified Berlese funnel separator. *J. Econ. Entomol.* **52**, 1085–91.

Doğramaci, M., DeBano, S.J., Kimoto, C., & Wooster, D.E. (2011) A backpack-mounted suction apparatus for collecting arthropods from various habitats and vegetation. *Entomol. Exp. Appl.* **139**(1), 86–90.

Dondale, C.D., Nicholls, C.F., Redner, J.H., Semple, R.B., & Turnbull, A.L. (1971) An improved Berlese-Tullgren funnel and a flotation separator for extracting grassland arthropods. *Can. Entomol.* **103**, 1549–52.

Doxon, E.D., Davis, C.A., & Fuhlendorf, S.D. (2011) Comparison of two methods for sampling invertebrates: vacuum and sweep-net sampling. *J. Field Ornithol.* **82**, 60–7.

Drummond, R.O. (1957) Observations on fluctuations of acarine populations from nests of *Peromyscus leucopus. Ecol. Monogr.* **27**, 137–52.

Dudley, C.O. (1971) A sampling design for the egg and first instar larval populations of the western pine beetle, *Dendroctonus brevicomis* (Coleoptera: Scolytidae). *Can. Entomol.* **103**, 1291–313.

Duffey, E. (1980) The efficiency of the Dietrick vacuum sampler (D-vac) for invertebrate population studies in different types of grassland. *Bull. Ecol.* **11**, 421–31.

Ellis, L.L. (1955) A survey of the ectoparasites of certain mammals in Oklahoma. *Ecology* **36**, 12–18.

Emden, H.F. van (1963) A technique for the marking and recovery of turf samples in stem borer investigations. *Entomol. Exp. Appl.* **6**, 194–8.

Erwin, T.L. (1983) Beetles and other insects of tropical rainforest canopies at Manaus, Brazil, sampled by insecticidal fogging. In: Sutton, S.L., Whitmore, T.C., & Chadwick, A.C. (eds), *Tropical Rain Forest Ecology and Management*. Blackwell, Oxford, pp. 59–75.

Erwin, T.L. (1989) Canopy arthropod biodiversity: a chronology of sampling techniques and results. *Rev. Peruana de Entomol.* **32**, 71–7.

Fowler, J.A. & Cohen, S. (1983) A method for the quantitative collection of ectoparasites from birds. *Ringing and Migration* **4**, 185–9.

Furniss, M.M. (1962) A circular punch for cutting samples of bark infested with beetles. *Can. Entomol.* **94**, 959–63.

Gara, R.I. & Vité, J.P. (1962) Studies on the flight patterns of bark beetles (Coleoptera: Scolytidae) in second growth Ponderosa pine forests. *Contrib. Boyce Thompson Inst.* **21**, 275–90.

Gibbs, D.G., Pickett, A.D., & Leston, D. (1968) Seasonal population changes in cocoa capsids (Hemiptera, Miridae) in Ghana. *Bull. Entomol. Res.* **58**, 279–93.

Gibson, R.W. (1975) Measurement of eriophyid mite populations on ryegrass using ultrasonic radiation. *Trans. R. Entomol. Soc. Lond.* **127**, 31–2.

Gladney, W.J., Ernst, S.E., & Drummond, R.O. (1974) Chlordimeform: a detachment-stimulating chemical for three-host ticks. *J. Med. Entomol.* **11**, 569–72.

Goodenough, J.L., Jank, P.C., Carroll, L.E., Sterling, W.L., Redman, E.J., & Witz, J.A. (1983) Collecting and preserving airborne arthropods in liquid at timed intervals with a Johnson–Taylor-type suction trap. *J. Econ. Entomol.* **76**(4), 960–3.

Goodey, J.B. (1957) *Laboratory Methods for Work with Plant and Soil Nematodes*, 3rd edn. Technical Bulletin 2, Ministry of Agriculture, HMSO, London.

Goosens, H.J. (1949) A method for staining insect egg plugs in wheat. *Cereal Chem.* **26**(5), 419–20.

Graham, H.M., Robertson, O.T., & Martin, D.F. (1964) Radiographic detection of pink bollworm larvae in cottonseed. *J. Econ. Entomol.* **57**, 419–20.

Gray, K.W. & Schuh, J. (1941) A method and contrivance for sampling pea aphid Populations. *J. Econ. Entomol.* **34**, 411–15.

Greig-Smith, P. (1964) *Quantitative Plant Ecology*. Butterworths, London.

Gwiazdowicz, D.J., Coulson, S.J., Grytnes, J.-A., & Pilskog, H.E. (2012) The bird ectoparasite *Dermanyssus hirundinis* (Acari, Mesostigmata) in the High Arctic; a new parasitic mite to Spitsbergen, Svalbard. *Acta Parasitol.* **57**(4), 378–84.

Hagstrum, D.W., Flinn, P.W., & Shuman, D. (1996) Automated monitoring using acoustical sensors for insects in farm-stored wheat. *J. Econ. Entomol.* **89**(1), 211–17.

Hand, S.C. (1986) The capture efficiency of the Dietrick vacuum insect net for aphids on grasses and cereals. *Ann. Appl. Biol.* **108**, 233.

Harper, C.A. & Guynn, D.C., Jr (1998) A terrestrial vacuum sampler for macroinvertebrates. *Wildlife Soc. Bull.* **26**, 302–6.

Harris, J.W.E., Collis, D.G., & Magar, K.M. (1972) Evaluation of the tree-beating method for sampling defoliating forest insects. *Can. Entomol.* **104**, 723–9.

Harris, M. (1971) Sampling pear foliage for nymphs of the pear Psylla, using the Berlese–Tullgren funnel. *J. Econ. Entomol.* **64**, 1317–18.

Harris, P. (1960) Number of *Rhyacionia buoliana* per pine shoot as a population index, with a rapid determination method of this index at low population levels. *Can. J. Zool.* **38**, 475–8.

Harshbarger, J.C. & Raffensperger, E.M. (1959) A method for collecting and counting populations of the shaft louse. *J. Econ. Entomol.* **52**, 1215–16.

Havelka, J., Žurovcová, M., Rychlý, S., & Starý, P. (2014) Russian Wheat Aphid, *Diuraphis noxia* in the Czech Republic – cause of the significant population decrease. *J. Appl. Entomol.* **138**, 273–80.

Heikinheimo, O. & Raatikainen, M. (1962) Comparison of suction and netting methods in population investigations concerning the fauna of grass leys and cereal fields, particularly in those concerning the leafhopper, *Calligypona pellucida (F.). Valt. Maatalousk. Julk. Helsingfors.* **191**, 31 pp.

Helson, G.A.H. (1958) Aphid populations: ecology and methods of sampling aphids Myzus-persicae (Sulz.) and Aulacorthum solani (Kltb). *N. Z. Entomologist* **2**, 20–3.

Henderson, C.F. & McBurnie, H.V. (1943) Sampling technique for determining populations of the citrus red mite and its predators. *USDA Circ.* **671**, 11 pp.

Henderson, I.F. & Whitaker, T.M. (1976) The efficiency of an insect sampler in grassland. *Ecol. Entomol.* **2**, 57–60.

Henry, L.G. & McKeever, S. (1971) A modification of the washing technique for quantitative evaluation of the ectoparasite load of small mammals. *J. Med. Entomol.* **8**, 504–5.

Hobbs, S.E. & Hodges, G. (1993) An optical method for automatic classification and recording of a suction trap catch. *Bull. Entomol. Res.* **83**(1), 47–51.

Hoerner, J.L. (1947) A separator for onion thrips. *J. Econ. Entomol.* **40**, 755.

Hopkins, G. (1949) The host associations of the lice of mammals. *Proc. Zool. Soc. Lond.* **119**, 387–604.

Hower, A.A. & Ferguson, W. (1972) A square-foot device for use in vacuum sampling alfalfa insects. *J. Econ. Entomol.* **65**, 1742–3.

Hughes, R.D. (1963) Population dynamics of the cabbage aphid *Brevicoryne brassicae* (L.). *J. Anim. Ecol.* **32**, 393–424.

Ives, W.G.H. (1959) A technique for estimating tamarack foliage production, a basis for detailed population studies of the larch sawfly. *Can. Entomol.* **91**, 513–19.

Jackson, J.A. (1985) On the control of parasites in nest boxes and the use of pesticides near birds. *Sialia* **7**, 17–25.

Jacobson, M. (1966) Chemical insect attractants and repellents. *Annu. Rev. Entomol.* **11**, 403–22.

Janion, S.M. (1960) Quantitative dynamics in fleas (Aphaniptera) infesting mice of Puszcza Kampinoska Forest. *Bull. Acad. Pol. Sci. II,* **8**(5), 213–18.

Janzen, D.H. (1963) Observations on populations of adult beaver-beetles, *Platypsyllus castoris* (Platypsyilidae: Coieoptera). *Pan. Pacif. Entomol.* **32**, 215–28.

Jennings, J.T. & Austin, A.D. (2011) Novel use of a micro-computed tomography scanner to trace larvae of wood boring insects. *Aust. J. Entomol.* **50**(2), 160–3.

Johnson, C.G. (1950) A suction trap for small airborne insects which automatically segregates the catch into successive hourly samples. *Ann. Appl. Biol.* **37**, 80–91.

Johnson, C.G. (1957) The distribution of insects in the air and the empirical relation of density to height. *J. Anim. Ecol.* **26**, 479–94.

Johnson, C.G. & Mellanby, K. (1942) The parasitology of human scabies. *Parasitology* **34**, 285–90.

Johnson, C.G., Southwood, T.R.E., & Entwistle, H.M. (1957) A new method of extracting arthropods and molluscs from grassland and herbage with a suction apparatus. *Bull. Entomol. Res.* **48**, 211–18.

Johnson, C.G. & Taylor, L.R. (1955a) The development of large suction traps for airborne insects. *Ann. Appl. Biol.* **43**, 51–61.

Johnson, C.G. & Taylor, L.R. (1955b) The measurement of insect density in the air. *Lab. Pract.* **4**, 187–92, 235–9.

Johnson, C.G., Taylor, L.R., & Southwood, T.R.E. (1962) High altitude migration of *Oscinella frit* L. (Diptera: Chioropidae). *J. Anim. Ecol.* **31**, 373–83.

Jones, B.M. (1950) A new method for studying the distribution and bionomics of trombiculid mites (Acarina: Trombidiidae). *Parasitology* **40**, 1–13.

Juillet, J.A. (1963) A comparison of four types of traps used for capturing flying insects. *Can. J. Zool.* **41**, 219–23.

Kalamarz, E. (1963) Badania nad biologia Mallophaga IV. Nowe methody zbierania ektopasozytów. *Ekol. Polska* **B 9**, 321–5.

Katz, R, Farrell, E.P., & Milner, M. (1954) The separation of grain by projection I. *Cereal Chem.* **31**, 316–25.

Kennard, W.C. & Spencer, J.L. (1955) A mechanical insect collector with high manoeuvrability. *J. Econ. Entomol.* **48**, 478–9.

Kershaw, W.J.S. (1964) Aphid sampling in sugar beet. *Plant. Path.* **13**, 101–6.

Kirk, I.W. & Bottrell, D.G. (1969) A mechanical sampler for estimating boll weevil populations. *J. Econ. Entomol.* **62**, 1250–1.

Kobayashi, F. & Murai, M. (1965) Methods for estimating the number of the cryrtomeria red mite, especially with the removal by solutions. *Res. Pop. Ecol.* **7**, 35–42.

Koop, J.A. & Clayton, D.H. (2013) Evaluation of two methods for quantifying passeriform lice. *J. Field Ornithol.* **84**(2), 210–15.

Koura, A. (1958) A new Transparency method for detecting internal infestation in grains. *Agric. Res. Rev.* **36**, 110–13.

Krantz, G.W. (1978) *A Manual of Acarology*. Oregon State University Book Stores, Corallis, USA.

Kretzschmar, G.P. (1948) Soy bean insects in Minnesota with special reference to sampling techniques. *J. Econ. Entomol.* **41**, 586–91.

La Croix, E.A.S. (1961) Observations on the ecology of the cotton-flea-beetles in the Sudan Gezira and the effect of sowing date on the level of population in cotton. *Bull. Entomol. Res.* **52**, 773–83.

Laster, M.L. & Furr, R.E. (1962) A simple technique for recovering insects from sorghum heads in insecticide tests. *J. Econ. Entomol.* **55**, 798.

Le Blanc, J.P.R., Hill, S.B., & Paradis, R.O. (1984) Oviposition in scout-apples by plum curculio, *Conotrachelus nenuphar* (Herbst)(Coleoptera: Curculionidae), and its relationship to subsequent damage. *Environ. Entomol.* **13**, 286–91.

Le Pelley, R.H. (1942) A new method of sampling thrips populations. *Bull. Entomol. Res.* **33**, 147–8.

Legner, E.F. & Oatman, E.R. (1962) Foliage-feeding Lepidoptera on young non-bearing apple trees in Wisconsin. *J. Econ. Entomol.* **55**, 552–4.

Leigh, T.F., Gonzalez, D., & Van Den Bosch, R. (1970) A sampling device for estimating absolute insect populations on cotton. *J. Econ. Entomol.* **63**, 1704–6.

Leos Martinez, J., Granovsky, T.A., Williams, H.J., Vinson, S.B., & Burkholder, W.E. (1986) Estimation of aerial density of the lesser grain borer (*Rhyzopertha dominica*) (Coleoptera: Bostrichidae) in a warehouse using dominicalure traps. *J. Econ. Entomol.* **79**(4), 1134–8.

Levings, S.C. & Windsor, D.M. (1985) Litter arthropod populations in a tropical deciduous forest: relationships between years and arthropod groups. *J. Anim. Ecol.* **54**, 61–9.

Levi, M.I., Chernov, S.G., Labunets, N.F., & Kosminskii, R.B. (1959) Aspiration method for the collection of fleas from rodents' nests. [In Russian.]. *Med. Parazitol.* **28**, 64–9.

Lewis, T. (1960) A method for collecting Thysanoptera from Gramineae. *Entomologist* **939**, 27–8.

Lewis, T. & Navas, D.E. (1962) Thysanopteran populations overwintering in hedge bottoms, grass litter and bark. *Ann. Appl. Biol.* **50**, 299–311.

Lewis, T. & Taylor, L.R. (1965) Diurnal periodicity of flight by insects. *Trans. R. Entomol. Soc. Lond.* **116**, 393–469.

Lipovsky, L.J. (1951) A washing method of ectoparasite recovery with particular reference to chiggers. *J. Kansas Entomol. Soc.* **24**, 151–6.

Locatelli, D.P. & Limonta, L. (1993) Catches of Psocoptera by a suction trap and notes for a list of Italian species. *Boll. Zool. Agrar. Bachicoltura* **25**(2), 131–1.

Locatelli, D.P. & Limonta, L. (1994) Psocoptera captured by using a suction trap in Valtellina during 1992–1993. *Boll. Zool. Agrar. Bachicoltura* **26**(2), 279–82.

Macaulay, E.D.M., Tatchell, G.M., & Taylor, L.R. (1988) The Rothamsted Insect Survey '12-meter' suction trap. *Bull. Entomol. Res.* **78**(1), 121–30.

McGroarty, D.L. & Dobson, R.C. (1974) Ectoparasite populations on house sparrows in northwestern Indiana. *Am. Mid. Natur.* **91**, 479–86.

MacLeod, J. (1948) The distribution and dynamics of ked populations, *Melophagus ovinus* Linn. *Parasitology* **39**, 61–8.

Macleod, I. & Donnelly, J. (1957) Some ecological relationships of natural populations of Calliphorine blowflies. *J. Anim. Ecol.* **26**, 135–70.

Maelzer, D.A. (1976) A photographic method and a ranking procedure for estimating numbers of the rose aphid, *Macrosiphum rosae* (L.) on rose buds. *Aust. J. Ecol.* **1**, 89–96.

Malcomson, R.O. (1960) Mallophaga from birds of North America. *Wilson Bull.* **72**, 182–97.

Mankin, R.W., Osbrink, W.L., Oi, F.M., & Anderson, J.B. (2002) Acoustic detection of termite infestations in urban trees. *J. Econ. Entomol.* **95**(5), 981–8.

Mankin, R.W., Hagstrum, D.W., Smith, M.T., Roda, A.L., & Kairo, M.T.K. (2011) Perspective and promise: a century of insect acoustic detection and monitoring. *Am. Entomol.* **57**(1), 30–44.

Maw, M.G. (1964) An effect of static electricity on captures in insect traps. *Can. Entomol.* **96**, 1482.

McCoy, J.R. & Lloyd, E.P. (1975) Evaluation of airflow systems for the collection of boll weevils from cotton. *Econ. Entomol.* **68**, 49–52.

McGovern, W.L., Leggett, J.E., Johnson, W.C., & Cross, W.H. (1973) Techniques for separating boll weevils and other small insects from samples taken with insect collecting machines. *J. Econ. Entomol.* **66**, 1332.

Milner, M., Barney, D.L., & Shellenberger, J.A. (1950) Use of selective fluorescent stains to detect egg plugs on grain kernels. *Science* **112**, 791–2.

Moore, N.W. (1964) Intra- and interspecific competition among dragonflies (Odonata). *J. Anim. Ecol.* **33**, 49–71.

Moran, V.C. & Southwood, T.R.E. (1982) The guild composition of arthropod communities in trees. *J. Anim. Ecol.* **51**, 289–306.

Morgan, C.V.G., Chant, D.A., Anderson, N.H., & Ayre, G.L. (1955) Methods for estimating orchard mite populations, especially with the mite brushing machine. *Can. Entomol.* **87**, 189–200.

Mosolov, L.P. (1959) A method of collecting the ectoparasites of rodents, without destroying the host population. [In Russian, Eng. *Summary.*] *Med. Parazitol.* **28**, 189–92.

Mueller-Dombois, D. & Ellenberg, H. (1974) *Aims and Methods of Vegetation Ecology.* J. Wiley & Sons, New York, 547 pp.

Murray, Y.F. (1957) An ecological appraisal of host–ectoparasite relationships in a zone of epizootic plague in central California. *Am. J. Trop. Med. Hyg.* **6**, 1068–86.

Nelson, W.A., Slen. S.B., & Banky, E.C. (1957) Evaluation of methods of estimating populations of the sheep ked, *Melophagus ovinus* (L.) (Diptera: Hippoboscidae), on mature ewes and young lambs. *Can. J. Anim. Sci.* **37**, 8–13.

Newell, I.M. (1947) Quantitative methods in biological and control studies of orchard mites. *J. Econ. Entomol.* **40**, 683–9.

Nicholls, C.F. (1960) A portable mechanical insect trap. *Can. Entomol.* **92**, 48–51.

Otvos, I.S & Bryant, D.G. (1972) An extraction method for rapid sampling of eastern hemlock looper eggs, *Lambdina fiscellaria* (Lepidoptera: Geometridae). *Can. Entomol.* **104**, 1511–14.

Paarmann, W. & Kerck, K. (1997) Advances in using canopy fogging technique to collect living arthropods from tree-crowns. In: Stork, N.E., Adis, J., & Didham, R.K. (eds), *Canopy Arthropods*. Chapman & Hall, London, pp. 53–66.

Pedersen, J.R. & Brown, R.A. (1960) X-ray microscope to study behaviour of internal infesting grain insects. *J. Econ. Entomol.* **53**, 678–9.

Perfect, T.J., Cook, A.G., Ferrer, E.R., Soriano, J., & Kenmore, P.E. (1989) Population sampling for planthoppers, leafhoppers (Hemiptera: Delphacidae & Cicadellidae) and their predators in flooded rice. *Bull. Entomol. Res.* **73**, 345.

Pielou, D.P. (1961) Note on a volumetric method for the determination of numbers of apple aphid, *Aphis pomi* Deg., on samples of apple foliage. *Can. J. Plant Sci.* **41**, 442–3.

Prescott, H.W. & Newton, R.C. (1963) Flight study of the clover root Curculio. *J. Econ. Entomol.* **56**, 368–70.

Richards, O.W. & Waloff, N. (1954) Studies on the biology and population dynamics of British grasshoppers. *Anti-Locust Bull.* **17**, 1–184.

Richards, O.W. & Waloff, N. (1961) A study of a natural population of *Phytodecta olivacea* (Forster) (Colcoptera, Chrysomeloidea). *Philos. Trans. Roy. Soc.* **244**, 204–57.

Richmond, C.A. & Graham, H.M. (1969) Two methods of operating a vacuum sampler to sample populations of the Cotton Fleahopper on wild hosts. *J. Econ. Entomol.* **62**, 525–6.

Ristich, S. & Lockard, D. (1953) An aspirator modified for sampling large populations. *J. Econ. Entomol.* **46**, 711–12.

Romney, V.E. (1945) The effect of physical factors upon catch of the beet leafhopper (*Eutettix tenellus* (Bak.) by a cylinder and two sweep methods. *Ecology* **26**, 135–47.

Rudd, W.G. & Jensen, R.L. (1977) Sweep net and ground cloth sampling for insects in soybeans. *J. Econ. Entomol.* **70**, 301–4.

Ru-Mei, X.U. (1985) Dynamics of within-leaf spatial distribution patterns of greenhouse whiteflies and the biological interpretations. *J. Appl. Ecol.* **22**, 63–72.

Santa, H. (1961) A method for sampling plant and leaf-hopper density in winter and early Spring. [In Japanese.]. *Plant Protection, Tokyo* **8**, 353–5.

Satchell, J.E. & Mountford, M.D. (1962) A method of assessing caterpillar populations on large forest trees, using a systemic insecticide. *Ann. Appl. Biol.* **50**, 443–50.

Schaefer, G.W., Bent, G.A., & Allsopp, K. (1985) Radar and opto-electronic measurements of the effectiveness of Rothamsted insect survey suction traps. *Bull. Entomol. Res.* **75**(4), 701–16.

Sealander, J.A. & Hoffman, C.E. (1956) A modified Berlese funnel for collecting mammalian and avian ectoparasites. *Southwest. Nat., Dallas* **1**, 134–6.

Sedivy, J. (1993) Variation in the population density of pollen beetles (*Meligethes aeneus* F.) in winter rape. *Ochrana Rostlin* **29**(1), 9–15.

Shands, W.A., Simpson, G.W., & Reed, L.B. (1954) Subunits of sample for estimating aphid abundance on potatoes. *J. Econ. Entomol.* **47**, 1024–7.

Shepard, M., Sterling, W., & Walker, J.K. (1972) Abundance of beneficial arthropods on cotton genotypes. *Environ. Entomol.* **1**, 117–21.

Simak, M. (1955) Insect damage on seeds of Norway spruce determined by X-ray photography. *Medd. Stat. Skogsforsknings-Inst., Uppsala* **41**, 299–310.

Skuhravy, V. & Novák, V. (1961) The study of field crop entomocenoses. [In Russian]. *Entomol. Obozr.* **41**, 807–14. (Transl. *Entomol. Rev.* **41**, 454–8.)

Skuhravy, V., Novák, K., & Stary, P. (1959) Entomofauna jetele (*Trifolium pratense* L.) a jeji vyvoj. *Rozpr. C sl. Akad. Ved.* **69**, 3–82.

Smith, J.W., Stadelbacher, E.A., & Gantt, C.W. (1976) A comparison of techniques for sampling beneficial arthropod populations associated with cotton. *Environ. Entomol.* **5**, 435.

Smith, B.D. (1960) Population studies of the black current gall mite (*Phytoptus ribis* Nal.). *Report, Agricultural and Horticultural Research Station Bristol* **1960**, 120–4.

Smith, L.B. (1977) Efficiency of Berlese–Tullgren funnels for removal of the rusty grain beetle, *Cryptolestes ferrugineus*, from wheat samples. *Can. Entomol.* **109**, 503–9.

Sonenshine, D.E. (1993) *Biology of Ticks*. Oxford University Press, Oxford, England.

Southwood, T.R.E., Jepson, W.F., & van Emden, H.F. (1961) Studies on the behaviour of *Oscinella frit* L. (Diptera) adults of the panicle generation. *Entomol. Exp. Appl.* **4**, 196–210.

Southwood, T.R.E., Moran, V.C., & Kennedy, C.E.J. (1982) The assessment of arboreal insect fauna: comparisons of knockdown sampling and faunal lists. *Ecol. Entomol.* **7**, 331–40.

Spencer, G.J. (1956) Some records of ectoparasites from flying squirrels. *Proc. Entomol. Soc. B. C.* **52**, 32–4.

Stage, H.H., Gjullin, C.M., & Yates, W.W. (1952) Mosquitoes of the North-western states. *USDA, Agric. Handb.* **46**, 95.

Stark, H.E. & Kinney, A.R. (1962) Abandonment of disturbed hosts by their fleas. *Pan. Pacif. Entomol.* **38**, 249–51.

Steiner, H. (1962) Methoden zur untersuchung der Populationsdynamik in Obstanlagen (ine. Musternahme und Sammeln). *Entomophaga* **7**, 207–14.

Stewart, A.J.A. & Wright, A.F. (1995) A new inexpensive suction apparatus for sampling arthropods in grassland. *Ecol. Entomol.* **20**(1), 98–102.

Street, M.W. (1971) A method for aural monitoring of in-kernel insect activity. *J. Georgia Entomol. Soc.* **6**, 72–5.

Strickland, A.H. (1954) An aphid counting grid. *Plant Path.* **3**, 73–5.

Stork, N.E. (1987) Guild structure of Bornean rainforest trees. *Ecol. Entomol.* **12**, 69–80.

Stork, N.E. & Hammond, P.M. (1997) Sampling arthropods from tree-crowns by fogging with knockdown insecticides: lessons from studies of oak tree beetle assemblages in Richmond Park (UK). In: Stock, N.E., Adis, J., & Didham, R.K. (eds), *Canopy Arthropods*. Chapman & Hall, London, 567 pp.

Stork, N.E., Adis, J., Didham, R.K. (eds) (1997) *Canopy Arthropods*. Chapman & Hall, London, 567 pp.

Summers, F.M. & Baker, G.A. (1952) A procedure for determining relative densities of brown almond mite populations on almond trees. *Hilgardia* **21**, 369–82.

Tarver, M.R., Shade, R.E., Tarver, R.D., *et al.* (2006) Use of micro-CAT scans to understand cowpea seed resistance to *Callosobruchus maculatus*. *Entomol. Exp. Appl.* **118**, 33–9.

Taylor, E.A. & Smith, F.F. (1955) Three methods for extracting thrips and other insects from rose flowers. *J. Econ. Entomol.* **48**, 767–8.

Taylor, L.R. (1951) An improved suction trap for insects. *Ann. Appl. Biol.* **38**, 582–91.

Taylor, L.R. (1955) The standardization of air flow in insect suction traps. *Ann. Appl. Biol.* **43**, 390–408.

Taylor, L.R. (1958) Aphid dispersal and diurnal periodicity. *Proc. Linn. Soc. Lond.* **1699**, 67–73.

Taylor, L.R. (1962a) The absolute efficiency of insect suction traps. *Ann. Appl. Biol.* **50**, 405–421.

Taylor, L.R. (1962b) The efficiency of cylindrical sticky insect traps and suspended nets. *Ann. Appl. Biol.* **50**, 681–5.

Taylor, L.R. (1974) Monitoring change in the distribution and abundance of insects. *Rep. Rothamsted Exp. Stn.* **1973**, 202–39.

Taylor, L.R. & Palmer, J.M.P. (1972) Aerial sampling. In: van Emden, H. (ed.), *Aphid Technology*. Academic Press, London, New York.

Topping, C.J., Sunderland, K.D., & Bewsey, J. (1992) A large improved rotary trap for sampling aerial invertebrates. *Ann. Appl. Biol.* **121**(3), 707–14.

Trumble, J.T. (1994) Sampling arthropod pests in vegetables. In: Pedigo, L.P. & Buntin, G.D. (eds), *Handbook of Sampling Methods for Arthropods in Agriculture*. CRC Press, Boca Raton, pp. 603–26.

Turnbull, A.L. & Nicholls, C.F. (1966) A 'quick trap' for area sampling or arthropods in grassland communities. *J. Econ. Entomol.* **59**, 1100–4.

Venables, E.P. & Dennys, A.A. (1941) A new method of counting orchard mites. *J. Econ. Entomol.* **34**, 324.

Vité, J.P. & Gara, R.I. (1961) A field method for observation on olfactory responses of bark beetles (Scolytidae) to volatile materials. *Contrib. Boyce Thompson Inst.* **219**, 175–82.

Wainhouse, D. (1980) A portable suction trap for sampling small insects. *Bull. Entomol. Res.* **70**(3), 491–4.

Walker, A.L., Cate, J.R., Pair, S.D., & Bottrell, D.G. (1972) A volumetric method for estimating populations of the Greenbug on grain sorghum. *J. Econ. Entomol.* **65**, 422–3.

Walther, B.A. & Clayton, D. H. (1997) Dust ruffling: a simple method for quantifying the ectoparasite loads of live birds. *J. Field Ornithol.* **68**, 509–18.

Wardlow, L.R. & Jackson, A.W. (1984) Comparison of laboratory methods for assessing numbers of apple rust mite (*Aculus schlechtendali*) overwintering on apple. *Plant Pathol.* **33**, 57.

Wasylik, A. (1963) Metoda analizy ciaglej roztoczy gniazd ptasich. *Ekol. Polska B* **9**, 219–24.

Way, M.T. & Heathcote, G.D. (1966) Interactions of crop density of field beans, abundance of *Aphis fabae* scop, virus incidence and aphid control by chemicals. *Ann. Appl. Biol.* **57**, 409–23.

Wickman, B.E. (1964) A comparison of radiographic and dissection methods for measuring siricid populations in wood. *Can. Entomol.* **96**, 508–10.

Wiegert, R.G. (1961) A simple apparatus for measuring density of insect population. *Ann. Entomol. Soc. Am.* **54**, 926–7.

White, T.C.R. (1975) A quantitative method of beating for sampling larvae of *Selidosema suavis* (Lepidoptera: Geometridae) in plantations in New Zealand. *Can. Entomol.* **107**, 403–12.

Williams, C.B. & Milne, P.S. (1935) A mechanical insect trap. *Bull. Entomol. Res.* **26**, 543–51.

Williamson, K. (1954) The Fair Isle apparatus for collecting bird ectoparasites. *Br. Birds* **47**, 234–5.

Wilson, L.F. (1962) A portable device for mass-collecting or sampling foliage inhabiting arthropods. *J. Econ. Entomol.* **55**, 807–8.

Wolfs, P.H., Lesna, I.K., Sabelis, M.W., & Komdeur, J. (2012) Trophic structure of arthropods in Starling nests matter to blood parasites and thereby to nestling development. *J. Ornithol.* **153**, 913–19.

Wood, G.W. & Small, D.N. (1970) A method of sampling for adults of *Chlamisus cribripennis*. *J. Econ. Entomol.* **63**, 1361–2.

Woodroffe, G.E. (1953) An ecological study of the insects and mites in the nests of certain birds in Britain. *Bull. Entomol. Res.* **44**, 739–72.

Wright, A.F. & Stuart, A.J.A. (1992) A study of the efficiency of a new inexpensive type of suction apparatus in quantitative sampling of grassland invertebrate populations. *Bull. Br. Ecol. Soc.* **23**, 116–20.

Yanoviak, S.P., Nadkarni, N.M., & Gering, J.C. (2003) Arthropods in epiphytes: a diversity component that is not effectively sampled by canopy fogging. *Biodivers. Conserv.* **12**(4), 731–41.

5   Absolute Population Estimates by Sampling a Unit of Aquatic Habitat

Many of the methods described in this chapter have comparatively low efficiencies, so that the emphasis must be on the use of a unit of habitat, rather than on the absoluteness of the estimate. Sampling costs and difficulties frequently dictate the use of a low-efficiency method that must be calibrated by a subsidiary study to obtain an efficiency conversion factor by which the results can be adjusted to give an absolute estimate. The choice of method is determined by the nature of the habitat, the size and mobility of the animals and the scale of the study. Marine studies frequently require larger, heavier, more complex samplers than those needed for freshwaters, but similar methodological approaches are applicable. As the primary determinant of the method is the type of habitat, methods will be classified under the following headings: (i) Open water: inhabited by surface dwellers, swimming macrofauna and planktonic animals; (ii) Vegetation: animals living on or around submerged or floating plants; and (iii) Bottom or benthic fauna: animals living on or in the substrate.

Methods for the study of the microfauna of inland or marine waters, which are not covered here, are introduced in works such as Edmondson & Winberg (1971), Baker & Wolff (1987), Pepper *et al.* (1995) and Ford (1993). In contrast to terrestrial habitats, major difficulties in making absolute estimates of the fauna lie in actually taking a sample of known size. As the problems of extraction, where this further operation is needed, are similar to those encountered with terrestrial samples these are discussed together in Chapter 6.

5.1   Open water

5.1.1   Nets

5.1.1.1   Plankton

Nets allow quantitative sampling for average density, and other methods (p. 187) must be used to study small-scale variations in plankton density. The development of quantitative zooplankton sampling methods from the 1800s to the present is reviewed by Wiebe & Benfield (2003). Typical plankton nets comprise a nylon gauze cone attached to a metal frame (Fig. 5.1; see also Plate 2). The mouth may be fitted with a collar or mouth-reducing cone and the cod-end should have a sampling bucket. A flow meter

Ecological Methods, Fourth Edition. P. A. Henderson and T. R. E. Southwood.
© 2016 John Wiley & Sons, Ltd. Published 2016 by John Wiley & Sons, Ltd.
Companion Website: www.wiley.com/go/henderson/ecologicalmethods

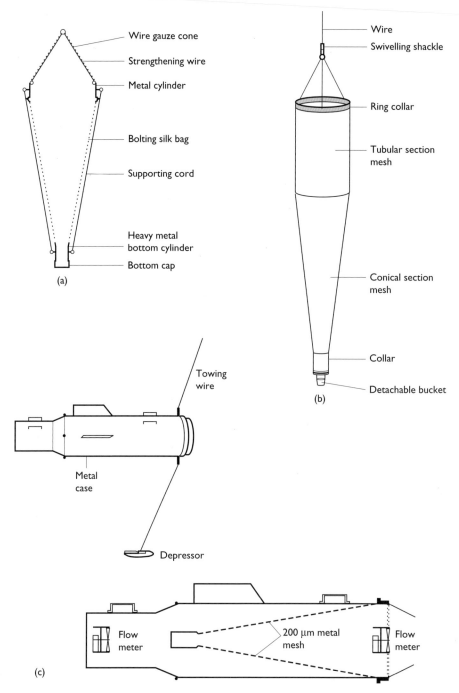

Figure 5.1   Typical plankton nets. (a) A Birge cone tow-net suitable for fresh waters with macro-phyte beds; (b) A modified working party 2 net used for vertical tows from depths of less than 200 m to the surface; (c) A Gulf III net used in marine environments for horizontal or oblique tows at speeds up to 3 m s^{-1}.

is attached in front of the mouth to measure the volume of water sampled. The main causes of sampling error are: (i) mesh penetration of animals; (ii) net avoidance; and (iii) changes in filtration efficiency. These errors are minimised by the correct choice of net type, mesh size and sampling speed. Care must be taken to ensure that the net is not moved too fast so that it 'pushes aside' some water and the animals in it; this is particularly important with fine-mesh nets (Ricker, 1938; Fujita, 1956); however, if the net is moved too slowly the more agile animals with good vision may be able to avoid it. Filtration efficiency is defined as the volume swept out by the net divided by the volume of water filtered. It is advisable to use a net with a mesh 75% of the minimum dimension of the animal to be sampled (Vannucci, 1968). Fine nets are easily clogged and Omori & Ikeda (1984) advise the use of a mesh sizes greater than 100 μm and 200 μm in ocean and coastal waters, respectively. For meso and macro plankton, a 330 μm mesh is commonly used. The *open area ratio* of a net is given by

$$R = \frac{aP}{A},$$

where a is the surface area of the net, A the mouth area and P the mesh porosity. Tranter & Smith (1968) found a 330 μm, 3.2 *open area ratio* net towed at 0.7–1.0 m s^{-1} had a filter efficiency >0.85 until it had filtered 49 m^3 of coastal water with abundant plankton. The volume of water sampled at > 0.85 efficiency increased to 300 m^3 when the open area ratio was increased to 6.4. Omari & Ikeda (1984) state that a net should have an open area ratio >3.5, and preferably >6.0 if it is to be used to sample high plankton densities.

A plankton net may be towed vertically, horizontally or obliquely (Fig. 5.2; see also Plate 2). If a sample is required from a set depth the net must be fitted with a closing or opening–closing mechanism. An example is the Clarke–Bumpus sampler (Clark & Bumpus, 1950; Tonolli, 1971) which is a normal cone-shaped plankton net in the mouth of which is a metal cylinder with a propeller blade (which records the throughput of water), two stabilising vanes and a shutter mechanism which is controlled by a messenger. Bongo (McGowan & Brown, 1966), MTD (Motoda, 1971) and

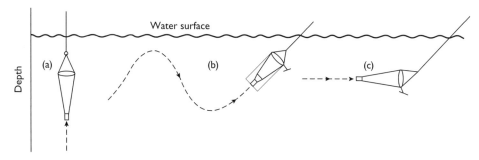

Figure 5.2   Sampling profiles for plankton nets. (a) Vertical; (b) Vertical–oblique; (c) Horizontal.

ORI nets have also been fitted with opening and closing mechanisms. Bongo and other marine plankton nets and closing nets can be purchased from Aquatic Research Instruments, 620 Wellington Place, Hope, ID 83836-0098; http://www.aquaticresearch .com/default.htm: General Oceanics, Florida; http://www.generaloceanics.com/ home.php: and SPARTEL, Totnes UK; http://www.spartel.u-net.com/index.htm).

Net avoidance is a serious problem when sampling larger plankton which swim well, and has been much discussed with respect to larval fish which can dart clear of the mouth of a net. The only solution is to use a large-mouthed net which can be trawled at high speed. The Isaacs–Kidd midwater trawl (Isaacs & Kidd, 1953) which can be towed at 2.5 m s^{-1} has been widely used. High-speed samplers such as the Gulf III (Gehringer, 1952), MTD underway plankton sampler V (Motoda, 1959), Miller sampler (Miller, 1961) and jet net (Clarke, 1964) have been developed for deploying from ships moving at up to 7 m s^{-1}. While, for some plankton, net avoidance is decreased using high-speed nets, the small mouth diameters of these nets can increase avoidance in swift-swimming groups. The design of the Gulf VII/PRO-NET and MAFF/Guildline high-speed samplers are described by Nash *et al.* (1998).

Specialist nets have been developed to sample close to the water surface (e.g. David, 1965; Matsuo *et al.*, 1976; Ellertson, 1977) and from sea or lake bed (Wickstead, 1953; Omori, 1969; Rothlisberg & Pearcy, 1977). In freshwaters where nets can become quickly clogged the Birge cone net, as modified by Wolcott (1901) (see Fig. 5.1), is a convenient tow-net; the anterior wire-mesh cone ensures that water-weed and other large debris do not enter the net, which is easily emptied by the removal of the bottom cap, conveniently made from a metal screw cap.

5.1.1.2   Particular methods for fish

The near shore open beach is one of the few habitats where netting can quantitatively sample open water fish. A beach seine comprises a curtain of net stretching from the surface to the bed is used to encircle a body of water and trap its fish, which are landed by pulling the net onto a beach. To be successful the water depth must be less than two-thirds of the depth of the net, otherwise fish will escape under the lead line. The method is only applicable on beaches that shelve gently, lack underwater obstructions such as fallen trees, and have little plant growth. The area swept by a seine can be increased by allowing it to drift with the flow parallel to the bank, though it is unlikely that the sample will reflect the total area swept. Some fish such as mullet escape by jumping the float line; this loss can be avoided by simultaneously fishing two seines one inside the other and kept sufficiently close together so that fish which jump the first net have insufficient space to accelerate to jump the second. Beach seines range in size from only 2 m long (as used to sample young fish) to large commercial nets 200 m long or larger. On gently sloping beaches drop or lift nets have been used to estimate fish densities. While these methods are some of the few which can give absolute estimates of species such as young flatfish, their main use has been to calibrate more easily used techniques such as beam trawling.

Electric fishing (Plate 3), which is normally undertaken within a limited stretch of a freshwater stream defined by stop nets, and uses a removal trapping approach, is described in Chapter 7 (p. 268). The use of beam and other trawls is discussed on p. 201.

5.1.2 Pumps

Pumps are often used when the objective is to sample plankton from a point locality or defined depth in shallow water. They are also advantageous in turbulent water, where the flow meter in a plankton net will no longer work. The disadvantages of this technique are that the precise depth from which the water is drawn is unknown, some animals may react to the current and so avoid capture, and delicate animals may suffer mechanical damage and become lost. Miller & Judkins (1981) discuss pumping-system design, which can be classified into systems which use either submersible or deck pumps.

Marine plankton samples can be collected using standard trash pumps such as the Flygt BS 2640 MT, which is able to lift water at a rate of $25\,l\,s^{-1}$ against a large and varying head. These pumps will not generally damage plankton, including delicate fish. The ability to operate over a wide head range is important in marine environments when the water is rising and falling with the tide. It is, of course unimportant when deployed from a boat or in freshwaters. To maintain a constant flow through the sampling equipment, water is pumped to a header tank with an overflow. This provides a constant head for the sampling equipment. Pumps do require a power supply; the Flygt BS 2640 needs a 415 V, three-phase supply rated 10.1 A (66.8 A peak starting current). When using a pump it is important to design the system to filter the water to ensure the plankton is not destroyed by the force of water hitting the net. A typical solution is to place the net in a large drum of water. When using a pump sampler it is possible to automate sample collection to collect standardised samples every hour over a 24-h period by using values to switch flow between a series of filters. A photograph of a typical system is shown in Plate 2).

5.1.3 Water-sampling bottles

A known volume of surface water may be collected with a bucket. For sub-surface water one of the most frequently used samplers is the Van Dorn bottle (Van Dorn, 1957), which will collect 5- to 20-l samples from a selected depth (Fig. 5.3). The device consists of a plastic cylinder, the ends of which can be sealed by stoppers. The sampler is lowered on a cable with the stoppers open so that water can flow through the cylinder. When correctly positioned, sending a messenger down the cable closes the stoppers and the sample is hauled to the surface. It is normal to deploy a number of bottles on a single cable. The low volume of the samples limits its usefulness for population estimation.

5.1.4 The Patalas–Schindler volume sampler

This sampler, which is particularly useful for small-scale limnological studies (e.g. Hardy, 1992), is a hybrid between a water sampler and a plankton net. The main body

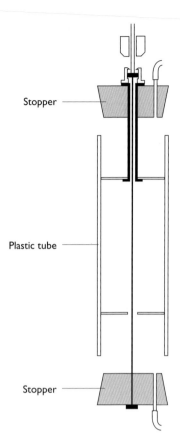

Stopper

Plastic tube

Stopper

Figure 5.3   A Van Dorn water-sampling bottle.

is a Perspex box with a capacity of 12 l with a large flap valve on the lower surface and a net attached to a side wall (Fig. 5.4). The sampler is lowered to the desired depth and then raised swiftly so that the valve closes, retaining the water sample. At the surface the box is turned on its side and its water drained via the side net. This procedure can be repeated a number of times to filter a larger volume before the sample is collected from the net bucket.

5.1.5   Particular methods for insects

No absolute quantitative method has been devised for estimating surface-dwelling insects. The larger forms may be counted directly *in situ*; their numbers and dispersion can also be studied by photography and possibly nearest-neighbour techniques used in their estimation (p. 367). Pond dipping with a long-handled ladle is usually regarded as a relative method (p. 259), but Croset *et al.* (1976) found that provided the ladle was large enough (so that no more than 5% of samples were empty), this method could be regarded as sampling from a unit of the habitat (the volume of the ladle). Such

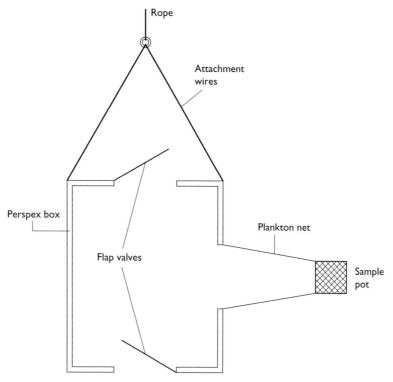

Figure 5.4   A cross-section of the Patalas–Schindler sampler. The sampler comprises a clear Perspex box, with valves on the upper and lower surfaces and a plankton net attached to the side. The sampler is lowered into the water, causing the valves to open. Once at the desired depth, the sampler is pulled swiftly to the surface, causing the valves to close. Once at the surface, the water in the sampler is filtered by running it through the attached plankton net and the sample removed.

samples provided population estimates of culicid larvae that were comparable to those from mark and recapture or removal trapping.

Because mosquito larvae dive to the bottom of a pond when disturbed and then gradually make their way up to the surface again, sampling by dipping with a net or strainer will give variable results, depending on the skill of the collector. Welch & James (1960) have, however, used this habit in the Belleville mosquito sampler (Fig. 5.5). The sampler consists of a cylinder, a cone, a concentrator and a bucket. The cylinder is placed in the pond, its base firmly pushed on to the substrate, the cone slipped inside, and the apparatus left for 20 min. The concentrator and bucket are then fixed to the top and the whole is rapidly reversed. Most of the larvae are collected in the bucket, as the water drains out through it and the concentrator; a few may remain stuck on the sides, so the apparatus should be rinsed.

The angle of the cone may have to be adjusted for different species; for *Aedes*, Welch & James (1960) found the minimum was 33°. The greater the angle of the cone, the greater its height; as the water must always be deeper than the height of the cone, this value

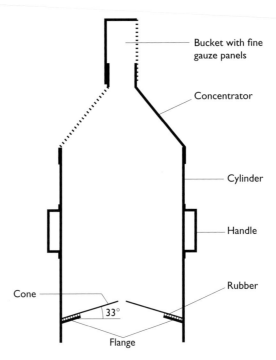

Bucket with fine gauze panels

Concentrator

Cylinder

Handle

Rubber

Cone

33°

Flange

Figure 5.5 Cross-section of the Belleville mosquito sampler.

will limit the depth of water in which the sampler can be used. Since the samplers need to be left for some time, a practical set of equipment for fieldwork would consist of ten cylinders and cones with a single concentrator and bucket. Laboratory tests gave disappointingly low efficiencies – in the region of 30–40% – and increasing the 'rising time' from 5 to 20 min did not markedly increase the number of larvae extracted. Welch & James suggest that the sampler be standardised by comparing the catch for a 24-h period with that for a shorter time. The series of catches may then be corrected to give an absolute estimate. A number of samples from a restricted habitat could be treated as a removal trapping experiment to obtain an absolute population estimate (p. 268). With small bodies of water, such as mosquito breeding pools, it may be possible to empty the complete pond, carefully separating the animals by a series of sieves, the contents of each sieve being sorted in a large pan. As developed by Christie (1954), the method allows the mosquito larvae to be returned to the pond after enumeration.

5.2 Vegetation

Besides the more general situation presented by rooted, but submerged, vegetation, particular solutions to the sampling problem have been developed for use in emergent rooted vegetation (e.g. reed beds) and with floating vegetation. The methods described below are loosely grouped according to habitat, but may be applicable in other situations, depending on the nature of the organisms being sampled. The available sampling methods are reviewed by Downing (1984). Sampling invertebrates from highly vegetated habitats such as marshland is demanding and frequently a suit of complementary methods will need to be deployed (Turner & Trexler, 1997).

5.2.1   *Floating vegetation*

One of the simplest devices is the sampling cylinder described by Hess (1941), which consists of a stout galvanised cylinder with copper mesh screening (Fig. 5.6). This is lowered under the vegetation, moved into position, raised, and any plant stems crossing its edge cut by striking the edges of the cylinder with a wooden paddle. If necessary this could be aided by sharpening the upper edges of the cylinder. The cylinder is then raised from the water and the plants and animals retained within it. The McCauley sampler (see below) may be used with floating vegetation, but such methods are only practical when the layer of vegetation is thin and easily cut.

Floating vegetation can form thick mats in some habitats, such as tropical lakes. The fish and macro crustaceans living in floating grass, *Paspalum repens*, meadows of Amazonian floodplains can be quantitatively sampled using a seine net.

Floating meadows, composed of grasses, are particularly extensive in tropical South America. In many parts of the world still waters develop floating meadows of water

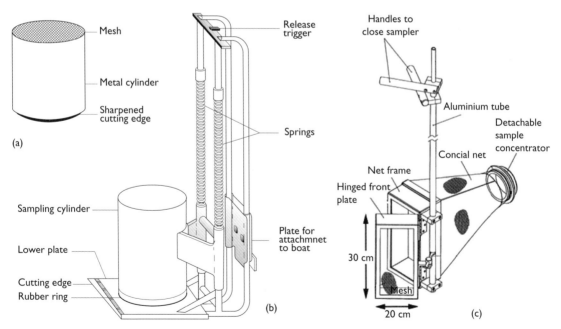

Figure 5.6   (a) Hess vegetation sampling cylinder; **(b)** McCauley's sampler for submerged and floating vegetation (after McCauley, 1975); **(c)** Kornijów's sampler for epifauna on submerged vegetation to depths of 2 m.

hyacinth, *Lemna* and other species. Animals among floating vegetation are frequently sampled by dipping with a pond net or a strainer, but such methods usually give only relative estimates of population density. For absolute estimates it is necessary to enclose a unit volume of the vegetation and associated water. When this has been strained the organisms are usually separated by hand-sorting; however, some of the methods described in Chapter 4 might be found useful (e.g. clearing and staining for eggs in water plants). Extraction by spraying the material with a fine mist, as used in plant nematode work, might also be found applicable (p. 247).

Henderson & Hamilton (1995) and Henderson & Crampton (1997) sampled the fish within floating vegetation mats using a 23 m sand eel seine net (see Plate 3). The net was made up of five panels of differing mesh size. The outer wing panels were of 1 cm stretched mesh, the next panels were 0.5 cm, and the central panel which acted as the bag was 0.25 cm. The vegetation was sampled by encircling an area of 2 to 12 m² with the net deployed from a boat. The lead line was then pulled under the root mat and the mass of plants pulled towards the boat. When possible, the entire mass was shipped and the fish sorted from the debris. When the weight was too heavy to be lifted, grass was removed from the net at the side of the boat. Bulla *et al.* (2011) took a different approach and sampled fish from floating macrophytes using a floating sieve 4 m long, 2 m wide, 0.6 m high, with a 2-mm mesh which was pushed under the floating mat and allowed to surface and capture the mat. Once the sieve was in place, aquatic plants were identified to species level and removed from the sieve. The fish in the sieve were then captured.

For insects and other invertebrates in the same habitat, Junk (1973) used a device attached to the front of a boat which could be used to take 'bites' from the outer edge of a floating meadow. A hand net of 0.1 m² mouth area with a powerful frame steel so that it could be thrust under the roots of floating plants and levered swiftly to the surface was used for insects, crustaceans and small fish by Henderson & Crampton (1997).

5.2.2   *Emergent vegetation*

Perhaps the simplest practicable sampling device for zooplankton in vegetated littoral areas is the littoral sampling tube of Pennak (1962). This consists of a length of 64 mm-diameter lightweight tubing (Pennak used a tube made of rubber-impregnated cloth with spiral reinforcing to maintain a constant diameter). To one end is attached some plankton net mesh which can be removed, and to the open end a recovery line (Fig. 5.7). The sample is taken by lowering the open end of the tube vertically to the chosen depth, ensuring that the mesh end remains in the air. Holding the mesh end steady the recovery line is used to raise the open end to the surface, when the mesh end is released. The tube is then raised and the plankton retained on the mesh washed into a sample holder. Tubes up to 9 m long can be constructed and the method is applicable to habitats other than amongst aquatic vegetation, for example, from under ice.

A number of simple devices have been used to sample amongst flooded rushes and grasses. Swanson (1978) used a column sampler constructed from a 135 cm-long, 6.5 cm-diameter acrylic tube with a disc cemented at one end with a 1.5 cm hole for a rubber bung. The open end was sharpened to give a cutting edge. Handles were

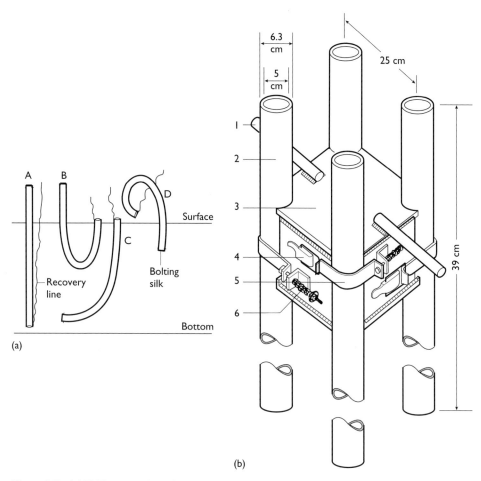

(a)

(b)

Figure 5.7   (a) Taking samples with a littoral sampling tube; (b) Schematic view of the multiple sampling device: 1. Handle; 2. Clear sampling tube; 3. Tube spacer; 4.Tube clamp; 5. Clamp strap; 6. Spring.

attached near the centre of the tube. Working in water <1 m deep, samples were collected by first pushing the tube into the substrate; the tube was then closed with a bung and the water column raised. A hand was placed over the open end as soon as possible and the tube up-ended. The multiple tube sampler (Fig. 5.7) described by Euliss *et al.* (1992) for collecting both water column and benthic invertebrates in shallow water operates on a similar principal. This simple multiple corer was used because it could sample both the benthic and water column fauna simultaneously. This allowed superior estimates of invertebrate density from those obtained when using separate devices for the benthos and water column which both tended to catch animals such as chironomids and oligochaetes which live at the sediment–water interface and thus when added together gave a biased estimate.

The Wisconsin trap and Gerking sampler (see below) may also be used with emergent vegetation.

5.2.3   *Submerged vegetation*

Sampling cages may be used to enclose a column of water of a known volume (James & Nicholls, 1961). It is a screen cage that is pushed down into the mud, the enclosed water is hand-sieved, and then the substrate dredged up and sorted on sorting trays attached to the sides of the cage. There is always the danger with this method that some animals will be missed in the hand-sieving. Bates (1941) and Goodwin & Eyles (1942) describe earlier versions of this sampling device.

5.2.3.1   Wisconsin trap

As described by Welch (1948) this is simply a canvas and gauze net with a closable mouth (Fig. 5.8c). The trap is lowered over the vegetation; the jaws of the trap are closed just above the substrate, the plants being uprooted or cut off. The whole is then raised from the water and drained.

5.2.3.2   Kornijów's 2014 sampler

A more complex design of net with a hinged mouth closing mechanism (see Fig. 5.6c) capable of sampling epiphytic communities at depths of up to 2m is described by Kornijów (2014). The sampler is attached to a manipulator pole used to place the sampler underwater and also close the jaws. Kornijów (2014) tested the sampler on

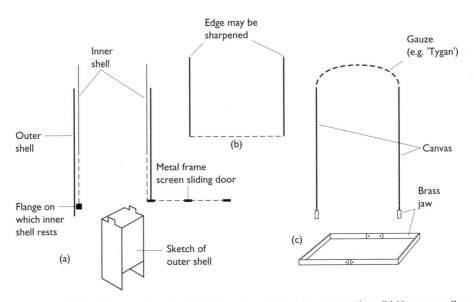

Figure 5.8   (a) Gerking sampler: sketch of the outer shell and cross-section; (b) Hess sampling cylinder for floating vegetation; (c) Wisconsin trap: sketch of jaws and cross-section.

Potamogeton perfoliatus beds at a depth of 0.5 m and found the sampler had a similar efficiency when sampling prawns, chironomids and amphipods to other devices not capable of sampling at up to 2 m deep.

5.2.3.3   The Gerking and Gates *et al.* samplers

Designed by Gerking (1957) for the sampling of littoral macrofauna, this equipment would seem to be satisfactory – apart from the labour involved – for obtaining absolute samples of all but the most active animals. It consists of two galvanised iron shells, each with a square cross-section; they may be nearly 1 m high if necessary (Fig. 5.8a). The two shells of the sampler are placed in position, the lower edges of the outer shell being forced into the mud. One side of the base of the outer shell is open (see sketch, Fig. 5.8a) and through this opening the stems of the water plants are quickly cut with a pair of grass shears. The metal and screen sliding door is then inserted and this effectively closes the inner shell, which may be slowly raised. The outer cylinder is left in position and Gerking used an Ekman dredge to remove the substrate from within the same area; alternatively, it could be scooped out. If it is not desired to sample the substrate simultaneously then only a single shell need be used; this should correspond in shape to the outer shell, but have a number of gauze panels.

Gates *et al.* (1987) describe a sampler with two components: an upper box sampler, fitted with opposable macrophyte-cutting jaws, and a lower Ekman grab sampler. Both, the macrophyte cutting and the Ekman grab jaws are fired simultaneously, allowing instantaneous compartmentalised sample collection from the vegetation and the pond substrate.

5.2.3.4   McCauley's samplers

McCauley (1975) described two samplers for the macro-invertebrates on aquatic vegetation. The general model (see Fig. 5.6b) for submerged and floating vegetation can be conveniently operated from the back of a boat. The lower plate of the sampler has a cutting edge, it is moved into position as close to the bottom as possible. When this has been done the sampling cylinder is released (triggered) and the pair of powerful springs force it down on to the lower plate. The plants are removed and the contents pumped out, using a hand-operated plastic bilge pump, and filtered through a fine mesh sieve. The sampler may then be removed and the inside washed down to remove any remaining animals.

McCauley's sampler for rushes consists of a length of acrylic tubing, which encloses the sample, and a sharpened brass plate, that is released by a trigger and cuts the bases of the rushes, rather like the sliding plate of the Gerking sampler.

5.3   Bottom fauna

A large number of methods have been developed for sampling the bottom fauna aquatic habitats; one early reviewer has commented that the number of samplers is nearly proportional to the number of investigators (Cummins, 1962)! This fact arises because five variables affect the choice of sampler: (i) the animal; (ii) the nature of the bottom substrate, whether soft or hard; (iii) the current; (iv) the depth of the water; and (v) the object of the study, whether a survey for pollution, an assessment of the food potentially available for fish, or an intensive ecological investigation for the development of a life-table for a single species.

A few generalisations are possible. Fast-flowing water has the advantage that the current may be used to carry animals, disturbed from the substrate, into a sampler (as with the Surber sampler, p. 197); however, it limits the use of devices that enclose a unit area, for as these are lowered the increased current immediately beneath them may scour the organisms from the very area that is to be sampled.

After the sample has been taken, the problem remains of separating the animals from the substrate. If the sample is too large or contains too many organisms, it may be divided and subsampled (see also p. 151). The material needs to be thoroughly agitated for satisfactory subsampling. Huckley (1975) describes a sample-splitter that uses compressed air for mixing. Hand-sorting is the most widely used method (Frost, 1971); its accuracy may be checked by staining any remaining animals with rhodamine B and viewing under ultraviolet light (Eckblad, 1973). Methods for sorting animals from sediment are discussed in Chapter 6.

As benthic sampling is often undertaken in open water with no absolute visual points of reference, it is important at the outset to consider how the position at which the samples are taken is to be determined. If repeat sampling is planned then the accuracy and repeatability of the method to get on station must also considered. GPS systems have made the determination of absolute position to ± 100 m generally available. Older methods for position fixing, and also those for depth determination, are reviewed by Holme & Willerton (1984). When sampling in the littoral zone it is particularly important to find the level as this determines the time the site will be exposed. Benthic communities are partially determined by sediment type, which is often collected as part of a benthic sample. Methods for sediment analysis are reviewed by Buchanan (1984).

Eleftheriou (2013) reviews methods for marine benthic sampling, and Elliott & Tullet (1978) present a bibliography of benthic sampling methods for freshwaters. Sampling efficiencies for grabs and air-lifts are assessed by Elliot & Drake (1981) and Drake & Elliot (1982), respectively, while sampling in stony streams has been reviewed by Macan (1958). Baker *et al.* (1987) discuss the type of sampling programme and pattern. Because benthic animals are usually particular in their substrate requirements it is important to characterise the substrate at the sampling locality. Kajak (1963) discusses the problems of sample unit size and the number of samples (see Chapter 2). While some marine habitats comprising a single sediment type can show little spatial variability in their fauna, other habitats such as streams have tremendous variability of microhabitats and the variance of a series of samples is often extremely large.

5.3.1 *Hand net sampling of forest litter*

In shallow waters the most practical way to lift a portion of a substrate composed of plants or forest litter may be with a hand net. Henderson & Walker (1986) used shallow hand nets of 0.1 m² area to quantitatively sample the insect and fish infaunal community of Amazonian steam litter banks. The net was rapidly thrust into the side of a submerged litter bank and raised rapidly vertically to the surface.

5.3.2 Sampling from under stones

The majority of the animals in fast-flowing streams will be underneath the stones on the bottom. Scott & Rushforth (1959) have investigated from the mathematical angle the influence of stone size and spacing on the area of the stream bed covered by stones. They proposed the symbol C, for this parameter and discussed its value. An estimate of the absolute population of animals may be obtained by picking up the stones from an area of the bottom; a net should be held on the downstream side to catch those animals washed off (Macan, 1958; Ulfstrand, 1968). If an estimate of population intensity (p. 2) is to be made, it will be necessary to make an estimate of the effective surface area of the stones. Calow (1972) showed that this could be calculated if longest perimeter of the stone (x) was found, then

$$\text{surface area} = 2.22 + 0.26x.$$

Some animals may be especially difficult to remove from the stones, and Britt (1955) found that these would become active if placed in a very weak acid alcohol solution (2–5% alcohol, 0.03–0.06% hydrochloric acid). A flotation solution (p. 229) might serve the same purpose and also separate the animals.

The Doeg and Lake sampler (Doeg & Lake, 1981; termed the DL sampler) and U net (Scrimgeour et al., 1993) sampler have been designed to sample invertebrates from individual stones (Fig. 5.9). Scrimgeour et al. (1993) concluded that the lighter and easier to deploy U net sampler gave less-biased invertebrate densities, while the DL sampler tending to over-estimate densities because the deployment produced a greater disturbance.

Alternatively, a net may be fixed in position, the stones and remainder of the substrate are disturbed from a unit area upstream, and the animals caught in the net. The simplest approach is the 'kicking technique', when the stones immediately upstream are 'kicked'. Frost et al. (1971) found that the first kick disturbed about 60% of the fauna yielded by 10 kicks, this is essentially a catch per unit effort (p. 272). The same principle, but with a more precise delimitation of the area, is used in the Surber sampler (Surber, 1936) (Fig. 5.10; see also Plate 3). The variability of this technique, with operator and stream conditions, and its efficiency has been tested by Needham & Usinger (1956). As some stones will lie partly inside and partly outside the square-foot frame, the selection and rejection of these is a matter of personal judgement. Indeed, Usinger & Needham found that one operator continually sampled half as much area again as the other four operators, but the latter were consistent. Comparison of the results from the Surber sampler with absolute counts of animals from buried trays (see below) showed that it caught only about a quarter of the population. The exact proportion will vary from site to site, but it is obvious that the Surber sampler cannot be used for absolute

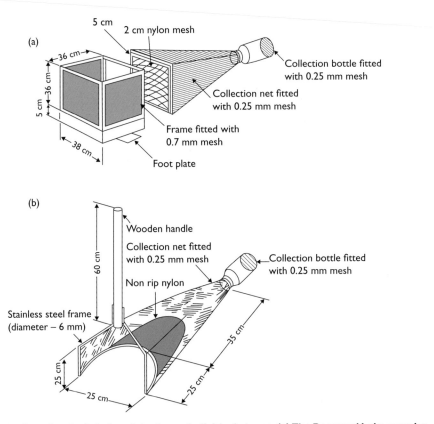

(a)

5 cm
2 cm nylon mesh

36 cm

36 cm

5 cm

38 cm

Collection bottle fitted
with 0.25 mm mesh

Collection net fitted
with 0.25 mm mesh

Frame fitted with
0.7 mm mesh

Foot plate

(b)

Wooden handle

Collection net fitted
with 0.25 mm mesh

Collection bottle fitted
with 0.25 mm mesh

60 cm

Non rip nylon

Stainless steel frame
(diameter – 6 mm)

25 cm

25 cm

35 cm

25 cm

Figure 5.9   Samplers for lotic invertebrates on individual stones. (a) The Doeg and Lake sampler has two components, a metal frame with metal wire on three sides. The back plate has a net attached, is covered with coarse mesh, and is detached after the stone has been removed from the substrate; (b) A U-shaped net sampler. The dimensions shown are for a sampler designed for large cobbles (maximum length and width 17 × 7 cm). Smaller nets should be used for smaller stones. (Adapted from Scrimgeour et al., 1993.)

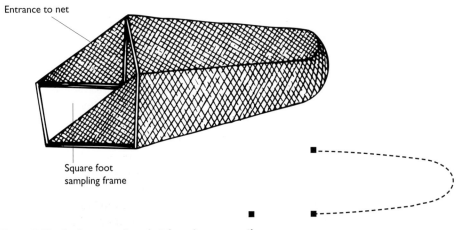

Entrance to net

Square foot
sampling frame

Figure 5.10   Surber sampler: sketch and cross-section.

population estimates from a shallow stream without a careful test of its efficiency to determine what correction factor should be applied (Kroger, 1972, 1974).

5.3.3 *The planting of removable portions of the substrate*

One of the most accurate methods of sampling the bottom fauna is to place a bag, tray or box in the stream or pond bed and either replace the substrate or allow the sediment to accumulate naturally (Usinger & Needham, 1956; Ford, 1962). An elaborate apparatus of this type is the box of Ford. This has two fixed wooden sides and a bottom, and is placed in a hole in the bed of the stream, the two sides being parallel to the course of the water. After a suitable time (Ford left it for 6 weeks) the other two sides, which are made of 'Perspex', are slid into position; the box is then made watertight and may be lifted from the stream bed with the sample undisturbed. In Ford's model it was possible to divide the sample horizontally so that each stratum could be separately analysed: one of the wooden sides had a series of slots in it; egress of mud and water through these was prevented by rubber flaps on the inside, the other sides were grooved at the corresponding level; 'Perspex' sheets could then be pushed through these slots dividing the sample. Simpler 'basket samplers' have been widely used and are generally found to give satisfactory assessments of the macro-invertebrate fauna (Mason *et al.*, 1973; Crossman & Cairns, 1974). A larger-framed net, which may be rapidly pulled to the surface, has been designed for use in the shallow waters of the Florida Everglades (Higer & Kolipinski, 1967).

The fauna of rocks or concrete substrates may be studied by placing easily removable blocks or plates of similar composition. Britt (1955) placed blocks of concrete, heavily scored on the undersurface, on a rubble and gravel bottom in deep water. The blocks can be easily located and raised with the help of the buoy and cord, but some animals may be lost as the block is raised. This loss is overcome when using the Mundie (1956) method, devised for the study of Chironomidae on the sloping slides of artificial reservoirs. The artificial substrate is a plate of asbestos cement composition (Fig. 5.11); attached to the centre of the plate is a guide wire which must be attached to a line above the water surface. The plates are left for the period of time necessary for them to be indistinguishable from the surrounding substrate (1–3 months) and then retrieved. Mundie's ingenious method of retrieval was as follows: a retrieving cone (Fig. 5.11) is slid down the guide line; this cone has flap-valves on the upper surface and these open as it descends, ensuring that there is no surge of water to disturb the sample as the cone settles. Rubber flanges around these valves and the base of the cone protect the sample from disturbance as the plate is gradually hauled up by the guide wire. Before the sample breaks the surface, a gauze-bottomed bucket (Fig. 5.11) is placed below it; this retains any organisms washed out as the water in the sampler drains through the flap valves of the cone.

Completely artificial substrates may also be used, but these give only relative estimates as settlement on them will be different to that on the natural substrate. These standardised artificial substrate units (SASUs) have been found particularly useful for estimating relative changes in the numbers of blackflies (Simulidae) (Wolfe & Peterson, 1958; Williasms & Obeng, 1962). Disney (1972) points out that the numbers obtained

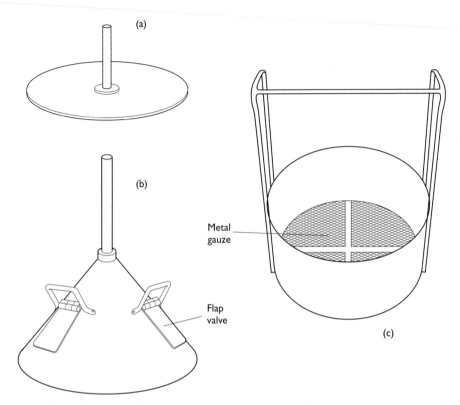

Figure 5.11 Artificial substrate and sampling apparatus. (Adapted from Mundie, 1956.) (a) Artificial substrate plate with central pipe, to which wire is attached; (b) Retrieving cone; (c) Gauze-bottomed landing bucket.

will depend on: (i) the absolute population of the blackflies; (ii) the relative area of other suitable substrata; (iii) the length of time the artificial substrate units are exposed; (iv) the intensity and nature of the factors that cause the larval blackflies to move from the substrate where they are already established; and (v) the nature of the substrate and its acceptability. The latter effect will vary with the animal and the substrate (Glime & Clemons, 1972; Mason *et al.*, 1973; Benfield *et al.*, 1974). Hester–Dendy multiplate artificial substrate samplers are approved by the USGS and the EPA in the USA and are frequently deployed. For example, Gillespie (2013) estimated the relative density of invertebrate prey of a salamander using Hester–Dendy artificial substrate samplers.

5.3.4 Cylinders and boxes for delimiting an area

As pointed out above, when a box or cylinder is lowered into flowing water, as it approaches the bottom it is likely to cause this to be scoured. This problem is least serious with the Hess circular square-foot sampler (Hess, 1941) (Fig. 5.12), which is made of a fairly coarse mesh and may be rapidly turned so as to sink a little way into the bottom. The smaller organisms may be lost through the coarse grid. If a finer grid is used, scouring becomes progressively more serious, but on the other hand drifting

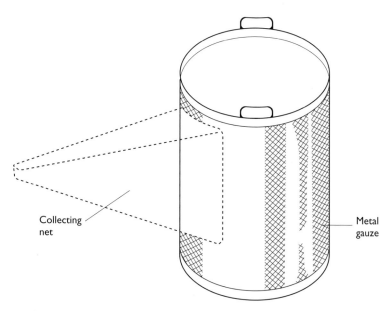

Figure 5.12   Hess circular square-foot sampler.

animals may pass through a coarse net and add cumulatively to the sample. Therefore, Waters & Knapp (1961) designed a circular sampler with fine-mesh screening; the emptying of the collecting bag was facilitated by attaching it to the sleeve with a zip fastener. The use of these samplers is restricted to streams, as a current is necessary to carry the organisms into the collecting bag. The Wilding sampler (Wilding, 1940; see also Welch, 1948) may be used in still or moving water. This consists of two cylinders. The outer cylinder has finely perforated sides and a band of saw teeth along the lower edge to aid penetration into the bed of the stream or lake (these are not essential). This cylinder delimits the area and the larger organisms, rocks and other debris are removed; the whole is then stirred and the inner cylinder (of slightly smaller diameter) lowered inside. The inner cylinder has a rotary valve at the bottom through which virtually all the water in the outer cylinder enters it (Fig. 5.13b and c). When it is resting on the bottom, the rotary valve is closed and the inner cylinder, containing all the water and smaller animals, is removed. Its contents can then be passed through a nest of sieves by opening the valve slowly.

The Gerking sampler and cages described above (p. 195) may also be used for delimiting an area of bed for sampling, and simple metal frames and cylinders were used for this purpose by Scott (1958) and Dunn (1961).

5.3.5   *Trawls, bottom sledges and dredges*

Marine epifauna may be quantitatively sampled by beam, Agissiz and otter trawls (Fig. 5.14), but their efficiency is generally low and they are species-selective (Mason *et al.*, 1979) and so are probably best considered qualitative methods. Numerous adaptations have been made to increase efficiency, including the use of additions such

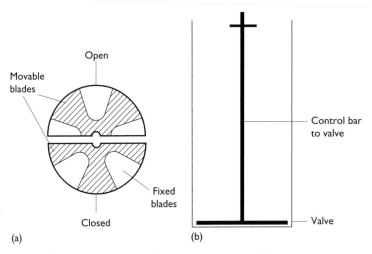

Figure 5.13 Simplified diagram of the Wilding sampler. (a) Diagram of rotary valve, showing open (upper half) and closed (lower half) positions; (b) Section of cylinder.

as tickler chains that scare animals burrowing into the substrate up into the mouth of the net (see Plate 2). For mobile animals, efficiency can change greatly between day and night. An additional problem is the measurement of the area swept for which the commonest solution is to fit a measuring wheel to the trawl or dredge frame (see Eleftheriou & Holme, 1984).

Recently, commercial fishermen have started to use pulse trawling in which the tickler chains of a conventional beam trawl are replaced with a series of electrical drag wires mounted on the net. An electrical pulse sent down these wires shocks fish off the sea bed and into the net. On soft sediments where beam trawling is difficult a notable increase in catch efficiency has been reported. This approach as not, as yet, been widely applied to research sampling.

Dredges are heavy devices designed to scrape organisms off hard surfaces while not digging deeply into the substrate (Fig. 5.14). Examples include the naturalists or rectangular dredge, rock dredge and sand dredge. Like trawls, dredges have a low sampling efficiency. Dickie (1955) found a specialised scallop dredge to range in efficiency from 5% to 12%, while for epifauna in general Richards & Riley (1967) consider efficiency to be about 10%. Given their low efficiencies, trawls and dredges are at best semi-quantitative; they are however, extremely useful for preliminary surveys to determine substrate and community composition.

Bottom sledges have been designed to glide over the substrate while sampling the epibenthic animals and proximo-benthic plankton. They often include opening and closing mechanisms to control when the sample is taken, as well as water flow meters. A typical example is the 'supra-benthic' sampler (Brunel *et al.*, 1978) which is a modification of the sledge design used by Macer (1967). Colman & Segrove (1955) describe a hand-towed sledge net for shallow water sampling.

In freshwaters, the use of the Surber sampler and the various sampling cylinders is limited to shallow waters (the depth of an arm's length or less); in deeper waters either

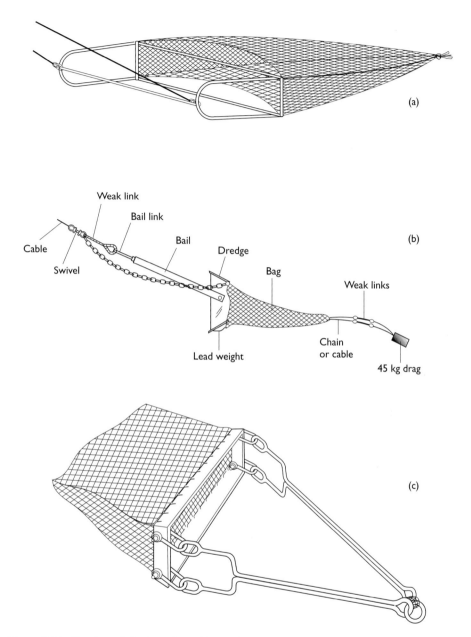

**Figure 5.14 Trawls and dredges for epifauna. (a) Agassiz trawl; (b) Rock dredge (from Nalwalk
et al., 1962); (c) Naturalist's or rectangular dredge. Note the weak link of twine joining one arm
to the ring.**

a moving net (drag, scoop, shovel) or a metal sampling box (dredge, grab, etc., see
below) must be used. Several nets that may be pushed or pulled have been described
(see Welch, 1948; Macan, 1958; Albrecht, 1959). One of the most robust models is that
devised by Usinger & Needham (1954, 1956) (Fig. 5.15) for sampling stone or gravel

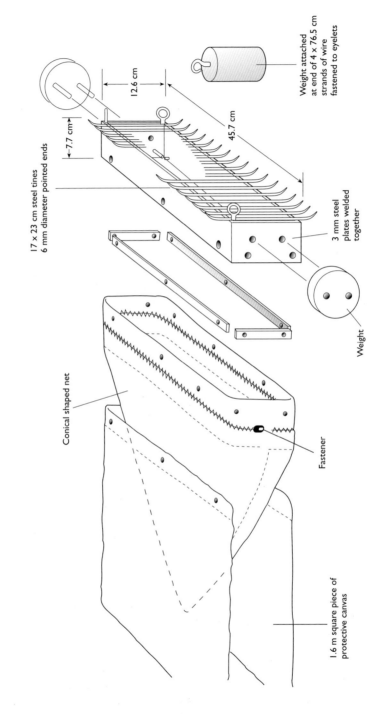

17 × 23 cm steel tines
6 mm diameter pointed ends

12.6 cm

7.7 cm

45.7 cm

Weight attached
at end of 4 × 76.5 cm
strands of wire
fastened to eyelets

3 mm steel
plates welded
together

Weight

Conical shaped net

Fastener

1.6 m square piece of
protective canvas

Figure 5.15   Usinger and Needham's drag for freshwater sampling. (Adapted from Usinger & Needham, 1956.)

bottoms (none of these movable nets is really suitable for soft mud). The tines on the mouth disturb the substrate, but prevent large stones and debris from entering the bag, while the weights and the heavy steel frame ensure that the tines really scour the bottom. The bag is attached by means of a brass zip-fastener and protected from tearing by a canvas sleeve. Usinger & Needham (1954) found that, under the conditions they were working, this drag caught about one-quarter of the animals from the area it traversed. Therefore, unless standardised this is only a relative method. A modification of this drag, particularly suitable for sampling unionid mussels, has the tines removed, an adjustable blade attached to the lower edge and a vertical handle (in addition to the tow-rope) fixed to the top leading edge (Negus, 1966).

5.3.6 Grabs

Samplers with closing jaws are called 'grabs'. They are often the method of choice for quantitative infaunal sampling. The basic principle is to lower the grab to the substrate surface where the grab is activated so that the jaws cut into and enclose a sample of the substrate which can then be raised to the surface (Fig. 5.16). The fundamental problems in grab design are to ensure first, that it rests correctly on the substrate so that when the jaws are closed the designed volume of substrate is taken; and second, that the jaws fully close to retain the sample during recovery. Wigley (1967) used motion pictures to determine the effectiveness of closing of various designs of Petersen's grab. There are ISO 16665 guidelines for quantitative sampling and sample processing of marine soft-bottom macrofauna.

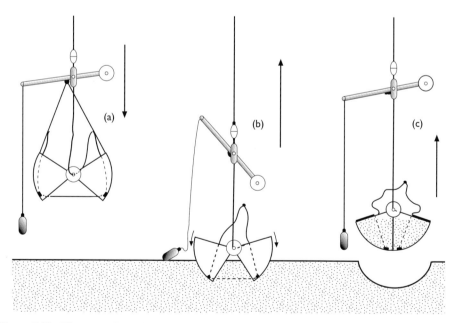

Figure 5.16   The operation of a Petersen grab, showing a method of closure that uses a counterpoised weight. (Adapted from Lisitsyn & Udintsev, 1955.)

More than 60 types of grab have been described (Elliott & Tullett, 1978). These tend to be variants of five basic types: Peterson, Van Veen, Birge-Ekmann, Smith-McIntyre, and orange-peel (Fig. 5.17; see Plate 1). The choice of grab will depend on depth, flow, substrate, weather (sea conditions) and the type of craft or platform from which it will be operated. Eleftheriou & Holme (1984) and Elliot & Tullet (1978) give extensive bibliographies of comparative sampling efficiency. Efficiency of capture of animals is

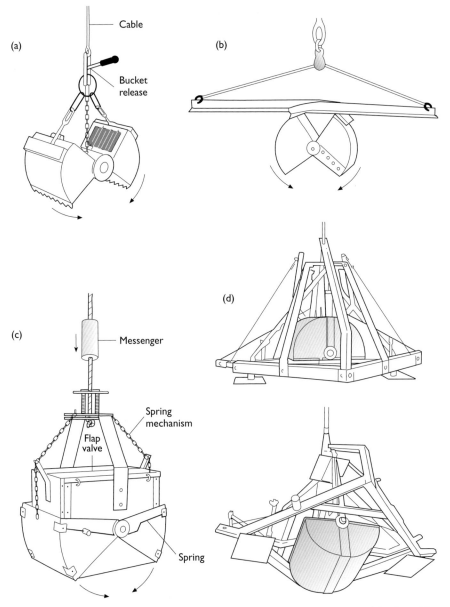

Figure 5.17 The basic types of grab design. (a) Peterson; (b) van Veen; (c) Birge–Ekman; (d) Smith–McIntyre.

only one criterion that needs to be considered. Others include substrate penetration characteristics, sample size, reliability of operation and ease of handling.

In marine habitats with large boats and mechanical winches, heavy grabs such as Petersen (Petersen & Boysen Jensen, 1911); Okean (Lisitsyn & Udintsev, 1955); van Veen (Van Veen, 1933); Ponar (Powers & Robertson, 1967); Hunter (Hunter & Simpson, 1976); Smith–McIntyre (Smith & McIntryre, 1954); Day; orange-peel (Brida & Reys, 1966); Baird (Baird, 1959); Hamon & Holme (Holme, 1949) can be used. Grab sampling at sea requires experience, and specialist advice should be sought. A general introduction to the subject is given in Eleftheriou & Holme (1984). Marine grabs can rarely penetrate deeper than 10–15 cm and thus will not sample deep-burrowing animals that can be sampled with a Foster anchor dredge, box-corer or suction sampler. Further, they will rarely capture faster-moving epibenthic animals such as crabs, which should be quantified using photographic, video or trawl methods.

Lighter grabs for use from small craft in ponds, rivers and lakes include: Van Veen (21 kg), Ponar (weighted 23 kg, unweighted 14 kg), Birge–Ekman (7 kg), Friedinger (8 kg), Dietz–La Fond (21 kg) and Allan (8 kg). In water up to 3 m deep the Birge–Ekmann and Allen grabs can be pole-operated. Elliott & Drake (1981) compared the performance of all these grabs for freshwater work and concluded that the Ponar grabs performed best with mud or fine gravel bottoms, and were the only design to sample at >3 cm depth when small stones (8–16 mm largest dimension) were present. When stones >16 mm (largest dimension) were present only the weighted Ponar sampled to a depth of 2 cm. No grab performed well when the substrate was predominately stones. Ponar, Birge–Ekmann and Allen grabs performed well in muddy substrates (particle size 0.004–0.06 mm). Freshwater grab capture efficiency for non-mobile objects in a fine gravel substrate is listed in Table 5.1. Caires & Chandra (2011) present efficiency conversion factors for Shipek, petite Ponar, standard Ekman, and large Ekman benthic grab samplers when used to sample invertebrates in lake substrates. The Shipek was the least efficient, while the large Ekman was the most efficient for silt substrates but performed poorly with sand. The petite ponar and small Ekman had similar efficiencies in both silt and sand.

The Ekman grab has been compared with various corers for sampling chironomid larvae in mud: Milbrink & Wiederholm (1973) found little difference, but Karlsson *et al.* (1976) found that coring followed by separation by flotation (p. 227) recovered more larvae.

Table 5.1 Efficiency of small grabs suitable for freshwater work measured as the percentage of pellets collected from a fine gravel substrate. (Adapted from Elliot & Drake, 1981.)

Grab	Efficiency at the surface	Efficiency at c. 3 cm	Area sampled per sample
Ponar	100	70	0.056
Van Veen	87	56	0.1
Birge–Ekmann	73	37	0.023
Allan	51	36	0.035
Friedinger	59	7	0.03
Dietz–Le Frond	22	26	0.016

5.3.7 *Dendy inverting sampler*

Freshwater substrates comprising mud with fine gravel or small sticks may be sampled with the Dendy (1944) inverting sampler (Fig. 5.18). Sticks and twigs are a particular problem for grabs as they stop the jaws from closing. The Dendy apparatus consists of a brass cylinder of about 8 cm diameter, the top of which is covered with a piece of brass gauze. The cylinder is on the end of a long handle (this limits the depth of operation),

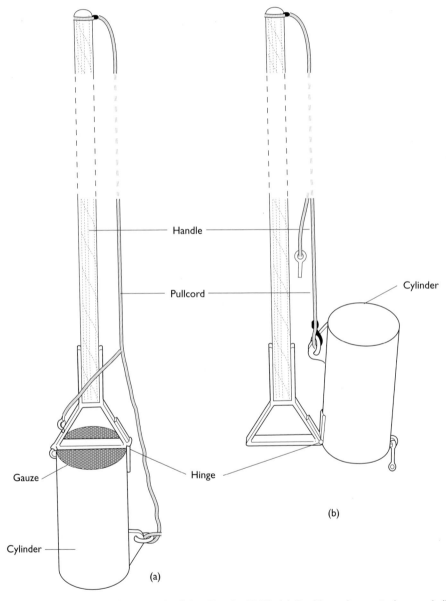

Figure 5.18   Dendy inverting sampler (after Dendy, 1944). (a) Position when entering mud; (b) Inverted, ready to raise to the surface.

which is used to drive it into the substrate; the pull cord is then tugged, and the cylinder inverted and lifted to the surface. Only material that can pass out through the gauze is lost in the ascent (apart from a small amount from the very bottom of the sample). The small size of this sampler relative to the Ekman grab may indicate its use in preference to the latter from cost and sampling pattern considerations (see p. 21). However, it does have the disadvantage of disturbing the light layer on the surface of the bottom, and for this reason Kajak (1971) recommends against its use.

5.3.8 Box samplers and corers

Samples from sand or mud beaches are obtained by driving a high-sided square or round cross-section frame into the substrate. Typical areas for square-sided frames are 0.1–0.25 m^2, and the sample is taken by digging out the substrate to the desired depth. Round cross-section samplers, which can simply be a length of pipe with the rim sharpened at one end, usually take a smaller sample which can be lifted undisturbed, although Grange & Anderson (1976) used a corer of 0.1 m^2 cross-section to a depth of 0.1 m on some beaches. To aid the retention of the sample as the corer is withdrawn from the substrate the tube may be fitted with a bung. Divers can also use corers (Gamble, 1984).

In large-scale marine studies, box corers (Reineck, 1963; Bouma & Marshall, 1964) are designed to first push a corer into the substrate; the corer is then closed prior to retrieval. For macrofaunal studies the corers must be large, heavy devices (400–750 kg), and can only be operated from large vessels. When available and practicable they are considered amongst the most reliable and efficient samplers of soft sediments. Large-diameter geological corers may be suitable for biological sampling but have been little used (Eleftheriou & Holme, 1984).

Piston-type samplers have been developed for small-scale, generally freshwater use. These enable a vertical core to be taken from the bottom and are ideal for the study of stratification of micro-organisms and for the assessment of the physical and chemical properties of the various layers. One of the earliest was the Jenkin surface mud sampler (Jenkin & Mortimer, 1938; Mortimer, 1942; Goulder, 1971; Milbrink, 1971) (Fig. 5.19).

Elgmork (1962) modified the Freidinger water sampler to take samples of soft mud; after sampling a piston was inserted and the sample slowly pushed out. The stratification of *Chaoborus punctipennis* was studied with this apparatus. The Kajak–Brinkhurst (KB) sampler (Kajak, 1971) is very suitable for soft sediments, the lid may be closed by a messenger after the sample is taken; it has been found to be nearly as efficient as using a diver (Holopainen & Sarvala, 1975). The KB sampler is available from Rickly Hydrological Company (http://www.rickly.com/index.htm). A multiple corer has been developed by Hamilton *et al.* (1970).

Livingston (1955) described a piston sampler that is particularly well adapted for obtaining the lower layers of sediment from a muddy bottom; Vallentyne (1955) and Rowley & Dahl (1956) introduced modifications that made the construction simpler and less costly, and Brown (1956) has modified it to sample the mud–water interface and, by using a clear plastic tube, to allow inspection of the intact core.

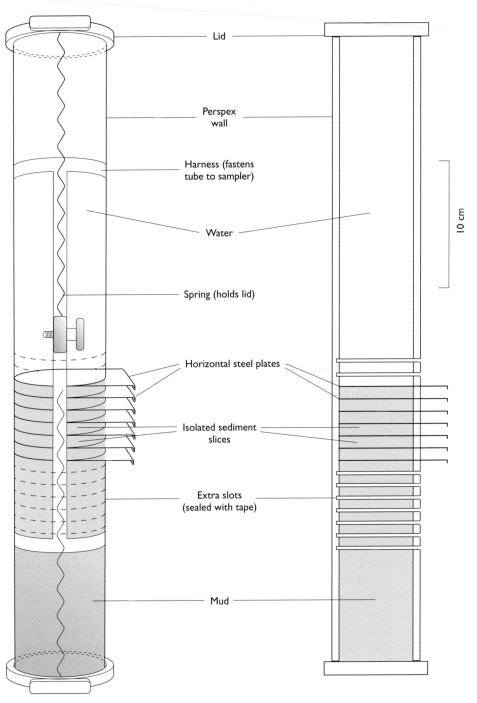

Figure 5.19 Modified Jenkin surface-mud sampler (adapted from Goulder, 1971), with horizontal steel plates in position to isolate slices of sediment.

A different approach is that of Shapiro (1958), who surrounded the sampling tube (corer) with a tapered jacket. Just before lowering, the jacket is filled with crushed solid carbon dioxide and *n*-butyl alcohol. It is then lowered quickly through the water, allowed to settle, left for about 5 min, and retrieved. The freezing mixture is replaced with water; the frozen core may soon be slid out and divided and handled in the solid state As pointed out by Kajak (1971), soil corers – especially O'Connor's (p. 223) – are very useful for hard bottoms in shallow water.

5.3.9 *Air-lift and suction devices*

Air-lifts, which are used in marine archaeology and mining, use the principle that when compressed air is discharged under water, an air–water mixture is formed which rises to the surface. A typical application is the diver-operated airlift sampler described by Barnett & Hardy (1967). This works by sucking material up a long vertical tube into a collecting sieve by the injection into the tube of compressed air (Fig. 5.20). This air-lift can be used to remove the sediment from within a core manually pushed into the substrate. Mackey (1972) describes a simple apparatus, essentially for taking core samples, whereby compressed air is fed to the centre of the base of a length of plastic drain pipe, the top of the pipe is angled and opens in a collecting net or sieve. A more elaborate device, in which the sampling area is delimited by a rectangular frame and the compressed air is liberated through a series of openings (Fig. 5.21), was developed by Pearson *et al.* (1973). They compared its efficiency with the Suber sampler and the Allen grab; overall there were not great differences, but particular animals were sampled in greater numbers by one or other method. The air-lift sampled significantly larger numbers of *Polycelis*, *Ephemerella*, *Sphaerium* and *Caenis*. Drake & Elliot (1982) have reviewed three airlift samplers for use in rivers.

A number of samplers employ suction to either force a coring tube into the substrate or draw the sediment into a collector. The Knudsen sampler (Knudsen, 1927; Barnett, 1969), which operates by sucking a coring tube into the substrate using a cable-operated pump, was the first sampler of this type to be designed. For shallow waters, Kaplan *et al.* (1974) and Thayer *et al.* (1975) have described samplers which work on the same principle. More recently, a number of samplers have used pumped water. The Benthic Suction Sampler of True *et al.* (1968) sucks a 0.1 m² core into the substrate while the substrate is drawn up into a collecting basket. Many other suction samplers have been developed (e.g. Emig & Lienhart, 1971; Mulder & Arkel, 1980; Grussendorf, 1981). Suction samplers have also been designed to be slowly towed along the bottom. Allen &Hudson (1970) used a vacuum sledge to quantify buried pink shrimp.

5.4 Poisons and anaesthetics used for sampling fish in rock pools and small ponds

Chemofishing has been long practised by mankind, and has been effectively used by ecologists seeking quantitative estimates of fish numbers in small isolated water bodies such as rock pools (e.g. Griffiths, 2000) and small ponds. In the past, rotenone – which almost inevitably killed the fish – was widely used but more recently fish anaesthetics,

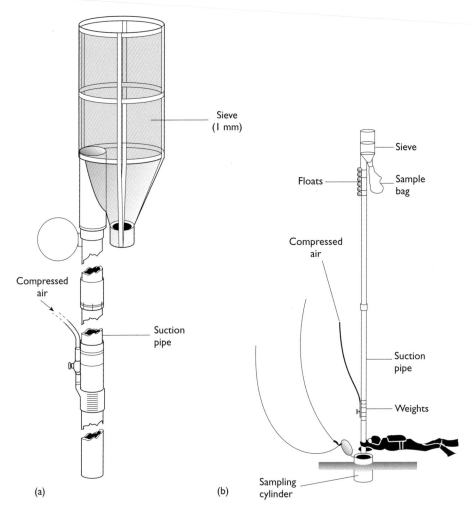

Figure 5.20   Air-lift suction sampler. (Adapted from Barnett & Hardy, 1967). (a) Diagram of the apparatus; (b) Method of use, with a diver extracting the contents of a sampling cylinder pushed into the substrate.

such as MS 222, have been the method of choice. Clove oil (eugenol) has recently been favoured (Griffiths, 2000) as, unlike quinaldine, MS-222, 2-phenoxyethanol and benzocaine it is not viewed as harmful to humans. While, in theory, anaesthetics should allow the fish to recover when placed in clean water, this is far harder to achieve in practice. For example, it can be difficult to capture fish after exposure as they may not float to the surface, resulting in them receiving a prolonged and lethal exposure. Further, recovered fish returned to the wild are often highly vulnerable to predation from birds and fish. Consideration should also be given to the effects of anaesthetics on non-target organisms. For example, Boyer *et al.* (2009) found that clove oil when used in coral reef environments reduced the growth of exposed coral.

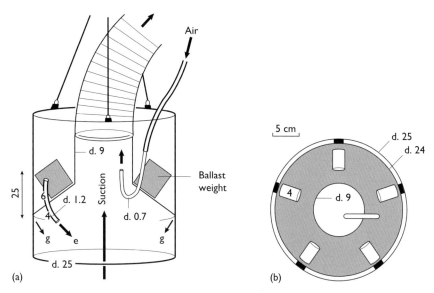

Figure 5.21 Suction sampler of Emig (1977). (a) In profile; (b) From below. Air introduced through the small tube produces suction in the central tube, through which sediment fauna is drawn up. The five compensating tubes (e) and the gap between cylinder and cone (g) are provided not only to enable the tube to dig into the substratum but also to help bring the sediment into suspension so that it is more easily collected. Diameters of various tubes (in cm) are shown.

References

Albrecht, M.-L. (1959) Die quantitative Untersuchung der Bodenfauna fliessender Gewiisser Untersuchungsmethoden und Arbeitsergebnisse). *Z. Fisch. (N.F.)* **8**, 481–550.

Allen, D.M. & Hudson, J.H. (1970) A sled-mounted suction sampler for benthic organisms., United States Fish and Wildlife Service Special Report – Fisheries.

Baird, R.H. (1959) A preliminary report on a new half square metre bottom sampler, International Council for the Exploration of the Sea, Shellfish Committee, C.M. 1958/70.

Baker, J.M., Hartley, J.P., & Dicks, B. (1987) Planning biological surveys. In: Baker, J.M. & Wolff, W.J. (eds), *Biological Surveys of Estuaries and Coasts.* Cambridge University Press, Cambridge.

Baker, J.M. & Wolff, W.J. (eds) (1987) *Biological Surveys of Estuaries and Coasts.* Cambridge, Cambridge University Press.

Barnett, P.R.O. (1969) A stabilizing framework for the Knudsen bottom sampler. *Limnol. Oceanogr.* **14**, 648–9.

Barnett, P.R.O. & Hardy, B.L.S. (1967) A diver-operated quantitative bottom sampler for sand macrofaunas. *Helgolander Wissensch. Meeresuntersuch.* **15**, 390–8.

Bates, M. (1941) Field studies of the anopheline mosquitoes of Albania. *Proc. Entomol. Soc. Wash.* **43**, 37–58.

Benfield, E.F., Hendricks, A.C., & Cairns, J. (1974) Proficiencies of two artificial substrates in collecting stream macroinvertebrates. *Hydrobiologia* **45**, 431–40.

Bouma, A.H. & Marshall, N.F. (1964) A method for obtaining and analysing undisturbed oceanic bed samples. *Mar. Geol.* **2**, 81–99.

Boyer, S.E., White, J.S., Stier, A.C., and Osenberg, C.W. (2009) Effects of the fish anesthetic, clove oil (eugenol), on coral health and growth. *J. Exp. Mar. Biol. Ecol.* **369**(1), 53–7.

Brida, C. & Reys, J.P. (1966) Modifications d'une benne 'orange peel' pour les prelevements quantitatifs du benthos de substrats meubles. *Recueil des Travaux de la Station Marine d'Endoume* **41**, 117–21.

Britt, N.W. (1955) New methods of collecting bottom fauna from shoals or rubble bottoms of lakes and streams. *Ecology* **36**, 524–5.

Brown, S.R. (1956) A piston sampler for surface sediments of lake deposits. *Ecology* **37**, 611–13.

Brunel, P., Besner, M., Messier, D., Poirier, L., Granger, D., & Weinstein, M. (1978) Le traineau suprabenthique Macer-GIROQ; appareil ameliore pour l'echantillonage quantitatif etage de la petite faune nageuse au voisinage du fond. *Int. Rev. Gasamten Hydrobiol.* **63**, 815–29.

Buchanan, J.B. (1984) Sediment analysis. In: Holme, N.A. & McIntyre, A.D. (eds), *Methods for the Study of Marine Benthos*. Blackwell Scientific Publications, Oxford.

Bulla, C.K., Gomes, L.C., Miranda, L.E., & Agostinho, A.A. (2011) The ichthyofauna of drifting macrophyte mats in the Ivinhema River, upper Paraná River basin, Brazil. *Neotrop. Ichthyol.* **9**(2), 403–9.

Caires, A.M. & Chandra, S. (2012) Conversion factors as determined by relative macroinvertebrate sampling efficiencies of four common benthic grab samplers. *J. Freshwater Ecol.* **27**(1), 97–109.

Calow, P. (1972) A method for determining the surface area of stones to enable quantitative density estimates of littoral stone dwelling organisms to be made. *Hydrobiologia* **40**, 37–50.

Christie, M. (1954) A method for the numerical study of larval populations of *Anopheles gambiae* and other pool-breeding mosquitoes. *Ann. Trop. Med. Parasitol., Liverpool* **48**, 271–6.

Clark, G.L. & Bumpus, D.F. (1950) The plankton sampler – an instrument for quantitative plankton investigations. *Spec. Publ. Am. Soc. Limnol. Oceanogr.* **5**, 1–8.

Clarke, W.D. (1964) The jet net, a new high speed plankton sampler. *J. Mar. Res.* **22**, 284–7.

Colman, J.S. & Segrove, F. (1955) The tidal plankton over Stoupe Beck Sands, Robin Hood's Bay (Yorkshire, North Riding). *J. Anim. Ecol.* **24**, 445–62.

Croset, H., Papierok, B., Rioux, J.A., Gabinaud, A., Cousserans, I., & Arnaud, D. (1976) Absolute estimates of larval populations of culicid mosquitoes: comparison of capture–recapture removal and dipping methods. *Ecol. Entomol.* **1**, 251–6.

Crossman, I.S. & Cairns, J. (1974) A comparative study between two different artificial substrate samplers and regular sampling techniques. *Hydrobiologia* **44**, 517–22.

Cummins, K.W. (1962) An evaluation of some techniques for the collection and analysis of benthic samples with special emphasis on lotic waters. *Am. Midl. Nat.* **67**, 477–503.

David, P.M. (1965) A Neuston net. A device for sampling the surface fauna of the oceans. *J. Mar. Biol. Assoc. UK* **45**, 313–20.

Dendy, J.S. (1944) The fate of animals in stream drift when carried into lakes. *Ecol. Monogr.* **14**, 333–57.

Dickie, L.M. (1955) Fluctuations in abundance of the giant scallop *Placopecten magellanicus* (Gmelin) in the Digby area of the Bay of Fundy. *J. Fish. Res. Bd. Can.* **12**, 797–857.

Disney, R.H.L. (1972) Observations on sampling pre-imaginal populations of blackflies (Dipt., Simuliidae) in West Cameroon. *Bull. Entomol. Res.* **61**, 485–503.

Doeg, T. & Lake, P.S. (1981) A technique for assessing the composition and density of the macroinvertebrate fauna of large stones in streams. *Hydrobiologia* **80**(1), 3–6.

Downing J.A. (1984) Sampling the benthos of standing waters. In: Downing, J.A., Rigler, M. (eds), *A Manual on Methods for the Assessment of Secondary Productivity in Fresh Waters*. Blackwell Scientific Publications, Oxford, pp. 112–30.

Drake, C.M. & Elliot, J.M. (1982) A comparative study of three air-lift samplers used for sampling benthic macro-invertebrates in rivers. *Freshwater Biol.* **12**, 511–33.

Dunn, D.R. (1961) The bottom fauna of Llyn Tegid (Lake Bala), Merionethshire. *J. Anim. Ecol.* **31**, 267–81.

Eckblad, J.W. (1973) Population studies of three aquatic gastropods in an intermittent backwater. *Hydrobiologia* **41**, 199–219.

Edmondson, W.T. & Winberg, G.G. (eds) (1971) *A Manual on Methods for the Assessment of Secondary Productivity in Freshwaters*. I.B.P. Handbook 17. Blackwell, Oxford, London.

Eleftheriou, A. (ed.) (2013) *Methods for the Study of Marine Benthos*. John Wiley & Sons.

Eleftheriou, A. & Holme, N.A. (1984) Macrofauna techniques. In: Holme, N.A. & McIntyre, A.D. (eds), *Methods for the Study of Marine Benthos*. Blackwell Scientific Publications, Oxford, pp. 140–216

Elgmork, K. (1962) A bottom sampler for soft mud. *Hydrobiologia* **20**, 167–72.

Ellertson, B.A. (1977) A new apparatus for sampling surface fauna. *Sarsia* **63**, 113–14.

Elliot, J.M. & Drake, C.M. (1981) A comparative study of seven grabs used for sampling benthic macroinvertebrates in rivers. *Freshwater Biol.* **11**, 99–120.

Elliott, J.M. & Tullet, P.A. (1978) *A Bibliography of Samplers for Benthic Invertebrates*. Freshwater Biological Association, Occasional Publication **4**, 61 pp.

Emig, C.C. & Lienhart, R. (1971) Principe de l'aspirateur sous-marin automatique pour sediments meubles. *Vie et Milieu* **22**, 573–8.

Euliss, N.H., Swanson, G.A., & Mackay, J. (1992) Multiple tube sampler for benthic and pelagic invertebrates in shallow wetlands. *J. Wildlife Manag.* **56**, 186–91.

Ford, I.B. (1962) The vertical distribution of larvae Chironomidae (Dipt.) *in the mud of a stream. Hydrobiologia* **19**, 262–72.

Ford, T.E. (1993) *Aquatic Microbiology: An Ecological Approach*. Blackwell Science, Oxford.

Frost, S. (1971) Evaluation of a technique for sorting and counting stream invertebrates. *Can. J. Zool.* **49**, 878–83.

Frost, S., Huni, A., & Kershaw, W.E. (1971) Evaluation of a kicking technique for sampling stream bottom fauna. *Can. J. Zool.* **49**, 167–73.

Fujita, H. (1956) The collection efficiency of a plankton net. *Res. Popul. Ecol.* **3**, 8–15.

Gamble, J.C. (1984) Diving. In: Holme, N.A. & McIntyre, A.D. (eds), *Methods for the Study of Marine Benthos*. Blackwell Scientific Publications, Oxford.

Gates, T.E., Baird, D.J., Wrona, F.J., & Davies, R.W. (1987) A device for sampling macroinvertebrates in weedy ponds. *J. North Am. Benthol. Soc.* **6**, 133–9.

Gehringer, J.W. (1952) An all metal plankton sampler (model Gulf III). *Spec. U.S. Fish. Wildlife Serv. Fish.* **88**, 7–12.

Gerking, S.D. (1957) A method of sampling the littoral macrofauna and its application. *Ecology* **38**, 219–26.

Griffiths, S.P. (2000) The use of clove oil as an anaesthetic and method for sampling intertidal rockpool fishes. *J. Fish Biol.* **57**(6), 1453–64.

Gillespie, J.H. (2013) Application of stable isotope analysis to study temporal changes in foraging ecology in a highly endangered amphibian. *PloS One* **8**(1), e53041.

Glime, J.M. & Clemons, R.M. (1972) Species diversity of stream insects on *Fontinalis* spp. compared to diversity on artificial substrates. *Ecology* **53**, 458–64.

Goodwin, M.H. & Eyles, D.E. (1942) Measurements of larval populations of *Anopheles quadrimaculatus, Say. Ecology* **23**, 376.

Goulder, R. (1971) Vertical distribution of some ciliated protozoa in two freshwater sediments. *Oikos* **22**, 199–203.

Grange, K.R. & Anderson, P.W. (1976) A soft-sediment sampler for the collection of biological specimens. *Records N. Z. Oceanogr. Inst.* **3**, 9–13.

Grussendorf, M.J. (1981) A flushing-coring device for collecting deep-burrowing infaunal bivalves in intertidal sand. *Fishery Bulletin, National Fisheries Service, NOAA, Seattle* **79**, 383–5.

Hamilton, A.L., Burton, W., & Flannagan, J.F. (1970) A multiple corer for sampling profundal benthos. *J. Fish. Res. Bd Canada* **27**, 1867–9.

Hardy, E.R. (1992) Changes in species composition of cladocera and food availability in a floodplain lake, Lago Jacaretinga, central Amazon. *Amazoniana* **12**, 155–68.

Henderson, P.A. & Crampton, W.G.R. (1997) A comparison of fish diversity and abundance between nutrient rich and poor lakes in the Upper Amazon. *J. Trop. Ecol.* **13**, 175–98.

Henderson, P.A. & Hamilton, H.R. (1995) Standing crop and distribution of fish in drifting and attached floating meadow within an Upper Amazonian varzea lake. *J. Fish Biol.* **47**, 266–76.

Henderson, P.A. & Walker, I. (1986) On the leaf-litter community of the Amazonian blackwater stream Tarumazinho. *J. Trop. Ecol.* **2**, 1–17.

Hess, A.D. (1941) New limnological sampling equipment. *Limnol. Soc. Am. Spec. Publ.* **6**, 1–15.

Higer, A.L. & Kolipinski, M.C. (1967) Pull-up trap: a quantitative device for sampling shallow-water animals. *Ecology* **48**, 1008–9.

Holme, N.A. (1949) A new bottom-sampler. *J. Mar. Biol. Assoc. UK* **28**, 323–32.

Holme, N.A. & Willerton, P.F. (1984) Position fixing of ship and gear. In: Holme, N.A. & McIntyre, A.D. (eds), *Methods for the Study of Marine Benthos*. Blackwell Scientific Publications, Oxford.

Holopainen, I.I. & Sarvala, J. (1975) Efficiencies of two corers in sampling soft-bottom invertebrates. *Ann. Zool. Fenn* **12**, 280–4.

Huckley, P. (1975) An apparatus for subdividing benthos samples. *Oikos* **26**, 92–6.

Hunter, W. & Simpson, A.E. (1976) A benthic grab designed for easy operation and durability. *J. Mar Biol. Assoc. UK* **56**, 951–7.

Isaacs, J.D. & Kidd, L.W. (1953) Isaacs–Kidd midwater trawl. Final report. *Scripps Institute of Oceanography.*

James, H.G. & Nicholls, C.F. (1961) A sampling cage for aquatic insects. *Can. Entomol.* **939**, 1053–5.

Jenkin, B.M. & Mortimer, C.H. (1938) Sampling lake deposits. *Nature* **142**, 834.

Junk, W.J. (1973) Investigations on the ecology and production-biology of the floating meadows (Paspalum-Echinochloa) on the Middle Amazon. *Amazoniana* **4**, 9–102.

Kajak, Z. (1963) Analysis of quantitative benthic methods. *Ekologia Polska A* **11**, 1–56.

Kajak, Z. (1971) Benthos of standing water. In: Edmondson, W.T. & Winberg, G.G. (eds), *I.B.P. Handbook 17*. Blackwell, Oxford, London.

Kaplan, E.H., Welker, J.R., & Krause, M.G. (1974) A shallow-water system for sampling macrobenthic infauna. *Limnol. Oceanogr.* **19**, 346–50.

Karlsson, M., Bohlin, T., & Stenson, J. (1976) Core sampling and flotation: two methods to reduce costs of a chironomid population study. *Oikos* **27**, 336–8.

Knudsen, M. (1927) A bottom sampler for hard bottom. *Meddelelser fra Komissionen for Havundersogelser, ser Fisk* **8**, 1–4.

Kornijów, R. (2014) A quantitative sampler for collecting invertebrates associated with deep submerged vegetation. *Aquat. Ecol.* **48**, 417–22.

Kroger, R.L. (1972) Underestimation of standing crop by the Surber sampler. *Limnol. Oceanogr.* **17**, 475–8.

Kroger, R.L. (1974) Invertebrate drift in the Snake River, Wyoming. *Hydrobiologia* **44**, 369–80.

Lisitsyn, A.P. & Udintsev, G.B. (1955) New model dredges. [In Russian.]. *Trudy vses. gidrobiol. Obsch.* **6**, 217–22.

Livingston, D.A. (1955) A lightweight piston sampler for lake deposits. *Ecology* **36**, 137–9.

Macan, T.T. (1958) Methods of sampling the bottom fauna in stony streams. *Mitt. int. Verein. theor. angew. Limnol.* **8**, 1–21.

Macer, T.C. (1967) A new bottom-plankton sampler. *Journal du Conseil Permanent International pour l'Exploration de la Mer* **31**, 158–63.

Mackey, A.P. (1972) An air-lift for sampling freshwater benthos. *Oikos* **23**, 413–15.

Mason, J., Chapman, C.J., & Kinnear, J.A.M. (1979) Population abundance and dredge efficiency studies on the scallop, *Pecten maximus* (L.). *Rapports et Proces Verbaux des Reunions. Conseil pour l'Exploration de la Mer* **175**, 91–6.

Mason, W.T., Weber, C.I., Lewis, P.A., & Julian, E.C. (1973) Factors affecting the performance of basket and multiplate macro-invertebrate samplers. *Freshwater Biol.* **3**, 409–36.

Matsuo, Y., Nemoto, T., & Marumo, R.A. (1976) A convertible Neuston net for zooplankton. *Bull. Plankton Soc. Japan* **23**, 26–30.

McCauley, V.J.E. (1975) Two new quantitative samplers for aquatic phytomacrofauna. *Hydrobiologia* **47**, 81–9.

McGowan, J.A. & Brown, D.M. (1966) A new opening-closing paired zooplankton net. *Univ. Calif. Scripps Inst. Oceanogr. Ref.*, 66-23.

Milbrink, G. (1971) A simplified tube bottom sampler. *Oikos* **22**, 260–3.

Milbrink, G. & Wiederholm, T. (1973) Sampling efficiency of four types of mud bottom sampler. *Oikos* **24**, 479–82.

Miller, D.A. (1961) A modification of the small Hardy plankton sampler for simultaneous high-speed plankton hauls. *Bull. Mar. Ecol.* **5**, 165–72.

Miller, D.A. & Judkins, D.C. (1981) Design of pumping systems for sampling zooplankton, with descriptions of two high-capacity samplers for coastal studies. *Biol. Oceangr.* **1**, 29–56.

Mortimer, C.H. (1942) The exchange of dissolved substances between mud and water in lakes. III and IV. *J. Ecol.* **30**, 147–201.

Motoda, S. (1959) Devices of simple plankton sampling. *Mem. Fac. Fish. Hokkaido Univ.* **7**, 73–94.

Motoda, S. (1971) Devices of simple plankton apparatus, V. *Bull. Fac. Fish. Hokkaido Univ.* **22**, 101–6.

Mulder, M. & Arkel, M.A.V. (1980) An improved system for quantitative sampling of benthos in shallow water using the flushing technique. *Neth. J. Sea Res.* **14**, 119–22.

Mundie, J.H. (1956) A bottom sampler for inclined rock surfaces in lakes. *J. Anim. Ecol.* **259**, 429–32.

Nash, R.D.M., Dickey-Collas, M., & Milligan, S.P. (1998) Descriptions of the Gulf VH/PRO-NET and MAFF/Guildline unencased high-speed plankton samplers. *J. Plankton Res.* **20**(10), 1915–26.

Needham, P.R. & Usinger, R.L. (1956) Variability in macrofauna of a single riffle in Prosser creek, California, as indicated by the Surber sampler. *Hilgardia* **24**, 383–409.

Negus, C.L. (1966) A quantitative study of growth and production of unionid mussels in the River Thames at Reading. *J. Anim. Ecol.* **35**, 513–32.

Omori, M. (1969) A bottom-net to collect zooplankton living close to the sea-floor. *J. Oceangr. Soc. Japan* **25**, 291–4.

Omori, M. & Ikeda, T. (1984) *Methods in Marine Zooplankton Ecology*. John Wiley & Sons, New York.

Pearson, R.G., Litterick, M.R., & Jones, N.V. (1973) An air-lift for quantitative sampling of the benthos. *Freshwater Biol.* **3**, 309–15.

Pennak, R.W. (1962) Quantitative zooplankton sampling in littoral vegetation areas. *Limnol. Oceanogr.* **7**, 487–9.

Pepper, I.L., Gerba, C.P., & Brendecke, J.W. (eds) (1995) *Environmental Microbiology: A Laboratory Manual*. Academic Press, London.

Petersen, C.G.J. & Boysen Jensen, P. (1911) Valuation of the sea. I. Animal life of the sea bottom, its food and quantity. *Report from the Danish Biological Station* **20**, 81 pp.

Powers, C.F. & Robertson, A. (1967) Design and evaluation of an all-purpose benthos sampler. *Great Lakes Research Division, Special Report* **30**, 126–31.

Reineck, H.E. (1963) Der Kastengreifer. *Natur und Museum* **93**, 102–8.

Richards, S.W. & Riley, G.A. (1967) The benthic epifauna of Long Island Sound. *Bull. Bingham Oceanogr. Collection* **19**, 89–135.

Ricker, W.E. (1938) On adequate quantitative sampling of the pelagic net plankton of a lake. *J. Fish. Res. Bd Can.* **4**, 19–32.

Rothlisberg, P.C. & Pearcy, W.G. (1977) An epibenthic sampler used to study the ontogeny of vertical migration of *Pandalus jordani* (Decapoda, Caridea). *Fishery Bull. US* **74**, 994–7.

Rowley, J.R. & Dahl, A. (1956) Modifications in design and use of the Livingstone piston sampler. *Ecology* **37**, 849–51.

Scrimgeour, G.J., Culp, J.M., & Glozier, N.E. (1993) An improved technique for sampling lotic invertebrates. *Hydrobiologia* **254**, 65–71.

Scott, D. (1958) Ecological studies on the Trichoptera of the River Dean, Cheshire. *Arch. Hydrobiol.* **54**, 340–92.

Scott, D. & Rushforth, J.M. (1959) Cover on river bottoms. *Nature* **183**, 836–7.

Shapiro, J. (1958) The core-freezer – a new sampler for lake sediments. *Ecology* **39**, 758.

Smith, W. & McIntryre, A.D. (1954) A spring-loaded bottom sampler. *J. Mar. Biol. Assoc. UK* **33**, 257–64.

Surber, E.W. (1936) Rainbow trout and bottom fauna production in one mile of stream. *Trans. Am. Fish. Soc.* **66**, 193–202.

Swanson, G.A. (1978) A water column sampler for invertebrates in shallow wetlands. *J. Wildlife Manag.* **42**, 670–1.

Thayer, G.W., Williams, R.B., Price, T.J., & Colby, D.R. (1975) A large corer for quantitatively sampling benthos in shallow water. *Limnol. Oceanogr.* **20**, 474–80.

Tonolli, V. (1971) Methods of collection. Zooplankton. In: Edmondson, W.T. & Winberg, G.G. (eds), *I.B.P. Handbook 17*. Blackwell, Oxford, London.

Tranter, D.J. and Smith, P.E. (1968) Loss of organisms through meshes. In: Tranter, D.J. & Fraser, J.H. (eds), *Zooplankton sampling. Monographs on Oceanographic Methodology*. UNESCO, Paris.

True, M.A., Reys, J.-P., & Delauze, H. (1968) Progress in sampling the benthos: the benthic suction sampler. *Deep Sea Res.* **15**, 239–42.

Turner, A.M. & Trexler, J.C. (1997) Sampling aquatic invertebrates from marshes: evaluating the options. *J. North Am. Benthol. Soc.* **16**(3), 694–709.

Ulfstrand, S. (1968) Benthic animal communities in Lapland streams. *Oikos (Suppl.)* **105**, 1–120.

Usinger, R.L. & Needham, P.R. (1954) A plan for the biological phases of the periodic stream sampling program (Mimeographed), Final Report to Calif. St. Water Pollution Coni. Bd.

Usinger, R.L. & Needham, P.R. (1956) C. *Progressive Fish. Cult.* **18**, 42–4.

Vallentyne, J.R. (1955) A modification of the Livingstone piston sampler for lake deposits. *Ecology* **36**, 139–41.

Van Dorn, W.G. (1957) Large volume water sampler. *Trans. Am. Geophys. Union* **37**, 682–4.

Van Veen, J. (1933) Onderzoek naar het zandtransport von rivieren. *De Ingenieur* **48**, 151–9.

Vannucci, M. (1968) Loss of organisms through meshes. In: Tranter, D.J. & Fraser, J.H. (eds), *Zooplankton Sampling. Monographs on Oceanographic Methodology.* UNESCO, Paris.

Waters, T.F. & Knapp, R.J. (1961) An improved stream bottom fauna sampler. *Trans. Am. Fish. Soc.* **90**, 225–6.

Welch, H.E. & James, H.G. (1960) The Belleville trap for quantitative samples of mosquito larvae. *Mosquito News* **20**, 23–6.

Welch, P.S. (1948) *Limnological Methods.* McGraw-Hill, New York.

Wickstead, J. (1953) A new apparatus for the collection of bottom plankton. *J. Mar. Biol. Assoc. UK* **32**, 347–55.

Wiebe, P.H. & Benfield, M.C. (2003) From the Hensen net toward four-dimensional biological oceanography. *Prog. Oceanogr.* **56**, 7–136.

Wigley, R.L. (1967) Comparative efficiencies of van Veen and Smith–McIntyre grab samplers as revealed by motion pictures. *Ecology* **48**, 168–9.

Wilding, J.L. (1940) A new square-foot aquatic sampler. *Limnol. Soc. Am. Spec. Publ.* **4**, 1–4.

Williams, T.R. & Obeng, L. (1962) A comparison of two methods of estimating changes in Simulium larval populations, with a description of a new method. *Ann. Trop. Med. Parasitol., Liverpool* **56**, 359–61.

Wolcott, R.H. (1901) A modification of the Birge collecting net. *Joppi. Microsc. Lab. Meth.* **4**, 1407–9.

Wolfe, L.S. & Peterson, D.G. (1958) A new method to estimate levels of infestations of black-fly larvae (Diptera: Simulidae). *Can. J. Zool.* **36**, 863–7.

6 Absolute Population Estimates by Sampling a Unit of Soil or Litter Habitat: Extraction Techniques

The extraction methods described here may be used not only with soils and aquatic sediments but also with plant and animal debris, litter and dung, plant material collected by suction apparatus or other means and the nests of vertebrates. The actual methods for obtaining the samples from these other habitats are discussed in Chapters 4 and 5. In general, the matrix holding the animals will be termed soil if of terrestrial origin, and sediment if of aquatic or marine sample origin.

Much of the work in soil zoology was originally aimed at the extraction of a large segment of the fauna by a single method. However, most workers have now concluded that a method that will give an almost absolute estimate of one species or group will give, at the most, a rather poor relative estimate for another. Not only does the efficiency of extraction vary with the animal, but also with the soil (see Table 6.1), the physical properties and size distribution of its particles, its water content, and the amount of vegetable matter in it. Therefore, although with certain animals under certain conditions each of these methods will give absolute population estimates, none of them will provide such data under all conditions. Further information on ecological methods in soil zoology is given in Phillipson (1970), Dindal (1990) and Gorny & Grum (1993). Techniques for sorting animals from benthic samples are reviewed by Eleftheriou & Holme (1984); Somerfield et al. (2005) and Hartley et al. (1987).

Bees, cicindelid and Bledius larvae, certain crabs and other comparatively large animals that make holes or casts in bare ground or the seabed may be counted directly in situ. In most studies, however, it is necessary both to take a sample and to extract the animals.

6.1 Sampling

The number and size of the samples and the sampling pattern in relation to statistical considerations have been discussed in Chapter 2. The soil samples are usually taken with a corer (Plate 4); golf-hole borers, metal tubing or plastic piping sharpened at one end make simple corers. It has been suggested that some animals may be killed by compression when the core is forced from such 'instruments', and furthermore it is highly desirable to keep the core undisturbed (especially for extraction by behavioural

Ecological Methods, Fourth Edition. P. A. Henderson and T. R. E. Southwood.
© 2016 John Wiley & Sons, Ltd. Published 2016 by John Wiley & Sons, Ltd.
Companion Website: www.wiley.com/go/henderson/ecologicalmethods

Table 6.1   **Recommended methods for extraction of soil arthropods. (Adapted from Edwards, 1991.) All, all methods suitable; A, Tullgren funnel; B, Kempson extractor; C, air-conditioned funnel; D, Salt & Hollick flotation; E, other flotation; F, grease film extractor; W, woodland; P, pasture; A, arable.**

Soil type/system									
	Peat			Clay/Loam			Sand		
Group	W	P	A	W	P	A	W	P	A
Isopoda	AC	ABC	ABC	AC	ABC	ABC	All	All	All
Pauropoda	ABC	ABC	ABC	ABC	ABC	All	All	All	All
Symphyla	ABC	ABC	ABC	All	ABC	All	All	All	All
Diplopoda	ABC	ABC	ABC	ABC	ABC	All	All	All	All
Chilopoda	ABC	ABC	ABC	All	ABC	All	All	All	All
*Acarina	ABC	ABC	ABC	ABCF	ABCDE	ACD	ABCF	ABC	ABC
Pseudoscorpionida	ABC	ABC	ABC	ABC	ABC	ABC	All	All	All
Araneae	ABC	ABC	ABC	ABC	ABC	ABC	ABC	ABC	All
*Collembola	ABC	ABC	ABC	All	ABCD	ABCD	ABCD	ABCD	ABCD
Protura	ABC	ABC	ABC	ABC	All	All	All	All	All
Psocoptera	ABC	ABC	ABC	All	All	All	All	All	All
Thysanoptera	ABC	ABC	ABC	DE	DE	All	All	All	All
Hemiptera	ABC	ABC	ABC	All	All	All	All	All	All
Hymenoptera	ABC	ABC	ABC	All	All	All	All	All	All
Coleoptera	ABC	ABC	ABC	All	All	CDEF	CDEF	All	All
Diptera	ABC	ABC	ABC	All	All	All	All	CDEF	CDEF
Eggs/pupae	DEF	DEF	DEF	DEF	DEF	DEF	DEF	DEF	DEF

methods), so more elaborate corers have been developed. In general, the larger the animal and the sparser its population, the bigger the sample; for example, Frick (1962) took 8 in-diameter cores for *Nomia* bees. The depth to which it is necessary to sample varies with the animal and the condition of the soil; it will be particularly deep in areas with marked dry seasons (Price, 1975); many soil animals have seasonal and diel vertical migrations (Erman, 1973). Termites and ants present special problems, due to their patchy contribution and the difficulty of taking adequately deep samples before the disturbance has caused them to escape (Brian, 1970; Sands, 1972).

With the O'Connor (1957) split corer (Fig. 6.1) the risk of compressing the sample by forcing it out of the corer is avoided. After the core has been taken the clamping band can be loosened, the two aluminium halves of the cover separated, and the sample exposed. Furthermore the sample can then be easily divided into the different soil layers: litter, humus, upper 2 cm, and so on. Plastic or metal rings may be inserted inside the metal sheath of the corer, just behind the cutting ring (MacFadyen, 1961; Dhillon & Gibson, 1962; Vannier & Vidal, 1965) (Fig. 6.1); these will enable the core to be extracted with its natural structure intact (see p. 237).

In order to penetrate hard tropical or frozen tundra soils it may be necessary to have equipment of the types described by Belfield (1956) and Potzger (1955). In contrast, in soft humus-rich situations, such as manure heaps, it is difficult to take an undisturbed sample. Törne (1962b) has devised a sampler for these habitats consisting of two concentric tubes each with cutting teeth. The inner tube is continuously pushed down

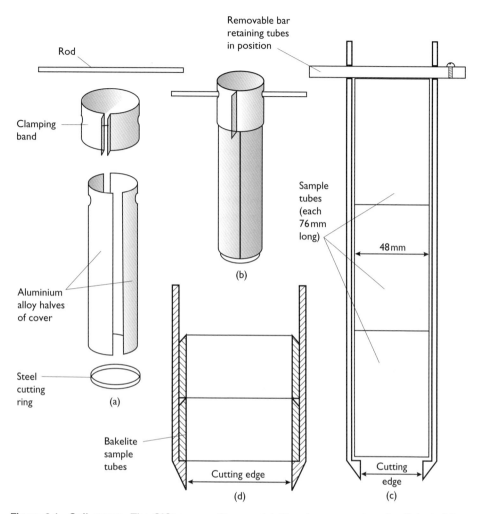

Figure 6.1   Soil corers. The O'Connor split corer. (a) Showing compartments. (Adapted from O'Connor, 1957); (b) Assembled; (c) Soil corer with sample tubes. (Adapted from Dhillon & Gibson, 1962); (d) Soil corer for the canister extractor. (Adapted from Macfadyen, 1961.)

firmly and held still, thereby protecting the sample, while the outer tube is rotated and cuts through the compost. Tanton (1969) described the construction of a corer of this type, with a series of internal plastic rings that can separate the sample into subsamples. Vannier & Alpern (1968) describe another corer that allows subsamples to be obtained from different depths.

Semi-fluid substances, such as the contents of water-filled tree holes, may be sampled with a device that includes an auger ('Worm bit') inside a normal corer; the turning of the auger takes the sample into the corer and retains it (Kitching, 1971).

Many pests of field crops lay their eggs on the bases of the plants and on the soil; the number of eggs usually falls off very rapidly with distance away from the plant (e.g. Lincoln & Palm, 1941; Abu Yaman, 1960). The scissor type of sampler described by Webley (1957) can take suitable samples of young plants and the soil around them.

Fallen leaves and other debris are usually sampled with quadrats, such as a metal box with top and bottom missing and the lower edge sharpened (Gabbutt, 1959).

6.2 Bulk staining

The staining of samples to facilitate sorting is occasionally used particularly with marine and freshwater benthic samples. Suitable stains include rose bengal, rhodamine B, eosin or Lugol's iodine. It is common practice to preserve benthic samples using formalin with $4 \, g \, l^{-1}$ rose bengal so that the organisms become stained during storage. For some animals rose bengal may obscure diagnostic features. After sieving, the animals appear bright red in the sorting tray and are easily picked from retained detritus. Rhodamine B-stained organisms fluoresce under UV light (Hamilton, 1969). For samples with a high detritus content, Williams & Williams (1974) advocate a counterstaining method where a high contrast is achieved by using rose bengal or Lugol's iodine and chlorazol black E.

6.3 Mechanical methods of extraction

Mechanical methods have the advantages that, in theory, they extract all stages – mobile and sedentary – and are not dependent on the behaviour of the animal or the condition of the substrate. Samples for mechanical extraction may be stored frozen for long periods before use. Their disadvantages are that, compared with behavioural methods, the operator must expend a great deal of time and energy on each sample, that sometimes they damage the animals and that, as mobile and immobile animals are extracted, it may be difficult to distinguish animals that were dead at the time of sampling from those that were alive.

There are a number of distinct mechanical processes that can be used to separate animals from the soil and vegetable material, including sieving, flotation, sedimentation, elutriation and differential wetting. The following account, however, is intended to be functional, rather than classificatory or historical. Only the main types of processes will be outlined, and it must be stressed that although there are already many different combinations and variants, others will need to be developed for particular animals and substrates.

6.3.1 Dry sieving

This method may be of use for the separation of fairly large (and occasionally small) animals, from friable soil or fallen leaves. Its disadvantages are that small specimens are often lost and a considerable amount of time needs to be spent in hand-sorting the sieved material. The Reitter sifter (see Kevan, 1962) is a simple device, but this is really more a collector's tool than a means of estimating populations. Lane & Shirck (1928) made the first mechanical sieve with a to-and-fro motion although it was worked by hand. Motorised sieve shakers are commercially available.

Molluscs have been extracted by dry-sieving methods (Jacot, 1935), but these have been shown to be inaccurate, with some snails remaining in the leaves (Williamson, 1959). Dry sieving has also been used to separate mosquito eggs from mud (Stage *et al.*, 1952). A modified grain drier is the basis of this method; the samples are dried until almost dusty, passed through a mesh sieve and then into the top of four shaker sieves of the grain drier. Coarser particles are removed by the first three sieves, but the eggs and similar-sized soil particles are held on the last one (80-mesh per inch); they then fall through an opening onto a 60-mesh per inch roll screen; a carefully adjusted air current blows away the lighter particles during the fall. Because of their spindle shape the eggs, with only a very small quantity of soil, eventually pass 'end-on' through the roll screen into the 'catch pan'. Such a degree of mechanisation is possible because only a single organism with a constant and regular shape and a uniform weight was required.

The active red-legged earth mite, *Halotydeus destructor*, lives amongst the grass and plant debris in Australian pastures; Wallace (1956) has shown that it can be very accurately sampled by taking shallow cores which are then inverted and rotated, still within the corer, over a sieve in a funnel. The sides of the metal corer are tapped a number of times and the mites fall through the sieve and may be collected in tubes below the funnel.

6.3.2 *Wet sieving*

This technique is most often used in conjunction with other methods (see below); on its own it is of particular value where the organisms are much smaller than the particles of the substrate (e.g. small snails amongst freshly fallen leaves) or much larger (e.g. many mud-dwellers), and hence separation by size alone is sufficient. The sieves may be of three basic designs: flat; revolving; or 'three-dimensional'. The method is particularly important for marine benthic samples.

Flat sieves may be built into a tower (see Fig. 6.2), and a series of sieves was the basis of early uses of this method by N.A. Cobb and H.M. Morris. Sieving is often used in

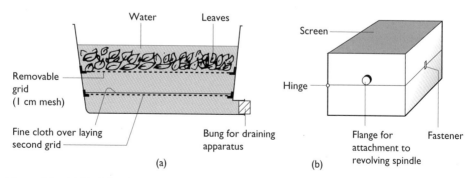

Figure 6.2 (a) Tank for the separation of molluscs from fallen leaves by wet sieving, based on Williamson's design; (b) Simple sieve box for wet-sieving sawfly cocoons, based on McLeod's design.

studies on the bottom animals of aquatic habitats for the separation of the mud from the organisms and stones that are retained by the sieve (Eleftheriou, 2013). The mesh size to be used is determined by the size of the organism being sampled. In marine studies, macrofauna is usually defined as animals retained by a 0.5–1.0 mm mesh. The exact mesh size to be used will be dictated by the particle size distribution, as the aim is to pass the majority of the sediment.

Gastropod molluscs and possibly other small animals may be efficiently separated from fresh leaf litter by wet sieving, but here it is the organisms that pass through the sieve and the unwanted material that is retained (Williamson, 1959). The leaves are placed on a coarse sieve (mesh size 1 cm), above a fine cloth-covered grid, the whole is immersed in a vessel of water (Fig. 6.2a), and the leaves are stirred from time to time. After about 15 min the vessel can be drained from below and the molluscs will be found on the cloth sheet, which is then inspected under the microscope. Clearly, this method is not satisfactory for molluscs amongst soil or humus as much of these materials will pass through the upper sieve; here a combination of wet sieving and floatation may be used (see Section 6.3.4)

Simple revolving sieves were used by McLeod (1961) for the extraction of sawfly cocoons. These consisted of wooden boxes, hinged in the middle and with screen tops and bottoms (Fig. 6.2b); they were revolved once or twice while a jet of water from a hose was played onto them. More elaborate revolving models were designed by Horsfall (1956) and Read (1958); these are discussed below.

The so-called 'three-dimensional' or rod sieve was designed by Stewart (1974) to facilitate the separation, undamaged, of tipulid larvae or other large animals from vegetable material in grassy soil samples, prior to flotation. Stewart's apparatus consists of 13 tiers of parallel galvanised steel rods set in an open-ended box. The rods are progressively closer in each tier (2.5 cm closing to 1.3 cm apart); the side of the box may be removed (for cleaning, a process assisted by a sliding panel). The sample is placed on the top row of rods and the box moved up and down in a tank of water; the inside of the tank is lined with a normal screen sieve. The vegetation will pass down between the rods, freeing insects and mineral material, which remain in the tank. The three-dimensional sieve is removed (and if necessary cleaned) and the sieve inside the tank removed and the normal flotation procedure followed (see below).

6.3.3   *Soil washing and flotation*

When the organisms to be extracted and the rest of the sample are of different particle-sizes, sieving alone is sufficient for extraction; however, in most situations after sieving the animals required remain mixed with a mass of similar-sized mineral and vegetable matter. As the specific gravity of mineral and biotic material is frequently different the extraction may be taken a stage further by flotation of the animals, but unfortunately plant material usually floats equally well. The combination of flotation, devised by A. Berlese many years before, with soil washing is usually associated with Ladell (1936). This approach, much modified and improved by Salt & Hollick (1944) and Raw (1962), is the basis of one of the most widely used, versatile and efficient techniques; it consists of four stages, three of which will be discussed under this heading. In a study of macrofauna in household compost, Gayatri (2011) found the only method to be more efficient was careful hand sorting.

Pretreatment

In order to disperse the soil particles, which is particularly important if there is a high clay content, the cores may be soaked in water and deep-frozen. Chemical dispersion may also be used: the core is soaked in solutions of sodium citrate (d'Aguilar et al., 1957) or sodium oxalate (Seinhorst, 1962). For the heavy clay soils it may be necessary to combine chemical and physical methods: the core is gently crumbled into a plastic container and covered with a solution of sodium hexametaphosphate (50 g) and sodium carbonate (20 g) in 1 litre of water; the whole is then placed under reduced pressure in a vacuum desiccator for a time. After the restoration of atmospheric pressure the sample is frozen for at least 48 h and may be stored in this condition.

Soil washing

The sample is placed in the upper (and coarsest) sieve of the washing apparatus (Fig. 6.3a); it may be washed through by slow jets of water or single sieves 'dunked' in the settling can. The material retained in the sieves must be carefully teased apart, and thoroughly washed; a few large animals may be removed at this stage. The mesh of the lowest sieve is such as to allow the animals required to pass through into the settling can, which is pivoted and is then tipped into the 'Ladell can'. A 'three-dimensional' sieve may be necessary for the above process if there is a lot of grass in the sample and large animals are being separated (see above). The Ladell, which resembles an inverted bottomless paraffin can, has a fine phosphor-bronze sieve and its lower opening immersed in the drainage tank; this tank should be arranged so that the level of water maintained in it, and hence in the Ladell, is slightly above the sieve of the latter – this minimises blockage of the sieve. The Ladell is allowed to drain – an often tedious process which may be aided by tapping the side of the can with the hand from time to time. The standard Ladell has a fine sieve (mesh size ca. 0.2 mm); for many animals this may be much finer than is necessary, and to overcome this Stephenson (1962) has designed a Ladell, with a set of interchangeable sieve plates of various mesh sizes. In conclusion, it must be warned that soil washing is an invariably wet operation and is best carried out in a room with a concrete floor and the operator wearing rubber boots and suitable apparel.

Flotation

The Ladell is removed from the tank to a stand (Fig. 6.3), allowed to drain completely, and then the process of flotation is begun. The lower opening of the Ladell is closed with a bung and the flotation liquid introduced until the Ladell is about two-thirds full (Fig. 6.3b). Concentrated magnesium sulphate solution (specific gravity ca. 1.2) is the most usual flotation liquid, but solutions of sodium chloride (Golightly, 1952), potassium bromide (d'Aguilar et al., 1957) or zinc chloride (Sellmer, 1956) may be used. Air is

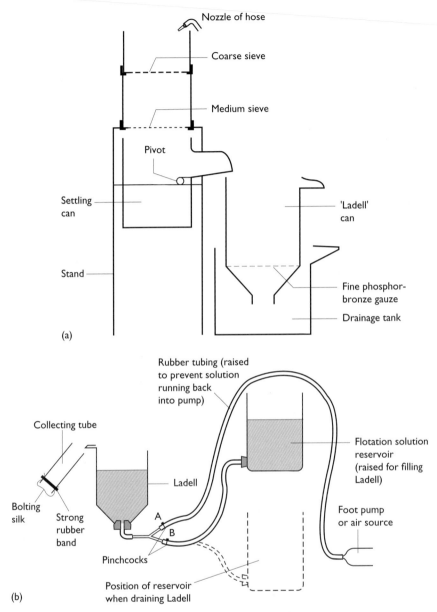

Figure 6.3 **(a) Soil washing apparatus. (Adapted from Salt & Hollick, 1944); (b) Ladell can and associated equipment during the agitation phase of floatation.**

then bubbled up from the bottom of the Ladell; this agitation is continued for 2–3 min and serves to free any animal matter that may have been trapped on the sieve. More of the flotation solution is introduced from below until the liquid and 'float' passes over the lip and the latter is retained on the collecting tube, which usually consists of a glass tube with a piece of bolting silk held in place with a rubber band. A wash bottle may be used to help direct the float out of the Ladell. When the surface and sides of the Ladell are completely clean, the animal and plant material will all be in the collecting tube and the process is complete. The flotation liquid is then drained from the Ladell, for further use, by appropriate manipulations of the reservoir and pinchclips (see Fig. 6.3).

The animals may now be separated from the plant material by direct examination under the microscope, or it may be necessary to carry out a further separation based on differential wetting and/or centrifuging (see below). However, before these are discussed other washing and flotation techniques must be reviewed.

The apparatus of d'Aguilar *et al.* (1957) first removes the smallest silt particles by agitating the sample with a water current in a cylinder with a fine gauze side; subsequently, the material is passed through a series of sieves, followed by flotation. Edwards *et al.* (1970) have developed a fully mechanised soil washing process that in general is as efficient as the Salt–Hollick process described above.

The eggs and puparia of the cabbage root fly (*Erioischia brassicae*) may be separated from the soil by sieving and flotation in water (Abu Yaman, 1960; Hughes, 1960); with such insects therefore there is no need for the Ladell – the final sieve should be fine enough to retain them, and this is immersed in water such that the insects float to the surface. A certain amount of foam often develops in soil washing and hinders examination of the float; it may be dispersed with a small quantity of caprylic acid (Abu Yaman, 1960). To separate relatively large insects (root maggots), Read (1958) used a cylindrical aluminium screen sieve that was sprayed with water and rotated and half-submerged in a tank; all the fine soil passed out of the sieve, which could then be opened and the floating insects removed. Gerard (1967) found a combination of wet sieving and simple magnesium sulphate flotation efficient for earthworm sampling, although studies by Nordstrom & Rundgren (1972) suggested that the additional labour involved, compared with chemical extraction, is not always justified (see p. 247).

A revolving sieve was also used by Horsfall (1956) for mosquito eggs. Here, the function of the drum, which consisted of three concentric sieves, was to disintegrate the sample and retain the larger materials, the eggs passing out and being collected in the finest of a series of sieves.

6.3.4 *Flotation separation of plankton, meiofauna and other small animals*

If the organisms differ in specific weight from the material in which they are dispersed, then separation by floatation may be possible. The normal approach is to mix the sample into a liquid of specific gravity sufficient allow the animals to float and the mineral particles to sink. Because organic debris also floats, such methods are inappropriate for samples of sediment rich in organic detritus. A large variety of liquids have been used including magnesium sulphate (Lydell, 1936), sodium chloride (salt) (Lyman, 1943), carbon tetrachloride (Birkett, 1957; Whitehouse & Lewis, 1966), heptane (Walter *et al.*, 1987), sugar solutions (Lackey & May, 1971; Higgins, 1977) and zinc chloride (Sellmer, 1956). At least for meiofaunal samples these solutions have been largely replaced by 'Ludox', a colloidal silica polymer. De Jonge & Bouwman (1977) extracted nematodes by mixing sediment samples with a 25% v/v solution of Ludox^{TM} (specific gravity 1.39 g ml^{-1}, density 1.31 g cm^{-3}). The surface of the mixture was then covered with water to prevent gel formation and left for 16 h, after which the surface layer was sieved and the organisms collected. Lugol can also be used with centrifugation (see p. 233). Eleftheriou and Holme (1984) state that this method has also been used successfully for marine macrofauna. With organic solvents, such as carbon tetrachloride, care must be taken over ventilation. Sellmer (1956) used a 75% solution of zinc chloride (specific gravity 2.1 g ml^{-1}) to float *Gemma*, a small bivalve mollusc, from marine sediment. Laurence (1954) separated insect larvae from dung by merely stirring the samples in a 25% solution of magnesium sulphate, while Iversen (1971) obtained 75–92% of mosquito eggs in samples from temporary woodland pools by this method, though Service (1968) added centrifugal flotation (see below). The efficiencies of this method, for the extraction of all stages of the midge, *Leptoconops*, from sand, were determined by Davies & Linley (1966) as: eggs 18%; 1st instar larvae 29%; 2nd instar larvae 72%; 3rd and 4th instar larvae 95%; pupae 85%. Sodium chloride solution or brine has been widely used for arthropods (e.g. Dondale *et al.*, 1971) and calcium chloride for larval and pupal midges (Sugimoto, 1967), while sucrose solution (specific gravity 1.12 g ml^{-1}) has been found effective for stream animals (Pask & Costa, 1971).

For semi-arid soils, where behavioural extractors can be inefficient, Walter *et al.* (1987) recommend heptane flotation. The basic protocol is as follows (Walter *et al.*, 1987; Kethley, 1991):

1. Pour the sample into a stoppered measuring cylinder and add 1 litre of 50% ethyl alcohol and about 10 ml of heptane.
2. Replace the stopper, without shaking invert the cylinder and allow the heptane to rise. Repeat this step twice.
3. Allow the cylinder to stand until the fine sediment has settled – minimum 4 h.
4. With minimal disturbance of sediment, decant the heptane layer into a sieve.
5. Rinse the sieve with 95% ethyl alcohol to remove the heptane and wash the sample into a sorting dish.

Slugs can be separated by wet sieving and flotation (Pinder, 1974). If the sample is first dried, flotation alone may be used for snails (Vagvolgyi, 1953) and nematode cysts (Goodey, 1957). Vagvolgyi's method has been modified by Mason (1970), who found it 84% efficient. When the sample is put into hot water and stirred, the dead shells and litter float and are removed; badly broken shells, soil and the originally live snails, which now are killed by heat, will sink. The sediment is then dried at 120 °C for about 16 h, during which time the contents of the originally live snails contract. Thus, when the sample is carefully crumbled into a weak solution of detergent these now float and may be removed by careful searching. Badly broken shells, soil and other debris again sink. With nematode cysts drying, dry sieving and flotation in water is the usual sequence; various details and modifications are given by Goodey (1957) and in reports from Murphy (1962a).

A flotation technique for separating fine plant material from insect eggs is described by King (1975). After rotary sieving, the mixture of eggs, fine plant debris and sand was added to water containing one part per hundred of 6% hydrogen peroxide. Oxygen, produced from the hydrogen peroxide when it comes into contact with the plant debris, causes this to float to the surface. After debris removal the eggs may be floated off in magnesium sulphate and glycerine solution (specific gravity 1.2 g ml^{-1}).

6.3.5 *Separation of plant and insects by differential wetting*

A mixture of animal and plant material results from soil washing and flotation and also from various methods of sampling animals on vegetation (see Chapter 4).

There are three possible approaches to their separation: one approach depends on the flotation effect of oxygen derived from decomposing hydrogen peroxide (see above), while two approaches depend on the lipoid and waterproof cuticle of arthropods. Either the arthropod cuticle can be wetted by a hydrocarbon 'oil', or the plant material can be waterlogged so that it will sink in aqueous solutions.

Wetting the arthropod cuticle

If arthropods and plant material are shaken up in a mixture of petrol or other hydrocarbon and water and then allowed to settle, the arthropods, the cuticles of which are wetted by the petrol, will lie in the petrol layer above the water and the plant material in the water. As Murphy (1962b) has pointed out, this separation will be imperfect if the plant material contains much air or if the specific gravities of the two phases are either too dissimilar or too close. The former may be overcome by boiling the suspension in water first (Cockbill *et al.*, 1945); the second by the choice of appropriate solutions. Ethyl alcohol (60%) may be used for the aqueous phase with a light oil. This concept was originally introduced into soil faunal studies by Salt & Hollick (1944). The float from the collection tube (see above) is placed in a wide-necked vessel with a little water, benzene is added, the whole shaken vigorously and allowed to settle. The arthropods will be in the benzene phase, which may be washed over into an outer vessel by adding further water below the surface from a pipette or wash bottle. Raw (1955) introduced the idea of freezing the benzene (m.p. 5.5 °C); the plug plus animals may then be removed and the benzene evaporated in a sintered glass crucible. (The fumes should be carried away by an exhaust fan as they are an accumulative poison.) A combination of decahydronophalene ('Dekalin') + carbon tetrachloride (specific gravity 1.2 g ml^{-1}) and zinc sulphate solution (specific gravity 1.3 g ml^{-1}) has been found particularly effective, the arthropods floating off in the organic solvent mixture and the organic debris in the zinc sulphate solution run off in a separating funnel (Heath, 1965). Xylene and paraffin have also been used for the oil phase and gelatine, that solidifies on cooling, for the aqueous phase; in the latter case, it is the plant material that is removed in a plug. If the cuticle is not easily wetted by the benzene, extraction is improved if a solution with a specific gravity of about 1.3 g ml^{-1} is used instead of water and the initial mixture exposed to a negative pressure (Sugimoto, 1967).

Waterlogging the plant material

This process, by boiling under reduced pressure (Hale, 1964) or vacuum extraction, may be used as part of the above technique. Repeated freezing and thawing also serves to impregnate the vegetable matter with water. Danthanarayan (1966) has devised a method for the separation of eggs of the weevil, *Sitona*, from the float by waterlogging the plant material by repeated freezing, filtering and transferring the whole float to a tube containing saturated sodium chloride solution; this is then centrifuged for five minutes (1500 rpm), whereupon the eggs and other animal matter come to the surface whilst the plant material sinks. The eggs of some insects (e.g. the wireworm, *Ctenicera destructor*) sink rather than float, in sodium chloride (salt) solution because they are covered with small soil particles that have adhered to the coating from the female's glands. Doane (1969) found that this could be overcome by centrifuging in 20% alcohol solution, cleaning in 5% bleach (sodium hypochlorite) solution, followed by flotation and centrifuging in salt solution.

Grease-film separation

Using the same principle as above, namely the resistance of arthropod cuticle to wetting by water and its lipophilic nature, Aucamp (1967) devised a 'grease-film machine', in which greased glass slides are revolved in an aqueous suspension of the litter or soil sample: the arthropods adhere to the grease. However, small stones remove the grease, thereby impairing the efficiency of the greased surface and providing an alternative surface. Shaw (1970) devised an apparatus to overcome this problem, in which the soil sample is enclosed in a cylinder of coarse mesh in the centre of a plastic box; the lid of the box is greased with anhydrous lanolin and the box filled with water containing a little sodium hexametaphosphate. Several such boxes may be revolved on a wheel, after which the grease and arthropods can be washed from the lid into a dish with hot soapy water or other solvent. Fragments and many whole arthropods may be recovered on a large scale from litter by using a Speight (1973) 'greased-belt machine' (Fig. 6.4). The litter is poured in an aqueous suspension onto a moving (9–12 m min^{-1}) fine nylon bolting cloth belt (350 × 15 cm; 1.6 mm mesh), coated with petroleum jelly slightly thinned with liquid paraffin. At its lowest point the belt passes through a water trough, when mineral and plant debris falls off. Subsequently, the belt – with the arthropod material

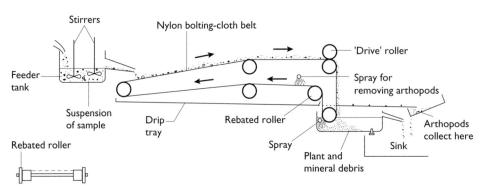

Figure 6.4   **Speight's greased belt machine for the separation of arthropods from litter. (Adapted from Speight, 1973.)**

now on its undersurface – passes over a special rebated roller under a powerful jet of water which washes the arthropods off into a tray. Speight found that extraction efficiency for whole arthropods or fragments was 70–90%, except for whole organisms with diameters of less than 0.75 mm or greater than 10.0 mm, for which it is unsatisfactory. About 1 litre of sample material can be extracted in 1 h, and the belt can run for 10 h before 'regreasing' is necessary. The apparatus is particularly appropriate for dead organisms and fragments. To remove fine organic matter, Speight treated all samples with formalin, before wet sieving (1 mm mesh) and adding to the feeder tank.

6.3.6 Centrifugation

This technique is little used for microfaunal studies because only small samples can be treated at any time. Müller (1962) found that Acarina and Collembola could be more efficiently extracted from soil by centrifugal flotation in saturated salt (sodium chloride) solution than by certain funnels and, after initially separating mosquito eggs from soil by washing and flotation in magnesium sulphate (see above), Service (1968) completed the extraction by this method. A similar technique has been used to separate nematode cysts from plant debris. Murphy (1962a) gives a useful summary, in tabular form, of early work using centrifugal flotation to separate animals from foodstuffs, soil and excreta.

For the extraction of marine meiofauna McIntyre & Warwick (1984) describe a technique developed at the Laboratoria voor Morfologie en Systematiek, Ghent which comprises the following steps:

1. Remove large particles and fine silt by decanting and sieve separation (see below).
2. Wash the contents of the sieve into a centrifuge tube.
3. Add about a heaped teaspoon of kaolin to every 250 ml of sample, shake vigorously and spin for 7 min at 6000 rpm.
4. Pour off supernatant, add Ludox-TM (specific gravity 1.15) and resuspend sediment.
5. Spin again at 6000 rpm for 7 min.
6. Pour supernatant into a sieve and wash with tap water before placing in a sorting dish.
7. Repeat steps 4 to 6 twice more.

6.3.7 Sedimentation

Sedimentation is of little value for terrestrial and marine macrofaunal studies, but live marine meiofauna, comprising groups such as nematodes, copepods, ostracods and gastrotrichs which will pass through a 1 mm sieve can be separated by decantation (McIntyre & Warwick, 1984). The basic procedure has the following steps:

1. Wash the sample into a stoppered measuring cylinder and make up to 800 ml. The size of the sample should give a sedimentation depth of about 30 cm.
2. Invert the cylinder several times and then leave to settle for 60 s.

3. Decant the supernatant through a sieve of pore size ≤ 63 µm.
4. Repeat steps 1 to 3 three times;
5. Wash material from the sieve for sorting and counting.

To stop the animals attaching to sediment particles and sinking, magnesium chloride or alcohol can be added to the sample. This technique, while ideally applied to live samples, can also be used for dead material. For groups such as nematodes, elutriation (see below) may be more efficient.

Seinhorst (1962) describes a sedimentation method ('two Erlenmeyer method') for soil nematodes in which the flask containing the suspension was moved across a series of collecting vessels, the first of which was also inverted and the sediment in it further precipitated. The efficiency of the method was, however, only 60–75%. Another sedimentation method for nematodes is described by Whitehead & Hemming (1965), but their 'tray method' (p. 244) was generally more efficient.

6.3.8 Elutriation

Elutriation is the separation of organisms by washing a sample in a constant stream of water, adjusted so that the animals are carried away and the heavier sediment remains behind. The technique is particularly appropriate for marine benthic samples from sand habitats. In some methods both water and air jets are used (Lauff et al., 1961; Pauly, 1973; Worswick & Barbour, 1974). For macrofauna, an apparatus which is easily constructed is Barnett's fluidised sand bath (Fig. 6.5) for which efficiencies of 98–100% have been claimed. In this case, water is forced vertically via a sintered sheet into a chamber holding sand that becomes fluidised. The sample is tipped into this chamber and the animals rise to the water surface where they are retained by a filter on the overflow. This apparatus can be used for samples as large as 20 litres and separation can be completed in 10 min. For meiofauna, an elutriation system can be easily built in the laboratory from standard glassware. The method was first used for meiofauna by Boissean (1957). Hockin (1981) gives a simple design using 'Quickfit' glassware. Small terrestrial arthropods such as Pauropoda and some Collembola that float in water may be extracted using von Törne's (1962a) elutriator and sieving process.

Elutriation is incorporated in two widely used methods for the extraction of soil nematodes: the Oostenbrink and Seinhorst elutriators. In the Oostenbrink model (Fig. 6.6a), the nematodes are washed out of the sample and carried in suspension by the opposing water currents out of the overflow and through a series of sieves (Oostenbrink, 1954; Goffart, 1959; Murphy, 1962a).

The Seinhorst elutriator is also part of a combined elutriation and sieving process; the soil is first passed through a coarse sieve and the suspension retained in a flask. This is inverted on top of the elutriator and the cork removed in situ (Fig. 6.6b); the upward flow of water (45 ml min^{-1}), coupled with the narrow sections A_2 and B_2, ensures that the nematodes are retained in sections A and B. After about 30 min the flask and the sections A and B may be drained through stopcocks A_3 and B_3. This material, and that collected from the overflow, is then wet-sieved. A few large worms may pass into the soil-collecting container and this should be re-elutriated, but only for a brief period so

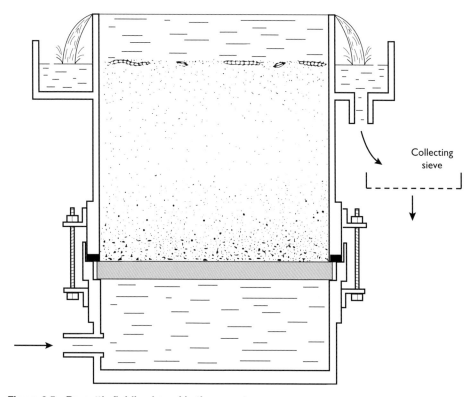

Collecting
sieve

Figure 6.5   Barnett's fluidised sand bath apparatus.

that they will still be retained in vessel A (Goodey, 1957; Seinhorst, 1962). This method
might well be adapted for single species studies on other organisms: those of a comparatively uniform size and mass (e.g. eggs) would be easily separated at a single level,
whereas healthy, parasitised and dead individuals having different masses would separate at different levels.

In a comparative study of methods for the extraction of nematodes from leaf litter,
Schouten & Kmarp (1991) found the blender–elutriation–cottonwool method to be the
most efficient. The sample of litter is first homogenised in a blender, after which the
nematodes are extracted using a funnel elutriator (as described above). The trapped
suspension is then sieved through a 40 μm sieve and finally the sieve washings are
filtered through a double cotton wool filter.

6.3.9   Sectioning

Small animals may be examined and counted in soil sections. Haarlov & Weis-Fogh
(1953, 1955) devised a method of fixing and freezing a soil core and then impregnating
it with agar and sectioning. Gelatine is probably a more suitable substance for impreg-
nation (Minderman, 1957; Anderson & Healey, 1970) and an automatic method for this
is described by Vannier & Vidal (1964). Although the method is valuable for giving

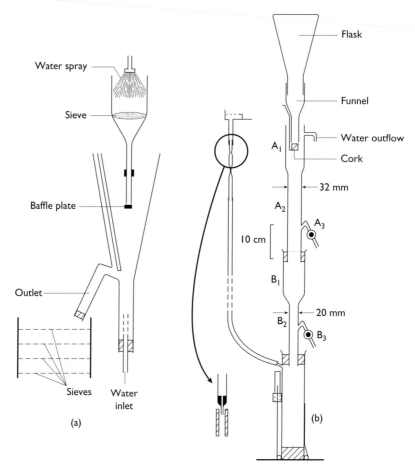

Figure 6.6   Elutriators. (a) Oostenbrink's model; (b) Seinhorst's model. (Adapted from Murphy, 1962a.)

qualitative information on feeding sites and microdistribution, it is less suitable than other methods for population studies (Pande & Berthet, 1973).

6.3.10   Aeration

Some aquatic organisms, such as smooth shelled ostracods, can be separated by aeration of the sediment (Szlauer-Łukaszewska & Radziejewska, 2013). Ostracods attach to the rising air bubbles and can be spotted and removed from the water surface.

6.4   Behavioural or dynamic methods

In these methods the animals are made to leave the substrate under a stimulus, such as heat, moisture (lack or excess) or a chemical. Their great advantage is that, unlike the mechanical methods, once the extraction has been set up it may usually be left, virtually unattended, and thus large quantities of material may be extracted simultaneously in batteries of extractors. Another important advantage is the ability to extract animals from substrates containing a large amount of vegetable material. The disadvantage is that, being based on the animal's behaviour, the extraction efficiency will vary with the condition of the animals and be influenced by changes in climate, water content, and so on, experienced before and after sampling as well as by variations in these conditions in the apparatus itself. If samples have to be retained for several days, polythene bags may be suitable containers (Rapoport & Oros, 1969); leaf litter may be held in cloth bags (e.g. cotton pillowcases) for up to 4–5 days as long as they are turned frequently and not exposed to excess moisture or extreme temperatures. Obviously, eggs and other immobile stages cannot be extracted by this method.

The exact behavioural mechanism of extraction is not fully understood. Temperature gradients are established from the start, but in some types of funnel the humidity gradient is not clearly marked in the lower part of the sample (Brady, 1969).

The pattern of egress of the mites is irregular and there is often a marked 'flush' towards the end of the extraction (Brady, 1969); this has been associated with the moisture content reaching about 20%, which triggers positive geotaxis (Nef, 1970, 1971). It seems that Oribatid mites are positively geotactic in dry conditions and negatively geotactic when the soil is moist; the adaptive significance of such responses is obvious and they are probably an important part of the mechanism of the vertical movements of soil fauna (p. 222). Whether all extractors act in this way, by the animal responding to a definite stimulus rather than moving along a gradient, is uncertain. Many workers have sought to maximise the gradients (e.g. MacFadyen, 1961; Kempson *et al.*, 1963) and these extractors have been found efficient (e.g. Edwards & Fletcher, 1970). There can be no doubt that, for certain groups, it is essential that the stimuli (usually heating and drying) are not applied too quickly (Lasebikan, 1971), but with the diversity of soil organisms it is not surprising that it is impossible to construct an extractor that is equally efficient for all groups. Therefore, although dry funnels, in particular, have been used extensively for community studies, their efficiency varies with soil type and animal group (MacFadyen, 1961; Satchell & Nelson, 1962; Block, 1966; Edwards and Fletcher, 1970). Edwards (1991) has reviewed most modifications of behavioural extractors and provides details of preferred methods for different soil types.

6.4.1 Dry extractors

The basic laboratory apparatus is the Berlese–Tullgren funnel, a combination of the heated copper funnel designed at the turn of the century by the Italian entomologist, A. Berlese, and subsequently modified by the Swede, A. Tullgren, who used a light-bulb as a heat source. The funnel has been considerably modified and improved by many workers, as described in the reviews of MacFadyen (1962), Murphy (1962a) and Edwards (1991). Some of the most important innovations have been the demonstration by Hammer (1944) that when soil is being extracted the core should be retained intact and inverted, thereby enabling the animals to leave the sample by the natural passageways, and the discovery by Haarlov (1947) that serious losses could result from the animals becoming trapped in condensation from the core on the sides of the funnel. Haarlov (1947) recommended that the core should never touch the sides of the funnel, and subsequent workers have sometimes referred to the space between the core and the sides of the funnel as the 'Haarlov passage'. When litter is being extracted and the 'passage' is difficult to maintain, it may be helpful to increase air circulation with wide plastic piping (Paris & Pitelka, 1962). Large numbers of small funnels were first grouped together by Ford (1937), and this approach has been much developed and improved by MacFadyen (1953, 1955, 1961, 1962), who also introduced the concepts of steepening the heat gradient and arranging a humidity gradient. Murphy (1955), Newell (1955), Dietrick *et al.* (1959) and Kempson *et al.* (1963) have all introduced devices to reduce the fall of soil into the sample, thereby ensuring a cleaner extraction. A folding Berlese funnel for use on expeditions has been developed by Saunders (1959) and a simpler model, in which the funnels are constructed of oil cloth and the animals 'forced down' by the vapour from naphthalene held in cheese-cloth bags above the sample, has been devised by Brown (1973). For large volumes of litter, Norton & Kethley (1988) describe a lightweight, collapsible and easily transportable Berlese–Tullgren funnel that is simple to construct.

There are many variants of the dry funnel in use, but most of them approximate to one of the following types.

6.4.1.1   Large Berlese funnel

This is used for extracting large arthropods such as Isopoda and Coleoptera, from bulky soil or litter samples (MacFadyen, 1961), and also for the extraction of insects from suction apparatus and other samples containing much vegetation (Clark *et al.*, 1959; Dietrick *et al.*, 1959; Dondale *et al.*, 1971). They can also be used to extract microarthropods from materials such as moss (Smrz, 1992). Desirable features are an air circulation system, introduced by MacFadyen, which may be opened to ensure a rapid drying for the extraction of desiccation-resistant animals, such as beetles and ants, or partly closed for a slower 'wet regime' (but avoid condensation) for beetle larvae, Campodea and other animals that are susceptible to desiccation. MacFadyen recommends testing the humidity below the sample by cobalt thiocyanate papers; for the resistant animals humidities down to 70% are tolerable, but for the others it should not fall below 90%. As phototactic animals (e.g. Halticine beetles) are often attracted upwards towards the light-bulb of the funnel and so fail to be extracted, this is replaced as a heat source by a Nichrome or similar wire-grid heating element – this also heats the sample more uniformly (Clark *et al.*, 1959). The extraction of positively phototactic animals has been further improved by Dietrick *et al.* (1959), who placed a 75 W spotlight below the collecting jar; this was switched on intermittently. An apparatus incorporating these and other features is illustrated in Fig. 6.7. As MacFadyen and others have frequently stressed, for most animals the heat should be applied slowly and therefore it may be useful to be able to control the voltage of the heating element via a rheostat (Dietrick *et al.*, 1959). Optimal temperature and humidity levels over the course of an extraction can be achieved using a computerised control system. Extraction time will vary from a few days to over a month (e.g. Park & Auerbach, 1954) depending on the substrate and the animal.

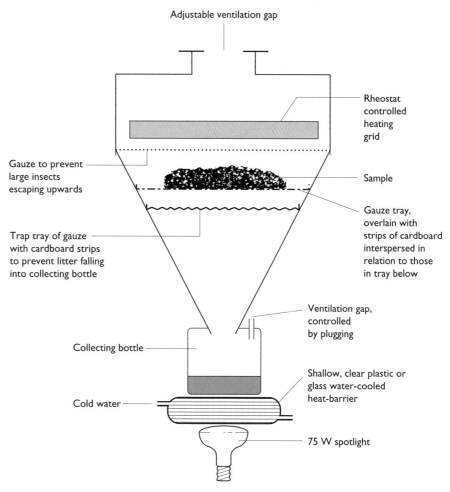

Figure 6.7 **A large Berlese funnel, with modification.**

Most recent enhancements to the Berlese funnels have been designed to produce steep temperature and humidity gradients. Crossley & Blair (1991) describe a low-cost design that can be used in a cool room with Christmas tree lights to produce a 20 °C gradient.

6.4.1.2 Horizontal extractor

Designed by Duffey (1962) for extracting spiders from grass samples, this extractor (Fig. 6.8) would probably serve equally well for any rapidly moving litter animal. It has the advantages that steep gradients are built up and that debris cannot fall into the collecting trough; it is also relatively compact and built in paired units. Heat is provided by a 500-W domestic heating element. Air passes in through the ventilation holes just above the aqueous solution of the collecting gutters; this helps to prevent a

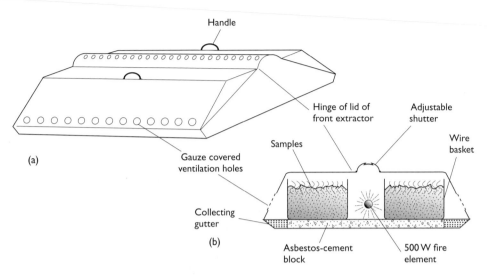

Figure 6.8 A pair of horizontal extractors, based on Duffey's (1962) design. (a) Sketch; (b) Sectional view.

too-rapid desiccation. Duffey recommended a 0.001% solution of phenylmercuric acetate (a fungicide and bactericide) with a few drops of a surface-active agent (detergent). The samples are held in wire trays 6 in (10 cm) wide. The extraction takes only a day or two.

6.4.1.3   High-gradient (multiple canister) extractor

Collembola, mites and other small animals may be obtained from small compact soil samples in this apparatus, designed by MacFadyen (1961). The samples are taken with the special corer (see Figs 6.1d and 6.9a), the samples removed, surrounded by the plastic ring, and a sieve plate is fitted to each ring. The collecting canisters are cooled in a water bath and 95% humidity is maintained below the samples during extraction (Fig. 6.9b). Heat is provided by two 160-W elements giving 10 W per sample, and the temperature of the upper surface of the sample reaches 35 °C; the temperature of the lower surface is determined by the water bath (e.g. 15 °C), so that a very steep gradient is set up over the 3 cm-deep sample. Extraction may be completed in 5 days, the maximum heat being applied after the first 24 h. MacFadyen recommended a 0.3% solution of the fungicide 'Nipagin' (which contains streptomycin and terramycin) be used in the collecting canisters, but the tests of Kempson *et al.* (1963) (see below) indicate that a saturated aqueous solution of picric acid with a little photographic wetting agent might be preferable. Constructional details are given by MacFadyen (1961), who has also (MacFadyen, 1962) described variants suitable for use under expedition conditions, the energy source then being kerosene or bottled gas, and the effect of various modifications on the physical conditions in the extractor (MacFadyen, 1968). Block (1966) found that this apparatus extracted 76% of the acarina from mineral soils.

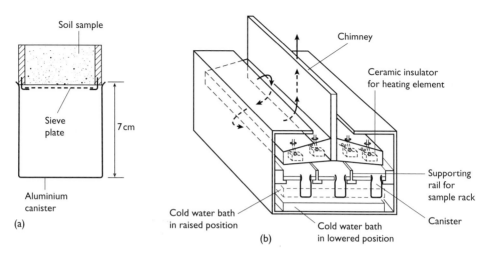

Figure 6.9   High-gradient extractor (after MacFadyen, 1961). (a) Canister, core and sieve plate; (b) Whole apparatus.

6.4.1.4   The Kempson bowl extractor (Fig. 6.10)

Developed by Kempson *et al.* (1963), this method is suitable for the extraction of mites, Collembola, Isopoda and many other arthropods in woodland litter. Extraction rates of 90–100% are recorded for groups other than the larvae of holometabolous insects,

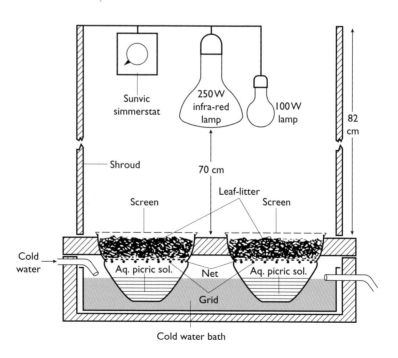

Figure 6.10   Kempson bowl extractor. (Adapted from Kempson, Lloyd & Ghelardi, 1963.)

for which it is not efficient. The apparatus consists of a box or shrouded chamber, containing an ordinary light-bulb and an infra-red lamp (250 W) that is switched on in pulses, at first only for a few seconds at a time and gradually increased until it is on for about a third of the time. The pulsing of the lamp is controlled by a time switch. The extraction bowls are sunk into the floor of the chamber. Each bowl contains a preservative fluid and is immersed in a cold water bath. A saturated aqueous solution of picric acid, with some wetting agent, was used by Kempson et al. (1963), but Sunderland et al. (1976) have found a 80 g l^{-1} solution of tri-sodium orthophosphate ($Na_3PO_4 \cdot 12H_2O$) a cheap, safe and suitable alternative. The litter is placed in a plastic tray above the bowl and is supported by two pieces of cotton fillet net laid on a coarse plastic grid. The top of the tray is covered by a fine black nylon screen. Kempson et al. (1963) give full instructions for the construction of this apparatus. Its great advantages, besides its comparatively simple construction, are the high humidity maintained on the lower surface of the sample, the lack of debris among the extracted material (the fillet net does not appear to present any barrier to the animals), the prevention by the black screen of animals escaping upwards, and the simple mechanism of gradually increasing the amount of heat to which the sample is subjected. Extractions with this apparatus generally take about a week. Adis (1987) found that a modified Kempson extractor had completed the removal of all arthropod groups from a tropical soil after 13 days.

6.4.1.5   The Winkler method

The Winkler sample extraction method is frequently used for ants and other macro-invertebrates in leaf litter. Because the equipment is light, unbreakable and does not require a power source, it is the method of choice for extractions from leaf litter in the field. Winkler extractors are available from Firma Hildegard Winkler; http://www.entowinkler.at/index.php/en/utensilien-supplies/geraete-und-utensilien, or B & S Entomological Services; http://entomology.org.uk/products.htm

The method comprises the following steps:

1. Sift samples of leaf litter and rotten wood by agitating them vigorously in a bag above a coarse mesh screen. A typical mesh size is 10 mm.
2. Litter arthropods are concentrated in the finer 'siftate' that passes through the screen.
3. The animals are then extracted from the siftate by a passive extraction method. The siftate is placed in a thin mesh sack and then suspended and enclosed within an outer cloth 'Winkler bag'. The Winkler bag tapers to a cup of ethanol. After loading the sample the Winkler bag is closed at the top and suspended in a sheltered location.
4. For some animals the siftate in the sac is agitated at regular, typically daily, intervals. This is because some animals move towards the centre of the sample as it dries out rather than falling through the mesh. However, this is best avoided if possible as it may result in loss of animals.
5. Arthropods fall from the litter and accumulate in the ethanol, and the sample is taken off after a known standardised time period which is often 3 days.

The efficiency of the Winkler method for the extraction of ants and other Hymenoptera, adult and larval beetles, earthworms, adult and larval Diptera, Lepidoptera larvae, Hemiptera, Arachnida, Chilopoda, Diplopoda, Mollusca and Isopoda was reported by Krell *et al.* (2005), who undertook extractions on series of leaf litter samples from both temperate and tropical forest. Given sufficient time, Krell *et al.* concluded that the method extracts more than 90% of the total fauna for most macro-invertebrate groups. The exceptions were the Isopoda, Diplopoda and Mollusca. The extraction times required for the maximum extraction varied between groups from 15 days for ants to 49 days for coleopteran larvae. However, for ants (typically the commonly most abundant group) 3 days was sufficient to obtain 70% of the individuals and nearly all species. For ants at least, Winkler extractors are useful field samplers, although for other groups the samples will probably need to be returned to the laboratory to allow extended extraction periods. Sabu & Shiju (2010), in a study in Indian moist decidu-ous forest, concluded that the Winkler method produced levels of extraction inferior to that obtained by Berlese funnel extraction. Similarly, Sakchoowong *et al.* (2007) con-cluded that Winkler extractors were inferior to Tullgren extractors for tropical leaf litter macroarthropods.

6.4.2 *Wet extractors*

The principle of this method is similar to that of the dry extractors, the animals being driven out of their natural substrate under the influence of a stimulus, possibly heat or, as the observations of Williams (1960a) would suggest for some cases, reduced oxygen tension. As the substrate is flooded with water in all variants of the method there is no risk of desiccation, and the method is particularly successful with those groups that are poorly extracted by a Berlese–Tullgren funnel – that is, nematode and enchytraeid worms, insect larvae and various aquatic groups (Williams, 1960a). The method, which is faster than dry extraction, seems to have been discovered independently at least three times.

6.4.2.1 Baermann funnel

As originally designed this consisted of a glass funnel with a piece of metal gauze or screen resting in the funnel and a piece of rubber tubing with a pinchcock on the stem (Fig. 6.11a); the sample, which must not be very deep, is contained in a piece of muslin and partly flooded with warm water. Nematode worms leave the sample, and fall to the bottom of the funnel where they collect in the stem. They may be drawn off, in a little water, by opening the pinchcock (Peters, 1955). A number of refinements of this method have been introduced.

Overgaard Nielsen (1947) ran a battery of Baermann funnels within a box with a lamp above; this heated the surface of the water in the funnels which was initially cold. Nematodes and rotifers could be extracted from soil and moss by this method, but it was not satisfactory for tardigrades. In order to extract Enchytraeidae, O'Connor (1957, 1962) used larger polythene funnels, spread the sample directly on to the gauze, without muslin, and submerged it completely; a powerful shrouded lamp heated the

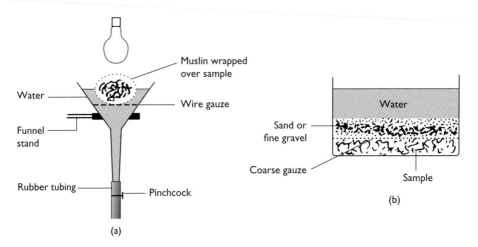

Figure 6.11 **(a) Simple heated Baermann funnel; (b) Sand extractor.**

surface of the water to 45 °C in 3 h; dry samples must be moistened before being placed in the funnel, and if the soil temperature is low the samples must be gradually warmed to room temperature for a day. Extraction usually takes about 3 h. In a series of tests O'Connor (1962) and Peachey (1962) have shown this method to be as efficient as the Nielsen inverted extractor (see below) for grassland soils, and more efficient for peat and woodland soils. It is also very efficient for second and later instar tipulid larvae in moorland soils (Hadley, 1971), though Kiritani & Matuzaki (1969) found only 10% of the larvae of the cotton root knot eelworm, *Meloidogyne incognita* to be extracted after a week. Didden *et al.* (1995) compared the efficiency of extraction of Enchytraeidae by wet funnel techniques and concluded that all of the methods tried gave good estimates. However, the omission of heating and a long extraction time gave the highest rates of extraction. These results suggest that, given sufficient time, heat should not be used to extract enchytraeids.

Whitehead and Hemming (1965) introduced a modification that has been termed the 'Whitehead tray'. The soil sample is placed in a small (23 × 33 cm) shallow tray of 8 mesh per cm phosphor-bronze wire cloth, the inside of the tray being lined with paper tissue and the outside supported by a plastic-coated wire basket. This stands in a plastic photographic tray or similar vessel, and is flooded with water until the surface reaches the underside of the soil (i.e. it is not inundated as much as in the Baermann funnel; Fig. 6.11a). After 24 h at room temperature the wire basket is carefully lifted out and the water in the tray agitated and poured into vessels resembling separating funnels. After about 4 h the nematodes collect at the tapered part of the funnel, which may be separated off by a rubber bung or diaphragm on a handle; the stop-cock is opened and the nematodes drawn off. The process may be repeated to further concentrate them. In comparative tests with clay, loam and sand and different nematodes, Whitehead & Hemming showed that this method usually extracted more efficiently than Seinhorst's 'Two Erlenmeyer method', elutriation or a sedimentation method they developed.

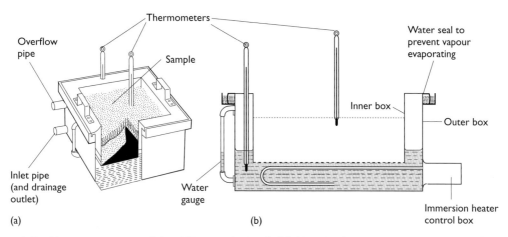

Figure 6.12   Hot-water extractor (after Milne *et al.*, 1958). (a) Sketch with segment 'removed'; (b) Sectional diagram.

6.4.2.2   Hot-water extractors

These have been developed by Nielsen (1952–53 for Enchytraeidae, and by Milne *et al.* (1958) for tipulid larvae. Nielsen's apparatus involved the heating of the lower surface of the sample, contained in an earthenware vessel, by hot water; the animals moved up and entered a layer of sand, kept cool by a cold water coil that was placed above the sample. The only advantage of this method over O'Connor's funnel (see above) is that it seems to extract more of the young worms (Peachey, 1962); under conditions with a large amount of humus it is less efficient (see above).

The larvae of many insects, notably of some Coleoptera and Diptera, earthworms, molluscs, and most motile animals seem to be efficiently and rapidly extracted from soil samples if these are heated from below. Schjotz-Christensen (1957) recorded 85% extraction of Elateridae using Nielsen's apparatus, modified by heating more strongly and omitting the cooling coil. Such an apparatus is not very different from that of Milne *et al.* (1958) (Fig. 6.12), which was found to be virtually 100% efficient for active larvae and pupae. Essentially, their device consists of two galvanised boxes, one within the other; the space between is a water bath heated by a thermostatically controlled immersion heater. The water temperature should not rise above 90 °C. The inner box has a wire gauze base and holds the sample. The initial mode of operation is to fill the water bath until the level is just above the base of the sample, the heater is switched on until the temperature registered by the thermometer in the centre of the turf is 40 °C; the water level is then raised to the level where it would just, eventually, flood the surface of the sample.

Heating and observation is continued, the insects being picked up as they leave the turf (but not before or they will retreat). The light above the extraction apparatus should not be too bright; Milne and co-workers recommended as a maximum, a 4-ft 40-W natural fluorescent tube held 3 ft (91.5 cm) above the sample; however, other species of insects might be more or less sensitive. Extraction times varied from 1 h in a light soil to nearly 3 h in heavy peat-covered clay. The chief disadvantage of this method as at present designed is that the apparatus must be watched continuously and the animals removed as they appear; however, equivalent mechanical methods usually take longer.

6.4.2.3   Sand extractors

Extremely simple in design, consisting of a metal can, a piece of wire gauze and some sand (Fig. 6.11b), these devices seem to be remarkably efficient for extracting aquatic and semi-aquatic insect larvae and other animals from mud and debris. The method was described by Bidlingmayer (1957) for the extraction of ceratopogonid midge larvae from salt marshes, and has been improved and extended by Williams (1960b), who found that the following freshwater invertebrates could be extracted from their soil substrate by this technique: Coelenterata, Turbellaria, Nematoda, Oligochaeta, Ostracoda, Acarina, Dipterous larvae, Gastropoda and Plecopoda. The sample, about 5 cm thick, is cut so as to fill the container completely: it is then covered with 5 cm of dry clean sand and the whole is flooded with water. After 24 h the sand may be scooped off and will contain large numbers of animals, but for 100% extraction of *Culicoides* larvae it should be left for 40 h (Williams, 1960b) and for other groups tests would need to be made. The animals may be separated from the sand, as both Bidlingmayer and Williams recommend, by stirring in a black photographic tray when the pale moving objects easily show up against the dark background. If the promise of this method is justified it would clearly repay further development. The final separation might be facilitated by the adoption of a sedimentation, elutriation or flotation method, aided perhaps by the replacement of the sand by fine washed gravel as used by Nielsen (1952–53 in his extractor (see above) that really combines this method with the hot-water process. The chief disadvantages of the method in its present form is that it is difficult to make a clean separation of the sand from the soil after extraction; the placing of a piece of wide-mesh wire gauze between the two substrates might aid this operation, or if the extraction was done in a square vessel a plate could be pushed across between the layers at the end of extraction and the top layer poured off.

6.4.2.4   Cold-water extractor

South (1964) found that slugs could be efficiently extracted from 1 ft (30.5 cm) -deep turves by slowly immersing these in water. Initially, they are stood in 1 in (2.5 cm) of water; after about 17 h the water level is raised to half the depth of the turf (i.e. 6 in, 15.2 cm); after 2 days the water level is again raised, this time in several stages, until at the end of another 24 h it is within a half-inch (1.2 cm) of the top of the turf. This approach has also been used to extract the collembolan, *Hyopogastrura* (Kraan, 1973) and overwintering rice water weevils, *Lissorhoptrus* (Gifford & Trahan, 1969). The latter authors used a technique basically similar to the 'van Emden biscuit tin' (see Fig. 10.4, p. 381), but constructed from plastic bleach bottles, in which the sample is flooded. Such a technique is, of course, only suitable for robust insects.

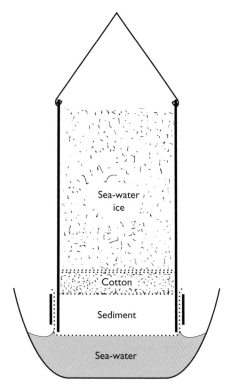

Figure 6.13   A cold-water extractor for marine interstitial fauna (after Uhlig, 1966, 1968).

6.4.2.5   Low-salinity extractor

Uhlig (1966, 1968) developed a method for marine meiofauna using melting seawater ice. The sample is placed in a tube with a gauze base. A layer of cotton wool, followed by crushed sea ice, is placed above it and the tube suspended in a dish of seawater (Fig. 6.13). As the ice melts the flow of cold, low-salinity water forces the fauna into the dish. Uhlig considered that it was the salinity rather than temperature gradient that forced the animals to move. This method can only be applied to sandy substrates.

6.4.2.6   Mistifier

This is essentially a method of extracting active nematodes from plant tissue. The plant material is spread on a tray, broken up and continually sprayed with very fine jets of water ('mists') (Webster, 1962). Combined with a cotton wool filter to collect the nematodes, this method can be much more efficient than other techniques for certain eelworms (Guiran, 1966; Oostenbrink, 1970).

6.4.2.7   Chemical extraction

Chemical fumes may be used to drive animals from vegetation (p. 149); the extension of such methods to soil and litter animals has been demonstrated successfully only with aphids and thrips by MacFadyen (1953), who used dimethylphthalate and 2-cyclohexyl-4,6-dinitrophenol. Chemicals in a liquid form, notably potassium permanganate solution, *ortho*-dichlorobenzene and formalin (often with a 'wetter') have been used for the extraction of insect larvae and earthworms (Svendesn, 1955; Milne *et al.*, 1958; Raw, 1959). Formalin should be applied four times, with 15- min intervals, at a rate of 75 ml of 4% formaldehyde per 3.5 1itre water per square metre of ground (Raw, 1959; Nordstrom & Rundgren, 1972). Many workers have found these methods to be inefficient for earthworms (Satchell, 1955; Svendsen, 1955); for tipulids in the grasslands studied by Milne *et al.* (1958) the efficiency was 85%. However, the method can be efficient, depending on the type of soil, the species of animal and the season (which will influence the activity level of the animal). The more porous the soil (Boyd, 1958) and the more active the earthworm, the more satisfactory the method. Among earthworms it is as efficient as hand-sorting for *Lumbricus terrestris* during active periods; on sandy soils it seems suitable for *Lirabellus and Dendrobaena octaedra*, but is seldom reliable for *Allobophora* spp.; where it is efficient it is more reliable for small worms than hand-sorting (Nordstrom & Rundgren, 1972). Daniel *et al.* (1992) found little difference in the efficiency of extraction of juvenile *Lumbricus terrestris* using formaldehyde and chloroacetophenone. Earthworms can be sampled using the 'hot mustard method' (Lawrence & Bowers, 2002; Burtis *et al.*, 2014), where a mixture of 80 g of powdered mustard seed in 8 litres of water is poured into a 50 × 50 cm area. The worms are collected with forceps as they leave the soil to escape the irritating effect of the mustard.

6.4.2.8 Electrical extraction

By discharging a current from a water-cooled electrode driven into the soil, Satchell (1955) was able to expel large numbers of earthworms; however, as he has pointed out, this method suffers from the disadvantage that the exact limits of the volume of soil treated are unknown (Nelson & Satchell, 1962) and the efficiency of extraction seems directly related to the soil pH (Edwards & Lofty, 1975). This method has been used to collect earthworms for pollution studies (Pizl & Josens, 1995).

6.5 Summary of the applicability of the methods

It has already been pointed out that the efficiency of the different methods varies greatly with the animal group and the substrate, as is shown for example by the studies of Nef (1960), Raw (1959, 1962), MacFadyen (1961), O'Connor (1962) and Satchell & Nelson (1962). The separation methods recommended by Edwards (1991) for different soil arthropod groups collected from woodland, pasture and arable land are listed in Table 6.1. The comments made here should only be taken as general indications; more specific information is given under each method.

Substrate type

Because of the difficulties of separating plant and animal material, the mechanical methods have generally been considered unsuitable for litter and for soils, pond mud and other substrates containing a large amount of organic matter. For these media, behavioural methods are likely to be the most satisfactory, and the hot-water and sand extraction techniques may well repay further development. There are, however, some exceptions; flotation is suitable for dung-dwelling insects (Laurence, 1954), Williamson's wet sieving (which may contain a behavioural element) works well for snails in fresh litter, and Satchell & Nelson (1962) found that flotation was more efficient than dry funnels for the extraction of Scutacarid mites from a moder soil.

With friable and sandy soils or aquatic sediments mechanical methods are generally to be preferred, and techniques as simple as wet-sieving or flotation, perhaps using a heavy organic solvent, may be satisfactory for the study of a single species. Marine benthic macrofauna are usually preserved in the field, where a stain such as rose bengal can be added later and they can be sorted by wet-sieving followed by manual picking of individuals from the debris. It has been common practice to use buffered 10% formalin solution as a preservative. While, exposure to formalin fumes should be avoided, there is no adequate substitute preservative available that will penetrate marine soft sediment samples. Occasionally, floatation or elutriation is used, possibly after initial sieving.

Animal type

Obviously, immobile stages can only be extracted by mechanical methods. Comparatively large and robust animals can often be estimated simply by sieving. When the specific gravity of the animal is very different from that of the substrate particles, possible techniques include elutriation and centrifugal separation, as well as the more widely used flotation. The two first named methods might also be useful for the separation of parasitised and unhealthy individuals from healthy ones, and for the extraction of the eggs of a single species. Broadly speaking, Acarina and Collembola seem to be more efficiently extracted by an appropriate funnel technique and insects, except on peat soils, by a flotation method (Edwards & Fletcher, 1970). Marine benthic meiofauna in coarse sediments can often be separated by simple decantation or floatation and from muddy substrates by sieving and centrifugation.

Behavioural methods must naturally be adopted to fit the behaviour of the animal; fast-moving animals that are comparatively resistant to desiccation need to be extracted by the horizontal extractor or the large Berlese funnel; the last named, when modified as described here, is also the only method really suitable for positively phototactic species such as Lygaeid bugs and phytophagous beetles. Groups such as Collembola and most mites need a slower extraction in the Kempson or multiple-canister extractors. Those animals that are very sensitive to desiccation (e.g. nematode and enchytraeid worms, dipterous and probably many coleopterous larvae) should be separated in wet extractors, which is also, of course, the behavioural method appropriate for animals in aquatic sediments. The degree of complexity for a single group is well illustrated by the earthworms, for while *Lumbricus terrestris*, but not other species, may be efficiently extracted from orchard soils by a surface application of formalin and several species from shallow soils by wet sieving and flotation, in other habitats even hand-sorting fails (Raw, 1959, 1960; Nelson & Satchell, 1962).

Cost

The time required to undertake an extraction is the most important factor determining the total cost of the procedure. Methods differ greatly in labour and time required. When using behavioural methods there is often a direct relationship between cost (time) and extraction efficiency. It has generally been found that behavioural methods work more quickly when the samples are exposed to higher gradients of environmental conditions such as temperature, salinity or humidity. However, using weaker gradients and taking longer over the extraction often – but by no means always – produces higher rates of extraction.

References

Abu Yaman, I.K. (1960) Natural control in cabbage root fly populations and influence of chemicals. *Meded. LandbHoogesch. Wageningen* **60**, 1–57.

Adis, J. (1987) Extraction of arthropods from neotropical soils with a modified Kempson apparatus. *J. Trop. Ecol.*, **3**(2), 131–8.

Anderson, J.M. & Healey, I.N. (1970) Improvements in the gelatine-embedding technique for woodlands soil and litter samples. *Pedobiologia* **8**, 108–20.

Aucamp, J.L. (1967) Efficiency of the grease film extraction technique in soil micro-arthropod survey. In: Graff, O. & Satchell, J.E. (eds), *Progress in Soil Biology*. North-Holland, Amsterdam.

Belfield, W. (1956) The arthropoda of the soil in a west African pasture. *J. Anim. Ecol.* **259**, 275–87.

Bidlingmayer, W.L. (1957) Studies on Culicoidesfurens (Poey) at Vero Beach. *Mosquito News* **17**, 292–4.

Birkett, L. (1957) Flotation technique for sorting grab samples. *J. Comm. Perm. Int. Explor. Mer.* **22**, 289–92.

Block, W. (1966) Some characteristics of the MacFadyen high gradient extraction for soil micro-arthropods. *Oikos* **17**, 1–9.

Boisseau, J-P. (1957) Technique pour l'étude quantitative de la faune interstitielle des sables. *Comptes Rendus du Congrés des Societes Savantes de Paris et des Départements* **1957**, 117–19.

Boyd, J.M. (1958) Ecology of earthworms in cattle-grazed machair in Tiree, Argyll. *J. Anim. Ecol.* **27**, 147–57.

Brady, J. (1969) Some physical gradients set up in Tullgren funnels during the extraction of mites from poultry litter. *J. Appl. Ecol.* **6**, 391–402.

Brian, M. V. (1970) Measuring population and energy flow in ants and termites. In: Phillipson, J. (ed.), *Methods of Study in Soil Ecology*. UNESCO, Paris.

Brown, R.D. (1973) Funnel for extraction of leaf litter organisms. *Ann. Entomol. Soc. Am.* **66**, 485–6.

Burtis, J.C., Fahey, T.J., & Yavitt, J.B. (2014) Impact of invasive earthworms on *Ixodes scapularis* and other litter-dwelling arthropods in hardwood forests, central New York state, USA. *Appl. Soil Ecol.* **84**, 148–57.

Clark, E.W., Williamson, A.L., & Richmond, C.A. (1959) A collecting technique for pink bollworms and other insects using a Berlese funnel with an improved heater. *J. Econ. Entomol.* **52**, 1010–12.

Cockbill, G.F., Henderson, V.E., Ross, D.M., & Stapley, J.H. (1945) Wire-worm populations in relation to crop production. I. A large-scale flotation method for extracting wireworms from soil samples and results from a survey of 600 fields. *Ann. Appl. Biol.* **32**, 148–63.

Crossley, D.A. & Blair, J.M. (1991) A high efficiency, 'low technology' Tullgren-type extractor for soil extractor for soil microarthropods. *Agricult. Ecosyst. Environ.* **34**, 187–92.

D'Aguilar, J., Bernard, R., & Bessard, A. (1957) Une méthode de lavage pour l'extraction des arthropodes terricoles. *Ann. Epiphytol. C* **8**, 91–9.

Daniel, O., Jager, P., Cuendet, G., & Bieri, M. (1992) Sampling of *Lumbricus terrestris* (Oligocheata, Lumbricidae). *Pedobiologia* **36**, 213–20.

Danthanarayan, A.W. (1966) Extraction of arthropod eggs from soil. *Entomol. Exp. Appl.* **9**, 124–5.

Davies, J.B. & Linley, J.B. (1966) A standardised flotation method for separating Leptoconops (Diptera: Ceratopogonidae) and other larvae from sand samples. *Mosquito News* **26**, 440.

De Jonge, V.N. & Bouwman, L.A. (1977) A simple density separation technique for quantitative isolation of meiobenthos using colloidal silica Lodox-TM. *Mar. Biol.* **82**, 379–84.

Dhillon, B.S. & Gibson, N.H.E. (1962) A study of the Acarina and Collembola of agricultural soils. 1. Numbers and distribution in undisturbed grassland. *Pedobiologia* **1**, 189–209.

Didden, W., Born, H., Domm, H., Graefe, U., Heck, M., Kuhle, J., Mellin, A., & Rombke, J. (1995) The relative efficiency of wet funnel techniques for the extraction of Enchytraeidae. *Pedobiologia* **39**, 52–7.

Dietrick, E.J., Schlinger, E.I., & Van Den Bosch, R. (1959) A new method for sampling arthropods using a suction collecting machine and modified Berlese funnel separator. *J. Econ. Entomol.* **52**, 1085–91.

Dindal, D.L. (1990) *Soil Biology Guide*. Wiley Interscience, New York.

Doane, I.F. (1969) A method for separating the eggs of the prairie grain wireworm, *Ctenicera destructor*, from soil. *Can. Entomol.* **101**, 1002–4.

Dondale, C.D., Nicholls, C.F., Redner, J.H., Semple, R.B., & Turnbull, A.L. (1971) An improved Berlese–Tullgren funnel and a flotation separator for extracting grassland arthropods. *Can. Entomol.* **103**, 1549–52.

Duffey, E. (1962) A population study of spiders in limestone grassland. Description of study area, sampling methods and population characteristics. *J. Anim. Ecol.* **31**, 571–99.

Edwards, C.A. (1991) The assessment of populations of soil-inhabiting invertebrates. *Agricult. Ecosyst. Environ.* **34**, 145–76.

Edwards, C.A. & Fletcher, K.E. (1970) Assessment of terrestrial invertebrate populations. In: Phillipson, J. (ed.), *Methods of Study in Soil Ecology*. UNESCO, Paris.

Edwards, C.A. & Lofty, J.R. (1975) The invertebrate fauna of the Park Grass Plots. 1. Soil fauna. *Rep. Rothamsted Exp. Sta. 1974, Part 2*, 133–54.

Edwards, C.A., Whiting, A.E., & Heath, G.W. (1970) A mechanized washing method for separation of invertebrates from soil. *Pedobiologia* **10**, 141–8.

Eleftheriou, A. (ed.) (2013) *Methods for the Study of Marine Benthos*. John Wiley & Sons.

Eleftheriou, A. & Holme, N.A. (1984) Macrofauna techniques. In: Holme, N.A. & McIntyre, A.D. (eds), *Methods for the Study of Marine Benthos*. Blackwell Scientific Publications, Oxford, pp. 140–216

Erman, D.C. (1973) Invertebrate movements and some diel and seasonal changes in a Sierra Nevada peatland. *Oikos* **24**, 85–93.

Ford, J. (1937) Fluctuations in natural populations of Collembola and Acarina. *J. Anim. Ecol.* **6**, 98–111.

Frick, K.E. (1962) Ecological studies on the alkali bee, *Nomia melanderi*, and its Bombyliid parasite, *Heterostylum robustum* in Washington. *Ann. Entomol. Soc. Am.* **559**, 5–15.

Gabbutt, P.D. (1959) The bionomics of the wood cricket, *Nemobius sylrestris* (Orthoptera: Gryllidae). *J. Anim. Ecol.* **28**, 15–42.

Gayatri, R.G. (2011) Efficiency of different extraction methods for the macro-fauna of household biocompost. *J. Ecobiol.* **28**, 371–4.

Gerard, B.M. (1967) Factors affecting earthworms in pastures. *J. Anim. Ecol.* **36**, 235–52.

Gifford, J.R. & Trahan, G.B. (1969) Apparatus for removing overwintering adult rice water weevils from bunch grass. *J. Econ. Entomol.* **62**, 752–4.

Goffart, H. (1959) Methoden zur Bodenuntersuchung auf nichtzystenbildende Nematoden. *Nachr Bl. Dtsch. PflSchDienst, Stuttgart* **11**, 49–54.

Golightly, W.H. (1952) Soil sampling for wheat-blossom midges. *Ann. Appl. Biol.* **39**, 379–84.

Goodey, J.B. (1957) *Laboratory methods for work with plant and soil nematodes.* H.M.S.O., London.

Gorny, M. & Grum, L. (eds) (1993) *Methods in Soil Zoology.* Amsterdam, Elsevier.

Guiran, G.D. (1966) Infestation actuelle et infestation potentielle du sol par nematodes phytoparasites du genre *Meloidogyne. C. R. Lebd. Seanc. Acad. Sci. (Paris)* **262**, 1754–6.

Haarlov, N. (1947) A new modification of the Tullgren apparatus. *J. Anim. Ecol.* **16**, 115–21.

Haarlov, N. & Weis-Fogh, T. (1953) A microscopical technique for studying the undisturbed texture of soils. *Oikos* **4**, 44–7.

Haarlov, N. & Weis-Fogh, T. (1955) A microscopical technique for studying the undisturbed texture of soils. In: Kevan, D.K.M. (ed.), *Soil Zoology.* University of Nottingham School of Agriculture.

Hadley, M. (1971) Aspects of the larval ecology and population dynamics of *Niolophilus ater* Meigen (Diptera: Tipulidae) on Pennine Moorland. *J. Anim. Ecol.* **40**, 445–66.

Hale, W.G. (1964) A flotation method for extracting Collembola from organic soils. *J. Anim. Ecol.* **33**, 363–9.

Hamilton, A.C. (1969) A method of separating invertebrates from sediments using longwave ultraviolet light and fluorescent dyes. *J. Fish. Res. Bd Can.* **26**, 1667–72.

Hammer, M. (1944) Studies on the Oribatids and Collemboles of Greenland. *Medd. Gronland* **141**, 1–210.

Hartley, J.P., Dicks, B., & Wolff, W.J. (1987) Processing sediment macrofauna samples. In: *Biological Surveys of Estuaries and Coasts.* Cambridge University Press, Cambridge, pp. 131–9.

Heath, G.W. (1965) An improved method for separating arthropods from soil samples. *Lab. Pract. April 1965.* **14**, 14–23.

Higgins, R.P. (1977) Two new species of Echinodered (Kinorhyncha) from South Carolina. *Trans. Am. Microsc. Soc.* **96**, 340–54.

Hockin, D.C. (1981) A simple elutriator for extracting meiofauna from sediment matrices. *Mar. Ecol. Prog. Series* **4**, 241–2.

Horsfall, W.R. (1956) A method for making a survey of floodwater mosquitoes. *Mosquito News* **16**, 66–71.

Hughes, R.D. (1960) A method of estimating the numbers of cabbage root fly pupae in the soil. *Plant Pathol.* **9**, 15–17.

Iversen, T.M. (1971) The ecology of the mosquito population (*Aedes communis*) in a temporary pool in a Danish beech wood. *Arch. Hydrobiol.* **69**, 309–32.

Jacot, A.P. (1935) Molluscan populations of old growth forests and rewooded fields in the Asheville basin of N. Carolina. *Ecology* **16**, 603–5.

Kempson, D., Lloyd, M., & Ghelardi, R. (1963) A new extractor for woodland litter. *Pedobiologia* **3**, 1–21.

Kethley, J. (1991) A procedure for extraction of microarthropods from bulk soil samples with emphasis on inactive stages. *Agricult. Ecosyst. Environ.* **34**, 193–200.

Kevan, D.K.M. (1962) *Soil Animals.* Witherby, London.

King, A.B.S. (1975) The extraction, distribution and sampling of the eggs of the sugarcane froghopper, *Aeneolamia varia saccharine* (Dist.) (Homoptera, Cercopidae). *Bull. Entomol. Res.* **65**, 157–64.

Kiritani, K. & Matuzaki, T. (1969) On the extraction rate of nematodes by the Baermann funnel. *Bull. Kochi Inst. Agr. For. Sci.* **2**, 25–30.

Kitching, R.L. (1971) A core sampler for semi-fluid substances. *Hydrobiologia* **37**, 205–9.

Kraan, C.V.D. (1973) Populations okologische untersuchungen an *Hypogastrura vintica* Tullb. 1872 (Collembola) anf Schiermonnikoog. *Faun-okol. Mitt.* **4**, 197–206.

Krell, F.T., Chung, A.Y., DeBoise, E., Eggleton, P., Giusti, A., Inward, K., & Krell-Westerwalbesloh, S. (2005) Quantitative extraction of macro-invertebrates from temperate and tropical leaf litter and soil: efficiency and time-dependent taxonomic biases of the Winkler extraction. *Pedobiologia* **49**(2), 175–86.

Lackey, R.T. & May, B.E. (1971) Use of sugar flotation and dye to sort benthic samples. *Trans. Am. Fish. Soc.* **100**, 794–7.

Ladell, W.R.S. (1936) A new apparatus for separating insects and other arthropods from the soil. *Ann. Appl. Biol.* **23**, 862–79.

Lane, M. & Shirck, E. (1928) A soil sifter for subterranean insect investigations. *J. Econ. Entomol.* **21**, 934–6.

Lasebikan, B.A. (1971) The relationship between temperature and humidity and the efficient extraction of Collembola by a dynamic-type method. *Rev. Ecol. Biol. Sol.* **8**, 287–93.

Lauff, G.M., Cummins, K.W., Enksen, C.H., & Parker, M. (1961) A method for sorting bottom fauna samples by elutriation. *Limnol. Oceanogr.* **6**, 462–6.

Laurence, B.R. (1954) The larval inhabitants of cow pats. *J. Anim. Ecol.* **23**, 234–60.

Lawrence, A.P. & Bowers, M.A. (2002) A test of the 'hot' mustard extraction method of sampling earthworms. *Soil Biol. Biochem.* **34**, 549–52.

Lincoln, C. & Palm, C.E. (1941) Biology and ecology of the Alfalfa Snout beetle. *Mem. Cornell Univ. Agric. Exp. Sta.* **236**, 3–45.

Lydell, W.R.S. (1936) A new apparatus for separating insects and other arthropods from the soil. *Ann. Appl. Biol.* **23**, 862–79.

Lyman, F.E. (1943) A pre-impoundment bottom fauna study of Watts Bar Reservoir area (Tennessee). *Trans. Am. Fish. Soc.* **72**, 52–62.

MacFadyen, A. (1953) Notes on methods for the extraction of small soil arthropods. *J. Anim. Ecol.* **22**, 65–78.

MacFadyen, A. (1955) A comparison of methods for extracting soil arthropods. In Kevan, D.K.M. (ed.), *Soil Zoology.* University of Nottingham School of Agriculture.

MacFadyen, A. (1961) Improved funnel-type extractors for soil arthropods. *J. Anim. Ecol.* **30**, 171–84.

MacFadyen, A. (1962) Soil arthropod sampling. *Adv. Ecol. Res.* **1**, 1–34.

MacFadyen, A. (1968) Notes on methods for the extraction of small soil arthropods by the high gradient apparatus. *Pedobiologia* **8**, 401–6.

Mason, C.F. (1970) Snail populations, beech litter production and the role of snails in litter decomposition. *Oecologia (Berl.)* **5**, 215–39.

McIntyre, A.D. & Warwick, R.M. (1984) Meiofauna techniques. In: Holme, N.A. & McIntyre, A.D. (eds), *Methods for the Study of Marine Benthos*, 2nd edn. Blackwell Scientific Publications, Oxford, pp. 217–44.

McLeod, J.M. (1961) A technique for the extraction of cocoons from soil samples during population studies of the Swaine sawfly, *Neodiprion swainei* Midd. (Hymenoptera: Diprionidae). *Can. Entomol.* **91**, 888–90.

Milne, A., Coggins, R.E., & Laughlin, R. (1958) The determination of the number of leather-jackets in sample turves. *J. Anim. Ecol.* **27**, 125–45.

Minderman, G. (1957) The preparation of microtome sections of unaltered soil for the study of soil organisms in situ. *Plant Soil* **8**, 42–8.

Müller, G. (1962) A centrifugal–flotation extraction technique and its comparison with two funnel extractors. In: Murphy, P.W. (ed.), *Progress in Soil Zoology*. Butterworths, London, pp. 207–11.

Murphy, P.W. (1955) Notes on processes used in sampling, extraction and assessment of the meiofauna of heathland. In: Kevan, D.K.M. (ed.), *Soil Zoology*. University of Nottingham School of Agriculture.

Murphy, P.W. (1962a) Extraction methods for soil animals. II. Mechanical methods. In: Murphy, P.W. (ed.) *Progress in Soil Zoology*. Butterworth, London.

Murphy, P.W. (ed.) (1962b) *Progress in Soil Zoology*. Papers from a colloquium on Research methods organised by the Soil Zoology Committee of the International Society of' Soil Science held at Rothamsted Experimental Station Hertfordshire 4th July, 1958. Butterworth, London.

Nef, L. (1960) Comparaison de l'efficacité de différentes variantes de l'appareil de Berlese–Tullgren. *Z. Agnew. Entomol.* **46**, 178–99.

Nef, L. (1970) Reactions des acariens a une dessication lente de la litiere. *Rel. Ecol. Biol. Soil* **7**, 381–92.

Nef, L. (1971) Influence de l'humidite sur le geotactisme des Oribates (Acarina) dans l'entracteur de Berlese–Tullgren. *Pedobiologia* **11**, 433–45.

Nelson, J.M. & Satchell, J.E. (1962) The extraction of Lumbricidae from soil with special reference to the hand-sorting method. In: Murphy, P.W. (ed.), *Progress in Soil Zoology*. Butterworth, London.

Newell, I. (1955) An autosegregator for use in collecting soil-inhabiting arthropods. *Trans. Am. Microsc. Soc.* **74**, 389–92.

Nielsen, C.O. (1952–53) Studies on Enchytraeidae 1. A technique for extracting Enchytracidae from soil samples. *Oikos* **4**, 187–96.

Nordstrom, S. & Rundgren, S. (1972) Methods of sampling lumbricids. *Oikos* **239**, 344–52.

Norton, R.A. & Kethley, J.B. (1988) A collapsible, full-sized Berlese-funnel system. *Entomol. News* **99**, 41–7.

O'Connor, F.B. (1957) An ecological study of the Enchytraeid worm population of a coniferous forest soil. *Oikos* **8**, 162–99.

O'Connor, F.B. (1962) The extraction of Enchytraeidae from soil. In: Murphy, P.W. (ed.), *Progress in Soil Zoology*. Butterworth, London.

Oostenbrink, M. (1954) Een doelrnatige methode voor het toetsen van aaltjesbestrijdingsmiddelen in grond met Hopiolaimus uniformis als proefdier. *Meded. Landb. Hoogesch. Gent.* **19**, 377–408.

Oostenbrink, M. (1970) Comparison of techniques for population estimation of soil and plant nematodes. In: Phillipson, J. (ed.), *Methods of Study in Soil Ecology*. UNESCO, Paris.

Overgaard Nielsen, C. (1947–48) An apparatus for quantitative extraction of nematodes and rotifers from soil and moss. *Naturajutl.* **1**, 271–8.

Pande, Y.D. & Berthet, P. (1973) Comparison of the Tullgren funnel and soil section methods for surveying Oribatid populations. *Oikos* **24**, 273–7.

Paris, O.H. & Pitelka, F.A. (1962) Population characteristics of the terrestrial isopod *Armadillidium vulgare* in California grassland. *Ecology* **43**, 229–48.

Park, O. & Auerbach, S. (1954) Further study of the tree-hole complex with emphasis on quantitative aspects of the fauna. *Ecology* **35**, 208–22.

Pauly, D. (1973) Uber ein Gerat zur Vorsortierung von Benthosproben. *Ber. Deutsch. Wissenschaft. Komm. Meeresforsch.* **22**, 458–60.

Pask, W.M. & Costa, R.M. (1971) Efficiency of sucrose flotation in recovering insect larvae from benthic steam samples. *Can. Entomol.* **103**, 1649–52.

Peachey, J.E. (1962) A comparison of two techniques for extracting Enchytraeidae from moorland soils. In: Murphy, P.W. (ed.), *Progress in Soil Zoology*. Butterworth, London.

Peters, B.G. (1955) A note on simple methods of recovering nematodes from soil. In: Kevan, D.K.M. (ed.), *Soil Zoology*. University of Nottingham School of Agriculture.

Phillipson, J. (ed.) (1970) *Methods of Study in Soil Ecology*. UNESCO, Paris.

Pinder, L.C.V. (1974) The ecology of slugs in potato crops, with special reference to the differential susceptibility of potato cultivars to slug damage. *J. Appl. Ecol.* **11**, 439–51.

Pizl, V. & Josens, G. (1995) The influence of traffic pollution on earthworms and their heavy metal contents in an urban ecosystem. *Pediobiologia* **39**, 442–53.

Potzger, J.E. (1955) A borer for sampling in permafrost. *Ecology* **36**, 161.

Price, D.W. (1975) Vertical distribution of small arthropods in a Californian Pine Forest soil. *Ann. Entomol. Soc. Am.* **68**, 174–80.

Rapoport, E.H. & Oros, E. (1969) Transporte y manipuleo de las muestras de suclo y su efecto sobre la micro y mesofauna. *Rev. Ecol. Biol. Soc.* **6**, 31–9.

Raw, F. (1955) A flotation extraction process for soil micro-arthropods. In: Kevan, D.K.M. (ed.), *Soil Zoology*. School of Agriculture, University of Nottingham, Nottingham, pp. 341–6.

Raw, F. (1959) Estimating earthworm populations by using formalin. *Nature* **184**, 1661–2.

Raw, F. (1962) Flotation methods for extracting soil arthropods. In: Murphy, P.W. (ed.), *Progress in Soil Zoology*. Butterworth, London.

Read, D.C. (1958) Note on a flotation apparatus for removing insects from soil. *Can. J. Plant Sci.* **38**, 511–14.

Sabu, T.K. & Shiju, R.T. (2010) Efficacy of pitfall trapping, Winkler and Berlese extraction methods for measuring ground-dwelling arthropods in moist-deciduous forests in the Western Ghats. *J. Insect Sci.* **10**(98), 1–17.

Salt, G. & Hollick, F.S.J. (1944) Studies of wireworm populations. 1. A census of wireworms in pasture. *Ann. Appl. Biol.* **31**, 53–64.

Sakchoowong, W., Nomura, S., Ogata, K., & Chanpaisaeng, J. (2007) Comparison of extraction efficiency between Winkler and Tullgren extractors for tropical leaf litter macroarthropods. *Thai J. Agric. Sci.* **40**(3-4), 97–105.

Sands, W.A. (1972) Problems in attempting to sample tropical subterranean termite populations. *Ekologia Polska* **20**, 23–31.

Satchell, J.E. (1955) An electrical method of sampling earthworm populations. In: Kevan, D.K.M. (ed.), *Soil Zoology*. University of Nottingham School of Agriculture.

Satchell, J.E. & Nelson, J.M. (1962) A comparison of the Tullgren-funnel and flotation methods of extracting acarina from woodland soil. In: Murphy, P.W. (ed.), *Progress in Soil Zoology*. Butterworth, London.

Saunders, L.G. (1959) Methods for studying Forcipomyia midges, with special reference to cacao-pollinating species (Diptera, Ceratopogonidae). *Can. J. Zool.* **27**, 33–51.

Schjotz-Christensen, B. (1957) The beetle fauna of the Corynephoretum in the ground of the Mols Laboratory, with special reference to *Cardiophorus asellus* Er. (Elateridae). *Natura Jutl.* **6-7**, 1–120.

Schouten, A.J. & Kmarp, K. (1991) A comparative study in the efficiency of extraction methods for nematodes from different forest litters. *Pedobiologia* **35**, 393–400.

Seinhorst, J.W. (1962) Extraction methods for nematodes inhabiting soil. In: Murphy, P.W. (ed.), *Progress in Soil Zoology*. Butterworth, London.

Sellmer, G.P. (1956) A method for the separation of small bivalve molluscs from sediments. *Ecology* **37**, 206.

Service, M.W. (1968) A method for extracting mosquito eggs from soil samples taken from oviposition sites. *Ann. Trop. Med. Parasitol.* **62**, 478–80.

Shaw, G.G. (1970) Grease-film extraction of an arthropod: a modification for organic soils. *J. Econ. Entomol.* **63**, 1323–4.

Smrz, J. (1992) The ecology of the microarthropod community inhabiting the moss cover of roofs. *Pedobiologia* **36**, 331–40.

Somerfield, P.J., Warwick, R.M., & Moens, M. (2005) Meiofauna techniques. *Methods for the Study of Marine Benthos*, Third edition, pp. 229–72.

South, A. (1964) Estimation of slug populations. *Ann. Appl. Biol.* **53**, 251–8.

Speight, M.C.D. (1973) A greased-belt technique for the extraction of arthropods from organic debris. *Pedobiologia* **13**, 99–106.

Stage, H.H., Gjullin, C.M., & Yates, W.W. (1952) *Mosquitoes of the Northwestern States*, Agricultural Handbook **46**, USDA, 95 pp.

Stephenson, J.W. (1962) An improved final sieve for use with the Salt and Hollick soil washing apparatus. In: Murphy, P.W. (ed.), *Progress in Soil Zoology*. Butterworth, London.

Stewart, K.M. (1974) A three-dimensional net sieve for extracting Tipulidae (Diptera) larvae from pasture soil. *J. Appl. Ecol.* **11**, 427–30.

Sugimoto, T. (1967) Application of oil-water flotation method to the extraction of larvae and pupae of aquatic midges from a soil sample. *Jap. J. Ecol.* **17**, 179–82.

Sunderland, K.D., Hassell, M., & Sutton, S.L. (1976) The population dynamics of *Philoscia muscorum* (Crustacea, Oniscoidea) in a dune grassland ecosystem. *J. Anim. Ecol.* **45**, 487–506.

Svendsen, J.A. (1955) Earthworm population studies: a comparison of sampling methods. *Nature* **175**, 864.

Szlauer-Łukaszewska, A. & Radziejewska, T. (2013) Two techniques of ostracod (Ostracoda, Crustacea) extraction from organic detritus-rich sediments. *Limnol. Ecol. Manag. Inland Waters* **43**(4), 272–6.

Tanton, M.T. (1969) A corer for sampling soil and litter arthropods. *Ecology* **5**, 134–5.

Törne, E. Von (1962a) An elutriation and sieving apparatus for extracting microarthropods from soil. In: Murphy, P.W. (ed.) *Progress in Soil Zoology*, Butterworth, London.

Törne, E. Von (1962b) A cylindrical tool for sampling manure and compost. In: Murphy, P.W. (ed.), *Progress in Soil Zoology*. Butterworth, London.

Uhlig, G. (1966) Untersuchungen zur Extraktion der vagilen Mikrofauna aus marinen Sedimenten. *Zoologischer Anzeiger* **29**, 151–7.

Uhlig, G. (1968) Quantitative methods in the study of interstitial fauna. *Trans. Am. Microsc. Soc.* **87**, 226–32.

Vagvolgyi, I. (1953) A new sorting method for snails, applicable also for quantitative researches. *Ann. Hist. Nat. Mus. Nat. Hung.* **44** (N.S. 3), 101–4.

Vannier, G. & Alpern, I. (1968) Techniques de prélévements pour l'étude des distribution horizontales et verticales des microarthropodes du sol. *Rev. Ecol. Biol. Sol.* **5**, 225–35.

Vannier, G. & Vidal, P. (1964) Construction d'unappareil automatique pour couper les inclusions de sol dans la gelatine. *Rev. Ecol. Biol. Sol.* **1**, 575–86.

Vannier, G. & Vidal, P. (1965) Sonde pédologique pour l'échantillonnage des microarthropodes. *Rev. Ecol. Biol. Sol.* **2**, 333–7.

Wallace, M.M.H. (1956) A rapid method of sampling small free-living pasture insects and mites. *J. Aust. Inst. Agric. Sci.* **22**, 283–4.

Walter, D.E., Kethley, J., & Moore, J.C. (1987) A heptane flotation method for recovering microarthropods from semiarid soils, with comparison to the Merchant–Crossley high-gradient extraction method and estimates of microarthropod biomass. *Pediobiologia* **30**, 221–32.

Webley, D. (1957) A method of estimating the density of frit fly eggs in the field. *Plant Pathol.* **6**, 49–51.

Webster, J.M. (1962) The quantitative extraction of *Ditylenchus dipsaci* (Kuhn) from plant tissues by a modified Seinhorst mistifier. *Nematologica* **8**, 245–51.

Whitehead, A.G. & Hemming, J.R. (1965) A comparison of some quantitative methods of extracting small vermiform nematodes from soil. *Ann. Appl. Biol.* **55**, 25–38.

Whitehouse, J.W. & Lewis, B.G. (1966) The separation of benthos from stream samples by flotation with carbon tetrachloride. *Limnol. Oceanogr.* **11**, 124–6.

Williams, D.D. & Williams, N.E. (1974) A counterstaining technique for use in sorting benthic samples. *Limnol. Oceanogr.* **19**, 152–4.

Williams, R.W. (1960a) A new and simple method for the isolation of fresh water invertebrates from soil samples. *Ecology* **41**, 573–4.

Williams, R.W. (1960b) Quantitative studies on populations of biting midge larvae in saturated soil from two types of Michigan bogs (Diptera: Ceratopogonidae). *J. Parasitol.* **46**, 565–6.

Williamson, M.H. (1959) The separation of molluscs from woodland leaf-litter. *J. Anim. Ecol.* **28**, 153–5.

Worswick, J.M. & Barbour, M.T. (1974) An elutriation apparatus for macroinvertebrates. *Limnol. Oceanogr.* **19**, 538–40.

7 Relative Methods of Population Measurement and the Derivation of Absolute Estimates

Most relative methods require comparatively simple equipment and, as they often serve to concentrate the animals, they provide impressive collections of data from situations where few animals will be found by absolute methods. This makes them particularly appropriate for initial faunal surveys, and from a statistical viewpoint plentiful data will appear to be preferable to the hard-won, often scant, information obtained from unit area sampling. Most traps will collect specimens continuously, providing a relatively large return for the amount of time spent working with them; the cost (see p. 19) of the data is low. With all these apparent advantages it is hardly surprising that these methods have been extensively used and developed. There are probably more accounts of their design and use in the literature than references to all the other topics in this book (therefore the list at the end of this chapter is highly selective).

7.1 Factors affecting the size of relative estimates

The biological interpretation of relative population estimates (p. 3) is extremely difficult. Their size is influenced by the majority or all of the following factors:

1. Changes in actual numbers: population changes.
2. Changes in the numbers of animals in a particular 'phase'.
3. Changes in activity following some change in the environment.
4. Sexual or species specific trap response.
5. Changes in the efficiency of the traps or the searching method.

It is clear, therefore, that the estimation of absolute population by relative methods is difficult; what one is really obtaining is the number of members of the population that were vulnerable to the trap under the prevailing climatic conditions and the current level of trapping efficiency. The influence of factors 2–5 on these relative methods must be considered further.

Theoretically, a trap may catch a sufficiently large proportion of the population to appreciably reduce the population being assessed. In practice, however, this is rarely the case. The failure of early attempts to control insect populations by light traps

Ecological Methods, Fourth Edition. P. A. Henderson and T. R. E. Southwood.
© 2016 John Wiley & Sons, Ltd. Published 2016 by John Wiley & Sons, Ltd.
Companion Website: www.wiley.com/go/henderson/ecologicalmethods

(e.g. Stahl, 1954) confirm the conclusions of Williams *et al.* (1955) that in general the previous nights' trapping has little effect. However, in isolated communities, such as small oceanic islands, ponds and streams or when the trap contains a powerful attractant (Petruska, 1968), a high proportion of the population may be captured.

7.1.1   *The 'phase' of the animal*

The susceptibility of an animal to capture or observation will alter with age if behavioural attributes or responses are age-dependent. Many relative methods rely to some extent on the animals' movements. Southwood (1962) divided insect movements into migratory or trivial, a classification that can be also applied to other animals. For many terrestrial species migratory movements occur mainly early in adult life, when insects, for example, have the power of flight, or between reproductive periods (Johnson, 1960, Johnson, 1963; Kennedy, 1961). The situation is often different for aquatic organisms such as fish where dispersal and migration often occurs early in life during the planktonic stage; an example is the migration of European eel, *Anguilla anguilla*, leptocephalus larvae from the Sargasso Sea to European waters.

For insects, trivial movements, during which they will be especially responsive to bait, occur mostly in later life. The effects of these phenomena on insect trap catches had in fact been recorded before they themselves were recognised: Geier (1960) showed with the codling moth that the majority of the females taken in light-traps were in the pre-reproductive phase, whilst bait-trap catches were predominantly mature (egg-laying) or post-reproductive females. Similarly, Spradbery & Vogt (1993) showed age-dependent trapping bias in the Old World screwworm fly, *Chrysomya bezziana*, to vary with the stage of ovarian development. Another example is the fall-off in numbers of *Culicoides* midges caught on sticky traps, which occurs before the actual population starts to decline (Nielsen, 1963). The reaction of animals to stimuli, important in many trapping techniques, also varies greatly with phase: this is particularly true of blood-sucking insects (Gillies, 1974).

Small mammals often vary in their vulnerability to trapping because of changes in their exploratory behaviour and response to novel objects, which are in turn influenced by factors such as sex, age, and social status (e.g. Carley & Knowlton, 1968; Myllymaki *et al.*, 1971; Adler & Lambert, 1997). Noting that feral house mouse populations have low recapture rates (0–20%) in live-trapping studies, Krebs *et al.* (1994) studied the movement of radio-collared mice. While low recapture rates during the breeding season were due to low trappability, during the non-breeding period this was caused by nomadic movements away from the study area.

Fish often vary in their vulnerability to light traps with age (often inferred from size distributions). Following light-trap sampling for coral reef fish, Choat *et al.* (1993) caught almost no larvae <5 mm long, and Thorrold (1992) reported that the pomacentrids, lethrinids, clupeids, mullids, and scombrids were exclusively late-stage larvae and pelagic juveniles. In a study of North American freshwater fish, Gregory & Powles (1985) found that Common carp and bluntnose minnow entered traps almost exclusively as yolk-sac larvae, while Iowa darter and pumpkinseed were taken only as yolk-sac and post-larvae and yellow perch as post-larvae from 5 to 33 mm total length. There are no published studies that have examined individual differences in the vulnerability of adult fish to traps, through Bagenal (1972) suggested that the reproductive status of perch, *Perca fluviatilis*, might be influencing catch rate. Fish traps such as fyke nets and salmon butts are usually positioned along the banks of channels where they intercept fish undertaking seasonal, tidal or migratory movements and give catches biased with respect to age, sex and reproductive condition.

7.1.2 *The activity of the animal*

The level of activity of an animal follows a diurnal cycle; for example, some insects fly by day, others at night, fish and mammals may forage at dawn and dusk others by night or day. The expression of this activity will be conditioned by the prevailing climatic conditions. The separation of changes in trap catch due to climate from that reflecting population change has long exercised entomologists. Williams (1940) approached the problem by taking running-means of the catches; the variation in these running-means (i.e. the long-term variation) reflected population changes and the departures of the actual catch from them reflected the influence of climate. Working with groups of species, such as the larger Lepidoptera, Williams was able to demonstrate both the long-term effects of climate through population change and the short-term effects on activity. When studying the populations of airborne aphids, Johnson (1952) found that the running-means reflected current climatic conditions and the deviations population trends; this was due to the relatively short period of time any given aphid spends in the air. Thus, the running-mean technique is unsafe on biological grounds, being influenced by the relative frequency of population and climatic change; furthermore, its use places severe restrictions on the number of degrees of freedom available for the calculation of significance levels. Even if no attempt is made to separate population trend from activity, when the regression of the actual catch of a group of animals on temperature is calculated, a highly significant result is frequently obtained, thereby emphasising the role of temperature in the determination of catch size (e.g. Williams, 1940: Southwood, 1960).

The activity of terrestrial vertebrates is greatly influenced by weather. Small mammals show changes of activity with respect to variables such as temperature, cloud cover and rain which change the number trapped (Lockart & Owings, 1974; Vickery & Bider, 1978; Mystowska & Sidorowicz, 1961; Plesner Jensen & Honess, 1995). Reptile activity is greatly influenced by temperature. Changes in physical factors such as water temperature, flow, light, tide or oxygen concentration greatly alter the activity of fish, although little quantification has been undertaken of their influence on the efficiency of capture. However, Bagenal (1972) found that captures of perch, *Perca fluviatilis*, declined when water temperature rose from 10 to 20 °C.

Occasionally it is possible to obtain a significant regression of catch size on temperature for a single species, particularly for groups such as fish which increase their swimming activity with temperature. However, activity may be determined by thresholds; for example, insects may have upper and lower temperature thresholds for flight (Taylor, 1963). If activity is known to occur between thresholds, then once these have been determined fluctuations in numbers between the lower and upper thresholds may

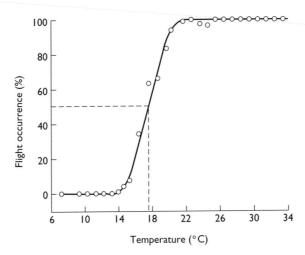

Figure 7.1 The graphical determination of the flight threshold of a species. (Adapted from Taylor, 1963.)

be considered as due to other causes. The thresholds may be determined by classifying each trapping period as either 1, when one or more insects of the particular species were caught or 0, none were caught. The trapping periods are then grouped according to the prevailing temperature and the percentage of occasions with flight plotted against temperature (Fig. 7.1). For example, if there were 25 trapping periods when the temperature was 16 °C and one or more specimens were collected on ten occasions, a point would be entered at the 40% flight occurrence (Fig. 7.1). Thresholds for light and other physical conditions may be determined in the same way. The transition from 0% flight occurrence to 100% is not sharp, presumably because of the variation in the individual animals and the microclimate of the sites from which they have flown. Thresholds for flight may appear to vary with the trapping method: Esbjerg (1987) found turnip moth, *Agrotis segetum*, would fly to sex traps in wind speeds that none would be caught in a light trap.

Taylor (1963) demonstrates elegantly that, when mixed populations of several species are considered, the series of thresholds will lead to an apparent regression of activity on temperature of the type demonstrated by Williams and others. Therefore, Taylor concludes that regression analysis as a means of interpreting the effect of temperature on insect flight should be limited to multispecific problems. Flight thresholds for temperature and wind velocity have been demonstrated in water beetles *Helophorus* (Landin, 1968; Landin & Stark, 1973) and for temperature and light in aphids (Halgren & Taylor, 1968; Dry & Taylor, 1970).

Southwood *et al.* (2003), using an extremely long time series of Heteroptera light trap-sampled between 1933 and 2000, showed that if days are grouped by their maximum daily temperature, the proportion in which flight occurred (at least one specimen was caught) gives a significant fit to a logistic regression (Fig. 7.2), and there is no step change that would indicate a threshold common for the whole group (Taylor, 1963). When a similar analysis is undertaken for the most abundant species,

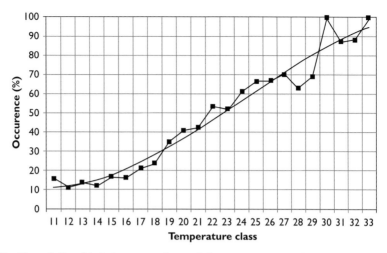

Figure 7.2 The relationship between maximum daily temperature and flight for Heteroptera from Rothamsted light-trap data. For each 1 °C temperature class the proportion of days on which at least one specimen of Heteroptera was caught was plotted against the maximum daily temperature. The calculations were made for all available years from 1933 to 2000 for the months of June to September. No clear temperature threshold was observed, and a third-order polynomial curve was fitted to the data by regression (after Southwood *et al.*, 2003).

Lygus rugulipennis Poppius, again there is no step change, so it seems that in practice this may not be a useful concept for trap catches of mirids.

7.1.3 *Differences in the response between species, sexes and individuals*

Our ability to observe or catch different species varies greatly, and this must be taken into account when choosing methods for faunal surveys. For example, small mammal species differ in their response to novel objects and those with a higher aversion are caught less frequently (Myllymaki *et al.*, 1971). Feldhamer & Maycroft (1992) found the mean number of captures per individual for golden mice, *Ochrotomys nuttalli*, were significantly less than those for white-footed mice, *Peromyscus leucopus*. Fish species vary greatly in their trappability. Pelagic species may avoid structures and thus never enter traps, while some species that live within floating meadows or weed beds may be so adept at navigating maze-like environments that they can escape. Species also differ greatly in their ability to detect and avoid gill nets which are particularly ineffective against electric fish and other elongate, non-spiny forms which do not use sight for navigation. Individual fish differ in their willingness to explore new objects and thus become trapped. There are reports that the presence of fish alarm substance, which is released from injured skin, will deter fish such as fathead minnows, *Pimephales promelas*, from entering traps (Mathis & Smith, 1992). While in this case the existence of pheromones has been questioned (Magurran *et al.*, 1996; Henderson *et al.*, 1997), the general principle that traps should be kept clean and give no signal of their prior use is sound practice. Eastop (1955) and Heathcote (1957a) have compared the ratios of the numbers of different species of aphids caught in yellow sticky traps to the numbers caught in suction traps, presumably the true population. Eastop, for example, found ratios varying from 31 to 0.5 and even within a genus (*Macrosiphum*), they ranged from 14.7 to 0.8. Analogous observations on the relative numbers of different species of mosquitoes caught in light-traps compared with the results from a rotary trap have been made by Love & Smith (1957), who found ratios (that they called the 'index of attraction') from over 7 down to 0.24.

In many groups significantly more of one sex or the other are caught in traps (for light traps, see Williams, 1939; Masaki, 1959). For example, mirid bug males make up the majority of light-trap catches and there is some evidence that male mirids do engage in significantly more trivial movement, 'flits', than females (Southwood, 1960; Waloff & Bakker, 1963). However, the excess of male Miridae is greater in ultraviolet than in tungsten filament light-traps (Southwood, 1960), so that at least a proportion of this predominance must be due to a selective effect of the traps between the sexes, rather than a real difference in flight activity. The positioning of traps can influence the sex ratio bias; pheromone traps for the oblique banded leafroller, *Choristoneura rosaceana*, placed in the tops of trees always caught higher numbers of male moths than those in lower- or middle-canopy positions (Agnello *et al.*, 1996). An interesting case of a difference between sexes is the large number of male relative to female tsetse flies usually taken on 'fly rounds' (Glasgow & Duffy, 1961). The biological interpretation of this seems to be that newly emerged female flies usually feed on moving prey and the early pairing, desirable in this species, is achieved by numbers of males following moving bait (Bursell, 1961). Sex differences in catchability have been reported for fish and crustaceans. For example, Myers *et al.* (2014) showed by tagging that capture of walleye, *Sander vitreus*, was both size- and sex-selective. For mammals, sex differences are also common; mature males are often caught more easily than females (e.g. Andreassen & Bondrup Nielsen, 1991; Drickamer *et al.*, 1995), but the reverse can be the case. For example, Adler and Lambert (1997) found that adult females of the Central American spiny rat, *Proechimys semispinosus*, were more trappable than males.

Individual differences in capture rates have been related to genetics. Gerard *et al.* (1994) found that heterozygous house mice, *Mus domesticus*, formed by the crossing of two chromosomal strains, had a lower trappability than the homozygous wild-types. Drickamer *et al.* (1995) also reported lower captures of heterozygous females. Diseases and parasites can also affect the trap response and should always be considered as a potential source of bias in a survey of infection rates as it is frequently easier to trap sick animals. Webster *et al.* (1994) found that *Toxoplasma gondii* changed avoidance behaviour in wild brown rats, *Rattus norvegicus*, resulting in the more frequent capture of infected individuals. It appears to be common for parasites to manipulate intermediate host behaviour to increase the chance of transmission to a predatory or scavenger definitive host (Moore, 2002).

7.1.4 *The efficiency of the trap or searching method*

The efficiency of a method of population estimation is the percentage of the animals actually present that are recorded. The efficiency of a searching method depends both

Table 7.1   The percentage efficiency of a cylindrical sticky trap for different insects at different wind speeds. (Data from Taylor, 1962.)

	Wind speed (miles/h)				
	1	2	6	10	20
Small aphid (*Jacksonia*)	20	45	64	66	68
Small fly (*Drosophila*)	35	58	72	73	74
Housefly (*Musca*)	45	64	76	76	76
Bumble-bee (*Bombus*)	52	68	78	78	78

on the skill of the observer and the habitat; for example, any observer is likely to see tiger beetles (Cicindellidae) far more easily on lacustrine mud flats than on grass-covered downlands.

Several types of trap – sticky, water and flight – will catch insects that are carried into or onto them by the wind, and Johnson (1950) and Taylor (1962) have shown that the efficiency of such traps varies with wind speed for which the results can be corrected for the size of the insect (Table 7.1). Variables noted by Martin *et al.* (1994) to influence sweep-net and sticky-trap sampling efficiency for three species of black flies, *Simulium truncatum*, complex, *Prosimulium mixtum*, and *Stegopterna mutata*, were wind speed, light, temperature, saturation deficit, and time of day. A similar list would be found for many insects.

Light traps are also affected by wind (e.g. Gregg *et al.*, 1994) and generally catch fewer specimens of most insects on nights when the moon is full (e.g. Taylor, 1986). This is considered to be due to a fall in efficiency of the traps (Williams *et al.*, 1956; Miller *et al.*, 1970; Bowden & Church, 1973). However, from time series obtained from tower-mounted light traps in northern New South Wales, Gregg *et al.* (1994) found no correlation with the phase of the moon. Steinbauer *et al.* (2012) found that Orthoptera caught in Farrow light traps used to monitor locusts were influenced by wind direction, wind speed, temperature and lunar light. For Lepidoptera, the catch was influenced by wind direction, temperature and humidity.

The efficiency of baits varies from many causes: the ageing and fermentation of artificial or non-living baits (Kawai & Suenaga, 1960) affecting their 'attractiveness', whilst living baits may differ in unexplained ways; Saunders (1964) found that two apparently identical black zebu oxen trapped different numbers of tsetse flies. Field biologists often find colleagues differ greatly in their attractiveness to biting flies and some can be employed as a useful decoy. The effectiveness of a given bait may vary from habitat to habitat (Starr & Shaw, 1944).

Fish traps and gill nets vary in efficiency with changes in water turbidity, turbulence, light, flow, temperature and depth. Bottom trawl efficiency changes with the nature of the substrate, and open-water fish and crustaceans may be able to avoid open-water trawls and nets when visibility is good. Efficiency of almost all towed nets varies with speed and should be standardised. The avoidance response of temperate fish tends to be faster at higher temperatures so that nets and suction samplers are often more effective in winter than summer.

The efficiency of the various methods will be discussed further below; the examples already presented indicate that variation in efficiency is a limitation on the value of relative estimates, even for comparative purposes.

7.2 The uses of relative methods

It is apparent from the above section that the actual data from relative methods should be used and interpreted with far more caution than has often been shown. Comparisons of different species and different habitats are particularly fraught with dangers.

7.2.1 Measures of the availability

This is the most direct approach; the availability of the population of an animal is the result of the response to the stimuli, the activity and the abundance; that is the product of factors 1–3 and 5 in the list above (p. 259); it may be defined as the ratio of total catch to total effort. Thus, assuming the efficiency of the trap or search does not change, the raw data of catch per unit time or effort will provide a measure of availability, or what Heydemann (1961) terms 'Aktivitatsdicht'.

True availability is meaningful in many contexts: it is most easily interpreted with natural 'baits'; for example, the availability of bloodsucking insects to their normal prey (in a bait trap) is a measure of their 'biting level', and the availability of flying insects to colonise or oviposit on a trap host plant is a measure of these parameters. Extrapolations from more artificial situations can only be made in the light of additional biological knowledge. The availability of codling moth females to a light-trap gives a convenient indication of the magnitude and phenology of emergence and oviposition, and this may be used to time the application of control measures. But the peak of availability of the females to the bait traps indicates that the main wave of oviposition has passed (Geier, 1960), and therefore in warm climates, where the eggs will hatch quickly, this information may be too late for effective control measures.

In general, measures of availability may be used for the immediate assessment of the 'attacking' or colonisation potential of a population and its phenology. Over a long period of time the changes in the species composition of the catches of the same trap in the same position may be used to indicate changes in the diversity of the fauna (see p. 495). It is only under exceptional circumstances that the trapping seems to affect population size, that is, when the traps are powerful and the population restricted.

7.2.2 Indices of absolute population

When the efficiency of the trap and the responsiveness of the animal to it can be regarded as constant, and if the effects of activity can be corrected for, then the resulting value is an index of the size of the population in that particular phase. The effects of activity due to temperature or other physical factors on the catch size of sticky, flight, pitfall and similar traps may be eliminated by the determination of the thresholds using Taylor's (1963) method (see above). With net catches, the regression technique of Hughes (1955) serves the same purpose (p. 279). The animals' diel periodicity cycle may need to be known before these corrections can be made and the index derived. Such an index may be used in the place of actual absolute population estimates in damage assessment and in studies on the efficiency of pest control measures. The value of independent estimates of population size has been stressed (p. 4), and a series of such indices may be compared with actual population estimates: if the ratio of one to the other is more or less constant the reliability of the estimates has been confirmed. Indeed, these indices may be used in place of absolute population estimates for any comparative purpose, but are of course of no value in life-table construction.

7.2.3   *Estimates of absolute population*

It is possible to derive estimates of absolute population from relative estimates by three approaches: (i) 'calibration' with absolute estimates; (ii) by estimating efficiency; and (iii) from the rate by which trapping reduces the sizes of successive samples (removal trapping). Distance sampling methods also provide absolute estimates, this is achieved by fitting a function which describes the decreasing efficiency of sighting an animal or its sign (e.g. nest) with distance from the observer. Such methods (as described in Chapter 9) are only applicable to large or easily spotted animals.

7.2.3.1   'Calibration' by comparison with absolute estimates

When a series of indices of absolute population have been obtained simultaneously with estimates of absolute population by another method, the regression of the index on absolute population may be calculated; Zoebisch *et al.* (1993) used a linear regression to relate sticky trap catches of the fly, *Liriomyza trifolii*, to population density. This can then be used to give estimates of population directly from the indices, but such 'corrections' should only be made under the same conditions that held during the initial series of comparisons.

7.2.3.2   Correcting the catch to allow for variations in trap efficiency

This approach is really a refinement of that above; trap catches and absolute estimates are made simultaneously under a variety of conditions that are known to affect trap efficiency. A table of the correction terms for each condition can then be drawn up and will give much greater precision than a regression coefficient for all conditions. A valuable approach along these lines has been made by Taylor (1962), who used the data from Johnson's (1950) experiment to determine the efficiency for sticky traps and townets using a calibrated suction trap (see p. 139), and these gave a direct measure of the absolute density of the aerial population. Taylor was therefore able to construct a table of correction terms (Table 7.2); these terms are of course only applicable to the particular traps that Johnson used, and it is important to note that the sticky trap was white, not yellow. Taylor also computed the expected fall in efficiency of the traps allowing for the different volumes of air passing at different winds and the effect of this on the impaction of the insects on the sticky trap, assuming that they behave as inert particles;

Table 7.2 The efficiency of a cylindrical sticky trap for different insects at different wind speeds. (Data from Taylor, 1962.)

Wind speed (m h^{-1})	1	2	6	10	20
Small aphid (*Jacksonia*)	20	45	64	66	68
Small fly (*Drosophila*)	35	58	72	73	74
House fly (*Musca*)	45	64	76	76	76
Bumble bee (*Bombus*)	52	68	78	78	78

he found that a large part of the fall in efficiency at low wind speeds could be accounted for by these factors.

7.2.4 Removal trapping or collecting

The principle of removal trapping or collecting is that the number of animals in a closed, finite, population will decline, as will the catch per unit effort, if captured animals are removed each time the population is sampled. Although Le Pelley (1935) first demonstrated that this fall of catch is geometric in observations on the hand-picking of the coffee bug, *Antestiopsis*, the method has been developed by mammalologists, starting in 1914 (Seber, 1982). Examples of studies on mammals are Rickart *et al.* (1991), Woodman *et al.* (1995) and Christensen (1996). These methods are particularly appropriate to small, isolated populations for which the probability of capturing each individual on each trapping occasion can be made >0.2. Removal methods are also applicable to the estimation of fish populations in streams and ponds which are sampled by electric fishing (see p. 274).

If this approach is to be used in life-table studies, it is necessary to use a method of catching that does not kill the animals, they will have to be kept captive and then released at the end of the estimation.

The basic procedure is to undertake s distinct periods of trapping and to record the number caught for the first time on each occasion, u_s. The various approaches for calculating N, the total population size, are based on either maximum likelihood or regression methods. Given constant sampling effort (e.g. the same number of traps on each occasion), maximum likelihood methods are superior (White *et al.*, 1982; Seber, 1986) and will be described in detail below. These estimates of N for a constant probability of capture model were first published by Moran (1951) and developed by Zippin (1956), who showed how to compute the maximum likelihood solution by iteration and gave tables for rapid estimation. Subsequently, Carle and Strub (1978) gave a modification of Zippin's method which produced a lower bias and variance by weighting the likelihood function with a beta distribution; probably the most important improvement is that their method did not fail for any sequence of catches. A more general model allowing variable probabilities of capture was presented by Otis *et al.* (1978). The suggestion (Seber, 1986) that the jackknife estimator for the M_{bh} model presented by Pollock & Otto (1983) should be tried has received little attention, probably because most ecologists find the maximum likelihood methods satisfactory. Regression methods are still used in studies where the sampling effort is variable, not being under the control of the researcher. For example, in fisheries research the number of fish landed varies with the activity of the fishing fleet. Methods for fisheries research are given in Ricker (1975) and Seber (1982), and will not be considered further here. When designing a removal trapping experiment every effort should be made to maintain constant sampling effort.

7.2.4.1 Assumptions underlying Zippin's and Carle and Strub's methods

For the application of these maximum likelihood methods the following conditions must be satisfied (Moran, 1951):

1. The catching or trapping procedure must not lower (or increase) the probability of an animal being caught. For example, the method will not be applicable if the insects are being caught by the sweep net and after the first collection the insects drop from the tops of the vegetation and remain around the bases of the plants, or if the animals are being searched for and, as is likely, the most conspicuous ones are removed first (Kono, 1953).
2. The population must remain stable during the trapping or catching period; there must not be any significant natality, mortality (other than by the trapping) or migration. The experimental procedure must not disturb the animals so that they flee from the area. As Glasgow (1953) has shown, if the trapping is extended over a period of time, immigration is likely to become progressively more significant as the population falls.
3. The population must not be so large that the catching of one member interferes with the catching of another. This is seldom likely to be a problem with insects where each trap can take many individuals, but may be significant in vertebrate populations where a trap can only hold one animal.
4. The chance of being caught must be equal for all animals. This is the most serious limitation in practice. Some individuals of an insect population, perhaps those of a certain age, may never visit the tops of the vegetation and so will not be exposed to collection by a sweep net. In vertebrates, 'trap-shyness' may be exhibited by part of the population. In electric fishing, where the method is used extensively, smaller individuals are more difficult to stun and individuals occupying territories under banks or other obstructions may be particularly difficult to catch.

Zippin (1956, 1958) has considered some of the specific effects of failures in the above assumptions. If the probability of capture falls off with time the population will be underestimated, but if the animals become progressively more susceptible to capture, the population will be overestimated. Changes in susceptibility to capture will arise not only from the effect of the experiment on the animal, but also from changes in behaviour associated with weather conditions or a diel periodicity cycle.

It has been shown by Zippin (1956, 1958) that a comparatively large proportion of the population must be caught to obtain reasonably precise estimates. His conclusions are presented in Table 7.3 from which it may be seen that, to obtain a coefficient of variation (CV = Estimate/Standard error × 100) of 30%, more than half the animals would have to be removed from a population of less than 200.

Table 7.3   Proportion of total population required to be trapped for specified coefficient of variation of *N*. (Data from Zippin, 1956.)

	Coefficient of variation (%)			
	30	20	10	5
200	0.55	0.60	0.75	0.90
300	0.50	0.60	0.75	0.85
500	0.45	0.55	0.70	0.80
1 000	0.40	0.45	0.60	0.75
10 000	0.20	0.25	0.35	0.50
100 000	0.10	0.15	0.20	0.30

For this reason. Turner (1962) found it impractical for estimating populations of insects caught in pitfall traps; the proportion of the population caught was too low.

7.2.4.2   Software for the computation of population size using removal sampling

Numerical iterative techniques are required to find the maximum likelihood value for the estimated population number, and are now undertaken on a computer using programs such as Removal Sampling (available at: http://www.pisces-conservation.com/index.html?softremoval.html$softwaremenu.html). A considerably less user friendly programme is Unmarked an R package (see http://cran.r-project .org/web/packages/unmarked/vignettes/unmarked.pdf). For the majority of researchers wishing to use R, the FSA: Fisheries Stock Assessment package (https://fishr.wordpress.com/packages/) is preferable to Unmarked because it is simple to use. The *removal* function in the FSA package calculates population estimates using a variety of methods, together with confidence intervals. The code listed below shows a simple application using the example data shown in Table 7.4 which is entered into a vector, ct (the 1st sample caught 65 individuals, the 2nd 43...). These data gave a population estimate of 195 with a standard error of 9.64.

```
library(FSA) #Load the FSA package see: https://fishr.wordpress.com/fsa/
ct <- c(65,43,34,18,12) #Load the observed captures on samples (passes) 1 to 5
ct                 #print data
(k <- length(ct))      #print number of removals
(T <- sum(ct))         #print total catch
                   #Zippin's method
Population.estimate <- removal(ct,method="Zippin") #run Zippin's method
summary(Population.estimate)                #Print results
confint(Population.estimate)                #print confidence intervals
#Carl & Strub method
Population.estimate.CS <- removal(ct,method="CarleStrub") #Run the Carl & Strub
                                       method
summary(Population.estimate.CS)             #Print results
```

Table 7.4   Example data from a removal trapping experiment.

Sampling occasion	1st (u_1)	2nd (u_2)	3rd (u_3)	4th (u_4)	5th (u_5)	Total
Number caught	65	43	34	18	12	172

Removal trapping methods require the number captured to fall with increasing sampling. If this does not occur the method may fail.

7.2.4.3   Simplified calculations with two or three sampling occasions

Simplified procedures for when at most three sampling occasions have been undertaken are given in Seber (1982). For two samples a good approximation to the maximum likelihood estimator is:

$$\hat{N} = \frac{u_1}{1 - \left(u_2/u_1\right)},$$

where u_1 are caught on the 1st sample and u_2 in the 2nd sample.

Population estimation based on two samples should only be used for p > 0.9. Generally, three or more sampling occasions should be undertaken so that the assumed constant probability of capture can be tested.

7.2.4.4   Graphical and regression methods with constant probability of capture

Because graphical and regression methods have been so widely used and are still frequently applied we briefly introduce them here. These methods are inferior to the maximum likelihood approach and do not give valid estimates of the standard error of the estimated population size. If the number caught on the ith occasion are plotted against the total catch up to occasion ($i-1$) a straight line may be fitted by eye (Menhinick, 1963) or a linear regression line calculated (Zippin, 1956). The point where this line cuts the x-axis gives the estimated population size. Fitting by eye is acceptable only when the points lie fairly close to a straight line. Fitting by regression is not really acceptable, as the values are not independent.

7.2.4.5   Dealing with variable probabilities of capture and the general maximum likelihood model

It will often be observed that the probability of capture changes during the course of a study. For example, larger fish are more easily caught when electric fishing; thus, the first sweep along a reach of a stream will tend to have the highest rate of capture. Generally, the best way to deal with this problem is to divide the population into a number of groups each of which can be independently estimated with a constant probability model. In an electric fish survey of a stream, it might be appropriate to undertake independent analyses for eel, trout and minnow, and to further subdivide the trout into size (age) classes. However, the ability to avoid capture may even vary between individuals of a single age or size class, and if this effect is large the mean probability of capture will change in what the observer perceives as a homogeneous population. When a constant probability model has been rejected and it is not possible to subdivide the population, the generalised removal method of Otis *et al.* (1978) is available.

A removal experiment does not generate sufficient information to allow the estimation of separate probabilities of capture for each sampling occasion. Otis *et al.* (1978) noted that as animals with a higher probability of capture tend to be caught first, the mean probability of capture (p_k) on sampling occasion k, tends to decline with successive samples so that after say m samples p can be assumed constant. A family of models can therefore be constructed in which the first 1,2,3 … $m-2$ samples are assumed to have different values for p and all later samples a constant p. The approach is to work upwards from $m = 1$ and stop when a χ^2 test shows that the model can no longer be rejected. The complicated procedure for calculating maximum likelihood estimates of the p_ks is given in outline in Otis *et al.* (1978). The method is available in computer programs such as CAPTURE (http://www.mbr-pwrc.usgs.gov/software/capture.shtml).

7.2.5 *Collecting*

Relative methods may be used simply as collecting methods, e.g. for animals for mark and recapture and for age determination in the construction of time-specific life-tables, provided the chances of all age and sex groups being captured are equal (e.g. Davies *et al.*, 1971), who found that parous rates in mosquitoes collected in light traps and at baits were only slightly different). They may also be used to determine the proportion of forms in a population (e.g. Bishop, 1972).

7.3 Relative methods: catch per unit effort

Methods grouped here are those in which the movement or action that results in the capture or observation of the insect is made by the observer.

7.3.1 *Observation by radar*

The use of radar in bird migration studies started in the 1960s and developed into a powerful tool for the study of insect movement in the upper air during the 1970s (Schaeffer, 1979). It has been found most valuable for studying large-scale night-time migrations in both birds (Bruderer, 1997a, b) and insects (Riley & Reynolds, 1990; Wolf *et al.*, 1990; Smith *et al.*, 1993). Because of the need to study bird and bat movements in relation to industrial developments such as wind farms, commercial bat and bird radars are available (e.g. Merlin™, http://www.detect-inc.com/avian.html).

Radar techniques, like the analogous sonar systems used for aquatic life, suffer from the general problem that they may not give sufficient taxonomic information to allow specific identification without supporting information. The obvious exceptions to this are large single species aggregations such as a desert locust swarm. However, the periodic change in signal strength of a bird linked to the flapping of wings, plus the speed or height of flight, may all give information that can lead to the identification of the species. The initial impetus for the use of radar in entomological research was the study of flying insect pests such as locusts. The techniques have developed with the advance in radar technology and recent innovations are reviewed in Drake & Reynolds (2012), while background information can be obtained from the Radar Entomology web site, http://www.pems.adfa.edu.au/~adrake/trews/ww_re_hp.htm. The detection of a flying object is related to the wavelength used. The upper part of the C band (3.8–7.5 cm) and the lower part of the S band (7.5–15 cm) are optimal for small birds, while the L band (15–30 cm) will not detect small passerines (Bruderer, 1997b). The reflectivity of birds seems to be related to their water content as the reflectivity of feathers is negligible, whereas for insects reflectivity is related to the amount of chitin present. Rainey & Joyce (1990) state that a radar with the operating parameters of 3.18 cm wavelength, peak power of 25 kW, pulse length 0.1 μS and pulse repetition rate of 1760 Hz could detect a single spruce budworm moth of 1 cm body length and 35 mg in weight at 750 m, and a desert locust 5 cm long at a considerably greater distance. Hobbs & Wolf (1989) describe an airborne radar system capable of detecting a moth with a radar cross-section of 1 cm^2, at a range of about 1 km. Individual detection efficiency is altered by the size of the animal, the direction of movement relative to the radar, the position within the radar beam, interference from rain or other airborne material, the terrain and the degree of flocking or swarming of the animal. Absolute counts of animals can rarely be obtained, as Bruderer (1997a) states: '…many ornithologists erroneously assume that counting echoes provides direct data on the number and distribution of birds, neglecting the difficulties ….'. These difficulties are even greater for insects where radar can only realistically give indices of density at different heights. The abundance of an animal can be expressed in terms of 'movement rates', which is the number of a species per minute as was done by Reynolds *et al.* (1997) in a study of Hawaiian sea birds and a bat. Radar has contributed most to the study of the timing, size and temporal variability of migrations, such as bird migration over Israel (Alfiya, 1995; Leshem & Yom Tov, 1996), passarine migration across the straits of Gibraltar (Hilgerloh, 1989, 1991) and seasonal insect flight activity in Texas (Beerwinkle *et al.*, 1994, 1995).

Radar remains a relatively expensive technique but has gradually become more important since the 1980s. Recently, high-resolution meteorological Doppler radars have been used to study insect migration (e.g. aphids in Finland; Leskinen *et al.*, 2011). Networks of weather radars have also been used to study bird migratory flight (e.g. Dokter *et al.*, 2011; Liechti *et al.*, 2013). An example of the use of a low-cost, hand-held Doppler shift radar is the study of Blake *et al.* (1990) to measure the flying speed of the barn swallow, *Hirundo rustica*.

Xu *et al.* (1996) described the use of ground-penetrating radar to detect termite nets.

7.3.2 *Hydroacoustic methods*

Sonar systems send out pulses of sound and detect echoes from underwater objects such as fish, crustaceans and cephalopods, and can even detect aggregations of planktonic crustaceans or the boundary between waters of different density. Acoustic methods can only be used for quantitative surveys of species which live in the water column and are only practicable in waters not dominated by aquatic macrophytes, logs or drifting leaves. When these conditions are met, dual or split-beam systems are able to give quantitative estimates of fish number and size distribution within a number of depth bands. For example, Kubecka *et al.* (1994) estimated brown trout *Salmo trutta* populations in freshwater lochs, and Duncan & Kubecka (1996) studied fish density and distribution in the River Thames. Acoustic surveys are extensively used for marine surveys of commercial species where normally the method gives an index of abundance that can be used to detect temporal and spatial changes. The tendency for fish and other aquatic organisms to form tight shoals can create problems for quantitative estimation (Appenzeller & Leggett, 1992). Absolute density estimates can be obtained for shoaling species using echo integration techniques, although errors of estimation can be large. An example of the use of acoustic surveys for marine crustaceans is the study of Brierley & Watkins (1996) on Antarctic krill *Euphausia superba*, and zooplankton. As in the case of radar, a weakness of the method is that it requires supporting information to distinguish between species.

While fish-finding sonars have become cheap and easy to use, quantitative fish-counting systems are expensive. Recently, high-resolution, imaging sonar has been used increasingly for fisheries monitoring, particularly migratory salmon. In shallow freshwaters waters the multi-beam system produced by Sound Metrics, and known by the acronym of DIDSON (Dual -frequency IDentification SONar), is the market leader (http://www.soundmetrics.com/products/didson-sonars). This device can produce high-resolution near-video quality images of fish only centimetres in length in low-visibility water. The detection range can be greater than 40 m. The resolution is often sufficient to allow species identification.

7.3.3   *Fish counters*

Automatic fish counters are widely used in rivers for counting migratory species such as salmon, *Salmo salar*. The detectors, which often work by measuring the change in conductivity caused by a passing fish, are usually placed into narrow channels. Fish counters give false counts by both not detecting passing fish, particularly if they pass as a group, and also counting drifting material. Accuracy can be tested by checking the record of fish detections against video recordings. Such counters are useful for studying seasonality of migration and fish movement in relation to factors such as flow or temperature and to produce annual indices of abundance. They generally cannot distinguish between species.

7.3.4   *Electric fishing*

Electric fishing is the method of choice for estimating fish populations in small streams and ponds where absolute estimates of population number can be made (Kennedy & Stange, 1988). Electric fishing is placed under 'relative methods' because it is never 100% efficient and population size is best estimated using removal trapping methods. Electric fishing is usually undertaken with a pulsed DC current that draws fish towards the anode and then stuns them. The power is supplied either from an electric generator or batteries in the case of back-pack units. Electric fishing is dangerous if undertaken by untrained personnel. An important early book on the method is Vibert (1967), and more recent developments are discussed in Cowx (1990).

The normal procedure for estimating population size in streams is to isolate a reach with stop nets then work down stream with a hand-held anode, removing any stunned fish (Plate 3). This procedure is repeated three or more times (the exact number depending on sampling efficiency) and the population number estimated by removal methods (p. 268). This is not possible in larger water bodies, where Persat & Copp (1990) argue that good estimates can be obtained by sampling at a large number of individual points, provided that the area sampled at each point can be estimated. However, on larger waters where boat-mounted gear is used it is normal to express the number caught as catch per unit effort, such as fish caught per 100 m^2 area swept.

Electric fishing efficiency is altered by:

1. The conductivity of the water. In low-conductivity freshwaters, as are normal in areas with non-calcareous bedrock such as the Amazon basin, the efficiency is too low for the method to be used.
2. Turbidity and flow, which affect the ability to spot and collect stunned fish.
3. Variation in water depth can have a large effect and electric fishing is rarely useful in waters >2 m deep.
4. The skill and tiredness of the operator
5. The size and species of fish. Large fish are more easily stunned than small fish.

7.3.5 *Aural detection*

Strictly speaking, sounds are 'products of animals' and this paragraph should come in the next chapter; however, this method has mostly been used with birds and the approach is similar to distance sampling methods with a 100% detection assumed out to a known distance beyond which none are heard. A simple estimator, due to W.H. Petrabough, is:

$$\hat{D} = \overline{h}\Big/_{L\pi r_a^2}$$

where h = average number of songs heard per stop, L = the total number of listening stops made, and r_a = the radius of audibility. The estimation of r_a presents a major source of error.

A further problem is that birds and insects have a definite periodicity of calling; Gates & Smith (1972) developed a time-dependent model to allow for this. Bird surveys often use the response of birds to tape-recorded calls to detect the presence of species that are difficult to spot. Such a technique is particularly useful in forests.

With many insects one 'caller' stimulates others, so that the probability of hearing a song might be better described by a skewed rather than a Poisson distribution with time. This method might be used with some insects (male cicadas, acridids) as a second or third method to compare estimates obtained with capture–recapture or other tech-niques. Indices of abundance of cetaceans and bats can be obtained by recording and quantifying their navigational calls (e.g. Verheggen, 1994).

7.3.6   *Exposure by plough*

The surveying of large, but aggregated, soil animals by conventional methods (Chapter 6) is often extremely laborious. Roberts & Smith (1972) found that a rapid survey of scarabaeid larvae could be made by ploughing a transect across grassland and counting those insects exposed in the furrow or on the overturned turf. Such transects would also detect patterns of aggregation; in general, there was a linear relationship between numbers per unit plough transect and absolute population as determined from core samples. Relatively more are, however, exposed in dense populations, because they cause the soil to break up.

7.3.7   *Collecting with a net or similar device*

A number of approaches are included under this heading: in aquatic habitats dipping and netting are widely practised relative methods; insects and other small animals on terrestrial vegetation may be collected with a sweep net or, for those on trees, with a beating tray. The latter approaches an absolute method and is discussed in Chapter 4 (p. 153).

7.3.7.1   Aquatic habitats

A wide variety of nets including, seine, beam trawl and push nets are used to sam-ple fish and crustaceans (see Fig. 7.3). Most nets are so inefficient that even when it is possible to estimate the volume or area swept out, the catch cannot be used to esti-mate population density. A probable exception is quantitative sampling with a seine net (as discussed in Chapter 5). Beam trawls are used extensively for sampling benthic fish and crustaceans. Because their efficiency is low and changes with substrate and aquatic conditions such as turbidity, they are difficult to calibrate and thus can rarely be used to estimate absolute density.

Plankton nets (Chapter 5; see also Plate 1) may also produce indices of abundance under conditions where it is not possible to estimate the volume of water filtered, as is normally the case in highly turbulent water. In freshwaters, drag nets can be used for sampling aquatic invertebrates in an analogous fashion to a sweep net.

7.3.7.2   Terrestrial insects

Aerial insects may be collected by random strokes through the air with a light net (Parker, 1949; Linsley *et al.*, 1952; Nielsen, 1963) or with net or gauze cones on a car (Almand *et al.*, 1974; Roberts & Kumar, 1994), a ship (Yoshimoto & Gressitt, 1959; Yoshi-moto *et al.*, 1962), or an aeroplane (Glick, 1939; Odintsov, 1960; Gressitt *et al.*, 1961). Because of the impedance to airflow due to the gauze, which becomes more severe at

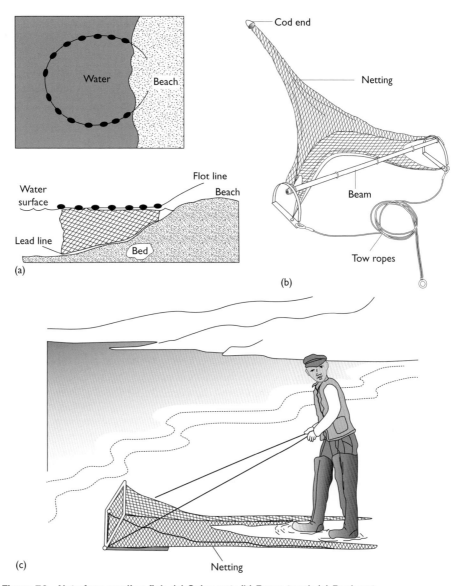

Figure 7.3   Nets for sampling fish. (a) Seine net; (b) Beam trawl; (c) Push net.

higher speeds, it is difficult – but not impossible – to obtain direct measures of aerial density from such sampling cones.

The sweep net is perhaps the most widely used piece of equipment for sampling insects from vegetation (Plate 4); its advantages are simplicity, speed and high return for little cost, and it will collect comparatively sparsely dispersed species. However, only those individuals on the top of the vegetation that do not fall off or fly away on the approach of the collector are caught. The influence of these behavioural patterns on the efficiency of the method has been investigated by the comparison of sweep net sam-ples with those from cylinders or suction apparatus (p. 155) (Beall, 1935; Romney, 1945; Johnson, Southwood & Entwistle, 1957; Race, 1960; Heikinheimo & Raatikainen, 1962; Rudd & Jensen, 1977; Harper *et al.*, 1993), sweep net samples with capture–recapture estimates (Nakamura *et al.*, 1967), by comparison with pit-falls (Cherrill & Sanderson, 1994) or by a long series of sweeps in the same habitat (Hughes, 1955; Fewkes, 1961; Saugstad *et al.*, 1967). Changes in efficiency may be due to:

1. changes in the habitat;
2. changes in species composition;
3. changes in the vertical distribution of the species being studied;
4. variation in the weather conditions; and
5. the influence of the diel cycle on vertical movements.

A sweep net cannot be used on very short vegetation. Once plants become more than about 30 cm tall, further increases in height mean that the net will be sampling progres-sively smaller proportions of any insect population whose vertical distribution is more or less random.

Even related species may differ in their availability for sweeping (Johnson *et al.*, 1957; Heikinheimo & Raatikainen, 1962), and therefore the methods are unsuitable for synecological work; these same workers also showed with a nabid bug and a leafhop-per that the vertical distribution of the various larval instars and the adults differed (Fig. 7.4). Within the adult stage the vertical distribution in vegetation may alter with

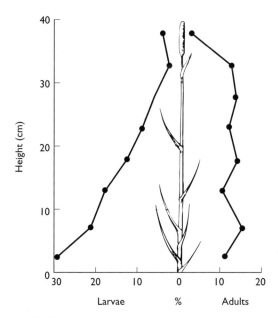

Figure 7.4 The vertical distribution of larval instars IV and V and adults of *Javesella pellucida* on Timothy grass. (Adapted from Heikinheimo & Raatikainen, 1962.)

age; ovipositing females of the grass mirid, *Leptopterna dolabrata*, are relatively unavailable to the sweep net; the same appears to be true of mature healthy males and females of the leafhopper, *Javesella* (= *Calligypona) pellucida*, in cereals, although parasitised individuals do not change their behaviour with age and so the percentage of parasitism would be overestimated by sweeping (Heikinheimo & Raatikainen, 1962) (see also p. 384)

Weather factors profoundly affect the vertical distribution, and hence the availability of insects to the sweep net. These have been studied by Romney (1945), Hughes (1955) and Saugstad *et al.* (1967); Hughes found with the chloropid fly, *Meromyza*, that the regressions of numbers swept against the various factors gave the following coefficients: wind speed –0.1774, time since saturation –0.0815, air temperature +0.0048, radiation intensity +0.0010 and radiation penetration –0.0367. Thus, the major influences on the catches of this fly are the first two factors listed, maximum efficiency being achieved at low wind speeds and immediately after a shower. Many arthropods have a diel periodicity of movement up and down the grass, the maximum numbers being on the upper parts a few hours after sunset (Romney, 1945; Fewkes, 1961; Benedek *et al.*, 1972).

The bags of sweep nets are usually made of linen, thick cotton or some synthetic fibre; the mouth is most often round; a square-mouthed net does not give more consistent results (Beall, 1935), but a D-shaped mouth is useful for collecting from short vegetation, especially young crops. There may be considerable variation in the efficiency of different collectors; usually, the more rapidly the net is moved through the vegetation the larger the catch (Balogh & Loksa, 1956).

The number of sweeps necessary to obtain a mean that is within 25% of the true value has been investigated by Gray & Treloar (1933) for collections from lucerne (alfalfa), and was found to vary from taxon to taxon, with an average of 26 units each of 25 sweeps. These authors suggested that this high level of variability was due to heterogeneity in the insect's spatial distribution, and much of this may have been attributable to the diurnal periodicity demonstrated by Fewkes (1961). In contrast, Luczak & Wierzbowska (1959) considered 10 units of 25 sweeps adequate for grassland spiders and Banks & Brown (1962) found that sweep net catches of the shield bug, *Eurygaster integriceps*, on wheat had sampling errors of only 10% and reflected absolute population differences determined by other methods.

7.3.8   *Visual searching and pooting*

Methods based on simple counts of the number of animals present are described in Chapter 9. The aspirator or pooter (Fig. 7.5) is convenient for rapidly collecting small insects in fixed time estimations When using a pooter, care should be taken to ensure that the operator cannot accidentally inhale damaging particles. Evans (1975) describes two pooter designs that work on the 'Venturi' principle, with the operator blowing. The one shown in Fig. 7.5b is emptied by carefully removing the top cork and the gauze filter; this operation and the re-alignment of the tubes before re-use must be carried out carefully. While searching a fixed area, if completely efficient, provides an absolute

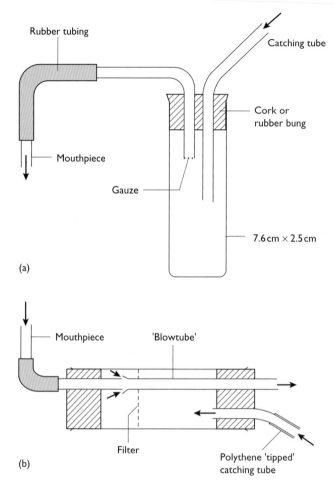

Figure 7.5 The aspirator or pooter. (a) A simple model; (b) A blow model. (Adapted from Evans, 1975.)

estimate of population size, but this is rarely possible. Sampling efficiency tends to vary with the weather and time of day.

7.4 Relative methods: trapping

The methods described in this section, in contrast to the previous section, are those in which it is the animal rather than the observer that makes the action that leads to its enumeration. Basically, traps may be divided into those that catch animals randomly and those that attract them in some way. (The word 'attract' is used in its widest sense without any connotation of desire: the studies of Verheijen (1960), for example, have shown that animals are trapped by artificial light through interference with the normal photic orientation and not strictly because of attraction.) It is important to distinguish between these two types as those that are based on attraction allow the possibility of a further source of error. But a strict division is impossible as some traps, more particularly water and sticky traps, are intermediate in position.

7.4.1 Interception traps

These are traps that intercept the animals, more or less randomly as they move through the habitat: air, water or land (Plate 5). Indices of absolute population may theoretically be obtained more easily from this type of trap than from others, as there is no variation due to attraction.

7.4.1.1 Air: flight traps

Here, we are concerned with stationary flight traps that are not believed to attract the insects; those that have some measure of attraction are discussed below (p. 304); others described elsewhere are moving (rotary) nets which may give absolute samples (p. 143) or, if the quantity of air they filter is uncertain, they are regarded as aerial sweeps (p. 277).

One of the simplest type is the suspended cone net as used by Johnson (1950); this may have a wind vane attachment to ensure that it swivels around to face into the wind; Taylor (1962) has shown that its efficiency at different wind speeds may be calculated (p. 144). Nets are particularly useful for weak flyers such as aphids (Davis & Landis, 1949), or at heights or in situations where the wind speed is always fairly high (Gressitt *et al.*, 1960), so that the insects cannot crawl out after capture. Similar traps have been used on moving ships (p. 276).

The Malaise trap is more elaborate; it consists basically of an open-fronted tent of cotton or nylon net, black or green in colour; the 'roof' slopes upwards to the innermost corner at which there is an aperture leading to a trap (Plate 5). It was developed by Malaise (1937) as a collector's tool; modified designs have been described by Gressitt & Gressitt (1962), Townes (1962) and Butler (1965), and a basically similar trap by Leech (1955). Gressitt's (Fig. 7.6) and Butler's models are much simpler to construct and transport than that of Townes. Gressitt's is large (7 m long and 3.6 m high). Butler's trap is smaller, being made from a bed mosquito-net by cutting out part of one side and a hole in the roof into which the collecting trap, a metal cylinder and a polythene bag, is placed. Townes (1962) gives very full instructions for the construction of his model, which is more durable and traps insects from all directions. When maximum catches are desired (for collecting) Malaise traps should be placed across 'flight paths' such as woodland paths, but in windy situations they cannot be used. The studies of Juillet (1963), who compared a Malaise trap with others – including the rotary, which is believed to give unbiased catches – suggest that for the larger Hymenoptera and some Diptera this trap is unbiased, but it is unsatisfactory for Coleoptera and Hemiptera. However, Roberts (1970, 1972) found that even for Diptera the form and colour of the trap would greatly influence the catch. Hansen (1988) found black traps to catch more specimens and species in both burnt and unburnt areas of open range, although the differences were not statistically significant.

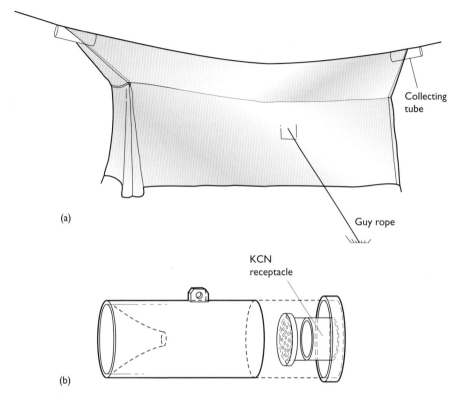

(a)

Collecting
tube

Guy rope

KCN
receptacle

(b)

Figure 7.6   Malaise trap. (a) Sketch of the Gressitt type; (b) Plastic collecting tube (after Gressitt & Gressitt, 1962).

A very large 'net-tent' (Filet-tente) extending in an arc for 37 m was designed by Aubert (1969) for trapping migratory insects in mountainous areas. An 'electrified net' has been used, sometimes with bait, for tsetse flies (Rogers & Smith, 1977).Walker & Lenczewski (1989) and Walker & Whitesell (1993) describe portable traps for catching migrating butterflies.

Flying Coleoptera and other insects that fall on hitting an obstacle during flight may be sampled with a window trap, which is basically a large sheet of glass or plexiglass held vertically with a trough, containing water with a wetting agent and a little preservative below it (Chapman & Kinghorn, 1955; Van Huizen, 1977) (Fig. 7.7), or with a layer of soy bean oil (Flamm et al., 1993). Living insects may be collected by replacing the trough with opaque cylinders whose ends have transparent collecting tubes: insects move towards the light and accumulate in the tubes (Nijholt & Chapman, 1968). A modified flight interception trap for use in tree canopies is described by Springate & Basset (1996).

One of the most important uses of stationary flight traps is to determine the direction of flight (see p. 431). Nielsen (1960) has observed that migratory butterflies will enter stationary nets and remain trapped, whereas those engaging in trivial movement will fly out again. The direction of migration can be determined by a number of stationary

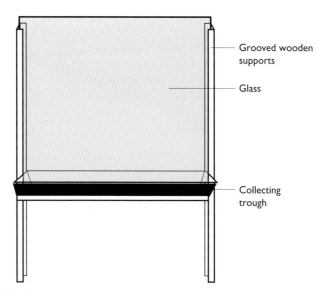

Figure 7.7 Window trap.

flight traps arranged so as to sample from four fixed directions. Stationary nets may be used (Nielsen, 1960) or a Malaise trap modified to collect the insects from each side separately (Roos, 1957). A more extensive trap is the 'robot observer', which was used in work on blowflies and consisted of a zigzag wall of net, with 10 V-shaped bays each with a 2 m² aperture and ending in a fly trap; the catches from the 10 opposing bays on each side are directly comparable (J. Macleod, pers. commun.).

Special traps have been designed to collect and distinguish between the insects entering and leaving the 'homes' of vertebrates, principally burrows and houses (Myers, 1956; Service, 1963, 1993; Muirhead-Thomson, 1963). Exit traps are usually net cones that are inverted above the exit hole. Entrance-traps have to be inconspicuous from the outside, and a metal gauze cylinder with a dark canvas 'skirt' that could be spread out to prevent passage round it was used in rabbit burrows by Myers, who concluded, however, that the entrance trap did not sample as randomly as the exit-trap. In houses, ingress and egress may be measured by identical traps, but Service's (1963) results show that in this habitat the former is measured less efficiently. The traps consist of a net cone leading into a net or net and 'Perspex' box; Saliternik (1960) trapped the mosquitoes leaving cesspits with a similar trap. This could be a measure of emergence (see Chapter 10).

7.4.1.2 Water: drift samplers and fish traps

7.4.1.2.1 *Drift samplers*

Waters (1969) and Elliott (1970) have reviewed methods of studying organisms drifting in streams. The simplest type is based on Waters' (1962) design (Fig. 7.8), a 1 m-long 'Nitex' net with a square mouth and a board sunk flush with the stream bed in front

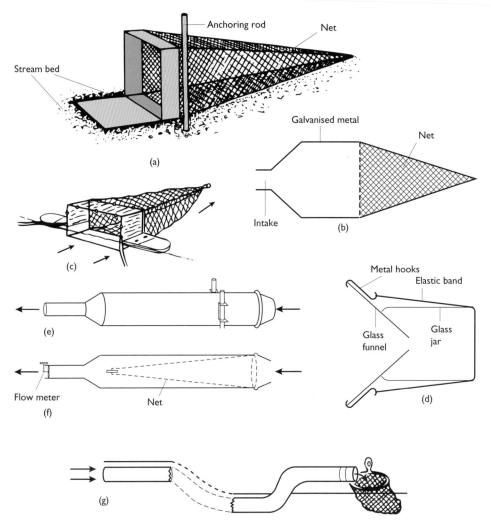

Figure 7.8   Aquatic interception traps. (a) For drifting invertebrates in streams. (b) Cushing-type trap for drifting invertebrates in streams. (c) Elliott floating trap (after Elliott, 1970). (d) Piezynski trap for water mites. (e) and (f) External and sectional views of Müller-type tubular trap with flowmeter (after Elliott, 1970); (g) a pipe trap latter Elliott, 1970): the dotted portion represents a length of pipe sufficient for a drop in level.

of the mouth. Nets of this type may become clogged, but if the mouth is narrowed (see Fig. 7.8b) this is alleviated and hence backflow is prevented and filtration increased (Cushing, 1964: Mundie, 1964; Anderson, 1967). A collecting vessel may be placed at the end of the net.

Elliott (1967) devised a floating trap (Fig 7.8b), similar to that of Waters, but having the advantage for 'drift sampling' that crawling forms were definitely excluded.

If the volume of water passing through a drift sampler is measured it then provides an absolute population estimate, analogous to a suction trap. Elliott (1970) was able to measure the water flow through his floating net using a current meter. The flow meter may be incorporated into the sampler (Fig. 7.8e and f); originally designed by K. Millier, they have been modified by Elliott (1967, 1970) and Cloud and Stewart (1974). A third approach to the measurement of drift is to pipe a portion over a sufficient distance so that it can be brought above the surface and filtered (Fig. 7.8g) (Elliott, 1970). Such methods are also particularly appropriate by weirs and falls. Anderson & Lehmkuhl (1968) funnelled the entire stream-flow through their trap. Sampling period is important because drift often shows a diel periodicity (Elliott, 1969), which is influenced by moonlight and other variations in light intensity (Anderson, 1966).

Pieczynski (1961) has found that what is in effect an unbaited glass lobster pot (Fig. 7.8d) will trap numbers of water mites and other free-swimming animals. Usinger & Kellen (1955) used a similar trap constructed from a plastic screen rather than glass and caught large numbers of aquatic insects; their trap is shown in Usinger (1963; p. 55). A modification with a collecting jar reached through a narrower entrance on the roof has been found particularly useful for water beetles (James & Redner, 1965). These traps are the aquatic equivalents of the pitfall trap.

7.4.1.2.2 *Fish and crustacean traps*

Mankind has trapped fish for thousands of years, and a great many methods are still currently used (for a review of methods, see Brant, 1984). Most familiar is the lobster pot which is used throughout the world. The scientist new to a region will find it useful to study local methods, which are often designed to perform well under local conditions of water movement and animal behaviour. Most fish traps operate on the principle that it is easier for a fish to enter than leave, and thus the catch need not be all the fish that have entered. Commonly used types of fish trap (see Fig. 7.9) include fish weirs, fyke nets, pot traps, salmon putts and minnow traps. All fish traps are highly selective. Aquatic traps can be used with an attractant such as light (see p. 302) or a bait.

A type of trap that is widely used for river, estuarine and coastal studies are the filter screens of industrial cooling water intakes at power stations, oil refineries and chemical works. These cooling water systems pump huge volumes of water (about 30 m³ s⁻¹ for a 1000 MW nuclear-powered station) and thus act as powerful samplers. The intakes are often placed on the seabed and any fish or crustacean swimming or crawling close to the intake is sucked in. Examples of studies using power station intake data are van den Broek (1979), Henderson (1989) and Henderson & Seaby (1994).

7.4.1.2.3 *Gill nets*

Gill nets are designed to work by catching swimming fish which attempt to pass through the net, but when they find the mesh is too small they cannot reverse out, generally because their gill covers catch on the net (Plate 2). Fish are, however, caught for a variety of reasons; for example, some catfish entangle their spiny pectoral fins. The size range of fish caught is related to mesh size, and it is common practice during general fisheries surveys to lay banks of nets with a range of mesh sizes. The efficiency is related to the ability of fish to detect the net. Fish with good vision are caught more often when light levels are low. However, this does not imply that they are always more effective in the dark as they will only catch swimming fish and many fish are inactive at night

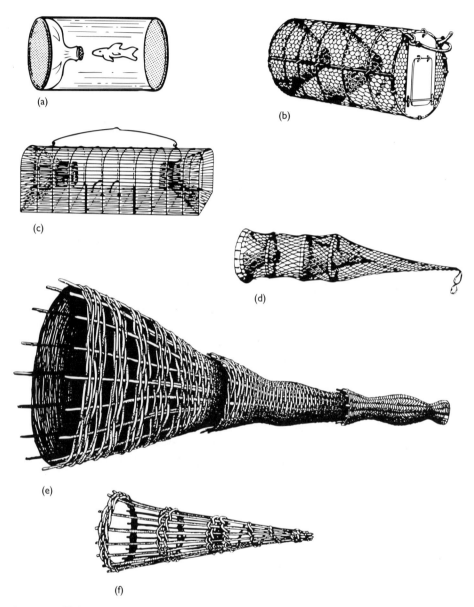

Figure 7.9 Fish traps.

(while other species become active). Mono-filament nets are more difficult than multi-filament nets for fish to detect and thus tend to be more efficient. However, they are more easily damaged by predators such piranhas, crabs or caiman, which are attracted by the catch and thus may be less practicable.

Fish abundance is normally expressed as catch per unit effort where effort is given in units of net length × time. This expression might suggest that the catch in a net increases linearly with time, which is not true (Minns & Hurley, 1988).

7.4.1.3   Land-pitfall and other traps

Like the lobster pot, the pitfall trap (Plate 4) was an adaptation by the ecologist of the technique of the hunter; basically, it consists of a smooth-sided hole into which the hunted animal may fall but cannot escape. Often, a glass, plastic or metal container is sunk into the soil so that the mouth is level with the soil surface. Pitfalls can be used to capture arthropods, amphibians, snakes and mammals. Insects may be emptied with a hand-operated or mechanical suction apparatus (p. 155), thereby avoiding the disturbance to the surroundings that would result from continued removal and re-sinking. Pitfalls can be arranged as trapping webs or grids for the estimation of population parameters (see p. 399), but the experimental results of Parmenter *et al.* (1989) on a captive population of darkling beetles *Eleodes* sp. at known density give little confidence that trapping webs will reliably estimate population parameters.

While pitfalls are used most frequently for arthropods they can be effective with vertebrates. When used to sample amphibians and lizards, pitfalls may be used in conjunction with a drift fence (Heyer *et al.*, 1994; Friend *et al.*, 1989; Morton *et al.*, 1988). An advantage of pitfalls over snap or live traps such as the Longworth is that they can produce multiple captures, although a captured animal may influence the probability of further captures. A description of an effective pitfall for small mammals is given in Walters (1989), and the relative merits of pit falls over other types of traps for mammals are considered by Innes & Bendell (1988).

Pitfall traps have been used extensively for studies on surface dwellers such as spiders (Koponen, 1992; Bultman, 1992; Bauchhenss, 1995), Collembola (Budaeva, 1993), centipedes, millipedes (Kurnik, 1988), ants (Abensperg *et al.*, 1995), beetles, especially Carabidae (e.g. Kowalski, 1976; Uetz, 1977; Epstein & Kulman, 1990; Cameron & Reeves, 1990; Togashi *et al.*, 1990; Braman & Pendley, 1993) and even crabs (Williams *et al.*, 1985). More elaborate traps have been designed to facilitate emptying (Rivard, 1962); for use under snow (Steigen, 1973), with rain-guards (Fichter, 1941; Steiner *et al.*, 1963) or with timing devices that allow the catch from each time period to be segregated (Blumberg & Crossley, 1988). Artificial (Stammer, 1949) or natural (Walsh, 1933) baits can be used in pitfall traps, and baited traps have been found useful for collecting beetles from the burrows of mammals (Welch, 1964). For example, Rieske & Raffa (1993) sampled pales weevil, *Hylobius pales*; pitch-eating weevil, *Pachylobius picivorus*; and pine root collar weevil, *H. radicis*, using pitfalls baited with ethanol and turpentine. However, the effects of baits are variable and in many cases, as their attractiveness will change with age, they may well introduce a further source of error, and simple traps are generally to be preferred in population studies (Greenslade & Greenslade, 1971).

Preservatives such as formalin or alcohol have sometimes been used in traps but, like baits, these may affect the catches of species differentially (Luff, 1968; Greenslade & Greenslade, 1971). While ethylene-glycol antifreeze has been found to work better (Clark & Blom, 1992) for the Carabidae, even this has been found to produce a different catch and sex ratio when compared with traps holding just water (Holopainen, 1992). The most suitable procedure for water-filled traps may be to add a small amount of detergent as a wetting agent.

Pitfall traps have many advantages: they are cheap (empty food or drink containers may be used), they are easy and quick to operate, and a grid of traps can provide
an impressive set of data (e.g. Gist & Crossley, 1973). However, many studies have
shown that catch size is influenced by a wide range of factors, apart from population
size (Greenslade, 1973; Petruska, 1969; Hayes, 1970; Luff, 1975). As pointed out by Topping & Sunderland (1992) in a study of the spider fauna of winter wheat, the catch
may not reflect either changes in abundance, relative abundance between species nor
the actual sex ratios. Pitfall traps must therefore be used with extreme caution but, as
Gist & Crossley (1973), Uetz & Unzicker (1976) and others have found, they may provide valuable information provided that proper attention is paid to potential sources
of variation. Pitfalls are frequently used in studies of community richness and for habitat assessment, but they sample only the surface-dwelling community and sweep nets
or suction sampling will reveal a different species pattern (Samu & Sarospataki, 1995).
Indeed, pitfalls are particularly appropriate for heavy ground-living beetles which are
poorly sampled by suction samplers such as the D-vac (Mommertz *et al.*, 1996). For
carabids, catches can reflect density (Hokkanen & Holopainen, 1986) but the capture
efficiency measured as the proportion of trap encounters resulting in capture can be
low. Halsall & Wratten (1988) found efficiencies ranging from zero to 40%, depending
on both species and conditions.

An interesting model of the action of pitfall traps has been developed by Jansen &
Metz (1979), who consider that the number of animals trapped depends on: (i) their
population density; (ii) their movement, assumed to be Brownian; (iii) the boundary
of the pitfall and its 'absorbitiveness'; and (iv) the outer boundary of the area and the
extent to which the animals penetrate it (the probability of absorption). Movement of
cursorial arthropods is affected by temperature, moisture and other weather conditions (Mitchell, 1963), food supply (Briggs, 1961), the characters of the habitat (e.g. the
amount of impedance by the vegetation) and the age, sex and condition of the individuals (Petruska, 1969; Hayes, 1970). If due account is taken of the thresholds, corrections
can be made for weather conditions (p. 261): the other variables need to be kept in view,
as they may or may not be significant in the particular study. The catches immediately
after a pitfall trap is placed in position are commonly found to be higher than those
subsequently achieved; Joosse (1975) and Greenslade (1973) termed these 'digging-in
effects'. Jansen & Metz (1979) obtained similar effects from their theoretical model due
to depletion of the individuals, moving in a Brownian manner, around the trap.

The boundary of the pitfall trap influences the catch both in respect of its magnitude
and its 'capture efficiency' (the probability of absorption). Traps are generally circular
and catch size will theoretically be related to the length of the perimeter (Morrill *et al.*,
1990); if approach, in the region of the trap, is effectively linear then rectangular or any
non-reflexed polygonal trap aperture will also catch in proportion to its diameter (Luff,
1975). However Luff's (1975) studies showed that the capture efficiency of trap boundaries varied both with respect to the nature of the trap (e.g. glass jar or tin) and the
species of ground beetle. The only generalisation that was possible was that smaller
aperture traps seemed to have higher efficiencies for the smaller beetles, whilst the
larger traps caught a higher proportion of the larger beetles that encountered the trap
boundary (aperture). Abensperg *et al.* (1995) concluded that larger traps caught more

ant species (especially those with larger individuals), however if results were expressed as catch per mouth area then the smallest (18 mm diameter) traps were the most efficient.

Gist & Crossley (1973) used exclusion barriers to isolate an area (provide a 'non-absorptive' *sensu* Jansen & Metz' outer boundary) and a grid of pitfalls 'trapped out' the area so that removal trapping methods could be used to estimate population size (p. 268). Such studies and the 'digging-in effect' suggest that, within a discrete habitat, a grid of pitfall traps could influence the population density of a highly mobile animal.

Two other factors may influence the efficiency of pitfall traps: the level of the trap lip and the retaining efficiency. Greenslade (1964) found that trap catches were qualitatively or quantitatively affected by the precise placement of the level of lip of the trap container, whether at the surface of the soil, or the surface of the litter. Retaining efficiency depends on escape; the smoother the side, the less likely this is. Glass and most plastic containers are therefore to be preferred to metal traps that quickly corrode, providing footholds. Small winged species may escape from large aperture traps by flight or they may be eaten by larger species: it is, of course, for these reasons that investigators often use preservative solutions.

Walking animals may also be caught on sticky traps, bands placed round tree trunks (see also p. 377) and glass plates lain on the ground. Mellanby (1962) found that the latter caught numbers of springtails and other animals; he suggested a gauze roof over the trap to prevent flying insects being caught. Details of the design of sticky traps are given below.

7.4.1.3.1 *Mammal traps*

A wide variety of traps for both large and small mammals are described by Bateman (1979). Most ecologists are only concerned with the live trapping of small mammals, which is normally accomplished with a trap that closes when the animal touches a lever or treadle. The most frequently used traps in Britain are Longworth traps (Penlon Ltd, Radley Road, Abingdon, Oxon, OX14 3PH; available from http://www.angleps.com/longworth_mammal_trap.php and http://www.nhbs.com/title/160222/longworth-small-mammal-trap) and in the USA the Sherman trap (H.B. Sherman Inc., P.O. Box 20267, Tallahassee, Florida 32316, USA; available from http://www.shermantraps.com/ and http://www.nhbs.com/title/183647/sherman-trap-small-folding-aluminium). Traps are often baited with food both to attract and later sustain the captive. References regarding the selectivity of these traps are given in Section 7.1.

7.4.1.4 Grain-probe traps

Commercial probe traps are available to detect beetle infestation in grain stores. The trap is a perforated cylinder, one end of which is sealed while the other has a collecting vial. The trap is placed within the grain and works rather like a pitfall in that it traps beetles which crawl through the perforations. White *et al.* (1990) discuss the design of probe traps, and Subramanyam *et al.* (1993) compare the capture efficiencies of the Storgard (manufactured by Trace, Salinas, CA, USA) and Grain Guard (manufactured by Grain Guard, Verona, WI, USA) commercial traps. The Storgard trap comprises a

45 cm-long, 3 cm internal diameter polyethylene tube with 4×2 mm rectangular perforations, while the Grain Guard version is a 37 cm polycarbonate cylinder of 2 cm internal diameter with 2.5 mm round, downward-pointing perforations.

Probe traps might also find application in studies of the infauna of habitats such as aquatic leaf litter banks where there is a lack of fine particulate matter that could clog or fill the trap.

7.4.2   *Flight traps combining interception and attraction*

The distinction between water and sticky traps and those flight traps described above (p. 281) should not be regarded as rigorous for, as Taylor (1962) has shown, a white sticky trap catches insects almost as if they are inert particles, whereas a net may sample selectively. The two approaches may be combined in the sticky net described by Provost (1960). However, many aphids and weevils are particularly attracted to yellow (Broadbent, 1948; Riley & Schuster, 1994), ceratopogonid midges to black (Hill, 1947); citrus midge to orange (Pena & Duncan, 1992); thrips to white (Lewis, 1959), blue (Brodsgaard, 1989) or yellow (Cho *et al.*, 1995), frit flies to blue (Mayer, 1961), and bark beetles to red (Entwistle, 1963). Therefore, the quality and magnitude of water and sticky trap catches will be greatly influenced by the colour of the trap, and this will be an additional source of variation in the efficiency of these traps. It is probable that the attraction of different colours to a given species may vary with both age and sex.

7.4.2.1   Sticky traps

These are an extension of the 'fly-paper'; the animal settles or impacts on the adhesive surface and is retained. A variety of adhesives may be used; those resins and greases developed for trapping moths ascending fruit trees have proved particularly useful; castor oil may be satisfactory for minute insects (Cameron *et al.*, 1995). One of the most commonly used sticky materials is polyisobutylene, and instructions for the removal of insects from this matrix are given by Murphy (1985). The insects are separated from most fruit-tree banding resins by warming and then scraping the resin plus insects into an organic solvent such as trichlorethylene or hot paraffin, from which they may be filtered. The separation from greases is easier; a mixture of benzene and isopropyl alcohol rapidly dissolves the adhesive. Thus, when possible a grease will be used in preference to a resin, but only weak insects will be trapped by a grease, notably mosquitoes (Provost, 1960), mites (Staples & Allington, 1959) and aphids (Close, 1959). Indeed, a grease may be more efficient for such insects, as the effective area of the trap is not reduced by the numbers of large insects that are also trapped if a powerful adhesive is used (Close, 1959). Greases may become too fluid at high temperatures.

Sticky traps are of several basic designs: Até strands, an African birdlime, were used by Golding (1941, 1946) in warehouses, and Provost (1960) used sticky nets, but most traps have been either large screens or small cylinders, boxes or plates. Cylindrical plastic traps are commonly used for small insects. Large screens consisting of a wooden lattice or series of boards have been used to measure movement at a range of heights of aphids and beetles (Dudley *et al.*, 1928; Taft & Jernigan, 1964). Individual boards,

Table 7.5   **The catches of some species of aphid on three types of trap expressed as a ratio of the number caught by a suction trap. (Data from Heathcote, 1957.)**

	Water	Cylindrical sticky	Flat sticky
Aphis fabae group	2.28	1.92	0.84
Tuberculoides annulatus	3.91	0.45	0.18
Cavariella aegopodii	0.36	2.00	0.27
Myzus persicae	1.34	0.34	0.78
Hyalopterus pruni	0.33	1.50	0.40
Drepanosiphum platanoidis	1.29	0.25	0.01
Brevicoryne brassicae	0.47	0.06	0.07
Sitobion spp.	0.08	0.17	0.06

white and yellow in colour, were used by Roesler (1953) and Wilde (1962) for studies of various flies and psyllids.

Ibbotson (1958) introduced the idea of glass plates for sticky traps. These were 8 inches square, with the upper surface coated with an adhesive and the under surface painted yellow (or as desired). Glass has the advantage that the adhesive is easily scraped off with a knife or the whole plate can be immersed in a solvent. Ibbotson used his trap for the frit fly; Staples & Allington (1959) used grease-coated microscope slides for trapping mites; Chiang (1973) used an oil-coated strip over a white disc for midges; Wakerley (1963) found Ibbotson's trap suitable for the carrot fly, *Psila rosae*, and Maxwell (1965) used a similar trap for the apple maggot fly, *Rhagoletis*. These plates may be exposed either vertically or horizontally; the exposure will affect the catch (Table 7.5) (Heathcote, 1957b); the influence of colour is less with vertical plates, where most insects are caught by wind impaction. That is, the catches are less biased, but horizontal plates catch more aphids in the landing phase (A' Brook, 1973) and gave high catches of *Adelges piceae* (Lambert & Franklin, 1967). However, fixed vertical plate traps will sample a different proportion of the passing air depending on the wind direction, and eddies will develop at the sides of the plates and around sticky boxes (Frohlich, 1956). Small glass plates may be mounted at right-angles to a wind vane and so present a constant exposure to the wind.

A revolving sticky trap was found most efficient for mosquitoes (Dow & Morris, 1972), but most *Stomoxys* were caught on the leeward side of white sticky traps (Williams, 1973). A good design, sampling at random from the passing air, is the cylinder sticky trap of Broadbent *et al.* (1948) (Fig. 7.10a), which consists basically of a piece of plastic material covering a length of stove-pipe. Taylor (1962) showed that if the wind speed was known, the catches of small insects on a white sticky trap of this type could be converted to a measure of aerial density (see Table 7.1, p. 265); however, yellow traps catch more aphids than white ones, and these catch more than black (Broadbent *et al.*, 1948; Taylor & Palmer, 1972; A'Brook, 1973). Care should therefore be exercised in applying these corrections to a species whose reaction to the colour of the trap is unknown. Colour will presumably have the greatest effect at low wind speeds.

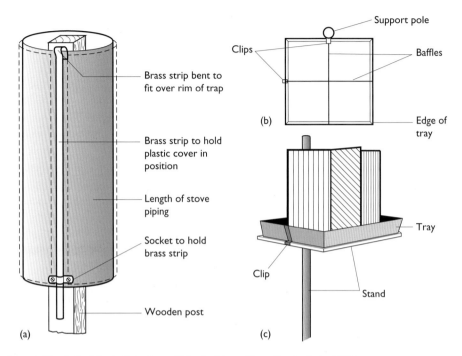

(a)

(b)

Clips

Support pole

Baffles

Edge of tray

Brass strip bent to fit over rim of trap

Brass strip to hold plastic cover in position

Length of stove piping

Socket to hold brass strip

Wooden post

Clip

Tray

Stand

(c)

Figure 7.10   Cylindrical sticky trap. (Adapted from Broadbent et al., 1948.)

Cylindrical sticky traps have been utilised in studies on the cacao mealy bug (Cornwell, 1960), while small ones (2.5 cm diameter) were used by Lewis (1959) for trapping thrips.

The relation of the size of the sticky trap to the catch has been investigated by Heathcote (1957a) and Staples & Allington (1959), who found that although catch usually increases with size, it is not proportional to size so that the smallest trap catches the largest number per unit area (Table 7.6).

Simpson & Berry (1973) exposed successive portions of a 'Tanglefoot' coated strip of plastic on the leading edge of a trap, with a vane, so that it always faced into the wind: in this way they were able to obtain a 24-h record of wind-borne aphids. A sticky trap which automatically changed the collecting surface is described by Rohitha & Stevenson (1987).

Mention was made above to the attraction of aphids and other insects in the landing phase to particular coloured horizontal plates. Prokopy (1968, 1977) and Kring

Table 7.6   The effect of trap size on the catch of black cylindrical sticky traps. (Data from Heathcote, 1957.)

Trap diameter	3 cm	6 cm	12 cm
Total catch	837	1473	2017
Catch/unit area (3-cm trap = 1)	837	736.5	504.2

(1970) have shown with fruit flies (Trypetidae = Tephritidae) that yellow panels and red spheres attract flies in different behavioural conditions: the yellow plate is a 'super leaf', the red sphere a fruit. The use of relative methods sensitive to behavioural phase to partition the population in this way is a powerful approach. Sticky trap catches are frequently biased towards one sex. For example, Kersting & Baspinar (1995), using yellow sticky traps to sample for the leafhopper *Circulifer haematoceps*, found >85% of the catch to be male, while D-VAC collections were 74% female. The sticky trap catch may also be biased towards females (e.g. Riley and Schuster, 1994), but it is often able to detect population change. For example, Palumbo *et al.* (1995) compared yellow sticky traps, direct counting and a suction sampler and found all three methods were able to show population trends in sweet-potato whitefly, *Bemisia tabaci*.

7.4.2.2 Water traps

These are simply glass, plastic or metal bowls or trays filled with water to which a small quantity of detergent and a preservative (usually a little formalin) have been added. Omission of the detergent will more than halve the catch (Harper & Story, 1962). The traps may be transparent or painted various colours and placed at any height. Yellow has long been recognised as one of the most effective colours. Yellow bowls were used by Moericke (1950) and others for trapping aphids, by Frohlich (1956) and Fritzsche (1956) for a weevil, and by D'Arcy-Burt & Blackshaw (1987) for bibionid flies. Fluorescent yellow trays caught twice as many cabbage-root flies, *Erioischia*, as yellow trays (Finch & Skinner, 1974). White trays were used by Southwood *et al.* (1961) for the frit fly, while Pollet & Grootaert (1987) found most species of Dolichopodid fly to be attracted to white, although soil-dwelling and arboreal species were most numerous in the red and the blue traps, respectively. A variety of colours were tested by Harper and Story (1962) for the sugar beet fly, *Tetanops myopaeformis:* the total numbers taken in the different coloured traps were as follows: yellow 330, white 264, black 202, red 107, blue 64, green 53.

For many purposes, sticky traps would be chosen in preference to water traps, because the relationship between wind speed and the catch of the water trap is likely to prove less simple than that for the sticky trap investigated by Taylor (1962). In addition, water traps must be frequently attended or they may overflow in heavy rain or dry out in the sun, although a model has been designed with a reservoir for automatically maintaining a constant level (Adlerz, 1971). On the other hand, water traps have certain advantages: the insects that are caught are in good condition for identification, as the catch is easily separated by straining or individual insects picked out with a pipette or forceps. Furthermore, when the population is sparse a water trap will, with aphids at least, make catches when a sticky trap of a manageable size would not (Heathcote, 1957b). The 'catching power' of a water trap can be further increased by standing two upright baffles of aluminium sheeting at right-angles to each other to form a cross in the tray (Coon & Rinicks, 1962) (Fig. 7.10b and c). These divisions enable one to separate the insect according to the quadrant in which it has been captured; this may be related to the direction of flight at the time of capture, although eddy effects could introduce error (see also p. 282). The results of Coon & Rinicks show that the effect of the baffle varies from species to species; in the four cereal aphids that they studied it increased the catch of *Rhopalosiphum maidis* tenfold, nearly doubled the very low catch of *Schizaphis agrostis* (= *Toxoptera graminum*), but had no effect on those of the two others, *Macrosiphum granarium* and *Rhopalosiphum fitchii*.

A floating water trap that collected the insects settling on the surface of ponds was devised by Grigarick (1959), the animals being drowned through the action of the detergent. The tray of his trap was surrounded by a wooden frame, therefore it would seem that the contrast presented by this surround might bias the catch and a more transparent float (e.g. an air-filled polythene tube) would be preferable.

7.4.3 *Light and other visual traps*

7.4.3.1 Mode of action and limitations

Light-traps are probably the most widely used insect traps and there are many thousands of references to them. The book by Muirhead-Thompson (2012) reviews light trap design and their capture efficiency for insects. Originally, paraffin and acetylene lamps were used by collectors (Frost, 1952); later, the tungsten filament electric light and, after its development by Robinson & Robinson (1950), the ultraviolet trap became widely used. The field continues to develop with the recent introduction of low-energy usage ultraviolet light-emitting diodes (LEDs). For example, Zheng *et al.* (2014) used different wavelength LED as attractants for adult, *Aleurodicus disperses* (Hemiptera: Aleyrodidae), while Hashiguchi *et al.* (2014) sampled the sand fly fauna (Diptera, Pcychodidae, Phlebotominae) using a mini-Shannon Trap with LED lights.

The uses of light-traps in ecology are subject to all the general limitations of relative methods outlined above (p. 259). However, the variation in efficiency of the trap from insect to insect, from night to night, and from site to site is more serious than in almost any other type of trap because light-traps are entirely artificial, relying on the disturbance of normal behaviour for their functioning. It is not justified to claim on a priori grounds, as Mulhern (1953) has done, that, being mechanical, light-traps are more reliable than hand collection per unit of habitat. Furthermore, in the discussion of the pros and cons of the many models of light-trap it is often implied that the bigger the catch, the 'better' the trap. Although it is true that the larger figures are often more acceptable for statistical analysis, it is unwise to assume that they are biologically more unbiased.

The exact mechanisms that lead to an insect's capture by a light trap are far from clear. Mikkola (1972) has shown that electroretinograms evoked by different types of light were at variance to the actual responses of the moths to traps. It seems reasonable to suppose that there is an area, sometimes called the radius of the trap or 'catchment area', within which the insect comes into the influence of the light. The brighter the trap the greater this area, and hence when there is bright moonlight the amount of contrast between the trap and its surroundings will be lessened, and the catchment area reduced (e.g. Miller *et al.*, 1970; Taylor, 1986; Nag & Nath, 1991). For example, a study of light-trap catches of rice green leafhoppers (*Nephotettix* spp.) showed that the effective radius of a 100 W tungsten varied from 245.20 m during the new moon to 16.72 m at full moon (Mukhopadhyay, 1991). Bowden & Church (1973) found that the extent of the reduction was consistent with the random, rather than fixed, capture of the insect within the catchment area. The behavioural studies of Verheijen (1960) and Mikkola (1972) suggest that the result may appear random because, even if there are zones of attraction corresponding perhaps to the insect seeking open flight routes through the vegetation, there are probably 'annuli of repulsion' and 'zones of dazzle' when the normal photic orientation is disturbed.

In a series of studies, Bowden has been concerned with the changes in the catchment area ('trap radius') with time – the lunar cycle; he has shown how the precise distribution of moonlight throughout the night needs to be considered in relation to the insects periodicity of flight (Bowden, 1973; Bowden & Church, 1973). Using this information, standardised catches may be calculated and these eliminate the variations due to the changes in the catchment area brought about by the effect of moonlight (Bowden & Morris, 1975). When this is done it is found that there are some insects that are more active on nights with full moon (Bidlingmayer, 1964; Anderson & Poorbaugh, 1964; Bowden & Gibbs, 1973).

Variation in the catchment area in space, with regard to the height of flight of the moths, the position and design of the trap and the role of wind in moving a column of air across the trap, have been studied by Taylor & Brown (1972) and Taylor & French (1974). They postulate a volume equivalent to the amount of air originally occupied by the catch, that is with 100% efficient extraction. This will, of course, be less than the catchment area, but one can visualise that the two approaches might be combined to make light traps a measure of absolute population. Catches generally fall with increasing wind speed, such as *Culicoides*, Edwards *et al.* (1987); turnip moth *Agrotis segetum*, Esbjerg (1987); migratory moths, Gregg *et al.* (1994). Other factors known to influence trap efficiency include temperature, relative humidity and rainfall.

Light traps (Plate 5) have been found useful in survey work, both for particular species (Geier, 1960; Otake & Kono, 1970; Grimm *et al.*, 1993; Bishop *et al.*, 1994; Zheng *et al.*, 2014) or on an extensive scale for a taxonomic group or region (Rings & Metzler, 1990; Richert & Huelbert, 1991; El Hag & El Meleigi, 1991; Kitching *et al.*, 2015).

7.4.3.2   The effect of trap design on catch

From the above we can conclude that catches in a light-trap will be influenced by the following design features:

1. The amount of contrast between the light source and surroundings: the greater the contrast, the greater the 'catchment area'.
2. Most animals will have a tendency to withdraw from the high light intensity immediately adjacent to the lamp.
3. The extent to which an animal may be able to change from approach to avoidance will depend on its flight speed; the faster, heavier flyers are unable to stop and change course quickly. Indeed, Stanley & Dominick (1970) found with a Pennsylvania-type trap with nested funnels that the large moths were trapped in the central smallest funnel, whilst coccinellids were most frequently caught in the widest outer funnel.

The effect of the illumination of the environment on the catch was utilised by Common (1959), who designed a transparent light-trap for Lepidoptera. The main advantage of this over the Robinson trap, of which it is a modification, is that the number of Coleoptera (especially Scarabaeidae) caught is greatly reduced; large numbers of beetles will damage many of the moths in a catch.

An increase in the intensity of the lamp, frequently brought about by the substitution of an ultraviolet lamp for an incandescent one, usually leads to an increased catch (e.g. Williams *et al.*, 1955; Barr *et al.*, 1963; Belton & Kempster, 1963). Because of the role of intensity in the mechanism of trapping, the actual size of the light source is of less importance (Belton & Kempster, 1963) and a small trap may often be substituted for a larger one without the catch being reduced in proportion to the change in light output (Smith *et al.*, 1959).

Many workers have considered that the differences in quality of the light, from incandescent lamps, from mercury-quartz lamps emitting ultraviolet and visible light, and from black (or 'blue') lamps emitting only or mainly ultraviolet, affect the catch; however, Mikkola's (1972) study showed this to be very complex. Whatever the mechanism, there are considerable differences; for example, many Diptera and Miridae are taken in the largest numbers in traps with incandescent (tungsten filament) lamps; Corixidae, many Lepidoptera (especially noctuids) are more abundant in 'UV traps' and are even better represented in 'black-light traps', as are Trichoptera (Frost, 1953; Williams *et al.*, 1955; Southwood, 1960; Tshernyshev, 1961).

Some of these differences are explained by the increased intensity of the ultraviolet traps leading to a repulsion of the insect when close to the lamp; slow flyers such as many Diptera, Nematocera and geometrid and pyralid moths stop their approach before they enter the trap, whilst the heavier, faster flyers, like noctuid and sphingid moths, pass straight into the trap. Suction fans have therefore been incorporated in light-traps to catch those insects that avoid capture at the 'last moment' (e.g. Glick *et al.*, 1956; Downey, 1962). The same phenomenon was demonstrated in the opposite way by Hollingsworth *et al.* (1961), who found that a suitably placed windbreak could increase the catch of trap: in the exposed situation many species would avoid trapping by being blown off course, but in the lee of the windbreak this would not occur and the larger moths would be caught particularly efficiently. The addition of baffles to the trap also serves to catch those insects whose repulsion by the high intensity close to the lamp would otherwise cause them to escape; Frost (1958a), by adding four intersecting baffles to a trap, almost doubled the numbers caught.

When differences due to flight speed and momentum have been eliminated, other specific differences in susceptibility to trapping remain; amongst the mosquitoes, *Aedes* seem especially susceptible to trapping (Love & Smith, 1957; Loomis, 1959b). Males are often taken in much greater numbers than females at light-traps, but this may be because males engage in more nocturnal flight activity than the females (e.g. Miridae; Southwood, 1960).

The height above the vegetation at which the trap is exposed will influence the catch. As the aerial density of most insects decreases with height, in general the higher the trap the smaller the catch (Frost, 1958a; Taylor & Brown, 1972). There are a few exceptions; Taylor & Carter (1961) have shown that some Lepidoptera may occur at a maximum density at as much as 10 m above the ground, while many ornithophilic mosquitoes and blackflies also have maxima some distance above the ground.

7.4.3.3   Techniques and types of trap

Many light-traps retain the insects alive; Merzheevskaya & Gerastevich (1962) did this by covering the collecting funnel of a Pennsylvania-type trap with a cloth bag and inserting another shallow cone in the neck. The Robinson trap retains the insects in the drum of the trap; but Belton and Kempster (1963) showed that unless the opening was under 1 in (2.5 cm) in diameter, many small moths, such as geometrids, would leave the trap in the morning, more especially if the sun shone on it; however, such a small cone opening will restrict the entry of other species.

A large number of killing agents have been used in light-traps; some of the chlorinated hydrocarbons such as tetrachloroethylene, trichloroethylene and tetrachloroethane are particularly useful. They may be poured onto a block of plaster of Paris at the bottom of the killing bottle, where their heavy vapours will be slowly evolved and remain relatively concentrated (Williams, 1958; Haddow & Corbet, 1960). However, it must be noted that the inhalation of their vapours over a long period may be dangerous to humans. Delicate insects such as Ephemeroptera or Psocoptera may be collected straight into alcohol, and chafer beetles into kerosene (Frost, 1964). Potassium and sodium cyanides are traditional, but dangerous, poisons; the last named is more suitable in moist climates (Frost, 1964).

White (1964) has designed a killing tin to preserve the light-trap catch in good condition (Fig. 7.11). It has the further advantage that only specimens too large to pass through the screen across the top of the funnel are retained; the mesh of this screen

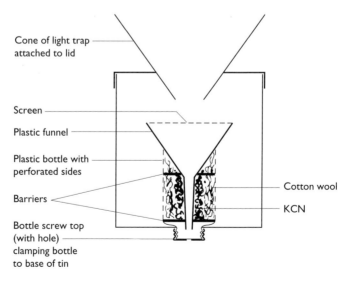

Figure 7.11   White's light-trap filling tin.

may be so fine as to retain the whole catch or, for example, it may be of such a size that the unwanted Diptera and small Trichoptera escape (E.G. White, pers. commun.). When only the smaller species are required, the larger ones may be excluded and so prevented from damaging the catch by covering the entrance to the trap with a coarse screen (e.g. Downey, 1962). If all specimens are to be captured a series of graded screens in the killing bottle will tend to separate the insects by size and aid in the subsequent sorting (Frost, 1964). In connection with the latter operation the sorting tray described by Gray (1955) may be useful.

7.4.3.4 The Rothamsted trap

The Rothamsted trap (Fig. 7.12a) was originally used by C. B. Williams as a research tool for studying the influence of climate on the numbers of flying insects. The trap takes moderate numbers of most groups, including nematocerous flies and the larger moths; it would seem to be less selective than most other light-traps and is useful in studies

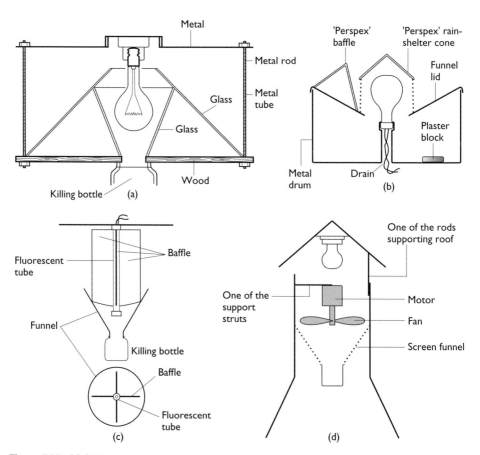

Figure 7.12 Light traps.

on the diversity of restricted groups. It is described in detail by Williams (1948): it has a roof, which allows its operation in all weather, but reduces the catch. The collections from different periods of the night may be separated by an automatic device. It may be purchased from the Burkard Co. (http://www.burkard.co.uk).

7.4.3.5   The Robinson trap

The Robinson trap (Fig. 7.12b) was the first trap using ultraviolet light, and was designed by Robinson and Robinson (1950) to make maximum catches of the larger Lepidoptera trap. It is therefore without a roof, and although the bulb is protected by a celluloid cone and there is a drainage hole in the bottom, it cannot be used in all weathers. The drum is partly filled with egg boxes or similar pieces of cardboard in which the insects can shelter. The catch may be retained alive or killed by placing blocks of plaster of Paris saturated with a killing agent on the floor. Large numbers of fast-flying nocturnal insects (e.g. Noctuidae, Sphingidae, Corixidae, Scarabaeidae) are trapped (Williams, 1951; Williams *et al.*, 1955); however, Common's (1959) transparent trap catches fewer beetles. The 'Muguga trap', widely used in eastern Africa, is a modification of this type of trap (Brown *et al.*, 1969; Siddorn & Brown, 1971).

7.4.3.6   The Pennsylvanian and Texas traps

These traps (Fig 7.12c) are basically similar, consisting of a central fluorescent tube surrounded by four baffles; below the trap is a metal funnel and a collecting jar (Frost, 1957; Hollingsworth *et al.*, 1963). The Pennsylvanian trap has a circular roof to prevent the entry of rain into the killing bottle, and it is presumably a reflection of the climates of the two States that the Texas trap is roofless. A wire mesh grid directly below the funnel in the Texas trap will allow rain and small insects to escape, whilst the larger specimens are deflected into a collecting container (Dickerson *et al.*, 1970). Frost's (1958c) experiments suggest that the largest catches would be obtained with a 15-W black-light fluorescent tube; Graham *et al.* (1961) used three argon-glow lamps in the Texas trap in survey work on the pink bollworm, *Platyedra gossypiella*, and found moths were trapped over a radius of 200 ft from the lamp. Fluorescent tubes may be run for many hours or even years from batteries (Clark & Curtis, 1973). Kitching *et al.* (2015), in a survey of forest moth assemblages, used Pennsylvania traps with an actinic tube powered by a 12 V battery held beneath the trap that allowed the unit to be suspended in the forest.

The Minnesota and Monks Wood light-traps are basically similar to the Pennsylvanian model, but the roof is cone-shaped rather than flat and this appears to reduce the catch (Frost, 1952; Frost, 1958b). The effects of many different modifications of this type of trap on the catches of small moths have been reported by Tedders & Edwards (1972).

7.4.3.7   The New Jersey trap

Developed by T. J. Headlee, the New Jersey trap (Fig. 7.12d) is primarily a trap for sampling mosquitoes; it combines light and suction – the lamp causes the insects to come into the vicinity of the trap and they are drawn in by the suction of the fan (Mulhern, 1942). The trap is therefore particularly useful for weak flyers that may fail to be caught by the conventional light-trap, as when close to the lamp they are repelled by its high intensity. Fitted with an ultraviolet lamp the New Jersey trap has been used to collect other groups of Nematocera, as well as mosquitoes (Zhogolev, 1959). Kovrov & Monchadskii (1963) found that if the trap was modified to emit polarised light, many insects – but not mosquitoes – were caught in larger numbers. Like the Rothamsted and Pennsylvanian traps, this trap may be fitted with an automatic interval collector. A tiny battery-operated model, the CDC (Centre for Disease Control) trap has been developed (Buckley & Stewart, 1970; Vavra *et al.*, 1974; Fontenille *et al.*, 1997). For comparative studies using different traps, it is important to standardise air flow and direction (Loomis, 1959a; Wilton & Fay, 1972) although, as has been pointed out several times, relative trapping methods, and especially light-traps, may sample different proportions of the population in different areas even when the traps are identical. Acuff (1976) found that light traps collected a wider range of mosquito species than bait and Malaise traps. For bloodmeal and arbovirus studies the insects should be collected into saline (7.5 g l^{-1}) with some antiseptic (Walker & Boreham, 1976).

7.4.3.8 The Haufe–Burgess visual trap

As Verheijen's (1960) studies have shown that the trapping effect of light depends on the degree of contrast between the lamp and the surroundings, it is not surprising that light-traps are ineffective in the twilight night of the arctic. Haufe & Burgess (1960) designed the present trap (Fig. 7.13a) for mosquitoes under these conditions, and it has been found to catch other groups as well. Basically, the trap is an exposed cone type of suction trap on the outside of which is a revolving cylinder painted with about 4 cm-wide black and white stripes. A horizontal disc extends out at the top of the cylinder and the insects are drawn through slits at the junction between the cylinder and the disc. Harwood (1961) retained the light-bulb above the trap (as in the New Jersey light-trap) and exposed various animals as bait below the trap, but the combination of so many attractants makes the interpretation of the results very difficult.

7.4.3.9 The Manitoba or canopy trap

Tabanidae are attracted to the highlights of a black or red sphere (Bracken *et al.*, 1962) and large numbers may be collected by suspending a sphere of these colours (e.g. a balloon) beneath a polythene collecting cone (Fig. 7.13b) (Thorsteinson *et al.*, 1965; Bracken & Thorsteinson, 1965). The collecting container may be of the type used on 'tents' (see Fig. 7.7) (Adkins *et al.*, 1972) and with modifications flies can be marked (see p. 103).

7.4.3.10 The pit-light trap

The pit-light trap (Fig. 7.14), as described by Heap (1988), is designed to capture ground-living insects and combines features of a light-trap and a pitfall trap. It is

Plate 1   Marine samplers. A, inshore trawler. B, midwater trawl for swiming fish. C, trawl cod end with catch. D, Day grab for benthic invertebrates. E, MIK net for elvers and larval fish. F, Beam trawling for epibenthic fish. (Photographs A, B and C by P. A. Henderson. Photographs D, E and F courtesy of Pisces Conservation Ltd.)

Ecological Methods, Fourth Edition. P. A. Henderson and T. R. E. Southwood.
© 2016 John Wiley & Sons, Ltd. Published 2016 by John Wiley & Sons, Ltd.
Companion Website: www.wiley.com/go/henderson/ecologicalmethods

Plate 2   Inshore marine samplers. A, 30 cm bongo plankton nets, with flow meter. B, A plankton net used with a pump sampler; the net is in a large water-filled bin to reduce sample damage. C, Push netting for small fish. D, A standard shallow water push net. E, A gill net used intertidally; a flounder is being removed. Gill nets can also be floated or set on the sea bed. (Photographs A, B and C by P. A. Henderson. Photographs D, E and F courtesy of Pisces Conservation Ltd.)

Plate 3 Freshwater sampling. A, Freshwater sampling nets, from the left, (i) plankton net, (ii) standard FBA net and (iii) Surber kick net sampler. B, Kich net sampling a small stream for invertebrates. C, Using a seine net to sample fish and invertebrates in an Amazonian floating meadow. D, Electric fishing gear comprising electrodes, control box and generator. E, Electric fishing a smale stream, note the stop net placed across the stream width. A second net was placed at the bottom of the reach. The method is used to sample fish and estimate density using removal sampling methods. (Photographs A and B courtesy of Pisces Conservation Ltd. Photographs C, D and E by P. A. Henderson.)

Plate 4 Sampling invertebrates in grassland and low vegetation. A, A selection of standard nets for insects and other terrestrial invertebrates. B, A univac sampler used with a defined area. C, sweep net sampling. D, Simple coring equipement. E, A pitfall trap made using a plastic cup and fitted with a cover to stop rain flooding the trap. (Photographs A, C, D and E by P. A. Henderson. Photograph B by Clive Hambler; reproduced with permission.)

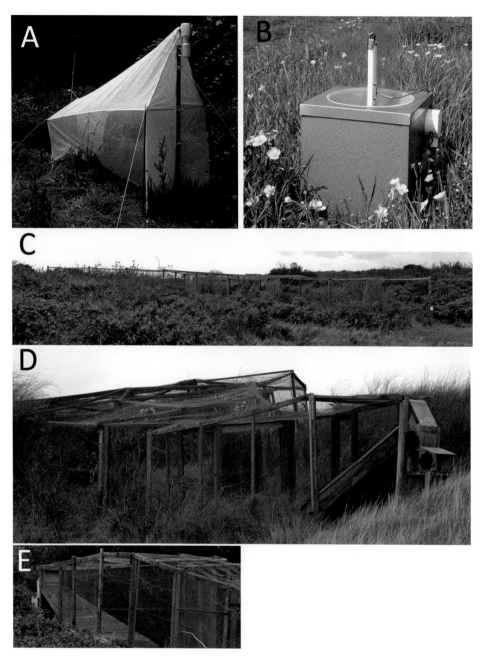

Plate 5 Terrestrial traps. A, Malaise trap. B, Small portable moth trap. C, D & E, Views of a large Heligoland trap for birds. (Photograph A courtesy of Pisces Conservation Ltd. Photographs B, C, D and E by P. A. Henderson.)

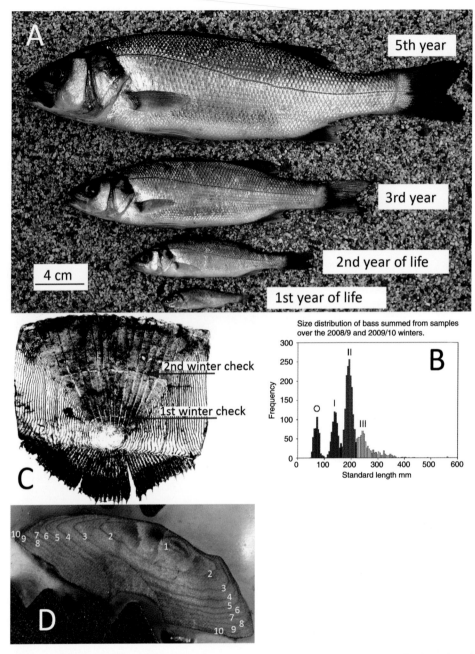

Plate 6 Ageing methods, using bass, *Dicentrarchus lobrax,* as an example. A, The lengths of young age groups. B, Frequency histogram showing modes for first 4 age classes, older age classes indistinct. C, Photograph of a scale showing two winter growth checks. D, Cross-section of an otolith of an old fish in the 11th year of life. The winter checks are labelled and have been stained with neutral red. (All photographs by P. A. Henderson. Plate B courtesy of P. A. Henderson.)

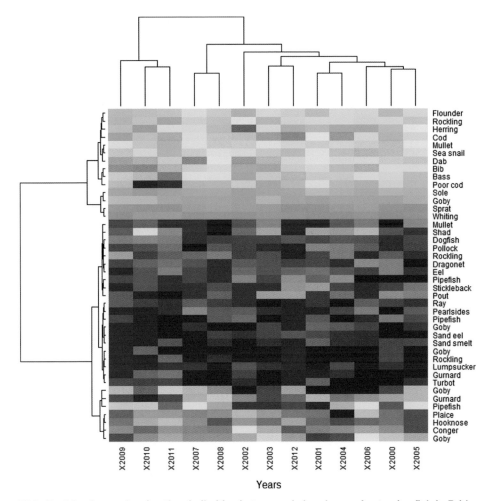

Plate 7 A heat map showing the similarities in temporal abundance of estuarine fish in Bridg-water Bay, England. The highest fish abundances are shown in green and the lowest in red. The analysis has arranged the fish with the most abundant towards the top. At the bottom are species which occasionally show bursts of abundance. The years are arranged in terms of their similarity in species composition, for example, 2003 and 2012 had similar species compositions. The heat map was generated using the R code given on p 527. (Courtesy of P. A. Henderson.)

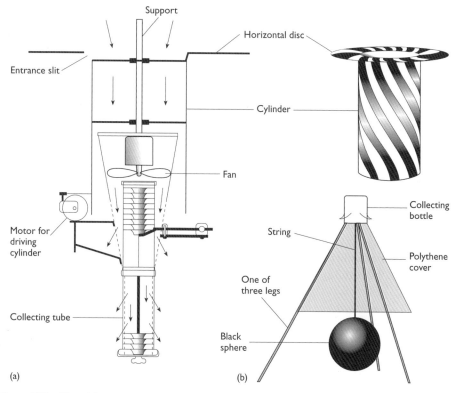

Figure 7.13   Visual traps.

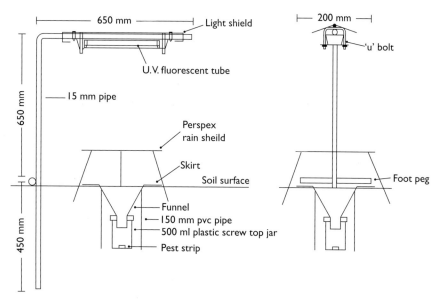

Figure 7.14   Pit-light trap (Heap, 1988).

claimed to capture a wide range of insects with a greater number of specimens and range of soil-dwelling species than can be captured using non-baited pitfall traps.

7.4.3.11   Aquatic light-traps

There are two types of aquatic light-trap, namely floating and submerged (Fig. 7.15). At present, light-traps are little used for sampling either invertebrates or fish but reports of their use are increasing in number. Light trapping for young fish is particularly appropriate amongst underwater obstructions such as fallen wood and macrophytes where netting becomes inefficient or impossible (Conrow et al., 1990; Dewey & Jennings, 1992). Neal et al. (2012) found night-time light trapping as the most effective sampling method for larval fish in tropical streams.

Baylor & Smith (1953) have described an ingenious floating trap, principally for Cladocera, which utilises their 'attraction' to yellow to get them into the trap, and their 'repulsion' to blue to drive them into the collecting net (Fig. 7.15). It is important that the blue light is not visible laterally; thus, the level of the blue 'Cellophane' must not extend above the lip of the funnel. Baylor & Smith suggest that the trap could be improved with refinements, such as the polarisation of the yellow light that would increase the 'fishing area' without increasing the intensity (see also Kovrov & Monchadskii, 1963). A floating light-trap, especially suitable for emerging insects, may be made out of a plastic bucket with a black light inside the lid and a funnel replacing the base (Carlson, 1971). A model for use in shallow water is described by Apperson & Yows (1976).

Subaquatic light-traps can be used in marine and freshwater environments and take a variety of forms. They are most frequently used to sample post-larval and juvenile fish when a common form is a plexiglass box with slits (Fig. 7.15). Both, electric lights and chemical light sticks may be used as a light source. Gehrke (1994) found that when larval golden perch, *Macquaria ambigua*, and silver perch, *Bidyanus bidyanus*, were exposed to light gradients the silver perch displayed stronger phototactic behaviour than golden perch, and both species were most responsive to light in the 601 nm waveband. The intensity of phototactic responses in both species was greater at higher irradiance levels. Traps fitted with 12-h yellow lightsticks attracted more golden perch larvae than traps with blue, green, orange, red lights. The greater efficacy of yellow lightsticks may be due to reduced attenuation; however, yellow lightsticks also emit a greater intensity of light over a longer time than other colours tested. Ponton (1994) tested an improved version of the quatrefoil light trap (Fig. 7.15) and concluded that it appeared useful to sample in particular freshwater microhabitats of several neotropical young fish. Kissick (1993) compared the attraction of larval and juvenile fish to traps lighted by either electric bulbs or photochemical light sticks. No significant differences in collection rates of bluegills, *Lepomis macrochirus*, and brook silversides, *Labidesthes sicculus*, were found. Significantly more threadfin shad, *Dorosoma petenense*, occurred in chemical light traps. Lightstick output will vary with temperature and is of only limited duration, but has the advantage over an electric light that it is less likely to fail. Examples of studies using lights to sample juvenile fish are Gregory & Powles (1985, 1988), Zigler & Dewey (1995), Thorrold (1992), Doherty et al. (1994), and Choat et al. (1993). Light traps can also be used to sample squid (Thorrold, 1992). Larval fish and aquatic

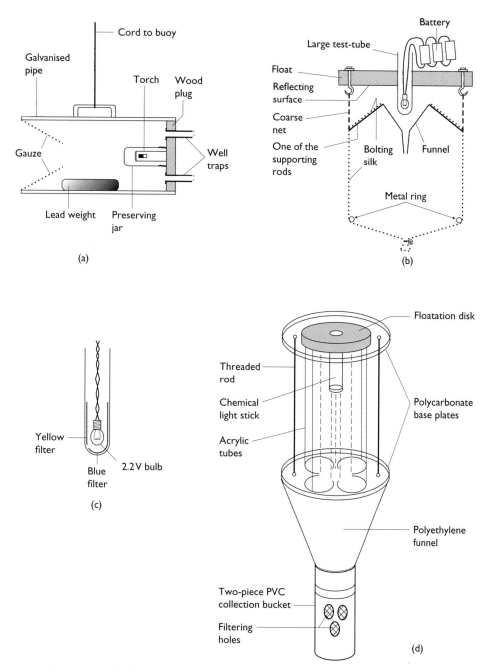

Figure 7.15   Aquatic light traps.

invertebrate light traps are available from Aquatic research instruments (web address: http://www.aquaticresearch.co./larval.htm)

7.5   Traps that attract animals by some natural stimulus or a substitute

When these traps are entirely natural they have the advantage over artificial ones in that the variations in efficiency reflect real changes in the properties of the population: the changes in the numbers of a phytophagous insect colonising a plant or of a bloodsucking species biting the host may not reflect population changes, but they do accurately reflect changes in the colonising or feeding rates of the species. That is, with the possible exception of artificial bait traps, these traps give measures of availability that are biologically meaningful.

7.5.1   Shelter traps

Cubical boxes (283 cm^3 in volume) with one open side and painted red have been used in mosquito surveys, and are often termed 'artificial resting units' (Goodwin, 1942; Goodwin & Love, 1957; Burbutis & Jobbins, 1958). As might be expected, only certain mosquitoes settle in these shelters, particularly species *Culex*, *Culiseta* and *Anopheles*. Loomis & Sherman (1959) found that visual counts of the resting mosquitoes with a torch were accurate to within 10 of the true value. Dales (1953) found that large numbers of *Tipula* would shelter in and be unable to escape from small metal truncated cones (15 cm aperture at top, 25 cm high and 28 cm at base), and Vale (1971) tested a range of artificial refuges for tsetse flies.

Ground dwellers (Cryptozoa) such as woodlice, centipedes and carabids that shelter beneath logs and stones, may be trapped by placing flat boards in the habitat. The 'cryptozoa boards' can be placed systematically in the habitat and used to study the dispersion of the animals (Cole, 1946); however, the proportion of the population that shelters beneath them will vary with soil moisture aspect, and other factors (Paris, 1965; Jenson, 1968).

Many insects shelter in cracks or under bark scales on tree trunks, for overwintering or as a mode of life. Artificial shelters consisting of several grooved boards bolted together have been placed in trees to trap earwigs (Chant & McLeod, 1952), in the field to trap overwintering *Bryobia* mites (Morgan & Anderson, 1958), and screwed onto the tree bark beside a fungus to trap the beetle, *Tetratoma fungorum* (Paviour-Smith, 1964). A variety of shelter traps (cryptic shelters) were used by Dahlsten *et al.* (1990) in a study of the Douglas fir tussock moth *Orgyia pseudosugata*; the most successful were wood blocks drilled with 2.5 cm holes. Fager (1968) used oak boxes filled with sawdust, with holes drilled in them, as 'synthetic logs' in studying the fauna of decaying wood.

Trap nests, made by drilling holes in woody stems (e.g. sumac) or in pieces of doweling which are split and then bound together, will be colonised by solitary Hymenoptera if the open ends are exposed in a board (Fig. 7.16); the binding may be removed and the nest contents easily exposed (Medler & Fye, 1956; Medler, 1964; Levin, 1957; Fye, 1965; Freeman & Jayasingh, 1975).

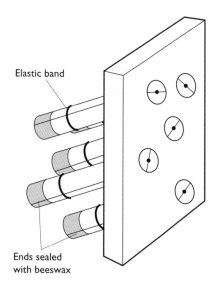

Elastic band

Ends sealed
with beeswax

Figure 7.16   Levins trap nests for solitary Hymenoptera.

7.5.2 *Trap host plants*

Insect-free potted host plants exposed in a habitat may be colonised by insects; these can be removed and counted and their numbers will be directly proportional to the colonisation potential of the population in that habitat. Trap host plants are therefore most useful for measuring emergence and colonisation (Fritzsche, 1956; Grigarick, 1959; Smith, 1962; Waloff & Bakker, 1963; Bucher & Cheng, 1970), and susceptible logs will record the same phenomena in bark beetles (Chapman, 1962). The actual adult insects or the resulting eggs or larvae may be counted. Michelbacher & Middlekauff (1954) studied population changes in *Drosophila* in tomato fields by placing as 'traps' ripe tomatoes that had been slit vertically on the sides and squeezed slightly; the number of eggs laid in the flesh on either side of the cuts could be easily ascertained in the laboratory. It is of course important to ensure that the plants, fruit or logs remain in the condition where they are attractive to the insects. Wireworms (Elateridae) are attracted to the CO_2 respired by germinating seeds, and Williams *et al.* (1992) used this behaviour to sample them by placing germinating seed into the soil for later retrieval.

7.5.3 *Baited traps*

The study of the behavioural responses of insects to scents from either their own species (pheromones) or their food sources (kairomones) remains an active and important field of entomological research. Such scents are used by man for a variety of purposes: for confusing the pest in its approach to the prey (plant) or in traps, either to measure or monitor the population or to reduce it by killing large numbers, either directly or by the incorporation of a sterilant into the bait. An enormous variety of traps have been described; the present account must be regarded as merely an outline of some of the types, with an emphasis on those most useful in ecological work. (Traps utilising vertebrates as bait are described in the previous section.)

7.5.3.1   Types of Trap

Basically traps can be divided into the two categories:

1. 'Lobster pot' traps (Fig. 7.17); these lure in insects through an inverted funnel when they are generally captured alive. Traps of this type have been used widely, with carbon dioxide as bait, for blood-sucking insects (Bellamy & Reeves, 1952; Wilson & Richardson, 1970); with carrion as bait for flies (MacLeod & Donnelly, 1956; Gillies, 1974), or with various extracts for bark beetles, when they are termed 'field olfactometers' (Chararas, 1977).
2. Impaction-traps (Fig. 7.17): in these the insect is held generally in an adhesive resin or grease or water trap in proximity to the lure. Designs of this type have been widely used in studies on chemical lures (Howell *et al.*, 1975; Lewis & Macaulay, 1976; Anderbrant *et al.*, 1989).

Pheromone traps are important monitoring tools for insect pests, and many different trap designs have been developed. The relative effectiveness of six commercially available pheromone traps, Multi-pher 1, 2, and 3 non-sticky traps and Scentry wing, Pherocon 1C, and Delta sticky traps for 6 orchard pest moths was compared by Knodel & Agnello (1990). Sticky traps generally captured more moths than non-sticky traps, regardless of species.

Williams *et al.* (2013) report a pheromone trap for oak processionary moth, *Thaumetopoea processionea*, which trialled pheromone lure source, trap design and height above the ground on capture rates. Two types of trap (Delta and funnel; Oecos, UK) containing one of three different commercially available pheromone lures were placed out in the lower (3–5 m), mid (5–10 m) and upper (10–15 m) canopy of 72 individual oak trees. Significantly more male moths were captured in traps positioned in the upper canopy (76.6%) compared with either mid-canopy (18.6%) or lower canopy (4.8%) positions. Further, funnel traps caught almost six times more male oak processionary moth than Delta traps. It is notable that this study found one of the three pheromone lures tested to be ineffective.

7.5.3.2   Baits or lures

The baits used have ranged from the actual biological material (e.g. virgin female insects or rotting fruit) to the extracted chemicals or their 'mimics'. Broadly, a distinction may be made between kairomones and pheromones, although some behavioural processes, such as the attraction of scolytids to a host tree, may involve both. Furthermore some of the lures have been discovered empirically and their classification is a matter of conjecture.

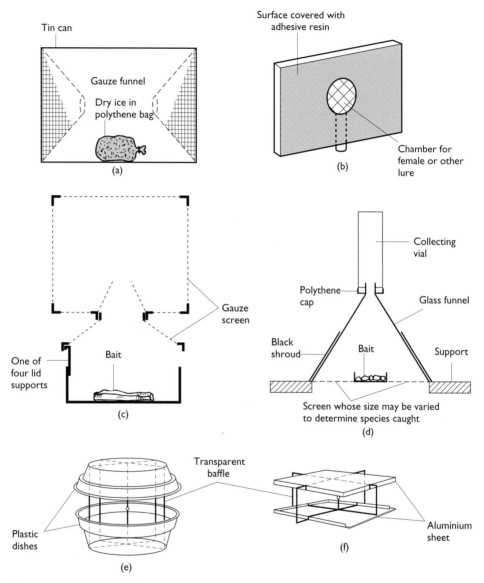

Figure 7.17   Bait traps. (a) Carbon dioxide trap; (b) Sticky trap (Coppel et al., 1960); (c) MacLeod's carrion trap; (d) Citrus Experimental Station general bait trap; (e) Water trap. (Adapted from Lewis & Macauley, 1976); (f) 'Lantern trap' (Adapted from Lewis & Macauley, 1976). (Chemical bait is exposed in the centre of the baffles that are covered with adhesive resin in the water and lantern traps.)

Kairomones

Carrion is used in trapping flies, especially blowflies (Dodge & Seaco, 1954); the freshness of the bait, the precise position of the trap and the presence of other flies have been found to affect the results (MacLeod & Donnelly, 1956; Fukuda, 1960; Kawai and Suenaga, 1960). Rotting fruit will attract certain butterflies (Sevastpulo, 1963), bananas or vinegar with yeast, *Drosophila* (Mason, 1963), hydrolysed proteins and similar compounds, fruit flies (McPhail, 1939; Neilson, 1960; Bateman, 1972) and proteinaceous materials, eye gnats, *Hippelates* (Mulla *et al.*, 1960). Various chemicals will act as lures; these are probably either the actual kairomone (as with carbon dioxide) or are related to it. Various fruit flies have been attracted by a variety of chemicals (Bateman, 1972), but complex synthetic lures such as 'Medlure' and 'Cuelure' are particularly effective (Beroza *et al.*, 1961; Fletcher, 1974). Many scolytids are attracted by turpinoids (Chararas, 1977), the cupesid beetle, *Priacma serrata*, by chlorine (Atkins, 1957), the cabbage rootfly, *Erioschia brassicae*, by allylisothiocyanata (Finch & Skinner, 1974) and the pine weevil, *Hylobius abietis* by the methyl ester of linoleic acid (Hesse *et al.*, 1955). Carbon dioxide, conveniently supplied as 'dry-ice', has been used as a lure for many blood-sucking insects: mosquitoes (Bellamy & Reeves, 1952; Gillies, 1974; Service, 1995), ticks (Garcia, 1962), tsetse flies (Rennison & Robertson, 1959), sandflies (Rioux & Golvan, 1969) and blackflies (Fallis & Smith, 1964). Carbon dioxide has been combined with a Malaise trap for tabanids (Blume *et al.*, 1972; Roberts, 1976), although Wilson & Richardson (1970) found *Chrysops* were less attracted to it than were *Tabanus*. Anderson & Yee (1995) trapped simulids on CO_2-baited three-dimensional horse/cow models with sticky ears and bodies. Tsetse fly can be attracted by baiting with natural and synthetic host odour (Hargrove *et al.*, 1995).

Pheromones

These may mostly be categorised as sex attractants, and the simplest approach is to expose a virgin female as the lure (Lewis *et al.*, 1971; Goodenough & Snow, 1973). The identification and synthesis of insect pheromones has been an important scientific achievement of the past 40 years. It has allowed the selective trapping of many moths and, more recently, of some Hemiptera and Coleoptera. There is an extensive literature on optimum trap design and some recent studies are: Anshelevich *et al.* (1993) and Anshelevich *et al.* (1994) on monitoring honeydew moth, *Cryptoblabes gnidiella*, and European vine moth, *Lobesia botrana*, respectively; Drapek *et al.* (1990) for the corn earworm, *Heliothis zea*, Kehat *et al.* (1994) for the peach twig borer, *Anarsia lineatella*, and Williams *et al.* (2013) for oak processionary moth, *Thaumetopoea processionea*. Particular progress has been made with codling moth (Reidl *et al.*, 1986), various tortricoid moths (Anshelevich *et al.*, 1994; Chang, 1995), noctuid moths (Witz *et al.*, 1992), pyralid moths (Anshelevich *et al.*, 1993) and bark beetles (Scolytidae) (Schlyter *et al.*, 1987).

Pheromone traps have been mostly used to detect and delimit outbreaks of insect pests. In some cases the number of individuals caught can be used to give an index of population size or impact. For example, Gage *et al.* (1990) found that returns from pheromone traps for gypsy moth were correlated with the amount of defoliation.

7.5.4 The use of vertebrate hosts or substitutes as bait for insects

The methods described in this section are used principally to determine the biting rate of intermittent ectoparasites under various conditions, the host range and the relative importance of different hosts and vectors and other measurements of significance in epidemiological studies. Some give unbiased indices of these behavioural characteristics, but in others the presence of the trap may repel some insects or attract others independently of the presence of the host. Dyce & Lee (1962) found that the presence of the drop-cone trap about 30 cm above the rabbit would deter certain species of mosquito, but not others, from biting it. In contrast, Colless (1959) showed that from one-third to one-half the catch of a Malayan trap is independent of the presence of the host. Comparisons of the responsiveness of tsetse flies to stationary and mobile baits showed that, whereas visual and olfactory stimuli are important with the former, visual cues predominate with moving bait (Vale, 1974). As there is often a marked diurnal periodicity in biting rate (Haddow, 1954) it is important that assessments should be made throughout the 24 hours; indeed, the 'biting cycle' is often longer (McClelland & Conway, 1971) so that if there is an unusual age distribution estimates need to be made over at least one full cycle. Human bait may be moving (see below) or stationary; vector landing rates being described in terms such as 'collector-nights' (Attenborough *et al.*, 1997).

These methods therefore give measures of availability and when attractiveness is constant indices of absolute population (i.e. relative abundance). Under certain circumstances they might be used to measure absolute population density by the removal technique (Roberts & Scanlon, 1975). One must note that the attractiveness of individual baits differs (Saunders, 1964; Khan *et al.*, 1970). Many special traps are reviewed by Service (1995).

Reference should also be made to the Manitoba or canopy trap (p. 300), the carbon dioxide trap (p. 307) and perhaps even the Haufe–Burgess mosquito trap (p. 300) which could have been included here on the grounds that they are substitutes for vertebrate hosts.

7.5.4.1 Moving baits

As ectoparasite incidence is often low and non-random, and as landing rate may vary in relation to the time of exposure in one spot (Fig. 7.18), a moving bait may give more extensive data than one stationary over a long period. The fly-round first developed by W.H. Potts has been used extensively in work on the ecology of the tsetse fly. In this, the catching party consisting of the human bait and the collectors would walk through the habitat, making a number of stops at which flies were caught. Ford *et al.* (1959) proposed a transect fly round that followed arbitrary straight lines with equidistantly placed halting sites. These transect rounds have advantages in the statistical analysis necessary for the determination of dispersion, but the catches are still variable and Glasgow (1961) concluded that a 7500-yard fly-round done once a week could not detect less than a fivefold change in the mean catch (i.e. in the index of absolute population). More males than females are usually (Glasgow & Duffy, 1961), but not always (Morris, 1960), caught on fly-rounds. This appears to be due to the fact that only teneral females feed on moving hosts, and mature females are therefore seldom taken; furthermore, the males follow the moving objects (even vehicles) not so much to feed (except those that

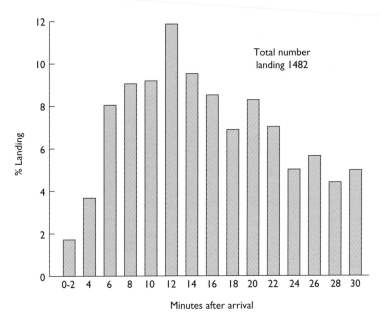

Figure 7.18   The variation in landing rate of Culicoides with time after arrival of human bait. (Adapted from Jamnback & Watthews, 1963.)

are teneral) as to be in the vicinity to pair with the young females (Bursell, 1961). Oxen and other animals may be used instead of humans as the bait on fly-rounds; the species of *Glossina* taken will be influenced by the bait animal (Jordan, 1962). Studying the biting habits of *Chrysops*, Duke (1955) found that biting rate per individual of a group of human baits was greater than that for a single individual working on his own. Single human baits were used by Murray (1963) in studies on *Aedes* and found to give a better picture of seasonal variation than light-traps.

Ticks and fleas sit on grass and other vegetation and seek to attach themselves to their hosts as they pass by, a behaviour termed 'questing'. Rothschild (1961) and Bates (1962) sampled fleas with what they termed artificial rabbits and birds: a skin with an internal heating device (a hot-water bottle) that was pulled across the ground. Ticks have been sampled by dragging blankets and pieces of cloth over the vegetation; Wilkinson (1961) has compared various methods and concluded that the 'hinged-flag sampler' a modification of Blagoveschenskii's (1957) technique, was most efficient and gave a good correlation with counts on cattle in the same enclosure. However, Randolph and Storey (pers. commun.) found that questing activity has a diurnal rhythm and is also influenced by the immediate microclimate, such that correction factors may be necessary. The hinged-flag sampler consists of a piece of plywood (50 × 50 cm), covered with cloth and with a handle strongly hinged to the centre of one side. The sampler is held as rigidly as possible with the lower edge making contact with the grass. In very rough country this cannot be used and cloth-covered leggings are the only available method.

Ticks may be removed from these samplers, retained and stored on transparent adhesive tape (e.g. 'Sellotape').

7.5.4.2   Stationary baits

The most natural situation is achieved when the host is freely exposed in the normal habit; when a human is the host this is of course possible, and it may be feasible for one individual to act as both bait and collector. In order to obtain a complete collection the bait may need to enclose him/herself periodically in a cage. Klock & Bidlingmayer (1953) devised one that was essentially an umbrella with blinds that could be released and would rapidly drop to the ground (the 'drop net'). Blagoveschenskii *et al.* (1943) dropped a bell-shaped cover over their human bait at intervals, and Myers (1956) extended this principle to the rabbit with the 'drop-cone trap', where a muslin cone is suspended by a rope over a pulley about 30 cm above a tethered bait (Fig. 7.19) and may be quickly lowered by an observer some distance away. But, as Dyce & Lee (1962) showed, the presence of the drop-cone trap just above the bait may affect the behaviour of some species. These authors also suggest that the effect of humans handling the traps in which other species are exposed may invalidate comparisons between the biting rate on humans and these other animals (Dyce & Lee, 1962).

Blackflies (Simuliidae) are less easily disturbed when feeding on their hosts, and Anderson & Defoliart (1961) investigated their host preferences by exposing various birds and mammals in net or wire cages for 15 min, and then covering these with a blackout box with a removable insect trap on top (Fig. 7.19b). The flies would soon be attracted to the light and move into the muslin trap. The blackout box was conveniently made of a strong cardboard box strengthened with wood. Disney (1972) lowered the box on a cantilever: it then becomes a type of 'drop-cone trap' (see above). Apart from any influence due to the trap, the catches at a bait may be influenced by the length of time it has been present (Fig. 7.18) (Jamnback & Watthews, 1963) and there are considerable local variations due to site (e.g. Saunders, 1964). Selection due to the influence of the trap itself will be greater with those methods where the animal is more or less enclosed.

Although traps of many different designs have been invented for flying blood-sucking insects, and often named after the inventor (Service, 1995), the majority can be classified into one of five types.

The Malayan or bed-net trap

This was originally designed by B.A.R. Garter, and apparently independently by R.C. Shannon. Basically, it consists of a gauze structure or modified mosquito net, operating rather like a Malaise trap over the bait, generally human. It is not surprising therefore that Colless (1959) found that an unbaited trap caught up to half as many mosquitoes as an unbaited one. Initially, the sides or part of the sides are open, but these flaps can be dropped by pulling a string, say every hour and the catch collected. Wharton *et al.* (1963) used a similar trap with monkeys, enclosed in a wire-netting cage, as bait, they may even be used with well tethered cattle (Mpofu & Masendu, 1986).

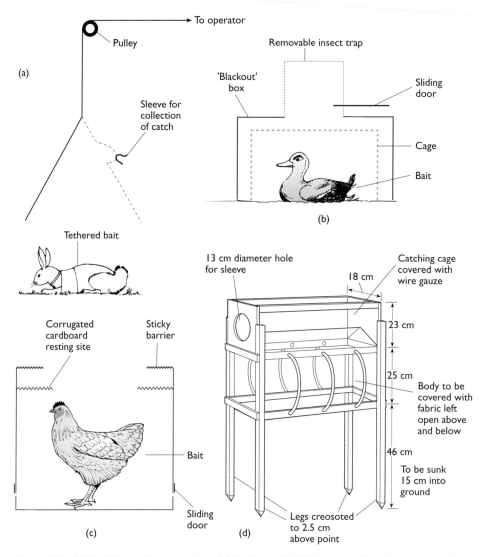

Figure 7.19   (a) The Myers drop-cone trap; (b) Anderson & Defoliart's method for trapping black-flies; (c) Harrison's trap for poultry mites; (d) The Morris trap.

The Stable trap

This was first described by E.H. Magoon and E.C. Earle, and consists of a fixed or portable sectional screen and wood shed into which mosquitoes and other biting flies can enter by horizontal slits in the lower walls. After feeding, most will fly upwards and are trapped (Bates, 1944). As the name suggests, these traps originally utilised cattle or horses as bait, but have also been widely used with dogs (e.g. Lewandowski *et al.*, 1980). A lightweight form, which can be hoisted into trees, uses birds as bait (Flemings, 1959). Trap huts are often of similar construction, but the slits are replaced by entry traps which protect the bait from being bitten.

The Lumsden trap

This was invented by W.H.R. Lumsden and combines bait with a weak suction device. It is constructed of a 'Perspex' hood leading to a suction fan, a gauze funnel and a collecting tube ('an inverted suction trap'). A bait animal is tethered, or placed in a cage, below the hood. The fan is switched on periodically and the resulting air current carries the insects feeding on the bait into the collecting tube. Portable, battery- or generator-operated versions were originally developed by Snow et al. (1960) and Minter (1961); these catch large numbers of *Culicoides* and *Phlebotomus*, but the airflow seems insufficient to capture many of the mosquitoes. A model with a more powerful suction device, used rabbits as bait and successfully trapped mosquitoes (Service, 1995).

Traps for small animals

These are generally open below (Worth & Jonkers, 1962; Turner, 1972), or the design may be like the carbon dioxide trap (see Fig. 7.17). The No. 10 Trinidad (Worth & Jonkers, 1962) and No. 17 Trinidad (Davies, 1971) traps are versatile and easily constructed gauze-covered models (Service, 1995). Another trap of this general type has been constructed by Harrison (1963) for studying the poultry red mite, *Dermanyssus*. It consists of a box containing the bait chicken with a number of holes near the base (Fig. 7.19c); these may be closed with a time switch or hand-operated barrier. The trap could be used for other ectoparasites that leave the host after feeding (e.g. Cimicidae).

Harris, Morris or tent traps

R.H.T.P. Harris devised a trap that apparently attracted and retained tsetse flies; the trap being the bait. His design was considerably modified and improved by Morris & Morris (1949) (Fig. 7.19d). A portable folding version has been described by Morris (1961). The flies enter the traps from below and only a few seem able to escape; however, if the traps were emptied less than once every 24 hours losses of trapped flies occurred due to predation by ants (Smith & Rennison, 1961). A high proportion of the flies caught in the Morris trap are females, and this is their great advantage compared with the fly-round; however, the dispersion pattern suggested by trap catches does not coincide with that found by fly-rounds or searching the resting sites (Glasgow & Duffy, 1961). Thus, their value as an index of absolute population must be doubted and the time of day of maximum catch in the Morris trap seems to differ from that at oxen (Smith & Rennison, 1961). Several species of *Glossina* and many tabanids are caught by Morris traps, but others are not taken even when they attack tethered oxen in the vicinity (Morris, 1960, 1961; Jordan, 1962; Vale, 1971). Black-coloured cloth on the trap seemed to attract more tabanids (Morris, 1961) and, in the wet season more tsetse flies, than did brown hessian (Morris, 1960). The variations in catch in these traps remain something of a mystery in spite of extensive tests (Hargrove, 1977). Fredeen (1961) used basically similar traps (termed silhouette traps) to collect Simuliidae, which, as Wenk & Schlorer (1963) have shown, are attracted to the silhouette of their hosts. Rodríguez-Pérez et al. (2013) tested a number of trap designs for black flies of the *Simulium ochraceum* complex. These included a human Silhouette coated with Tangle-Trap Insect Trap Coating® (The Tanglefoot Company, Grand Rapids, MI, USA) and baited with BG-Lure™ attractant (a mixture of compounds found in human skin secretions [Biogents AG, Regensburg,

Germany]). They found that the silhouette trap performed relatively poorly compared with a BG-Sentinel trap (Biogents AG) baited with either BG-Lure™ or 1-octen-3-ol lure (AgriSense Ltd, UK) attractant or a new design they called an Esperanza window trap.

7.6 Using Sound

Many insects respond to sounds of their own and other species: recordings could be used as the lure in traps (Belton, 1962; Cade, 1975). Kanda *et al.* (1988) used dry ice, a hamster and a speaker emitting sound at 500 Hz (which is the wing beat frequency) to attract female *Mansonia*, while Ikeshoji & Yap (1987) and co-workers have developed a number of devices that attract male mosquitoes and trap them on cylindrical or other shaped sticky traps (Service, 1995).

References

A'Brook, J. (1973) Observations on different methods of aphid trapping. *Ann. Appl. Biol.* **74**, 263–77.

Abensperg-Traun, M. & Steven, D. (1995) The effects of pitfall trap diameter on ant species richness (Hymenoptera: Formicidae) and species composition of the catch in a semi-arid eucalypt woodland. *Aust. J. Ecol.* **20**(2), 282–7.

Acuff, V.R. (1976) Trap biases influencing mosquito collecting. *Mosquito News* **36**, 173–6.

Adkins, T.R., Ezell, W.G., Sheppard, D.C., & Askey, M.M. (1972) A modified canopy trap for collecting Tabanidae (Diptera). *J. Med. Entomol.* **9**, 183–5.

Adler, G.H. & Lambert, T.D. (1997) Ecological correlates of trap response of a Neotropical forest rodent, *Proechimys semispinosus. J. Trop. Ecol.* **13**(1), 59–68.

Adlerz, W.C. (1971) A reservoir-equipped Moericke trap for collecting aphids. *J. Econ. Entomol.* **64**, 966–7.

Agnello, A.M., Reissig, W.H., Spangler, S.M., Charlton, R.E., & Kain, D.P. (1996) Trap response and fruit damage by oblique banded leafroller (Lepidoptera: Tortricidae) in pheromone-treated apple orchards in New York. *Environ. Entomol.* **25**(2), 268–82.

Alfiya, H. (1995) Surveillance radar data on nocturnal bird migration over Israel, 1989–1993. *Israel J. Zool.* **41**(3), 517–22.

Almand, L.K., Sterling, W.L., & Green, C.L. (1974) A collapsible truck-mounted aerial net for insect sampling. *Texas Agric. Exp. Sta. Mise Pub.* **1189**, 1–4.

Anderbrant, O., Lofqvist, J., Jonsson, J., & Marling, E. (1989) Effects of pheromone trap type, position and color on the catch of pine sawfly, *Neodiprion sertifer* (Geoff.) (Hymenoptera, Diprionidae). *J. Appl. Entomol.* **107**(4), 365–9.

Anderson, J.R. & Defoliart, G.R. (1961) Feeding behaviour and host preferences, of some black flies (Diptera: Simulidae). *Ann. Entomol. Soc. Am.* **54**, 716–29.

Anderson, J.R. & Poorbaugh, J.H. (1964) Observation on the ethology and ecology of various Diptera associated with northern California poultry ranches. *J. Med. Entomol.* **1**, 131–47.

Anderson, J.R. & Yee, W.L. (1995) Trapping black flies (Diptera: Simuliidae) in northern California: I. Species composition and seasonal abundance on horses, host models, and in insect flight traps. *J. Vector Ecol.* **20**, 7–25.

Anderson, N.H. (1966) Depressant effect of moonlight on activity of aquatic insects. *Nature* **209**, 319–20.

Anderson, N.H. (1967) Biology and downstream drift of some Oregon Trichoptera. *Can. Entomol.* **99**, 507–21.

Anderson, N.H. & Lehmkuhl, D.M. (1968) Catastrophic drift of insects in a woodland stream. *Ecology* **49**, 198–206.

Andreassen, H.P. & Bondrup Nielsen, S. (1991) Home range size and activity of the wood lemming, *Myopus schisticolor. Holarctic Ecology* **14**(2), 138–41.

Anshelevich, L., Kehat, M., Dunkelblum, E., & Greenberg, S. (1993) Sex pheromone traps for monitoring the honeydew moth, *Cryptoblabes gnidiella*: Effect of pheromone components, pheromone dose, field ageing of dispenser, and type of trap on male captures. *Phytoparasitica* **21**(3), 189–98.

Anshelevich, L., Kehat, M., Dunkelblum, E., & Greenberg, S. (1994) Sex pheromone traps for monitoring the European vine moth, *Lobesia botrana*: Effect of dispenser type, pheromone dose, field aging of dispenser, and type of trap on male captures. *Phytoparasitica* **22**(4), 281–90.

Appenzeller, A.R. & Leggett, W.C. (1992) Bias in hydroacoustic estimates of fish abundance due to acoustic shadowing: Evidence from day-night surveys of vertically migrating fish. *Can. J. Fish. Aquat. Sci.* **49**(10), 2179–89.

Apperson, C.S. & Yows, D.G. (1976) A light trap for collecting aquatic organisms. *Mosquito News* **36**, 205–6.

Atkins, M.D. (1957) An interesting attractant for *Priacma serrata* (Lec.) (Cupesidae: Coleoptera). *Can. Entomol.* **89**, 214–19.

Attenborough, R.D., Burkot, T.R., & Gardner, D.S. (1997) Altitude and risk of bites from mosquitoes infected with malaria and filariosis among the Mianmia people of Papua New Guinea. *Trans. R. Soc. Trop. Med. Hyg.* **9**, 8–10.

Aubert, J. (1969) Un appareil de capture de grandes dimensions destiné au marquage d'insectes migrateurs. *Mitt. Schweiz. Entomol. Gesell.* **42**, 135–9.

Bagenal, T.B. (1972) The variability in the number of perch, *Perca fluviatilis* L., caught in traps. *Freshwater Biol.* **2**, 27–36.

Balogh, J. & Loksa, I. (1956) Untersuchunger über die Zoozönose des Luzernenfeldes. *Acta Zool. Acad. Sci. Hung.* **2**(1-3), 17–114.

Banks, C.J. & Brown, E.S. (1962) A comparison of methods of estimating population density of adult sunn pest *Eurygaster integriceps* Put. (Hemiptera: Scutelleridae) in wheat fields. *Entomol. Exp. Appl.* **5**, 255–60.

Barr, A.R., Smith, T.A., Boreham, M.M., & White, K.E. (1963) Evaluation of some factors affecting the efficiency of light traps in collecting mosquitoes. *J. Econ. Entomol.* **569**, 123–7.

Bateman, J. (1979) *Trapping: A Practical Guide*. David and Charles, Newton Abbot.

Bateman, M.A. (1972) The ecology of fruit flies. *Annu. Rev. Entomol.* **17**, 493–518.

Bates, J.K. (1962) Field studies on the behaviour of bird fleas. 1. Behaviour of the adults of three species of bird flea in the field. *Parasitology* **52**, 113–32.

Bates, M. (1944) Notes on the construction and use of stable traps for mosquito studies. *J. Natn. Malar. Soc.* **3**, 135–45.

Bauchhenss, E. (1995) Ground-living spiders of sandy soil areas in Northern Bavaria (Germany) (Arachnida: Araneae). *Zoologische Beitraege* **36**(2), 221–50.

Baylor, E.R. & Smith, F.E. (1953) A physiological light trap. *Ecology* **34**, 223–4.

Beerwinkle, K.R., Lopez, J.D., Jr, Schleider, P.G., & Lingren, P.D. (1995) Annual patterns of aerial insect densities at altitudes from 500 to 2400 meters in east-central Texas indicated by continuously-operating vertically-oriented radar. *Southwestern Entomologist Supplement* **18**, 63–79.

Beerwinkle, K.R., Lopez, J.D., Jr, Witz, J.A., Schleider, P.G., Eyster, R.S., & Lingren, P.D. (1994) Seasonal radar and meteorological observations associated with nocturnal insect flight at altitudes to 900 meters. *Environ. Entomol.* **23**(3), 676–83.

Beall, G. (1935) Study of arthropod populations by the method of sweeping. *Ecology* **16**, 216–25.

Bellamy, R.W. & Reeves, W.C. (1952) A portable mosquito bait trap. *Mosquito News* **129**, 256–8.

Belton, P. (1962) Effects of sound on insect behaviour. *Proc. Entomol. Soc. Manitoba* **18**, 1–9.

Belton, P. & Kempster, R.H. (1963) Some factors affecting the catches of Lepidoptera in light traps. *Can. Entomol.* **95**, 832–7.

Benedek, P., Erdelyi, C., & Fesus, I. (1972) General aspects of diel vertical movements of some arthropods in flowering lucerne stands and conclusions for the integrated pest control of lucerne. *Acta Phytopathol. Hung.* **7**, 235–49.

Beroza, M., Green, N., Gertler, S.I., Steiner, L.R., & Miyashita, D.H. (1961) New attractants for the Mediterranean fruit fly. *J. Agric. Food Chem.* **9**, 361–5.

Bidlingmayer, W.L. (1964) The effect of moonlight on the flight activity of mosquitoes. *Ecology* **45**, 87–94.

Bishop, A.L., McKenzie, H.J., Spohr, L.J., & Barchia, I.M. (1994) *Culicoides brevitarsis* Kieffer (Diptera: Ceratopogonidae) in different farm habitats. *Austr. J. Zool.* **42**(3), 379–84.

Bishop, J.A. (1972) An experimental study of the cline of industrial melanism in *Biston betularia* (L.) (Lepidoptera) between urban Liverpool and rural North Wales. *J. Anim. Ecol.* **41**, 209–43.

Blagoveschenskii, D.I. (1957) Biological principles of the control of ixodid ticks. [In Russian]. *Ent. Obozr.* **36**, 125–33.

Blagoveschenskii, D.I., Sregetova, N.G., & Monchadskii, A.S. (1943) Activity in mosquito attacks under natural conditions and its diurnal periodicity. [In Russian]. *Zool. Zhur.* **22**, 138–53.

Blake, R.W., Kolotylo, R., & De La Cueva, H. (1990) Flight speeds of the barn swallow, *Hirundo rustica*. *Can J. Zool.* **68**(1), 1–5.

Blumberg, A.Y. & Crossley, D.A.J. (1988) Diurnal activity of soil-surface arthropods in agroecosystems: Design for an inexpensive time-sorting pitfall trap. *Agricult. Ecosyst. Environ.* **20**(3), 159–64.

Blume, R.R., Miller, J.A., Eschle, J.L., Matter, J.I., & Pickens, M.O. (1972) Trapping tabanids with modified Malaise traps baited with CO_2. *Mosquito News* **32**, 90–5.

Bowden, J. (1973) The influence of moonlight on catches of insects in light-traps in Africa. Part 1. The moon and moonlight. *Bull. Entomol. Res.* **63**, 113–28.

Bowden, J. & Church, B.M. (1973) The influence of moonlight on catches of insects in light traps in Africa. Part II. The effect of moon phase on light trap catches. *Bull. Entomol. Res.* **63**, 129–42.

Bowden, J. & Gibbs, D.G. (1973) Light-trap and suction-trap catches of insects in the northern Gezira, Sudan, in the season of southward movement of the InterTropical Front. *Bull. Entomol. Res.* **62**, 571–96.

Bowden, J. & Morris, M. (1975) The influence of moonlight on catches of insects in light-traps in Africa. Ill. The effective radius of a mercury-vapour light trap rid the analysis of catches using effective radius. *Bull. Entomol. Res.* **65**, 303–48.

Bracken, G.K., Hanec, W., & Thorsteinson, A.J. (1962) The orientation of horseflies and deer flies (Tabanidae: Diptera). *Can. J. Zool.* **40**, 685–95.

Bracken, G.K. & Thorsteinson, A.J. (1965) Orientation behaviour of horse flies; influence of some physical modifications of visual decoys on orientation of horse flies. *Entomol. Exp. Appl.* **8**, 314–18.

Braman, S.K. & Pendley, A.F. (1993) Activity patterns of Carabidae and Staphylinidae in centipede grass in Georgia. *J. Entomol. Sci.* **28**(3), 299–307.

Brant, A.V. (1984) *Fish Catching Methods of the World.* Fishing Newsbooks, Farnham, UK.

Brierley, A.S. & Watkins, J.L. (1996) Acoustic targets at South Georgia and the South Orkney Islands during a season of krill scarcity. *Mar. Ecol. Prog. Ser.* **138**(1-3), 51–61.

Briggs, J.B. (1961) A comparison of pitfall trapping and soil sampling in assessing populations of two species of ground beetles (Col.: Carabidae). *Rep. East Malling Res. Sta.* **1960**, 108–12.

Broadbent, L. (1948) Aphis migration and the efficiency of the trapping method. *Ann. Appl. Biol.* **35**, 379–94.

Broadbent, L., Doncaster, J., Hull, R., & Watson, M. (1948) Equipment used for trapping and identifying alate aphids. *Proc. R. Entomol. Soc. Lond. A* **23**, 57–8.

Brodsgaard, H.F. (1989) Colored sticky traps for *Frankliniella occidentalis* (Pergande) (Thysanoptera, Thripidae) in glasshouses. *J. Appl. Entomol.* **107**(2), 136–40.

Brown, E.S., Betts, E., & Rainey, R.C. (1969) Seasonal changes in the distribution of the African armyworm, *Spodoptera exempta* (Wlk.) (Lep., Noctuidae) with special reference to eastern Africa. *Bull. Entomol. Res.* **58**, 661–728.

Bruderer, B. (1997a) The study of bird migration by radar. Part 2: Major achievements. *Naturwissenschaften* **84**(2), 45–54.

Bruderer, B. (1997b). The study of bird migration by radar: Part 1: The technical basis. *Naturwissenschaften* **84**(1), 1–8.

Bucher, G.E. & Cheng, H.H. (1970) Use of trap plants for attracting cutworm larvae. *Can. Entomol.* **102**, 797–8.

Buckley, D.I. & Stewart, W.W.A. (1970) A light-activated switch for controlling battery operated light traps. *Can. Entomol.* **102**, 911–12.

Budaeva, L.I. (1993) Peculiarities of surface dwelling collembolan communities (Collembola) in Khakasia steppe. *Zoologicheskii Zhurnal* **72**(3), 45–52.

Bultman, T.L. (1992) Abundance and association of cursorial spiders from calcareous fens in southern Missouri. *J. Arachnol.* **20**(3), 165–72.

Burbutis, P.P. & Jobbins, D.M. (1958) Studies on the use of adiurnal resting box for the collection of *Culiseta melanura* (Coquillet). *Bull. Brooklyn Entomol. Soc. (N.S.)* **53**, 53–8.

Bursell, E. (1961) The behaviour of tsetse flies (*Glossina swynnesioni* Austen) in relation to problems of sampling. *Proc. R. Entomol. Soc. Lond. A* **36**, 9–20.

Butler, G.D. (1965) A modified Malaise insect trap. *Pan. Pacif. Entomol.* **41**(1), 51–3.

Cade, W. (1975) Acoustically orienting parasitoids: fly phototaxis to cricket song. *Science* **190**, 1312–13.

Cameron, E.A. & Reeves, R.M. (1990) Carabidae (Coleoptera) associated with gypsy moth, *Lymantria dispar* (L.) (Lepidoptera: Lymantriidae), populations subjected to *Bacillus thuringiensis* Berliner treatments in Pennsylvania (USA). *Can. Entomol.* **122**(1–2), 123–30.

Cameron, M.M., Pessoa, F.A.C., Vasconcelos, A.W., & Ward, R.D. (1995) Sugar meal sources for the phlebotomine sandfly *Lutzomyia longipalpis* in Ceara State, Brazil. *Med. Vet. Entomol.* **9**(3), 263–72.

Carle, F.L. & Strub, M.R. (1978) A new method for estimating population size from removal data. *Biometrics* **34**, 621–30.

Carley, C.J. & Knowlton, F.F. (1968) Trapping wood rats: effectiveness of several techniques and differential catch by sex and age. *Texas J. Sci.* **19**, 248–51.

Carlson, D. (1971) A method for sampling larval and emerging insects using an aquatic black light trap. *Can. Entomol.* **103**, 1365–9.

Chang, V.C.S. (1995) Trapping *Cryptophlebia illepida* and *C. ombrodelta* (Lepidoptera: Tortricidae) in macadamia in Hawaii. *Int. J. Pest Manag.* **41**(2), 104–8.

Chant, D.A. & McLeod, J.H. (1952) Effects of certain climatic factors on the daily abundance of the European earwig, *Forficula auricularia* L. (Dermaptera: Forficulidae), in Vancouver, British Columbia. *Can. Entomol.* **84**, 174–80.

Chapman, J.A. (1962) Field studies on attack flight and log selection by the Ambrosia beetle, *Trypodendron lineatum* (Oliv.) (Coleoptera: Scolytidae). *Can. Entomol.* **94**, 74–92.

Chapman, J.A. & Kinghorn, J.M. (1995) Window-trap for flying insects. *Can. Entomol.* **82**, 46–7.

Chararas, C. (1977) Attraction chimique exercée sur certains Scolytidae par les pinacees et les cupressaées. *Comp. Insect. Milieu Trop., Coll. Int. C.N.R.S.* **265**, 165–86.

Cherrill, A.J. & Sanderson, R.A. (1994) Comparison of sweep-net and pitfall trap samples of moorland Hemiptera: Evidence for vertical stratification within vegetation. *Entomologist* **113**(1), 70–81.

Chiang, H.C. (1973) A simple trap for certain minute flying insects. *Ann. Entomol. Soc. Am.* **669**, 704–5.

Cho, K., Eckel, C.S., Walgenbach, J.F., & Kennedy, G.G. (1995) Comparison of colored sticky traps for monitoring thrips populations (Thysanoptera: Thripidae) in staked tomato fields. *J. Entomol. Sci.* **30**(2), 176–90.

Choat, J.H., Doherty, P.J., Kerrigan, B.A., & Leis, J.M. (1993) A comparison of towed nets, purse seine, and light-aggregation devices for sampling larvae and pelagic juveniles of coral reef fishes. *U S National Marine Fisheries Service Fishery Bulletin* **91**(2), 195–209.

Christensen, J.T. (1996) Home range and abundance of *Mastomys natalensis* (Smith, 1834) in habitats affected by cultivation. *African J. Ecol.* **34**(3), 298–311.

Clark, J.D. & Curtis, C.E. (1973) A battery-powered light trap giving two years' continuous operation. *J. Econ. Entomol.* **66**, 393–6.

Clark, W.H. & Blom, P.E. (1992) An efficient and inexpensive pitfall trap system. *Entomol. News* **103**(2), 55–9.

Close, R. (1959) Sticky traps for winged aphids. *N. Z. J. Agric. Res.* **2**, 375–9.

Cloud, T.J. & Stewart, K.W. (1974) Seasonal fluctuations and periodicity in the drift of caddis fly larvae (Trichoptera) in the Brazos River, Texas. *Ann. Entomol. Soc. Am.* **67**, 805–11.

Cole, L.C. (1946) A study of the Cryptozoa of an Illinois woodland. *Ecol. Monogr.* **16**, 49–86.

Colless, D.H. (1959) Notes on the culicine mosquitoes of Singapore. VI. Observations on catches made with baited and unbaited trap-nets. *Ann. Trop. Med. Parasitol., Liverpool* **53**, 251–8.

Common, I.F.B. (1959) A transparent light trap for the field collection of Lepidoptera. *J. Lepid. Soc.* **13**, 57–61.

Conrow, R., Zale, A.V., & Gregory, R.W. (1990) Distributions and abundances of early life stages of fishes in a Florida (USA) lake dominated by aquatic macrophytes. *Trans. Am. Fish. Soc.* **119**(3), 521–8.

Coon, B.F. & Rinicks, H.B. (1962) Cereal aphid capture in yellow baffle trays. *J. Econ. Entomol.* **55**, 407–8.

Coppel, H.C., Casida, J.E., & Dauterman, W.C. (1960) Evidence for a potent sex attractant in the introduced pine sawfly, *Diprion similis* (Hymenoptera: Diprionidae). *Ann. Entomol. Soc. Am.* **53**, 510–12.

Cornwell, P.B. (1960) Movements of the vectors of virus diseases of cacao in Ghana. *Bull. Entomol. Res.* **51**, 175–201.

Cowx, I.G. (ed.) (1990) *Developments in Electric Fishing*. Oxford, UK, Fishing News Books, Blackwell Scientific Publications.

Cushing, C.E. (1964) An apparatus for sampling drifting organisms in streams. *J. Wildlife Manag.* **28**, 592–4.

Dales, R.P. (1953) A simple trap for Tipulids (Dipt.). *Entomol. Monthly Mag.* **89**, 304.

D'Arcy-Burt, S. & Blackshaw, R.P. (1987) Effects of trap design on catches of grassland Bibionidae (Diptera: Nematocera). *Bull. Entomol. Res.* **77**(2), 309–16.

Dahlsten, D.L., Rowney, D.L., Copper, W.A., & Wenz, J.M. (1992) Comparison of artificial pupation shelters and other monitoring methods for endemic populations of Douglas-fir tussock moth, *Orgyia pseudotsugata (McDunnough) (Lepidoptera, Lymantriidae). Can. Entomol.* **124**, 259–369.

Davies, J.B., Corbet, P.S., Gillies, M.T., & McCrae, A.W.R. (1971) Parous rates in some Amazonian mosquitoes collected by three different methods. *Bull. Entomol. Res.* **61**, 125–32.

Davis, E.W. & Landis, B.J. (1949) An improved trap for collecting aphids. *USDA Bur. Entomol. Plant Quar. ET* **278**, 3 pp.

Dewey, M.R. & Jennings, C.A. (1992) Habitat use by larval fishes in a backwater lake of the upper Mississippi River. *J. Freshwater Ecol.* **7**(4), 363–72.

Dickerson, W.A., Gentry, C.R., & Mitchell, W.G. (1970) A rainfree collecting container that separates desired Lepidoptera from small undesired insects in light traps. *J. Econ. Entomol.* **63**, 1371.

Disney, R.H.L. (1972) Observations on chicken-biting blackflies in Cameroon with a discussion of parous rates of *Simulium damnosum. Ann. Trop. Med. Parasitol.* **66**, 149–58.

Dokter, A.M., Liechti, F., Stark, H., Delobbe, L., Tabary, P., & Holleman, I. (2011) Bird migration flight altitudes studied by a network of operational weather radars. *J. Royal Soc. Interface*, **8**(54), 30–43.

Dodge, H.R. & Seaco, J.M. (1954) Sarcophagidae and other Diptera taken by trap and net on Georgia mountain summits in 1952. *Ecology* **35**, 50–9.

Doherty, P.J., Fowler, A.J., Samoilys, M.A., & Harris, D.A. (1994) Monitoring the replenishment of coral trout (Pisces: Serranidae) populations. *Bull. Mar. Sci.* **54**(1), 343–55.

Dow, R.P. & Morris, C.D. (1972) Wind factors in the operation of a cylindrical bait trap for mosquitoes. *J. Med. Entomol.* **9**, 60–6.

Downey, J.E. (1962) Mosquito catches in New Jersey Mosquito traps and ultra-violet light traps. *Bull. Brooklyn Entomol. Soc.* **57**, 61–3.

Drake, V.A. & Reynolds, D.R. (2012) *Radar Entomology: Observing Insect Flight and Migration*. CABI. 512 pp.

Drapek, R.J., Coop, L.B., Croft, B.A., & Fisher, G.C. (1990) *Heliothis zea* pheromone trapping: Studies of trap and lure combinations and field placement in sweet corn. *Southwestern Entomologist* **15**(1), 63–70.

Drickamer, L.C., Lenington, S., Erhart, M., & Robinson, A.S. (1995) Trappability of wild house mice (*Mus domesticus*) in large outdoor pens: Implication for models of t-complex gene frequency. *Am. Mid. Nat.* **133**(2), 283–9.

Dry, W.W. & Taylor, L.R. (1970) Light and temperature thresholds for take-off by aphids. *J. Anim. Ecol.* **39**, 493–504.

Dudley, J.E., Searles, E.M., & Weed, A. (1928) Pea aphid investigations. *Trans. IV Int. Congr. Entomol.* **2**, 608–21.

Duke, B.O.L. (1955) Studies on the biting habits of Chrysops. III. *Ann. Trop. Med. Parasitol. Liverpool* **49**, 362–7.

Duncan, A. & Kubecka, J. (1996) Patchiness of longitudinal fish distributions in a river as revealed by a continuous hydroacoustic survey. *ICES J. Mar. Sci.* **53**(2), 161–5.

Dyce, A.L. & Lee, D.J. (1962) Blood-sucking flies (Diptera) and myxomatosis transmission in a mountain environment in New South Wales. 11. Comparison of the use of man and rabbit as bait animals in evaluating vectors of myxomatosis. *Aust. J. Zool.* **10**, 84–94.

Eastop, V. (1955) Selection of aphid species by different kinds of insect traps. *Nature* **1769**, 936.

Edwards, P.B., Kettle, D.S., & Barnes, A. (1987) Factors affecting the numbers of Culicoides (Diptera: Ceratopogonidae) in traps in coastal South-East Queensland (Australia) with particular reference to collections of *Culicoides subimmaculatus* in light traps. *Austr. J. Zool.* **35**(5), 469–86.

El Hag, E.T.A. & El Meleigi, M.A. (1991) Insect pests of spring wheat in central Saudi Arabia. *Crop Protection* **10**(1), 65–9.

Elliott, J.M. (1967) Invertebrate drift in a Dartmoor stream. *Arch. Hydrobiol.* **63**, 202–37.

Elliott, J.M. (1969) Diel periodicity in invertebrate drift and the effect of different sampling periods. *Oikos* **20**, 524–8.

Elliott, J.M. (1970) Methods of sampling invertebrate drift in running water. *Ann. Limnol.* **6**, 133–59.

Entwistle, P.F. (1963) Some evidence for a colour sensitive phase in the flight period of Scolytidae and Platypodidae. *Entomol. Exp. Appl.* **6**, 143–8.

Epstein, M.E. & Kulman, H.M. (1990) Habitat distribution and seasonal occurrence of carabid beetles in East-central Minnesota (USA). *Am. Mid. Nat.* **123**(2), 209–25.

Esbjerg, P. (1987) The influence of diurnal time and weather on sex trap catches of the turnip moth (*Agrotis segetum* Schiff.) (Lepidoptera, Noctuidae). *J. Appl. Entomol.* **103**(2), 177–84.

Evans, L.J. (1975) An improved aspirator (pooter) for collecting small insects. *Proc. Br. Entomol. Nat. Hist. Soc.* **8**, 8–11.

Fager, E.W. (1968) The community of invertebrates in decaying oak wood. *J. Anim. Ecol.* **37**, 121–42.

Fallis, A.M. & Smith, S.M. (1964) Ether extracts from birds and CO_2 as attractants for some ornithophilic simuliids. *Can. J. Zool.* **42**, 723–30.

Feldhamer, G.A. & Maycroft, K.A. (1992) Unequal capture response of sympatric golden mice and white-footed mice. *Am. Mid. Nat.* **128**(2), 407–10.

Fewkes, D.W. (1961) Diel vertical movements in some grassland (Nabidae Heteroptera). *Entomol. Monthly Mag.* **97**, 128–30.

Fichter, E. (1941) Apparatus for the comparison of soil surface arthropod populations. *Ecology* **22**, 338–9.

Finch, S. & Skinner, G. (1974) Some factors affecting the efficiency of water-traps for capturing cabbage root flies. *Ann. Appl. Biol.* **77**, 213–26.

Flamm, R.O., Pulley, P.E., & Coulson, R.N. (1993) Colonization of disturbed trees by the southern pine bark beetle guild (Coleoptera: Scolytidae). *Environ. Entomol.* **22**, 62–70.

Flemings, M.B. (1959) An altitude biting study of *Culex tritaeniorhynchus* (Giles) and other associated mosquitoes in Japan. *J. Econ. Entomol.* **52**, 490–2.

Fletcher, B.S. (1974) The ecology of a natural population of the Queensland fruit fly, *Dacus tryoni* IV. *Aust. J. Zool.* **21**, 541–65.

Fontenille, D., Diallo, M., Mondo, M., Ndiaye, M., & Thonnon, J. (1997) First evidence of natural vertical transmission of yellow fever virus in *Aedes aegypti* its epidemic vector. *Trans. R. Soc. Trop. Med. Hyg.* **91**, 533–5.

Ford, J., Glasgow, J.P., Johns, D.L., & Welch, J.R. (1959) Transect fly-rounds in field studies of *Glossina. Bull. Entomol. Res.* **50**, 275–85.

Fredeen, P.J.H. (1961) A trap for studying the attacking behaviour of black flies *Simulium arcticum* Mall. *Can. Entomol.* **93**, 73–8.

Freeman, B.E. & Jayasingh, D.B. (1975) Population dynamics of *Pachodynerus nasidens* (Hymenoptera) in Jamaica. *Oikos* **26**, 86–91.

Friend, G.R., Smith, G.T., Mitchell, D.S., & Dickman, C.R. (1989) Influence of pitfall and drift fence design on capture rates of small vertebrates in semi-arid habitats of Western Australia. *Aust. Wildlife Res.* **16**, 1–10.

Fritzsche, R. (1956) Untersuchungen zur Bekampfung der Rapssch;idlinge. IV. Beitrdge zur Okologie und Bekampfung des Grossen Rapsstengelriisslers (*Ceuthorrhynchus napi* Gyll.). *NachrBl. dt. Pfl. Schntzdienst, Berl.* **10**, 97–105.

Frohlich, G. (1956) Methoden Zur Bestimmung der Befalls-bzw. Bekdampfungster-mine verschiedener Rapsschddlinge, insebesondere des Rapsstengeiriisslers (*Ceuthorrhynchus napi* Gyll.). *NachrBl. dt. Pfl. Schutzdienst, Berl.* **10**, 48–53.

Frost, S.W. (1952) Light traps for insect collection, survey and control. *Bull. PA Agric. Exp. Sta.* **550**, 1–32.

Frost, S.W. (1953) Response of insects to black and white light. *J. Econ. Entomol.* **46**, 376–7.

Frost, S.W. (1957) The Pennsylvania insect light trap. *J. Econ. Entomol.* **50**, 287–92.

Frost, S.W. (1958a) Insects attracted to light traps placed at different heights. *J. Econ. Entomol.* **51**, 550–1.

Frost, S.W. (1958b) Insects captured in light traps with and without baffles. *Can. Entomol.* **909**, 566–7.

Frost, S.W. (1958c) Traps and lights to catch night-flying insects. *Proc. X Int. Congr. Entomol.* **2**, 583–7.

Frost, S.W. (1964) Killing agents and containers for use with insect light traps. *Entomol. News* **75**, 163–6.

Fukuda, M. (1960) On the effect of physical condition of setting place upon the number of flies collected by fish baited traps. *Endemic Dis. Bull. Nagasaki Univ.* **2**, 222–8.

Fye, R.E. (1965) The biology of the Vespidae, Pompilidae and Sphecidae (Hymenoptera) from trap nets in Northwestern Ontario. *Can. Entomol.* **97**, 716–44.

Gage, S.H., Wirth, T.M., & Simmons, G.A. (1990) Predicting regional gypsy moth (Lymantriidae) population trends in an expanding population using pheromone trap catch and spatial analysis. *Environ. Entomol.* **19**, 370–7.

Garcia, R. (1962) Carbon dioxide as an attractant for certain ticks (Acarina: Argasidae and lxodidae). *Ann. Entomol. Soc. Am.* **55**, 605–6.

Gates, C.E. & Smith, W.B. (1972) Estimation of density of mourning doves from aural information. *Biometrics* **28**, 345–59.

Gehrke, P.C. (1994) Influence of light intensity and wavelength on phototactic behaviour of larval silver perch *Bidyanus bidyanus* and golden perch *Macquaria ambigua* and the effectiveness of light traps. *J. Fish Biol.* **44**(5), 741–51.

Geier, P.W. (1960) Physiological age of codling moth females (*Cydia pomonella* L.) caught in bait and light traps. *Nature* **185**, 709.

Gerard, D., Bauchau, V., & Smets, S. (1994) Reduced trappability in wild mice, *Mus musculus domesticus*, heterozygous for Robertsonian translocations. *Anim. Behav.* **47**(4), 877–83.

Gillies, M.T. (1974) Methods for assessing the density and survival of blood sucking Diptera. *Annu. Rev. Entomol.* **19**, 345–62.

Gist, C.S. & Crossley, D.A. (1973) A method for quantifying pitfall trapping. *Environ. Entomol.* **2**, 951–2.

Glasgow, J.P. (1953) The extermination of animal populations by artificial predation and the estimation of populations. *J. Anim. Ecol.* **22**, 32–46.

Glasgow, J.P. (1961) The variability of fly-round catches in field studies of Glossina. *Bull. Entomol. Res.* **51**, 781–8.

Glasgow, J.P. & Duffy, B.J. (1961) Traps in field studies of *Glossina pallidipes* Austen. *Bull. Entomol. Res.* **52**, 795–814.

Glick, P.A. (1939) *The Distribution of Insects, Spiders and Mites in the Air. Technical Bulletin* **673**, USDA, 150 pp.

Glick, P.A., Hollingsworth, J.P., & Eitel, W. (1956) Further studies on the attraction of pink bollworm moths to ultra-violet and visible radiation. *J. Econ. Entomol.* **49**, 158–61.

Golding, F.D. (1941) Two new methods of trapping the cacao moth (*Ephestia cautella*). *Bull. Entomol. Res.* **32**, 123–32.

Golding, F.D. (1946) A new method of trapping flies. *Bull. Entomol. Res.* **37**, 143–54.

Goodenough, J.L. & Snow, J.W. (1973) Tobacco budworms: nocturnal activity of adult males as indexed by attraction to live virgin females in electric grid traps. *J. Econ. Entomol.* **66**, 543–4.

Goodwin, M.H. (1942) Studies on artificial resting places of *Anopheles quadrimaculatus* Say. *J. Nat. Malar. Soc.* **1**, 93–9.

Goodwin, M.H. & Love, G.J. (1957) Factors influencing variations in populations of *Anopheles quadrimaculatus* in southwestern Georgia. *Ecology* **38**, 561–70.

Graham, H.M., Glick, P.A., & Hollingsworth, J.P. (1961) Effective range of argon glow lamp survey traps for pink bollworm adults. *J. Econ. Entomol.* **54**, 788–9.

Gray, P.H.H. (1955) An apparatus for the rapid sorting of small insects. *Entomologist* **889**, 92–3.

Gray, H. & Treloar, A. (1933) On the enumeration of insect populations by the method of net collection. *Ecology* **14**, 356–67.

Greenslade, P.I.M. (1964) Pitfall trapping as a method for studying populations of Carabidae (Coleoptera). *J. Anim. Ecol.* **33**, 301–10.

Greenslade, P. (1973) Sampling ants with pitfall traps: digging-in effects. *Insectes Soc.* **20**, 343–53.

Greenslade, P. & Greenslade, P.J.M. (1971) The use of baits and preservatives in pitfall traps. *J. Aust. Entomol. Soc.* **10**, 253–60.

Gregg, P.C., Fitt, G.P., Coombs, M., & Henderson, G.S. (1994) Migrating moths collected in tower-mounted light traps in northern New South Wales, Australia: Influence of local and synoptic weather. *Bull. Entomol. Res.* **84**(1), 17–30.

Gregory, R.S. & Powles, P.M. (1985) Chronology, distribution, and sizes of larval fish sampled by light traps in macrophytic Chemung Lake (Ontario, Canada). *Can. J. Zool.* **63**(11), 2569–77.

Gregory, R.S. & Powles, P.M. (1988) Relative selectivities of Miller high-speed samplers and light traps for collecting ichthyoplankton. *Can. J. Fish. Aquat. Sci.* **45**(6), 993–8.

Gressitt, J.L. & Gressitt, M.K. (1962) An improved Malaise trap. *Pacific Insects* **4**, 87–90.

Gressitt, J.L., Leech, R.E., & O'Brien, C.W. (1960) Trapping of air-borne insects in the Antarctic area. *Pacific Insects* **2**, 245–50.

Grigarick, A.A. (1959) Bionomics of the rice leafminer, *Hydrellia griseola* (Fallen.), in California (Diptera: Ephydridae). *Hilgardia* **29**, 1–80.

Grimm, F., Gessler, M., & Jenni, L. (1993) Aspects of sandfly biology in southern Switzerland. *Med. Vet. Entomol.* **7**(2), 170–6.

Haddow, A.I. & Corbet, P.S. (1960) Observations on nocturnal activity in some African Tabanidae (Diptera). *Proc. R. Entomol. Soc. Lond. A* **35**, 1–5.

Haddow, A.J. (1954) Studies on the biting habits of African mosquitoes. An appraisal of methods employed, with special reference to the twenty-four-hour catch. *Bull. Entomol. Res.* **45**, 199–242.

Halgren, L.A. & Taylor, L.R. (1968) Factors affecting flight responses of alienicolne of *Aphis fabae* Scop. and *Schizaphis graminum* Rondeni (Homoptera: Aphididae). *J. Anim. Ecol.* **37**, 583–93.

Halsall, N.B. & Wratten, S.D. (1988) The efficiency of pitfall trapping for polyphagous predatory Carabidae. *Ecol. Entomol.* **13**, 293–9.

Hansen, J.D. (1988) Trapping methods for rangeland insects in burned and unburned sites: A comparison. *Great Basin Naturalist* **48**(3), 383–7.

Hargrove, J.W. (1977) Some advances in the trapping of tsetse (*Glossina* spp.) and other flies. *Ecol. Entomol.* **2**, 123–37.

Hargrove, J.W., Holloway, M.T.P., Vale, G.A., Gough, A.J.E., & Hall, D.R. (1995) Catches of tsetse (*Glossina* spp.) (Diptera: Glossinidae) from traps and targets baited with large doses of natural and synthetic host odour. *Bull. Entomol. Res* **85**(2), 215–27.

Harper, A.M., Schaber, B.D., Entz, T., & Story, T.P. (1993) Assessment of sweepnet and suction sampling for evaluating pest insect populations in hay alfalfa. *J. Entomol. Soc. British Columbia* **90**(0), 66–76.

Harper, A.M. & Story, T.P. (1962) Reliability of trapping in determining the emergence period and sex ratio of the sugar-beet root maggot *Tetanops myopaeformis* (Rbder) (Diptera: Otitidae). *Can. Entomol.* **94**, 268–71.

Harrison, I.R. (1963) Population studies on the poultry red mite *Dermanyssus gallinae* (Deg.). *Bull. Entomol. Res.* **53**, 657–64.

Harwood, R.F. (1961) A mobile trap for studying the behaviour of flying bloodsucking insects. *Mosquito News* **21**, 35–9.

Hashiguchi, K., Velez, L., Kato, H., Criollo, H., Romero, D., & Gomez, E. (2014) Sand fly fauna (Diptera, Pcychodidae, Phlebotominae) in different leishmaniasis-endemic areas of Ecuador, surveyed using a newly named Mini-Shannon trap. *Trop. Med. Health*, **42**, 163.

Haufe, W.C. & Burgess, L. (1960) Design and efficiency of mosquito traps based on visual response to patterns. *Can. Entomol.* **92**, 124–40.

Hayes, W.B. (1970) The accuracy of pitfall trapping for the sand-beach isopod *Tylos punctatus*. *Ecology* **51**, 514–16.

Heap, M.A. (1988) The pit-light, a new trap for soil-dwelling insects. *Aust. J. Entomol.* **27**, 239–40.

Heathcote, G.D. (1957a) The comparison of yellow cylindrical, flat and water traps and of Johnson suction traps, for sampling aphids. *Ann. Appl. Biol.* **45**, 133–9.

Heathcote, G.D. (1957b) The optimum size of sticky aphid traps. *Plant Pathol.* **6**, 104–7.

Heikinheimo, O. & Raatikainen, M. (1962) Comparison of suction and netting methods in population investigations concerning the fauna of grass leys and cereal fields, particularly in those concerning the leafhopper, *Calligypona pellucida* (F.). *Valt. Maatalourk. Julk. Helsingfors* **191**, 31 pp.

Henderson, P.A. (1989) On the structure of the inshore fish community of England and Wales. *J. Mar. Biol. Assoc. UK* **69**, 145–63.

Henderson, P.A., Irving, P.W., & Magurran, A.E. (1997) Fish pheromones and evolutionary enigmas: a reply to Smith. *Proc. R. Soc. Lond. B.* **264**, 451–3.

Henderson, P.A. & Seaby, R.M.H. (1994) On the factors influencing juvenile flatfish abundance in the lower Severn Estuary. *Neth. J. Sea Res.* **32**, 321–30.

Hesse, G., Kauth, H. & Wachter, R. (1955). Frasslockstoffe beim Fichtenriisselkäfer *Hylobius abietis. Z. Angew. Entomol.* **37**, 239–244.

Heydemann, B. (1961) Untersuchungen iber die Aktivatdts und Besiedlungsdichte bei epigaischen spinnen. *Verh. di. zoot. Ges.*, 538–56.

Heyer, W.R., Donnelly, M.A., McDiarmid, R.W., Hayek, L.A., & Foster, M.S. (eds) (1994) *Measuring and Monitoring Biological Diversity. Standard Methods for Amphibians.* Smithsonian Institution Press, Washington DC,.

Hilgerloh, G. (1989) Autumn migration of trans-Saharan migrating passerines in the Straits of Gibraltar (Mediterranean Sea). *Auk* **106**(2), 233–9.

Hilgerloh, G. (1991) Spring migration of passerine trans-Saharan migrants across the Straits of Gibraltar. *Ardea* **79**(1), 57–62.

Hill, M.A. (1947) The life-cycle and habits of *Culicoides impuctatus* Goet. and *C. obsoletus* Mg., together with some observations on the life-cycle of *Culicoides odibilis* Aust., *Culicoides pallidicornis Kief. Culicoides cubitalis Edw. and Culicoides chiopterus Mg. Ann. Trop. Med. Parasitol.* **41**, 55–115.

Hobbs, S.E. & Wolf, W.W. (1989) An airborne radar technique for studying insect migration. *Bull. Entomol. Res.* **79**(4), 693–704.

Hokkanen, H. & Holopainen, J.K. (1986) Carabid species and activity densities in biologically and conventionally managed cabbage fields. *J. Appl. Ent.* **36**, 257–61.

Hollingsworth, J.P., Briggs, C.P., Glick, P.A., & Graham, H.M. (1961) Some factors influencing light trap collections. *J. Econ. Entomol.* **54**, 305–8.

Hollingsworth, J.P., Hartsock, J.G., & Stanley, J.M. (1963) *Electrical insect traps for survey purposes.* USDA Agric. Res. Series. (ARS), 10 pp.

Holopainen, J.K. (1992) Catch and sex ratio of Carabidae (Coleoptera) in pitfall traps filled with ethylene glycol or water. *Pedobiologia* **36**, 257–61.

Howell, J.F., Cheikh, M., & Harris, E.J. (1975) Comparison of the efficiency of three traps for the Mediterranean Fruit Fly baited with minimum amounts of trimedlure. *J. Econ. Entomol.* **68**, 277–9.

Hughes, R.D. (1955) The influence of the prevailing weather conditions on the numbers of *Meromyza variegata* Meigèn (Diptera: Chloropidae) caught with a sweep net. *J. Anim. Ecol.* **24**, 324–35.

Ibbotson, A. (1958) The behaviour of frit fly in Northumberland. *Ann. Appl. Biol.* **46**, 474–9.

Ikeshoji, T. & Yap, H.H. (1987) Monitoring and chemosterilization of a mosquito population, *Culex quinquefasciatus* (Diptera, Culicidae) by sound traps. *Appl. Entomol. Zool.* **22**, 474–81.

Innes, D.G.L. & Bendell, J.F. (1988) Sampling of small mammals by different types of traps in northern Ontario, Canada. *Acta Theriol.* **33**(26–43), 443–50.

James, H.G. & Redner, R.L. (1965) An aquatic trap for sampling mosquito predators. *Mosquito News* **25**, 35–7.

Jamnback, H. & Watthews, T. (1963) Studies of populations of adult and immature *Culicoides sanguisuga* (Diptera: Ceratopogonidae). *Ann. Entomol. Soc. Am.* **56**, 728–32.

Jansen, M.J.W. & Metz, J.A.J. (1979) How many victims will be pitfall make? *Acta Biotheoretica* **28**, 98–122.

Jenson, P. (1968) Changes in Cryptozoan numbers due to systematic variation of covering boards. *Ecology* **49**, 409–18.

Johnson, C.C. (1952) The role of population level, flight periodicity and climate in the dispersal of aphids. *Trans. IX Int. Congr. Entomol.* **1**, 429–31.

Johnson, C.G. (1950) The comparison of suction trap, sticky trap and tow-net for the quantitative sampling of small airborne insects. *Ann. Appl. Biol.* **37**, 268–85.

Johnson, C.G. (1960) A basis for a general system of insect migration and dispersal flight. *Nature* **186**, 348–50.

Johnson, C.G. (1963) Physiological factors in insect migration by flight. *Nature* **198**, 423–7.

Johnson, C.G., Southwood, T.R.E., & Entwistle, H.M. (1957) A new method of extracting arthropods and molluscs from grassland and herbage with a suction apparatus. *Bull. Entomol. Res.* **48**, 211–18.

Joosse, E.N.G. (1975) Pitfall-trapping as a method for studying surface dwelling Collembola. *Z. Morph. Oekol. Tiere* **55**, 587–96.

Jordan, A.M. (1962) The ecology of the jiisca group of tsetse flies (*Glossina*) in Southern Nigeria. *Bull. Entomol. Res.* **53**, 355–85.

Juillet, J.A. (1963) A comparison of four types of traps used for capturing flying insects. *Can. J. Zool.* **41**, 219–23.

Kanda, T., Loong, K.P., Chiang, G.L., Cheong, W.H., & Lim, T.W. (1988) Field study on sound trapping and the development of trapping method for both sexes of Mansonia in Malaysia. *Trop. Biomed.* **5**(1), 37–42.

Kawai, S. & Suenaga, O. (1960) Studies of the methods of collecting flies. III. On the effect of putrefaction of baits (fish). [In Japanese, Eng. summary.]. *Endemic Dis. Bull. Nagasaki Univ.* **2**, 61–6.

Kehat, M., Anshelevich, L., Dunkelblum, E., & Greenberg, S. (1994) Sex pheromone traps for monitoring the peach twig borer, *Anarsia lineatella* Zeller: Effect of pheromone components, pheromone dose, field aging of dispenser, and type of trap on male captures. *Phytoparasitica* **22**, 291–8.

Kennedy, G.J.A. & Stange, C.D. (1988) Efficiency of electric fishing for salmonids in relation to river width. *Fish. Manag.* **12**, 55–60.

Kennedy, J.S. (1961) A turning point in the study of insect migration. *Nature* **189**, 785–91.

Kersting, U. & Baspinar, H. (1995) Seasonal and diurnal flight activity of *Circulifer haematoceps* (Hom., Cicadellidae), an important leafhopper vector in the Mediterranean area and the Near East. *J. Appl. Entomol.* **119**(8), 533–7.

Khan, A.A., Maibach, H.I., Strauss, W.G., & Fisher, J.L. (1970) Differential attraction of the yellow fever mosquito to vertebrate hosts. *Mosquito News* **30**, 43–7.

Kitching, R.L., Nakamura, A., Yasuda, M., Hughes, A.C., & Min, C. (2015) Environmental determinism of community structure across trophic levels: moth assemblages and substrate type in the rain forests of south-western China. *J. Trop. Ecol.* **31**, 81–9.

Kissick, L.A. (1993) Comparison of traps lighted by photochemicals or electric bulbs for sampling warmwater populations of young fish. *N. Am. J. Fish. Manag.* **13**(4), 864–7.

Klock, J.W. & Bidlingmayer, W.L. (1953) An adult mosquito sampler. *Mosquito News* **13**, 157–9.

Knodel, J.J. & Agnello, A.M. (1990) Field comparison of nonsticky and sticky pheromone traps for monitoring fruit pests in western New York (USA). *J. Econ. Entomol.* **83**(1), 197–204.

Kono, T. (1953) On the estimation of insect population by time unit collecting. [In Japanese.]. *Res. Popul. Ecol.* **2**, 85–94.

Koponen, S. (1992) Spider fauna (Araneae) of the low Arctic Belcher Islands, Hudson Bay. *Arctic* **45**(4), 358–62.

Kovrov, B.G. & Monchadskii, A.S. (1963) The possibility of using polarized light to attract insects. [In Russian.]. *Ent. Obo-r.* **42**, 49–55 (transl. *Ent. Rev.* **42**, 25–8).

Kowalski, R. (1976) Obtaining valid population indices from pitfall traps. *Bull. Acad. Pol. Sci. Ser. Sci. Biol. II* **23**, 799–803.

Krebs, C.J., Singleton, G.R., & Kenney, A.J. (1994) Six reasons why feral house mouse populations might have low recapture rates. *Wildlife Res.* **21**(5), 559–67.

Kring, J.B. (1970) Red spheres and yellow panels combined to attract apple maggot flies. *J. Econ. Entomol.* **63**, 466–9.

Kubecka, J., Duncan, A., Duncan, W.M., Sinclair, D., & Butterworth, A.J. (1994) Brown trout populations of three Scottish lochs estimated by horizontal sonar and multimesh gill nets. *Fish. Res.* **20**(1), 29–48.

Kurnik, I. (1988) Millipedes from xerothermic and agricultural sites in South Tyrol (Austria). *Berichte Des Naturwissenschaftlich Medizinischen Vereins In Innsbruck* **75**(0), 109–114.

Lambert, H.L. & Franklin, R.T. (1967) Tanglefoot traps for detection of the balsam woolly aphid. *J. Econ. Entomol.* **60**, 1525–9.

Landin, J. (1968) Weather and diurnal periodicity of flight by *Helophorus bretipalpis* Bedel (Col.: Hydrophilidae). *Opusc. Entomol.* **33**, 28–36.

Landin, J. & Stark, E. (1973) On flight thresholds for temperature and wind velocity, 24-hour flight periodicity and migration of the beetle, *Helophorus brevipalpis* Bedel (Col. Hydrophilidae). *Zoon. Suppl.* **1**, 105–14.

Leech, H.B. (1955) Cheesecloth flight trap for insects. *Can. Entomol.* **85**, 200.

Le Pelley, R.H. (1935) Observations on the control of insects by hand-collection. *Bull. Entomol. Res.* **26**, 533–41.

Leskinen, M., Markkula, I., Koistenen, J., Pylkko, P., Ooperi, S., Siljamo, P., Ojanen, H., Raisko, S., & Tiilikkala, K. (2011) Pest insect immigration warning by an atmospheric dispersion model, weather radars and traps. *J. Appl. Entomol.* **135**, 55–67.

Leshem, Y. & Yom Tov, Y. (1996) The magnitude and timing of migration by soaring raptors, pelicans and storks over Israel. *Ibis* **138**(2), 188–203.

Levin, M.D. (1957) Artificial nesting burrows for *Osmia lignaria* Say. *J. Econ. Entomol.* **50**, 506–7.

Lewandowski Jr, H.B., Hooper, G.R., & Newson, H.D. (1980) Determination of some important natural potential vectors of dog heartworm in central Michigan. *Mosquito News* **40**(1), 73–9.

Lewis, T. (1959) A comparison of water traps, cylindrical sticky traps and suction traps for sampling Thysanopteran populations at different levels. *Entomol. Exp. Appl.* **2**, 204–15.

Lewis, T. & Macaulay, E.D.M. (1976) Design and elevation of sex-attractant traps for pea moth, *Cydia nigricana* (Steph.) and the effect of plume shape on catches. *Ecol. Entomol.* **1**, 175–87.

Lewis, W.J., Snow, J.W., & Jones, R.L. (1971) A pheromone trap for studying populations of *Cardiochiles nigricaps*, a parasite of *Heliothis virescens*. *J. Econ. Entomol.* **64**, 1417–21.

Liechti, F., Guélat, J., & Komenda-Zehnder, S. (2013) Modelling the spatial concentrations of bird migration to assess conflicts with wind turbines. *Biol. Conserv.* **162**, 24–32.

Linsley, E.G., MacSwain, J.W., & Smith, R.F. (1952) Outline for ecological life histories of solitary and semi-social bees. *Ecology* **33**, 558–67.

Lockart, R.B. & Owings, D.H. (1974) Seasonal changes in the activity pattern of *Dipodomys spectabilis*. *J. Mammol.* **55**, 291–7.

Loomis, E.C. (1959a) A method for more accurate determination of air volume displacement of light traps. *J. Econ. Entomol.* **52**, 343–5.

Loomis, E.C. (1959b). Selective response of *Aedes nigronaculis* (Ludlow) to the Minnesota light trap. *Mosquito News* **19**, 260–3.

Loomis, E.C. & Sherman, E.J. (1959) Comparison of artificial shelter and light traps for measurements of *Culex tarsalis* and *Anopheles freeborni* populations. *Mosquito News* **19**, 232–7.

Love, G.I. & Smith, W.W. (1957) Preliminary observations on the relation of light trap collections to mechanical sweep net collections in sampling mosquito populations. *Mosquito News* **17**, 9–14.

Luczak, J. & Wierzbowska, T. (1959) Analysis of likelihood in relation to the length of a series in the sweep method. *Bull. Acad. Pol. Sci. Ser. Sci. Biol.* **7**, 313–18.

Luff, M.L. (1968) Some effects of formalin on the numbers of Coleoptera caught in pitfall traps. *Entomol. Monthly Mag.* **104**, 115–16.

Luff, M.L. (1975) Some features influencing the efficiency of pitfall traps. *Oecologia* **19**, 345–57.

Lumsden, W.H.R. (1958) A trap for insects biting small vertebrates. *Nature* **181**, 819–20.

MacLeod, J. & Donnelly, J. (1956) Methods for the study of Blowfly populations. 1. Bait trapping. Significance of limits for comparative sampling. *Ann. Appl. Biol.* **44**, 80–104.

Magurran, A.E., Irving, P.W., & Henderson, P.A. (1996) Why is fish alarm 'pheromone' ineffective in the wild? *Proc. R. Soc. Lond B.* **263**, 1551–6.

Malaise, R. (1937) A new insect-trap. *Entomol. Tidskr.* **58**, 148–60.

Martin, F.R., McCreadie, J.W., & Colbo, M.H. (1994) Effect of trap site, time of day, and meteorological factors on abundance of host-seeking mammalophilic black flies (Diptera: Simuliidae). *Can. Entomol.* **126**(02), 283–9.

Masaki, J. (1959) Studies on rice crane fly (*Tipula aino* Alexander, Tipulidae, Diptera) with special reference to the ecology and its protection. [In Japanese.]. *J. KantoTosan Agric. Exp. Sta.* **13**, 1–195.

Mason, H.C. (1963) Baited traps for sampling *Drosophila* populations in tomato field plots. *J. Econ. Entomol.* **56**, 897–8.

Mathis, A. & Smith, R.J.F. (1992) Avoidance of areas marked with a chemical alarm substance by fathead minnows (*Pimephales promelas*) in a natural habitat. *Can. J. Zool.* **70**, 1473–6.

Maxwell, C.W. (1965) Tanglefoot traps as indicators of apple maggot fly activities. *Can. Entomol.* **97**, 1–10.

Mayer, K. (1961) Untersuchungen ilber das Wahlverhalten der Fritfiiege (*Oscinelta frit* L.) beim Anflug von Kulturpflanzen im Feldversuch mit der Fangschalenmethode. *Mitt. biol. Bund. Anst. Ld- u. Forstw., Berlin* **106**, 1–47.

McClelland, G.A.H. & Conway, G.R. (1971) Frequency of blood-feeding in the mosquito *Aedes aegypti. Nature* **232**, 485–6.

McPhail, M. (1939) Protein lures for fruit flies. *J. Econ. Entomol.* **32**, 758–61.

Medler, J.T. (1964) Biology of *Rygchium foraminatum* in trap-nests in Wisconsin (Hymbroptera: Vespidae). *Ann. Entomol. Soc. Am.* **57**, 56–60.

Medler, J.T. & Fye, R.E. (1956) Biology of *Ancistrocerus antilope* (Panzer) in trap nests in Wisconsin. *Ann. Entomol. Soc. Am.* **49**, 97–102.

Mellanby, K. (1962) Sticky traps for the study of animals inhabiting the soil surface. In: Murphy, P.W. (ed.), *Progress in Soil Zoology*. Butterworth, London.

Menhinick, E.F. (1963) Estimation of insect population density in herbaceous vegetation with emphasis on removal sweeping. *Ecology* **44**, 617–21.

Merzheevskaya, O.I. & Gerastevich, E.A. (1962) A method of collecting living insects at light. [In Russian.]. *Zool. Zhur.* **41**, 1741–3.

Michelbacher, A.E. & Middlekauff, W.W. (1954) Vinegar fly investigations in Northern California. *J. Econ. Entomol.* **47**, 917–22.

Mikkola, K. (1972) Behavioural and electrophysiological responses of night-flying insects, especially Lepidoptera, to near-ultraviolet visible light. *Ann. Zool. Fennici* **99**, 225–54.

Miller, T.A., Stryker, R.G., Wilkinson, R.N., & Esah, S. (1970) The influence of moonlight and other environmental factors on the abundance of certain mosquito species in light-trap collections in Thailand. *J. Med. Entomol.* **7**, 555–61.

Minns, C.K. & Hurley, D.A. (1988) Effects of net length and set time on fish catches in gill nets. *North Am. J. Fish. Manag.* **8**, 216–23.

Minter, D.M. (1961) A modified Lumsden suction-trap for biting insects. *Bull. Entomol. Res.* **529**, 233–8.

Mitchell, B. (1963) Ecology of two carabid beetles, *Bembidion lampros* (Herbst.) and *Trechus quadristriatus* (Schrank). *J. Anim. Ecol.* **32**, 377–92.

Moericke, V. (1950) Ober den Farbensinn der Pfirsichblattlaus Wy, odespersicae Sulz. *Z. Tierpsychol.* **7**, 265–74.

Mommertz, S., Schauer, C., Koesters, N., Lang, A., & Filser, J. (1996) A comparison of D-Vac suction, fenced and unfenced pitfall trap sampling of epigeal arthropods in agro-ecosystems. *Ann. Zool. Fennici* **33**, 117–24.

Moore, J. (2002) *Parasites and the Behavior of Animals*. Oxford University Press.

Moran, P.A.P. (1951) A mathematical theory of animal trapping. *Biometrica* **38**, 307–11.

Morgan, C.V.G. & Anderson, N.H. (1958) Techniques for biological studies of tetranychid mites, especially *Bryobia arborea* M. & A. and *B. praetiosa* Koch. (Acarina: Tetranychidae). *Can. Entomol.* **90**, 212–15.

Morrill, W.L., Lester, D.G., & Wrona, A. E. (1990) Factors affecting efficacy of pitfall traps for beetles (Coleoptera: Carabidae and Tenebrionidae). *J. Entomol. Sci.* **25**(2), 284–93.

Morris, K.R.S. (1960) Trapping as a means of studying the game tsetse, *Glossina pallidipes Aust. Bull. Entomol. Res.* **51**, 533–57.

Morris, K.R.S. (1961) Effectiveness of traps in tsetse surveys in the Liberian rain forest. *Am. J. Trop. Med. Hyg.* **10**, 905–13.

Morris, K.R.S. & Morris, M.G. (1949) The use of traps against tsetse in West Africa. *Bull. Entomol. Res.* **39**, 491–523.

Morton, S.R., Gillam, M.W., Jones, K.R., & Fleming, M.R. (1988) Relative efficiency of different pit-trap systems for sampling reptiles in spinifex grasslands. *Aust. Wildife. Res.* **15**, 571–7.

Muirhead-Thomson, R.C. (ed.) (1963) *Practical entomology in malaria eradication, Part 1, Chapter . MHO/PA/62.63*, WHO, mimeographed.

Muirhead-Thompson, R.C. (2012) *Trap Responses of Flying Insects: The Influence of Trap Design on Capture Efficiency*. Academic Press.

Mukhopadhyay, S. (1991) Lunation-induced variation in catchment areas of light-traps to monitor rice green leafhoppers (*Nephotettix* spp.) in West Bengal. *Indian J. Agric. Sci.* **61**(5), 337–40.

Mulhern, T.D. (1942) New Jersey mechanical trap for mosquito surveys. *N. J. Agric. Exp. Sta. Circ.* **421**, 1–8.

Mulhern, T.D. (1953) Better results with mosquito light traps through standardizing mechanical performance. *Mosquito News* **13**, 130–3.

Mulla, M.S., Georchiou, G.P., & Dorner, R.W. (1960) Effect of ageing and concentration on the attractancy of proteinaceous materials to *Hippelates* gnats. *Ann. Entomol. Soc. Am.* **53**, 835–41.

Mundie, J.H. (1964) A sampler for catching emerging insects and drifting materials in streams. *Limnol. Oceanogr.* **9**, 456–9.

Murphy, W.L. (1985) Procedure for the removal of insect specimens from sticky-trap material. *Ann. Entomol. Soc. Am.* **78**(6), 881.

Murray, W.D. (1963) Measuring adult populations of the pasture mosquito, *Aedes migromaculis* (Ludlow). Proceedings, 27th Conference Californian Mosquito Control Association, 1959.

Myers, K. (1956) Methods of sampling winged insects feeding on the rabbit *Oryctolagus cuniculus* (L.). *Aust. C.S.I.R.O. Wildlife Res.* **1**, 45–58.

Myers, R.A., Smith, M.W., Hoenig, J.M., Kmiecik, N., Luehring, M.A., Drake, M.T., Schmalz, P.J., & Sass, G.G. (2014). Size-and sex-specific capture and harvest selectivity of walleyes from tagging studies. *Trans. Am. Fish. Soc.* **143**, 438–50.

Mpofu S. M., & Masendu H.T. (1986) Description of a baited trap for sampling mosquitoes. *J. Am. Mosq. Control Assoc.* **2**, 363–65.

Myllymaki, A., Paasikallio, A., Pankakoski, E., & Kanervo, V. (1971) Removal experiments on small quadrates as a means of rapid assessment of abundance of small mammals. *Ann. Zool. Fennici* **8**, 177–85.

Mystowska, E.T. & Sidorowicz, J. (1961) Influence of the weather on captures of micromammalia II. *Insectivora Acta Theriologica* **5**, 263–73.

Nag, A. & Nath, P. (1991) Effect of moon light and lunar periodicity on the light trap catches of cutworm *Agrotis ipsilon* (Hufn.) moths. *J. Appl. Entomol.* **111**(4), 358–60.

Nakamura, K., Ito, Y., Miyashita, K., & Takai, A. (1967) The estimation of population density of the green rice leafhopper, *Nephotettix cincticeps* Uhler. in spring field by the capture–recapture method. *Res. Popul. Ecol.* **9**, 113–29.

Neal, J.W., Adelsberger, C.M., & Lochmann, S.E. (2012) A comparison of larval fish sampling methods for tropical streams. *Mar. Coastal Fish.* **4**, 23–9.

Neilson, W.T.A. (1960) Field tests of some hydrolysed proteins as lures for the apple maggot, *Rhagoletis pomonella* (Walsh). *Can. Entomol.* **92**, 464–7.

Nielsen, B. (1963) The biting midges of *Lyngby aamose* (Culicoides: Ceratopogonidae). *Naturajutl.* **10**, 1–48.

Nijholt, W.W. & Chapman, J.A. (1968) A flight trap for collecting living insects. *Can. Entomol.* **100**, 1151–3.

Odintsov, V.S. (1960) Air catch of insects as a method of study upon entomofauna of vast territories. [In Russian.] *Entomol. Obozr.* **39**, 227–30.

Otake, A. & Kono, T. (1970) Regional characteristics in population trends of smaller brown planthopper, *Laodelphax striatellus* (Fallen) (Hemiptera: Delphacidae), a vector of rice stripe disease: an analytical study of light trap records. *Bull. Shikoku Exp. Sta.* **21**, 127–47.

Otis, D.L., Burnham, K.P., White, G.C., & Anderson, D.R. (1978) Statistical inference from capture data on closed animal populations. *Wildife. Monogr.* **62**, 1–135.

Palumbo, J.C., Tonhasca, A., Jr, & Byrne, D.N. (1995) Evaluation of three sampling methods for estimating adult sweetpotato whitefly (Homoptera: Aleyrodidae) abundance on cantaloupes. *J. Econ. Entomol.* **88**(5), 1393–400.

Paris, O.H. (1965) The vagility of P32-labelled isopods in grassland. *Ecology* **46**, 635–48.

Parker, A.H. (1949) Observations on the seasonal and daily incidence of certain biting midges (Culicoides Latreille – Diptera, Ceratopogonidae) in Scotland. *Trans. Roy. Entomol. Soc. Lond.* **100**, 179–90.

Parmenter, R.R., Macmahon, J.A., & Anderson, D.R. (1989) Animal density estimation using a trapping web design: Field validation experiments. *Ecology* **70**(1), 169–79.

Paviour-Smith, K. (1964) The life history of *Tetratoma fungorum* F. (Col., Tetratomidae) in relation to habitat requirements, with an account of eggs and larval stages. *Entomol. Monthly Mag.* **100**, 118–34.

Pena, J.E. & Duncan, R. (1992) Sampling methods for *Prodiplosis longifila* (Diptera: Cecidomyidae) in limes. *Environ. Entomol.* **21**, 996–1001.

Persat, H. & Copp, G.H. (1990) Electric fishing and point abundance sampling for the ichthyology of large rivers. In: Cowx, I.G. (ed.), *Developments in Electric Fishing*. Fishing News Books, Blackwell Scientific Publications, Oxford, UK.

Petruska, F. (1968) Members of the group Silhini as a component part of the insects fauna of sugar beet fields in the Unicov Plain. *Acta Univ. Palackianae Olomucensis Fac. Rerum Natur.* **38**, 189–200.

Petruska, F. (1969) On the possibility of escape of various components of the epigenic fauna of the fields from the pitfall traps containing formalin (Coeoptera) (in Czech, English summary). *Acta Univ. Palackianae Olomucensis Fac. Rerum Natur.* **31**, 99–124.

Piecynski, E. (1961) The trap method of capturing water mites (Hydracarina). *Ekol. Poll B,* **7**, 111–15.

Plesner Jensen, S. & Honess, P. (1995) The influence of moonlight on vegetation height preference and trappability of small mammals. *Mammalia* **59**(1), 35–42.

Pollet, M. & Grootaert, P. (1987) Ecological data on Dolichopodidae (Diptera) from a woodland ecosystem: I. Color preference, detailed distribution and comparison of different sampling techniques. *Bulletin De L'Institut Royal Des Sci. Naturelles De Belgique Entomol.* **57**(0), 173–87.

Pollock, K.H. & Otto, M.C. (1983) Robust estimation of population size in closed animal populations from capture–recapture experiments. *Biometrics* **39**, 1035–49.

Ponton, D. (1994) Sampling neotropical young and small fishes in their microhabitats: An improvement of the quatrefoil light-trap. *Archiv fur Hydrobiologie* **131**(4), 495–502.

Prokopy, R.J. (1968) Sticky spheres for estimating Apple Maggot adult abundance. *J. Econ. Entomol.* **61**, 1082–5.

Prokopy, R.J. (1977) Host plant influences on the reproductive biology of Tephritidae. In: Labeyrie, V. (ed.), *Comportement des Insectes et Milieu Trophique.* Coll. Int. C.N.R.S.

Provost, M.W. (1960) The dispersal *Aedes taeniorhynchus.* III. Study methods for migratory exodus. *Mosquito News* **20**, 148–61.

Race, S.R. (1960) A comparison of two sampling techniques for lygus bugs and stink bugs on cotton. *J. Econ. Entomol.* **53**, 689–90.

Rainey, R.C. & Joyce, R.J.V. (1990) An airborne radar system for desert locust control. *Philos. Trans. Roy. Soc. B* **328**, 585–606.

Reidl, H., Howell, J.F., McNally, P.S., & Westigard, P.H. (1986) Codling moth management: use and standardisation of pheromone trapping systems. *Univ. Calif. Agric. Exp. Sta. Bull.* **1918**.

Rennison, B.D. & Robertson, D.H.H. (1959) The use of carbon dioxide as an attractant for catching tsetse. *Rep. E. Afr. Trypan. Res. Organ.* **1958**, 26.

Reynolds, M.H., Cooper, B.A., & Day, R.H. (1997) Radar study of seabirds and bats on Windward Hawaii. *Pacific Sci.* **51**(1), 97–106.

Richert, A. & Huelbert, D. (1991) Evaluation of catches made by light traps over five years (1984 to 1988) on the phenological basis of Hohenfinow (district Eberswalde-Finow) (Germany) for the faunistics of Lepidoptera. *Beitraege Zur Entomologie* **41**(1), 251–64.

Rickart, E.A., Heaney, L.R., & Utzurrum, R.C.B. (1991) Distribution and ecology of small mammals along an elevational transect in southeastern Luzon, Philippines. *J. Mammal.* **72**(3), 458–69.

Ricker, W.E. (1975) *Computation and interpretation of biological statistics of fish populations.* Blackburn Press, Ottawa.

Rieske, L.K. & Raffa, K.F. (1993) Potential use of baited pitfall traps in monitoring pine root weevil, *Hylobius pales, Pachylobius picivorus*, and *Hylobius radicis* (Coleoptera: Curculionidae) populations and infestation levels. *J. Econ. Entomol.* **86**(2), 475–85.

Riley, D.G. & Schuster, D.J. (1994) Pepper weevil adult response to colored sticky traps in pepper fields. *Southwestern Entomol.* **19**(2), 93–107.

Riley, J.R. & Reynolds, D.R. (1990) Nocturnal grasshopper migration in West Africa: Transport and concentration by the wind, and the implications for air-to-air control. *Philos. Trans. Ro. Soc. Lond. B* **328**(1251), 655–72.

Rings, R.W. & Metzler, E.H. (1990) The Lepidoptera of Fowler Woods State Nature Preserve, Richland County, Ohio (USA). *Great Lakes Ent.* **23**(1), 43–56.

Rioux, J.A. & Golvan, Y.I. (1969) Epidemiologie des Leishmanioses dans le sud de la France. *Monogr. Inst. Nat. Sante Recherche Med.* **37**, 1–220.

Rivard, I. (1962) Un piège à fosse améliore pour la capture d'insectes actifs à la surface du sol. *Can. Entomol.* **94**, 270–1.

Roberts, D. & Kumar, S. (1994) Using vehicle-mounted nets for studying activity of Arabian sand flies (Diptera: Psychodidae). *J. Med. Entomol.* **31**(3), 388–93.

Roberts, D.R. & Scanlon, J.E. (1975) The ecology and behavior of *Aedes atlanticus* D. & K. and other species with reference to Keystone virus in the Houston area. *Texas. J. Med. Entomol.*, **12**(5), 537–46.

Roberts, R.H. (1970) Colour of Malaise trap and collection of Tabanidae. *Mosquito News* **30**, 567–71.

Roberts, R.H. (1972) The effectiveness of several types of Malaise traps for the collection of Tabanidae and Culicidae. *Mosquito News* **32**, 542–7.

Roberts, R.H. (1976) The comparative efficiency of six trap types for the collection of Tabanidae (Diptera). *Mosquito News* **36**, 530–7.

Roberts, R.I. & Smith, T.J.R. (1972) A plough technique for sampling soil insects. *J. Appl. Ecol.* **9**, 427–30.

Robinson, H.S. & Robinson, P.J.M. (1950) Some notes on the observed behaviour of Lepidoptera in flight in the vicinity of light-sources together with a description of a light-trap designed to take entomology samples. *Entomol. Gaz.* **1**, 3–15.

Rodríguez-Pérez, M.A., Adeleke, M.A., Burkett-Cadena, N.D., Garza-Hernández, J.A., Reyes-Villanueva, F., Cupp, E.W., Toe, L., & Unnasch, T.R. (2013) Development of a novel trap for the collection of black flies of the *Simulium ochraceum* complex. *PloS One* **8**, e76814.

Roesler, R. (1953) Uber eine Methode zur Feststellung der Flugzeit schddlicher Fliegenarten (Kirschfliege, Kohlfliege, Zwiebelfiiege). *Mitt. Biol. Reichsant. (Zent Anst.)Ld- u. Forstw., Berlin* **75**, 97–9.

Rogers, D.J. & Smith, D.T. (1977) A new electric trap for tsetse flies. *Bull. Entomol. Res.* **67**, 153–9.

Rohitha, B.H. & Stevenson, B.E. (1987) An automatic sticky trap for aphids (Hemiptera: Aphididae) that segregates the catch daily. *Bull. Entomol. Res.* **77**, 67–72.

Romney, V.E. (1945) The effect of physical factors upon catch of the beet leafhopper (*Eutettix tenellus* (Bak.)) by a cylinder and two sweep-net methods. *Ecology* **26**, 135–48.

Roos, T. (1957) Studies on upstream migration in adult stream-dwelling insects. *Inst. Freshw. Res. Drottringholm Rep.* **38**, 167–93.

Rothschild, M. (1961) Observations and speculations concerning the flea vector of myxomatosis in Britain. *Entomol. Monthly Mag.* **96**, 106–9.

Rudd, W.G. & Jensen, R.L. (1977) Sweep net and ground cloth sampling for insects in soybeans. *J. Econ. Entomol.* **70**, 301–4.

Saliternik, Z. (1960) A mosquito light trap for use on cesspits. *Mosquito News* **20**, 295–6.

Samu, F. & Sarospataki, M. (1995) Design and use of a hand-hold suction sampler, and its comparison with sweep net and pitfall trap sampling. *Fol. Entomol. Hung.* **56**(0), 195–203.

Saugstad, E.S., Bram, R.A., & Nyquist, W.E. (1967) Factors influencing sweep-net sampling of alfalfa. *J. Econ. Entomol.* **60**, 421–6.

Saunders, D.S. (1964) The effect of site and sampling and method on the size and composition of catches of tsetse flies (Glossina) and Tabanidae (Diptera). *Bull. Entomol. Res.* **55**, 483.

Schaeffer, G.W. (1979) An airborne radar technique for investigation and control of migrating insects. *Philos. Trans. R. Soc. Lond.* **287**, 459–65.

Schlyter, F., Lofqvist, J., & Byers, J. A. (1987) Behavioral sequence in the attraction of the bark beetle *Ips typographus* to pheromone sources. *Physiol. Entomol.* **12**(2), 185–96.

Seber, G.A.F. (1982) *Estimation of Animal Abundance and Related Parameters*. Griffin, London.

Seber, G.A.F. (1986) A review of estimating animal abundance. *Biometrics* **42**, 267–292.

Seber, G.A.F. & Whale, J.F. (1970) The removal method for two and three samples. *Biometrics* **26**, 393–400.

Service, M.W. (1963) The ecology of the mosquitoes of the northern Guinea savannah of Nigeria. *Bull. Entomol. Res.* **54**, 601–32.

Service, M.W. (1995) *Mosquito Ecology Field Sampling Methods*, 2nd edn. Chapman & Hall, 988 pp.

Sevastpulo, D.G. (1963) Field notes from East Africa – Part XI. *Entomologist* **96**, 162–5.

Siddorn, J.W. & Brown, E.S. (1971) A Robinson light trap modified for segregating samples at predetermined time intervals, with notes on the effect of moonlight on the periodicity of catches of insects. *J. Appl. Ecol.* **8**, 69–75.

Simpson, R.G. & Berry, R.E. (1973) A twenty-four hour directional aphid trap. *J. Econ. Entomol.* **66**, 291–2.

Smith, A.D., Riley, J.R., & Gregory, R.D. (1993) A method for routine monitoring of the aerial migration of insects by using a vertical-looking radar. *Philos. Trans. R. Soc. Lond. B* **340**(1294), 393–404.

Smith, B.D. (1962) The behaviour and control of the blackcurrant gall mite *Phytoptus ribis* (Nal.). *Ann. Appl. Biol.* **50**, 327–34.

Smith, I.M. & Rennison, B.D. (1961) Studies of the sampling of *Glossina pallidipes* Aust. I, II. *Bull. Entomol. Res.* **52**, 165–89.

Smith, P.W., Taylor, J.C., & Apple, J.W. (1959) A comparison of insect traps equipped with 6- and 15-watt black light lamps. *J. Econ. Entomol.* **52**, 1212–14.

Snow, W.E., Pickard, E., & Sparkman, R.E. (1960) A fan trap for collecting biting insects attacking avian hosts. *Mosquito News* **20**, 315–16.

Southwood, T.R.E. (1960) The flight activity of Heteroptera. *Trans. R. Entomol. Soc. Lond.* **112**, 173–220.

Southwood, T.R.E. (1962) Migration of terrestrial arthropods in relation to habitat. *Biol. Rev.* **37**, 171–214.

Southwood, T.R.E., Jepson, W.F., & Van Emden, H.F. (1961) Studies on the behaviour of *Oscinella frit* L. (Diptera) adults of the panicle generation. *Entomol. Exp. Appl.* **4**, 196–210.

Southwood, T.R.E., Henderson, P.A., & Woiwod, I.P. (2003) Stability and change over 67 years – the community of Heteroptera as caught in a light-trap at Rothamsted, UK. *Eur. J. Entomol.* **100**(4), 557–62.

Spradbery, J.P. & Vogt, W.G. (1993) Mean life expectancy of Old World screwworm fly, *Chrysomya bezziana*, inferred from the reproductive age-structure of native females caught on wormlure-baited sticky traps. *Med. Vet. Entomol.* **7**(2), 147–54.

Springate, N.D. & Basset, Y. (1996) Diel activity of arboreal arthropods associated with Papua New Guinean trees. *J. Nat. Hist.* **30**(1), 101–12.

Stahl, C. (1954) Trapping hornworm moths. *J. Econ. Entomol.* **47**, 879–82.

Stammer, H.J. (1949) Die Bedeutung der Athylenglycolfallen für tier6kologische und phanologische Untersuchungen. *Verh. Dtsch. Zoologen. Kiel* **(1948)**, 387–91.

Stanley, J.M. & Dominick, C.B. (1970) Funnel size and lamp wattage influence on light-trap performance. *J. Econ. Entomol.* **63**, 1423–6.

Staples, R. & Allington, W.B. (1959) The efficiency of sticky traps in sampling epidemic populations of the Eriophyid mite *Aceria tulipae* (K.), vector of wheat streak mosaic virus. *Ann. Entomol. Soc. Am.* **52**, 159–64.

Starr, D.F. & Shaw, J.C. (1944) Pyridine as an attractant for the Mexican fruit fly. *J. Econ. Entomol.* **37**, 760–3.

Steinbauer, M.J., Haslem, A., & Edwards, E.D. (2012) Using meteorological and lunar information to explain catch variability of Orthoptera and Lepidoptera from 250 W Farrow light traps. *Insect Conserv. Divers.* **5**(5), 367–80.

Steigen, A.L. (1973) Sampling invertebrates active below a snow cover. *Oikos* **24**, 373–6.

Steiner, P., Wenzel, F., & Baumert, D. (1963) Zur Beeinflussung der Arthropodenfauna nordwestdeutscher Kartoffelfelder durch die Anwendung synthetischer Kontaktinsektizide. *Mitt. Biol. BundAnst. Ld- u. ForstK., Berlin* **109**, 1–38.

Subramanyam, B., Hagstrum, D.W., & Schenk, T.C. (1993) Sampling adult beetles (Coleoptera) associated with stored grain: comparing detection and mean trap catch efficiency of two types of probe traps. *Environ. Entomol.* **22**, 33–42.

Taft, H.M. & Jernigan, C.E. (1964) Elevated screens for collecting boll weevils flying between hibernation sites and cottonfields. *J. Econ. Entomol.* **57**, 773–5.

Taylor, L.R. (1962) The efficiency of cylindrical sticky insect traps and suspended nets. *Ann. Appl. Biol.* **50**, 681–5.

Taylor, L.R. (1963) Analysis of the effect of temperature on insects in flight. *J. Anim. Ecol.* **32**, 99–112.

Taylor, L.R. & Brown, E.S. (1972) Effects of light-trap design and illumination on samples of moths in the Kenya highlands. *Bull. Entomol. Res.* **62**, 91–112.

Taylor, L.R. & Carter, C.I. (1961) The analysis of numbers and distribution in an aerial population of Macrolepidoptera. *Trans. R. Entomol. Soc. Lond.* **113**, 369–86.

Taylor, L.R. & French, R.A. (1974) Effects of light-trap design and illumination on samples of moths in an English woodland. *Bull. Entomol. Res.* **63**, 583–94.

Taylor, L.R. & Palmer, J.M.P. (1972) Aerial sampling. In: van Emden, H.F. (ed.), *Aphid Technology*. Academic Press, London.

Taylor, R.A.J. (1986) Time series analysis of numbers of Lepidoptera caught at light traps in East Africa, and the effect of moonlight on trap efficiency. *Bull. Entomol. Res.* **76**(4), 593–606.

Tedders, W.L. & Edwards, G.W. (1972) Effects of black light trap design and placement on catch of adult Hickory Shuckworms. *J. Econ. Entomol.* **65**, 1624–7.

Thorrold, S.R. (1992) Evaluating the performance of light traps for sampling small fish and squid in open waters of the central Great Barrier Reef lagoon. *Mar. Ecol. Progress Series* **89**(2–3), 277–85.

Thorsteinson, A.J., Bracken, G.K., & Hanec, W. (1965) The orientation behaviour of horse flies and deer flies (Tabanidae, Diptera). 111. The use of traps in the study of orientation of Tabanids in the field. *Entomol. Exp. Appl.* **8**, 189–92.

Togashi, I., Nakata, K., Sugie, Y., Hashimoto, M., & Tanaka, A. (1990) Beetles belonging to the subtribe Carabina captured by pitfall traps in Ishikawa Prefecture (Japan). *Bull. Biogeograph. Soc. Japan* **45**(1–22), 103–6.

Topping, C.J. & Sunderland, K.D. (1992) Limitations to the use of pitfall traps in ecological studies exemplified by a study of spiders in a field of winter wheat. *J. Appl. Ecol.* **29**(2), 485–91.

Townes, H. (1962) Design for a Malaise trap. *Proc. Entomol. Soc. Wash.* **64**, 253–62.

Tshernyshev, W.B. (1961) Comparison of field responses of insects to the light of a mercury-quartz lamp and clear ultra-violet radiation of the same lamp. [In Russian.]. *Ent. Obozar.* **40**, 568–70 (transl. *Ent. Rev.* **40**, 308–9).

Turner, E.C. (1972) An animal-baited trap for the collection of Culicoides spp. (Diptera:Ceratopogonidae). *Mosquito News* **32**, 527–30.

Turner, F.B. (1962) Some sampling characteristics of plants and arthropods of the Arizona desert. *Ecology* **43**, 567–71.

Uetz, G.W. (1977) Co-existence in a guild of wandering spiders. *J. Anim. Ecol.* **46**, 531–41.

Uetz, G.W. & Unzicker, J.D. (1976) Pit fall trapping in ecological studies of wandering spiders. *J. Arachnol.* **3**, 101–11.

Usinger, R.L. (ed.) (1963) *Aquatic Insects of California.* University of California Press, Berkeley, Los Angeles, 508 pp.

Usinger, R.L. & Kellen, W.R. (1955) The role of insects in sewage disposal beds. *Hilgardia* **23**, 263–321.

Vale, G.A. (1971) Artificial refuges for tsetse flies (*Glossina* spp.). *Bull. Entomol. Res.* **61**, 331–50.

Vale, G.A. (1974) The responses of tsetse flies (Diptera, Glossinidae) to mobile and stationary baits. *Bull. Entomol. Res.* **64**, 545–88.

van den Broek, W.L.F. (1979) A seasonal survey of fish populations in the Lower Medway Estuary, Kent, based on power station screen sampling. *Estuar. Coastal Mar. Sci.* **9**, 1–15.

Van Huizen, T.H.P. (1977) The significance of flight activity in the life cycle of *Amara plebeja* Gyll. (Coleoptera, Carabidae). *Oecologia* **29**, 27–41.

Vavra, I.R., Carestia, R.R., Frommer, R.L., & Gerberg, E.J. (1974) Field evaluation of alternative light sources as mosquito attractants in the Panama Canal Zone. *Mosquito News* **34**, 382–4.

Verheggen, L. (1994) Monitoring van territorial vleermuizen in de paartijd. *Nieusbrief, Vleermuiswerkgroep Nederland* **17**, 5–8.

Verheijen, F.I. (1960) The mechanisms of the trapping effect of artificial light sources upon animals. *Arch. Néerland. Zool.* **13**, 1–107.

Vibert, R. (ed.) (1967) *Fishing with Electricity. Its Application to Biology and Management.* Fishing News (Books) Ltd, Farnham, UK.

Vickery, W.L. & Bider, J.R. (1978) The effect of weather on *Sorex cinereus* activity. *Can. J. Zool.* **56**, 291–7.

Wakerley, S.B. (1963) Weather and behaviour in carrot fly (*Psila rosae* Fab. Dipt. Psilidae) with particular reference to oviposition. *Entomol. Exp. Appl.* **6**, 268–78.

Walker, A.R. & Boreham, P.F.L. (1976) Saline, as a collecting medium for Culicoides (Diptera: Ceratopogonidae) in blood-feeding and other studies. *Mosquito News* **36**, 18–20.

Walker, T.J. & Lenczewski, B. (1989) An inexpensive portable trap for monitoring butterfly migration. *J. Lepidopt. Soc.* **43**(4), 289–98.

Walker, T.J. & Whitesell, J.J. (1993) A superior trap for migrating butterflies. *J. Lepidopt. Soc.* **47**(2), 140–9.

Waloff, N. & Bakker, K. (1963) The flight activity of Miridae (Heteroptera) living on broom, *Sarothamnus scoparius* (L.). *J. Anim. Ecol.* **32**, 461–80.

Walsh, G.B. (1933) Studies in the British necrophagous Coleoptera. II. The attractive powers of various natural baits. *Entomol. Monthly Mag.* **69**, 28–32.

Walters, B.B. (1989) Differential capture of deer mice with pitfalls and live traps. *Acta Theriologica* **34**(29–43), 643–7.

Waters, T.F. (1962) Diurnal periodicity in the drift of stream invertebrates. *Ecology* **43**, 316–20.

Webster, J.P., Brunton, C.F.A., & MacDonald, D.W. (1994) Effect of *Toxoplasma gondii* upon neophobic behaviour in wild brown rats, *Rattus norvegicus. Parasitology* **109**(1), 37–43.

Welch, R.C. (1964) A simple method of collecting insects from rabbit burrows. *Entomol. Monthly Mag.* **100**, 99–100.

Wenk, P. & Schlorer, G. (1963) Wirtsorientierung und Kopulation bei blutsaugenden Simuliiden (Diptera). *Zeit. Trop. Med. Parasitol.* **14**, 177–91.

Wharton, R.H., Eyles, D.E., & Warren, M.C.W. (1963) The development of methods for trapping the vectors of Monkey malaria. *Ann. Trop. Med. Parasitol., Liverpool* **57**, 32–46.

White, E.G. (1964) A design for the effective killing of insects caught in light traps. *N. Z. Entomol.* **3**, 25–7.

White, G.C., Anderson, D.R., Burnham, K.P., & Otis, D.L. (1982) *Capture–Recapture and Removal Methods for Sampling Closed Populations*. Los Alamos National Laboratory, Los Alamos, New Mexico.

White, N.D.E., Arbogast, R.T., Fields, P.G., Hillman, R.C., Loschiavo, S.R., Subramanyam, B., Throne, J.E., & Wright, V.F. (1990) The development and use of pitfall and probe traps for capturing insects in stored grain. *J. Kansas Entomol. Soc.* **63**, 506–25.

Wilde, W.H.A. (1962) Bionomics of the pear psylla, *Psylla pyricola* Foerster in pear orchards of the Kootenay valley of British Columbia, 1960. *Can. Entomol.* **94**, 845–9.

Wilkinson, P.R. (1961) The use of sampling methods in studies of the distribution of larvae of *Boophilus microplus* on pastures. *Aust. J. Zool.* **9**, 752–83.

Williams, B.G., Naylor, E., & Chatterton, T.D. (1985) The activity patterns of New Zealand mud crabs under field and laboratory conditions. *J. Exp. Mar. Biol. Ecol.* **89**(2-3), 269–82.

Williams, C.B. (1939) An analysis of four years captures of insects in a light trap. Part I. General survey; sex proportion; phonology and time of flight. *Trans. R. Entomol. Soc. Lond.* **89**, 79–132.

Williams, C.B. (1940) An analysis of four years captures of insects in a light trap. Part II. The effect of weather conditions on insect activity; and the estimation and forecasting of changes in the insect population. *Trans. R. Entomol. Soc. Lond.* **90**, 228–306.

Williams, C.B. (1948) The Rothamsted light trap. *Proc. R. Entomol Soc. Lond. A* **23**, 80–85.

Williams, C.B. (1951) Comparing the efficiency of insect traps. *Bull. Entomol. Res.* **42**, 513–17.

Williams, C.B., French, R.A., & Hosni, M.M. (1955) A second experiment on testing the relative efficiency of insect traps. *Bull. Entomol. Res.* **46**, 193–204.

Williams, C.B., Sincih, B.P., & El Ziady, S. (1956) An investigation into the possible effects of moonlight on the activity of insects in the field. *Proc. R. Entomol. Soc. Lond. A* **319**, 135–44.

Williams, D.F. (1973) Sticky traps for sampling populations of *Stomoxys calcitrans. J. Econ. Entomol.* **66**, 1279–80.

Williams, G. (1958) Mechanical time-sorting of pitfall captures. *J. Anim. Ecol.* **27**, 27–35.

Williams, D.T., Straw, N., Townsend, M., Wilkinson, A.S., & Mullins, A. (2013) Monitoring oak processionary moth *Thaumetopoea processionea* L. using pheromone traps: the influence of pheromone lure source, trap design and height above the ground on capture rates. *Agric. Forest Entomol.* **15**, 126–34.

Williams, L., Schotzko, D.J., & McCaffrey, J.P. (1992) Geostatistical description of the spatial distribution of *Limonius californicus* (Coleoptera: Elateridae) wireworms in the Northwestern United States with comments on sampling. *Environ. Entomol.* **21**, 983–95.

Wilson, B.H. & Richardson, C.C. (1970) Attraction of deer flies (Chrysops) (Diptera: Tabanidae) to traps baited with dry ice under field conditions in Louisiana. *J. Med. Entomol.* **7**, 625.

Wilton, D.P. & Fay, R.W. (1972) Air-flow direction and velocity in light trap design. *Entomol. Exp. Appl.* **15**, 377–86.

Witz, J.A., Lopez, J.D.J., & Latheef, M.A. (1992) Field density estimates of *Heliothis virescens* (Lepidoptera: Noctuidae) from catches in sex pheromone-baited traps. *Bull. Entomol. Res.* **82**(2), 281–6.

Wolf, W.W., Westbrook, J.K., Raulston, J., Pair, S.D., & Hobbs, S.E. (1990) Recent airborne radar observation of migrant pests in the USA. *Philos. Trans. R. Soc. Lond. B* **328**(1251), 619–30.

Woodman, N., Slade, N.A., Timm, R.M., & Schmidt, C.A. (1995) Mammalian community structure in lowland, tropical Peru, as determined by removal trapping. *Zool. J. Linn. Soc.* **113**(1), 1–20.

Worth, C.B. & Jonkers, A.H. (1962) Two traps for mosquitoes attracted to small vertebrate animals. *Mosquito News* **22**, 18–21.

Xu, X., Wu, J., Wu, X., Li, D., & Rao, Q. (1996) A study on ground-penetrating radar exploration of subterranean termites nests in dykes and dams. *Acta Entomol. Sinica* **39**(1), 46–52.

Yoshimoto, C.M. & Gressitt, J.L. (1959) Trapping of airborne insects on ships on the Pacific (Part II). *Proc. Hawaiian Entomol. Soc.* **17**(1), 150–5.

Yoshimoto, C.M., Gressitt, J.L., & Wolff, T. (1962) Airborne insects from the Galathea expedition. *Pacific Insects* **4**, 269–91.

Zheng, L.X., Zheng, Y., Wu, W.J., & Fu, Y.G. (2014) Field evaluation of different wavelengths light-emitting diodes as attractants for adult *Aleurodicus dispersus* Russell (Hemiptera: Aleyrodidae). *Neotrop. Entomol.* **43**, 409–14.

Zhogolev, D.T. (1959) Light-traps as a method for collecting and studying the insect vectors of disease organisms. [In Russian.]. *Ent. Obozr.* **38**, 766–73.

Zigler, S.J. & Dewey, M.R. (1995) Phototaxis of larval and juvenile northern pike. *N. Am. J. Fish. Manag.* **15**(3), 651–3.

Zippin, C. (1956) An evaluation of the removal method of estimating animal populations. *Biometrics* **12**, 163–89.

Zippin, C. (1958) The removal method of population estimation. *J. Wildlife Manag.* **22**, 82–90.

Zoebisch, T.G., Stimac, J.L., & Schuster, D.J. (1993) Methods for estimating adult densities of *Liriomyza trifolii* (Diptera: Agromyzidae) in staked tomato fields. *J. Econ. Entomol.* **86**(2), 523–8.

8 Estimates of Species Richness and Population Size Based on Signs, Products and Effects

Comparative surveys of species richness for some animal groups can be undertaken by surveying signs or products such as footprints, faeces, nests, burrows or cast skins. Measures of the size of populations based on the magnitude of their products or effects are often referred to as population indices. The relationship of these indices to the absolute population varies from equivalence when, for example, the number of exuviae are counted, to no more than an approximate correlation, when the index is obtained from general measures of damage.

8.1 Arthropod products

8.1.1 Exuviae

The larval or pupal exuviae of insects with aquatic larval stages are often left in conspicuous positions around the edges of water bodies, and where it is possible to gather these they will provide a measure of range, species diversity, emergence rate and of the absolute population of newly emerged adults. The method is most easily applied to large insects such as dragonflies (Corbet, 1957; Wiesmath, 1989; Hartnoll & Bryant, 1990; Pollard & Berrill, 1992; Kielb et al., 1996). Raebel et al. (2010) consider dragonfly exuviae surveys the best method to assess the habitat quality of ponds and their odonate biodiversity as they demonstrate the completion of the life-cycle at the site. Bried et al. (2012) noted that dragonfly exuviae surveys for biodiversity tend to be biased low because of inefficient detection.

Surveys of chironomid communities in lakes, streams and rivers have frequently been undertaken by collecting floating pupal exuviae (Fend & Carter, 1995; Kownacki, 1995; Franquet & Pont, 1996; Cranston et al., 1997; Ruse, 2010). In some habitats, such as the Everglades National Park, identification keys for chironomid exuviae have been prepared (Jacobsen, 2008). Collection from the surface gives data on recent emergence. To produce a species list integrated over space and time, surface sediment sampling for fossil chironomid remains can be undertaken (Luoto, 2011).

The exuviae of the last immature, subterranean stage of some insects are also often conspicuous, and with cicadas Strandine (1940) and Dybas & Lloyd (1962, 1974) found

Ecological Methods, Fourth Edition. P. A. Henderson and T. R. E. Southwood.
© 2016 John Wiley & Sons, Ltd. Published 2016 by John Wiley & Sons, Ltd.
Companion Website: www.wiley.com/go/henderson/ecologicalmethods

that the number of larval cases per unit area was correlated with the number of emergence holes and provided a useful index of population; different species could be recognised from the exuviae. The spatiotemporal distribution and biodiversity of seven cicadas in tropical East Asian forest and plantations was studied using exuviae in trees (Lee *et al.*, 2010)

Paramonov (1959) drew attention to the possibility of obtaining an index of the population of arboreal insects from their exuviae, more particularly from the head capsules of lepidopterous larvae that may be collected in the same way as frass (see below). He found that the head capsules could be separated from the other debris by flotation. In most species, the stage of the larva can be determined from the size of the capsule, and if all the capsules could be collected an elegant measurement of absolute population would be obtained. However, a significant, and probably fluctuating, number fails to be recovered and this variable needs to be carefully investigated in a given situation before the method is employed to measure either absolute population or even an index of it. Higashiura (1987) estimated absolute larval density of each stage of a larval population of the gypsy moth, *Lymantria dispar* using head-capsule collections sorted by hand and produced an age-specific life table. He concluded that the results were similar to those based on frass collection. Pupal exuviae on the branches and trunks of peach trees were used to provide relative estimates of the moth, *Synanthedon*, which accorded with estimates made with pheromone traps. Yonce *et al.* (1977) and Floater (1996) used exuviae and frass of processionary caterpillars, *Ochrogaster lunifer*, left at the foot of trees to estimate the number of individuals.

8.1.2   Frass

The faeces of insects are generally referred to as frass; as this is based on the incorrect use of a German word, the term feculae has been suggested (Frost, 1928), but seldom used. (The German word frass = insect damage or food, whilst *kot* = faeces.) Weiss (2006) reviews insect defecation behaviour and the use of frass by ecologists.

The frass-drop, the number of frass pellets falling to the ground, was first used as an index of both population and insect damage by a number of forest entomologists in Germany (Rhumbler,1929; Gösswald, 1935). The falling frass is collected in cloth or wooden trays or funnels under the trees (Fig. 8.1). In order for such collections to be of maximum value for population estimation, one should be able to identify the species and the developmental stage. Information is also required on the quantity of frass produced per individual per unit of time and the proportion of this that falls to the ground and is collected.

8.1.2.1   Identification

Although the frass of early instars of several species may be confused, in the later instars it is generally distinctive and keys for the identification of frass pellets have been prepared (Morris, 1942; Weiss & Boyd, 1950, 1952; Hodson & Brooks, 1956; Solomon, 1977). With the nun moth, *Lymantria monacha*, Eckstein (1938) concluded the frass size was a more reliable indicator of larval instar than the width of the head capsule; in the sawfly

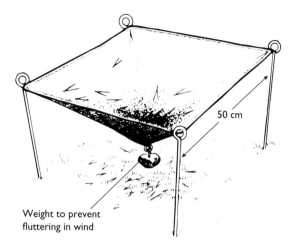

Figure 8.1   Frass collector, constructed of cloth. (Adapted from Tinbergen, 1960.)

Diprion hercyniae (Morris, 1949), moth *Hypsipyla robusta* (Mo & Tanton, 1995) and the armyworm *Pseudaletia unipuncta* (Pond, 1961). the instar can easily be determined from pellet size. Water-soluble constituents are washed out when the pellets are exposed to rain, and so pellet weight may fall by as much as 30%, though in these species the volume is not affected. Bean (1959) considered that in the spruce budworm, *Choristoneura fumiferana*, the width of the pellet rather than its volume was closely correlated with larval instar; furthermore, this author also found that larvae, feeding on pollen, ejected frass pellets that were considerably larger than those produced by the same instar larvae feeding on foliage.

8.1.2.2   The rate of frass production

This can be measured in terms of dry weight or number of pellets, but the two may not vary in direct proportion (Fig. 8.2a). Whatever units are used, the rate of frass production is affected by many factors: the phase and generation (Fridén, 1958; Iwao, 1962), the temperature and humidity (Morris, 1949; Green & De Freitas, 1955; Pond, 1961; Liebhold & Elkinton, 1988; Kamata & Yanbe, 1994), the available food, photoperiod (Kamata & Yanbe, 1994), both the plant species and its condition (Morris, 1949; Green & de Freitas, 1955; Pond, 1961; Dadd, 1960; Waldbauer, 1964), the diel periodicity of larval feeding (Liebhold & Elkinton, 1988), the developmental rate as determined by the phonology of the season (Tinbergen, 1960), and the presence of adult parasites (Green & de Freitas, 1955) (Fig. 8.2). Therefore, it is not possible to assume that the number of pellets produced in different areas or different seasons reflects very precisely changes in the actual populations. Direct comparisons should be made with other independent measurements of the population. Tinbergen (1960) found that in six out of seven years frass-drop provided a reliable index of the absolute population. Zandt (1994) compared frass production with direct counting from sampled branches and water basin traps as

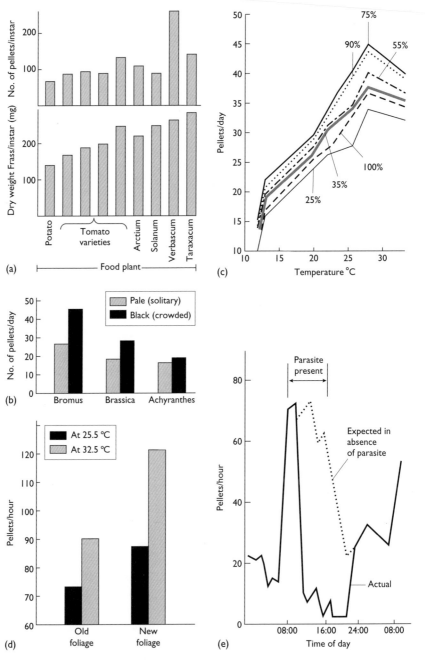

Figure 8.2 Variations in frass-drop rate with different factors. (a) Influence of food plant on the weight and number of pellets per instar from *Protoparce sexta* (Sphingidae). (Data from Waldbauer, 1964); (b) Influence of phase and foodplant on the number of pellets per day from *Leucania separate* (Noctuidae). (Data from Iwao, 1962); (c) Influence of temperature and humidity on the number of pellets per day from *Dendrolimus pini* (Lasiocampidae). (Data from Gösswald, 1935); (d) Influence of temperature and condition of foliage on the number of pellets per hour from 20 larvae of the sawfly, *Neodiprion lecontei* (Diprionidae). (Data from Green & de Freitas, 1955); (e) Influence of the presence of a dipterous parasite on the number of pellets per hour from 20 larvae of the sawfly, *N. lecontei*. (Data from Green & de Freitas, 1955.)

techniques for measuring population size of sawfly larvae. He concluded that, while frass production could successfully estimate mean caterpillar intensity per tree, the sampling error was too great to estimate the population intensity in a particular tree. When methods are found to give different estimates this must be investigated further to determine whether the absolute method has become inaccurate or if the frass-drop rate per individual has changed. The latter can be measured by the use of a 'coprometer' that collects the frass voided each hour into a separate compartment (Green & Henson, 1953; Green & de Freitas, 1955).

8.1.2.3 Efficiency of collection

Another variable is the proportion of the frass that actually reaches the ground and is collected. As Morris (1949) points out, a greater proportion of the frass will be retained, on the foliage, in calm weather than under more windy conditions, and where the young larvae produce webbing the frass will tend to be caught up with this and its fall delayed (Bean, 1959).

Most of the above work has referred to Lepidoptera and sawflies; the frass of Coleoptera may also be identified, and Eckstein (1939) and Campbell (1960) used frass-drop as an index of population in studies of stick insects, Phasmida.

It can be concluded that frass-drop measurements may, with caution, be used as a second method to check trends established by another technique (see p. 4). If comparisons are limited to a certain area where the insect has but a single host plant, the chances that a good relative estimate will be obtained are maximal.

8.1.2.4 Other products

Populations of web-building spiders can be estimated by counting the number of webs and the visibility of the webs increased by dusting with lycopodium powder (Cherrett, 1964), or by spraying with a fine mist of water from a knapsack sprayer. Webs over a large area can be rapidly assessed using the latter method.

Indices of the populations of colonial nest-building caterpillars have been obtained by counting the number of nests, rather than the caterpillars; the great advantage of this approach lies in extensive work, as it enables the population level to be measured over many acres (Tothill, 1922; Legner & Oatman, 1962; Morris & Bennett, 1967; Ito *et al.*, 1970). Some mites also produce webbing and this may be used as an index (Newcomer, 1943). Cercopid larvae may be estimated by counting the spittle masses, but this is less accurate for the first instars (Whittaker, 1971).

8.2 Vertebrate products and effects

Bird and reptile and mammal abundance can be estimated from counting nests or burrows. The main problem is to ensure that the structure is occupied. Nest counting is particularly appropriate for sea birds and other colonial breeders that synchronise their reproduction. The census techniques to be used will vary with the habitat, and for sea birds are discussed by Walsh *et al.* (1995).

Bird and mammal numbers may be estimated by counting faeces. For many mammals their dung is much more conspicuous than the animal. An interesting example is the Amazonian manatee, which is extremely shy and difficult to spot at the water surface. However, it produces floating dung that is characteristic and easily spotted. Generally, a simple count of the droppings in a known area will give an index of abundance over the indeterminate period over which the droppings have remained intact. As shown by Kuehl *et al.* (2007) for gorilla dung, the decay rate can vary enormously between locations. Rainfall and other physical variables will greatly change this time period, as will biotic factors such as the abundance of dung beetles. There are two ways to overcome this problem, either remove all the droppings from the study area prior to starting the census, or to mark some fresh dung and subsequently never count anything which looks older. Animal droppings are often highly clumped in their distribution and may conform to a negative binomial distribution (p. 28); statistical methods to compare densities are discussed by White & Eberhardt (1980).

The presence of mammals such as squirrels can be detected by the characteristic tooth marks they leave on the remains of their food, such as nuts. A wide variety of feeding signs have been used in mammalian studies. An unusual example is the use by Turner (1975) of bite marks on cattle as an index of vampire bat, *Desmodus rotundus*, abundance.

Mammal density has also been monitored using hair samplers which snag hair from a passing animal. These can work in similar fashion to the way a barbed wire fence catches the wool of sheep, or the hair may be retained on sticky tape placed inside a tube placed where the animals will use it as a runway (Suckling, 1978). These devices are probably only useful for demonstrating the presence of a species at a locality as the quantity gathered will reflect many variables other than population size. DNA identification techniques now make it practicable to identify an individual from a single hair.

Footprints and runways may also be used. Mooty & Karns (1984) showed that the density of white-tailed deer inferred from tracks was correlated with the density inferred from pellet counts. Whatever type of sign is used, the data can be analysed in two principal ways. The density can be expressed as counts per area of search, or alternatively signs can be used instead of direct animal observations in distance sampling methods (see Chapter 9). In Russia, a relative index of abundance for game species is generated by counting the tracks in snow which intersect standard transects. These relative indices have been used to estimate absolute density using the Formozov–Malyshev–Pereleshin formula (Stephens *et al.*, 2006). The precision of this method is known to vary with animal density and sampling effort, and it is known to produce poor results in low-density populations.

8.3 Effects due to an individual insect

These are of great value to the ecologist, as they are immediately convertible into an absolute estimate. When, at a given stage of the life-cycle, each animal has some unique effect (for example, the commencement of a leaf mine), a count of these – after all the animals have passed this stage – will provide a precise measure of the total number passing through that stage. Counts of the actual animals would need to be integrated through time to provide such a total (see p. 430). These effects are therefore measures of the number of animals entering a stage, the estimation of which is discussed in Chapter 11.

Cicada larvae construct conspicuous and distinctive turreted emergence holes which may be counted to measure the total emerging population (Dybas & Davis, 1962). Populations of solitary Hymenoptera can be assessed from the number of nest holes (Bohart & Lieberman, 1949); this will give the number of reproducing females, but in some species a female will construct more than one nest.

Measures of absolute population can be obtained for plant feeders when they enter the plant. Leaf-miners allow particularly elegant studies, since at the end of a generation it is possible to determine the number of larvae that commenced mining, the number that completed development, and often some indication of the age and cause of death in those that failed (Askew, 1968; Pottinger & Le Roux, 1971; Leroi, 1972, 1974; Payne *et al.*, 1972). When the larva cuts characteristic cases at different ages even more information may be gathered (De Gryse, 1934) and, because of the 'historical' record of spatial position precise studies of interspecific competition are possible (Murai, 1974). Leaf-miner abundance has been studied at large spatial scales; for example, Kozlov *et al.* (2013) studied latitudinal variation in birch-feeding leaf miners in Northern Europe. The web site of W. N. Ellis, Zoölogisch Museum Amsterdam, entitled 'Leafminers and plant galls of Europe' (at http://www.bladmineerders.nl/index.htm) gives extensive information on leaf miners and galls and their identification.

Galls are easily counted, but many harbour a variable number of insects and it is usually impossible to determine externally whether these have been parasitised or not (Bess & Haramoto, 1959; Nakamura *et al.*, 1964; Howse & Dimond, 1965; Redfern, 1968; Redfern & Cameron, 1993).

Some stem-borers cause the growing shoot to die; when multiple invasion is sufficiently rare to be overlooked estimates of these 'dead hearts' may be taken as equivalent to the total number of larvae invading, and the same approach can be applied to insects in grains or seeds (Jepson & Southwood, 1958; Gomez & Bernardo, 1974). Oviposition punctures have been found to provide an accurate index of the number of eggs per plant unit for the weevil, *Hypera postica* (Harcourt *et al.*, 1974).

8.4 General effects: plant damage

The effect of an insect population on a plant stand is the product of two opposing processes: the eating rate of the insects, and the growing rate of the plant. The eating rate of the insects will depend on their age, the temperature conditions and the other factors that affect frass production (see above), as well as on population size. An example of the complex interaction of these processes for various cruciferous plants and two different pests is given by Taylor & Bardner (1968a,b). Therefore, measures of damage are only approximate indices of population size, although they may have advantages over other methods (e.g. Johnson & Burge, 1971). Damage *per se* is, of course, an important parameter for many entomologists; an outline of its assessment will be given.

8.4.1 Criteria

8.4.1.1 Economic damage

This measure is the one least related to the population size and describes the effects of the animals in economic terms; it is of considerable importance to the agriculturist and forester as it indicates the need for control measures, which become desirable once the economic threshold is reached. The economic threshold is defined by Stern *et al.* (1959) as '…the density at which control measures should be determined to prevent an increasing pest population from reaching the economic injury level'. Changing economic conditions will alter the economic injury level from season to season, and its relation to the size of the insect population will vary greatly from species to species. Pests that cause blemishes to fruits or destroy the leading shoots of growing trees cause damage whose economic level is very high relative to the amount of plant material destroyed (Smith, 1969; Southwood & Norton, 1973; Stern, 1973).

8.4.1.2 Loss of yield

This is a measure of the extent to which the weight, volume or number of the marketable parts of the plant is changed; normally reduced, but occasionally increased (Harris, 1974). Although more of a biological measure than economic damage, variations in yield may be caused by many factors other than pest numbers (Möllerström, 1963). Furthermore, the same number of insects attacking a plant can have very different effects depending on the timing of the attack in relation to the growth stages of the plant. For example, the larva of the frit fly invades and kills a young shoot of the oat plant; early in the season the plant will have time to develop another tiller that will contribute towards the yield of grain, and so the latter is not affected. Later in the season any tillers destroyed would have contributed towards the crop, but there is insufficient growing time for the tillers that replace them to ripen grain; this is the time of the maximum effect of attack on yield. At a still later time, any tillers that are young enough to be susceptible to the frit fly are so far behind the bulk of the crop that their contribution to the eventual yield would be insignificant. If the frit fly attacks such tillers, yield will not be reduced; it might even be increased. Prasad (1961) studied the reduction in weight of cabbage heads following the artificial colonisation by newly emerged larvae of the small white cabbage butterfly, *Pieris rapae*, at various dates after transplantation. It was found (Fig. 8.3) that the greatest reduction in yield occurred when the larvae were introduced seven weeks after transplantation, as the leaves damaged at this time contribute to the head. On the other hand, attacks soon after transplanting affect the growth of the plant.

8.4.1.3 The amount of plant consumed

This is perhaps the most meaningful biological measure and could be expressed precisely in kilojoules as the amount of the primary production consumed by the insect

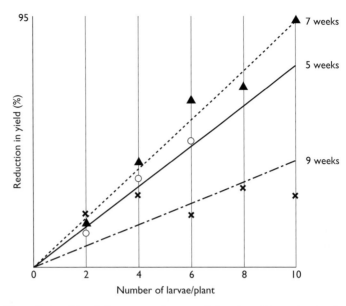

Figure 8.3   The relationship of damage to the age of the plant. The reduction in yield of cabbage plants when infested with different numbers of *Peris rapae* larvae at 5, 7 and 9 weeks after transplantation. (Data from Prasad, 1961.)

(see Chapter 14). The actual amount of damage may be assessed in terms of dry weight reduction (e.g. Ortman & Painter, 1960; Buntin, 1991) or leaf area destroyed (e.g. Abbott *et al.*, 1993). The area of the leaf may be measured electronically (Hatfield *et al.*, 1976) by using a flatbed scanner or digital photography with a computer; photocopying provides a rapid permanent record. The area of leaf blotching due to the removal of cell contents (by leafhoppers and mirids) may be estimated by photography or by colour scanning, followed by image analysis in which the area occupied by different colours is calculated by a computer. For example, Baur *et al.* (1990) quantified leaf area reduction in the grey alder, *Alnus incana*, by initially photographing leaves in the field against a white background. Subsequently, on the photographic prints the missing outlines of the leaves were re-constructed and these images entered into a computer using a video camera for analysis by an image-processing program. The authors estimated measurement error as <2% of the total leaf area and considered this method suitable for quantifying leaf mine size and necrotic tissue area. Mathur *et al.* (2013) quantified *Brassica juncea* leaf area using scanned images analysed with the public domain image processing program ImageJ, which can be downloaded from http://imagej.nih.gov/ij/.

It is often easier to measure some index of the amount of plant destroyed rather than the actual quantity. For example, Coaker (1957) used various degrees of battering of the leaf margin of cotton as an index of mirid damage; Dimond & Allen (1974) found the length of pine needles closely correlated with the numbers of first instar adelgid, *Pineus* larvae; and Coombs (1963) assessed the level of damage in stored grain from the weight of fine dust present in a given volume of grain. Nuckols &

Connor (1995) visually examined each sampled leaf for damage caused by chewing herbivores, skeletonisers, sap-feeders, leaf-miners and gall formers and scored the effect in terms of the percentage of the leaf area affected. The degree of defoliation can also be measured by the amount of light passing through the canopy (Higley, 1992).

Estimates of the quantity of leaf consumed may be confused by the subsequent enlargement of the hole by the growth of the leaf. Reichle *et al.* (1973) allowed for this by studying the expansion of standard holes punched in young leaves, and Lowman (1987) showed for a range of tropical forest trees that leaf holes expand proportionately with leaf growth.

Remote sensing techniques (see Chapter 15) using optical and thermal imaging from aircraft provides a method of surveying insect damage over large areas, generally in forests (Wear *et al.*, 1964; Klein, 1973; Riley, 1989), but also in orchards and crops (Stern, 1973; Harris *et al.*, 1976; Wallen *et al.*, 1976; Riley, 1989). Infestations are often detected by their secondary effects; for example, corn leaf aphids, *Rhopalosiphum maidis*, are detected photographically by the sooty mould which grows on the honeydew (Wallen *et al.*, 1976).

Satellite images have recently started to be applied for the surveying of plant damage by insects. Meigs *et al.* (2011) discuss the use of Landsat time series in conjunction with aerial surveys and field measurements to characterise the impacts on North American conifer forest of bark beetles and insect defoliators. Bjerke *et al.* (2014) report on plant damage in the Nordic Arctic linked to climate and pest outbreaks which were quantified using moderate resolution imaging spectroradiometer data from the MODIS Terra satellite.

Insect defoliation of trees affects the growth rate and is reflected in the annual growth rings (Lessard & Buffam, 1976; Creber, 1977), whilst aphids may also effect the time of leaf fall and other characters (Dixon, 1971). Considering more particularly the effects of defoliates on trees, Henson & Stark (1959) proposed that insect populations could be defined as:

1. Tolerable – Populations that do not utilise the entire excess biological productivity of the host; that is, the insect and the host plant populations could continue at this level indefinitely.
2. Critical – 'Populations that utilise more than the excess biological productivity of the hosts, but less than the total productivity'. Such population levels cannot be continued indefinitely and Henson & Stark (1959) (Fig. 8.4) and Churchill *et al.* (1964) show the long-term effects of insect attack on perennial hosts.
3. Intolerable – 'Populations that are depleting the host at a rate greater than the current rate of production'.

The condition of the host tree (expressed as the number of leaf needles per branch tip) and the history of attack in previous years will interact with the intensity of larval population (larvae per tip) in the determination of the appropriate description of population in these terms (Fig. 8.4).

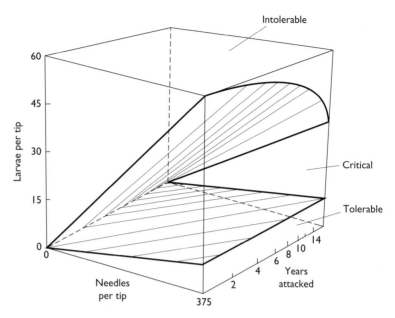

Figure 8.4 The relationship of tree condition, insect population intensity, and previous history to intolerable, critical and tolerable populations. (Data from Henson & Stark, 1959.)

8.5 Determining the relationship between damage and insect populations

As has been indicated above, the level of damage a plant suffers is influenced by many factors, such as soil, climate, age and health of the plant, as well as the size of the insect population. Therefore, attempts to correlate insect numbers to damage when these variables are measured in different areas and seasons are seldom satisfactory. However, it may be possible to separate the roles of the different factors by a carefully planned sampling programme emphasising the comparison of nearby fields (Sen & Chakrabarty, 1964). The comparisons will be even more exact if they are made in the same field, as then climatic and soil effects will be identical. Theoretically, there are four techniques for making such comparisons:

1. The introduction of a known number of animals on to pest-free plants (e.g. Andrzejewska & Wojcik, 1970; Ogunlana & Pedigo, 1974; Harrison & Maron, 1995).
2. The exclusion of the animals from certain plots by mechanical barriers.
3. The exclusion of the animals or the reduction of their populations in some plots by the use of pesticides.
4. A comparison of the growth of individual plants attacked by the pest compared with unattached plants growing in the same crop (Bardner, 1968).

Methods 1 and 2 both suffer from the fact that the barriers erected to exclude or retain the animals will severely modify the climate and growth conditions of the plant; the exclusion technique is of course widely used in studies on the effects of grazing by vertebrates.

The third method has been used successfully by Dahms & Wood (1957) and Jepson (1959). Working on the frit fly, *Oscinella frit*, the last-named author found that a detailed knowledge of the periods of oviposition of the pest allowed frequent 'blanket spraying' to be replaced by a limited number of timed sprays or granular applications (Jepson & Mathias, 1960).

References

Abbott, I., Van Heurck, P., Burbidge, T., & Williams, M. (1993) Damage caused by insects and fungi to eucalypt foliage: Spatial and temporal patterns in Mediterranean forest of Western Australia. *Forest Ecol. Manag.* **58**(1-2), 85–110.

Andrzejewska, L. & Wojcik, Z. (1970) The influence of Acridoidea on the primary production of a meadow (field experiment). *Ekol. Polska* **18**, 89–109.

Askew, R.R. (1968) A survey of leaf-miners and their parasites on laburnum. *Trans. Roy. Entomol. Soc. Lond.* **120**, 1–37.

Bardner, R. (1968) Wheat bulb fly, *Leptohylemyia coarctata* Fall. and its effect on the growth and yield of wheat. *Ann. Appl. Biol.* **61**, 1–11.

Baur, R., Fritsche, A., Camenzind, R., & Benz, G. (1990) Quantitative analysis of leaf area loss caused by insects: A method combining photography and electronic image processing. *J. Appl. Entomol.* **109**(2), 182–8.

Bean, J.L. (1959) Frass size as an indicator of spruce budworm larval instars. *Ann. Entomol. Soc. Am.* **52**, 605–8.

Bess, H.A. & Haramoto, F.H. (1959) Biological control of Pamakani, *Eupatorium adenophorum*, in Hawaii by a tephritid gall fly, *Proc. ecidochares utilis*. 2. Population studies of the weed, the fly and the parasites of the fly. *Ecology* **40**, 244–9.

Bjerke, J.W., Karlsen, S.R., Høgda, K.A., Malnes, E., Jepsen, J.U., Lovibond, S., Vikhamar-Schuler, D., & Tømmervik, H. (2014) Record-low primary productivity and high plant damage in the Nordic Arctic Region in 2012 caused by multiple weather events and pest outbreaks. *Environ. Res. Lett.* **9**, 084006.

Bohart, G.E. & Lieberman, F.V. (1949) Effect of an experimental field application of DDT dust on *Nomia melanderi*. *J. Econ. Entomol.* **42**, 519–22.

Bried, J.T., D'Amico, F., & Samways, M.J. (2012) A critique of the dragonfly delusion hypothesis: why sampling exuviae does not avoid bias. *Insect Conserv. Divers.* **5**, 398–402.

Buntin, G.D. (1991) Effect of insect damage on the growth, yield, and quality of sericea lespedeza forage. *J. Econ. Entomol.* **84**(1), 277–84.

Campbell, K.G. (1960) Preliminary studies on population estimation of two species of stick insects (Phasmatidae) occurring in plague numbers in Highland Forest areas of southeastern Australia. *Proc. Linn. Soc. N. S. W.* **85**, 121–37.

Cherrett, J.M. (1964) The distribution of spiders on the Moor House National Nature Reserve, Westmorland. *J. Anim. Ecol.* **33**, 27–48.

Churchill, G.B., John, H.H., Duncan, D.P. & Hodson, A.C. (1964) Long-term effects of defoliation of aspen by the forest tent caterpillar. *Ecology* **45**, 630–3.

Coaker, T.H. (1957) Studies of crop loss following insect attack on cotton in East Africa. II. Further experiments in Uganda. *Bull. Entomol. Res.* **48**, 851–66.

Coombs, C.W. (1963) A method of assessing the physical condition of insect-damaged grain and its application to a faunistic survey. *Bull. Entomol. Res.* **54**, 23–35.

Corbet, P.S. (1957) The life-history of the Emperor dragonfly, *Anax imperator* Leach (Odonata: Aeshnidae). *J. Anim. Ecol.* **26**, 1–69.

Cranston, P.S., Cooper, P.D., Hardwick, R.A., Humphrey, C.L., & Dostine, P.L. (1997) Tropical acid streams: The chironomid (Diptera) response in northern Australia. *Freshwater Biol.* **37**(2), 473–83.

Creber, G.T. (1977) Tree rings: a natural data storage system. *Biol. Rev.* **52**, 349–82.

Dadd, R.H. (1960) Observations on the palatability and utilisation of food by locusts, with particular reference to the interpretation of performances in growth trials using synthetic diets. *Entomol. Exp. Appl.* **3**, 283–304.

Dahms, R.G. & Wood, E.D. (1957) Evaluation of green bug damage to small grains. *J. Econ. Entomol.* **50**, 443–6.

De Gryse, J.J. (1934) Quantitative methods in the study of forest insects. *Sci. Agr.* **14**, 477–95.

Dimond, J.B. & Allen, D.C. (1974) Sampling populations of pine leaf chermid *Pineus pinifoliae* (Homoptera: Chermidae). III. Neosisterites on white pine. *Can. Entomol.* **106**, 509–18.

Dixon, A.F.G. (1971) The role of aphids in wood formation II. The effect of the lime aphid, *Eucallipterus tiliae* L. (Aphididae), on the growth of lime, *Tilia* × *vulgaris* Hayne. *J. Appl. Ecol.* **8**, 393–9.

Dybas, H.S. & Davis, D.D. (1962) A population census of seventeen-year periodical Cicadas (Homoptera: Cicadidae: Magicicada). *Ecology* **43**, 432–59.

Dybas, H.S. & Lloyd, M. (1962) Isolation by habitat in two synchronized species of periodical cicadas (Homoptera: Cicadidae: Magicicada). *Ecology* **43**, 444–59.

Dybas, H.S. & Lloyd, M. (1974) The habitats of 17-year periodical cicadas (Homoptera: Cicadidae: Magicicada spp.). *Ecol. Monogr.* **44**, 279–324.

Eckstein, K. (1938) Die Bewertung des Kotes der Nonnenraupe, *Psilura monacha* L., als Grundlage fiir die Festellung ihres Auftretens und zu ergreifenden Massregeln. *Allgem. Forst. Jagdztg.* **114**, 132–48.

Eckstein, K. (1939) Das Bohrmehl des Waldgärtners, *Myelophilus pimperda* L., nebst Bemerkungen über den 'Frass' der Borkenkäfer und anderen Insekten. *Arb. Physiol. Angew. Entomol.* **6**, 32–41.

Fend, S.V. & Carter, J.L. (1995) The relationship of habitat characteristics to the distribution of Chironomidae (Diptera) as measured by pupal exuviae collections in a large river system. *J. Freshwater Ecol.* **10**(4), 343–59.

Floater, G.J. (1996) Estimating movement of the processionary caterpillar Ochrogaster lunifer Herrich-Schaffer (Lepidoptera: Thaumetopoeidae) between discrete resource patches. *Australian J. Entomol.* **35**(3), 279–83.

Franquet, E. & Pont, D. (1996) Pupal exuviae as descriptors of the chironomid (Diptera: Nematocera) communities of large rivers. *Archiv. fur Hydrobiologie* **138**(1), 77–98.

Fridén, F. (1958) *Frass-drop frequency in Lepidoptera.* (Almqvist & Wiksells Boktryckeri), Uppsala.

Frost, S.-W. (1928) Insect scatology. *Ann. Entomol. Soc. Am.* **21**, 35–46.

Gomez, K.-A. & Bernardo, R.C. (1974) Estimation of stem borer damage in rice fields. *J. Econ. Entomol.* **67**, 509–16.

Gösswald, K. (1935) Über die Frasstätigkeit von Forstschädlingen unter verschiedener Temperatur und Luftfeuch tigkeit und ihre praktische und physiologische Bedeutung. *I. Z. Angew. Entomol.* **21**, 183–7.

Green, G.W. & De Freitas, A.S. (1955) Frass drop studies of larvae of *Neodiprion americanus banksianae*, Rob. and *Neodiprion lecontii*, Fitch. *Can. Entomol.* **87**, 427–40.

Green, G.W. & Henson, W.R. (1953) A new type of coprometer for laboratory and field use. *Can. Entomol.* **85**, 227–30.

Harcourt, D.G., Mukerji, M.K., & Guppy, J.C. (1974) Estimation of egg populations of the alfalfa weevil, *Hypera postica* (Colcoptera: Curculionidae). *Can. Entomol.* **106**, 337–47.

Harris, M.K., Hart, W.G., Davis, M.R., Ingle, S.I., & Van Cleave, H.W. (1976) Aerial photography shows caterpillar infestation. *Pecan Q.* **10**, 12–18.

Harris, P. (1974) A possible explanation of plant yield increases following insect damage. *Agro-ecosystems* **1**, 219–25.

Harrison, S. & Maron, J.L. (1995) Impacts of defoliation by tussock moths (*Orgyia vetusta*) on the growth and reproduction of bush lupine (*Lapinus arboreus*). *Ecol. Entomol.* **20**, 223–9.

Hartnoll, R.G. & Bryant, A.D. (1990) Size-frequency distributions in decapod Crustacea: The quick, the dead, and the cast-offs. *J. Crustacean Biol.* **10**, 14–19.

Hatfield, J.L., Stanley, C.D., & Carlton, R.E. (1976) Evaluation of an electronic foliometer to measure leaf area in corn and soybeans. *Agron. J.* **68**, 434–6.

Henson, W.R. & Stark, R.W. (1959) The description of insect numbers. *J. Econ. Entomol.* **529**, 847–50.

Higashiura, Y. (1987) Larval densities and a life-table for the gypsy moth, *Lymantria dispar*, estimated using the head-capsule collection method. *Ecol. Entomol.* **12**(1), 25–30.

Higley, L.G. (1992) New understandings of soybean defoliation and their implications for pest management. In: Copping, L.G., Green, M.B., & Rees, T.R. (eds) *Pest Management of Soybean*. Elsevier, Amsterdam.

Hodson, A.C. & Brooks, M.A. (1956) The frass of certain defoliators of forest trees in the north central United States and Canada. *Can. Entomol.* **88**, 62–8.

Howse, G.M. & Dimond, J.B. (1965) Sampling populations of Pine leaf adelgid, *Pineus pinifoliae* (Fitch). I. The gall and associated insects. *Can. Entomol.* **97**, 952–61.

Ito, Y., Shibazaki, A., & Iwahashi, O. (1970) Biology of *Hyphantria cunea* Drury (Lepidoptera: Arctiidae) in Japan. XI. Results of road-survey. *Appl. Entomol. Zool.* **5**, 133–44.

Iwao, S. (1962) Studies on the phase variation and related phenomena in some Lepidopterous insects. *Mem. Coll. Agric. Kyoto Univ. (Ent. no. 12)* **84**, 1–80.

Jacobsen, R.E. (2008) *A Key to the Pupal Exuviae of the Midges (Diptera: Chironomidae) of Everglades National Park, Florida*. US Geological Survey.

Jepson, W.F. (1959) The effects of spray treatments on the infestation of the oat crop by the frit fly (*Oscinella frit* L.). *Ann. Appl. Biol.* **47**, 463–74.

Jepson, W.F. & Mathias, P. (1960) The control of frit fly, *Oscinella frit* (L.) in sweet corn (*Zea mays*) by Thirpet (*O,O*-diethyl *S*-ethylthio-methyl phosphorodithioate). *Bull. Entomol. Res.* **51**, 427–33.

Jepson, W.F. & Southwood, T.R.E. (1958) Population studies on *Oscinella frit* L. *Ann. Appl. Biol.* **46**, 465–74.

Johnson, C.G. & Burge, G.A. (1971) Field trials of anti-capsid insecticides on farmers cocoa in Ghana, 1956–60. 2. Effects of different insecticides compared by counting capsids, and capsid counting compared with counting the percentage of newly damaged trees. *Ghana. J. Agric. Sci.* **4**, 33–8.

Kamata, N. & Yanbe, T. (1994) Frass production of last instar larvae of the beech caterpillar, *Quadricalcarifera punctatella* (Motschulsky) (Lep., Notodontidae), and method of estimating their density. *J. Appl. Entomol.* **117**(1), 84–91.

Kielb, M.A., Bright, E., & O'Brien, M. F. (1996) Range extension of *Stylogomphus ablistylus* (Odonata: Gomphidae) for the Upper Peninsula of Michigan. *Great Lakes Entomologist* **29**(2), 87–8.

Klein, W.H. (1973) Beetle killed pine estimates. *Photogrammetric Eng.* **39**, 385–8.

Kownacki, A. (1995) The use of chironomid pupal exuviae for ecological characterization of the upper Vistula (southern Poland). *Acta Hydrobiol.* **37**(1), 41–50.

Kozlov, M.V., van Nieukerken, E.J., Zverev, V., & Zvereva, E.L. (2013) Abundance and diversity of birch-feeding leafminers along latitudinal gradients in northern Europe. *Ecography* **36**, 1138–49.

Kuehl, H.S., Todd, A., Boesch, C., & Walsh, P.D. (2007) Manipulating decay time for efficient large-mammal density estimation: gorillas and dung height. *Ecol. Appl.* **17**, 2403–14.

Lee, Y.F., Lin, Y.H., & Wu, S.H. (2010) Spatiotemporal variation in cicada diversity and distribution, and tree use by exuviating nymphs, in East Asian tropical reef-karst forests and forestry plantations. *Ann. Entomol. Soc. Am.* **103**, 216–26.

Legner, E.F. & Oatman, E.R. (1962) Sampling and distribution of summer eyespotted bud moth *Spilonota ocellana* (D. & S.) larvae and nests on apple trees. *Can. Entomol.* **94**, 1187–9.

Leroi, B. (1972) A study of natural populations of the celery leaf-miner *Philophylla heraclei* L. (Diptera Tephritidae). 1. Methods of counting of larval populations. *Res. Popul. Ecol.* **13**, 201–15.

Leroi, B. (1974) A study of natural populations of the celery leaf-miner, *Philophylla heraclei* 1. (Diptera, Tephritidae). 11. Importance of changes of mines for larval populations. *Res. Popul. Ecol.* **15**, 163–82.

Lessard, G. & Buffam, P.E. (1976) Effects of *Rhyacionia neomexicana* on height and radial growth in Ponderosa Pine reproduction. *J. Econ. Entomol.* **69**, 755–60.

Liebhold, A.M. & Elkinton, J.S. (1988) Estimating the density of larval gypsy moth, *Lymantria dispar* (Lepidoptera: Lymantriidae), using frass drop and frass production measurements: Sources of variation and sample size. *Environ. Entomol.* **17**(2), 385–90.

Lowman, M.D. (1987) Relationships between leaf growth and holes caused by herbivores. *Aust. J. Ecol.* **12**, 189–91.

Luoto, T.P. (2011) The relationship between water quality and chironomid distribution in Finland – a new assemblage-based tool for assessments of long-term nutrient dynamics. *Ecol. Indicators* **11**, 255–62.

Mathur, V., Tytgat, T.O., de Graaf, R.M., Kalia, V., Reddy, A.S., Vet, L.E., & van Dam, N.M. (2013) Dealing with double trouble: consequences of single and double herbivory in *Brassica juncea*. *Chemoecology*, **23**, 71–82.

Meigs, G.W., Kennedy, R.E., & Cohen, W.B. (2011) A Landsat time series approach to characterize bark beetle and defoliator impacts on tree mortality and surface fuels in conifer forests. *Remote Sensing of Environment* **115**, 3707–18.

Mo, J. & Tanton, M.T. (1995) Estimation of larval instars of *Hypsipyla robusta* Moore (Lepidoptera: Pyralidae) by larval frass widths. *Aust. Entomol.* **22**(2), 59–62.

Möllerström, G. (1963) Different kinds of injury to leaves of the sugar beets and their effect on yield. *Medd. Växtskyddsanst.* **12**, 299–309.

Mooty, J.J. & Karns, P.D. (1984) The relationship between white-tailed deer track counts and pellet-group surveys. *J. Wildlife Manag.* **48**, 275–9.

Morris, R.F. (1942) The use of frass in the identification of forest insect damage. *Can. Entomol.* **74**, 164–7.

Morris, R.F. (1949) Frass-drop measurement in studies of the European spruce sawfly. *Univ. Michigan Sch. Forestry and Conserv. Bull.* **12**, 1–58.

Morris, R.F. & Bennett, C.W. (1967) Seasonal population trends and census methods for *Hyphantria cunea*. *Can. Entomol.* **99**, 9–17.

Murai, M. (1974) Studies on the interference among larvae of the citrus leaf miner, *Phyllocristis citrella* Stainton (Lepidoptera: Phyllocristidae). *Res. Popul. Ecol.* **16**, 80–111.

Nakamura, M., Kondo, M., Ito, Y., Miyashita, K., & Nakamura, K. (1964) Population dynamics of the chestnut gall-wasp, *Dryocosmus buriphilus*. 1. Description of the survey station and the life histories of the gall wasp and its parasites. *Jap. J. Appl. Entomol. Zool.* **8**, 149–58.

Newcomer, E.J. (1943) Apparent control of the Pacific mite with xanthone. *J. Econ. Entomol.* **369**, 344–5.

Nuckols, M.S. & Connor, E.F. (1995) Do trees in urban or ornamental plantings receive more damage by insects than trees in natural forests? *Ecol. Entomol* **20**, 253–60.

Ogunlana, M.O. & Pedigo, P. (1974) Economic injury levels of the potato leafhopper on Soybeans in Iowa. *J. Econ. Entomol.* **67**, 29–32.

Ortman, E.E. & Painter, R.H. (1960) Quantitative measurements of damage by the greenbug, *Toxoptera graminum* to four wheat varieties. *J. Econ. Entomol.* **53**, 798–802.

Paramonov, A. (1959) A possible method of estimating larval numbers in tree crowns. *Entomol. Manag. Mag.* **95**, 82–3.

Payne, J.A., Tedders, W.L., Cosgrove, G.E., & Foard, D. (1972) Larval mine characteristics of four species of leaf-mining Lepidoptera in Pecan. *Ann. Entomol. Soc. Am.* **65**, 74–81.

Pollard, J.B. & Berrill, M. (1992) The distribution of dragonfly nymphs across a pH gradient in south-central Ontario lakes. *Can. J. Zool.* **70**(5), 878–85.

Pond, D.D. (1961) Frass studies of the armyworm, *Pseudalctia unipuncta*. *Ann. Entomol. Soc. Am.* **54**, 133–40.

Pottinger, R.P. & Le Roux, E.J. (1971) The biology and dynamics of *Lithocolletis blancardella* (Lepidoptera: Gracillariidae) on apple in Quebec. *Mem. Entomol. Soc. Canada* **77**, 1–437.

Prasad, S.K. (1961) Quantitative estimation of damage to cabbage by cabbage worm, *Pieris rapae* (Linn.). *Indian J. Entomol.* **23**, 54–61.

Raebel, E.M., Merckx, T., Riordan, P., Macdonald, D.W., & Thompson, D.J. (2010) The dragonfly delusion: why it is essential to sample exuviae to avoid biased surveys. *J. Insect Conserv.* **14**, 523–33.

Redfern, M. (1968) The natural history of spear thistle-heads. *Field Studies* **2**, 669–717.

Redfern, M. & Cameron, R.A.D. (1993) Population dynamics of the yew gall *Taxomyia taxi* and its chalcid parasitoids: a 24 year study. *Ecol. Entomol.* **18**, 365–78.

Reichle, D.E., Goldstein, R.A., Van Hook, R.I., & Dobson, G.J. (1973) Analysis of insect consumption in a forest canopy. *Ecology* **54**, 1076–84.

Rhumbler, L. (1929) Zur Begiftung des Kiefernspanners (*Bupalus piniarius* L.) in der Oberfbrsterei Hersfeld-Ost 1926. *Z. Anger. Entomol.* **15**, 137–58.

Riley, J.R. (1989) Remote sensing in entomology. *Annu. Rev. Entomol.* **34**, 1–247.

Sen, A.R. & Chakrabarty, R.P. (1964) Estimation of loss of crop from pests and diseases of tea from sample surveys. *Biometrics* **20**, 492–504.

Ruse, L. (2010) Classification of nutrient impact on lakes using the chironomid pupal exuvial technique. *Ecol. Indicators* **10**, 594–601.

Smith, R.F. (1969) The importance of economic injury levels in the development of integrated pest control programs. *Qual. Plant. Mater. Veg.* **17**, 81–92.

Solomon, J.D. (1977) Frass characteristics for identifying insect borers (Lepidoptera: Cossidae and Sesiidae; Coleoptera: Cerambyciidae) in living hardwoods. *Can. Entomol.* **109**, 295–303.

Southwood, T.R.E. & Norton, C.A. (1973) Economic aspects of pest management strategies and decisions. In: Geier, P.W. *et al.* (eds) *Insects: Studies in Population Management*. Ecological Society of Australia, Canberra, 168–184.

Stern, V.M. (1973) Economic thresholds. *Ann. Rev. Entomol.* **18**, 259–80.

Stern, V.M., Smith, R.F., Van Den Bosch, R., & Hagen, K.S. (1959) The integrated control concept. *Hilgardia* **29**, 81–101.

Strandine, E.J. (1940) A quantitative study of the periodical cicada with respect to soil of three forests. *Am. Midl. Nat.* **24**, 177–83.

Stephens, P.A., Zaumyslova, O.Y., Miquelle, D.G., Myslenkov, A.I., & Hayward, G.D. (2006) Estimating population density from indirect sign: track counts and the Formozov–Malyshev–Pereleshin formula. *Animal Conserv.* **9**, 339–48.

Suckling, G.C. (1978) A hair sampling tube for the detection of small mammals in trees. *Aust. Wildlife Res.* **5**, 249–52.

Taylor, W.E. & Bardner, R. (1968a) Effects of feeding by larvae of *Phaedon cochleariae* (F.) and *Plutella maulipennis* (Curt.) on the yield of radish and turnip plants. *Ann. Appl. Biol.* **62**, 249–54.

Taylor, W.E. & Bardner, R. (1968b) Leaf injury and food consumption by larvae of *Phaedon cochleariae* (Col. Chrysomelidae) and *Plutella maculipennis* (Lep. Plutellidae) feeding on turnip and radish. *Entomol. Exp. Appl.* **11**, 177–84.

Tinbergen, L. (1960) The natural control of insects in pinewoods. 1. Factors influencing the intensity of predation by song birds. *Arch. Neérl. Zool.* **13**, 266–343.

Tothill, J.D. (1922) The natural control of the fall webworm (*Hyphentia cunea* Drury) in Canada. *Bull. Can. Dept. Agric. 3 (n.s) (Ent. Bull.* **19**), 1–107.

Turner, D.C. (1975) *The Vampire Bat*. John Hopkins Press, Baltimore, Maryland.

Waldbauer, G.P. (1964) Quantitative relationships between the numbers of fecal pellets, fecal weights and the weight of food eaten by tobacco hornworms, *Protoparce sexta* (Johan.) (Lepidoptera: Sphingidae). *Entomol. Exp. Appl.* **7**, 310–14.

Wallen, V.R., Jackson, H.R., & MacDiarmid, S.W. (1976) Remote sensing of corn aphid infestation, 1974 (Hemiptera: Aphididae). *Can. Entomol.* **108**, 751–4.

Walsh, P.M., Halley, D.J., Harris, M.P., del Nevo, A., Sim, I.M., & Tasker, M.L. (1995) *Seabird Monitoring handbook for Britain and Ireland*. JNCC, Peterborough.

Wear, J.F., Pope, R.B., & Lauterbach, P.G. (1964) Estimating beetle-killed Douglas fir by photo and field plots. *J. Forest.* **62**, 309–15.

Weiss, H.B. & Boyd, W.M. (1950) Insect feculae I. *J. New York Entomol. Soc.* **58**, 154–68.

Weiss, H.B. & Boyd, W.M. (1952) Insect feculae II. *J. New York Entomol. Soc.* **60**, 25–30.

Weiss, M.R. (2006) Defecation behavior and ecology of insects. *Annu. Rev. Entomol.* **51**, 635–61.

White, G.C. & Eberhardt, L.E. (1980) Statistical analysis of deer and elk pellet group data. *J. Wildlife Manag.* **44**, 121–31.

Whittaker, J.B. (1971) Population changes in *Neophilaenus lineatus* (L.) (Homoptera: Cercopidae) in different parts of its range. *J. Anim. Ecol.* **40**, 425–43.

Wiesmath, I. (1989) Faunistic and ecological studies of damselflies and dragonflies (Odonata: Zygoptera and Anisoptera) on bodies of water in the Tubingen area (West Germany). *Jahreshefte Der Gesellschaft Fuer Naturkunde In Wuerttemberg* **144**(0), 297–314.

Yonce, C.E., Gentry, C.R., Tumlinson, J.H., Doolittle, R.E., Mitchell, E.R., & McLaughlin, I.R. (1977) Seasonal distribution of the lesser peach tree borer in Central Georgia as monitored by pupal skin counts and pheromone trapping techniques. *Environ. Entomol.* **6**, 203–6.

Zandt, H.S. (1994) A comparison of three sampling techniques to estimate the population size of caterpillars in trees. *Oecologia* **97**(3), 399–406.

9 Wildlife Population Estimates by Census and Distance Measuring Techniques

Counting the number of sightings forms the basis for estimating density for many animal groups. This is particularly the case for large or easily seen animals such as birds, large grassland mammals, whales, crocodillians and large, active insects such as butterflies. While it may be possible to count animals from a suitable vantage point or while moving along a transect, the count can only be converted to a density estimate if the area scanned can be estimated. This simple approach (see Chapter 4) is often difficult to undertake for two reasons: first, it may not be possible to estimate sufficiently accurately the area scanned; and second, not all of the animals present may have been spotted. Distance sampling methods have been developed to allow for these problems by assuming that the likelihood an individual will be observed will decline with distance in a mathematically definable way. The methods discussed here are also useful for small, sessile or slow-moving organisms such as barnacles, corals and molluscs, and some can also be applied to data collected non-visually. For example, a bird census may be based on bird song, or an electric fish survey on the detection of distinctive electrical signals. One problem, as can be the case with a visual survey, is to ensure that the same individual is counted only once. Distance sampling and nearest-neighbour methods give absolute estimates of animal density. These represent the fourth and final approach to absolute population estimation the others are mark–recapture (Chapter 3), removal trapping (Chapter 7) and direct counting of a unit volume or area of habitat (Chapters 4 to 6). We briefly introduce wildlife census methods here rather than with other direct counting methods because they are the natural starting point from which to introduce distance sampling methods.

By convention, a *census* is defined as the counting of *all* the individuals belonging to the group of interest within a defined area, and a *survey* when only a *proportion* are counted. When the study area can be divided into quadrates which can vary in size or shape – for example, a rectangular 10 ha area of forest or a 0.1 m² core or grab sample – then a census can be made of a randomly selected set and standard statistical theory used to estimate average density and its variance for the entire area. Survey methods such as distance sampling and nearest-neighbour aim to estimate density using observations on the distance between animals or from a selected line or point to the animals, and thus do not require the worker to accurately map out or define the sampling area. They are thus particularly appropriate for the estimation of population density

Ecological Methods, Fourth Edition. P. A. Henderson and T. R. E. Southwood.
© 2016 John Wiley & Sons, Ltd. Published 2016 by John Wiley & Sons, Ltd.
Companion Website: www.wiley.com/go/henderson/ecologicalmethods

for large animals living at low density in difficult-to-traverse habitat. For example, distance sampling using a Fourier series is frequently the method of choice for estimating population size of primates in neotropical forest (Buckland *et al.*, 2010). The animals are counted while walking along specially cut forest trails. The high density of the forest and the low density and mobility of the monkeys would make it futile to try to census a number of predefined quadrates

9.1 Census methods

If it proves possible to count all of the individuals, n, within a known area, a, then this is termed a census and the estimated density, D, is simply:

$$\hat{D} = \frac{n}{a}. \tag{9.1}$$

Counting often requires the observer to move over the census area, and thus favours the use of strip transects (long, thin quadrats) of length, L, and width, $2w$, along which the observer moves in a straight line. Strips have been surveyed by foot, car, boat, aircraft, drone, remote-operated vehicle or submarine (e.g. Starr *et al.*, 1996). The strip width must be determined at the outset; if this cannot be done, or animals may be missed, then line transect methods (see p. 360) should be used.

Aerial census methods, as described in detail by Norton-Griffiths (1978), are frequently used for wildlife surveys (e.g. Pinder, 1996). Remote sensing methods in general are discussed further in Chapter 15. The counts obtained suffer from error, which varies with the counting rate and bias because of the tendency of observers to undercount. LeResche & Rausch (1974), from a study of bias during aerial surveys of moose, *Alces alces*, concluded that this bias was sufficient to invalidate the method as a means of absolute population estimation. A survey of 17 studies on large mammals by Caughley (1974) found that the proportion of the population counted varied from 23% to 89%. Similarly, Redfern *et al.* (2002) considered visibility bias to be the primary source of error during aerial surveys of large herbivores in the Kruger National Park. One possible means to reduce or assess the bias is to undertake counts from photographs or video film. However, as Harris & Lloyd (1977) show for a study of sea bird colony size by aerial photography, estimates may still vary substantially between observers, and a tendency to undercount remains because not all nests are visible. Ways of handling the bias inherent in aerial surveys are discussed by Caughley (1974) and Caughley *et al.* (1976). A comparison of census methods for bird nest estimation was made by Dodd & Murphy (1995).

Some animals concentrate at particular seasons or time of the day when a census becomes possible. For example, Allsteadt & Vaughan (1992) were able to census caiman during the dry season by counting the reflective eyes by torchlight at night. Technological solutions to improve detection rates can also make census methods feasible; Naugle *et al.* (1996) were able to detect white-tailed deer using an aerial infra-red imaging system. Local concentrations of conspicuous insects can also allow a direct census (see Chapter 4).

9.2 Point and line survey methods

9.2.1 Indices of abundance using transects

These methods are frequently used in conservation studies where minimising cost is important, and when the objective is to detect any change in abundance rather than absolute magnitude. A typical example applied to insects is the study by Thomas (1983) on a number of species of British butterflies. The protocol was:

1. Undertake a preliminary inspection to define the extent of the butterfly colony: this is termed the flight area, A.
2. Mark out a zig-zag trail of length L m, which should either exceed 1000 m in length or be long enough to record 40 individuals.
3. Walk the trail and count individuals, N, seen within an imaginary box from 0 to 5 m ahead with a fixed width which varies with species, but is typically 4 to 6 m.
4. Calculate the index as:

$$p = \frac{NA}{L}. \tag{9.2}$$

Thomas (1983) showed that this abundance index was correlated to the absolute population density as determined by mark–recapture, and could be converted to an estimate of absolute abundance by regression analysis.

An example of a major monitoring program based on transect walking is the UK Butterfly Monitoring Scheme (http://www.ukbms.org), which has been used to monitor butterfly abundance since 1976. Recorders walk transects in more than 1500 separate sites across Britain and have now counted over 16.4 million butterflies.

9.2.2 Methods based on flushing

The simplest approach is to estimate the number disturbed from a known width of habitat. Under certain circumstances, such as sea bird colonies, all individuals may be flushed and the problem is simply obtaining an accurate count (Bibby *et al.*, 1992), but usually only a proportion is disturbed. If this is constant then the numbers themselves give an index of absolute population and, by incorporating the proportion and the area covered, an estimate of absolute population density. If the proportion flushed is unknown or varies, only a relative measure of availability is given.

This method has been used for the assessment of populations of the red locust, *Nomadacris septemfasciata*, in its outbreak areas. The observer moved across the area in a motor vehicle (Land Rover) and counted the insects disturbed ahead (Scheepers & Dunn, 1958). Although the response of a locust to flushing depends on its condition (Nickerson, 1963), this method appears to have a fairly constant efficiency – about 75% of the locusts within the path of the vehicle rise (Symmons *et al.*, 1963). In an attempt to cover a larger area more readily, flushing by a low-flying aircraft was tried, but the efficiency of flushing, even when increased by spraying a noxious chemical, was low in sparse populations. It was suggested that a siren might be used to disturb a higher proportion of locusts (Symmons *et al.*, 1963), and such an approach has been found to be effective with nesting sea birds (Bullock & Gomersall, 1981; Bibby *et al.*, 1992). In a review of sampling methods for orthoptera in grasslands, Gardiner *et al.* (2005) concluded that methods involving flushing grasshoppers from the sward are fairly accurate in short, open swards (<50 cm sward height) when grasshoppers are in low densities (<2 adults per m²). At higher population densities, methods which require the capture of grasshoppers such as box quadrats and sweep netting (Chapter 7) may be more appropriate.

Hayne (1949) developed the first estimator of density for a flushing experiment that could claim to be robust for studies on bird populations. This simple method was designed to estimate the numbers of grouse which it was assumed would flush as the observer came within a certain radius, r, of the bird. The observer moves along a transect noting the radial distance at which each bird is flushed. The density estimate, D_H, is then given by:

$$\hat{D}_H = \frac{n}{2L}\left(\frac{1}{n}\sum\frac{1}{r_i}\right),$$ (9.3)

where n is the total number of animals counted, L is the transect length, and r_i is the sighting distance to the ith animal.

The approximate variance of this estimate is:

$$\text{var}(\hat{D}_H) \approx D_H^2\left[\frac{\text{var}(n)}{n^2} + \frac{\sum\left(\frac{1}{r_i}-R\right)^2}{R^2 n\,(n-1)}\right],$$ (9.4)

where R is the mean of the reciprocals of the sighting distances.

Burnham & Anderson (1976) point out that an underlying assumption is that the average sighting angle, θ, is 32.7°. To test this assumption, calculate

$$z = \frac{\sqrt{n}(\theta - 32.7)}{21.56},$$ (9.5)

and reject the assumption if the absolute value of z is greater than 1.96.

If this assumption is rejected, then the following adjustment can be applied (Burnham and Anderson, 1976):

$$\hat{D}_{MH} = \hat{c}\frac{n}{2l}\left[\frac{1}{n}\sum_{i=1}^{n}\frac{1}{r_i}\right],$$ (9.6)

where $\hat{c} = 1.9661 - 0.0295\,(\theta)$.

More than six other estimators have been developed (Gates *et al.*, 1968; Gates, 1969); an unbiased estimator with minimal variance is considered to be that of Gates (1969) and Kovner & Patil (1974), although some others may be more robust. This estimator is:

$$\hat{D} = [2n - 12L\bar{r}], \qquad (9.7)$$

where n = number of animals sighted, L = length of the line transect and $\bar{r}$ = average radial distance over which the observer encountered animals. The assumptions are made that encounter distances (r) follow a negative exponential distribution, and that the flushing of one animal will not affect that of another. The effect of varying the two latter assumptions has been explored by Sen *et al.* (1974), who produced several sets of data for birds and concluded that, although r often followed a negative exponential function, a Pearson type III distribution was more general. They also concluded that the flushing of one animal was not independent of the flushing of another, and the best fit was obtained if cluster size was considered to follow a Poisson distribution (i.e. the distribution of 'flushings' would fit the negative binomial). The estimation of each r can, of course, be a major source of error.

The observer should seek to fix the position where the animal was seen and then measure the distance; a range-finder may be helpful (Bibby *et al.*, 1992). If, because of the risk of disturbing other animals, r has to be estimated, the observer should regularly check the accuracy of the estimation, at another site if necessary.

The transect route may simply follow an existing path and this may be satisfactory for relative estimates. A more acceptable basis for absolute population estimates will be provided if the single route is replaced with a number of shorter straight, transepts each with starting point and direction determined randomly. If the habitat has different zones each should be traversed separately, a stratified sampling system (see Chapter 2) (Bibby *et al.*, 1992).

Line transect theory normally assumes that the animals (objects under observation) are immobile prior to detection. This assumption will not cause serious error provided that the movement relative to that of the observer is random and slow. For cases where there is substantial movement, a different estimator was developed by Yapp (1956):

$$\hat{D} = \frac{E}{2rV} \qquad (9.8)$$

where E = number of encounters per unit time (i.e. n/t), and V = average velocity of the organism relative to the observer, which is given by:

$$V^2 = \bar{u}^2 + \bar{w}^2 \qquad (9.9)$$

where u = average velocity of the observer and w = average velocity of the animal.

This estimator is based on the kinetic theory of gases, and assumes that the observer walks in a straight line at a constant speed, the animals move at random at a constant speed, and there is a fixed and known radius r within which the observer can recognise the animals.

The applicability of this approach has been investigated by Skellam (1958), who concluded that Equation (9.8) was valid, but that the derivation of V contained an approximation, the effect of which would be most serious when the speed of the observer and the animal were equal and when almost negligible when the two were very dissimilar. Skellam also showed that Equation (9.9) could be regarded as a Poisson variable, provided that the animals did not move back on their tracks or move in groups; under these conditions the variance would equal the mean. Aggregation will increase the variance. The major practical difficulty in the application of this method – for the estimation of populations of butterflies, dragonflies, birds and other animals for which it would seem to be appropriate – is the determination of the average speed of the animal. Care must of course be taken with the units: if E is expressed as the number of encounters per hour and r in metres, the speed, must be expressed in metres per hour. Burnham *et al.* (1980) concluded that the inherent assumptions greatly restricted the valid use of this method.

9.2.3 *Line transect methods: the Fourier series estimator*

Line transect methods have been developed for situations when it is not possible to count all the animals within a strip transect. The methods are based on the idea that only animals lying on the centre line of the strip transect along which the observer moves will be certain to be detected, and that the probability of detection will fall with perpendicular distance from this line. This was also the case with the Haynes and Gates estimators discussed above, but they made strong assumptions about the shape of the detection function (rectangular and negative exponential respectively) which will often not hold. The techniques presented here have been reviewed in detail by Buckland *et al.* (1993) and more recently by Thomas *et al.* (2010). A bibliography of distance sampling literature is available at http://distancesampling.org/dbib.html. The calculations can be undertaken using the free program Distance (http://distancesampling.org/). All of the methods offered by Distance are also available as R packages (http://distancesampling.org/R/index.html). These are discussed below, together with an example R code.

For many animals, the observer walks the transect, but counts can also be undertaken from motor vehicles, ships and remote-operated vehicles.

For these methods it is assumed that :

1. Objects on the line are always detected.
2. The observer does not influence the recorded positions. For mobile animals, the position must be that prior to any response to the presence of the observer. The theory has been developed under the assumption that the objects are immobile, but slow movement in relation to the observer creates little inaccuracy.
3. Distances and/or angles are measured accurately.
4. The objects are correctly identified

The basic field procedure is for the transect route to be a straight line of length, L, randomly placed with respect to the animals or objects to be counted, and the perpendicular distance to each detected object of interest, x, recorded. In practice a number of lines, arranged as a regular grid, and randomly placed in the study may be used.

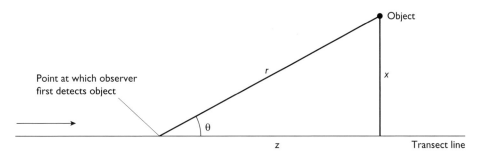

Figure 9.1 **The relationship between sighting distance, r, sighting angle, θ, and the perpendicular distance, x, from the transect line.**

Further, it is usually easier to record the sighting distance, r, which is the distance from the observer to the object, and the sighting angle, θ, which is the angle of the object from the transect line, and to calculate x rather than measure x directly (see Fig. 9.1). Methods of population estimation based on r and θ were reviewed by Hayes & Buckland (1983) but are not presented here as they are considered inferior to those based on the perpendicular distance (Buckland *et al.*, 1993).

If all n objects in a strip of length, L, and width, $2w$ are counted then the estimated density is:

$$\hat{D} = \frac{n}{2wL}. \tag{9.10}$$

If a proportion, P, of the animals present are detected, then the equation becomes:

$$\hat{D} = \frac{n}{2wLP}. \tag{9.11}$$

Line transect methods use the distribution of the perpendicular detection distances to estimate P.

Figure 9.2 shows a typical histogram of the decline in the number of observations with perpendicular distance. A detection function, $g(x)$, defined as the probability of detecting an object at distance, x, can be fitted to these data. Given assumption (1) above, $g(0)$, the probability of detecting an object lying on the line, is assumed to equal 1.

The probability of detecting an object within a strip of area $2wL$, P, is:

$$P = \frac{\int_0^w g(x)dx}{w} \tag{9.12}$$

which on substitution into Equation (9.11) gives

$$\hat{D} = \frac{n}{2L \int_0^w \hat{g}(x)dx}. \tag{9.13}$$

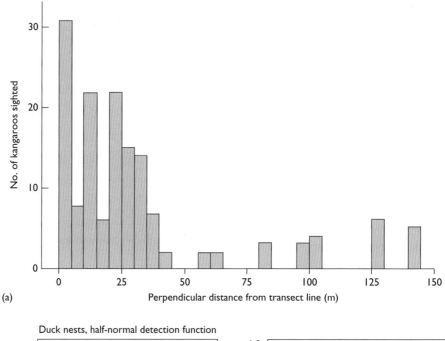

(a)

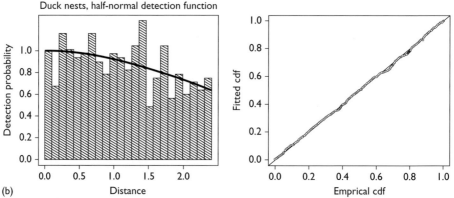

(b)

Figure 9.2 (a) A typical histogram of the decline in the number of observations with perpendicular distance from the transect line. (Data from Coulson & Raines, 1985); (b) An example of a fitted half-normal detection function to the duck nest data set used in Section 9.3, together with the fitting diagnostic plot. For the R code, see p. 366.

As it is assumed that $g(0) = 1$, the probability density function (pdf) evaluated at $x = 0$ is

$$f(0) = \frac{1}{\displaystyle\int_0^w g(x)dx} \qquad (9.14)$$

and thus the general estimator of density is often expressed as:

$$\hat{D} = \frac{n\hat{f}(0)}{2L}. \qquad (9.15)$$

Density is expressed as numbers per square unit of length where length of the transect, and distance to the objects are expressed in the same length units such as metres or kilometres. A flexible function with the desired properties to use as a detection function is a cosine series. This is the Fourier model of Crain *et al.* (1979), and is a general model which has been shown to give good results for a wide variety of data.

When perpendicular distance has been measured as a continuous variable and not grouped into size classes, the Fourier series $f(0)$ is given by:

$$\hat{f}(0) = \frac{1}{w^*} + \sum_{k=1}^{m} \hat{a}_k,$$ (9.16)

where

$$\hat{a}_k = \frac{2}{nw^*} \left[\sum_{i=1}^{n} \cos\left(\frac{k\pi x_i}{w^*}\right) \right]$$ (9.17)

and w^* is the transect half-width (without truncation this is the largest perpendicular distance observed); x_i is the perpendicular distance of the ith animal, n is the number of observations and m is the number of cosine terms determined by the stopping rule. You choose the first value of m such that

$$\frac{1}{w^*} \left(\frac{2}{n+1}\right)^{1/2} \geq |\hat{a}_{m+1}|.$$ (9.18)

Experience suggests that m should be < 7.

Having calculated $\hat{f}(0)$, Equation (9.14) is then used to estimate D. The sampling variance of D requires considerable computation. First, calculate the variance and covariances of the parameters a_k

$$\text{var}(\hat{a}_k) = \frac{1}{n-1} \left[\frac{1}{w^*} \left(a_{2k} + \frac{2}{w^*} \right) - a_k^2 \right],$$ (9.19)

$$\text{cov}(\hat{a}_k, \hat{a}_j) = \frac{1}{n-1} \left[\frac{1}{w^*} (a_{k+j} + a_{k-j}) - a_k a_j \right].$$ (9.20)

Then sum the variance–covariance matrix to give

$$\text{var}\left(\frac{1}{\hat{a}}\right) = \sum_{j=1}^{m} \sum_{k=1}^{m} \text{cov}(\hat{a}_j, \hat{a}_k)$$ (9.21)

Finally, the estimated variance is calculated using

$$\text{var}(\hat{D}) = \hat{D}^2 \left[\frac{\text{var}(n)}{n} + \frac{\text{var}\left(1/\hat{a}\right)}{\left(1/\hat{a}\right)^2} \right].$$ (9.22)

It is normally assumed that the variance of the number of animals counted, $var(n)$ is a Poisson variable so that it is equal to n. As this need not be so, Burnham *et al.* (1980) suggest that a sounder approach is to undertake a number of replicate transects, calculate D for each, and then find the variance of these replicates.

Buckland *et al.* (1993) argue that extreme observations of perpendicular distance should be removed from the data set prior to calculating population density. As a rough rule, they suggest that 5% of the data should be truncated.

The Fourier series method can also be applied to grouped data. Such data are generated when it is impossible to assign observations to accurately measured distances, so they are allocated to distance bands such as 0–5, 5–10, 10–15 and 15–20 m. It may also be advantageous to group data when bias in favour of certain distances is suspected. A description of these methods is beyond the scope of the present text, but they are offered in specialised computer programs such as Distance, which is available free (see below). If possible grouping data should be avoided as it reduces accuracy (Southwell & Weaver, 1993).

From the above Fourier model it can be seen that the key computational issue is the estimation of $g(x)$ or $f(0)$ and a variety of other functions could have been used to relate the number of detections to perpendicular distance. The approach of Buckland *et al.* (1993) is to estimate $g(x)$ from a small group of models chosen for their robustness, a shape criterion and computational efficiency. Buckland (1992) argues that the formulation of the model can be a two-stage process. First, a key function is selected, two initial candidates are a uniform or a half-normal function because they require just zero and 1 estimated parameters, respectively. Second, a series expansion to improve the fit so that the function is represented as:

$$g(x) = key(x)\left[1 + series(x)\right]. \tag{9.23}$$

Buckland *et al.* (1993) consider the key functions and series expansions listed in Table 9.1 useful.

Examples of the application of distance sampling methods are given in Buckland *et al.* (1993), while Ensign *et al.* (1995) apply the technique to benthic fish in streams, and Kelley (1996) to wood duck.

Table 9.1 Key functions and series expansions useful for fitting the detection function for distance sampling (Buckland *et al.*, 1993).

Key function	
Uniform, $1/w$	Cosine, $\sum\limits_{j=1}^{m} a_j \cos\left(\frac{j\pi y}{w}\right)$
Uniform, $1/w$	simple polynomial, $\sum\limits_{j=1}^{m} a_j \left(\frac{y}{w}\right)^{2j}$
Half-normal, $\exp\left(\frac{-y^2}{2\sigma^2}\right)$	Cosine, $\sum\limits_{j=1}^{m} a_j \cos\left(\frac{j\pi y}{w}\right)$
Half-normal, $\exp\left(\frac{-y^2}{2\sigma^2}\right)$	Hermite polynomial, $\sum\limits_{j=2}^{m} a_j H_{2j}(y_s)$
Hazard-rate, $1 - \exp\left(-\left(\frac{y}{\sigma}\right)^{-b}\right)$	Cosine, $\sum\limits_{j=1}^{m} a_j \cos\left(\frac{j\pi y}{w}\right)$

Clemente-Sánchez *et al.* (2013) assessed the accuracy of walked and driven line transect methods for the estimation of pronghorn, *Antilocapra americana*, numbers in the Chihuahuan Desert rangelands of New Mexico, USA. A total of 16 transects each 5 km long were walked, and the observed pronghorn were positioned using Leopard RXB-IV laser rangefinder binoculars with electronic compass. For the driving survey a total of 14 transects each 5 km long were surveyed with a pickup and spotlight at night, and each pronghorn sighting position recorded. The accuracy of both methods was compared with the true population size for the area obtained from aerial survey. It was concluded that the walking transect method overestimated population size. In contrast, the driving surveys gave estimates not significantly different from the actual population present. The population density estimates were made using the program Distance (http://distancesampling.org/).

9.2.4 *Point transects*

Instead of traversing a transect the observer may move to a number of fixed points and record the distance, r, to individual animals. These methods are almost only used for bird surveys, where the patchy suitability of the habitat to the birds may make transects inappropriate because they cut across a number of habitat types. Point transects are often easier to undertake because the observer needs only estimate distance and markers may be placed in advance to aid the estimation of distance. Population density is given by:

$$\hat{D} = \frac{nh(0)}{2\pi k} \tag{9.24}$$

where n is the number of animals observed, k the number of point transects undertaken, and $h(0)$ the slope of the probability density function of detection distances evaluated at zero distance. In similar fashion to what was described for the line transect method, the central problem is to estimate $h(0)$. This is normally accomplished using programs such as Distance (http://distancesampling.org/).

For the case of a half-normal detection function the maximum likelihood estimator for density has the particularly simple form (Buckland *et al.*, 1993):

$$\hat{D} = \frac{n^2}{2\pi k \sum_{i=1}^{n} r_i^2} \tag{9.25}$$

9.3 Distance sampling software in R

A wide range of R packages are available to analyze distance sampling data (see http://distancesampling.org/R/index.html). The main R packages are as follows:

- Mrds: to fit detection functions to point and line transect distance sampling survey data.
- Distance: analysis of single observer distance sampling surveys with a simpler interface to that of mrds.

- Dsm: fits density surface models to spatially referenced distance sampling data. Count data are corrected using detection function models fitted using mrds or Distance. Spatial models are constructed using generalised additive models.
- DSsim: a package for simulating distance sampling surveys.
- Mads: deals with unidentified sightings, covariate uncertainty and model uncertainty in Distance sampling.

The following example R code listing is based on the vignette available from http://distancesampling.org/R/vignettes/ducknests.html, which presents an analysis of duck nest data. It used the ds function in the Distance package. The data set comprises a simulated data set of 20 transects each 128.8 km long. The perpendicular distances from each transect are given in metres. Data sets are simple lists stored as csv files that can be organised using a spreadsheet such as Excel. Each observation is a single row comprising the following columns: (1) a name label; (2) Area which can be 0 if no area was measured; (3) a transect or sample identifier; (4) Transect length (effort); and (5) sighting distance. Care needs to be taken with units; note that in this example the transect lengths are in kilometres and the sighting distances in metres. The plotted output for the fitted detection function is shown in Fig. 9.2b.

```
library (Distance) # load the distance library
library (knitr) # Library for tabulated output
nests = read.csv("C:\\Users\\ R working\\ducknests.csv") # read csv file
head(nests) # Prints out the first few lines of the data set to check it is OK

# transect length in kilometres,perpendicular distance in metres,
# therefore a conversion factor of 0.001 converts perpendicular distance
of metres to kilometres.
# The calculated nest density is nests per square kilometres.

#First fit a half normal detection function with cosine adjustment
halfnorm.ducks <- ds(nests, key="hn", adjustment="cos", convert.units = 0.001)

#Second fit a uniform detection function with cosine adjustments
unifcos.ducks <- ds(nests, key="unif", adjustment="cos", mono="strict",
convert.units = 0.001)

#Third fit a hazard-rate with polynomial adjustments
hazard.ducks <- ds(nests, key="hr", adjustment="poly", convert.units = 0.001)

#Plot detection function and show goodness of fit
par(mfrow=c(1,2))
plot(halfnorm.ducks, main="Duck nests, half-normal detection function")
fit.test <- ddf.gof(halfnorm.ducks$ddf)
par(mfrow=c(1,1))

# Present table giving encounter rate - nests per kilometer surveyed
kable(halfnorm.ducks$dht$individuals$summary,format="markdown")

# Present table giving estimated density, std. error & conf. intervals
kable(halfnorm.ducks$dht$individuals$D,format="markdown")
```

9.4 Spatial distribution and plotless density estimators

In their simplest form these methods are based on the premise that, if the individuals of a population are randomly distributed, the population density can be estimated from either the distance between individuals (nearest-neighbour methods) or from randomly chosen points to the nearest individual (closest individual methods). If these 'basic distance' methods are to be applied in their simpler forms, tests of randomness must be carried out. For populations sampled in quadrates, with a mean greater than one, the Poisson Index of Dispersion (see Chapter 2) is a satisfactory test of randomness. Seber (1982) gives other test procedures.

If the density is known from some other method, these methods may themselves be used as a test of randomness (Waloff & Blackith, 1962). Pielou's (1969) Index (I_α) is particularly useful for this purpose:

$$I_\alpha = \bar{a}_i^2 mn \qquad (9.26)$$

where m = density per unit area calculated from another method, and $\bar{a}$ is the mean distance from a random point to the closest individual (see below); then $I_\alpha = 1$ for random distributions, >1 for contagious, and <1 for regular, suggesting competition. Underwood (1976) has used this approach in a study of intertidal molluscs.

Although widely used with vegetation (especially trees), these approaches have not been much unused by zoologists. In the main this is undoubtedly due to the difficulty, and high cost in time, of finding the exact positions of animals. Furthermore, habitat heterogeneity is often fine-grained so that interpretations of pattern in terms of intra-specific interactions (for example by using Pielou's Index) may be invalid, more especially if the second or third closest neighbour is used (Waloff & Blackith, 1962). They are, however, particularly appropriate for animals living or nesting in fairly uniform, two-dimensional habitats, such as muddy or sandy littoral zones or on the soil surface of arable crops. The practical difficulties are reduced if the positions of the animals, nests or other artefact can be recorded photographically.

Whilst the non-randomness of many populations may preclude using the basic equations for exact population estimation, they do have the potential to provide an order of magnitude check (p. 4) on estimates obtained by other methods. Like mark and recapture, they are independent of sample size.

A complete review of plotless density estimators is provided by Engeman et al. (1994), who used simulated data to assess the utility and robustness of the various estimators with different spatial-patterns and different sample sizes. Their major conclusions are incorporated in the accounts below.

9.4.1 Closest individual or distance method

A point is first selected at random, after which searching round it in concentric rings the closest individual is found and its distance (X) from the point measured. The process should be continued to find the second, third, …ith closest individual. Referred to by Engeman et al. (1994) as an '…ordered distance estimator', they recommend using the third as being the most practical, as did Keuls *et al.* (1963) who used it for estimating the density of the snail *Limnaea truncatula*. The estimate of density, $\hat{D}$, is (Moristita, 1957):

$$\hat{D} = sr - 1 \bigg/ \pi \sum_{i=1}^{s} X_i \tag{9.27}$$

where r = the rank of the individual in distance from the randomly selected point, e.g. for the nearest-neighbour $r = 1$, for the second nearest 2 …; X = distance between the randomly selected point and the individual, and s is the number of random sampling points.

The variance of D is given by Seber (1982) as:

$$\text{var}[\hat{D}] = \hat{D}^2 \bigg/ (sr - 2). \tag{9.28}$$

A somewhat different approach, the variable area transect, was developed by Parker (1979) and considered by Engeman *et al.* (1994) to be the simplest and most practical. The search area is a fixed-width strip of width, w. Commencing from a random point, the area is searched until the ith individual is found. When i is the number of individuals searched for from each random point and l is the length searched from the random point to the ith individual, then:

$$\hat{D} = \frac{(si - 1)}{w \sum l} \tag{9.29}$$

This method assumes the organisms are Poisson randomly distributed with a constant density in the study area.

Dobrowski & Murphy (2006) noted that the variable area transect method had received little attention despite robust estimation properties. They undertook simulations which indicated that transect width affected the quality of the estimates and should be as narrow as practicable.

9.4.2   *Nearest-neighbour methods*

In these methods, an animal is selected at random and searching is continued until another – the nearest neighbour – is encountered. The distance between the two (r) is then measured. As with the closest individual method there are variants that use the 2nd to ith nearest neighbour. Strictly speaking, this is not a random method (Pielou, 1969), as a truly random method would involve numbering all the animals in advance and following random numbers for the selection of the first animal. Also, in its simplest form the method is not robust if the distribution is non-random (Engeman *et al.*, 1994). The basic expression is due to Clark & Evans (1954):

$$\hat{D} = \frac{1}{4\bar{r}^2} \tag{9.30}$$

where $\hat{D}$ = density per unit area and $\bar{r}$ = mean distance between nearest neighbours.

A number of other methods have been devised that are more robust if the distribution is non-random. These are based on the separate consideration of each quadrate in the circle surrounding, the original animal (angle order estimators), or on a combination of information on the closest individual and nearest neighbour (methods of M.G. Kendal and P.A.P. Moran). However, the algorithms for their calculation are extremely complex (Engeman *et al.*, 1994) and have not been much used by zoologists.

References

Allsteadt, J. & Vaughan, C. (1992) Population status of *Caiman crocodilus* (Crocodylia: Alligatoridae) in Cano Negro, Costa Rica. *Brenesia* **0**(38), 57–64.

Bibby, C.J., Burgess, N.D., & Hill, D.A. (1992) *Bird Census Techniques.* Academic Press, San Diego. 257 pp.

Buckland, S.T. (1992) Fitting density functions using polynomials. *Appl. Statist.* **41**, 63.

Buckland, S.T., Anderson, D.R., Burnham, K.P., & Laake, J. L. (1993) *Distance Sampling: Estimating abundance of biological populations.* Chapman & Hall, London.

Buckland, S.T., Plumptre, A.J., Thomas, L., & Rexstad, E.A. (2010) Design and analysis of line transect surveys for primates. *Int. J. Primatol.* **31**, 833–47.

Bullock, I.D. & Gomersall, C.H. (1981) The breeding populations of terns in Orkney and Shetland in 1980. *Bird Study* **28**, 187–200.

Burnham, K.P. & Anderson, D.R. (1976) Mathematical models for non-parametric inferences from line transect data. *Biometrics* **32**, 325–36.

Burnham, K.P., Anderson, D.R., & Laake, J.L. (1980) Estimation of Density from Line Transect Data. *Wildlife Monograph* **72**.

Caughley, G. (1974) Bias in aerial survey. *J. Wildlife Manag.* **38**, 921–33.

Caughley G., Sinclair, R. & Scott, D. (1976) Experiments in aerial survey. *J. Wildlife Manag.* **40**, 290–300.

Clark, P.I. & Evans, F.C. (1954) Distance to nearest neighbor as a measure of spatial relationships in populations. *Ecology* **35**, 445–53.

Clemente-Sánchez, F., Holechek, J.L., Valdez, R., Mendoza-Martínez, G.D., Rosas-Rosas, O.C., & Tarango-Arámbula, L.A. (2013) Accuracy of two techniques used to estimate pronghorn (*Antilocapra americana*) numbers in Chihuahuan Desert rangelands. *J. Appl. Animal Res.* **41**, 149–55.

Coulson, G.M. & Raines, J.A. (1985) Methods for small-scale surveys of grey kangaroo populations. *Aust. Wildlife Res.* **12**, 119–25.

Crain, B.R., Burnham, K.P., Anderson, D.R., & Laake, J.L. (1979) Nonparametric estimation of population density for line transect sampling using Fourier series. *Biometric. J.* **21**, 731–48.

Dobrowski, S.Z. & Murphy, S.K. (2006) A practical look at the variable area transect. *Ecology,* **87**, 1856–60.

Dodd, M.G. & Murphy, T.M. (1995) Accuracy and precision of techniques for counting great blue heron nests. *J. Wildlife Manag.* **59**(4), 667–73.

Engeman, R.M., Sugihara, R.T., Pank, L.F., & Dusenberry, W.E. (1994) A comparison of plotless density estimators using Monte Carlo simulation. *Ecology* **75**(6), 1769–79.

Ensign, W.E., Angermeier, P.L., & Dolloff, C.A. (1995) Use of line transect methods of estimate abundance of benthic stream fishes. *Can. J. Fish. Aquat. Sci.* **52**(1), 213–22.

Gardiner, T., Hill, J., & Chesmore, D. (2005) Review of the methods frequently used to estimate the abundance of Orthoptera in grassland ecosystems. *J. Insect Conserv.* **9**, 151–73.

Gates, C.E. (1969) Simulation study of estimators for the line transect sampling method. *Biometrics* **25**, 317–28.

Gates, C.E., Marshall, W.H., & Olson, D.P. (1968) Line transect method of estimating grouse population densities. *Biometrics* **24**, 135–45.

Harris, M.P. & Lloyd, C.S. (1977) Variations in counts of seabirds from photographs. *Bird Study* **34**, 187–90.

Hayes, R.J. & Buckland, S.T. (1983) Radial-distance models for the line-transect method. *Biometrics* **39**, 29–42.

Hayne, D.W. (1949) Two methods of estimating populations from trapping records. *J. Mammal.* **30**, 399–411.

Kelley, J.R., Jr (1996) Line-transect sampling for estimating breeding wood duck density in forested wetlands. *Wildlife Soc. Bull.* **24**(1), 32–6.

Keuls, M., Over, H.I., & De Wit, C.T. (1963) The distance method for estimating densities. *Statist. Neerland.* **17**, 71–91.

Kovner, J.L. & Patil, S.A. (1974) Properties of estimators of wildlife population density for the line transect method. *Biometrics* **30**, 225–30.

LeResche, R.E. & Rausch, R.A. (1974) Accuracy and precision of aerial moose censusing. *J. Wildlife Manag.* **38**, 175–82.

Moristita, M. (1957) Estimation of population density by spacing method. *Mem. Fac. Sci. Kyushu Univ. E* **1**, 187–97.

Naugle, D.E., Jenks, J.A., & Kernohan, B.J. (1996) Use of thermal infrared sensing to estimate density of white-tailed deer. *Wildlife Soc. Bull.* **24**(1), 37–43.

Nickerson, B. (1963) An experimental study of the effect of changing light intensity on the activity of adult locusts. *Entomol. Monthly Mag.* **99**, 139–40.

Norton-Griffiths, M. (1978) *Counting Animals*. African Wildlife Leadership Foundation, Nairobi.

Parker, K.R. (1979) Density estimation by variable area transect. *J. Wildlife Manag.* **43**, 484–92.

Pielou, E.C. (1969) *An Introduction to Mathematical Ecology*. Wiley-Interscience, New York, London.

Pinder, L. (1996) Marsh deer *Blastocerus dichotomus* population estimate in the Parana River, Brazil. *Biol. Conserv.* **75**(1), 87–91.

Redfern, J.V., Viljoen, P.C., Kruger, J.M., & Getz, W.M. (2002) Biases in estimating population size from an aerial census: a case study in the Kruger National Park, South Africa: Starfield Festschrift. *South Afr. J. Sci.* **98**, 455.

Scheepers, C. & Dunn, D.L. (1958) Enumerating populations of adults of the red locust, *Nomadacris septemjasciata* (Serville), in its outbreak areas in East and Central Africa. *Bull. Entomol. Res.* **49**, 273–85.

Seber, G.A.F. (1982) *The Estimation of Animal Abundance and Related Parameters*. Griffin, London.

Sen, A.R., Tourigny, I., & Smith, G.E.J. (1974) On the line transect sampling method. *Biometrics* **30**, 329–40.

Skellam, J.G. (1958) The mathematical foundations underlying the use of line transects in animal ecology. *Biometrics* **14**, 385–400.

Southwell, C. & Weaver, K. (1993) Evaluation of analytical procedures for density estimation from line-transect data: Data grouping, data truncation and the unit of analysis. *Wildlife Res.* **20**(4), 433–44.

Starr, R.M., Fox, D.S., Hixon, M.A., Tissot, B.N., Johnson, G.E., & Barss, W.H. (1996) Comparison of submersible-survey and hydroacoustic-survey estimates of fish density on a rocky bank. *U. S. National Mar. Fish. Service Fish. Bull.* **94**(1), 113–23.

Symmons, P.M., Dean, G.J.W., & Stortenbere, R.C.W. (1963) The assessment of the size of populations of adults of the red locust. *Nomadacris septemfasciata* (Serville), in an outbreak area. *Bull. Entomol. Res.* **54**, 549–69.

Thomas, J.A. (1983) A quick method for estimating butterfly numbers during surveys. *Biol. Conserv.* **27**, 195–211.

Thomas, L., Buckland, S.T., Rexstad, E.A., Laake, J.L., Strindberg, S., Hedley, S.L., & Burnham, K.P. (2010) Distance software: design and analysis of distance sampling surveys for estimating population size. *J. Appl. Ecol.* **47**, 5–14.

Underwood, A.J. (1976) Nearest neighbour analysis of spatial dispersion of intertidal prosobranch gastropods within two substrata. *Oecologia* **26**, 257–66.

Waloff, N. & Blackith, R.E. (1962) The growth and distribution of the mounds of *Lasius flavus* (Fabricius) (Hym.: Formicidae) in Silwood Park, Berkshire. *J. Anim. Ecol.* **31**, 421–37.

Yapp, W.B. (1956) The theory of line transects. *Bird Study* **3**, 93–104.

10 Observational and Experimental Methods for the Estimation of Natality, Mortality and Dispersal

In this chapter are discussed the methods of directly quantifying the various processes that produce population change. The magnitude of these different 'pathways' may also be obtained by the subtraction or integration of census figures in a budget: methods of calculation and of analysis of budgets are discussed in the next chapter, but there is no hard and fast distinction between the contents of the two chapters. As is indicated below, in the appropriate sections, the terms 'natality' and 'dispersal' are used in their widest sense.

10.1 Natality

Natality is the number of births. For an egg layer, such as birds, most insects or fish, this is, strictly speaking, the number of living eggs laid; however, from the practical point of view of constructing a population budget the number of individuals entering any post-ovarian stage (i.e. a larval instar, pupa, fledgling or adult) can be considered as the 'natality' of that stage.

10.1.1 Fertility

Fertility is the number of viable eggs laid by a female and fecundity is a measure of the total egg production; the latter is often easier to measure. In those animals where all the eggs are mature on emergence, the *total potential fecundity* may be *estimated by examining the ovaries* as Davidson (1956) did with sub-imagines of a mayfly, *Cloeon*. The ovaries were removed and lightly stained in methylene blue, the eggs were separated by sieving through bolting silk, and then counted in a Sedwick–Rafter plankton-counting cell. For fish and other animals that produce many eggs, fecundity is frequently estimated by a gravimetric method (Bagenal & Braum, 1978). A sample of eggs is counted and the total number of eggs estimated from the weight of the ovaries. Some fish are batch spawners so that the ovary comprises a population of eggs at a number of stages of development. For these species, the fecundity of a single batch can be estimated by only counting the eggs in the largest size group, but total annual fecundity is difficult to estimate, as there is no direct method to estimate the number of batches produced during

Ecological Methods, Fourth Edition. P. A. Henderson and T. R. E. Southwood.
© 2016 John Wiley & Sons, Ltd. Published 2016 by John Wiley & Sons, Ltd.
Companion Website: www.wiley.com/go/henderson/ecologicalmethods

the spawning season. Many crustaceans, including copepods, prawns and crabs and certain spiders and cockroaches, carry the eggs externally where they can be counted, as may the eggs in birds' nests. Crustaceans may remove dead eggs from the egg mass, and therefore the developmental stage of the eggs should be noted as the number of eggs tends to decline with age. Other crustaceans such as mysids, darwinulid ostracods and cladocerans carry their eggs and early stages internally, where they can often be counted under a microscope without dissection.

When eggs are matured throughout much of adult life, fecundity can be measured directly by keeping females captive under as natural conditions as possible and recording the total number of eggs laid (e.g. Huffaker & Spitzer, 1950; Spiller, 1964). If viable eggs can be distinguished from non-viable ones, usually by the onset of development, then fertility may be measured (e.g. Fewkes, 1964). Egg cannibalism can occur with many animals, particularly following disturbance; for insects, the influence of this behaviour may be calculated from the loss of a proportion of a number of marked eggs (Rich, 1956).

It is often found that fecundity is proportional to weight. Honěk (1993) reviewed the relationship between intra-specific variation in female body size and potential fecundity using published literature on 57 oviparous species of Coleoptera, Diptera, Ephemeroptera, Heteroptera, Homoptera, Hymenoptera, Lepidoptera, and Trichoptera, and 11 species of larviparous Aphidina and Diptera. He found that the common regression for oviparous and larviparous species predicts a 0.95% increase in median fecundity for each 1% increase in dry body weight. The number of ovarioles (in 10 species of Coleoptera, Diptera, Hymenoptera and Orthoptera) also increased with body weight. The general relationship predicted a 0.81% increase in ovariole number for each 1% increase in dry body weight.

For insects, the fecundity–weight relationship has been used to estimate female fecundity (e.g. Prebble, 1941; Richards & Waloff, 1954; Waloff & Richards, 1958; Colless & Chellapah, 1960; Lozinsky, 1961; Murdie, 1969; Taylor, 1975). As female weight is directly related to size, measures of size such as wing length (Gregor, 1960) or pupal length (Miller, 1957), or both (Hard, 1976), may be substituted for weight. Once the relation between size and fecundity has been established (regression analysis is a convenient method) it may be applied to estimate the natality in field populations by measuring wild females or female pupae – the length of the empty pupal cases is a particularly suitable criterion (Miller, 1957). The same approach can be used for fish and crustaceans.

The size (age)–fecundity relationship is frequently required for fisheries population models. Following a comparison of a range of possible relationships, Zivkov & Petrova (1993) considered a logistic curve to be the best model of the length–fecundity relationship over the full adult size range, although other models were equally appropriate over reduced size ranges.

Within a given population of insects, the rate of oviposition will be influenced by temperature, and the extent to which the potential fecundity is realised is influenced by the longevity of the females. The influence of temperature can be studied in the laboratory and incorporated into the regression equation; the observations of Richards & Waloff (1954) show that it may be justified to apply data from the laboratory to populations in the field. Information on longevity may be obtained from wild populations by marking and recapture. An equation incorporating all these variables will be in the form (Richards & Waloff, 1954):

$$\text{No. of eggs} = \text{regression coeff.} \times \text{weight} \pm \text{reg. coeff.} \times \text{temp.}$$
$$+ \text{reg. coeff.} \times \text{longevity} + \text{constant}$$

Such equations cannot, however, be applied outside the populations for which they were derived. In fish, for example, there is often considerable zoogeographical variation in size-specific fecundity, as shown by Henderson (1998) for the dab, *Limanda limanda*, in the north-east Atlantic. Further, females that have developed under different regimes of nutrition and crowding, will have different potential fecundities (Blais, 1953; Clark, 1963). Therefore, for example, it is necessary to have different equations for different stages in an insect pest outbreak (Miller, 1957). Larval density of course affects size and longevity directly (Miller & Thomas, 1958), but the fact that new equations are necessary shows that the effects on fecundity are not just reflections of changes in these properties. Evidence is accumulating that changes may occur in the genetic constitution of fluctuating populations and the fecundity of different forms of a species may be quite different. Therefore, laboratory measurements of fertility should be continually checked, for it is important in any population study to establish not only the potential fertility, but any variation in it, for this may be the essence of a population regulation mechanism (p. 453).

When large numbers of eggs are laid together in a group the weight of the individual female will fall sharply after each oviposition and slowly rise again until the next group is deposited. Where individual animals can be marked it may be possible by frequently recapturing and weighing to establish the time of oviposition of actual egg batches; Richards & Waloff (1954) were able to do this with a field population of grasshoppers. It is sometimes possible to mark some insects with a radionuclide (e.g. zinc-65 [^{65}Zn]) and then estimate egg production by measuring the loss of isotope (Mason & McGraw, 1973). The actual number of eggs per egg mass may vary with size or age of the female (Richards & Waloff, 1954) or from generation to generation (Iwao, 1956).

Age- and size-specific variation in eggs per batch and the individual size of eggs is commonly observed in fish and crustaceans. There is an extensive literature on the effects of egg size on juvenile growth and survival (e.g. Segers & Taborsky, 2011).

10.1.2 *Numbers entering a stage*

When an animal can be trapped as it passes from one part of its habitat to another part at a specific stage of its life-history, the summation of the catches provides a convenient measure of the total population entering the next stage. Such traps are commonly referred to in entomological research as *emergence traps*. The particular design will depend on the insect and its habitat; a number are described by Peterson (1934), Nicholls (1963) and Ives *et al.* (1968). Basically, the traps consist of a metal or cloth box that covers a known area of soil and glass collecting vials with 'lobster-pot'-type baffles inserted in some of the corners (Fig. 10.1). A newly emerged insect being positively

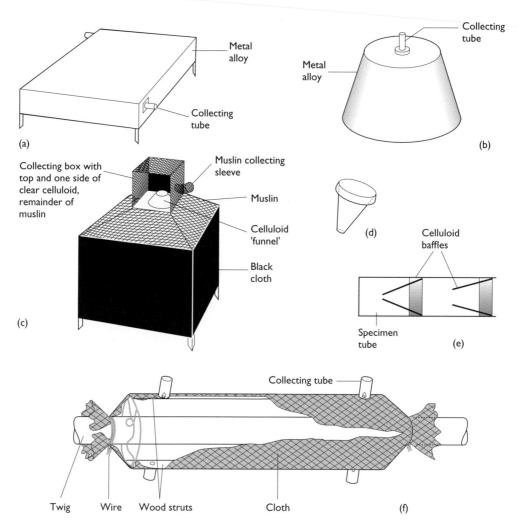

Figure 10.1   Emergence traps for terrestrial habitats. (a, b) Metal box and tub, respectively; (c) Cloth-covered Calnaido type; (d) Celluloid baffle of collecting tube; (e) Sectional view of collecting tube with pair of baffles designed to separate, partially at least, large and small insects; (f) Cloth trap for twigs. Panels (a, b, c) (adapted from Southwood & Siddorn, 1965); Panel (f) (adapted from Glen, 1976).

phototactic will make its way into the collection vials, which are emptied regularly; Turnock (1957) used an adhesive resin in the collecting containers of his trap so that they could be emptied at less-frequent intervals. Releasing a known number of newly emerged individuals into it can test the efficiency of a trap for a particular insect. The construction of the trap will influence its effect on the microclimate; all traps tend to reduce the daily temperature fluctuations (Fig. 10.2), and the deeper they are (the greater the insulating layer of air) the smaller the fluctuations for the same covering. Cloth-covered traps lose the heat less quickly than metal ones, and wood and screen traps do not heat the soil to the same degree as metal ones (Higley & Pedigo, 1985).

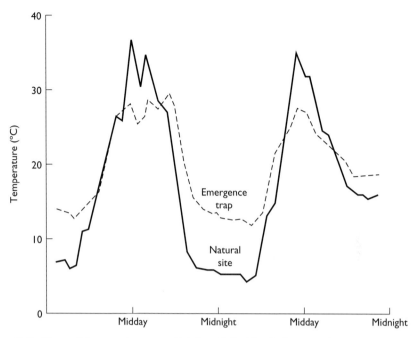

Figure 10.2   The soil temperature over 48 h in a Calnaido cloth-covered emergence trap compared with that from a natural site. (Adapted from Southwood & Siddorn, 1965.)

Although the daytime deficit may be approximately balanced by the greater warmth at night (Fig 10.2), over a period of several weeks these small daily excesses or deficits will accumulate to levels where they might influence the development rate of pupae (Southwood & Siddom, 1965). For this reason, caution should be exercised in using emergence trap data for phenology. Rice & Reynolds (1971) collected twice as many pink bollworms (*Platyedra*) from metal-screened emergence cages as from plastic-screened ones. Special traps have been described by many workers, including one for fleas (Bates, 1962) and others for ceratopogonid midges (Campbell & Pelham-Clinton, 1960; Neilsen, 1963; Davies, 1966; Braverman, 1970), for anthomyiid flies (Dinther, 1953), for cecidomyiid midges (Speyer & Waede, 1956; Nijveldt, 1959; Guennelon & Audemard, 1963) for beetles (Richards & Waloff, 1961; Polles & Payne, 1972; Boethel *et al.*, 1976) and for lepidopterous larvae emerging from maize 'ears' (Straub *et al.*, 1973).

A number of special traps have been designed for bark-dwelling insects. Moeck (1988) describes a trap constructed from a plastic pail and funnel that was found to be 97.5% efficient at collecting insects from lodgepole pine bolts. For bark-beetles and other insects occurring at high densities it may even be useful to subdivide the trap to record the dispersion pattern (Reid, 1963). Simpler traps have been developed by Nord & Lewis (1970) and Glen (1976), essentially consisting of a wire and wood frame, covered with black cloth (see Fig. 8.5): Glen found mean temperature was hardly affected.

Emergence traps (Fig. 10.3) have been used extensively in aquatic environments for measuring the quality, quantity and biomass of insects emerging from the habitat,

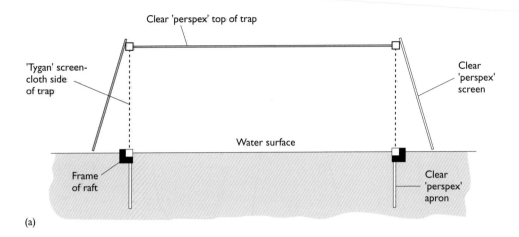

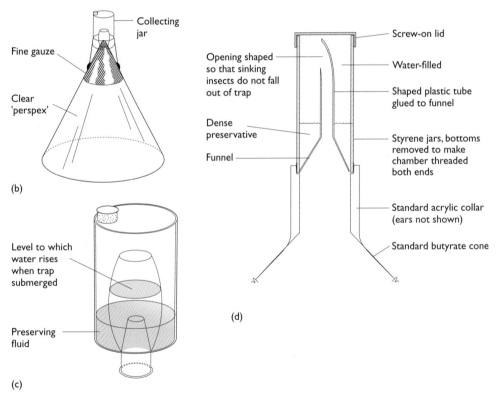

Figure 10.3 Emergence traps for aquatic habitats. (a) Floating box trap for use underexposed situations; the screen may be omitted in sheltered situations (adapted from Morgan *et al.*, 1963); (b) Submerged funnel trap, with details of collecting jar (c) (Adapted from Borutsky, 1955); (d) Submerged trap for chironomids. (Adapted from Welch *et al.*, 1988.)

and have been reviewed by Davies (1984). Examples of their application are Malison *et al.* (2010), Aagaard *et al.* (1997), Boettger & Rudow (1995), Speth (1995) and Dewalt *et al.* (1994). In shallow water, amongst emergent vegetation and in sheltered situations, floating box traps have been utilised (Adamstone & Harkness, 1923; Macan, 1949; Vallentyne, 1952; Sommerman *et al.*, 1955; Judge, 1957; Morgan & Waddell, 1961). In exposed situations submerged funnel traps have been used as these avoid damage from wind and rain (Grandilewskaja-Decksbach, 1935; Brundin, 1949; Jónasson, 1954; Palmén, 1955, 1962; Mundie, 1956; Darby, 1962; Mulla *et al.*, 1974; McCauley, 1976). A careful comparison of these two types of trap for emerging insects have been made by Morgan *et al.* (1963) and Kimerle & Anderson (1967), from which the following conclusions may be drawn. Funnel traps are generally much less efficient than floating box traps. A number of factors probably contribute to this: one factor is the decomposition of the catch in the small air space of the collecting jar of the funnel trap. Such losses will be proportionally greater the denser the population of emerging insects and the longer the intervals between emptying the trap, which should therefore be done daily. The ascending pupae and larvae are strongly phototactic and the slight shade produced by the gauze of the funnel trap will cause it to be avoided; furthermore some insects (e.g. Odonata) will crawl out again. It is impossible to exclude small predatory insects from the traps and they may consume much of the catch. Lastly, the effective trapping area is reduced if the trap becomes tipped. However, in very exposed situations these traps must be used, and Morgan *et al.* (1963) recommend a design similar to that shown in Fig. 10.3, to overcome as many of these faults as possible. Borutsky (1955) originally devised the collecting jar; it is very important that the 'Perspex' be kept clean as algal growth will soon render it opaque. Figure 10.3 shows the preserving funnel trap described by Welch *et al.* (1988), which is a modification of the basic design of Davies (1984). The sampling efficiency of these traps was tested with and without preservative, with transparent and opaque collecting tubes, and with and without the funnel which stops the insects falling back into the water. Traps holding preservative (normally phenoxyethanol, but diesel fuel, saltwater with formalin and Mistovan® were also tried) caught as many chironomids as those without. However, these chironomids were rarely in the adult form as they died before eclosion. It was found that the preservatives were unable to stop decay in traps left in place for 12 or 18 days. Contrary to the general finding that opaque traps catch fewer animals (Davies, 1984), no significant difference was found by Welch *et al.* (1988), although they did find that most chironomids remained as pupae in the opaque traps. Funnels are clearly important, as those without them achieved only 17% of the pupal and 58% of the adult catch of chironomids obtained by those with funnels. Further information on the choice of aquatic emergence trap and their relative efficiencies are given in Davies (1984), LeSage & Harrison (1979), Morgan (1971) and Mundie (1971).

The most important causes of loss of efficiency of floating box traps are waves, which will swamp the catch, and the shading effect of the trap itself on the ascending larvae and pupae. Morgan *et al.* (1963) point out that as it is probably only those animals that ascend near the edges of the trap, that can take successful avoiding reaction, the larger the area of the trap the smaller this edge-effect relative to the size of the catch. The trap designed by Morgan *et al.* (Fig. 10.3a) is constructed mainly from clear Perspex;

the apron projecting in the water retains floating exuviae within the trap and helps to reduce wave damage; the latter function is also served by the lateral screens. The wooden frame should be as narrow as possible and painted white to minimise shadow; wire netting stretched under the frame reduces predation by fish. The trap needs to be emptied every two days, and this is a difficult process. A floating tray is inserted under the trap and the whole towed to a boathouse or similar site where the unidirectional light source ensures that the edge on the darkest side may be lifted, to allow the entrance of an entomological pooter (= aspirator; see Fig. 7.5) without any of the insects escaping. On dull days or at night, the trap may be tilted towards a lamp during emptying. In view of the time-consuming nature of this operation it would seem worthwhile to determine whether shrouding the traps for a short while in the day would drive a large proportion of the catch into a collecting tube inserted in the roof, which is the method of collection from tents (p. 162).

Kimerle & Anderson (1967) had a sliding base in their trap that greatly facilitated the retrieval of the whole catch, and Corbet (1966) arranged for the traps to be closed and emptied in the laboratory. Carlson (1971) increased catch size (and difficulties of interpretation regarding the catchment area) by incorporating a battery-operated light (see p. 294).

Langford & Daffern (1975) designed traps with floats for work in large, fast-flowing, rivers: they found these robust, but there was some evidence that the numbers of Trichoptera, but not Ephemeroptera, caught were affected by the presence of floats. Wrubleski & Rosenberg (1984) noted that polystyrene floats, as fitted to the 'week' model tent trap of LeSage & Harrison (1979), tended to become colonised by chironomids; a polyvinyl chloride (PVC) pipe, which is easily cleaned, would be preferable.

In very shallow fast-running streams it may be necessary to use a tent trap (Ide, 1940; Anderson & Wold, 1972); basically, this resembles a gauze tent that is attached to the substrate. A rather unsatisfactory feature of the design is that the observer has to enter the trap to remove the insects; not only (as Ide mentions) is he/she exposed to noxious insects (e.g. Simuliidae) but the resultant trampling of the substrate must affect future emergence. Possibly a smaller tent should be used – one that can be emptied from the exterior.

In flowing waters an emergence trap receives input from the insects which were living within the substrate below the trap, plus drifting animals brought in from upstream. To quantify the drifting component, Steffan (1997) describes an emergence trap fitted with an underwater sieve plate so that any drifting fauna enters the trap while those of benthic origin are excluded.

Using a using a trap similar in appearance to the submerged-funnel type, Lindeberg (1958) sampled Chironomids from very shallow rock pools. The funnel was made of glass and the whole upper part of the trap was airtight so that it could be kept full of water, even though it projected well above the level of water in the pool. Cook & Horn (1968) developed a sturdy trap for damselflies and similar insects emerging from rush in shallow water in areas frequented by cattle. This consisted of fencing wire, covered with fibreglass window screening.

Material containing the resting stage of the animal may be collected and exposed under field conditions or in the laboratory. Bark beetle emergence from logs may be

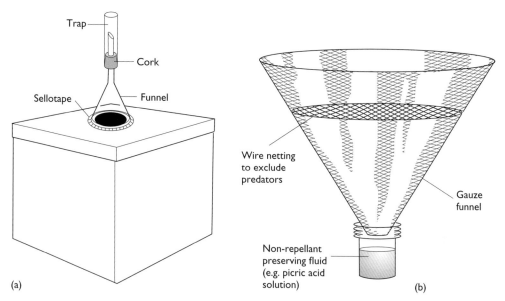

Figure 10.4   **(a) Emergence tin to determine in the laboratory the emergence of insects from samples. (Adapted from van Emden, 1962); (b) Funnel trap for collection of descending arboreal larvae.**

measured under natural conditions by placing the logs in screened cages (McMullen & Atkins, 1959; Clark & Osgood, 1964). Wikars *et al.* (2005) compared the efficiency of window traps, emergence traps and sieving to estimate the species richness of saproxylic beetles in logs and stumps of Norway spruce, and found that emergence trapping collected the greatest number of species over the whole season. The emergence of flies from heads of grass and corn was determined by placing these in muslin bags in the field (Southwood *et al.*, 1961). The emergence of mites from overwintered eggs on twigs or bark was measured by Morgan & Anderson (1958) by attaching the substrate to the centre of a white card and surrounding it with a circle of a fruit-tree banding resin. As the mites crawled out they were caught in the resin and, being red, were easily counted.

Unseasonable emergence may be forced in the laboratory (Wilbur & Fritz, 1939; Terrell, 1959; van Emden, 1962; McKnight, 1969; Gruber & Prieto, 1976). Commercial containers such as shoeboxes, ice-cream cartons and biscuit tins have been found useful in such studies (Fig. 10.4a).

Insects that are arboreal for part of their life may often be trapped on their upward or downward journeys. Wingless female moths and others moving up tree trunks may be trapped in sackcloth and other trap-bands (DeBach, 1949; Reiff, 1955; Otvos, 1974) or in inverted funnel traps (Varley & Gradwell, 1963; Ives *et al.*, 1968; Agassiz, 1977). Larvae descending to pupate may be caught in funnel traps (Fig. 10.4b) (Ohnesorge, 1957; Pilon *et al.*, 1964).

Emergence traps of the types used for aquatic insects can also be placed over gravel spawning substrate to quantify trout or salmon fry emergence (e.g. Weaver & Fraley, 1993). An assessment of the vertical movement of benthic marine copepods upwards from sea grass beds and their subsequent resettlement has been made by Bell *et al.*

(1989) and Kurdziel & Bell (1992), using emergence-type traps. Freshwater animals that migrate to different habitats at set developmental stages such as salmon, *Salmo salar*, can be counted by placing automatic counters in weirs (see p. 274).

10.1.3 *The birth-rate from mark and recapture data*

The details of marking animals and of the equations for estimating the birth-rate are given in Chapter 3. This birth-rate will include any immigration; it is thus more correctly referred to as the dilution rate.

10.2 Mortality

10.2.1 *Total*

Measurements of total mortality are also commonly obtained by the subtraction of population estimates for successive stages (e.g. Miller, 1958; Cook & Kettlewell, 1960; see also Chapter 11).

10.2.1.1 Successive observations on the same cohort

When it is possible to make these, mortality may be measured directly. Completely natural cohorts can in general only be followed in sedentary or relatively immobile animals (e.g. the egg and pupal stages of insects, scale insects, some aphids and Hemiptera, birds eggs, benthic worms, barnacles and other attached epibenthic animals). Colonies present in the field in accessible positions may be repeatedly examined, each colony being identified by labelling (e.g. MacLellan, 1962; van Emden, 1963); indeed, with an animal as large as the hornworm (Sphingidae), Lawson (1959) was able to follow the larvae as well as eggs. In other situations, it may be necessary to delimit the sample population or even to 'plant' it. Examples of the former are Ives & Prentice's (1959) and Turnock & Ives' (1962) studies, where they placed pupal-free moss-filled trays or blocks of peat with screen bottoms and sides under the host trees of certain sawflies at pupation. Later in the season, these natural traps may be removed and the pupae classified. More mobile animals will need to be retained by a more extensive enclosure (Vlijm *et al.*, 1968; Ashby, 1974) or 'field cage' (Dobson *et al.*, 1958; see p. 26). Hoddle (2006) followed cohorts of the hemipteran, *Tetraleurodes perseae*, held in fine mesh bags on the branches of avocado. It is important to remember, however, that the screens and cages are bound to alter – albeit only slightly – the situation one is trying to assess. The planting of known populations in the field and their subsequent re-examination gives a measure of the level of mortality. Sometimes, it is possible to recover the remains of the dead individuals in such experiments, and then further information may be gained about the cause of death (Graham, 1928; Morris, 1949; Buckner, 1959; Pavlov, 1961).

When an insect leaves a mark of its presence, as do many that bore into plant tissue, a single count at, for example, the pupal stage enables the still-living and the dead to be distinguished (see also p. 343). Furthermore, the latter may be divided into those whose burrows have been opened by predators and those killed by parasites and disease. Such a study, which of course is in effect population estimates of two successive stages (i.e. total young larvae = total burrows, and pupae), was made by Gibb (1958) for the Eucosmid moth, *Ernarmonia conicolana*, in pine cones.

10.2.1.2   The recovery of dead or unhealthy individuals

This gives another measure of mortality. Although Gary (1960) devised a trap for collecting dead and unhealthy honey-bees, it is unusual to be able to recover the non-survivors of a mobile stage unless they are marked with a tracking device (p. 120). However, unhatched eggs can often be examined to ascertain the cause of their death (e.g. Bess, 1961; Way & Banks, 1964). Matsuzawa *et al.* (2002), from an examination of the unhatched eggs of loggerhead sea turtle, *Caretta caretta*, together with temperature records from within the nest, concluded that pre-emergent mortality was linked to excess heat.

10.2.2   *The death-rate from mark and recapture data*

Details of these models and the calculation of the death-rate, which includes any emigration, are given in Chapter 3 (see also p. 405).

10.2.3   *Climatic factors*

Apart from direct observations on a known cohort, the main method of establishing the role of climate in the total mortality has been experimental. A known number of individuals may be exposed to field conditions (e.g. Lejeune *et al.*, 1955) or predictions may be made from laboratory experiments (MacPhee, 1961, 1964; Green, 1962; Sullivan & Green, 1964; Kensler, 1967; Carter, 1972; Abdel Rahman, 1974; Neuenschwander, 1975); in the latter case it is important to allow for the effect of acclimatisation. Cold hardiness may be influenced by the presence of free water in the surroundings and of food in the gut (Eguagie, 1974).

10.2.4   *Biotic factors*

As with climatic factors, knowledge of the role of various biotic factors is often obtained from successive observations on the same cohort (see above); this section is restricted to techniques for the recognition of the role of individual factors. It is important to remember that the effect of a parasite will vary with the host, as has been demonstrated by Loan & Holdaway (1961). For example, the braconid parasite of a weevil, which has no effect on the adult male, causes egg production by the female to fall off quickly. The quantitative evaluation of natural enemies has been reviewed by Kiritani & Dempster (1973) and Sunderland (1988): there are four methods of direct assessment, together with the indirect approach by laboratory experiment.

10.2.4.1 Direct observation of predation

Sometimes it is possible to estimate predation on the basis of sightings of predators consuming prey (e.g. Kiritani *et al.*, 1972; Kiritani & Dempster, 1973; Latham & Mills, 2009). The observed frequency of feeding (F) is determined by routine counts, together with the length of time captured prey are retained (R), and the diurnal rhythm of feeding to give the proportion of feeding (C) that occurs during the time interval when the observations are made. Now, if the probability of observing feeding (P_f) is retention time in hours (R) divided by 24, then an estimate of the total number taken is:

$$n = {FC}/{P_f} \tag{10.1}$$

A series of values of n may be plotted against time, and the area under the graph gives the total prey killed in that habitat. This 'sight–count' method was devised and used by Kiritani *et al.* (1972) when studying spider predation of leafhoppers. It depends on a high accuracy in observing all instances of predation at a given time, and on the values of C and R being fairly constant. Latham & Mills (2009) compared estimates of predation by aphid predators derived from field observation, laboratory areas and field cages. They concluded that, for highly mobile predators or predators of dispersed prey, field observation combined with observations of both the duration and pattern of feeding activity throughout the day, is the best option for quantifying daily per capita consumption.

As the young of many birds remain in the nest, it is possible to record their food. The simplest method is to observe feeding through the glass side of a nesting box (Tinbergen, 1960). However, identification and counting of the prey is often difficult, and Promptow & Lukina (1938) and Betts (1954, 1958) have shown that parent birds may be induced to put the whole, or at least part, of the food for the young into an artificial gape, from which it may be removed. Photography and video can also be used to record the parental visits, when it is often possible to identify the food carried.

When a single parasite larva emerges from each individual of an arboreal host and drops to the ground to pupate, these may be collected in trays or cone traps (see p. 339) to give an absolute measure of their own population, which is equivalent to the number of hosts they have killed. Such conditions are fulfilled, for example, in many tachinid parasites of Lepidoptera. When the percentage parasitism of the host is known accurately such figures may also be used to calculate the actual host population (Dowden *et al.*, 1953; Bean, 1958).

10.2.4.2 Examination of the prey (host)

The examination of vertebrates for exoparasites has been described on p. 166. However, a description of methods for the detection of internal vertebrate parasites is beyond the scope of this book. These methods include dissection, histological examination of tissues, forced removal from the gut, and examination of faeces.

Parasitic insects may often be detected in their hosts by dissection (e.g. Miller, 1955; Hafez, 1961; Evenhuis, 1962; Hughes *et al.*, 2003) or by breeding out the parasites from a sample of hosts (e.g. Richards, 1940; Sasaba & Kiritani, 1972; Core *et al.*, 2012). As the rate of parasitism varies throughout a generation of the host, a single assessment will not give an adequate degree of precision unless the hosts are closely synchronised. Parasites may sometimes be detected within their hosts without dissection by the use of soft X-rays (wavelength >0.25 Å) and fine-grain film (Holling, 1958). It is possible that elutriation (p. 234) could also be used to diagnose parasitism. The proportion of hosts attacked by a parasite is referred to as the *'apparent parasitism'*. The problems involved in the combination of a series of apparent mortalities and its interpretation are discussed in the next chapter.

Sometimes it is possible to find a good proportion of the corpses of insects killed by a predator, generally when the predator collects the food together in one place, such as spiders' webs (Turnbull, 1960), hunting wasps' nests (Rau & Rau, 1916; Richards & Hamm, 1939; Evans & Yoshimoto, 1962), thrushes' anvils (Goodhart, 1958) and shrikes' larders (Lefranc & Worfolk (1997). Ants provide unusual opportunities for this type of assessment because in many species the foragers return to the nest with virtually intact prey, along trails; special apparatus has been designed to sample returning foragers and their prey (Chauvin, 1966; Finnegan, 1969). The frequency of beak-marks on butterflies, arising from unsuccessful attacks by birds, has often been recorded. Its interpretation is difficult, but it may be used as evidence for seasonal variation in predation intensity (Shapiro, 1974). The food brought by their parents to nestling birds may be recorded photographically and subsequently identified to broad categories.

A semi-natural assessment of the role of predators may be made by 'planting' a known number of prey in natural situations. Buckner (1958) describes how sawfly cocoons were exposed in this way and the type of predator determined by the markings left on the opened cocoons.

The diagnosis and determination of the extent of infection by pathogens in an insect population is a complex and specialised problem that is outside the scope of this book; reference should be made to works such as Poinar & Thomas (1984), Lacey (2012) and Steinhaus (2012). It should be noted that some methods of collection (e.g. sweep net) may lead to an overestimate of the proportion of diseased insects in the population (Newman & Carner, 1975). It has also been frequently suggested that parasitised fish are more vulnerable to capture.

10.2.4.3 Examination of the predator

Analysis of faeces and pellets

For some (particularly vertebrate) predators the faeces can be examined for the remains of prey. If the number of prey individuals can be counted, then given the gut evacuation rate the number of prey consumed can be estimated. Unfortunately, for many poikilotherms the evacuation rate is affected by the size of the meal (Sunderland, 1988), although this may not be the case with fish (Elliott & Persson, 1978). For specialist predators quantification may even be possible if assimilation data are available by using the weight of faeces produced to estimate the number of prey consumed. Laboratory studies such as that of Hewitt & Robbins (1996) on the grizzly bear, *Ursus arctos*, need to be undertaken to relate the volume of food eaten to the faecal volume produced if the relative importance of different foods is to be estimated from their presence and quantity in the faeces. Captive studies were also undertaken by Cottrell *et al.* (1996) on harbour seals, *Phoca vitulina*, to assess the survival in recognisable form of fish remains such as eye lenses, scales, vertebrae and otoliths, which suggested that no one single structure such as an otolith should be used. Generally, care must be taken because the species list from stomach contents (or pellets) and the actual food eaten will often not correspond (Hartley, 1948. Cavallini & Volpi (1995) compared gut and faecal content analysis for the red fox, *Vulpes vulpes*, and concluded that there were significant differences with birds in particular, declining in dominance from stomach to faeces. Many predation studies have used faecal analysis, some examples are: Cavallini & Serafini (1995) on small Indian mongoose, *Herpestes auropunctatus*; Kohira & Rexstad (1997) on wolves, *Canis lupus*; Le Jacques & Lode (1994) on European genets *Genetta genetta*; Mukherjee *et al.* (1994) on Asiatic lion, *Panthera leo persica*, Santori *et al.* (1995) on the marsupials *Metachirus nudicaudatus* and *Didelphis aurita*, and Michalski *et al.* (2011) on insectivorous Blue tits, *Cyanistes caeruleus*, and Great tits, *Parus major*.

Some predators regurgitate pellets, comprising the indigestible parts of their meals from which their prey can be identified. For example, the remains of insects and other prey are found in the pellets of owls (Hartley, 1948; Miles, 1952; Southern, 1954; Jedrzejewski *et al.*, 1996; Avenant, 2005; Lloveras *et al.*, 2009), cormorants, *Phalacrocorax auritus* (Johnson *et al.*, 1997), brown skua, *Catharacta skua* (Moncorps *et al.*, 1998) and white stork, *Ciconia ciconia* (Barbraud & Barbraud, 1997). Jedrzejewski *et al.* (1996) quantified the numbers of prey of each species taken by owls from 1 ha of woodland by collecting the pellets under roost trees and identifying the prey eaten and their number in each pellet from skeletal remains and the presence of hair and feathers. This quantification required estimates of owl density, a coefficient to correct for the fact that some of the prey disappears during digestion, and the number of pellets produced per day. When both pellets and faeces are available for analysis the best results are likely to come from pellet analysis as the remains are more easily identified. Johnson & Ross (1996) examined fish remains in the pellets and faeces of the double-crested cormorant, *Phalacrocorax auritus*, and found that about 90% of all the diagnostic fish remains found were from pellets. Fish otoliths are particularly valuable as they can occasionally be identified to species or genus and are not easily digested. They can also be used to estimate the age and size of the fish.

Gut contents analysis

One of the most direct and frequently used methods for the identification of predator–prey linkages is the examination of gut contents. For small animals this normally requires the animals to be killed so that the gut can be dissected. It is generally undesirable in a study (quite apart from conservational and legal considerations) to kill large numbers of vertebrates, although this is considered acceptable in fisheries studies. For certain larger animals, such as fish or caiman (e.g. Laverty & Dobson, 2013) it is often possible to wash out the stomach contents without causing long-term harm to the animal. Similarly, the young of some birds will regurgitate the food if their necks are manipulated shortly after feeding (Errington, 1932; Lack & Owen, 1955); with others a neck ring needs to be used to prevent swallowing. Emetics may sometimes be administered to recover stomach contents (Radke & Frydendall, 1974). Even when the predator eats large prey, macroscopic examination alone may be insufficient to identify all the prey consumed: Douglas (1992) found that a microscopic examination of the gut contents of the grass snake, *Psammophylax rhombeatus*, improved both the detection and identification of both mammalian and lizard remains. The ability to detect the remains of different animals in invertebrates often requires skill and experience because the meal is often broken into small pieces. However, given experience the gut contents of even microscopic animals such as copepods and ostracods can be identified. By measuring the size of remains in the gut it may be possible to infer the size range of the prey. For example, using samples taken from the guts of large predatory fish, Smale (1986) identified and estimated the size of fish and squid from otoliths and 'beaks', respectively.

While many predatory insects are fluid feeders, skeletal remains can be found and identified from the guts of Dermoptera, coccinellids and carabids (Sunderland, 1988). In general, more refined methods of detection are needed for insects and these are discussed below.

Differential digestion is one of the problems in interpreting data from stomach and faecal contents (Buckner, 1966).

Chromatography

Generally of little application, but Knutsen & Vogt (1985) used gas chromatography to identify dietary components in the guts of the lobster, *Homarus gammarus*.

Electrophoresis

Electrophoresis can be used to separate prey enzymes such as esterases to produce banding patterns that are species-specific. The different enzymes (and other proteins) are separated because they migrate through a gel at different rates when an electrical field is applied. Specific stains detect the individual enzymes. For the method to be applied, reference gels must be constructed for potential prey, starved predators and predator guts holding known prey. Because of their diversity and high activity, and thus ease of staining, esterase banding patterns are frequently used although other enzymes are also potentially available. Giller (1986), for example, used esterase banding to detect the prey of notonectids. Sunderland (1988) suggests that this method is best used for predator–prey systems with few species as this enhances the likelihood that banding patterns for the various species will be unique. The method suffers from numerous pitfalls; perhaps one of the most important is that, if not all the prey is consumed, then the prey banding pattern from the gut may not contain all the bands seen on the reference gels. The method may also be unable to distinguish between some species. Diallo *et al.* (1997) describes a technique based on the electrophoresis of superoxide dismutase to distinguish between tsetse blood meals of human and non-human origin.

Serological methods

Serological techniques, originally developed by vertebrate immunologists, were once used extensively in entomology for gut analysis of species which suck fluids, particularly for the identification of blood meals of biting flies. Sunderland (1988) reviews the use of these techniques for the quantification of predation. These methods often only identify the family or order of the source of the blood meal and, while still used, they are becoming largely replaced by DNA-based methods which can give species level identification (see below).

DNA-based methods

Polymerase chain reaction (PCR)-based techniques for detecting prey remains in the guts, faeces and regurgitates of predators has rapidly developed during recent years, and is becoming the technique of choice when solid remains cannot be identified. A description of these techniques is beyond the scope of this book (see King *et al.*, 2008 for a review of DNA-based methodologies and best practice). The identification of blood meals by DNA-based methods is reviewed by Kent (2009).

Labelled prey

The prey may be labelled with a dye, a rare element or a radioactive isotope; it is important that the label does not modify the behaviour of the predator. The egg predators of a moth were detected by spraying the eggs with an alcoholic suspension of a powdered fluorescent dye. The dye could be detected, under UV light, in the gut of the predators; this was particularly easy if the gut was macerated and dried (Hawkes, 1972).

Feeding can also easily be established if the prey are marked with a rare element (see p. 119) or a radioactive isotope and the label is subsequently detected in the predator. Ito *et al.* (1972) used europium for this purpose, but found it to be rapidly excreted by insects. However, Hamann & Iwannek (1981) found europium to remain detectable for 3 weeks in tsetse flies, while lanthanum was detectable for only a few days. If each prey carried a similar burden of rare element or radionuclide, the level of these labels in the predator would be a measure of the prey consumed.

Historically most radioactive labelling studies involved prey tagged with ^{32}P and had as their objective the identification of the predators (Jenkins & Hassett, 1950; James, 1961, 1966; Jenkins, 1963; Walker *et al.*, 1991). Radiotracer studies are now rarely undertaken and are not discussed further here.

10.2.4.4   Exclusion techniques

These methods seek to demonstrate the effects of predators, parasites or competitors by artificially excluding them and measuring the increase in the prey's population under the new conditions; this increase represents the action of the predator (or parasite) in the exposed situation. Conceptually, the method is very attractive, but its weakness is that the techniques of excluding the predators often affect the microclimate and other aspects of the habitat, so that it is difficult to be certain that the observed effects are entirely due to the predator (Fleschner, 1950).

Mechanical or other barriers

Mammals and birds are relatively easily excluded by wire netting or nets, and it is unlikely that such methods have marked side effects on the prey species, unless of

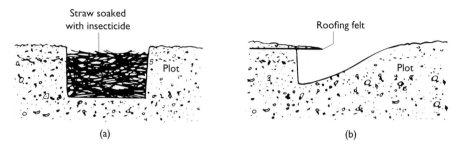

Figure 10.5   Barriers to carabid beetles. (a) Exclusion barrier of straw soaked with insecticide; (b) Emigration barrier that allows beetles to enter the plot but not leave. (Adapted from Wright *et al.*, 1960.)

course they also exclude herbivorous mammals whose activities have an important role on the vegetation. Buckner (1959) used mammal exclusion cages in his study of the predation of larch sawfly, *Pristiphora*, cocoons; in this case both the protected and exposed cocoons were 'planted' and therefore there was the risk that the resulting artificially high density might have acted as a bait to the mammals.

Predatory insects need to be excluded by smaller muslin cages of the sleeve-cage type (Smith & DeBach, 1942), but as Fleschner (1950) and Hunter and West (1990) have shown such cages may affect the population directly and therefore, even if some cages are left open for predator access as the cages affect the prey's rate of increase, the natural situation cannot always be measured (De Bach & Huffaker, 1971). Weseloh (1990) tethered Gipsy moth, *Lymantira*, larvae in cages with various-sized meshes to determine the role of predators of different sizes in different microhabitats.

Non-flying insects can be more easily excluded by mechanical or insecticidal barriers. Wright *et al.* (1960) and Coaker (1965) demonstrated the effects of carabid beetles on populations of the cabbage root fly, *Erioischia brassicae*, by means of trenches round the plots, filled with straw soaked with an insecticide (see Fig. 10.5). Spiders (Clarke & Grant, 1968) and ants can often be eliminated by similar barriers or bands on trees (p. 290) (e.g. Weseloh, 1993). Various combinations of exclusion cages may be used to separate the roles of vertebrates (wire cages), ground-dwelling invertebrates (mechanical barriers) and flying invertebrates (net cages – variations in net size can produce further separation) (Ashby, 1974; Eickwort, 1977).

Marine benthic ecologists have frequently used nets or cages to exclude predators (e.g. Posey *et al.*, 2006).

Elimination of predator or parasite

Vertebrates may be either shot (Dowden *et al.*, 1953) or trapped (Buckner, 1959) until their numbers are extremely low; as this must be done over a fairly large area to be effective it will lead to major disturbance in the habitat; this is undesirable in ecological studies and often on conservation grounds. Fleschner (1950) found hand-picking of invertebrate predators from a part of the tree was both a feasible and reliable method.

As some predatory insects are very susceptible to certain insecticides to which the prey is resistant, it is possible to assess their effects by an insecticidal check method (Debach, 1946; Debach & Huffaker, 1971). It is very important to ensure that the pesticide has no side effects on the prey species. Although this method was found reliable for studies on predators of *Aonidiella* scale insects (Debach, 1955), it was not satisfactory for various plant-feeding tetranycid mites (Fleschner, 1950).

A biological check method has also been described in which large populations of ants are built up; these attack and often prevent predator and parasite action (Fleschner, 1950). Although such observations do give a measure of predator importance, there are so many other factors involved that they cannot be considered as equivalent to the quantitative estimation of predation.

10.2.5   *Experimental assessment of natural enemies*

There is an extensive theoretical background, supported by field and laboratory studies, on the interactions of prey with their natural enemies; the subject is reviewed by Murdoch & Oaten (1975), Beddington *et al.* (1976), Hassell *et al.* (1976), Hassell (1976a,b), May (1976) and Hassell (1978). The interactions may be viewed in relation to two population variables (Hassell, 1976a; Hassell *et al.*, 1976; Beddington *et al.*, 1976):

1. The death rate of the prey.
2. The rate of increase of the predator.

The two are, of course, interrelated; the prey death rate depends on the numbers of predators and on their searching efficiency. To some extent this classification parallels that between functional and numerical responses (Solomon, 1949; Holling, 1959a,b, 1961, 1965, 1966), but is somewhat broader in that prey density is now only one of the independent variables against which predator efficiency or reproductive rate is viewed. The major relationships are summarised in Table 10.1. The following sections indicate some approaches to the measurement of these relationships. Once this has been done the best procedure, as almost invariably in ecology, is to plot the data; when this has been completed it will, in many cases, be possible to select the most appropriate model and determine precisely the value of the major component parameters. For some other relationships, clear models have yet to be developed.

10.2.5.1   **Death rate of prey at differing prey densities**

This is most elegantly measured in the laboratory with different prey densities per predator and the numbers of prey killed or hosts parasitised (for insect parasitoids) recorded.

The numbers of prey eaten per predator at different prey densities can also be obtained from field data, although it is rare that an adequate estimate of the density of searching predators is available. Such results, however, should only be interpreted as a functional response *sensu strictu* if the density is relatively constant. Wide fluctuations

Table 10.1   Some relationships whose measurement is important in assessing the role of natural enemies. Other variables: A = age of predator; B = age of other predators of same species; C = climate and other external drivers; D = other prey (switching); E = other animals (competition and mutualism).

Component	In relation to	Other variables	Affecting
Attack rate a Handling time T_h	prey density 'functional response'	A,C,D,E.	
Interference constant m Quest constant q or Time lost through encounters bt_w	predator density	A,B,C,D,E.	death rate of prey
Predator dispersion P_i (number/ith area) T_i (time/ith area)	prey dispersion N_i 'aggregative component of numerical response'	A,C,D,E.	
Predator fecundity Survival rate of predator of different stages Duration of different stages of predator	prey density 'breeding component of numerical response'	B,C,D,E.	rate of increase of predator

in predator density (and food choice – if polyphagous) may confuse the relationship by different levels of predator interference or cooperation occurring (see below).

There are two essential components to a functional response: an instantaneous attack rate (a) and a saturation term, often called the 'handling time' (T_h) but also including any effects of satiation. In addition the speed of movement of the predator (Glen, 1975), and its reactive distance (both of which may be influenced by habitat characteristics) and the proportion of attacks that are successful (Nelmes, 1974) and of prey that are fully consumed, are further variables. The number of prey dying and the number eaten may not be the same, either because the predator kills more than it consumes (Buckner, 1966; Kruuk, 1972; Toth & Chew, 1972) or because it disturbs the prey, driving them from the host plant, and exposing it to other mortalities. Nakasuji *et al.* (1973) showed this for spiders with lepidopterous larvae, and W. Milne (unpublished data) for birds with aphids. The type of habitat will influence predator–prey interactions through its effect on speed of moving and reactive distance, as well as through dispersion (Martin, 1969).

Many functional responses have been found to be of the 'Holling Type II form (Fig. 10.6). This has been traditionally described by Hollings disc equation:

$$N_a = \frac{aNTP}{1 + aT_hN} \tag{10.2}$$

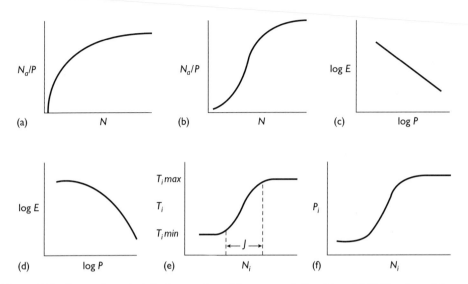

Figure 10.6 Some characteristic forms of predator–prey relationships. (a) Holling type II functional response; (b) Holling type III functional response; (c) Interference – Hassel & Varley model; (d) Interference – Rogers & Hassel or Beddington model; (e and f) Aggregative response measured in time and numbers.

where N_a = total number of prey attacked, a = attack rate, N = total number of prey, P = predator density, usually one in experimental systems, T_h = handling time and T = total time (searching time + $T_h N_a$, but excluding any regular sleeping time, i.e. sleeping not related to the amount of feeding).

The method of estimating a and T_h from functional response data using the above equation is described in Holling (1959b). This technique, however, does not allow for the removal of prey during the experiment (i.e. it assumes prey replaced as eaten or a systematically searching predator) and will lead to incorrect estimates if prey depletion is significant (Royama, 1971; Rogers, 1972) (c.f. removal trapping, p. 268). A more satisfactory method for attaining a and T_h is discussed by Rogers (1972), which assumes random exploitation of the available prey. Using this technique, it is necessary to distinguish between predators (that remove prey as eaten) and parasitoids (that can re-encounter hosts, thus involving more handling time):

Random predator:

$$N_a = N[1 - \exp\{-a(PT - N_a T_h)\}] \tag{10.3}$$

Random parasitoid:

$$N_a = N\left[1 - \exp\left\{-{TaP}\big/{(1 + aT_h N)}\right\}\right] \tag{10.4}$$

where P = number of predators (parasitoids).

Although the type II response of Holling is usually regarded as the form typical of invertebrate predators and parasitoids, there is increasing evidence that sigmoid type III responses are also widespread, at least amongst insects (Murdoch & Oaten, 1975;

Hassell, 1978), and are viewed as likely to occur with generalist predators that can switch prey. For example, Schenk & Bacher (2002) report a type III response in a study of paper wasp, *Polistes dominulus*, predation upon the shield beetle, *Cassida rubiginosa*. Experimental evidence suggests that these responses result from some component of *a*, or the time initially available for search (*T*), now being a rising function of prey density. The increased number of parameters involved in describing such responses makes their adequate estimation less straightforward than for type II responses, and requires a non-linear, least-squares method.

10.2.5.2   The death rate of prey varies with predator density

The presence of other predators may enhance (cooperation) or reduce (interference) the prey death rate due to a single predator. Instances of cooperation are particularly common amongst vertebrates and range from the group-hunting tactics of carnivorous mammals to the mere confusion of prey by several birds fishing in the same region. For example, Fryxell *et al.* (2007) developed group-dependent functional responses for lions and their prey in the Serengeti National Park. The defence mechanisms of, for example, a colony of aphids would probably be more easily disrupted by a number of coccinellid larvae than by a singleton, but this does not seem to have been quantitatively demonstrated; the more general result with invertebrates is interference (Hassell *et al.*, 1976). It is important to remember that, in laboratory experiments, predators or parasitoids are often prevented from dispersing when crowded; therefore unreal levels of interference and superparasitism may be observed (Hassell *et al.*, 1976). Interference effects, when plotted in a log scale, are commonly either linear (Fig. 10.6c) or curvilinear (Fig. 10.6d). When the relationship is linear then the component coefficients may be extracted using the model of Hassell & Varley (1969):

$$\log E = \log Q - m \log P \tag{10.5}$$

where P = predator density, m = interference constant, Q = quest constant (= E when $P = 1$) and E = searching efficiency of the predator. In general, E may be defined (Hassell, 1976b) as:

$$E = \frac{1}{P} \log_e \left[\frac{N}{N - N_{ha}} \right] \tag{10.6}$$

when N_{ha} = prey actually eaten or parasitised. Under the special circumstances when, every prey encountered is actually eaten and the number attacked is small compared with the total prey population, then

$$E = N_a / NP \tag{10.7}$$

The curvilinear relationships were discussed and modelled by Rogers & Hassell (1974) and also by Beddington (1975), who provided an expression for searching efficiency (*E*):

$$E = aT / [1 + bt_w(P - 1)] \tag{10.8}$$

where a = attack rate = N_a/NT_sP', T_s = total search time and P' = predator number, P = predator density, b = rate of encounter between parasites and t_w = time wasted per encounter.

Arditi & Ginzburg (1989) argued that the functional response when predator density has an effect can be modelled using the ratio of prey population size to predator population size rather than the absolute numbers of each species. The validity of this approach has been widely debated (see Abrams & Ginzburg, 2000).

10.2.5.3   The predator and prey form aggregations

Predators may aggregate in a region of high prey density. This aggregative or behavioural component of the numerical response has considerable significance in the stability of predator–prey systems (Hassell & May, 1974; Murdoch & Oaten, 1975). Observations may be made in the laboratory or in the field, measuring the number of predators found in regions of different prey density. When this can be done sequentially (e.g. by time-lapse photography) the series of numbers obtained may be used to calculate the proportion of time spent in each area. More rarely, continuous observations can be made to give data on actual searching time per area (Murdie & Hassell, 1973).

The results may be plotted as numbers of predators per unit area (Pi) or the time spent by predators per unit area (Ti) against prey density per unit area (Ni) (Fig. 10.6e and f). In their analysis of the significance of the form of the aggregative response for stability, Hassell & May (1974) showed that the greater the difference between the maximum and minimum times per unit area and the closer the region J (Fig. 10.6e) corresponded to the average range of the prey's density, the more stabilising the aggregative response.

10.2.5.4   Fecundity, developmental and survival rates of predators in relation to prey density

The importance of these relationships is reviewed by Beddington *et al.* (1976), who suggest certain expressions for them, but stress that the lack of documentation for all the components for one species '…represents a major gap in experimental ecological work'. The methodology for such studies is, of course, basically similar to that used for any animal (e.g. Turnbull, 1962).

10.2.5.5   The role of other prey

'Other prey' is often important in allowing the survival of natural enemies over a period when the prey species being considered is sparse. However, the few quantitative studies in this area have been concerned with 'switching', a sudden change in the predators' preference between various prey: switching is reviewed by Murdoch & Oaten (1975). The idea arose from L. Tinbergen's theory of a 'search image'; some supporting evidence is given by Murton (1971). Royama (1970) suggested however that the predator maximises the 'profitability' of its searching: the profitability of feeding on a particular prey is compounded of size, population density and ease of capture. Royama's own work, together with that of Bryant (1973), in which the prey of insectivorous birds were compared with the available prey, tends to support his concept, with larger prey items tending to be relatively favoured: a finding that also accords with that of Hespenheide (1975). However, even in these studies the prey spectrum was not markedly different from that available. No significant preference or switching has been discerned for trout, *Salmo trutta* (Elliott, 1970), coccinellid larvae (Murdoch & Marks, 1973), mites (Santos, 1976), or generally for predatory littoral snails (Murdoch, 1969). In a study of the predatory crab, *Cancer borealis*, Siddon & Witman (2004) found that crabs switched from feeding on urchins to mussels when available. The switching of a generalist predator from a native to an invasive alien prey may also occur. Jaworski *et al.* (2013) studied the prey preference of the generalist mirid predator *Macrolophus pygmaeus* when encountering simultaneously the local tomato pest *Bemisia tabaci* and the invasive alien pest *Tuta absoluta*. Prey switching did occur, with the predator favouring the most abundant prey.

A number of different indices have been used to detect preference, of which V. S. Ivlev's 'elective index' and various forage ratios are examples (Lawton *et al.*, 1974; Jacobs, 1974; Mustafa, 1976). These were reviewed by Cock (1978), who follows Lawton *et al.* (1974) in using the random Equations (10.4) and (10.5) to predict the numbers of each prey type (subscripts 1 and 2 in the notation) eaten:

$$N_{a1} = N_1 \left[1 - \exp\left\{ -a_1 P \left(T - T_{h2} N_{a1} - T_{h2} N_{a2} \right) \right\} \right] \tag{10.9}$$
$$N_{a2} = N_2 \left[1 - \exp\left\{ -a_{21} P \left(T - T_{h2} N_{a2} - T_{h1} N_{a1} \right) \right\} \right] \tag{10.10}$$

when the ratio of the two prey types becomes:

$$\frac{N_{a1}}{N_{a2}} = \frac{N_1 \left[1 - \exp\left(-a_1 P T_s \right) \right]}{N_2 \left[1 - \exp\left(-a_2 P T_s \right) \right]} \tag{10.11}$$

where $T_s = (T - T_{h1}, N_{a1} - T_{h2} N_{a2})$, T = total time available for searching and feeding, and T_h = handling time. A similar procedure based on the random parasitoid Equation (10.4) should be used with parasitoids.

Cock (1978) (see also Hassell, 1978) recommends the following procedure for detecting preference:

1. Carry out function response experiments using each prey separately.
2. Estimate a_1, a_2, T_{h1}, and T_{h2} from the random Equations (10.9) and (10.10) as appropriate (i.e. predator or parasitoid).
3. Any preference resulting from differences in the functional response parameters (i.e. $a_1 \neq a_2$ and/or $T_{h1} \neq T_{h2}$) can now be conveniently displayed in terms of N_{a1}/N_{a2} [calculated from Equation (10.10)] plotted against N_1/N_2 alternatively, as the proportion of one of the species in the total diet against the proportion available (e.g. $N_{a1}/(N_{a1} + N_{a2})$ against $N_1/(N_1 + N_2)$. Such innate preference will then be detected as a deviation from a slope of unity passing through the origin.
4. Carry out a further experiment in which various ratios of the two prey types are presented together, and so contrast predicted and observed results. Ideally, this procedure should then be repeated for a range of total prey densities that encompass those used in the functional response experiments (item 1). Any difference between

the predicted preference from item (3) and that observed will now be due either to an active rejection of one of the prey or some change in a_1, a_2, T_{h1} or T_{h2} as a result of the predator experiencing the two prey types together.

In the field situation the position may be further complicated because polyphagous predators will often have wider trivial ranges than their prey whose micro-habitat preferences may be more restricted. Thus, if a predator modifies its hunting pattern from one microhabitat to another in response to relative differences in prey availability, then the switching effect might will be greater than that estimated from preference alone when prey are randomly mixed. Jolicoeur & Brunel's (1966) observations on the cod, *Gadus morhus*, may be an illustration of this.

10.2.5.6 Changes during the development of the predator

This concept was formalised by Murdoch (1971) and termed the developmental response. Clearly, many components (T_h, a, m, etc.) will change with the age of the predator; Hassell *et al.* (1976) show how age structure of predator and prey may alter the parameter values of the functional response. Such data are most easily gained from laboratory experiments (Turnbull, 1962; Thompson, 1975; Evans, 1976; Wratten, 1976).

10.2.5.7 Changes due to other animals

Competition between predators is often recorded (e.g. between parasitoids, leading to multiparasitism). Mutualism between two predators or between a non-predator (buffalo) and a predator (egret) may also be important. Mougi and Kondoh (2012) found in modelling studies that a mixture of antagonistic and mutualistic interactions of moderate strength could aid the stability of complex model communities.

10.2.5.8 Climatic and similar effects

The appropriate experimental methods are referred to on p. 383.

10.3 Dispersal

The term dispersal covers any movement away from the initial locality. It is useful to define neighbourhood dispersal for the process by which individuals migrate into and via adjacent areas, and jump dispersal for longer distance dispersal where individuals are transported or purposively move quickly to a new area. Jump dispersal is often caused by humans; for example, the transportation of marine organisms by ships is regularly observed. Neighbourhood dispersal may be described by random walk or diffusion equations, and an example would be the movement of newly hatched larvae away from their egg mass. Given the wide range of process that are encompassed by jump dispersal – the transport of ostracods in the feet and feathers of ducks or the displacement of a fish by a flash flood are two examples – there cannot be a single mathematical approach. In expanding populations, neighbourhood dispersal progresses on a continuous front whereas jump dispersal creates isolated local populations. Natural populations will often exhibit both types of dispersal, or even exhibit a continuum running between these two extremes (Plank, 1967).

10.3.1 Detecting and quantifying jump dispersal

Often, the only practical methods to study long-distance movements involve the use of marked animals that are released in a known locality and subsequently caught or observed elsewhere. Methods for marking are given in Chapter 3. Such an approach is most applicable to large, long-lived animals such as fish, reptiles, birds or mammals, although they are also applicable to some insects such as butterflies (e.g. Urquhart, 1960). For some insects, long-distance aerial movements can be quantified using suction traps (see Chapter 4), and fish or crustacean movements via estuaries can be detected in a similar fashion using traps such as cooling water intakes (p. 285). Migratory birds can be counted by observers placed on known routes, or counted using radar (p. 272). It may be possible to determine if the rate of long-distance immigration between populations is of any significance by either a comparison of gene frequencies between populations or from the frequency of parasites on hosts. However, such methods cannot estimate the magnitude of the movements. Often the only approach that can be taken to study jump dispersal is through the analysis of historical records, both published and anecdotal, plus available museum specimens to reconstruct the invasion. This approach was used by Saurez *et al.* (2001) for argentine ants, *Linepithema humile*, in the USA.

10.3.2 Quantifying neighbourhood dispersal

When neighbourhood dispersal is modelled as a diffusion process then, as Fagan (1997) argues, there are three approaches that can be taken. The traditional, concentration curve, approach aims to study the spatial distribution of a large number of individuals away from the release point at known instants in time, and then estimating dispersal parameters by fitting an appropriate model to the observed change of numbers with distance (Fig. 10.7a). A second approach is to plot the movements of individuals and to calculate step-lengths and turning angles for a random-walk model, the parameters of which are then used to predict the population dispersal rate (Fig. 10.7b). The third method, termed by Fagan (1997) the boundary-flux approach, is to count through time the cumulative number of individuals which have moved out from a release point and then to use these data to estimate the dispersal coefficient (Fig. 10.7c). The approaches presented below assume that dispersal can be successfully modelled as a diffusion process, as has been found often to be so by Okubo (1980, 1989). All of the methods presented assume that the dispersal (diffusion) coefficient is constant over the period of observation and each animal behaves in a similar way. Neither of these assumptions can be strictly true, although they often lead to reasonable assumptions. When a model gives a poor fit it is generally because one or both of these assumptions is violated. Diffusion models tend to be most in error close to the release point shortly after release, and at later times close to the leading edge of the dispersing population. A number

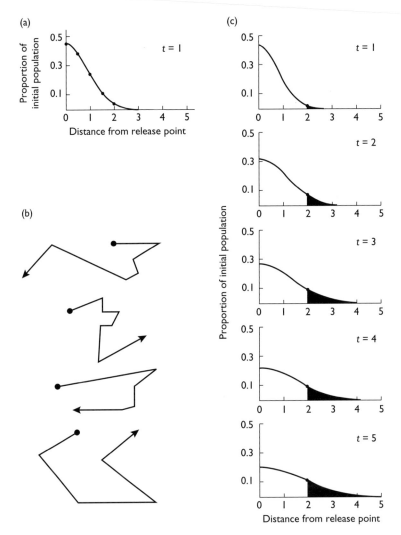

Figure 10.7 Possible approaches to the modelling of animal movement as a diffusion process. (Adapted from Fagan, 1997.) (a) The traditional concentration–curve approach; (b) Plotting the movements of individuals and calculating step lengths and turning angles for a random-walk model, the parameters of which are then used to predict the population dispersal rate; (c) The boundary-flux approach, in which counts through time of the cumulative number of individuals which have moved out from a release point are made.

of issues are raised by the requirement of these methods to release and/or recognise animals in the field, which will be discussed first.

10.3.2.1 The use of marked or introduced animals

When marked animals are released where they were found, or directly marked in the field (Greenberg & Bornstein, 1964), their subsequent movements may be altered by the experience. An alternative approach is to release unmarked animals in an 'empty' habitat. Some workers have used laboratory-bred material for these releases; although this may be acceptable (MacLeod & Donnelly, 1956; Fletcher & Economopoulos, 1976), it should not be adopted without a check that the behaviour is normal. The use of animals captured in the wild also poses problems: the level of migratory activity in insects is commonly related to age (e.g. Johnson, 1969) and the process of marking may so disturb them that dispersal is exceptional during the first few days after release (Clark, 1962; Greenslade, 1964). Dean (1973) studied the movements of aphids in an uninfested crop by placing pots of cereals, with dense colonies of aphids, at various points.

10.3.2.2   Methods based on a radial solution of the diffusion equation

The basic approach is for animals to be released at a known point and subsequently recaptured in traps placed at increasing distances (in annuli or in the form of a cross) from the release point (e.g. Doane, 1963). The influence of irregularities in environmental favourableness around different traps may be reduced by using the ratio of unmarked to marked animals in each trap (Gilmour *et al.*, 1946; MacLeod & Donnelly, 1963). As it is assumed that the animals can disperse equally in all directions, a one-dimensional model can be used where distance is expressed as the radial distance from the point of release.

First, it should be determined if there is drift or non-randomness in the direction of dispersal. One approach to this is to superimpose upon the map of the release and recapture sites a horizontal and vertical grid, and the mean and the variance are calculated for each day in terms of these grids (Clark, 1962). If the means differ significantly, then there was markedly more dispersal in one direction than another drift. If the means are similar, but the variances differ, then movement was non-random. Clark (1962) describes a method for testing whether the resulting spatial distribution is circular or elliptical.

Another approach is that of Paris (1965) who, in a series of experiments, determined the number of woodlice in different radii and compared the results for eight radii using Friedman's (1940) analysis of variance by ranks test, which is non-parametric.

Both of the above methods are limited to situations when a group of marked animals are released from a central point. When the members of a natural population have been marked individually it may be possible to use the approach of Frank (1964), who demonstrated that individual limpets, *Acmaea digitalis*, were not moving randomly and hence concluded that they had home ranges. Let the habitat be regarded as a grid of identical squares, and assume that the probability of movement from one square to the next over each time step is constant. The movement can then be modelled as a Markov process, for which the transition probability matrix at the end of a given number of time periods for the probabilities can be easily calculated. The actual movements may then be compared with the expected using the χ^2 test. Work on bird navigation has revealed some ability to orientate in a time-compensated manner to sun, moon and certain constellations and perhaps magnetic field; attention has therefore been directed at analysing the effect of a small bias on an otherwise random movement (Kendall, 1974; Matthews, 1974).

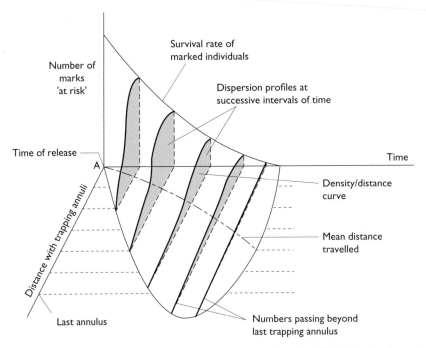

Figure 10.8 **A schematic representation of the dispersal of marked individuals released at a fixed point A, showing some of the variables that may be estimated. (Adapted from Inoue *et al.*, 1973.)**

Provided that the drift in one direction is not excessive, the dispersal of a marked population from a point source may be represented by Fig. 10.8, where the shaded panels represent the successive density/distance profiles (Inoue *et al.*, 1973). Five interrelated variables may be calculated to determine the properties of a dispersing population as revealed by such an experiment:

1. The extent to which the population is heterogeneous with some dispersing while others remain almost sedentary.
2. The numbers or relative densities at successive distances from the release point.
3. The fall-off of density with distance; i.e. the form of the leading edge of the shaded segments in Fig. 10.8.
4. The mean distance travelled (by the emigrating component if some are stationary) (Fig. 10.8).
5. The proportion, or number, of marked individuals that have passed beyond the area.

Account must also be taken of the density at the start because these variables may well be density modified (Taylor & Taylor, 1976). Furthermore, the tail of the distribution, representing the most mobile individuals, may not be found in small samples.

10.3.2.3 The detection of heterogeneity with respect to the rate of dispersal for the individuals of a population

In an experiment on the dispersal of marked *Drosophila* from a central point of a cross of traps, Dobzhansky & Wright (1943) demonstrated that the flies were heterogeneous with respect to the distance they travelled. The departure from normality of the curve of the frequency curve of numbers with distance from the point of release on a given day was detected by calculating the kurtosis by the formula:

$$K = \frac{R \sum_{i=1}^{y} x_i^4 n_i}{\left(\sum_{i=1}^{y} x_i^2 n_i\right)^2} \tag{10.12}$$

where R = total animals caught (recaptured) in all traps, x_i = distance of the recapture point (i) from the point of release, n_i = total number of animals caught in traps at the same distance (x_i) from the release point and y = the last equidistant set of recapture points (the first point will be at the centre, i.e. $x_i = 0$).

Heterogeneity of dispersal is an expected property of populations. Shaw (1970) has shown how even alate aphids may differ greatly in their potential for dispersal.

10.3.2.4 The numbers at various distances from the release point

The approach with the fewest assumptions is that of Fletcher (1974), who studied the Queensland fruit fly, *Dacus tryoni*, and calculated the proportion of the migrant flies that were in a particular annulus (F_i). Let n_i/g_i represent the total number of marked flies trapped (recaptured) divided by the number of traps at annulus i. Then

$$\hat{F}_i = \frac{n_i}{g_i}\left(x_{i+1}^2 - x_i^2\right) \bigg/ \sum_{i-1}^{y} \frac{n_i}{g_i}\left(x_{i+1}^2 - x_i^2\right) \tag{10.13}$$

where x_i = the distance the inner radius of the ith annulus is from the central release point, therefore x_{i+1} = the distance of the outer radius of the same annulus from the central point and y = total number of annuli.

This method cannot be used when significant numbers have migrated beyond the outer annulus; proportions are used because the catchment area of the traps could not be ascertained.

Crumpacker & Williams (1973), in their study of *Drosophila*, assumed that the zone of attraction of one trap would overlap the next, but the number of marked flies (M) in the annulus could be estimated for a particular day as:

$$\hat{M}_i = r_i\left(\frac{U_i}{u_i}\right)\left(\frac{A_i}{\alpha_i}\right), \tag{10.14}$$

where r_i = marked flies recaptured on the particular day in the ith annulus, u_i = the unmarked flies captured, U_i = the total unmarked flies, A_i = the total area of the annulus and α_i = attractive area of the annulus. On their assumption $A_i/\alpha_i = 1$; clearly M_i and U_i are both unknown, but these could easily be computed from a series of equations. The probability that a marked fly reached the ith annulus is estimated by:

$$\hat{q}_i = \frac{M_i}{\sum\limits_{i=1}^{y} M_i} \tag{10.15}$$

Given the assumption that $A_i = \alpha_i$ this may be expressed as:

$$\hat{q} = \frac{A_i r_i / u_i}{\sum\limits_{i=1}^{y} \frac{A_i r_i}{u_i}} \tag{10.16}$$

where the first annulus is numbered 1 and the last y.

Of course, these probabilities will be influenced by heterogeneity in the population discussed above (see also p. 405 for a method assuming random movement).

10.3.2.5 The fall-off of density with distance

This topic was comprehensively reviewed by Freeman (1977) and Taylor (1978); the latter shows that the equations belong to two 'families' (Table 10.2). The shapes of the curves generated vary significantly (Fig. 10.9), and both Taylor (1978) and Freeman (1977) found that the equations in family II give the best fits to most field data for insects; this implies that the lengths of individual movements are not random. Random dispersal would be better fitted by equations in family I. The general equation for family II is:

$$N = \exp(\eta + bX^c) \tag{10.17}$$

Taylor (1978) suggests that the parameter c is a measure of non-randomness: $c < 2$ indicates a tendency to aggregation, $c \approx 2$ random, and $c > 2$ repulsion leading to

Table 10.2   Equations for the dispersal of organisms from a release point following the classification. (Data from Taylor, 1978), where N = number dispersing to distance X and η, b and c are constants.

	General form for family	Equations	Author
I.	$N = \eta + b\,f(X)$	$N = \eta + c/X$	Paris, 1965
		$N = \eta + b.\log_e X$	Wolfenbarger, 1946
		$N = \eta + b.\log_e X + c/X$	Wolfenbarger, 1946
II.	$N = \exp(\eta + bX^c)$	$N = \exp(\eta + c/X)$	Taylor, 1978
		$N = \exp(\eta + b.\log_e X)$	MacLeod & Donnelly, 1963
		$N = \exp(\eta + b\sqrt{X})$	Hawkes, C., 1972
		$N = \exp(\eta + bX)$	Gregory & Read, 1949
		$N = \exp(\eta + bX^2)$	Dobzhansky & Wright, 1943

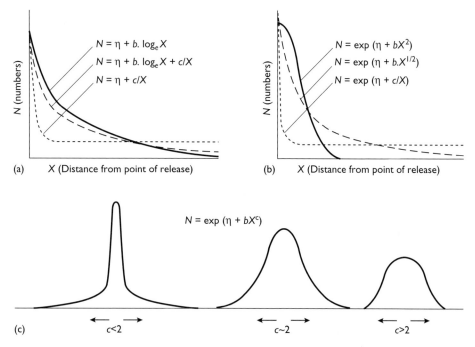

Figure 10.9 **Various forms of the dispersal–distance curves. (a and b) Half-distributions for various equations (see Table 10.2). (Adapted from Taylor, 1978 based on Dobzhansky & Wright's 1943 data.) (a) Family I type curves; (b) Family II type curves, which imply a density relationship; (c) Representations of the whole distributions for family II curves with different values of c in the general equation.**

regularity (see Fig. 2.4 and p. 48). Taylor & Taylor (1976, 1978) have taken this concept of density-dependence further in their general model (see below).

A simple test for random movement – a random-walk, Brownian motion or diffusion model – is to compare the fall-off of numbers with distance with half a normal distribution (Fletcher, 1974). Considered as a diffusion phenomenon, the density of marked insects per unit area (m') at a particular point is given by (Scotter *et al.*, 1971; Aikman & Hewitt, 1972):

$$m' = \frac{M}{4\pi Dt} \exp\left(\frac{-x^2}{4Dt} \right) \tag{10.18}$$

where M = effective (i.e. surviving and moving) number of marked animals released, x = the radial distance from the point of release, t = time, and D = dispersal rate coefficient. This equation may be solved by iteration, but D will be found to have two values: one as the density at a particular distance builds up, and one as it gradually falls as they disperse even further (as shown by Clark, 1962 in his study of grasshoppers).

It should be noted that the magnitude of the maximum x in relation to the scale of the animals habitat will strongly influence the type of frequency distribution found with distance: many species will show a different pattern within the population habitat and outside it.

10.3.2.6 The mean distance travelled and rate of dispersal

This is most easily calculated from data on the proportional frequency (F_i) (Fletcher, 1974) or actual numbers (P_i) (Crumpacker & Williams, 1973) of marked individuals estimated for successive annuli (see above). The estimate is:

$$\hat{d} = \sum_{i=0}^{y} F_i \frac{1}{2} (x_{i+1} + x_i)$$

(10.19)

where x_i is the distance the inner radius of the ith annulus is from the central point, x_{i+1} is the same for the outer radius and y is the outermost annulus. If numbers have been estimated, M_i is substituted for F_i in the above equation. These estimates are reliable only as long as no animals pass beyond the sampling annuli.

If a random walk can be assumed then a measure of the rate of dispersal is given by dispersal coefficient, D. In the experiment with *Drosophila* referred to above, Dobzhansky & Wright (1943, 1947) determined the speed of dispersal by comparing the change in variance on successive days. If the speed of dispersal is constant, the variance should change by a constant amount from day to day. They suggested that as the curve of numbers on distance was not normal, variance should be estimated:

$$s^2 = \frac{\pi \sum_{i=2}^{y} x_i^3 F_i}{\left(\sum_{i=2}^{y} x_i \bar{r}_i \right) + r_1}$$

(10.20)

where x_i = distance of recapture point from central release point, F_i = the mean number of animals captured per site at a given distance (x_i) from the release point, y = the equidistant set of traps or capture sites furthest from the release point, and r_i = the number of recaptures in the central trap (i.e. where $x = 0$). If animals are released at a central point it is possible, making the assumptions of random flight movements and constant speed, to calculate a theoretical curve for the fall-off in numbers of marked individuals with distance. By relating this curve to actual data of the ratio of marked to unmarked flies, Gilmour et al. (1946) were able to calculate the number of marked blowflies beyond the last ring of traps.

10.3.2.7 The number of marked animals that have left the area

Using a standard diffusion model with the dispersal rate (diffusion) coefficient (D) calculated as above, assuming trapping or searching methods are 100% efficient (i.e. $M_i = r_i$), Scotter et al. (1971) give the equation as:

$$\hat{M}_i = a \left[\exp \left(\frac{-x_i^2}{4Dt} \right) - \exp \left(\frac{-x_{i+1}^2}{4Dt} \right) \right]$$

(10.21)

where a = number of marked flies released, t = time, x_i = distance of the inner radius of the ith annulus from the central point and x_{i+1} = distance of outer radius from i^{th} annulus from point. Where $x_i = 0$, i.e. the innermost radius, there is a unique solution to D. Then the probability (P) that a particular insect travels more than a certain distance beyond the radius of the outer annulus (y) is:

$$P = \exp\left(\frac{-y^2}{4Dt}\right) \tag{10.22}$$

The actual number of animals migrating into or out of a population may also be determined from mark and recapture analysis (see Chapter 3). If there are no births or deaths, then the loss and dilution rates may be taken as equivalent to migration rates; alternatively, the two components may be separated by measuring natality and mortality separately (see pp. 373–389).

The simple Lincoln Index may be used to calculate the proportion of a population that has migrated if the total population is known from some other method and there are no births or deaths. A number of marked individuals are released; knowing this number, the total population and the size of subsequent samples, expected recapture values may be calculated. The proportion of the marked insects that have emigrated (a_e) is estimated by the ratio of actual to expected recaptures:

$$a_e = {r}/{r_e} \tag{10.23}$$

where r = actual recaptures r_e = expected recaptures = an/N (where a = number of marked animals released, n = total size of sample and N = total population known from some other estimate). Such an approach is of particular value for a highly mobile animal.

10.3.2.8   Methods based on a two-dimensional solution of the diffusion equation

10.3.2.8.1   *The use of quadrat counts of unmarked individuals*

Where the habitat is uniform and it is possible to assume random diffusion, movement may be separated from mortality by using the method of Dempster (1957), who solved the diffusion equation using a standard numerical approximation. The changes in the numbers of animals (f) with time are given by Skellam (1951):

$$\frac{df}{dt} = D\left(\frac{d^2f}{dx^2} + \frac{d^2f}{dy^2}\right) - \mu f \tag{10.24}$$

where D = the dispersal coefficient, x and y are the two dimensions in which dispersal occurs, and μ = mortality rate.

Now, if we consider a block of 9 quadrats (Fig. 10.10) for the central quadrat, f_{22}, the second derivatives can be approximated by:

$$\frac{d^2f}{dx^2} + \frac{d^2f}{dy^2} = \frac{1}{3}\left(2f_{11} - f_{12} + 2f_{13} - f_{21} - 4f_{22} - f_{23} + 2f_{31} - f_{32} + 2f_{33}\right) \tag{10.25}$$

This expression can be substituted into Equation (10.23), and if the number of animals in each quadrat is known at two points in time an expression with only two unknowns, D and μ is formed. Now imagine a situation with a number of central squares so that

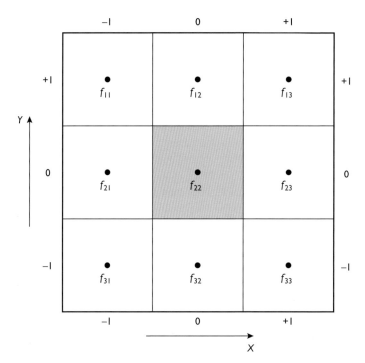

Figure 10.10   The populations (f$_{11}$, f$_{12}$....) along the two axes (x and y) across a central quadrat surrounded by quadrats of equal area. (Adapted from Dempster, 1957.)

a series of equations can be produced. While the absolute minimum number of equal-sized quadrats needed to solve for D and μ would be 15, arranged in a block of 3 × 5, a larger number would be preferable – Dempster used 18. Equations for the four middle squares can be calculated. The df/dt value being the changes in numbers – normally a decrease – between time periods. The equations are then solved by normal mathematical procedures.

For example (from Dempster, 1957): consider square B2 in Fig. 10.11. The number of insects in this square has changed from 50500 to 26475.

Therefore,

$$df/dt = 24024$$

and

$$\left(\frac{d^2f}{dx^2} + \frac{d^2f}{dy^2}\right) = \frac{1}{3}[(2x44230) - 45650 + (2x6478) - 47370 - (4x50500) - 16660$$

$$+ (2x50720) - 103900 + (2x89160)] = -11465$$

Substituting in Equation (10.25) we get $-24205 = -11465D - 50500\mu$.

At least three, preferably more similar, equations are obtained. The normal equation for μ is then found by multiplying each equation by the coefficient of μ in it and adding all equations. The normal equation for α is found in the same way. These two equations are then solved.

	A	B	C
1	44 230 14 600	45 650 16 145	6478 5840
2	47 370 22 580	50 500 26 475	16 650 6623
3	50 720 30 015	103 900 52 680	89 160 57 860

Figure 10.11 The estimated number of locusts entering the first (upper figure) and second instar (lower figure) in a number of quadrates. (Data from Dempster, 1957.)

10.3.2.8.2 *The boundary-flux approach*

This method solves the two-dimensional diffusion equation for the dispersal coefficient, D, and the mortality rate, μ, by observing the cumulative capture of animals dispersing outwards from a release point. Fagan (1997) illustrated the approach with a study of dispersal in two mantids. The experiments were undertaken in quadrats delimited by tanglefoot bands that would capture any mantid passing over them. The mantids were initially released in the centre of the quadrat and the tanglefoot bands at the periphery of the area were inspected daily for 3 weeks and the number of mantids captured was counted. As the mantids could not escape from the tanglefoot the bands acted as an absorbing boundary and the count is the flux of mantids at a known time and distance from the release point.

For a square plot, Fagan (1997) gives the cumulative number caught at the boundary as:

$$C(t) = \sum_{j=0}^{\infty}\sum_{k=0}^{\infty}\left\{\left(16(-1)^{(1+j+k)}D[(1+2j)^2+(1+2k)^2]\left[\exp\left(\frac{-t\Omega_{jk}}{L^2}\right)-1\right]\right)\Big/[(1+2j)(1+2k)\Omega_{jk}]\right\} \quad (10.26)$$

$$\Omega_{jk} = D\pi^2[(1+2j)^2+(1+2k)^2]+\mu L^2.$$

where t is time and L is the length of one side of the square. This large and ugly expression was solved numerically for D with the summations approximated as a double Fourier series. The mortality rate was estimated from other data.

10.3.2.9   The rate of population interchange between two areas

Richards & Waloff (1954) described a method for studying the movement between two grasshopper colonies; their basic assumption was that the survival rates in the two colonies were similar. Iwao (1963), who has derived equations that are applicable to populations where both the survival rates and sampling ratios differ, further developed their method. A series of three sets of observations on days 1 (t_1), 2 (t_2) and 3 (t_3) are necessary, as with Bailey's triple catch (p. 97). A number (a_1) of animals are marked and released in both areas on day 1; on day 2 a sample (n_2) is taken and the number of already marked individuals recorded, all the individuals (a_2) are then given a distinctive mark and released. On the third day, samples are again taken in both areas and the number of already marked individuals recorded together with the details of their marks.

Thus, the estimate of the emigration rate from area x to area y during the time interval from day 1 to day 2 is given by:

$$_{xy}\hat{e} = \frac{\left(\frac{_{yy}r_{31}\,_ya_2 + _{yy}r_{21}\,_{yy}r_{32}}{_{yy}r_{32}\,_ya_1} \right) _ya_1\,_{xy}r_{21}}{_xa_1\,_{yy}r_{21}} \tag{10.27}$$

where the notation has been adapted to conform with that in Chapter 3, the anterior subscripts being added to denote the areas; with both the anterior and posterior subscripts the symbol nearest the character represents the actual condition and that furthest away its previous history. Thus, $_{xy}r_{21}$ represents the recaptures in area y on day 2 that were marked in area x on day 1. To recapitulate the notation for the above equation:

$_xa_1$ = no. of marked individuals released in area x on day 1
$_ya_1$ = no. of marked individuals released in area y on day 1
$_ya_2$ = no. of marked individuals released in area y on day 2
$_{yy}r_{21}$ = recaptures in area y on day 2 marked in area y on day 1
$_{yy}r_{31}$ = recaptures in area y on day 3 marked in area y on day 1
$_{yy}r_{32}$ = recaptures in area y on day 3 marked in area y on day 2

The equivalent equation for the estimation of the emigration rate from y to x is:

$$_{yx}\hat{e} = \frac{\left(\frac{_{xx}r_{31}\,_ya_2 + _{xx}r_{21}\,_{xx}r_{32}}{_{xx}r_{32}\,_xa_1} \right) _xa_1\,_{yx}r_{21}}{_ya_1\,_{xx}r_{21}} \tag{10.28}$$

If the total population (N) has been estimated by capture–recapture or some other way, then the actual numbers that are estimated to have emigrated (E) are:

$$_{xy}\hat{E} = {}_xN_1\,_{xy}\hat{e}_1 \tag{10.29}$$

and:

$$_{yx}\hat{E} = {}_yN_1\,_{yx}\hat{e}_1 \tag{10.30}$$

Survival rates (ϕ) can also be calculated:

$$_x\phi_1 = {}_{xx}\mu_1 + {}_{xy}\hat{e}_1 \tag{10.31}$$

where

$$_{xx}\mu_1 = \frac{_{yy}r_{31}a_2 + {}_{yy}r_{21yy}r_{32}}{_{yy}r_{32y}a_1} \tag{10.32}$$

This expression is the bracketed term in Equation (10.29).

10.3.2.10   The description of population displacement in relation to its dispersion

Migration has long been recognised as one of the three pathways of population change; however, its formal incorporation into population mechanisms is more recent. Many models now envisage organisms as forming metapopulations composed of semi-isolated subpopulations, each of which may exchange members (Hanski, 1991; Hassel et al., 1991; Hastings & Harrison, 1994). Taylor & Taylor (1976, 1978) showed, on the basis of the extensive monitoring of adult Lepidoptera in Britain, that populations must be regarded as spatially fluid. Emigration and immigration are linked as two opposing density-related behaviour sets such that the coefficient of displacement is:

$$\Delta = Gp^p - Hp^q$$

where ρ = population density; p = the density-related moderator in relation to emigration, and q that for congregation. This may be written as:

$$\Delta = \Gamma\left[\left(\frac{p}{p_0}\right)^p - \left(\frac{p}{p^0}\right)^q\right]$$

where ρ_0 is the density at which emigration balances immigration and:

$$\Gamma = {}^{(p-q)}\sqrt{H^p g^{-q}} \tag{10.33}$$

is a scale factor. The net result of displacement is therefore congregation when $\rho < \rho_0$ and emigration when $\rho > \rho_0$. Taylor & Taylor fitted the above expression to various sets of density/distance frequency data for animals and calculated values for the parameters E, ρ_0, p and q. The significance of ρ_0, the equilibrium density, is most easily grasped, although of course habitat, as well as behavioural characters will influence its magnitude. Taylor & Taylor's model provides a description of displacement with its link to spatial pattern (dispersion) and so provides a dynamic description of the changes of the population of an animal with time.

10.4 The measurement and description of home range and territory

The determination of the home range or territory of an individual or, for social animals, a colony, is of value in the analysis of competition and density effects, the assessment of resources and similar problems (Brown & Onans, 1970). There are well-recognised problems with the most commonly used definition of home range as '…that area traversed by the individual in its normal activities of food gathering, mating and caring for young. Occasional sallies outside the area, perhaps exploratory in nature, should not be considered as in part of the home range.' (Burt, 1943). First, how does one recognise an occasional sally? Second, the intensity of use within the home range is not addressed (Kie *et al.*, 2010). The definition is still of value, particularly when traditional home range estimation methods are used.

Many studies have been made on the territories of vertebrates, and advances in telemetry and use of the Global Positioning System (GPS) have recently allowed this at high frequency and accuracy. GPS has revolutionised the tracking of large animals as it is accurate to 30 m or less, can record position at high frequencies down to 1 s or less, and provides accurate time stamping (Tomkiewicz *et al.*, 2010). A key limitation of GPS is signal obstruction, which may occur in dense canopy or when an animal goes undercover.

From the 1930s to 1980s relatively less work was undertaken on insects except for crickets (Alexander, 1961), dragonflies (Borror, 1934; Moore, 1952, 1957), some grasshoppers (Clark, 1962), and ants (Elton, 1932). Some insects may be marked and recaptured on a number of occasions (e.g. Borror, 1934; Green & Pointing, 1962; Greenslade, 1964; Dolný *et al.*, 2014); this type of experiment gives data suitable for the computation of the home range, using the methods of vertebrate ecologists. Ten recaptures are usually taken as sufficient to calculate the home range, but considerably more data would be required to use the method of Frank (1964) (p. 399) and ensure that one was not measuring an artefact.

Local foraging flight for insects large enough to carry transponders, including social bees and moths, is now studied using scanning harmonic radar (Chapman *et al.*, 2011) (see p. 117). Radiofrequency identification (RFID) tags to allow radar tracking weighing 3.8 mg (about 8% of the insect weight) have been attached to Chinese citrus fly, *Bactrocera minax*, with no visible impact on the ability of the flies to take off (Gui *et al.*, 2011). As Kissling *et al.* (2013) note, the fixed detection zone of a stationary radar unit (<1 km diameter) and the restricted detection distance of RFID tags (usually <1–5 m) place a major constraint on the use of radar. For insects with a body mass above 1 g, active battery power transmitters can be attached which offer radiotelemetry tracking over ranges of 100–500 m. However, these transmitters weigh more than RFID passive tags and are likely to inflict an appreciable energetic cost on the insect and alter behaviour.

The term 'territory implies the exclusion of at least certain other individuals. While this is frequently observed for vertebrates it is rarely observed for invertebrates, although clear examples are crickets and dragonflies (Alexander, 1961; Moore, 1952, 1957). Nevertheless, most invertebrates will have a home range over which they forage and search for mates and where they rest; that is, the area over which they engage in trivial movement (*sensu* Southwood, 1962). As Moore (1957) points out, the home range, especially for insects, should not be regarded as a hard and fast geographical area, but provided it is measured in a strictly comparative way it should be possible to detect changes due to inter- or intra-specific competition, to variation in the available resources, or to changes in the behaviour of the animals themselves (e.g. Wellington, 1964).

There are a number of methods for calculating the home range of vertebrates. Recent advances in radiotelemetry and GPS generate large amounts of data and have stimulated the development of statistical approaches such as kriging, the application of nonlinear generalised regression models and mechanistic home range models (Moorcroft & Lewis, 2013). In a review of recent developments, Kie *et al.* (2010) concluded that traditional methods such as kernel density estimators are likely to remain popular because of their ease of use. These traditional methods were reviewed by Worton (1987), who subsequently recommended kernel methods to estimate the utilisation distribution (the probability density function of the location) (Worton, 1989). Traditional methods will likely remain of value for the study of small animals that cannot carry a tag and thus do not generate large numbers of observed locations.

10.4.1   *The minimum convex polygon area method for estimating home range*

This is the simplest method and can be done by hand, although computer software, such as the R package adehabitatHR (Calenge, 2011), is available. The points of recapture are mapped and the outermost points joined up to enclose the area, which may be measured (Mohr, 1947; Odum & Kuenzler, 1955). This method has the advantage that no assumption is made about the shape of the range. However, with such a method the worker may feel it necessary to exclude 'incidental forays outside the area' (Jorgensen & Tanner, 1963). It is common practice to remove a set percentage of the most extreme outliers; in the R example below, the 5% furthest away from the centroid are excluded using the percent=95 option in the mcp function.

The following R code calculates the home range of 4 wild boars using the adehabitatHR package. The simple plots obtained and the tabulated results are shown in Fig. 10.12.

```
library(adehabitatHR)
#The Minimum Convex Polygon (Mohr, 1947)
data(puechabonsp) #load wild boar demo data
# The data comprises (x,y) coordinates for 4 boar named Jean, Chou, Calou and Brock
#default coordinates in metres
puechabonsp #print the data
#The home range areas are calculated with the
#removal of the 5% most extreme points (100 - the excluded percent)
cp <- mcp(puechabonsp$relocs[,1], percent=95)
as.data.frame(cp) #Prints the 4 home ranges in hectares (default)
plot(cp)#A simple plot of the 4 polygons
#Add the observed locations to the polygon plots
plot(puechabonsp$relocs, add=TRUE)
```

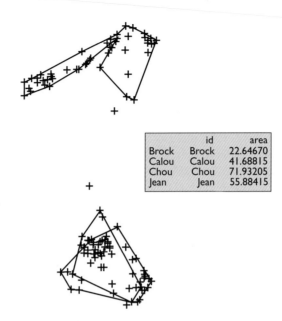

	id	area
Brock	Brock	22.64670
Calou	Calou	41.68815
Chou	Chou	71.93205
Jean	Jean	55.88415

Figure 10.12 The plotted output for the minimum polygon method generated by the R package adehabitatHR (Calenge, 2011). The four polygons shown are the estimated home ranges for four boars, which were found to have estimated home ranges varying from 22.6 to 71.9 km². The R code is given in Section 10.2.4.1.

All polygon methods have limitations, and estimates obtained are correlated to the number of data points plotted. The enclosed area may include areas which are not visited; they tend to produce overestimates.

10.4.2 *The kernel estimation method for home range*

Worton (1989) proposed the use of kernel methods and these are still in frequent use. The basic scheme is that a bivariate kernel function is placed over each reported location and the values of these functions are averaged together to generate the utilisation distribution. A wide variety of kernel functions have been proposed. Two common choices are the bivariate normal and Epanechnikov kernel functions. Although the Epanechnikov kernel has some mathematical advantages, the choice does not greatly influence the estimates obtained and the bivariate normal is a commonly used. More recently, the kernel approach has been extended to include a number of nonparametric methods such as the alpha-hull method (Burgman & Fox, 2003) and the local convex hull (LoCoh) method (Getz & Wilmers, 2004)

Where an animal's habitat is not homogeneous – for example, if it spans two types of ecosystem – the above methods are not appropriate for the calculation of home range. Van Winkle *et al.* (1973) developed a method for frogs living on the edge of a pond which could be applied to other animals in similar ecotones.

10.5   The rate of colonisation of a new habitat and artificial substrates

This property, which is related to migration, may be measured by planting virgin artificial habitats (e.g. new plants in a field or stones in a stream) and determining the rate of colonisation (Breymeyer & Pieczynski, 1963) (see also pp. 199 and 305).

The larval settlement of marine benthic invertebrates such as mussels and barnacles are typically studied using artificial substrates. For example, Hoffmann *et al.* (2012) used plastic mesh pot scourers to give a three-dimensional structure for mussels to colonise and polycarbonate plates coated in an anti-slip floor covering to monitor barnacle settlement.

10.6   The direction of migration

There are a number of traps (p. 281) that enable one to determine the direction an insect was flying at the time of capture. It is possible that the proportionality values obtained from a series of such traps round a habitat (Sylven, 1970) might be used in conjunction with measures of either the net change in population or a measurement of migration to determine values for immigration and emigration. Furthermore, it has been pointed out how the dilution and 'loss' rates obtained from multiple capture–recapture analysis are compounded of birth and immigration and death and emigration, respectively. Sometimes it is possible to separate the two components of the dilution rate from knowledge of natality or the numbers entering the stage. When 'births' are known, the proportion of emigration to immigration determined from a 'directional trap', might be used to give an indication of the amount of immigration and hence allow the partitioning of the death-rate into mortality and immigration. Caution would need to be used in such an approach, particularly to ensure an adequate distribution of traps around the habitat, and difficulty would be experienced in determining the numbers of insects leaving and arriving in vertical air currents, that is, only crossing the habitat boundaries at a considerable height.

Aquatic samplers can often give information on the direction of movement. For example, gill nets (p. 285) show the direction of movement of the captured fish and, when placed around the circumference of a lake, can show the diurnal pattern of movement from shore habitat to open water. Traps in flowing waters can often be arranged to only catch animals moving up or down stream.

References

Aagaard, K., Solem, J.O., Nost, T., & Hanssen, O. (1997) The macrobenthos of the pristine stream, Skiftesaa, Hoylandet, Norway. *Hydrobiologia* **348**, 81–94.

Abdel Rahman, I. (1974) The effect of extreme temperatures on Californian red scale, *Aonidiella aurantii* (Mask.) (Hemiptera: Diaspididae) and its natural enemies. *Aust. J. Zool.* **22**, 203–12.

Abrams, P.A. & Ginzburg, L.R. (2000) The nature of predation: prey dependent, ratio dependent or neither? *Trends Ecol. Evol.* **15**, 337–41.

Adamstone, F.B. & Harkness, W.J.K. (1923) The bottom organisms of Lake Nipigon. *Univ. Toronto Stud. Biol.* **22**, 121–70.

Agassiz, D. (1977) A trap for wingless female moths. *Proc. Br. Entomol. Nat. Hist. Soc.* **10**, 69–70.

Aikman, D. & Hewitt, G. (1972) An experimental investigation of the rate and form of dispersal in grasshoppers. *J. Appl. Ecol.* **9**, 807–17.

Alexander, R.D. (1961) Aggressiveness, territoriality and sexual behaviour in field crickets (Orthoptera: Gryllidae). *Behaviour* **17**, 130–223.

Anderson, N.H. & Wold, J.L., (1972) Emergence trap collections of Trichoptera from an Oregon stream. *Can. Entomol.* **104**, 189–201.

Arditi, R. & Ginzburg, L.R. (1989) Coupling in predator-prey dynamics: ratio-dependence. *J. Theoret. Biol.* **139**, 311–26.

Ashby, J.W. (1974) A study of arthropod predation of *Pieris rapae* L. using serological and exclusion techniques. *J. Appl. Ecol.* **11**, 419–25.

Avenant, N.L. (2005) Barn owl pellets: a useful tool for monitoring small mammal communities. *Belg. J. Zool.* **135**, 39–43.

Bagenal, T.B. & Braum, E. (1978) Eggs and early life history. In: Bagenal, T. (ed.), *Methods for Assessment of Fish Production in Fresh Waters*, 3rd edn. IBP Handbook No. 3, Blackwell Scientific Publications, Oxford, pp. 165–201.

Bates, J.K. (1962) Field studies on the behaviour of bird fleas. 1. Behaviour of the adults of three species of bird flea in the field. *Parasitology* **52**, 113–32.

Barbraud, C. & Barbraud, J.C. (1997) Diet of White Stork *Ciconia ciconia* chicks in 'Charente-Maritime', importance of insects. *Alauda* **65**, 259–62.

Bean, J.L. (1958) The use of larvaevorid maggot drop in measuring trends in spruce budworm populations. *Ann. Entomol. Soc. Am.* **51**, 400–3.

Beddington, J.R. (1975) Mutual interference between parasites or predators and its effect on searching efficiency. *J. Anim. Ecol.* **44**, 331–40.

Beddington, J.R., Hassell, M.P., & Lawton, J.H. (1976) The components of arthropod predation II. The predator rate of increase. *J. Anim. Ecol.* **45**, 165–85.

Bell, S.S., Hicks, G.R.F., & Walters, K. (1989) Experimental investigations of benthic re-entry by migrating meiobenthic copepods. *J. Exp. Mar. Biol. Ecol.* **130**, 291–304.

Bess, H.A. (1961) Population ecology of the gipsy moth, *Porthetria dispar* L. (Lepidoptera: Lymantridae). *Bull. Conn. Agric. Exp. Sta.* **646**, 43 pp.

Betts, M.M. (1954) Experiments with an artificial nestling. *Brit. Birds* **47**, 229–31.

Betts, M.M. (1958) Further experiments with an artificial nestling gape. *Brit. Birds* **49**, 213–15.

Blais, J.R. (1953) The effects of the destruction of the current year's foliage of balsam fir on the fecundity and habits of flight of the spruce budworm. *Can. Entomol.* **85**, 446–8.

Boethel, D.J., Morrison, R.D., & Eikenbary, R.D. (1976) Pecan weevil *Curculio caryae* (Coleoptera: Curculionidae). 2. Estimation of adult populations. *Can. Entomol.* **1089**, 19–22.

Boettger, K. & Rudow, A. (1995) The Chironomidae (Diptera, Nematocera) of the Kossau, a North German lowland stream using the emergence trap method: Limnological studies in the nature reserve Kossautal (Schleswig- Holstein): IV. *Limnologica* **25**, 49–60.

Borror, D.J. (1934) Ecological studies of *Agria moesta* Hogen (Odonata: Coenagrionidae) by means of marking. *Ohio J. Sci.* **34**, 97–108.

Borutsky, E.V. (1955) A new trap for the quantitative estimation of emerging chironomids.] [In Russian.] *Trudy vses. gidrobiol. Obshch.* **6**, 223–6.

Braverman, Y. (1970) An improved emergence trap for Culicoides. *J. Econ. Entomol.* **63**, 1674–5.

Breymeyer, A. & Pieczynski, E. (1963) Review of methods used in the Institute of Ecology, Polish Academy of Sciences, for investigating migration. [In Polish.] *Ekol. Polska* **B 9**, 129–44.

Brown, J.L. & Orians, G.H. (1970) Spacing patterns in mobile animals. *Annu. Rev. Ecol. Syst.* **1**, 239–62.

Brundin, L. (1949) Chironomiden und andere Bodentiere der Siidschwedischen Urgebirgsseen. *Rep. Inst. Freshw. Res. Drottringholm* **20**, 915 pp.

Bryant, D.M. (1973) The factors influencing the selection of food by the House Martin (*Delichon urbica* (L.). *J. Anim. Ecol.* **42**, 539–64.

Buckner, C.H. (1958) Mammalian predators of the larch sawfly in eastern Manitoba. *Proc. Xth Int. Congr. Entomol.* **49**, 353–61.

Buckner, C.H. (1959) The assessment of larch sawfly cocoon predation by small mammals. *Can. Entomol.* **91**, 275–82.

Buckner, C.H. (1966) The role of vertebrate predators in the biological control of forest insects. *Annu. Rev. Entomol.* **11**, 449–70.

Burgman, M.A. & Fox, J.C. (2003) Bias in species range estimates from minimum convex polygons: implications for conservation and options for improved planning. *Animal Conserv.* **6**, 19–28.

Burt, W.H. (1943) Territoriality and home range concepts as applied to mammals. *J. Mammal.* **24**, 346–52.

Calenge, C. (2011) Home range estimation in R: the adehabitatHR package. http://cran.r-project.org/web/packages/adehabitatHR/vignettes/adehabitatHR.pdf, April 2011.

Campbell. J.A. & Pelham-Clinton, E.C. (1960) A taxonomic review of the British species of 'Culicoides' Latreille (Diptera, Ceratopogonidae). *Proc. R. Soc. Edinb.* **B67**, 181–302.

Carlson, D. (1971) A method for sampling larval and emerging insects using an aquatic black light trap. *Can. Entomol.* **103**, 1365–9.

Carter, C.L. (1972) Winter temperature and survival of the green spruce aphid *Elatobium abietinum. Forestry Commission. Forest Records* No. **84**, 10 pp.

Cavallini, P. & Serafini, P. (1995) Winter diet of the small Indian mongoose, *Herpestes auropunctatus*, on an Adriatic Island. *J. Mammal.* **76**, 569–74.

Cavallini, P. & Volpi, T. (1995) Biases in the analysis of the diet of the red fox *Vulpes vulpes. Wildlife Biol.* **1**, 243–8.

Chapman, J.W., Drake, V.A., & Reynolds, D.R. (2011) Recent insights from radar studies of insect flight. *Annu. Rev. Entomol.* **56**, 337–56.

Chauvin, R. (1966) Un procédé pour recolter automatiquement les proies que les *Formica polyctena* rapportent au rid. *Insect. Soc.* **13**, 59–67.

Clark, D.P. (1962) An analysis of dispersal and movement in *Phaulacridium vittatum* (Sj6st.) (Acrididae). *Aust. J. Zool.* **10**, 382–99.

Clark, E.W. & Osgood, E.A. (1964) An emergence container for recovering southern pine beetles from infested bolts. *J. Econ. Entomol.* **57**, 783–4.

Clark, L.P. (1963) The influence of population density on the number of eggs laid by females of *Cardiaspina albitextura* (Psyllidae). *Aust. J. Zool.* **11**, 190–201.

Clarke, R.D. & Grant, P.R. (1968) An experimental study of the role of spiders as predators in a forest litter community. Part 1. *Ecology* **49**(6), 1152–4.

Coaker, T.H. (1965) Further experiments on the effect of beetle predators on the numbers of the cabbage root fly, *Erioischia brassicae* (Bouché), attacking brassica crops. *Ann. Appl. Biol.* **56**, 7–20.

Cock, M.I.W. (1978) The assessment of preference. *J. Anim. Ecol.* **47**, 805–16.

Colless, D.H. & Chellapah, W.T. (1960) Effects of body weight and size of bloodmeal upon egg production in *Aedes aegypti* (Linnaeus) (Diptera, Culicidae). *Ann. Trop. Med. Parasitol.* **54**, 475–82.

Cook, L.M. & Kettlewell, H.B.D. (1960) Radioactive labelling of lepidopterous larvae: a method of estimating late larval and pupal mortality in the wild. *Nature* **1879**, 301–2.

Cook, P.P. & Horn, H.S. (1968) A sturdy trap for sampling emergent Odonata. *Ann. Entomol. Soc. Am.* **61**, 1506–7.

Corbet, P.S. (1966) Diel periodicities of emergence and oviposition in riverine Trichoptera. *Can. Entomol.* **98**, 1025–34.

Core, A., Runckel, C., Ivers, J., Quock, C., Siapno, T., DeNault, S., Brown, B., DeRisi, J., Smith, C., & Hafernik, J. (2012) A new threat to honey bees, the parasitic phorid fly *Apocephalus borealis*. *Plos One* **7**, e29639.

Cottrell, P.E., Trites, A.W., & Miller, E.H. (1996) Assessing the use of hard parts in faeces to identify harbour seal prey: Results of captive-feeding trials. *Can. J. Zool.* **74**(5), 875–80.

Crumpacker, D.W. & Williams, J.S. (1973) Density, dispersion and population structure in *Drosophila pseudoobscura*. *Ecol. Monogr.* **43**(4), 499–538.

Custer, T.W. & Pitelka, F.A. (1974) Correction factors for digestion rates for prey taken by snow buntings (*Plectrophenax nivalis*). *Condor* **77**, 210–12.

Darby, R.E. (1962) Midges associated with California rice fields, with special reference to their ecology (Diptera: Chironomidae). *Hilgardia* **32**, 1–206.

Davidson, A. (1956) A method of counting Ephemeropteran eggs. *Entomol. Monthly Mag.* **92**, 109.

Davies, I.J. (1984) Sampling aquatic insect emergence. In: Downing, J.A. & Rigler, F.H. (eds), *A Manual of Methods for the Assessment of Secondary Productivity in Fresh Waters*. Blackwell Scientific Publications, Oxford.

Davies, J.B. (1966) An evaluation of the emergence or box trap for estimating sandfly (*Culicoides* spp. Heleidae) populations. *Mosquito News* **26**, 170–2.

Dean, G.T. (1973) Aphid colonization of spring cereals. *Ann. Appl. Biol.* **75**, 183–93.

Debach, P. (1946) An insecticidal check method for measuring the efficacy of entomophagous parasites. *J. Econ. Entomol.* **39**, 695–7.

Debach, P. (1949) Population studies of the long-tailed mealy bug and its natural enemies on citrus trees in Southern California, 1946. *Ecology* **30**, 14–25.

Debach, P. (1955) Validity of the insecticidal check method as a measure of the effectiveness of natural enemies of Diaspine scale insects. *J. Econ. Entomol.* **48**, 584–8.

Debach, P. & Huffaker, C.B. (1971) Experimental techniques for evaluation of the effectiveness of natural enemies. In: Huffaker, C.B. (ed.), *Biological Control*. Plenum Press, New York. pp. 113–40.

Dempster, J.P. (1957) The population dynamics of the Moroccan locust (*Dociostaurus maroccanus* Thunberg) in Cyprus. *Anti-Locust Bull.* **27**, 1–60.

Dewalt, R.E., Stewart, K.W., Moulton, S.R., & Kennedy, J.H. (1994) Summer emergence of mayflies, stoneflies and caddisflies from a Colorado mountain stream. *Southwestern Naturalist* **39**(3), 249–56.

Diallo, B.P., Truc, P., & Laveissiere, C. (1997) A new method for identifying blood meals of human origin in tsetse flies. *Acta Tropica* **63**(1), 61–4.

Dinther, J.B.M. Van (1953) Details about some flytraps and their application to biological research. *Entomol. Ber.* **14**, 201–4.

Doane, J.F. (1963) Dispersion on the soil surface of marked adult *Ctenicera destructor* and *Hypolithus bicolor* (Coleoptera: Elateridae), with notes on fight. *Ann. Entomol. Soc. Am.* **56**, 340–5.

Dobson, R.M., Stephenson. J.W., & Lofty, J.R. (1958) A quantitative study of a population of wheat bulb fly, *Leptohylemyia coarctala* (Fall.), in the field. *Bull. Entomol. Res* **49**, 95–111.

Dobzhansky, T. & Wright, S. (1943) Genetics of natural populations: X. Dispersion rates in *Drosophila pseudoobscura*. *Genetics* **28**, 304–40.

Dobzhansky, T. & Wright, S. (1947) Genetics of natural populations. XV. Rate of diffusion of a mutant gene through a population of *Drosophila pseudoobscura*. *Genetics* **32**, 303–24.

Dolný, A., Harabiš, F., & Mižičová, H. (2014) Home Range, Movement, and Distribution Patterns of the Threatened Dragonfly *Sympetrum depressiusculum* (Odonata: Libellulidae): A Thousand Times Greater Territory to Protect? *PloS One* **9**, e100408.

Douglas, R.M. (1992) Microscopic investigation of the digestive tract contents of spotted grass snakes, *Psammophylax rhombeatus rhombeatus* (Reptilia: Colubridae). *J. African Zool.* **106**(5), 401–11.

Dowden, P.B., Jaynes, H.A., & Carolin, V.M. (1953) The role of birds in a spruce budworm outbreak in Maine. *J. Econ. Entomol.* **46**, 307–12.

Eguagie, W.E. (1974) Cold hardiness of *Tingis ampliata* (Heteroptera: Tingidae). *Entomol. Exp. Appl.* **17**, 204–14.

Eickwort, K.R. (1977) Population dynamics of a relatively rare species of milkweed beetle, (Labidomera). *Ecology* **58**, 527–38.

Elliott, J.M. (1970) Diel changes in invertebrate drift and the food of trout (*Salmo trutta* L.) *J. Fish. Biol.* **2**, 161–5.

Elliott, J.M. & Persson, L. (1978) The estimation of daily rates of food consumption for fish. *J. Anim. Ecol.* **47**, 977–91.

Elton, C. (1932) Territory among wood ants (*Formica rufa* L.) at Picket Hill. *J. Anim. Ecol.* **1**, 69–76.

Emden, H.F. Van (1962) A preliminary study of insect numbers in field and hedgerow. *Entomol. Monthly. Mag.* **98**, 255–9.

Emden, H.F. Van (1963) A field technique for comparing the intensity of mortality factors acting on the cabbage aphid, *Brevicoryne brassicae* (L) (Hem., Aphididae) in different areas of a crop. *Entomol. Exp. Appl.* **6**, 53–62.

Errington, P. (1932) Technique of raptor food habits study. *Condor* **34**, 75–86.

Evans, H.E. & Yoshimoto, C.M. (1962) The ecology and nesting behaviour of the Pompilidae (Hymenoptera) of the Northeastern United States. *Misc. Pub. Entomol. Soc. Am.* **3**(3), 65–119.

Evans, H.F. (1976) The role of predator-prey size ratio in determining the efficiency of capture by *Anthrocoris nemorum* and the escape reactions of its prey, *Acyrthosiphon pisum*. *Ecol. Entomol.* **1**, 85–90.

Evenhuis, H.H. (1962) Methods to investigate the population dynamics of aphids and aphid parasites in orchards. *Entomophaga* **7**, 213–20.

Fagan, W.F. (1997) Introducing a 'boundary-flux' approach to quantifying insect diffusion rates. *Ecology* **78**(2), 579–87.

Fewkes, D.W. (1964) The fecundity and fertility of the Trinidad sugar-cane froghopper, *Aeneolamia varia saccharine* (Homoptera, Cercopidae). *Trop. Agricult. Trin.* **419**, 165–8.

Finnegan, R.J. (1969) Assessing predation by ants on insects. *Insectes Sociaux* **16**, 61–5.

Fleschner, C.A. (1950) Studies on searching capacity of the larvae of three predators of the citrus red mite. *Hilgardia* **20**(13), 233–65.

Fletcher, B.S. (1974) The ecology of a natural population of the Queensland fruit fly, *Dacus tryoni* V. The dispersal of adults. *Aust. J. Zool.* **22**, 189–202.

Fletcher, B.S. & Economopoulos, A.P. (1976) Dispersal of normal and irradiated laboratory strains and wild strains of the olive fly *Dacus oleae* in an olive grove. *Entomol. Exp. Appl.* **20**, 183–9.

Frank, P.W. (1964) On home range of limpets. *Am. Nat.* **98**, 99–100.

Freeman, G.H. (1977) A model relating numbers of dispersing insects to distance and time. *J. Appl. Ecol.* **14**, 477–87.

Friedman, M. (1940) A comparison of alternative tests of significance for the problem of m rankings. *Ann. Math. Stat.* **11**, 86–92.

Fryxell, J.M., Mosser, A., Sinclair, A.R., & Packer, C. (2007) Group formation stabilizes predator–prey dynamics. *Nature* **449**, 1041–3.

Gary, N.E. (1960) A trap to quantitatively recover dead and abnormal honey bees from the hive. *J. Econ. Entomol.* **53**, 782–5.

Getz, W.M. & Wilmers, C.C. (2004) A local nearest-neighbor convex-hull construction of home ranges and utilization distributions. *Ecography* **27**, 489–505.

Gibb, J.A. (1958) Predation by tits and squirrels on the Eucosmid *Ernarmonia conicolana* (Heyl.). *J. Anim. Ecol.* **27**, 375–96.

Giller, P.S. (1986) The natural diet of the Notonectidae: Field trials using electrophoresis. *Ecol. Entomol.* **11**(2), 163–72.

Gilmour, D., Waterhouse, D.F., & McIntyre, G.A. (1946) An account of experiments undertaken to determine the natural population density of the sheep blowfly, *Lucilia cuprina* Wied. *Bull. Coun. Sci. Indust. Res. Aust.* **195**, 1–39.

Glen. D.M. (1975) Searching behaviour and prey-density requirements of *Blephandopterus angulatus* (Fall.) (Heteroptera: Miridae) as a predator of the Lime Aphid, *Eucallipterus tiliae* (L.) and Leafhopper, *Ainetoidea aineti* (Dahlbom). *J. Anim. Ecol.* **44**, 116–34.

Glen, D.M. (1976) An emergence trap for bark-dwelling insects, its efficiency and effects on temperature. *Ecol. Entomol.* **1**, 91–4.

Goodhart, C.B. (1958) Thrush predation on the snail *Cepaea hortensis*. *J. Anim. Ecol.* **279**, 47–57.

Graham. S.A. (1928) The influence of small mammals and other factors upon larch sawfly survival. *J. Econ. Entomol.* **21**, 301–10.

Grandilewskaja-Decksbach, M.L. (1935) Materialien zur Chironomidenbiologicn verschiedener Becken. Zur Frage ilber die Schwankungen der Anzahl und der Biomasse der Chironomidenlarven. *Trudy Limnol. Sta. Kosine* **19**, 145–82.

Green, G.W. (1962) Low winter temperatures and the European pine shoot moth, *Rhyacionia buoiiana* (Schiff.) in Ontario. *Can. Entomol.* **94**, 314–36.

Green, G.W. & Pointing, P.J. (1962) Flight and dispersal of the European pine shoot moth, *Rhyacionia buoliana* (Schiff.) II. Natural dispersal of egg-laden females. *Can. Entomol.* **94**, 299–314.

Greenberg, B. & Bornstein, A.A. (1964) Fly dispersion from a rural Mexican slaughter house. *Am. J. Trop. Med. Hyg.* **13**(6), 881–6.

Greenslade, P.I.M. (1964) The distribution, dispersal and size of a population of *Nebria brevicollis* (F.) with comparative studies on three other carabidae. *J. Anim. Ecol.* **33**, 311–33.

Gregor, F. (1960) Zur Eiproduktion des Eichenwicklers (*Tortrix viridana* L.). *Zool. Listy.* **9**, 11–18.

Gruber, F. & Prieto, C.A. (1976) Collecting chamber suitable for recovery of insects from large quantities of host plant material. *Environ. Entomol.* **5**, 343–4.

Guennelon, G. & Audemard, M.H. (1963) Enseignements écologiques donnés par la méthode de captures par cuisses-éclosion de la cécidomyie des lavandes (*Thomasmiana lavandulae* Barnes). Critique de la méthode. Conclusions pratiques. *Ann. Epiphyt. C* **149**, 35–48.

Gui, L.Y., Xiu-Qin, H., Chuan-Ren, L., & Boiteau, G. (2011) Validation of harmonic radar tags to study movement of Chinese citrus fly. *Can. Entomol.* **143**, 415–22.

Hafez, M. (1961) Seasonal fluctuations of population density of the cabbage aphid, *Brevicoryne brassicae* (L.) in the Netherlands, and the role of its parasite, *Aphidius* (*Diaeretiella*) *rapae* (Curtis). *Tijdschr. Pl. Ziekt.* **67**, 445–548.

Hamann, H. & Iwannek, K.H. (1981) The use of the indicator activation method for the labelling of tsetse flies. *Z. Angew. Entomol.* **91**, 206–12.

Hanski, I. (1991) Single-species metapopulation dynamics: concepts, models and observations. *Biol. J. Linn. Soc.* **42**, 17–38.

Hard, J.S. (1976) Estimation of Hemlock sawfly (Hymenoptera: Diprionidae) fecundity. *Can. Entomol.* **108**, 961–6.

Hartley, P.H.T. (1948) The assessment of the food of birds. *Ibis* **90**, 361–81.

Hassell, M.P. (1976a) Arthropod predator–prey systems. In: May, R.M. (ed.), *Theoretical Ecology, Principles and Applications*. Blackwell Scientific Publications, Oxford, pp. 71–93.

Hassell, M.P. (1976b) *The Dynamics of Competition and Predation*. Edward Arnold, London.

Hassell, M.P. (1978) *The Dynamics of Arthropod Predator–Prey Relationships*. Princeton Monograph on Population Biology **13**, pp. 245. Princeton, New Jersey.

Hassell, M.P., Lawton. J.H., & Beddington, J.R. (1976) The components of arthropod predation. 1. The prey death rate. *J. Anim. Ecol.* **45**, 135–64.

Hassell, M.P. & May, R.M. (1974) Aggregation of predators and insect parasites and its effect on stability. *J. Anim. Ecol.* **43**, 567–94.

Hassell, M.P. & Varley C.C. (1969) A new inductive population model for insect parasites and its bearing on biological control. *Nature* **223**, 1133–7.

Hassel, M.P., Comins, H.N., & May, R.M. (1991) Spatial structure and chaos in insect population dynamics. *Nature* **353**, 255–8.

Hastings, A. & Harrison, S. (1994) Metapopulation dynamics and genetics. *Annu. Rev. Ecol. Syst.* **25**, 167–88.

Hawkes, R.B. (1972) A fluorescent dye technique for marking insect eggs in predation studies. *J. Econ. Entomol.* **65**, 1477–8.

Henderson, P.A. (1998) On the variation in dab *Limanda limanda* recruitment: a zoogeographic study. *J. Sea Res.* **40**, 131–42.

Hespenheide, H.A. (1975) Prey characteristics and predator niche width. In Cody, M.L. & Diamond, J.M. (eds), *Ecology and Evolution of Communities*. Harvard University Press, Cambridge, Mass., pp. 158–80.

Hewitt, D.G. & Robbins, C.T. (1996) Estimating grizzly bear food habits from fecal analysis. *Wildlife Soc. Bull.* **24**(3), 547–50.

Higley, L.G. & Pedigo, L.P. (1985) Examination of some adult sampling techniques for the seedcorn maggot (*Delia platura*). *J. Agricult. Entomol.* **2**(1), 52–60.

Hoddle, M.S. (2006) Phenology, life tables, and reproductive biology of *Tetraleurodes perseae* (Hemiptera: Aleyrodidae) on California avocados. *Ann. Entomol. Soc. Am.* **99**, 553–9.

Hoffmann, V., Pfaff, M.C., & Branch, G.M. (2012) Spatio-temporal patterns of larval supply and settlement of intertidal invertebrates reflect a combination of passive transport and larval behavior. *J. Exp. Mar. Biol. Ecol.* **418**, 83–90.

Holling, C.S. (1958) A radiographic technique to identify healthy, parasitised and diseased sawfly prepupae within cocoons. *Can. Entomol.* **90**, 59–61.

Holling, C.S. (1959a) The components of predation as revealed by a study of small mammal predation of the European pine sawfly. *Can. Entomol.* **91**, 293–320.

Holling, C.S. (1959b) Some characteristics of simple types of predation and parasitism. *Can. Entomol.* **91**, 385–98.

Holling, C.S. (1961) Principles of insect predation. *Annu. Rev. Entomol.* **6**, 163–82.

Holling, C.S. (1965) The functional response of predators to prey density and its role in mimicry and population regulation. *Mem. Entomol. Soc. Can.*, No. **45**, 60 pp.

Holling, C.S. (1966) The functional response of invertebrate predators to prey density. *Mem. Entomol. Soc. Can.*, No. **48**, 86 pp.

Honěk, A. (1993) Intraspecific variation in body size and fecundity in insects: a general relationship. *Oikos* **66**, 483–92.

Huffaker, C.B. & Spitzer, C.H. (1950) Some factors affecting red mite populations on pears in California. *J. Econ. Entomol.* **43**, 819–31.

Hughes, D.P., Beani, L., Turillazzi, S., & Kathirithamby, J. (2003) Prevalence of the parasite *Strepsiptera* in Polistes as detected by dissection of immatures. *Insectes Sociaux* **50**, 62–8.

Hunter, M.D. & West, C. (1990) Variation in the effects of spring defoliation on the late season phytophagous insects of *Quercus robur*. In: Watt, A.D., Leather, S.R., Hunter, M.D., & Kidd, N.A.C. (eds), *Population Dynamics of Forest Insects*. Intercept, Andover, UK, pp. 123–35.

Ide, F.P. (1940) Quantitative determination of the insect fauna of rapid water. *Univ. Toronto Stud. Biol. Ser.* **47** (Publ. Ontario Fish. Res. Lab. 59), 20 pp.

Inoue, T., Kamimmura, K., & Watanabe, M. (1973) A quantitative analysis of dispersal in a horse-fly *Tabanus iyoensis* Shiraki and its application to estimate the population size. *Res. Popul. Ecol.* **14**, 209–33.

Ito, Y., Yamanaka, H., Nakasuji, F., & Kiritani, K. (1972) Determination of predator–prey relationship with an activable tracer, Europium-151. *Kontyu* **40**, 278–83.

Ives, W.G.H. & Prentice, R.M. (1959) Estimation of parasitism of larch sawfly cocoons by *Bessa harveyi* Tnsd. in survey collections. *Can. Entomol.* **91**, 496–500.

Ives, W.G.H., Turnock, W.J., Buckner, J.H., Heron, R.I., & Muldrew, I.A. (1968) Larch sawfly population dynamics: techniques. *Manitoba Entomol.* **2**, 5–36.

Iwao, S. (1956) On the number of eggs per egg-mass of the paddy rice borer, *Schoenobius incertellus* Walker and the percentage of their parasitization. [In Japanese.] *Gensei (Kochi Konchu Dokokai)* **5**, 45–9.

Iwao, S. (1963) On a method for estimating the rate of population interchange between two areas. *Res. Popul. Ecol.* **5**, 44–50.

Jacobs, J. (1974) Quantitative measurement of food selection: a modification of the forage ratio and Iviev's selectivity index. *Oecologia* **14**, 413–17.

James, H.G. (1961) Some predators of *Aedes stimulans* (Walk) and *Aedes trichurus* (Dyar) (Diptera: Culicidae) in woodland pools. *Can. J. Zool.* **39**, 533–40.

James, H.G. (1966) Location of univoltine *Aedes* eggs in woodland pool areas and experimental exposure to predators. *Can. Entomol.* **98**, 550–5.

Jaworski, C.C., Bompard, A., Genies, L., Amiens-Desneux, E., & Desneux, N. (2013) Preference and prey switching in a generalist predator attacking local and invasive alien pests. *PloS One* **8**, e82231.

Jedrzejewski, W., Jedrzzejewska, B., Szymura, B., & Zub, K. (1996) Tawney owl (*Strix aluco*) predation in a pristine deciduous forest (Bialowieza National Park, Poland). *J. Anim. Ecol.* **65**, 105–20.

Jenkins, D.W. (1963) Use of radionuclides in ecological studies of insects. In: Schultz, V. & Klement, A.W. (eds), *Radioecology*. Rheinhold, New York, pp. 431–40.

Jenkins, D.W. & Hassett, C.C. (1950) Radioisotopes in entomology. *Nucleonics* **6**(3), 5–14.

Johnson, C.G. (1969) *Migration and Dispersal of Insects by Flight*. Methuen, London, 763 pp.

Johnson, J.H. & Ross, R.M. (1996) Pellets versus faeces: Their relative importance in describing the food habits of double-crested cormorants. *J. Great Lakes Res.* **22**(3), 795–8.

Johnson, J.H., Ross, R.M., & Smith, D.R. (1997) Evidence of secondary consumption of invertebrate prey by double-crested Cormorants. *Colonial Waterbirds* **20**(3), 547–51.

Jolicoeur, P. & Brunel, D. (1966) Application du diagramme hexagonal a l'étude de la sélection de ses proies par la morue. *Vie et Milieu* **17**, 419–33.

Jónasson, P.M. (1954) An improved funnel trap for capturing emerging aquatic insects, with some preliminary results. *Oikos* **5**, 179–88.

Jorgensen, C.D. & Tanner, W.W. (1963) The application of the density probability function to determine the home ranges of *Ulla stansburiana stansburiana* and *Cnemidophorus tigris tigris*. *Herpetologica* **19**, 105–15.

Judge, W.W. (1957) A study of the population of emerging and littoral insects trapped as adults from tributary waters of the Thames River at London, Ontario. *Am. Midl. Nat.* **58**, 394–412.

Kendall, D.G. (1974) Pole seeking, Brownian motion and bird navigation. *J. R. Stat. Soc. (B)* **36**, 365–417.

Kensler, C.B. (1967) Desiccation resistance of intertidal crevice species as a factor in their zonation. *J. Anim. Ecol.* **36**, 391–406.

Kent, R.J. (2009) Molecular methods for arthropod bloodmeal identification and applications to ecological and vector-borne disease studies. *Mol. Ecol. Resources* **9**, 4–18.

Kie, J.G., Matthiopoulos, J., Fieberg, J., Powell, R.A., Cagnacci, F., Mitchell, M.S., Gillard, J.M., & Moorcroft, P.R. (2010) The home-range concept: are traditional estimators still relevant with modern telemetry technology? *Philos. Trans. R. Soc. B: Biol. Sci.* **365**, 2221–31.

Kimerle, R.A. & Anderson, N.H. (1967) Evaluation of aquatic insect emergence traps. *J. Econ. Entomol.* **60**, 1255–9.

King, R.A., Read, D.S., Traugott, M., & Symondson, W.O.C. (2008) Invited review. Molecular analysis of predation: a review of best practice for DNA-based approaches. *Mol. Ecol.* **17**, 947–63.

Kiritani, K. & Dempster, J.P. (1973) Different approaches to the quantitative evaluation of natural enemies. *J. Appl. Ecol.* **10**, 323–30.

Kiritani, K., Kawahara, S., Sasaba, T., & Nakasuji, F. (1972) Quantitative evaluation of predation by spiders on the green rice leafhopper, *Nephotettix cincticeps* Uhler, by a sight-count method. *Res. Popul. Ecol.* **13**, 187–200.

Kissling, D.W., Pattemore, D.E., & Hagen, M. (2013) Challenges and prospects in the telemetry of insects. *Biol. Rev. Cambridge Philos. Soc.* **89**, 511–30.

Knutsen, H. & Vogt, N.B. (1985) An approach to identifying the feeding patterns of lobsters using chemical analysis and pattern recognition by the method of SIMCA. II Attempts at assigning stomach contents of lobsters, *Homarus gammarus* (L) to infauna and detritus. *J. Exp. Mar. Biol. Ecol.* **89**, 121–34.

Kohira, M. & Rexstad, E.A. (1997) Diets of wolves, *Canis lupus*, in logged and unlogged forests of southeastern Alaska. *Can. Field Naturalist* **111**(3), 429–35.

Kruuk, H. (1972) Surplus killing by carnivores. *J. Zool.* **166**, 233–44.

Kurdziel, J.P. & Bell, S.S. (1992) Emergence and dispersal of phytal-dwelling meiobenthic copepods. *J. Exp. Mar. Biol. Ecol.* **163**(1), 43–64.

Lack, D. & Owen, D.F. (1955) The food of the swift. *J. Anim. Ecol.* **24**, 120–36.

Lacey, L.A. (ed.) (2012) *Manual of Techniques in Invertebrate Pathology*. Academic Press.

Langford, T.E. & Daffern, J.R. (1975) The emergence of insects from a British river warmed by power station cooling-water. *Part. I. Hydrobiologia* **46**, 71–114.

Latham, D.R. & Mills, N.J. (2009) Quantifying insect predation: a comparison of three methods for estimating daily per capita consumption of two aphidophagous predators. *Environ. Entomol.* **38**, 1117–25.

Laverty, T.M. & Dobson, A.P. (2013) Dietary overlap between black caimans and spectacled caimans in the Peruvian Amazon. *Herpetologica* **69**, 91–101.

Lawson, F.R. (1959) The natural enemies of the hornworms on tobacco (Lepidoptera: Sphingidae). *Ann. Entomol. Soc. Am.* **52**, 741–55.

Lawton, J.H., Beddington, J.R., & Bonser, R. (1974) Switching in invertebrate predators. In: Usher, M.B. & Williamson, M.H. (eds), *Ecological Stability*, Chapman & Hall, London, pp. 141–58.

Le Jacques, D. & Lode, T. (1994) Feeding habits of European genets (*Genetta genetta* L., 1758) from a grove in western France. *Mammalia* **58**(3), 383–9.

Lefranc, N. & Worfolk, T. (1997) *Shrikes*. Pica Press, Sussex, UK.

LeJeune, R.R., Fell, W.H., & Burbidge, D.P. (1955) The effect of flooding on development and survival of the larch sawfly *Pristiphora erichsonii* (Tenthredinidae). *Ecology* **36**, 63–70.

LeSage, L. & Harrison, A.D. (1979) Improved traps and techniques for the study of emerging aquatic insects. *Entomol. News* **90**(2), 65–78.

Lindeberg, B. (1958) A new trap for collecting emerging insects from small rockpools, with some examples of the results obtained. *Suom. hyönt. Aikak. (Ann. Ent. Fenn.)* **24**, 186–91.

Lloveras, L., Moreno-García, M., & Nadal, J. (2009) The eagle owl (*Bubo bubo*) as a leporid remains accumulator: taphonomic analysis of modern rabbit remains recovered from nests of this predator. *Int. J. Osteoarchaeol.* **19**, 573–92.

Loan, C. & Holdaway, F.G. (1961) *Microctonus aethiops* (Nees) auctt. and *Perilitus rutilus* (Nees) (Hymenoptera: Braconidae). European parasites of *Sitona* weevils (Coleoptera: Curculionidae). *Can. Entomol.* **93**, 1057–78.

Lozinsky, V.A. (1961) On the correlation existing between the weight of pupae and the number and weight of eggs of *Lymantria dispar* L. [In Russian.] *Zool. Zh.* **40**, 1571–3.

Macan, T.T. (1949) Survey of a moorland fishpond. *J. Anim. Ecol.* **18**, 160–86.

Matsuzawa, Y., Sato, K., Sakamoto, W., & Bjorndal, K. (2002) Seasonal fluctuations in sand temperature: effects on the incubation period and mortality of loggerhead sea turtle (*Caretta caretta*) pre-emergent hatchlings in Minabe, Japan. *Mar. Biol.* **140**, 639–46.

McCauley, V.J.E. (1976) Efficiency of a trap for catching and retaining insects emerging from standing water. *Oikos* **27**, 339–45.

McKnight, M.E. (1969) Distribution of hibernating larvae of the Western Budworm, *Choristoneura occidentalis* on Douglas Fir in Colorado. *J. Econ. Entomol.* **62**, 139–42.

MacLellan, C.R. (1962) Mortality of codling moth eggs and young larvae in an integrated control orchard. *Can. Entomol.* **94**, 655–66.

MacLeod, I. & Donnelly, J. (1956) Methods for the study of blowfly populations. II. The use of laboratory-bred material. *Ann. Appl. Biol.* **44**, 643–8.

MacLeod, I. & Donnelly, J. (1963) Dispersal and interspersal of blowfly populations. *J. Anim. Ecol.* **32**, 1–32.

McMullen, L.H. & Atkins, M.D. (1959) A portable tent-cage for entomological field studies. *Proc. Entomol. Soc. B.C.* **56**, 67–8.

MacPhee, A.W. (1961) Mortality of winter eggs of the European red mite *Panonychus ulmi* (Koch), at low temperatures, and its ecological significance. *Can. J. Zool.* **39**, 229–43.

MacPhee, A.W. (1964) Cold-hardiness, habitat and winter survival of some orchard Arthropods in Nova Scotia. *Can. Entomol.* **96**, 617–36.

Malison, R.L., Benjamin, J.R., & Baxter, C.V. (2010) Measuring adult insect emergence from streams: the influence of trap placement and a comparison with benthic sampling. *J. North Am. Bentholog. Soc.* **29**, 647–56.

Martin, F.J. (1969) Searching success of predators in artificial leaflitter. *Am. Midl. Nat.* **819**, 218–27.

Mason, W.H. & McGraw, K.A. (1973) Relationship of ^{65}Zn excretion and egg production in *Trichoplusia ni* (Hubner). *Ecology* **54**, 214–16.

Matthews, G.V.T. (1974) On bird navigation with some statistical undertones. *J. R. Stat. Soc. (B)* **36**, 349–64.

May, R. M. (1976) Models for two interacting populations. In: May, R.M. (ed.), *Theoretical Ecology, Principles and Applications*, Blackwell Scientific Publications, Oxford, pp. 49–70.

Michalski, M., Nadolski, J., Marciniak, B., Loga, B., & Banbura, J. (2011) Faecal analysis as a method of nestling diet determination in insectivorous birds: a case study in Blue Tits *Cyanistes caeruleus* and Great Tits *Parus major*. *Acta Ornithol.* **46**, 164–72.

Miles, P.M. (1952) Entomology of Bird Pellets. *Amat. Entomol. Soc.* Leaflet **24**, 8 pp.

Miller, C.A. (1955) A technique for assessing larval mortality caused by parasites. *Can. J. Zool.* **33**, 5–17.

Miller, C.A. (1957) A technique for estimating the fecundity of natural populations of the spruce budworm. *Can. J. Zool.* **35**, 1–13.

Miller, C.A. (1958) The measurement of spruce budworm populations and mortality during the first and second larval instars. *Can. J. Zool.* **36**, 409–22.

Miller, R.S. & Thomas, J.L. (1958) The effects of larval crowding and body size on the longevity of adult *Drosophila melanogaster*. *Ecology* **39**, 118–25.

Moeck, H.A. (1988) A bucket emergence trap for corticolous insects. *Can. Entomol.* **120**(10), 855–8.

Mohr, C.O. (1947) Table of equivalent populations of North American small mammals. *Am. Midl. Nat.* **37**, 223–49.

Moncorps, S., Chapuis, J.L., Haubreux, D., & Bretagnolle, V. (1998) Diet of the brown skua *Catharacta skua loennbergi* on the Kerguelen archipelago: Comparisons between techniques and between islands. *Polar Biol.* **19**(1), 9–16.

Moorcroft, P.R. & Lewis, M.A. (2013) *Mechanistic Home Range Analysis*. (MPB-43). Princeton University Press.

Moore, N.W. (1952) On the so-called 'territories' of dragonflies (Odonata: Anisoptera). *Behaviour* **4**, 85–100.

Moore, N.W. (1957) Territory in dragonflies and birds. *Bird Study* **4**, 125–30.

Morgan, C.V.G. & Anderson, N.H. (1958) Techniques for biological studies of Tetranychid mites, especially *Bryobia arborea* M. and A. and *B. praetiosa* Koch (Acarina: Tetranychidae). *Can. Entomol.* **90**, 212–15.

Morgan, N.C. (1971) Factors in the design and selection of insect emergence traps. In: Edmondson, W.T. & Winberg, G.G. (eds), *A Manual on Methods for the Assessment of Secondary Productivity in Fresh Waters*. Blackwell Scientific Publications, Oxford.

Morgan, N.C. & Waddell, A.B. (1961) Insect emergence from a small trout loch, and its bearing on the food supply of fish. *Sci. Invest. Freshw. Fish. Scot.* **25**, 1–39.

Morgan, N.C., Waddell, A.B., & Hall, W.B. (1963) A comparison of the catches of emerging aquatic insects in floating box and submerged funnel traps. *J. Anim. Ecol.* **32**, 203–19.

Morris, R.F. (1949) Differentiation by small mammals between sound and empty cocoons of the European spruce sawfly. *Can. Entomol.* **81**, 114–20.

Mougi, A. & Kondoh, M. (2012) Diversity of interaction types and ecological community stability. *Science* **337**, 349–51.

Mukherjee, S., Goyal, S.P., & Chellam, R. (1994) Refined techniques for the analysis of Asiatic lion *Panthera leo persica* scats. *Acta Theriol.* **39**(4), 425–30.

Mulla, M.S., Norland, R., Ikeshoji, T., & Kramer, W.L. (1974) Insect growth regulators for the control of aquatic midges. *J. Econ. Entomol.* **67**(2), 165–70.

Mundie, J.H. (1956) Emergence traps for aquatic insects. *Mitt. Int. Verein. Theor. Angew. Limnol.* **7**, 1–13.

Mundie, J.H. (1971) Techniques for sampling emerging aquatic insects. In: Edmondson, W.T. & Winberg, G.G. (eds), *A Manual on Methods for the Assessment of Secondary Productivity in Fresh Waters.* Blackwell Scientific Publications, Oxford.

Murdie, G. (1969) The biological consequences of decreased size caused by crowding or rearing temperatures in apterae of the pea aphid, *Acyrthosiphon pisum* Harris. *Trans. R. Entomol. Soc. Lond.* **121**, 443–55.

Murdie, G. & Hassell, M.P. (1973) Food distribution, searching success and predator–prey models. In: Hioms, R.W. (ed.), *The Mathematical Theory of the Dynamics of Biological Populations.* Academic Press, London, pp. 87–101.

Murdoch, W.W. (1969) Switching in general predators: experiments on predators specificity and stability of prey populations. *Ecol. Monogr.* **39**, 335–54.

Murdoch, W.W. (1971) The developmental response of predators to changes in prey density. *Ecology* **52**, 132–7.

Murdoch, W.W. & Marks, J.P. (1973) Predation by coccinellid beetles: experiments on switching. *Ecology* **54**, 160–7.

Murdoch, W.W. & Oaten, A. (1975) Predation and population stability. *Adv. Ecol. Res.* **9**, 1–131.

Murton, R.K. (1971) The significance of a specific search image in the feeding behaviour of the wood pigeon. *Behaviour* **40**, 10–42.

Mustafa, S. (1976) Selective feeding behaviour of the common carp, *Esomus danricus* (Ham.) in its natural habitat. *Biol. J. Linn. Soc.* **8**, 279–84.

Nakasuji, R., Yamanaka, H., & Kiritani, K. (1973) The disturbing effect of microphantid spiders on the larval aggregation of the tobacco cutworm, *Spodoptera litura* (Lepidoptera: Noctuidae). *Kontyu* **41**, 220–7.

Neilson, M.M. (1963) Disease and the spruce budworrn. *Mem. Entomol. Soc. Can.* **31**, 272–88.

Nelmes, A.J. (1974) Evaluation of the feeding behaviour of *Prionchulus punctatus* (Corb.) a nematode predator. *J. Anim. Ecol.* **43**, 553–65.

Neuenschwander, P. (1975) Influence of temperature and humidity on the immature stages of *Hemerobius pacificus. Environ. Entomol.* **4**(2), 215–20.

Newman, G.G. & Carner, G.R. (1975) Disease incidence in soy bean loopers collected by two sampling methods. *Environ. Entomol.* **4**, 231–2.

Nicholls, C.F. (1963) Some entomological equipment. *Res. Inst. Can. Dept. Agric. Belleville, Inf. Bull.* **2**, 85 pp.

Nijveldt, W. (1959) Overhet gebruik van vangekegels bij het galmugonderzoek. *Tijdschr. PlZieki.* **65**, 56–9.

Nord, J.C. & Lewis, W.C. (1970) Two emergence traps for wood-boring insects. *J. Ga. Entomol. Soc.* **5**, 155–7.

Odum, E.P. & Kuenzler, E.J. (1955) Measurement of territory and home range size in birds. *Auk* **72**, 128–37.

Ohnesorge, B. (1957) Untersuchungen ilber die Populationsdynamik der kleinen Fichtenblattwespe, *Pristiphora abietina* (Christ) (Hym. Tenthr.). I. Teil. Fertilitdt und Mortalitdt. *Z. Angew. Entomol.* **40**, 443–93.

Okubo, A. (1980) *Diffusion and Ecological Problems: Mathematical Models.* Springer-Verlag, Berlin.

Okubo, A. (1989) Dynamical aspects of animal grouping: swarms, schools, flocks, and herds. *Adv. Biophys.* **22**, 1–94.

Otvos, I.S. (1974) A collecting method for pupae of *Lambderia fiscellaria* (Lepidoptera: Geometridae). *Can. Entomol.* **106**, 329–31.

Palmén, E. (1955) Diel periodicity of pupal emergence in natural populations of some chironomids (Diptera). *Ann. Zool. Soc. Vanamo* **17**(3), 1–30.

Palmén, E. (1962) Studies on the ecology and phonology of the Chironomids (Dipt.) of the Northern Baltic. *Ann. Entomol. Fenn.* **28**(4), 137–68.

Paris, O.H. (1965) The vagility of p32-labelled Isopods in grassland. *Ecology* **46**, 635–48.

Pavlov, I.F. (1961) Ecology of the stem moth *Ochsenheimeria vaculella* F.-R. (Lepidoptera Tineoidea). [In Russian.] *Ent. Obozr.* **40**, 818–27 (transl. *Entomol. Rev.* **40**, 461–6).

Peterson, A. (1934) *A Manual of Entomological Equipment and Methods*. Pt 1. Edwards Bros Inc., Ann Arbor.

Pilon, J.G., Tripp, H.A., McLeod, J.M., & Ilnitzkey, S.L. (1964) Influence of temperature on prespinning eonymphs of the Swainejack-pine sawfly, *Neodiprion swainei* Midd. (Hymenoptera: Diprionidae). *Can. Entomol.* **96**, 1450–7.

Plank, V. der (1967) Spread of plant pathogens in space and time. In: Gregory, P.H. & Monteith, J.L. (eds), *Airborne Microbes*. Cambridge University Press, Cambridge.

Poinar, G.O. & Thomas, G.M. (1984) *Laboratory Guide to Insect Pathogens and Parasites*. Plenum Press, New York, pp. 392.

Polles, S.G. & Payne, J.A. (1972) An improved emergence trap for adult Pecan weevils. *J. Econ. Entomol.* **65**, 1529.

Posey, M.H., Alphin, T.D., & Cahoon, L. (2006) Benthic community responses to nutrient enrichment and predator exclusion: influence of background nutrient concentrations and interactive effects. *J. Exp. Mar. Biol. Ecol.* **330**, 105–18.

Prebble, M.L. (1941) The diapause and related phenomena in *Gilpinia polytoma* (Hartig). IV. Influence of food and diapause on reproductive capacity. *Can. J. Res. D* **19**, 417–36.

Promptow, A.N. & Lukina, E.W. (1938) Die Experimente beim biologischen Studium und die Erndhrung der Kohlmeise (*Parus major* L.) in der Brutperiode. [In Polish.] *Zool. Zh.* **17**, 777–82.

Radke, W.I. & Frydendall, M.J. (1974) A survey of emetics for use in stomach contents recovery in the house sparrow. *Am. Midl. Nat.* **92**, 164–72.

Rau, P. & Rau, N. (1916) The biology of the mud-daubing wasps as revealed by the contents of their nests. *J. Anim. Behavior* **6**, 27–63.

Reid, R.W. (1963) Biology of the mountain pine beetle, *Dendrocionus monticoiae* Hopkins, in the East Kootenay Region of British Columbia. III. Interaction between the beetle and its host, with emphasis on brood mortality and survival. *Can. Entomol.* **95**, 225–38.

Reiff, M. (1955) Untersuchungen zum Lebenszyklus der Frostspanner *Cheimatobia* (*Operophthera*) *brumata* L. und *Hibernia defoliaria*. *Ch. Mitt. Schweiz. Ent. Ges.* **269**, 129–44.

Rice, R.E. & Reynolds, H.T. (1971) Seasonal emergence and population development of the Pink Bollworm in Southern California. *J. Econ. Entomol.* **64**, 1429–32.

Rich, E.R. (1956) Egg cannibalism and fecundity in *Tribolium*. *Ecology* **37**, 109–20.

Richards, O.W. (1940) The biology of the small white butterfly (*Pieris rapae*), with special reference to the factors controlling its abundance. *J. Anim. Ecol.* **9**, 243–88.

Richards, O.W. & Hamm, A.H. (1939) The biology of the British Pompilidae (Hymenoptera). *Trans. Soc. Br. Entomol.* **6**, 51–114.

Richards, O.W. & Waloff, N. (1954) Studies on the biology and population dynamics of British grasshoppers. *Anti-Locust Bull.* **17**, 184 pp.

Richards, O.W. & Waloff, N. (1961) A study of a natural population of *Phytodecta olivacea* (Forster) (Coleoptera: Chrysomelidae). *Phil. Trans. B* **244**, 205–57.

Rogers, D.J. (1972) Random search and insect population models. *J. Anim. Ecol.* **41**, 369–83.

Rogers, D.J. & Hassell, M.P. (1974) General models for insect parasite and predator searching behaviour: interference. *J. Anim. Ecol.* **43**, 239–53.

Royama, T. (1970) Factors governing the hunting behaviour and selection of food by the great tit (*Parus major* L.). *J. Anim. Ecol.* **39**, 619–68.

Royama, T. (1971) A comparative study of models for predation and parasitism. *Res. Pop. Ecol. Suppl.* **1**, 90 pp.

Santori, R.T., Moraes, D.A.D., & Cerqueira, R. (1995) Diet composition of *Metachirus nudicaudatus* and *Didelphis aurita* (Marsupialia, Didelphoidea) in southern Brazil. *Mammalia* **59**(4), 511–16.

Santos, M.A. (1976) Prey selectivity and switching response of *Zetzellia maki*. *Ecology* **579**, 390–4.

Sasaba, T. & Kiritani, K. (1972) Evaluation of mortality factors with special reference to parasitism of the Green rice leafhopper, *Nephotettix cinticeps* Uhler (Hemiptera: Deltocephalidae). *Appl. Entomol. Zool.* **7**, 83–93.

Schenk, D. & Bacher, S. (2002) Functional response of a generalist insect predator to one of its prey species in the field. *J. Animal Ecol.* **71**, 524–31.

Scotter, D.R., Lamb, K.P., & Hassan. E. (1971) An insect dispersal parameter. *Ecology* **52**, 174–77.

Segers, F.H. & Taborsky, B. (2011) Egg size and food abundance interactively affect juvenile growth and behaviour. *Funct. Ecol.* **25**, 166–76.

Shapiro, A.M. (1974) Beak-mark frequency as an index of seasonal predation intensity on common butterflies. *Am. Nat.* **108**, 229–32.

Shaw, M.J.P. (1970) Effects of population density on alienicolae of *Aphis fabae* Scop. *Ann. Appl. Biol.* **65**, 191–4, 197–203, 205–12.

Siddon, C.E. & Witman, J.D. (2004) Behavioral indirect interactions: multiple predator effects and prey switching in the rocky subtidal. *Ecology* **85**, 2938–45.

Skellam, J.G. (1951) Random dispersal in a theoretical population. *Biometrika* **38**, 196–218.

Smale, M.J. (1986) The feeding habits of six pelagic and predatory teleosts in eastern Cape coastal waters (South Africa). *J. Zool. Series B* **1**, 357–410.

Smith, H.S. & Debach, P. (1942) The measurement of the effect of entomophagous insects on population densities of the host. *J. Econ. Entomol.* **35**, 845–9.

Solomon, M.E. (1949) The natural control of animal population. *J. Anim. Ecol.* **18**, 1–35.

Sommerman, K.M., Sailer, R.I., & Esselbaugh, C.O. (1955) Biology of Alaskan black flies (Simuliidae, Diptera). *Ecol. Monogr.* **25**, 345–85.

Southern, H.N. (1954) Tawny owls and their prey. *Ibis* **96**, 384–410.

Southwood, T.R.E. (1962) Migration of terrestrial arthropods in relation to habitat. *Biol. Rev.* **37**, 171–214.

Southwood, T.R.E., Jepson, W.F., & Emden, H. F. Van (1961) Studies on the behaviour of *Oscinella frit* L. (Diptera) adults of the panicle generation. *Entomol. Exp. Appl.* **4**, 196–210.

Southwood, T.R.E. & Siddom, J.W. (1965) The temperature beneath insect emergence traps of various types. *J. Anim. Ecol.* **34**, 581–5.

Speth, S. (1995) The emergence of Ephemeroptera, Plecoptera and Trichoptera (Insecta) from two streams of the north German Lowland (Osterau and Rodenbek, Schleswig-Holstein). *Limnologica* **25**(3-4), 237–50.

Speyer, W. & Waede, M. (1956) Eine MethodezurVorhersagedes Weizengallmiicken-fluges. *Nachr bl. dt. PflSchDienst. Stuttg.* **8**, 113–21.

Spiller, D. (1964) Numbers of eggs laid by *Anobium punctatum* (Degeer). *Bull. Entomol. Res.* **559**, 305–11.

Steffan, A.W. (1997) Drift emergence-traps for the recording of hatching running water-insects (Ephemeroptera, Plecoptera, Trichoptera, Diptera). *Entomol. General.* **21**(4), 293–306.

Steinhaus, E. (ed.) (2012) *Insect Pathology V1: An Advanced Treatise (Vol.* **1***).* Elsevier.

Straub, R.W., Fairchild, L.M., & Keaster, A.J. (1973) Corn earworm: use of larval traps on corn ears as a method of evaluating corn lines of resistance. *J. Econ. Entomol.* **66**(4), 989–90.

Suarez, A.V., Holway, D.A., & Case, T.J. (2001) Patterns of spread in biological invasions dominated by long-distance jump dispersal: insights from Argentine ants. *Proc. Natl Acad. Sci. USA* **98**, 1095–100.

Sullivan, C.R. & Green, G.W. (1964) Freezing point determination in immature stages of insects. *Can. Entomol.* **96**, 158.

Sunderland, K.D. (1988) Quantitative methods for detecting invertebrate predation occurring in the field. *Ann. Appl. Biol.* **112**, 201–24.

Sylven, E. (1970) Field movement of radioactively labelled adults of *Dasyneura brassicae* Winn. (Dipt., Cecidomyiidae). *Entomol. Scand.* **1**, 161–87.

Taylor, L.R. (1975) Longevity, fecundity and size: control of reproductive potential in a polymorphic migrant, *Aphis fabae* Scop. *J. Anim. Ecol.* **44**, 135–63.

Taylor, L.R. & Taylor, R.A.J. (1976) Aggregation, migration and population mechanics. *Nature* **265**, 415–21.

Taylor, L.R. & Taylor, R.A.J. (1978) Dynamics of spatial behaviour. In: *Population Control by Social Behaviour.* Institute of Biology Symposium, pp. 181–212.

Taylor, R.A.J. (1978) The relation between density and distance of dispersing insects. *Ecol. Entomol.* **3**, 63–70.

Terrell, T.T. (1959) Sampling populations of overwintering spruce budworm in the Northern Rocky Mountain region. *Res. Notes Intermountain Forest Range Exp. Sta., Ogden, Utah* **61**, 8 pp.

Thompson, D.J. (1975) Towards a predator-prey model incorporating age structure, *J. Anim. Ecol.* **44**, 907–16.

Tinbergen, L. (1960) The natural control of insects in pinewoods. 1. Factors influencing the intensity of predation by songbirds. *Arch. Néerland. Zool.* **13**, 266–343.

Tomkiewicz, S.M., Fuller, M.R., Kie, J.G., & Bates, K.K. (2010) Global positioning system and associated technologies in animal behaviour and ecological research. *Philos. Trans. Royal Soc. B: Biol. Sci.* **365**, 2163–76.

Toth, R.S. & Chew, R.M. (1972) Development and energetics of *Notonecta undulata* during predation on *Culex tarsalis. Ann. Entomol. Soc. Am.* **5**, 1270–9.

Turnbull. A.L. (1960) The prey of the spider *Linyphia triangularis* (Clerck) (Araneac: Linyphiidae). *Can. J. Zool.* **38**, 859–73.

Turnbull, A.L. (1962) Quantitative studies of the food of *Linyphia triangularis* Clerck (Araneac: Linyphiidae). *Can. Entomol.* **94**, 1233–49.

Turnock, W.J. (1957) A trap for insects emerging from the soil. *Can. Entomol.* **89**, 455–6.

Turnock, W.I. & Ives, W.G.H. (1962) Evaluation of mortality during cocoon stage of the larch sawfly, *Pristiphora erichsonii* (Htg.). *Can. Entomol.* **94**, 897–902.

Urquhart, F.A. (1960) *The Monarch Butterfly.* University of Toronto Press, Toronto, 361 pp.

Vallentyne, J.R. (1952) Insect removal of nitrogen and phosphorus compounds from lakes. *Ecology* **33**, 573–7.

Varley, G.C. & Gradwell, G.R. (1963) The interpretation of insect population change. *Proc. Ceylon Assn Adv. Sci. (D) (1962)* **18**, 142–56.

Vlijm, L., Van Dijck, T.S., & Wijmans, S.Y. (1968) Ecological studies on Carabid beetles. III. Winter mortality in adult *Calathus melanocephalus* (Linn.) egg production and locomotory activity of the population which has hibernated. *Oecologia* **1**, 304–14.

Walker, I., Henderson, P.A., & Sterry, P. (1991) On the patterns of biomass transfer in the benthic fauna of an Amazonian black-water river, as evidenced by 32P label experiment. *Hydrobiologia* **215** (2), 153–62.

Waloff, N. & Richards, O.W. (1958) The biology of the Chrysomelid beetle, *Phytodecta olivacea* (Forster) (Coleoptera: Chrysomelidae). *Trans. R. Entomol. Soc. Lond.* **110**, 99–116.

Way, M.I. & Banks, C.J. (1964) Natural mortality of eggs of the black bean aphid, *Aphis fabae* (Scop.), on the spindle tree, *Euonymus europaeus*. *Ann. Appl. Biol.* **54**, 255–67.

Weaver, T.M. & Fraley, J.F. (1993) A method to measure emergence success of westslope cutthroat trout fry from varying substrate compositions in a natural stream channel. *N. Am. J. Fish. Manag.* **13**(4), 817–22.

Welch, H.E., Jorgenson, J.K., & Curtis, M.F. (1988) Measuring abundance of emerging Chironomidae (Diptera): experiments on trap size and design, set duration, and transparency. *Can. J. Fish. Aquat. Sci.* **45**, 738–41.

Wellington, W.G. (1964) Qualitative changes in populations in unstable environments. *Can. Entomol.* **96**, 436–51.

Weseloh, R.M. (1990) Gypsy moth predators: an example of generalist and specialist natural enemies. In: Watt, A.D., Leather, S.R., Hunter, M.D., & Kidd, N.A.C. (eds), *Population Dynamics of Forest Insects*. Intercept Limited.

Weseloh, R.M. (1993) Manipulation of forest ant (Hymenoptera: Formicidae) abundance and resulting impact on gypsy moth (Lepidoptera: Lymantriidae) populations. *Environ. Entomol.* **22**, 587–94.

Wikars, L.O., Sahlin, E., & Ranius, T. (2005) A comparison of three methods to estimate species richness of saproxylic beetles (Coleoptera) in logs and high stumps of Norway spruce. *Can. Entomol.* **137**, 304–24.

Wilbur, D.A. & Fritz, R. (1939) Use of shoebox emergence cages in the collection of insects inhabiting grasses. *J. Econ. Entomol.* **32**, 571–3.

Winkle, W. Van, Martin, D.C., & Sebetich, M.J. (1973) A home-range model for animals inhabiting an ecotone. *Ecology* **54**, 205–9.

Worton, B.J. (1987) A review of models of home range for animal movement. *Ecol. Model.* **38**, 277–98.

Worton, B.J. (1989) Kernel methods for estimating the utilization distribution in home-range studies. *Ecology* **70**, 164–8.

Wratten, S.D. (1976) Searching by *Adatea bipunctata* (L.) (Coleoptera: Coccinellidae) and escape behaviour of its aphid and cicadellid prey on lime (*Tilia* × *vulgaris* Hayne). *Ecol. Entomol.* **1**, 139–42.

Wright, D.W. Hughes, R.D., & Worrall, J. (1960) The effect of certain predators on the numbers of cabbage root fly (*Erioischia brassicae* (Bouché) and on the subsequent damage caused by the pest. *Ann. Appl. Biol.* **48**, 756–63.

Wrubleski, D.A. & Rosenberg, D.M. (1984) Overestimates of Chironomidae (Diptera) abundance from emergence traps with polystyrene floats. *Am. Midl. Nat.* **111**(1), 195–7.

Zivkov, M. & Petrova, G. (1993) On the pattern of correlation between the fecundity, length, weight and age of pikeperch *Stizostedion lucioperca*. *J. Fish Biol.* **43**, 173–82.

11 The Construction, Description and Analysis of Age-specific Life-tables

11.1 Types of life-table and the budget

The construction of a number of life-tables is an important component in the understanding of the population dynamics of a species. Although some animal ecologists, such as Richards (1940), had expressed their results showing the successive reductions in the population of an insect throughout a single generation, Deevey (1947) was really the first to focus attention on the importance of this approach. Life-tables have long been used by actuaries for determining the expectation of life of an applicant for insurance, and thus the column indicating the expectation of life at a given age (the e_x column) is an essential feature of human life-tables. However, the fundamental interests of the ecologist and, even more so, of the applied biologist are essentially different from those of the actuary, and it is a mistake to believe that the approaches and parameters of primary interest in the study of human populations are also those of greatest significance to the animal ecologist. Because many insects, other invertebrates and small fish have discrete generations and their populations are not stationary, the age-specific life-table is more widely applicable than the time specific life-table. The differences between these two types are as follows:

An *age-specific* (or *horizontal*) *life-table* is based on the fate of a real cohort; conveniently, the members of a population belonging to a single generation. The population may be stationary or fluctuating.

A *time-specific* (or *vertical*) *life-table* is based on the fate of an imaginary cohort found by determining the age structure, at one instant in time, of a sample of individuals from what is assumed to be a stationary population with considerable overlapping of generations – that is, a multi-stage population. Age determination is a prerequisite for time-specific life-tables (see Chapter 12). A modification of this approach is the *variable life-table* of Gilbert *et al.* (1976), which is an inductive strategic computer model of the population: this is varied until it provides a reasonable description of the population (see Chapter 12).

The data in age-specific life-tables may be corrected so as to start with a fixed number of individuals (e.g. 1000); however, this practice, which simplifies the calculation of life

Ecological Methods, Fourth Edition. P. A. Henderson and T. R. E. Southwood.
© 2016 John Wiley & Sons, Ltd. Published 2016 by John Wiley & Sons, Ltd.
Companion Website: www.wiley.com/go/henderson/ecologicalmethods

expectancy, causes the very important information on actual population size to be lost. It will be seen in this chapter that the variations in population size, from generation to generation, provide the frame of reference against which the roles of the various factors are analysed. Therefore, in much work on insect populations, the type of table required lists the actual absolute populations at different stages and records the action of mortality factors where these are known (see Table 11.4, p. 453). Such a table, giving just the observed data, is well described by the term *budget* proposed by Richards (1961), which also has the advantage that it emphasises the distinction between this approach and that of the actuary.

11.2   The construction of a budget

The ideal budget will contain absolute estimates of the total population of as many stages as possible. At some points in the generation it may be possible to determine the total number entering a stage directly, as described in Chapter 9 (see also p. 283). For other stages there will be a series of estimates, made on successive sampling dates, using methods based on the numbers per unit area (see Chapters 4, 5 and 6) or from mark and recapture (Chapter 3), nearest-neighbour (p. 368) or removal trapping and distance sampling (see Chapter 7) techniques. The problem now arises as to how to determine from these estimates the total number of individuals that pass through a particular stage in one generation.

The degree of synchronisation in the life-cycle is an important factor affecting the ease or difficulty of this step. The ideal situation is when there is a point of time when all the individuals of a generation are in a given stage; a census at this time will provide a 'peak estimate' that may be used in a life-table. As the overlap in time of successive stages increases, it will be necessary to integrate a number of estimates to obtain the total population; special techniques for doing this are described below. When there is complete overlap of all stages, methods based on the age structure of the population are most appropriate (see Chapter 12). Once the series of estimates of total population at each stage have been developed it can be assumed that the differences between these represent mortality and/or dispersal. It may be possible, and is an advantage, to check this assumption by direct measurement of these factors (Chapter 9).

A budget is to some extent self-checking; erroneous population measurements may be exposed by increases in numbers at a stage when natality or immigration is impossible, or by other inconsistencies. Thus, the more terms in a budget the greater the confidence that can be placed in it (Richards, 1959); this confidence is additional to that obtained from the statistical confidence limits of the individual estimates. The latter are based solely on the information gathered by the given method for that particular stage, and indicate that the true value is likely (to the extent of the chosen probability level) to lie within them. It is reasonable to claim that when there is agreement with other estimates, by other methods made simultaneously (p. 4) or sequentially, this substantially increases the probability that the true value lies close to the estimate.

11.3 Analysis of stage-frequency data

Stage-frequency data comprises counts of the individuals in different development stages in samples taken from a population over a period of time. The aim is usually to estimate: (1) the durations of the stages; (2) the numbers entering stages; and (3) survival rates.

Manly (1990) tabulates 23 methods which use stage-frequency data for estimating some combination of the numbers entering a stage, stage duration and stage-specific survival rates. These methods vary in their assumptions and requirements, and in determining which to use the first consideration must be the validity of the assumptions, followed by computational complexity and software availability. There has been little published on the relative merits of these methods, although Manly & Seyb (1989), Munholland *et al.* (1989) and Manly (1990) do compare some approaches. Of the early and computationally less-demanding methods, the graphical method is probably the simplest and most robust (Ruesink, 1975); that of Kiritani & Nakasuji (1967) as modified by Manly (1976) is also relatively simple and reliable (Manly, 1974c). From the beginning of the 1970s increased computer availability and power allowed the development of stochastic models for which the parameters can be estimated by maximum likelihood. The initial work of Ashford *et al.* (1970), who described a mathematically complex method based on the general theory of stochastic branching processes and using maximum likelihood estimates, had little direct impact as it was difficult for biologists to understand and implement. Birley's (1977) method, which was further developed by Bellows & Birley (1981), is flexible because survival can vary between the different stages; however, this can result in the desire to estimate more parameters than the observations will allow – this is termed 'over-parameterisation'. A theoretical framework for a class of widely applicable models was presented by Kempton (1979). and models of this type were compared by Manly & Seyb (1989). Manly (1990) considered the Kempton model as offering the most efficient means of estimating the parameters by maximum likelihood. Manly (1987) describes a multiple regression approach for the estimation of the numbers entering each stage and stage survival rates. A limitation is that the samples must be equally spaced through time.

As Manly (1997) points out, it is a difficult task to simultaneously estimate a complex pattern of recruitment to the first stage, stage-specific survival rates and stage duration. He shows, using simulated data, that, even under ideal conditions, estimates are subject to large biases. The quality of the estimates greatly improves if stage durations are known, indicating that it is wise to undertake stage duration observations or experiments whenever possible.

More recently, Knape *et al.* (2014) have developed a model for the analysis of stage-frequency data in which the same individuals are repeatedly sampled through time, as would be the case in a laboratory study of insect development. The supporting material appendices with pseudo code as well as R code implementing the method using mealybug data are available with this paper at the Biometrics website at the Wiley Online Library. Finally, there have been methods developed for populations followed over a number of generations or which have overlapping generations, but these will not discussed further here. When the population is sampled over a number of generations, parameters for the different stages may be estimated by fitting age-structured simulation models. Schneider & Ferris (1986) describe an algorithm for estimating the stage-specific duration and survival parameters for such a model. An age-structured population can also be envisaged as a surface obtained from a plot of age by time by population size. Wood (1994) shows how age- and time-specific mortality and birth rates can be estimated by fitting a smoothed population surface to age-structured data using spline interpolation.

Below, we describe in more detail the older, computationally less-demanding, methods.

11.3.1 Southwood's graphical method

This is the crudest and simplest of the methods of integration. It gives reasonable estimates of the numbers entering a stage if mortality, which may be heavy, occurs entirely at the end of the stage. If a constant mortality rate applies throughout the stage, then the estimate will overestimate the numbers at the median age of the stage. Successive estimates are plotted, most conveniently allowing one square per individual and per day (Fig. 11.1). The points are joined up and the number of squares under the line counted; this total is then divided by the mean developmental time under field conditions to give the estimate of the numbers reaching the median age for the stage. This technique has been found to be robust (Ruesink, 1975) and has been quite widely used (e.g. Southwood & Jepson, 1962; Helgeson & Haynes, 1972).

11.3.2 Richards & Waloff's first method

This method (Richards & Waloff, 1954) assumes a single impulse of recruitment and a constant mortality rate. It is therefore only applicable to a stage with a well-marked

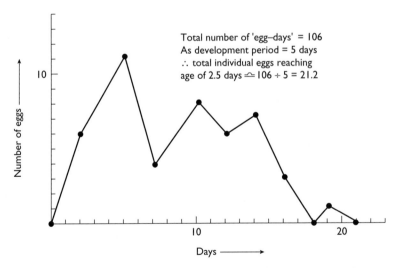

Figure 11.1 The determination of the total number of individuals in a stage from a series of estimates by graphical summation (hypothetical example with unrealistically small numbers for simplicity).

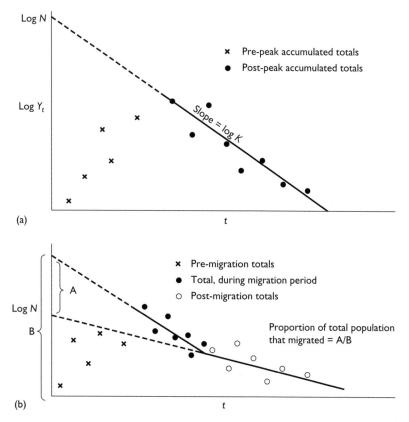

Figure 11.2 The estimation of the number of individuals entering a stage by Richards and Waloff's first method. (a) Simple case without migration. (b) When the animal may migrate.

peak in abundance. By plotting the regression of the fall-off of numbers with time after the peak, and then extending this back to the time when the stage was first found, the total number entering the stage can be estimated (Fig. 11.2).

A population exposed to a steady mortality may be expressed as:

$$Y_t = N_0 \phi^t \tag{11.1}$$

where Y_t = population on day t, N_0 is the peak population (ideally the number hatched) and ϕ = the fraction of the population which survives to the end of a unit of time (e.g. a day), the survival rate per unit time. Hence:

$$\log Y_t = \log N_0 + t \log \phi \tag{11.2}$$

and after the peak this will describe a straight line which is conveniently obtained from the regression of $\log Y_t$ on t: the regression coefficient will be $\log \phi$, the logarithm of the average, and assumed constant, survival rate. The number entering the stage will then be given by the value of Y_t, found by inserting into the equation a value of t corresponding to the start of the stage. Alternatively, if t is numbered from the start of the stage in question, $\log N_0$ will equal the number entering the stage.

More reliable estimates are obtained if Y_t is taken as the accumulated population. That is, after the peak of a particular instar, the values of Y_t are the total of that instar, plus any individuals that have passed through that instar and are in subsequent ones. For example, Y_t for the second instar would consist of the populations of second instar, third instar and fourth instar, and so on, estimated for day t, but Y_t for the third instar would consist only of the populations of the third and subsequent instars. Dempster (1956) found that, if Y_t was based solely on the numbers of the particular instar, the estimates of population were too high, the slope of the line being too steep due to the confusion of moulting with mortality. The use of accumulated totals avoids this.

Waloff & Bakker (1963) used a modification of this method to determine the total change in population due to migration. Basically, separate regressions were calculated for the migration and post-migration period and the slopes (log ϕ) compared. The value of log ϕ for the migration period represents migration and mortality, but for the post-migration period just mortality. Therefore, if the difference between these is expressed as a percentage of the larger, this will be the percentage of the initial population that was lost during the migratory period.

11.3.3 Manly's method

Manly's (1974a,b) method is essentially a modification of Richards & Waloff's First Method, but it avoids the need for a distinct peak. Recruitment is assumed to be normal, but Manly considered that as long as the entry was unimodal, then reasonable estimates would be provided. Simulation studies suggest that significant biases will arise if recruitment to the first stage is 'far from normal' (Manly, 1974c). The basic equation is:

$$Y_t = N_0 \int_{-\infty}^{t} \exp[-(t - x)f(x)(dx)] \tag{11.3}$$

where Y_t = the number of animals in a particular stage and later stages at time t, N_0 = total number entering the stage, $\exp(-\theta(t - x)$ = the probability that an animal that enters the population at time x is still alive at time t (θ being the constant age specific death rate) and $f(x)$ is the frequency of the stage at time x. Because of Manly's assumption of normality, the frequency function for this distribution can be inserted and Equation (11.3) becomes:

$$Y_t = N_0^* \exp(-\theta t) \int_{-\infty}^{t - \mu^*/\sigma} (2\pi)^{-\frac{1}{2}} \exp\left(-\frac{1}{2}x^2\right) dx \tag{11.4}$$

where

$$N_0^* = \exp\left[\theta(\mu + \frac{1}{2}\theta\sigma^2)\right] N_0$$

and

$$\mu^* = \mu + \theta\sigma^2$$

and μ is mean time of entry to the stage and σ is its standard deviation.

The four unknowns, N_0, μ, σ and θ may be estimated for a series of at least four values of Y_t using non-linear regression. Manly (1974a) also describes how the values can be estimated from graphical methods.

11.3.4 Ruesink's method

Ruesink's (1975) method, like that of Birley, allows the calculation of the stage survival rate. It does this by a comparison of the rate of recruitment into the stage in question with the rate of recruitment to the next stage. The calculations are most easily performed by computer and the basic equation is:

$$\phi_{jj+1} = \frac{C_{j+1}(t+n) - C_{j+1}(t+n)}{D_j(t+n) - D_j(t-n)}, \tag{11.5}$$

where ϕ_{jj+1} = the survival between the jth stage and the next stage, C_{j+1} = the total number of individuals that have entered stage $j+1$ in the periods $(t+n)$ and $(t-n)$, likewise D_j = the total number of individuals that have left stage j, n = an arbitrary time period. Ruesink (1975) points out that if there are large sampling errors, n must be large to provide meaningful estimates. The values of C_{j+1} and D_j are obtained either graphically or by computer. In the former method a continuous line is drawn through the points from field data for the numbers of that stage: this is in essence the approach of Richards & Waloff's First Method.

11.3.5 Dempster's method

This method requires an independent estimate of the total natality of the first stage, together with a series of population estimates, at least two more in number than the numbers of stages (Dempster, 1961). It is assumed that the mortality during each stage is constant. The number of any one stage dying between sampling days is given by the product of the number of that stage present, the fraction dying per day and the number of days.

For simplicity, the number present is estimated as the average of the numbers on the two sampling days, an approximation only valid if sampling is frequent. If then I_0 is the number of first instar larvae present on day 0, and I_t is the number present t days later, the number of first instar larvae dying during that time is

$$\frac{(I_0 + I_t)}{2} t\mu_1$$

where μ_1 is the average daily mortality of first-stage larvae.

Similar equations are constructed for second, third and following stages, giving a total change in the population between days 0 and t during the hatching period as:

$$Y_0 - Y_t = N\alpha_{0 \to t} - \frac{(I_0 + I_t)}{2} t\mu - \frac{(II_0 + II_t)}{2} t\mu_2 ... \frac{(Ad_0 + Ad_t)}{2} t\mu_a \tag{11.6}$$

where Y_0 and Y_t = total population at days 0 and t (successive sampling dates), N = the total number hatching (or emerging) which is found independently; $\alpha_{(0 \to t)}$ = the proportion of the total hatch that occurs between days 0 and t; I_0 and I_t = the total numbers of first instar larvae on the first and second sampling dates respectively, II_0 and II_t are the numbers of the second instar larvae; further terms are inserted, as appropriate, until the adult (Ad_0 and Ad_t) stage is reached; t is the time internal between day 0 and day t (the two sampling occasions and $\mu_1, \mu_2 \ldots \mu_a$ are the average daily mortality of first and second stage larvae and adults, respectively.

An equation of the form of (11.6) is generated for each time step. The unknowns, $\mu_1, \mu_2 \ldots \mu_a$ may be found by least squares and then solving a series of simultaneous equations. Alternatively, if sufficient observations have been taken, the unknowns can be found by multiple regression. However, the usual formulae for the standard errors do not apply as there are errors in the regressor variables. Where there is migration, a further term can be added to these equations. This is a method capable of wide application, but it is not robust. In order to obtain reliable estimates, a large number of accurate population samples are required (Manly, 1974c). Dempster (1961) uses as an example a *Locusta migratoria* population sampled daily for 40 days.

11.3.6 Richards & Waloff's Second Method

This method may be used when recruitment and mortality overlap widely. The total number (N_i) of the ith stage taken in all the daily samples will be given by:

$$N_t = N_0 \int_0^a \emptyset^t dt = \frac{N_0(\emptyset^a - 1)}{\ln \emptyset} \qquad (11.7)$$

where N_0 = the total number entering the stage, ϕ = the survival rate (for a unit of time) and a = the duration of the stage (Richards, 1959; Richards *et al.*, 1960; Richards & Waloff, 1961). Now, if the number of eggs laid (N_o) and duration of the egg stage (a) are known and N_i is obtained from samples, the equation can then be solved for ϕ. The percentage mortality is $100(1 - \phi^a)$, and this may be used to calculate the number surviving the egg stage and entering the first larval instar. Thus, N_0, a and N_i will again be known values in the equation for the first instar so that the percentage mortality for this instar and the numbers entering the next may be found. The process can be repeated to the end of the life-cycle.

In order to apply this method, one must have an accurate estimate for the number entering the stage N_0. Richards & Waloff (1961) obtained this from the population of adult females assessed independently and the oviposition rate measured under the field conditions. It is also necessary to determine the duration of the stage (a) experimentally and, as Richards *et al.* (1960) pointed out, differences, of as little as half a day, in the estimate of the duration of an instar can make a large difference in the estimated mortality. It is clear, therefore, that inaccuracies in these independent estimates of a and N_0, or discrepancies between the actual and assumed field conditions, which will influence the values assigned to a and N_0, are potential sources of error with this method.

11.3.7 *Kiritani, Nakasuji & Manly's method*

As originally developed (Kiritani & Nakasuji, 1967), this method required that samples be taken at regular intervals throughout the generation and, unless the number entering stage *i* are known independently, only survival rates can be estimated. Manly (1976) has modified the method to remove these restrictions. Samples are taken at irregular intervals ($h_1, h_2...$), the area under the frequency trend curve is estimated by the trapezoidal rule:

$$\hat{A}_i = \frac{1}{2} \sum_{l=1}^{n} (h_1 + h_{j+1})\hat{f}_{il} \tag{11.8}$$

where f_{il} = the number of the *i*th instar estimated from the samples taken on the *i*th occasion, which is at the end of the sampling intervals h_1, there are $n + 1$ sampling intervals, the last interval extending from the last occasion when the stage was present to the next sampling occasion (when it was found to be absent). It will be noted that these f_{il} values correspond to I, II ... A_d of Dempster's method, but whereas in his method two numbers are averaged over one sampling interval, in this method the numbers are 'spread' into the sampling intervals on either side, then the sum divided by two.

The survival rate of the *i*th stage is then estimated:

$$\hat{\phi}_i = |-A_i| \sum_{j=1}^{q} A_j \tag{11.9}$$

where *j* = the next stage and *q* = the last stage.

Manly (1976) shows how further estimates can be obtained if the median stage (instar value, which will not, of course, normally be a whole number) is plotted against time so that the time for a 'median' animal to pass from stage to another can be read off. Now, the daily survival rate raised to the power of the duration of the stage should equal the stage survival rate as determined above:

$$\hat{\phi}_i = \hat{\phi}_d^{\hat{a}_i} \tag{11.10}$$

where ϕ_d = daily survival rate and a_i = the duration of the *i*th stage as determined above. (This calculation can be done for two or more instars at once, which may give a better estimate of duration). Thus:

$$\log \hat{\phi}_d = \log \hat{\phi}_i / \hat{a}_i \tag{11.11}$$

If, on the other hand, time is taken over several instars (i.e. the divisor is, say $a_1 + a_2 + a_3$), then having found ϕ_i and ϕ_d the appropriate stage durations (*a*-values) can be found. Then the numbers entering each stage may be found:

$$\hat{N}_{0i} = -\ln \phi_d \left[\sum_{j=1}^{q} A_j \right] \tag{11.12}$$

where N_{0i} = the number entering the *i*th stage and the others as above.

It will be apparent that this method depends on the assumption of a constant daily survival rate for all stages. This method is, like the two following methods, able to model multi-cohort data.

The Kiritani–Nakasuji–Manly method requires the cohort to be followed until death. Yamamura (1998) describes a modification for situations when data are only available for a restricted number of stages which is simpler to apply, but does require the stage duration to be known.

11.3.8   Kempton's method

Kempton (1979) gave the probability that an individual will be in stage j at time t as:

$$p_j(t) = w(t) \int_0^t \{g_j(y) - g_{j+1}(y)\} dy, j < q$$

and

$$p_q(t) = w(t) \int_0^t g_q(y) dy \qquad (11.13)$$

where $w(t)$ is the probability of surviving to time t and $g_j(t)$ is the probability-density function of the time of entry to stage j. The integral term in Equation (11.13) calculates the probability that an individual alive at time t is at stage j. For example, the integral from 0 to t of $g_j(y)$ gives the probability that an individual has entered stage j by time t.

Kempton (1979) noted that a stage structured model required three components:

1. A description of stage survival. When constant survival rates cannot be assumed, a gamma, Weibull or lognormal distribution in life expectancy might be reasonable assumptions.
2. A model for the time of entry into stage 1. Conceptually, each individual can be considered to start in stage 0 and to enter stage 1 at a time that is either normally or gamma distributed.
3. A model for the distribution of the duration of each stage. It is advantageous to assume that this is either constant or distributed normally, gamma, or inverse normally.

The particular distributions chosen for components 1, 2 and 3 above will determine the functions $w(t)$ and $g_j(t)$ and the difficulty in estimating $p_j(t)$. Kempton (1979) in a worked example assumed survival per unit time to be constant and stage duration and time of entry to stage 1 to be gamma distributed with a common scale parameter. He further assumed that the number of individuals counted in stage j at time t was Poison distributed with mean $N_0 p_j(t_i)$ where N_0 is the total number of individuals at time 0. He was then able to give a likelihood function which he was able to maximise. While considered by Manly and Seyb (1989) to be one of the most efficient methods for modelling multi-cohort data, access to a computer program is generally required to use this method and there are no readily available programs.

11.3.9 The Bellows and Birley Method

Bellows and Birley (1981) refined the model originally developed by Birley (1977). The method allows each stage to have a different survival rate.

The number of individuals in stage j at time t, Y_t, is given by :

$$Y_j(t) = N_1 \sum_{i=0}^{t} P_j(i)\phi_j^{t-i}\{1 - H_j(t - i)\} \tag{11.14}$$

where N_1 is the total number of individuals entering stage 1 and $P_j(i)$ is the proportion of these individuals which have entered stage j at time t. The survival rate of stage j is ϕ_j and $H_j(t)$ is the probability that an individual remains for less than or equal to t time units in stage j.

The proportion of the population entering stage j at time t is a function of the numbers that have arrived at other times:

$$P_{j+1}(t) = \sum_{i=0}^{t} P_j(i)\phi_j^{t-i}h_j(t - i) \tag{11.15}$$

where $hj(t)$ denotes a duration of t time units in stage j.

The model requires assumptions about the distribution of stage duration times and the distribution of entry to stage 1. Bellows and Birley (1981) suggest that stage duration distribution is inverse normal because development time is normally distributed. Manly (1990) however, considered the Weibull distribution as advantageous because of its simple form and he also used it to model the time of entry to stage 1 so that:

$$H_j(t) = 1 - \exp\left\{-\left(\frac{t}{Q_j}\right)^{aj}\right\}, \tag{11.16}$$

and

$$P_1(t) = \exp[-\{(t - 1)/Q_0\}^{\alpha_0}] - \exp[-(t/Q_0)^{\alpha_0}] \tag{11.17}$$

where α and Q_j are parameters to be estimated.

While Bellows and Birley (1981) suggested using non-linear regression to estimate the parameters of the model, Manly (1990) noted that, if stage-frequency data could be assumed to be Poisson variates, then the general maximum likelihood algorithm MAXLIK can be used.

As in the case of Kempton's method, a computer program is required to undertake the calculations. A recent example of the application of this method is the study of González-Zamora & Moreno (2011) on the white fly, *Bemisia tabaci*, on sweet pepper.

11.4   The description of budgets and life-tables

11.4.1   Survivorship curves

The simplest description of a budget is the graphical representation of the fall-off of numbers with time – the survivorship curve. The number living at a given age (l_x) are plotted against the age (x); the shape of curve will describe the distribution of mortality with age. Slobodkin (1962) recognised four basic types of curve (Fig. 11.3): in type I, mortality acts most heavily on the old individuals; in type II (a straight line when the l_x scale is arithmetic) a constant number die per unit of time; in type III (a straight line when the l_x scale is logarithmic) the mortality rate is constant; and in type IV mortality acts most heavily on the young stages. Deevey (1947) also drew attention to these different types, but only recognised three and, in order to avoid confusion between his classification and Slobodkin's, it should be noted that Deevey plotted survivors (l_x) on a log scale and thus type II of Slobodkin was not recognised by him; type II of Deevey = type III of Slobodkin and type III of Deevey = type IV of Slobodkin.

Fish and many marine organisms with high fecundities show type IV curves. In insects, mortality often occurs in distinct stages so that survivorship curves show a number of distinct steps (Ito, 1961).

11.4.2   Stock–recruitment (Moran–Ricker) curves

The density-dependent relationship between the number of adults in a population and recruitment can be expressed by the stock–recruitment curve. This approach is particularly important in fisheries research where independent estimates of the number of adult and larval fish are available. The management objective in fisheries is to maintain the adult population at a level that will produce sufficient recruits to replace the adults killed by both natural causes and fishing. If recruitment is density-dependent

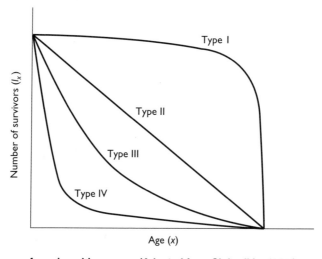

Figure 11.3   Types of survivorship curves. (Adapted from Slobodkin, 1962.)

then, at least in theory, it is possible to exploit a population without appreciably reducing recruitment. Ricker (1954) and Beverton & Holt (1957) originally derived the two basic stock recruitment relationships that are commonly used.

The curve, developed by W.E. Ricker, is expressed by the equation:

$$R = CNe^{-DN}. \tag{11.18}$$

where N is the size of the population, R is the number of recruits, C and D are constants, and e is the base of the natural logarithm.

The parameter D measures the tendency of the population to adjust to changes in population number, and C measures the number of individuals which would survive to adulthood if only density-independent mortality was operating. This curve reaches a maximum when $N = 1/D$ and $R = C/De$, and has been found to describe the stock relationships of some fish, including salmon (Gulland, 1983). Given estimates of N and R for a number of years, the parameters C and D are estimated by regression. In practice, it is difficult to obtain reliable estimates of C and D.

The Beverton–Holt curve is given by the equation

$$R = \frac{1}{A + \frac{B}{N}}. \tag{11.19}$$

where N and R are as above and A and B are constants to be estimated. This is an asymptotic curve with an asymptote of $R = 1/A$.

These stock–recruitment curves, examples of which are shown in Fig. 11.4, are particularly appropriate for the description of density-dependent relationships in fisheries studies because of the biological characteristics of fish populations. First, the extremely high fecundity of fish makes a Ricker-type relationship feasible. Second, adults and

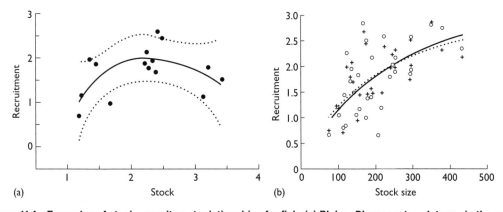

Figure 11.4   Examples of stock–recruitment relationships for fish. (a) Plaice, *Pleuronectes platessa*, in the western English Channel. (Adapted from Iles, 1994.) The fitted model is the Saila–Lorda equation with Scheffé confidence bands. (b) English sole, *Pleuronectes vetulus*. (Adapted from Iles & Beverton, 1998); o, original data, +, adjusted for Ekman transport; Beverton–Holt curves were fitted to the original data (solid line) and the adjusted data (dotted line).

young fish often differ greatly in life style and distribution so that different methods are used to estimate their numbers, making it natural to divide the adult stock from the recruits. Both, because of the difficulty of fitting appropriate curves to noisy data and because of their lack of biological realism, the utility of stock–recruitment relationships has been frequently questioned (e.g. Fletcher & Deriso, 1988).

An alternative use of this approach was developed by Rogers (1979), and depends on plotting the data on log/log scales. The finite capacity for increase (λ) then describes a straight line, parallel to the 45° line until density-dependent factors begin to operate (Fig. 11.5). Rogers termed this the Moran curve (after P.A.P. Moran, who proposed such methods of expressing population data). A long series of successive populations are plotted; the upper limit will be likely to delineate the point at which density dependence starts to operate – where the curve inflects. The vertical departures from

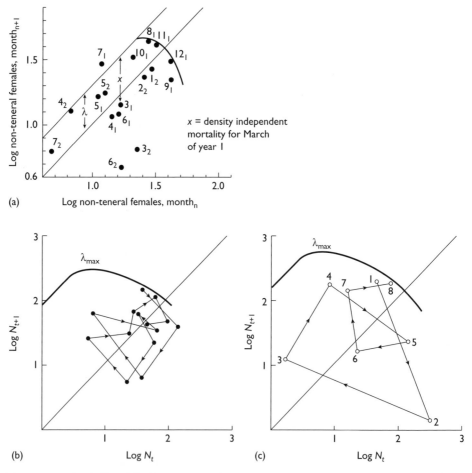

Figure 11.5 Examples of the Moran curve. (a) Moran plot for the tsetse fly, *Glossina morsitans submorsitans*, populations in the Yankori Game Reserve, Nigeria. (Data from Rogers, 1979); the numbers represent the month and the subscripts the year; (b) Sequential Moran plot for the pine moth, *Bupalus piniarius*, populations. (Data from Southwood, 1981); (c) Sequential Moran plot for the cinnabar moth, *Tyria jacobaeae*. (Data from Southwood, 1981.)

the line before the point of inflection will represent the action of density-dependent factors, departures after this point (the right-hand side of Fig. 11.5) represent the action of both density-dependent and -independent factors. The Moran curve may be fitted by eye to a long series of data (as in Fig. 11.5), when $\lambda = e^{rm}$. Alternatively, the straight portion may be calculated from a knowledge of maximal potential fertility (as in Fig. 11.5a) when $\lambda = e^{rp}$. Even when the Moran curve cannot be determined with certainty, the plotting of successive populations on a log-log scale allows the comparison of the population dynamics of different species (Southwood, 1981). An example is provided by the tight cycling of *Bupalus* (Fig. 11.5) compared with that of another moth, *Tyria* (Fig. 11.5), where density-dependent factors seem absent between the second and third generations.

Further approaches to the analysis of time-series including methods for the detection of density-dependence in long time series are given in Chapter 15 (see p. 608).

11.4.3 The life-table and life expectancy

For the further description of the data collected in the form of a budget or any type of life-table (Chapter 12) it is often most convenient if these are corrected so as to commence with a fixed number, usually 1000.

A table may then be constructed with the following columns (Deevey, 1947):

x the pivotal age for the age class in units of time (days, weeks, etc.)
l_x the number surviving at the beginning of age class x (out of a thousand originally born)
d_x the number dying during the age interval x
e_x the expectation of life remaining for individuals of age x

In practice the table may have two further columns (Table 11.1) to facilitate the calculation of the expectation of life as follows:

(i) The number of animals alive between age x and $x+1$ is found. This is calculated using:

$$L_x = \int 1_x d_x \approx \frac{l_x + 1_{x+1}}{2} \tag{11.20}$$

Table 11.1   A hypothetical life-table.

x	l_x	d_x	L_x	T_x	e_x	$1000q_x$
1	1000	300	850	2180	2.18	300
2	700	200	600	1330	1.90	286
3	500	200	400	730	1.46	400
4	300	200	200	330	1.10	667
5	100	50	75	130	1.30	500
6	50	30	35	55	1.10	600
7	20	10	15	20	1.00	500
8	10	10	5	5	0.50	1000

(ii) The total number of animal x age units beyond the age x, which is given by:

$$T_x = L_x + L_{x-1} + L_{x-2}....L_w,$$ (11.21)

where w = the last age. In practice it is found by summing the L_x column from the bottom upwards.

The expectation of life which is theoretically:

$$e_x = \frac{\int_x^w l_x d_x}{l_x}$$ (11.22)

and is therefore given by:

$$e_x = \frac{T_x}{l_x}$$ (11.23)

When the survivorship curve is of type I (Fig. 11.6), e_x will decrease with age; it will be constant for type II, and it will increase for types III and IV.

A further column is sometimes added to life-tables, the mortality rate per age interval (q_x); this is usually expressed as the rate per thousand alive at the start of the interval:

$$1000q_x = 1000\frac{d_x}{l_x}$$ (11.24)

11.4.4 *Life and fertility tables and the net reproductive rate*

In the two sections above we have been concerned solely with the description of one of the pathways of population change, namely mortality. In this section and the next, methods of describing natality and its interaction with mortality in the population will be discussed.

A life and fertility table (or fecundity schedule) may be constructed by preparing a life-table with x and l_x columns as before, except that the l_x column refers entirely to females and should represent the number of females alive, during a given age interval, as a fraction of an initial population of one. Or, expressed another way, the life expectancy at birth to age x as a fraction of one (Birch, 1948) (Table 11.2).

A new column is then added on the basis of observations, this is the m_x or age-specific fertility column (often termed the fecundity column but, as it refers to live births, fertility is more appropriate) that records the number of living females born per female in each age interval. In practice it is frequently necessary to assume a 50:50 sex ratio when $m_x = N_x/2$, N_x being the total natality per female of age x.

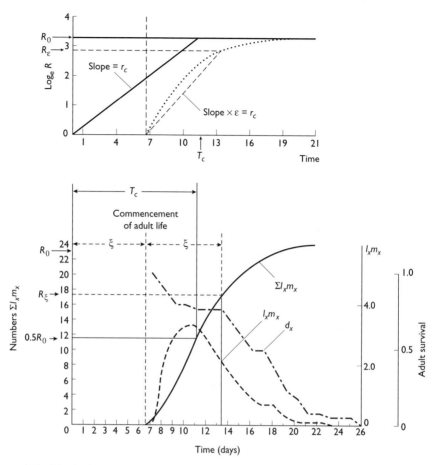

Figure 11.6 The intrinsic rate of natural population increase. Population growth data for the mite, *Tetranychus urticae*. (Data from Bengstron, 1979), displayed to illustrate the values used to make approximate estimates following the methods of Laughlin (1965) and Bengstron (1969) (*ri*) and Wyatt and White (1977) (*rc/e*) (see text for further explanation).

Table 11.2 Life and fertility table for the beetle, *Phyllopertha horticola*. (Data from Laughlin, 1965.)

x (in weeks)	l_x	m_x	$l_x m_x$ (V_x)
0	1.00	—	Immature
49	0.46	—	Immature
50	0.45	—	Immature
51	0.42	1.0	0.42
52	0.31	6.9	2.13
53	0.05	7.5	0.38
54	0.01	0.9	0.01

Columns l_x and m_x are then multiplied together to give the total number of female births (female eggs laid) in each age interval (the pivotal age being x); this is the $l_x m_x$ column.

The number of times a population will multiply per generation is described by the net reproductive rate R_0, which is:

$$R_0 = \int_0^\infty l_x m_x d_x = \sum l_x m_x \qquad (11.25)$$

Thus, from Table 11.2, $R_0 = 2.94$; this net reproductive rate may be expressed in another way as the ratio of individuals in a population at the start of one generation to the numbers at the beginning of the previous generation. Thus,

$$R_0 = \frac{N_{t+\tau}}{N_t} \qquad (11.26)$$

where τ = generation time.

Clearly, values of R_0 in excess of one imply an increasing population, and values of less than one a decreasing population; when $R_0 = 1$ the population will be stationary. Where it is difficult to define the generation time the value of R_0 as a description is limited and this led A.J. Lotka, a student of human demography, to propose the consideration of the growth rates of populations.

11.4.5 *Population growth rates*

As A.J. Lotka pointed out, the growth rate of a population is r in the equation:

$$\frac{dN}{dt} = rN, \qquad (11.27)$$

where N is the number of individuals at any given time (t); which on integration gives:

$$N_t = N_0 e^{rt} \qquad (11.28)$$

The parameter r in this equation describes population growth. Under conditions of an unlimited environment and with a stable age distribution this parameter becomes a constant.

The maximum value of the parameter r that is possible for the species under the given physical and biotic environment is denoted as r_m, and is variously termed the intrinsic rate of natural increase, the Malthusian parameter, the innate capacity for increase, or various combinations of these (Leslie & Ranson, 1940; Birch, 1948; Caughley & Birch, 1971), whilst $e^r = \lambda$ is the finite capacity for increase. It must be stressed that this value is the maximum under various natural conditions; that is, it allows for some mortality. The theoretical potential r_p with maximal fertility and zero mortality will be higher (Southwood, 1969). Under optimal physical conditions r_m will approach r_p, at the limits of a species environmental range r_m will approach zero, and beyond the boundary of the range r_m. will be negative. If the population is constrained by its environment (perhaps through intraspecific effects), another parameter r_s may be determined which is the rate at which the population would change (it may be positive or negative) if the age distribution was stable (Caughley & Birch, 1971). As environmental constraints (due to population density) approach zero, so r_s comes to equal r_m. The parameter r_s is of particular interest to students of vertebrate populations.

The intrinsic rate of natural increase (r_m) is of value as a means of describing the growth potential of a population under given climatic and food conditions (Messenger, 1964; Watson, 1964). It is an important parameter in inductive strategic and management models for insect pest populations and in fisheries work (Jensen, 1975).

When a stable age distribution has been achieved, but the population is still growing in an unlimited environment, and given l_x, m_x values from a life table, r_m may be calculated from the expression:

$$\sum e^{-r_m x} l_x m_x = 1 \tag{11.29}$$

However, ecologists are interested in r_m or the 'time r' in a broader context, its relationship to life history parameters such as generation size and fecundity (Smith, 1954; Fenchel, 1974; Southwood et al., 1974), whilst the establishment of a stable age distribution in an unlimited environment may be virtually impossible. Sometimes the instantaneous birth rate per individual ($\sim r_p$) has been used instead of the instantaneous rate of natural increase per individual, thereby eliminating the mortality pathway.

An approximation for r_m, the capacity for increase (r_c) has long been used by insect ecologists, and Laughlin (1965) presented a graph that showed its error over r_m in relation to the relative length of the reproductive period and the net reproductive rate. It is defined as:

$$r_c = \ln R_0 \big/ T_c, \tag{11.30}$$

where R_0 is as defined in Equations (11.25) and (11.26), and T_c = cohort generation time – the mean age of the females in the cohort at the birth of female offspring or the pivotal age where $l_x m_x = 0.5 R_0$ (Bengstron, 1969).

The simple and general relationship between r_m and r_c (and between τ and T_c) has been elucidated by May (1976):

$$r_c = r_m \left(1 - \frac{r\sigma^2}{2T_c} + ... \right) \tag{11.31}$$

where σ^2 the variance of the $1_x.m_x$ distribution defined as:

$$\sigma^2 = \left(\frac{\sum x^2 l_x m_x}{R_0} \right) - T_c^2 \tag{11.32}$$

The further correction terms, omitted from Equation (11.31), involve higher-order moments (skewness, etc.): if we can assume that $1_x m_x$ is normal (Gaussian), then the equation is exact. May (1976) rearranged the equation for r_c to give an expression of the relative error involved in calculating r_c rather than r_m:

$$\frac{r_m - r_c}{r_c} \approx \frac{1}{2}(\ln R_0)(CV)^2, \tag{11.33}$$

where $CV = \sigma/T_c$ the variation in the distribution of age of reproduction within a cohort. A consideration of Equation (11.33) immediately shows that, provided the $l_x m_x$ distribution is approximately normal, the relative error of r_c compared with r_m will be small, provided that:

1. $R_0 \approx 1$; that is, the population essentially just replaces itself, e.g. some human populations, other vertebrates with low fecundities (e.g. albatross, whales).
2. CV is small; that is, reproduction tends to occur in all individuals at about the same age and is not spread out over a long reproductive life. This is true of many arthropods, but not of long-lived, virtually iteroparous species such as the tropical rain forest butterflies, *Heliconius*.

CV is clearly related to Laughlin's (1965) comparison of reproductive period to total period (to oldest reproductive age).

Another parameter that is involved in expression of life table data is generation time (τ) which is defined as:

$$\tau = \frac{\ln R_0}{r_m}, \tag{11.34}$$

and has been considered to have little biological significance; however, May (1976) shows that:

$$\tau = T_c \left[1 - \frac{r_m \sigma^2}{2T_c} \right] \tag{11.35}$$

Thus, the difference is significant only if the $l_x m_x$ distribution has a high variance, particularly if this implies (as it generally will) a significant departure from the normal (i.e. additional terms are required in the above). Lefkovitch (1963) suggests that variation in the stored products beetle, *Lasioderma serricome*, is not normally distributed; Messenger (1964) also obtained various anomalous results, possibly for the same reason. The subject would repay further investigation in the light of May's (1976) paper, but such errors are not among the most substantial facing ecologists!

11.4.6 *The calculation of r*

Where a life-table can be constructed for a stable age distribution in an unlimited environment, then r_m may be calculated by numerical methods using Equation (11.29). For many insect species, where the conditions listed above are met, r_c is adequate and may be calculated from Equation (11.30).

Several species of aphid and spider mites have been found by Wyatt & White (1977) to have steep cumulative l_x,m_x curves early in adult life that flatten out at a time period that happens to equal the period of pre-reproductive life – that is, the development time (ξ) from birth to reproduction (Fig. 11.6). The value on the $\Sigma\, l_x,m_x$ curve at a particular time represents the total births until that time per number of original females. Wyatt & White (1977) point out that it is much easier to determine: (i) the pre-reproductive development period (ξ); and (ii) the total reproduction/original female (R_ξ) in the first part of the reproductive period of length $= \xi$, than the l_x,m_x values. The slope given by these values will be slightly steeper than r_c (see Fig. 11.6) so that

$$r = a(\log_e R_\xi)\big/_\xi \tag{11.36}$$

where a is a constant that corrects the slope of this line to correspond with r_c; Wyatt & White (1977) found $a = 0.74$ to hold for many aphids and mites, but when a whole series of determinations are to be made under different climatic or biotic conditions, the value of the constant should be confirmed because it is very dependent on the form of the $\Sigma\, l_x,m_x$ curve. This method additionally assumes, of course, that CV, as defined for Equation (11.33), is small.

11.5 The analysis of life-table data

In the previous section we were concerned with the description of a life-table by a single parameter, r_m. The present section will discuss the methods of 'taking the life-table apart', so as to determine the role of each factor. Several life-tables are necessary; ideally these will be a series for a number of generations of the same population, but some information can be obtained from life-tables from different populations or by measuring density, mortality and the associated factors in different parts of the habitat. Such analysis is not only of considerable theoretical interest, but will eventually provide a rational basis for prediction of animal populations: enabling, for example, the forecasting of climatically induced pest outbreaks and the making of prognoses of the effects of changes in cultural or control practices. The analysis of population on multiple sites has been developed by Dennis and Tapper (1994) and Dennis et al. (1998) (see Chapter 15).

11.5.1   The comparison of mortality factors within a generation (Table 11.3)

11.5.1.1   Apparent mortality

This is the measured mortality, the numbers dying as a percentage of the numbers entering that stage, i.e. d_x as a percentage of l_x (see also p. 431). Its main value is for simultaneous comparison either with independent factors or with the same factor in different parts of the habitat.

Table 11.3 Various measures for the comparison of mortality factors.

Measure	Stage			
	Eggs	Larvae	Pupae	Adults
l_x	1000	500	300	30
d_x	500	200	270	
% apparent mortality	50	40	90	
% real mortality	50	20	27	
% indispensible mortality	3	2	27	
Mortality/survivor ratio	1.00	0.66	9.00	
log population	3.00	2.70	2.48	1.48
k-values	0.30	0.22	1.00	

11.5.1.2 Real mortality

This is calculated on the basis of the population density at the beginning of the generation, i.e. $100 \times d_i/l_c$ = the deaths in the ith age interval and 1_c the size of the cohort at the commencement of the generation. The real mortality row in Table 11.3 is the only % row that is additive, and is useful for comparing the role of population factors within the same generation.

11.5.1.3 Indispensable (or irreplaceable) mortality

This is that part of the generation mortality that would not occur, should the mortality factor in question be removed from the life system, after allowance is made for the action of subsequent mortality factors. It is often assumed that these will still destroy the same percentage independent of the change in prey density; clearly, this assumption will not always be justified. To take an example of the calculation of the indispensable mortality from Table 11.3, consider the egg stage mortality: if there is no egg mortality, 1000 individuals enter larval stage where a 40% mortality leaves 600 survivors to pupate; in the pupal stage a 90% mortality leaves 60 survivors, that is 30 more than when egg mortality occurs, and thus its indispensable mortality $=30/1000 \times 100 = 3\%$. Pereira *et al.* (2007) present life tables for the coffee leafminer, *Leucoptera coffeella*, in the wet and dry seasons, which include the calculation of irreplaceable mortality.

When it is known that the subsequent mortalities are unrelated to density, the indispensable component of a factor may be used for assessing its value in control programmes (e.g. Huffaker & Kennet, 1965). If the exact density relationship of subsequent factors is known then a corrected, and more realistic, indispensable mortality can be calculated.

11.5.1.4 Mortality–survivor ratio

Introduced by Bess (1945), this measure represents the increase in population that would have occurred if the factor in question had been absent. If the final population is multiplied by this ratio, then the resulting value represents, in individuals, the indispensable mortality due to that factor.

11.5.1.5   The simple statistical relationship of population size to a factor

Originally, such relationships were investigated by straightforward regression analysis. However, such an approach ignores autocorrelation that is almost inevitable in time series data. Given time series of sufficient length more sophisticated approaches are available (see Chapter 15).

11.5.2   *Survival and life budget analysis*

Very significant advances in the analysis of population census data were made by Watt (1961, 1963, 1964), R.F. Morris (1959, 1963a & b) and their coworkers. Watt developed the mixed deductive–inductive model that blended field data and mathematical analysis in a way that allowed 'feedback' and correction in the model's development (Conway, 1973). Morris introduced the concept of key-factor analysis that aimed to determine the factor (or factors) that were of the greatest predictive value in forecasting future population trends; he also sought to estimate the magnitude of density dependence in the system. Subsequent studies (Hassell & Huffaker, 1969; Southwood, 1967; Maelzer, 1970; St Amant, 1970; Luck, 1971; Brockelman & Fagen, 1972; Benson, 1973) have shown that many difficulties arise in the interpretation of results from the Morris method. At this time, the method of Varley & Gradwell (1960), the development of which was stimulated by Morris's original paper, was to be preferred. However, Varley & Gradwell's method still has serious conceptual and statistical problems (Manly, 1990; Royama, 1992; Sibley & Smith, 1998). These are addressed by Brown *et al.* (1993) with a method they called 'structured demographic accounting', by Vickery (1991) who tested for bias and density dependence by randomisation and simulation, and subsequently by Sibley and Smith (1998) with the development of λ contribution analysis. Below, we first describe the traditional (Varley & Gradwell's) method because it has been widely used and is simpler to understand. The calculations required for λ contribution analysis are then described; once these have been undertaken, the graphical presentation and identification of a key factors is the same as for the traditional method.

Sibley and Smith (1998) have summarised the technical problems suffered by traditional key factor analysis. First, it does not correctly account for births as the k (mortality) values only measure changes in population caused by death and emigration. A way of including variation in births is by calculating birth-rate mortality defined as the difference between the maximum potential birth-rate and the observed birth-rate. Second, the number of individuals in each final age class declines with age, so that the estimation of k-values becomes progressively more unreliable with age. However, all k-values are given equal weight in the calculation of K, which can lead to serious errors. When applied to populations with overlapping generations, no account is taken of the different contributions made by the age classes present to population growth. Thus, the mortality experienced by older individuals with low reproductive potential is given equal weight with that of young adults (Sibley and Smith, 1998).

Royama (1996) points out that in any key factor analysis the potential importance of a factor that varies little through time is overlooked and that a key factor with a large effect may be arbitrarily created by the choice of stage division in a life-table.

11.5.2.1 Varley & Gradwell's method

This method, developed by Varley & Gradwell (1960, 1963a,b, 1965; Varley, 1963; Varley *et al.*, 1973) demands the data from a series of successive life-tables. The whole generation is considered; thus, it is immediately apparent in which age interval the density-dependent and key factors lie, and it provides a direct method of testing the role of changes in natality from generation to generation.

Varley & Gradwell's method may be outlined as follows:

1. The maximum potential natality is found by multiplying the number of females of reproductive age by the maximum mean fecundity female and this figure is entered in the budget (Table 11.4).
2. The values in the budget are converted to logarithms.
3. A convenient generation basis in this method is adult to adult; the base chosen will affect the recognition of key factors (Varley & Gradwell, 1965). The total generation

Table 11.4 A budget prepared for Varley & Gradwell's analysis.

	Numbers/10 m² (observed, unless marked*)	Log numbers/10 m²	k's
Maximum potential natality (no. reprod. ♀ × maximum natality	× 30 × 100 =1500	3.176	
k^0 (variation in natality)			0.076
Eggs laid	1260	3.100	
k^1 (egg loss)			0.171
Eggs hatching	850	2.929	
k^2 (predation, etc.)			0.498
3rd stage larvae	270	2.431	
k^3 (apparent parasitism of larvae)	100		0.201
[Larvae surviving parasitism]	170*	2.230	
k^4 (predation and other larval mortality)			0.276
Pupae	90	1.954	
k^5 (overwintering loss)			0.301
Adults emerging	45	1.653	
k^6 (dispersal, etc.)			0.352
Adults reproducing	20	1.301	
		K	= 1.875

'mortality' is given by subtracting the log of the population of adults entering the reproductive stage from the log maximum potential natality of the previous generation – this value is referred to as K^* (Table 11.4).

4. The series of age-specific mortalities are calculated by subtracting each log population from the previous one (Table 11.4); these are referred to as k-values, so that:

$$K = k_0 + k_1 + k_2 + k_3 + \ldots\ldots k_i, \tag{11.37}$$

5. Where precise estimates of mortality and migration (see Chapter 10) are available these are also incorporated into the equation. These series of k-values – one series for each generation – provide a complete picture of population changes. In the subsequent steps of this analysis the role of each k factor is examined separately, but it must be remembered that sampling errors are 'hidden' in each k and may be responsible for spurious results (Kuno, 1971). The identification of the k-value with a specific factor presents no problem when it is based on an apparent mortality, but when it is the difference of successive estimates it should, strictly, be referred to as 'overwintering loss' or 'loss of young adults'. It may be possible from other knowledge to indicate the major components of these losses; such assumptions could be checked by testing the correlation between the k-value and an independent measure of the loss factor, such as the abundance of a predator, over a number of generations. The difference between the log maximum potential natality and log actual natality, k_0 has a special significance, for it does not represent mortality in the strict sense, but the variation in natality. This is compounded of two separate causes: (i) death of the reproducing females before the end of the reproductive life; and (ii) variations in the fertility of the females. The two could be separated by determining the form of survival curve of the reproducing female (Chapter 12).

6. The next step involves recognition of the key factor for the index of population trend from adult to adult. An assessment may be made by visual correlation, K and k_0 to k_i are plotted against generation, and it may easily be seen which k is most closely correlated with K (Fig. 11.7). A quantitative evaluation of the roles of each k is provided by Podoler & Rogers' (1975) method in which the individual k-values are plotted (on the y-axis) against total mortality (K on the x-axis) and the regression calculated. The relative importance of each factor will be proportional to its regression coefficient. Because sampling errors incorporated in k will also appear in K this is not a precise statistical test for the importance of each mortality, but for the role of the estimates of each mortality in contributing to changes in the value of K.

7. The various ks are then tested for direct density-dependence. Each k is plotted against the number (N_t) entering the stage (age interval) on which it acts (Fig. 11.8) and if the regression is significant, then density dependence may be suspected. But the principal difficulty now arises: the variables are not independent (Varley & Gradwell, 1960; Bulmer, 1975), the regression is:

$$k = \log a + b \log N_t \tag{11.38}$$

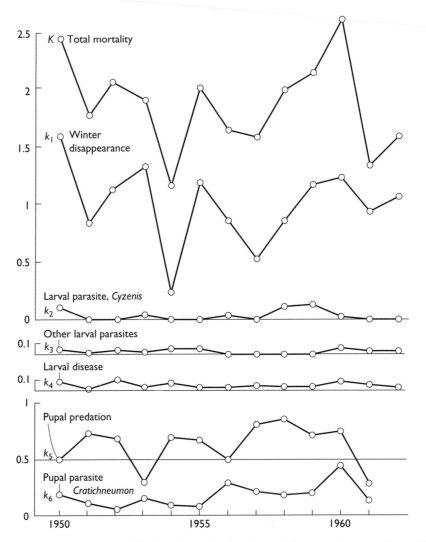

Figure 11.7   **The recognition of the key factor by visual correlation of various *k* values with *K*. (Data from Varley & Gradwell, 1960.)**

where b = the slope of the density-dependence (see also May *et al.*, 1974) and α a constant, but

$$k = \log N_t - \log N_t + 1. \tag{11.39}$$

Thus, the regression could be spurious, due to sampling errors. The second step therefore is to plot the log numbers entering the stage ($\log N_t$) against the log number of survivors ($\log N_{t+1}$). The regressions of $\log N_t$ on $\log N_{t+1}$ and of $\log N_{t+1}$ on $\log N_t$ should be calculated, and if both the regression coefficients depart

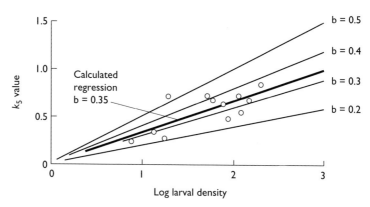

Figure 11.8 **The recognition of density dependence of a *k* factor by plotting the value of this factor against the log density at the start of the stage on which it acts. (Data from Varley & Gradwell, 1963a.)**

significantly from 1.0, then the density dependence may be taken as real. Bulmer (1975) has devised a more precise test based on the ratio:

$$R = \frac{\displaystyle\sum_{t=1}^{n} (\log N_t - \log \bar{N})^2}{\displaystyle\sum_{t=1}^{n-1} k^2} \qquad (11.40)$$

where N = the mean population level at time t over the years studied; R is the reciprocal of von Neumann's ratio which is known to have a certain distribution depending on the value of N. Bulmer (1975) shows by simulation that a value of R smaller than the lower limit of R (due to random variation as predicted from tables of von Neumann's ratio) is an indication of density dependence. However, as he points out, this test will eliminate any trend (i.e. delayed density dependence) and it is not strictly valid if the estimates of N_t and N_{t+1} contain sampling errors (as they normally do). However, Bulmer concludes that '...a series of 62 observations is only likely to reveal the existence of rather strong density-dependence and its existence may be masked by temporal trends (i.e. delayed effects) in the data'; so the practical utility of this test is unfortunately minimal (unless the organisms have very short generation times so that a series of about 100 generations may be accumulated).

8. If density dependence is believed to be real, attention may now be refocused on the plot of *k*, against the numbers entering the stage (Fig. 11.9). The slope of the line, the regression coefficient, should be determined as this will give a measure of how the factor will act; the closer the regression coefficient is to 1.0, the greater the stabilising effect of that regulatory factor. If the coefficient is exactly 1.0 the factor will compensate completely for any changes in density; if the coefficient is less than 1.0, the factor will be unable to compensate completely for the changes in density caused by other disturbing factors; whilst a coefficient of more than 1.0 implies

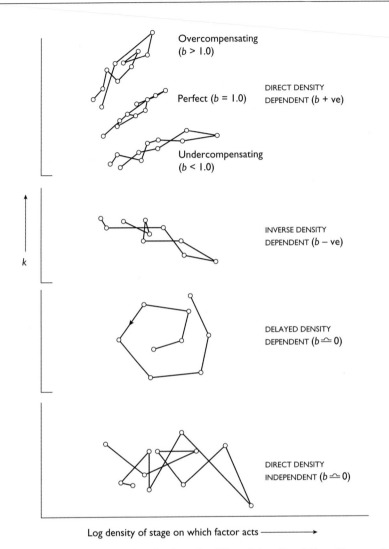

Log density of stage on which factor acts ⟶

Figure 11.9 Time-sequence plots showing how the different density relationships may be recognised from the patterns produced.

overcompensation. Using Equation (11.38), Stubbs (1977) was able to detect density dependence in 46 sets of life-budget data and draw significant conclusions. Hassell (1975) developed a more general model for density-dependence, for single species populations:

$$k = b \log (1 + aN_t), \qquad (11.41)$$

where k is defined in Equation (11.38), b is a measure of density dependence $a = 1/N_c$, and N_c = a critical density for the organism at which scramble competition starts to operate. Hassell (1975) was able to estimate the level of density dependence in populations of several species by fitting his model to published population budgets.

9. The detection of temporal trends (*sensu* Bulmer, 1975) or of delay in density dependence may be undertaken by plotting the *k*-value against log initial density (as in Fig. 11.9) and then joining the points up in a time sequence plot (Varley, 1947, 1953; Morris, 1959; Varley & Gradwell, 1965). The different types of factor will trace different patterns (Fig. 11.9): direct density-dependent factors will trace a more or less straight line or narrow band of points, delayed density factors circles or spirals; density-independent factors irregular or zigzag plots, whose amplitude reflects the extent to which they fluctuate. Solomon (1964) discusses how the order in which various factors act will influence the magnitude of their effect. This type of plot, the linking of consecutive points serially, is of particular value in investigating what Hassell (1966) has appropriately termed 'intergeneration relationships', and it is the only method by which a delayed relationship (a delayed density-dependent factor) can be recognised.

In conclusion, Varley & Gradwell's method allows:

1. The recognition of the key factor (or factors) or the period in which it acts.
2. The investigation of the role of natality in population dynamics.
3. The consideration of the role of mortality factors at every stage of the life-cycle, the recognition of the different density relationships of these factors and an indication of their mode of operation: direct density-dependent factors tend to stabilise, delayed density-dependent factors lead to oscillations, density-independent factors lead to fluctuations, and inverse density dependent factors will tend to accentuate the fluctuations.

Many life tables have now been analysed by these methods and comparative evaluations of the role of density-dependent and density-independent factors have been made (e.g. Stiling, 1987, 1988). Density-dependence, which may be hidden by stochastic variation (Hassel, 1987), can be studied on various spatial scales. The survival of whitefly, *Aleurotrachelus jelinekii*, was measured on individual leaves on two branches of *Viburnum tinus* (Southwood & Reader, 1976; Hassell *et al.*, 1987; Southwood *et al.*, 1989; Hassell, 1998). They found that the dynamics could be understood only with a long (16 generations) series of observations which allowed the recognitions of different density-dependent relationships in K_0 at the low and higher population densities, as measured on the bush and of additional density dependence in the survival from egg to adult on a leaf to leaf basis. From this detailed, but complex, picture, it was possible to develop a model with multiple equilibria as predicted by other studies (Southwood & Comins, 1979).

Working on data from individual farms, Randolph (1994, 1997) and Randolph & Rogers (1997) have constructed life budgets for the African tick, *Rhipicephalus appendiculatus*. This was done from a series of measurements of the population of different stages by subtracting the monthly $\log_{10}$ number + 1 of one stage of the tick from the

$\log_{10}$ number +1 of the preceding stage with the time interval equivalent to the development period. These k-values were then plotted against the density of the earlier stage and against climatic variables. Clear examples were found of density-dependent mortality and climate-related density-independent mortality in different stages.

Using several thousand time series of moths and aphids sampled in Britain by light and suction traps, Woiwod & Hanski (1992) found significant evidence of density dependence by Bulmer's methods in 79% of the moths and 88% of the aphid series greater than 20 years in length.

11.5.3   Sibley's λ contribution analysis

Whereas, traditional key factor analysis seeks to determine the influences on mortality, this method aims to identify those elements in the life-table that most influence the population growth rate measured as either λ or r ($\lambda = e^r$) (Sibley & Smith, 1998). Given an age-specific birth rate b_x, the rate of change of λ with respect to b_x is given by:

$$\frac{\partial \lambda}{\partial b_x} = \frac{\lambda^{-x} l_x}{\sum\limits_{i=1}^{M} i \lambda^{-i-1} b_i l_i} \tag{11.42}$$

where M is the age of last reproduction.

Similarly, the rate of change of λ with respect to k_x is given by:

$$\frac{\partial \lambda}{\partial k_x} = \frac{\sum\limits_{i=x+1}^{M} \lambda^{-i} b_i l_i}{\sum\limits_{i=1}^{M} i \lambda^{-i-1} b_i l_i}. \tag{11.43}$$

and the total population growth rate is the summation:

$$\lambda_{total} = \sum \frac{\partial \lambda}{\partial p_i} p_i \tag{11.44}$$

where p_i is a life history parameter b_x or k_x.

Note that Equation (11.44) is analogous to the traditional key factor equation

$$K = k_0 + k_1 + k_2 + k_3 + \dots \dots k_i.$$

The difference is that in Equation (11.44) each term is weighted by its contribution to the population growth rate and birth as well as mortality rates are included.

Having calculated the various terms:

$$\frac{\partial \lambda}{\partial b_x} b_x$$

and

$$\frac{\partial \lambda}{\partial k_x} k_x$$

for each year, they are plotted over the years on a graph with λ_{total} and a visual examination and regression analysis undertaken as for traditional key factor analysis to identify the key factors.

References

Ashford, J.R., Read, K.L.Q., and Vickers, G.G. (1970) A system of stochastic models applicable to studies of animal population dynamics. *J. Anim. Ecol.* **39**, 29–50.

Bellows, T.S. & Birley, M.H. (1981) Estimating developmental and mortality rates and stage recruitment from insect stage-frequency data. *Res. Pop. Ecol.* **23**, 232–44.

Bengstron, M. (1969) Estimating provisional values for intrinsic rate of natural increase in population growth studies. *Aust. J. Sci.* **32**, 24.

Benson. J.F. (1973) Some problems of testing for density-dependence in animal populations. *Oecologia* **13**, 183–90.

Bess, H.A. (1945) A measure of the influence of natural mortality factors on insect survival. *Ann. Entomol. Soc. Am.* **38**, 472–82.

Beverton, R.J.H. & Holt, S.J. (1957) On the dynamics of exploited fish populations. *Fisheries Investigations, Series 2*, **19**, 533 pp. Ministry of Agriculture, Fisheries & Food. HMSO, London.

Birch, L.C. (1948) The intrinsic rate of natural increase of an insect population. *J. Anim. Ecol.* **17**, 15–26.

Birley, M.H. (1977) The estimation of insect density and instar survivorship functions from census data. *J. Anim. Ecol.* **46**, 497–510.

Brockelman. W.Y. & Fagen, R.M. (1972) On modelling density-independent population change. *Ecology* **53**, 944–8.

Brown, D., Alexander, N.D.E., Marrs, R.W., & Albon, S. (1993) Structured accounting of variance of demographic change. *J. Anim. Ecol.* **62**, 490–502.

Bulmer, M.G. (1975) The statistical analysis of density-dependence. *Biometrics* **31**, 901–11.

Caughley, G. & Birch, L.C. (1971) Rate of increase. *J. Wildlife Manag.* **35**, 658–63.

Conway, G.R. (1973) Experience in insect pest modeling: a review of models, uses and future directions. *Mem. Aust. Ecol. Soc.* **1**, 103–30.

Deevey, E.S. (1947) Life tables for natural populations of animals. *Q. Rev. Biol.* **22**, 283–314.

Dempster. I.P. (1956) The estimation of the numbers of individuals entering each stage during the development of one generation of an insect population. *J. Anim. Ecol.* **25**, 91–5.

Dempster, J.P. (1961) The analysis of data obtained by regular sampling of an insect population. *J. Anim. Ecol.* **30**, 429–32.

Dennis, B. & Taper, M.L. (1994) Density dependence in time series observations: estimation and testing. *Ecol. Monogr.* **64**, 205–24.

Dennis, B., Kemp, W.P., & Taper, M.L. (1998) Joint density dependence. *Ecology* **79**, 426–41.

Fenchel, T. (1974) Intrinsic rate of natural increase: the relationship with body size. *Oecologia* **14**, 317–26.

Fletcher, R.I. & Deriso, R.B. (1988) Fishing in dangerous waters: remarks on a controversial appeal to spawner-recruit theory for long-term impact assessment. In: *Science, law, and Hudson River power plants: A case study in environmental impact assessment.*

Gilbert, N., Gutierrez, A.R., Frazer, B.D., & Jones, R.E. (1976) *Ecological Relationships.* Addison-Wesley, Reading & San Francisco, 157 pp.

González-Zamora, J.E. & Moreno, R. (2011) Model selection and averaging in the estimation of population parameters of *Bemisia tabaci* (Gennadius) from stage frequency data in sweet pepper plants. *J. Pest. Sci.* **84**, 165–77.

Gulland, J.A. (1983) *Fish Stock Assessment*. John Wiley & Sons, New York, pp. 223.

Hassell, M.P. (1966) Evaluation of parasite or predator responses. *J. Anim. Ecol.* **35**, 65–75.

Hassell, M.P. (1975) Density-dependence in single-species populations. *J. Anim. Ecol.* **44**, 283–95.

Hassell, M.P. (1987) Detecting regulation in patchily distributed animal populations. *J. Anim. Ecol.* **56**, 705–13.

Hassell, M.P. & Huffaker, C.B. (1969) The appraisal of delayed and direct density-dependence. *Can. Entomol.* **101**, 353–61.

Hassell, M.P., Southwood, T.R.E., & Reader, P.M. (1987) The dynamics of the viburnum whitefly (*Aleurotrachelus jelinkii*): a case study of population regulation. *J. Anim. Ecol.* **56**, 283–300.

Helgeson, R.G. & Haynes, D.L. (1972) Population dynamics of the cereal leaf beetle, *Aulema melanopus* (Coieoptera: Chrysomelidae): model for age-specific mortality. *Can. Entomol.* **104**, 797–814.

Huffaker, C.B. & Kennet, C.E. (1965) Ecological aspects of control of olive scale *Parlatonia oleae* (Colvee) by natural enemies in California. *Proc. XII. Int. Congr. Entomol.*, 585–6.

Iles, T.C. (1994) A review of stock–recruitment relationships with reference to flatfish populations. *Netherlands J. Sea Res.* **32**, 399–420.

Iles, T.C. & Beverton, R.J.H. (1998) Stock, recruitment and moderating processes in flatfish. *J. Sea Res.* **39**, 41–55.

Ito, Y. (1961) Factors that affect the fluctuations of animal numbers, with special reference to insect outbreaks. *Bull. Nat. Inst. Agric. Sci.* C **13**, 57–89.

Jensen, A.L. (1975) Comparison of logistic equations for population growth. *Biometrics* **31**, 853–62.

Kempton, R.A. (1979) Statistical analysis of frequency data obtained from sampling an insect population grouped by stages. In: Ord, J.K., Patil, G.P., & Taille, C. (eds), *Statistical Distributions in Scientific Work*. International Cooperative Publishing House, Maryland.

Kiritani, K. & Nakasuji, F. (1967) Estimation of the stage-specific survival rate in the insect population with overlapping stages. *Res. Popul. Ecol.* **9**, 143–52.

Knape, J., Daane, K.M., & de Valpine, P. (2014) Estimation of stage duration distributions and mortality under repeated cohort censuses. *Biometrics* **70**, 346–55.

Kuno, E. (1971) Sampling error as a misleading artifact in key factor analysis. *Res. Popul. Ecol.* **13**, 28–45.

Laughlin. R. (1965) Capacity for increase: a useful population statistic. *J. Anim. Ecol.* **349** 77–91.

Lefkovitch, L.P. (1963) Census studies on unrestricted populations of *Lasioderma serricorne* (F.) (Coleoptera: Anobiidae). *J. Anim. Ecol.* **32**, 221–31.

Leslie, P.H. & Ranson, R.M. (1940) The mortality, fertility and rate of natural increase of the vole (*Microtus agrestis*) as observed in the laboratory. *J. Anim. Ecol.* **9**, 27–52.

Luck, R.F. (1971) An appraisal of two methods of analysing insect life tables. *Can. Entomol.* **1039**, 1261–71.

Maelzer, D.A. (1970) The regression of log N n on log N, as a test of density dependence: an exercise with computer-constructed density-independent populations. *Ecology* **51**, 810–22.

Manly, B.F.J. (1974a) Estimation of stage-specific survival rates and other parameters for insect populations developing through several stages. *Oecologia* **15**, 277–85.

Manly, B.F.J. (1974b) A note on the Richards, Waloff & Spradbery method for estimating stage-specific mortality rates in insect populations. *Biometr. Z.* **17**, 77–83.

Manly, B.F.J. (1974c) A comparison of methods for the analysis of insect stage-frequency data. *Oecologia* **17**, 335–48.

Manly, B.F.J. (1976) Extensions to Kiritani and Nakasuji's method for analysing insect stage-frequency data. *Res. Popul. Ecol.* **17**, 191–9.

Manly, B.F. (1987) A multiple regression method for analysing stage-frequency data. *Res. Pop. Ecol.* **29**, 119–27.

Manly, B.F.J. (1990) *Stage-Structured Populations, Sampling, Analysis and Simulation.* Chapman & Hall, London.

Manly, B.F. (1997) A method for the estimation of parameters for natural stage-structured populations. *Res. Pop. Ecol.* **39**, 101–11.

Manly, B.F.J. & Seyb, A. (1989) A comparison of three maximum likelihood models for stage-frequency data. *Res. Pop. Ecol.* **31**, 367–80.

May, R.M. (1976) Estimating r: a pedagogical note. *Am. Nat.* **110**, 496–9.

May, R.M., Conway, G.R., Hassell, M.P., & Southwood, T.R.E. (1974) Time delays, density-dependence and single-species oscillations. *J. Anim. Ecol.* **43**, 747–70.

Messenger, P.S. (1964) Use of life tables in a bioclimatic study of an experimental aphid-braconid wasp host–parasite system. *Ecology* **45**, 119–31.

Morris, R.F. (1959) Single-factor analysis in population dynamics. *Ecology* **40**, 580–8.

Morris, R.F. (ed.) (1963a) The dynamics of epidemic spruce budworm populations. *Mem. Entomol. Soc. Can.* **31**, 1–332.

Morris, R.F. (1963b) The dynamics of epidemic spruce budworm populations. *Mem. Entomol. Soc. Can.* **31**, 30–7, 116–29.

Munholland, P.L., Kalbfleisch, J.D., & Dennis, B. (1989) A stochastic model for insect life history data. In: McDonald, L.L., Manly, B.F.J., Lockwood, J.A., & Logan, J.A. (eds), *Estimation and Analysis of Insect Populations.* Springer-Verlag, Berlin.

Pereira, E.J.G., Picanço, M.C., Bacci, L., Crespo, A.L.B., & Guedes, R.N.C. (2007) Seasonal mortality factors of the coffee leafminer, *Leucoptera coffeella. Bull. Entomol. Res.,* **97**, 421–32.

Podoler, H. & Rogers, D. (1975) A new method for the identification of key factors from life-table data. *J. Anim. Ecol.* **44**, 85–114.

Randolph, S.E. (1994) Population dynamics and density-dependent seasonal mortality indices of the tick *Rhipicephalus appendiculatus* in eastern and southern Africa. *Med. Vet. Entomol.* **8**, 351–68.

Randolph, S.E. (1997) Abiotic and biotic determinants of the seasonal dynamics of the tick *Rhipicephalus appendiculatus* in South Africa. *Med. Vet. Entomol.* **11**, 25–37.

Randolph, S.E. & Rogers, D.J. (1997) A generic population model for the African tick Rhipicephalus appendiculatus. *Parasitology* **115**, 265–79.

Richards, O.W. (1940) The biology of the small white butterfly (*Pieris rapae*), with special reference to the factors controlling abundance. *J. Anim. Ecol.* **9**, 243–88.

Richards, O.W. (1959) The study of natural populations of insects. *Proc. R. Entomol. Soc. Lond.* C **23**, 75–9.

Richards, O.W. (1961) The theoretical and practical study of natural insect populations. *Annu. Rev. Entomol.* **6**, 147–62.

Richards, O.W. & Waloff, N. (1954) Studies on the biology and population dynamics of British grasshoppers. *Anti-Locust Bull.* **17**, 182 pp.

Richards, O.W. & Waloff, N. (1961) A study of a natural population of *Phytodecta olivacea* (Forster) (Coleoptera: Chrysomeloidea). *Phil. Trans. B* **244**, 205–57.

Richards, O.W., Waloff, N., & Spradbery, J.P. (1960) The measurement of mortality in an insect population in which recruitment and mortality widely overlap. *Oikos* **11**, 306–10.

Ricker, W.E. (1954) Stock and recruitment. *J. Fisheries Res. Bd Canada* **11**, 559–623.

Rogers, D.J. (1979) Tsetse population dynamics and distribution: a new analytical approach. *J. Anim. Ecol.* **48**, 825–49.

Royama, T. (1992) *Analytical Population Dynamics*. Chapman & Hall, London.

Royama, T. (1996) A fundamental problem in key factor analysis. *Ecology* **77**, 87–93.

Ruesink, W.G. (1975) Estimating time-varying survival of arthropod life stages from population density. *Ecology* **56**, 244–7.

St Amant, J.L.S. (1970) The detection of regulation in animal populations. *Ecology* **51**, 823–8.

Schneider, S.M. & Ferris, H. (1986) Estimation of stage-specific developmental times and survivorship from stage-frequency data. *Res. Pop. Ecol.* **28**, 267–80.

Sibley, R.M. & Smith, R.H. (1998) Identifying key factors using lambda contribution analysis. *J. Anim. Ecol.* **67**, 17–24.

Slobodkin, L.B. (1962) *Growth and Regulation of Animal Populations*. Holt, Rinehart and Winston, New York, 184 pp.

Smith, F.E. (1954) Quantitative aspects of population growth. In: Boell, E. (ed.), *Dynamic of Growth Processes*. Princeton University Press, Princeton, New Jersey.

Solomon, M.E. (1964) Analysis of processes involved in the natural control of insects. *Adv. Ecol. Res.* **2**, 1–58.

Southwood, T.R.E. (1967) The interpretation of population change. *J. Anim. Ecol.* **36**, 519–29.

Southwood, T.R.E. (1969) Population studies of insects attacking sugar cane. In: Williams, J.R. *et al.* (eds), *Pests of Sugar Cane*. Elsevier, Amsterdam, pp. 427–59.

Southwood, T.R.E. (1981) Stability in field population of insects. In: Hiorns, R.W. (ed.), *Mathematical Theory of the Dynamics of Biological Populations II*. London Mathematical Society, London, pp. 31–46.

Southwood, T.R.E., Hassel, M.P., Reader, P.M., & Rogers, D.J. (1989) Population dynamics of the viburnum whitefly (*Aleurotrachelus jelinekii*). *J. Anim. Ecol.* **58**, 921–42.

Southwood, T.R.E. & Comins, H.N. (1979) A synoptic population model. *J. Anim. Ecol.* **45**, 949–65.

Southwood, T.R.E. & Jepson, W.F. (1962) Studies on the populations of *Oscinella frit* L. (Dipt.: Chloropidae) in the oat crop. *J. Anim. Ecol.* **31**, 481–95.

Southwood, T.R.E., May, R.M., Hassell, M.P., & Conway, G.R. (1974) Ecological strategies and population parameters. *Am. Nat.* **108**, 791–804.

Southwood, T.R.E. & Reader, P.M. (1976) Population census data and key factor analysis for the Viburnum whitefly, *Aleurotrachelus jelinekii* (Frauenf.) on three bushes. *J. Anim. Ecol.* **45**, 313–25.

Stiling, P.D. (1987) The frequency of density dependence in insect host-parasitoid systems. *Ecology* **68**, 844–56.

Stiling, P. (1988) Density-dependent processes and key factors in insect populations. *J. Anim. Ecol.* **57**, 581–93.

Stubbs, M. (1977) Density-dependence in the life-cycles of animals and its importance in K and r-strategies. *J. Anim. Ecol.* **46**, 677–88.

Varley, G.C. (1947) The natural control of population balance in the knapweed gallfly (*Urophora jaceana*). *J. Anim. Ecol.* **16**, 139–87.

Varley, G.C. (1953) Ecological aspects of population regulation. *Trans. IX Int. Congr. Entomol.* **2**, 210–14.

Varley, G.C. (1963) The interpretation of change and stability in insect populations. *Proc. R. Entomol. Soc. Lond.* C **27**, 52–7.

Varley, G.C. & Gradwell, G.R. (1960) Key factors in population studies. *J. Anim. Ecol.* **29**, 399–401.

Varley, G.C. & Gradwell, G.R., (1963a) The interpretation of insect population changes. *Proc. Ceylon Assoc. Adv. Sci.* **18** (D), 142–56.

Varley, G.C. & Gradwell, G.R. (1963b) Predatory insects as density dependent mortality factors. *Proc. XVI Int. Zoo. Congr.* **1**, 240.

Varley, G.C. & Gradwell, G.R. (1965) Interpreting winter moth population changes. *Proc. XII Int. Congr. Entomol.* 377–8.

Varley, G.C. Gradwell, G.R., & Hassell, M.P. (1973) *Insect Population Ecology: An Analytical Approach.* Blackwell, Oxford, 212 pp.

Vickery, W.L. (1991) An evaluation of bias in k factor analysis. *Oecologia* **85**, 413–19.

Waloff, N. & Bakker, K. (1963) The flight activity of Miridae (Heteroptera) living on broom, *Sarothamnus scoparius* (L.) Wimn. *J. Anim. Ecol.* **32**, 461–80.

Watson, T.F. (1964) Influence of host plant condition on population increase of *Tetranychus telarius* (Linnaeus) (Acarina: Tetranychidae). *Hilgardia* **35**(11), 273–322.

Watt, K.E.F. (1961) Mathematical models for use in insect pest control. *Can. Entomol.* **93**, (Suppl. 19), 62 pp.

Watt, K.E.F. (1963) Mathematical population models for five agricultural crop pests. *Mem. Entomol. Soc. Can.* **32**, 83–91.

Watt, K.E.F. (1964) Density dependence in population fluctuations. *Can. Entomol.* **96**, 1147–8.

Woiwod, I.P. & Hanski, I. (1992) Patterns of density dependence in moths and aphids. *J. Anim. Ecol.* **61**, 619–29.

Wood, S.N. (1994) Obtaining birth and mortality patterns from structured population trajectories. *Ecol. Monogr.* **64**, 23–44.

Wyatt, I.I. & White, P.F. (1977) Simple estimation of intrinsic increase rates for aphids and tetranychid mites. *J. Appl. Ecol.* **14**. 757–66.

Yamamura, K. (1998) A simple method to estimate insect mortality from field census data: a modification of the Kiritani-Nakasuji-Manly method. *Res. Popul. Ecol.* **40**, 335–40.

12 Age-grouping, Time-specific Life-tables and Predictive Population Models

This chapter is concerned with techniques for animals whose generations overlap widely; age-grouping is a prerequisite for these methods, which have been most widely applied to vertebrate populations. Analysis is easiest with two extreme types of population: the stationary and the exponentially expanding. If the population can be assumed to be stationary, then the fall-off in numbers in successive age groups will reflect the survivorship curve and thus a time-specific (or vertical) life-table can be constructed on this basis (p. 476). If the population is expanding and unconstrained by its environment, the age structure may become stabilised and then mortality can be estimated from the difference between expected and actual growth rates (p. 446). Age-grouping may also provide useful information on the fertility or potential fertility of the population (p. 373).

12.1 Age-grouping

Considerable research has been undertaken on estimating age, and thus this is only a selective introduction to the literature. In the following, taxonomically arranged sections, the methods available for age-grouping (or age-grading) both vertebrates and invertebrates are outlined. While a method may be generally applicable to the members of a taxon, the ease and accuracy with which it may be applied can vary greatly between species, so whenever possible, the advice of someone experienced with the group should be sought. There are still many animals such as leaches, flatworms and sea anemones that cannot be aged.

It is common practice, particularly in animals which reproduce in synchrony once a year, such as some fish, to age-group animals using a size-frequency histogram to identify the different cohorts (Plate 6). There are two major problems with this method: first, individual variation in growth can be considerable both within and between cohorts resulting in erroneous age assignment; and second, growth usually decelerates with age, making it difficult or impossible to discriminate between the older age classes. An extreme example of these pitfalls is shown by the study of Galinou Mitsoudi & Sinis (1995) on the long-lived bivalve, *Lithophaga lithophaga*, where it was found that individuals of 5.0 ± 0.2 cm in length ranged in age from 18 to 36 years. Size-frequency distributions can be used if they are supported by prior knowledge on the age–length relationship and individual variation (Goodyear, 1997).

Ecological Methods, Fourth Edition. P. A. Henderson and T. R. E. Southwood.
© 2016 John Wiley & Sons, Ltd. Published 2016 by John Wiley & Sons, Ltd.
Companion Website: www.wiley.com/go/henderson/ecologicalmethods

12.2 Aging young by developmental stage

Age cannot be inferred from developmental stage without reference to the environment. The speed of development may be temperature-dependent or influenced by factors such as oxygen and food availability. Rearing experiments may be required to understand the influence of environmental conditions on development. It is sometimes possible to relate development to the degree–days (the product of temperature and time) needed to reach a certain stage. However, it has been found that this measure can also vary with temperature. If allowance is made for the environment, developmental stage can be a useful method for ageing juveniles.

The immature stages of most insects and crustaceans are easily distinguished, larval instars being recognised by the diameter of the head or carapace, the length or presence of the appendages and other structural features, although even so basic a feature as the width of the larval head capsule may be misleading (Kishi, 1971). For the lobster, *Homarus gammarus*, where the number of moults is not fixed, Henocque (1987) has suggested that the number of segments in the antennules increases with each successive moult. It is even possible to age-group within the instars of some Heteropterous larvae as the number of eye facets increases for some time after moulting; this was noted in lace-bugs (Tingidae) (Southwood & Scudder, 1956) and in Corixidae, where it is even more marked, the interocular distance decreases throughout the instar (E.C. Young, unpublished results). In the last larval instar of Exopterygota the wing pads frequently darken shortly before the final moult.

Different ages of the insect pupal stage may often be determined by dissection, particularly in the Diptera, where various categories based on pigmentation can be recognised: for example, unpigmented, eyes of pharate adult pigmented, head-pigmented, pharate adult fully pigmented (Schneider & Vogel, 1950; Emden *et al.*, 1961). Using the falling phenolic content, measured by the reduction of a dye, Chai & Dixon (1971) were able to assess the age of sawfly cocoons.

In fisheries research, the eggs, larval and post-larval development of fish are divided into a number of distinct developmental stages for ageing. Typical features used for the egg include the presence of the embryonic shield, pigment and eye development, the number of myomeres, and the length of the embryo. Most fish hatch with a yolk sac and the larvae are termed post-larvae after yolk sac resorption. Features used to stage larval and post-larval fish include fin rays, scale formation and pigmentation. Daily growth bands are laid down by fish on scales and otoliths which can be used to age larval and juvenile stages (Antunes & Tesch, 1997; Narimatsu & Munehara, 1997; Sepulveda, 1994). Amphibian tadpoles also can be easily assigned to developmental stage (Brown, 1990).

For birds, it is often easy to distinguish juveniles or young adults from older individuals because of their plumage (e.g. Hohman & Cypher, 1986; Graves, 1997), and they are frequently divided into three age classes of unfledged, juvenile and adult.

For mammals, one of the most useful features for ageing juveniles is the pattern of tooth eruption (e.g. Anderson, 1986; Taylor, 1988; Bianchini & Delupi, 1990; Ancrenaz & Delhomme, 1997), although a wide range of other skeletal and developmental features have also been used.

12.3 Aging by using structures

12.3.1 Annelids

In temperate regions annual growth rings are formed in the jaws of the polychaetes *Harmothoe imbricata*, *H. derjugini*, *Halosydna nebulosa*, *Hermilepidonotus robustus*, *Lepidonotus squamatus* and *Eunoe* sp. (Britaev & Belov, 1993). An interesting example of how habitat clues can aid aging is the study by Nishi & Nishihira (1996), in which they used the growth rings of coral to age the polychaete Christmas tree worm, *Spirobranchus giganteus*, which lives embedded within the skeleton of the coral.

12.3.2 Crustaceans

Crustaceans, particularly the long-lived Malacostraca, are difficult to age. Sheehy *et al.* (1994, 1996) have shown that the amount of lipofuscin in the left olfactory lobe cell mass of the brain of Australian red-claw crayfish *Cherax quadricarinatus*, or in the eyestalk ganglia of European lobster *Homarus gammarus*, is positively correlated with age. The lipofuscin concentration was measured using confocal fluorescence microscopy and image analysis. Krill can also be aged by the amount of florescent pigment (Berman *et al.*, 1989)

The time since the last moult has been determined for spider crab, *Maja squinado* and European lobster, *H. gammarus*, by measuring the natural radionuclides activity ratio, $^{228}Th/^{228}Ra$ in the exoskeleton (Le Foll *et al.*, 1989).

While it was once believed that crustaceans lose all hard structures when they moult, it has recently been found that some structures such as the gastric mill and the endocuticle region of the eye stalk may be retained and exhibit growth bands in shrimps, crabs and lobsters (Kilada *et al.*, 2012).

12.3.3 Insects

Some of the types of criteria used are indicated below. Early methods for age-grouping of mosquitoes are reviewed by Hamon *et al.* (1961) and by Muirhead-Thomson (1963), and for arthropods of medical importance as a whole by Detinova (1962, 1968); Russian workers have contributed greatly to this subject. Most methods rely on criteria that are indications of some physiological process, particularly reproduction and excretion, or on general wear and tear; therefore, the precise chronological equivalent of a given condition may vary from habitat to habitat or even between individuals.

12.3.3.1 Biochemical markers

Thomas and Chen (1989, 1990) reported that fluorescent pteridines in the compound eyes of adult screwworms, *Cochliomyia hominivorax*, increased in concentration through time and could be used for age determination. Age was calculated using a regression model that also took account of the size of the head capsule. Similarly, Camin *et al.* (1991) reported that the homogenised head capsules of Mediterranean fruit flies, *Ceratitis capitat*, measured spectrofluorometrically showed significant differences in fluorescence from 0 to 28 days after emergence. This pteridine accumulation method has also been shown to work for melon fly, *Bactrocera cucurbitae* (Mochizuki *et al.*, 1993). For tsetse flies it is not possible to accurately age recently emerged flies using this method, but Msangi & Lehane (1991) showed that compounds in the abdomens of young *Glossina morsitans morsitans* displayed high levels of fluorescence which decreased linearly with time for the first five days post-emergence. Tomic-Carruthers *et al.* (1996), in a detailed study of the head capsules of adults of the fly, *Anastrepha ludens*, identified a range of pteridines and concluded that deoxysepiapterin, a pteridine with yellow fluorescence, accumulated sufficiently and over a sufficiently long period to be useful for aging. Fluorescence has been found not to work for aging the fly *Culicoides variipennis sonorensis* (Mullens & Lehane, 1995).

A considerable literature exists on the aging of necrofagous insects used in forensic entomology to estimate the post-mortem interval. For example, Xu *et al.* (2014) report cuticular hydrocarbons (CHs) as promising age indicators in some insect species, especially the larvae of necrophagous flies. They used gas chromatography (GC) and gas chromatography-mass spectrometry (GC-MS) to characterise age-dependent changes in CH abundance in larval *Aldrichina grahami* (Aldrich) (Diptera: Calliphoridae). The majority of low-molecular-weight alkanes ($\leq$C25) and almost all of the alkenes decreased in abundance with larval development. In contrast, high-molecular-weight alkanes of chain length greater than C25 gradually increased with age.

12.3.3.2   Cuticular bands

The demonstration by Neville (1963) that there were apparently daily growth layers present in some areas of the cuticle of certain insects opened up the possibility of an ageing criterion for insects, as precise as that available from fish scales or otoliths.

Further studies have shown that such cuticular banding is widespread, but care is needed in its interpretation. In some Heteroptera, food availability affects band deposition (Neville, 1970), but not in the blowfly, *Lucilia* (Tyndale-Biscoe & Kitching, 1974). The banding was thought to depend on a rhythm that was merely influenced by temperature, although in Coleoptera the number of bands may vary in different parts of the same animal (Neville, 1970); however, Tyndale-Biscoe & Kitching (1974) showed that in *Lucilia* they were dependent on temperature changes; precisely, a rise of 3.5 °C or more over a threshold of 15.5 °C. With the normal daily temperature rhythm, one band (a dark layer + a light one, seen most clearly under polarised light) is laid down each day: however, this commences at apolysis so the number of bands formed in the pharate adult must be determined before they can be used to estimate post-ecdysal adult age. In the Exopterygota, where there seems to be more rhythmicity in their deposition, bands may be seen by sectioning the hind tibia, and have been demonstrated in most orders (Neville, 1970). Similar rings have been noted in honey-bees and beetles. In the Diptera, Schlein & Gratz (1972, 1973) showed that the skeletal apodemes exhibited more bands than other regions of the cuticle: they describe a staining technique, but Tyndale-Biscoe & Kitching (1974) could count banding in *Lucilia* without polarised light or staining. Band formation appears to be limited to the early part of adult life, up to 10–13 days in

Anopheles mosquitoes (Schlein & Gratz, 1973), and as many as 63 have been observed in a large grasshopper (Neville, 1970). Zuk (1987) found that the field crickets, *Gryllus pennsylvanicus* and *Gryllus veletis*, added one growth ring per day until about day 25–30 when cuticle growth was completed and rings are no longer added. It practice, this was not a serious limitation on the method as few animals were found in the wild with more than 18 rings.

12.3.3.3 Sclerotisation and colour changes in the cuticle and wings

In all insects, the newly emerged adult is pale and the cuticle untanned – in most species the major part of tanning is completed during a short teneral period.

However, complete sclerotisation may not occur for a considerable time, especially if a period of diapause or aestivation intervenes; for example, Lagace & Van Den Bosch (1964) found in the weevil, *Hypera*, that the elytra remained thin and almost teneral for much of the summer diapause. During teneral development in the Corixidae and Notonectidae the cuticle is pigmented at different rates; the sequence of pigmentation of the mesotergum is remarkably constant, proceeding forwards, in a number of stages, from the posterior region (Young, 1965). The wings of young dragonflies are milky, becoming clear when mature (Corbet, 1962a,b).

Various excretory pigments are accumulated throughout adult life and especially during diapause; some of these occur in the cuticle and may cause progressive colour changes. In other instances cuticular colour change is a product of the sclerotisation process. Dunn (1951) showed that the region of the discal cell and the associated veins of the hind wing of the Colorado beetle (*Leptinotarsa*) change from yellowish, in the newly emerged individual, to reddish after hibernation, and the green shieldbug, *Palomena prasina*, becomes redder, often a dark bronze, during hibernation (Schiemenz, 1953); in contrast, the pink colouring of another pentatomid, *Piezodorus lituratus*, disappears while overwintering. Many Miridae change colour during or after hibernation and yellows on the forewings of various Heteroptera often darken with age becoming orange or even reddish (Southwood & Leston, 1959). The coloration of the bodies of adult dragonflies changes with age (Corbet, 1962a,b). Red pigments, believed to be pheomelanin and eumelanin, are contained in the wing veins of the glassy-winged sharpshooter, *Homalodisca vitripennis* (Hemiptera: Cicadellidae). This red pigment darkens with age and eventually becomes brown/black in colour. Timmons *et al.* (2011) aged *H. vitripennis* by taking high-resolution wing images and using analysis software to calculate the amount of red pigment present.

12.3.3.4 Developmental changes in the male genitalia

The males of many insects have a period of immaturity during which they may be recognised by the condition of the male genitalia (e.g. in *Dacus tryoni*, as described by Drew, 1969). In most Nematocera the male hypogydium rotates soon after emergence; Rosay (1961) found in mosquitoes that the time for rotation was similar in the species of *Aedes* and *Culex* studied, but was strongly influenced by temperature (at 17 °C this was 58 h, and at 28 °C it was 12 h). The aedaegus of the weevil, *Hypera*, remains relatively unsclerotised throughout the period of aestivation lasting several months (Lagace & Van Den Bosch, 1964). In carabid beetles, comparable changes and the colour of the accessory glands may be used to determine age class (Dijk, 1972).

12.3.3.5 Changes in the internal non-reproductive organs

The fat-body is perhaps the most variable of these; in some insects it is large at the start of adult life, gradually waning (e.g. in the moth *Argyroploce*; Waloff, 1958), and in others it is slowly built up prior to diapause. The availability and quality of food affect the condition of the fat-body and other organs (Fedetov, 1947, 1955, 1960; Haydak, 1957), as does parasitism.

The deposition of coloured excretory products in various internal organs, especially the malpighian tubules, provides another measure of age. Haydak (1957) found in worker honey-bees that the malpighian tubules are clear for the first three days of life, milky from three to ten days, and in older bees generally (not always) yellowish-green. Female *Culicoides* acquire a burgundy-red pigment in the abdominal wall during the development of the ovarian follicles and, as this remains, parous midges can easily be recognised without dissection (Dyce, 1969).

In newly emerged mosquitoes, bees, moths and other endopterygote insects, part of the gut is filled with the brightly coloured meconium; this is usually voided within a day or two, at the most (Haydak, 1957; Rosay, 1961).

The detailed studies of Haydak (1957) and others on the honeybee have shown that ranges of organs, especially various glands, change their appearance with age.

12.3.3.6 The condition of the ovaries and associated structures

Criteria associated with changes in the female reproductive system have been used widely in ageing studies; they are valuable for even when they do not provide a precise chronological age, as they do give information on the extent of egg laying, which is for some purposes more useful. The principal characters that have been used may be summarised under the following headings.

1. Egg rudiments. In those insects that do not develop additional egg follicles during adult life, a count of the number of egg rudiments will indicate the potential fecundity. At the start of adult life the count will be high, gradually falling off, though a large number of rudiments may still remain at the end of life (Waloff, 1958; Corbet, 1962b).

2. Follicular relics. After an egg has been laid, the relics of the follicle, which are often pigmented, will remain in the bases of the ovariole pedicels or in the oviducts for a variable period enabling a parous female to be recognised (Corbet, 1960, 1961; Hamon *et al.*, 1961; Saunders, 1962, 1964; Anderson, 1964; Vlijm & Van Dijk, 1967; Lewis *et al.*, 1970; van Dijk, 1972; Hoc & Charlwood, 1990; Hoc & Wilkes, 1995). In some insects the total amount of relics accumulates with an increasing number of cycles (Anderson, 1964), but in others they are eventually lost. V.P. Polovodova demonstrated in *Anopheles* that each time an egg develops it causes local stretching

of the stalk of the oviduct. These dilations contain the remains of the follicles, and if the ovaries are carefully stretched and examined (see Giglioli, 1963) the number of dilations may be counted. In certain mosquitoes the maximum number of bead-like dilations in any ovariole can be taken as equal to the number of ovarian cycles; with other species difficulties have been encountered (Muirhead-Thomson, 1963).

3. Granulation of the basal body. In the black fly, *Simulium woodi*, the basal body of each ovariole is a group of six to eight cells in the calyx wall enclosed by the end of the ovariolar sheath. The granulation of these cells was reported by Hoc and Wilkes (1995) to progressively increase in intensity with each ovulation.

4. Ovarian tracheoles. The tracheoles supplying the ovaries of nulliparous females are tightly coiled ('tracheal skeins'); as the eggs mature these become stretched, so that they do not resume their previous form, even in interovular periods. These changes have been observed and used for ageing in mosquitoes (Hamon *et al.*, 1961; Detinova, 1962; Kardos & Bellamy, 1962), in dragonflies (Corbet, 1961), and in calypterate flies (Anderson, 1964). A standard method of dissection for age-grading mosquitoes, by this method, is given by Davies *et al.* (1971).

5. Ovariole cycles and combined evidence. In many insects the ovaries pass through a series of cycles; the stage in any given cycle may be easily recognised by the size of the most mature egg rudiments. Kunitskaya (1960) recognised six stages in fleas; but five stages seem normal within the Diptera, as initially recognised in mosquitoes by Christophers (1960) (e.g. ceratopogonids; Linley, 1965); simulilds (Le Berre, 1966), anthomyiids (Jones, 1971). By combining these with information on follicular relics, Tyndale-Biscoe & Hughes (1969) were able to recognise 17 different stages in *Musca*, and Vogt *et al.* (1974) 16 stages in *Lucilia*. Taken in conjunction with other evidence that indicates whether the female is nulliparous or parous, the age may be determined over two ovarian cycles. Vogt & Morton (1991) assigned sheep blowfly *Lucilia cuprina* to age class on the basis of parity and stage of ovarian development. In viviparous insects and others (e.g. Schizopteridae), where one large egg is laid at a time, the ovaries develop alternatively and this may allow a larger number of reproductive cycles to be recognised (Saunders, 1962, 1964). In some insects the number of functional ovarioles decreases after the first two cycles, and this may be used as an index of age.

12.3.3.7 Indices of copulation

Although frequency of copulation is not a reliable index of chronological age, it may be 'calibrated' for a particular population and gives *per se* information of ecological significance. The presence of sperm in the spermatheca will provide evidence of pairing; in some mosquitoes a gelatinous mating plug remains in the oviduct for a short period (Muirhead-Thomson, 1963). In insects where the sperm are deposited in a spermatophore the remains of these in the spermatheca will give a cumulative measure of copulation (Waloff, 1958).

In some insects the appendages of the male will leave characteristic 'copulation marks'. In dragonflies these are on the compound eyes or occipital triangle of the female, while in Zygoptera some sticky secretion may remain on the sides of the female's thorax (Corbet, 1962a); in tsetse flies they are on the sixth sternum (Squire, 1952). In Cimicidae with a broad spermalege, every pairing will leave a characteristic groove (Usinger, 1966). The mated status of male calliphorine flies can be determined as traces of the accessory gland secretion remain in the lateral penis ducts for up to a week after mating (Pollock, 1969).

12.3.3.8 Changes in weight

Insects that feed little or not at all during adult life will become progressively lighter; Waloff (1958) showed that individual variation due to size differences could be reduced if a weight/length ratio was used. With Lepidoptera, weight/wing length is a convenient measure, and Waloff found in a number of species that this falls off strikingly with age. For example, male *Argyroploce* with an expected life of 10 days had a ratio of 1.93, but when the expected life was 2.5 days the ratio was 1.13.

In other insects, the adult weight will fluctuate, reflecting ovarian cycles (Waloff, 1958; see also p. 373) or changes associated with diapause.

12.3.3.9 'Wear and tear'

As an adult insect becomes older its cuticle and appendages become damaged by contact with the environment, and this 'wear and tear' may be used as an index of age. Kosminskii (1960) has found that fleas may be age-graded by the extent to which the ctenidal (genal comb) bristles are broken, whilst Michener *et al.* (1955) and Daly (1961) used the wear on the mandibles of wild bees. However, the most widely applicable index of this type is provided by the tearing or battering of the wings, e.g. in Lepidoptera, Diptera (Saunders, 1962) and Hymenoptera (Michener *et al.*, 1955). Although apparently crude, the number of tears ('nicks') in the wings has been found to correlate well with other indices of ageing (Michener & Lange, 1959; Saunders, 1962). Phoretic water mites die or fall off mosquitoes, and possibly other insects that emerge from water, with time: under certain conditions (Corbet, 1970), they may be used to indicate age (Gillett, 1957; Corbet, 1963).

12.3.4 Molluscs

Bivalve molluscs gradually increase the thickness of the shell so that growth checks can show as a series of bands in suitably prepared cross-sections. In temperate regions the horse mussel, *Modiolus modiolus*, has alternating patterns of light (summer) and dark (winter) growth lines in the middle nacreous layer of the shell (Anwar *et al.*, 1990), while *Mytilus galloprovincialis* in the Black Sea produces dark summer and light winter bands(Shurova & Zolotarev, 1988). By counting such annual bands, Galinou-Mitsoudi & Sinis (1995) were able to show that piddock, *Lithophaga lithophaga*, could have a longevity of over 54 years. For some species it may be possible to count annuli in thin cross-sections using transmitted light, alternatively the cross-section can be polished and etched to emphasise the bands and an acetate-peel impression of the surface taken for examination (Richardson *et al.*, 1993). For species such as the European flat oyster, *Ostrea edulis*, acetate peels are taken from a cross-section of the umbo (Richardson *et al.*, 1993) or chondrophore (Tian & Shimizu, 1997). While clear annual rings on bivalve shells are often present in some highly variable habits they may be unclear and confused by other growth checks. When this is the case Garcia (1993) found detailed examination using scanning electron microscopy useful.

Pulmonates may be aged by growth bands. Raboud (1986) aged *Arianta arbustorum* from the Swiss Alps using thin sections of the shell margins. In some habitats, harsh winter conditions can give the shells of gastropods such as *Hydrobia ulva* clear winter growth checks. Terrestrial helicid snails are known to generally show bands caused by winter growth checks in the northern temperate and summer growth checks in Australia; these can be used for ageing (Baker & Vogelzang, 1988).

Cross-sections of squid statoliths show daily growth rings (Rodhouse & Hatfield, 1990), and Raya *et al.* (1994) have used these to age the cuttlefish, *Sepia hierredda*. Larval gastropods have also been aged by this method (Grana-Raffucci & Appeldoorn, 1997). Daily growth rings have also been observed in young giant scallop, *Placopecten magellanicus* (Parson *et al.*, 1993). However, shell growth patterns and checks may be related to the tidal cycle (Richardson, 1987; Richardson *et al.*, 1990). The tiny (0.22 mm-diameter) statoliths in the whelk, *Nassarius reticulatus*, have also been shown to exhibit growth rings (Barroso *et al.*, 2005).

12.3.5 *Fish*

The allocation of fish to year class is normally accomplished by the examination of structures such as scales (Loubens & Panfili, 1992; Almeida *et al.*, 1995; Moreira *et al.*, 1991; Singh & Sharma, 1995; Vidalis & Tsimenidis, 1996), otoliths (Berg, 1985; Gordo, 1996; Hales & Belk, 1992; Hassager, 1991; Oxenford *et al.*, 1994; Worthington *et al.*, 1995), spines (Esteves *et al.*, 1995; Tucker, 1985), opercular bones (Achionye Nzeh, 1994), fin rays (Lai *et al.*, 1987) and vertebrae (Parsons, 1993; Officer *et al.*, 1996) which are marked with bands because of seasonal growth checks. In temperate regions, rapid summer and slow winter growth result in an annual pattern comprising a summer and a winter zone which differ in appearance. The annulus is usually defined as the winter zone. These methods may be inappropriate for tropical fish living in environments with little seasonality. Bands will be formed whenever the growth of a fish is checked, as could be caused by drought, famine, periods of low oxygen, reproduction or disease. For example, the scales of the small goby, *Pomatoschistus microps*, show one growth check per year in the Tagus estuary, Portugal, but two in the British Isles. This may be because growth checks are caused both by reproduction and winter and in Portugal the two checks coincide (Moreira *et al.*, 1991).

Computer-aided systems have been developed for the identification and counting of annuli (Cailliet *et al.*, 1996).

Scales must be taken from a part of the fish known to acquire scales early in life, where scale replacement is unlikely and where they exhibit complete and clear growth patterns. For example, bluefish, *Pomatomus saltatrix*, or black sea bass, *Centropristis striata*, scales should be removed from the area behind the pectoral fin. A number of scales should be removed and examined from each fish. Age determinations may be made from direct observation, scale impressions, or photographs. A microfiche viewer is convenient for scale examination. For thick scales which do not show a clear structure with transmitted light, impressions in laminated plastic film can be taken for examination (Dery, 1983). Frequently encountered problems include difficulty in identifying the first annulus (Mann & Steinmetz, 1985) and indistinct patterns towards the edge of the scale (Hesthagen, 1985).

Otoliths are removed by dissection from the head of the fish and may be stored dry prior to examination. Examination for annuli may be undertaken using whole, baked, broken or crossed-sectioned otoliths, depending on the species. Visibility of the annuli is enhanced by baking or burning, because the hyaline zones turn brown in contrast to the white opaque zones (e.g. Aprahamian, 1987). Richter & McDermott (1990) tested the use of histological stains to enhance otolith annulus visibility, and found best results with acidified Neutral Red for sole, turbot, brill and scad, with aqueous Aniline Blue for cod, hake, whiting, plaice and grey mullet, and Eosin Y for pelagic species with small, translucent otoliths. Worthington *et al.* (1995) suggested that the relationship between otolith weight and fish age may be used to age tropical reef fish, and found the method satisfactory provided that the weight–age relationship was frequently re-calibrated.

Transverse sections of a dorsal spine show annuli in species such as the sail fish, *Istiophorus platypterus* (Alvarado-Castillo & Felix-Uraga, 1996). For sharks and other elasmobranchs, growth rings in the vertebral centra and dorsal spines can be used for aging (Parsons, 1993; Officer *et al.*, 1996; Tucker, 1985). Age determination of elasmobranchs can be problematical; Polat & Gumus (1995), for example, could not age spiny dogfish, *Squalus acanthias*, using vertebrae but were successful using the annuli on the surface of the mantle and the basal diameter of the second dorsal spine. The visibility of growth bands in the vertebrae of blackmouth catsharks, *Galeus melastomus*, was enhanced by decalcification with 5% nitric acid solution (Correia & Figueiredo, 1997). Tropical catfish can be difficult to age, and Al Hassan & Al Sayab (1994) show that eye lens weight, which is often used for small mammals, can indicate age. The pectoral spines of catfish can also be used (Blouin & Hall, 1990). Brennan & Cailliet (1989) concluded that for white sturgeon, pectoral fin sections were the most practical ageing structure in terms of ease of collection, processing, legibility, and precision of interpretation.

12.3.6 *Lampreys*

Beamish & Medland (1988) estimated lamprey age using statolith banding. However, annuli were not recognisable in the statoliths of the southern brook lamprey, *Ichthyomyzon gagei*, because growth was even throughout the year so that length–frequency distribution analysis is the only method available.

12.3.7 *Reptiles and amphibians*

Examination of skeletal structures are commonly used for reptile and amphibian aging (Francillon *et al.*, 1984; Castanet, 1986; Smirina, 1994; Gharzi & Yari, 2013). Francillon Viellot *et al.* (1990) aged the newts *Triturus cristatus* and *Triturus marmoratus* by counting annual growth rings in cross-sections of femurs and phalanges, staining with Ehrlich haematoxylin. Hutton (1986) found that male and non-breeding female Nile crocodiles, *Crocodylus niloticus*, could be aged from laminae in osteoderms while Tucker (1997), in a study of the freshwater crocodile *Crocodylus johnstoni*, found that these growth lines could be reliably counted in unstained 60–80-μm sections viewed by Nomarski interference microscopy.

12.3.8 *Birds*

Reliable methods for aging adult birds are few, probably because of the widespread use of rings. Klomp & Furness (1992) found that the count of endosteal layers in the tibia could accurately determine age in fulmar, *Fulmarus glacialis*, shag, *Phalacrocorax aristotelis*, redshank, *Tringa tetanus*, and great skua, *Catharacta skua*. Chae & Fujimaki (1996) suggested that the extent of skull pneumatisation could be used to age Hazel grouse, *Bonasa bonasia*. Fornasari *et al.* (1994) considered it possible to age hawfinches, *Coccothraustes coccothraustes*, using a set of criteria which scored rectrix abrasion, the width of the outer web of the sixth primary, and sharpness of the greater alula feather.

12.3.9 *Mammals*

While skeletal development assessed from radiography can be used to assign individuals to age classes (Nassar Montoya *et al.*, 1992), the most widely applicable methods use the teeth. One of the most accurate estimates of age at death is the count of incremental lines in the cementum of teeth (Martin, 1996). Particular applications of this method are: red foxes, *Vulpes vulpes* (Goddard & Reynolds, 1993; Cavallini & Santini, 1995); coyotes, *Canis latrans* (Jean *et al.*, 1985); marmots (Mashkin & Kolesnikov, 1990) and beaver, *Castor fiber* (Stiefel & Piechock, 1986). This method is not always reliable; for example, in warthog the incremental lines are often indistinct (Mason, 1984) and in tropical ungulates more than one band per annum may be laid down in the cementum (Kay & Cant, 1988). Other tooth characteristics used include; incisor height in deer (Moore *et al.*, 1995); degree of root closure and dentine layers as observed using radiography in beaver, *Castor fiber*, and pine martens, *Martes martes* (Hartman, 1992; Helldin, 1997) and tooth wear (Hartman, 1995; Ohtaishi *et al.*, 1990). Tooth wear will vary with the amount of abrasive material in the diet (Fandos *et al.*, 1993).

Growth rings can sometimes be observed in bone. Iason (1988) found that polished cross-sections of the mandibles of mountain hares, *Lepus timidus*, from Scotland showed annual growth rings which are laid down in the autumn. Growth lines in bone have also been used to age Florida manatees, *Trichechus manatus latirostris* (Marmontel *et al.*, 1996).

Eye lens weight, which increases linearly in weight with age, is considered the best character with which to age small mammals from genera such as such as *Microtus*, *Clethrionomys, Peromyscus, Mus, Apodemus, Pitymys* and *Rattus* (Rowe *et al.*, 1985; Heise *et al.*, 1991; Quere & Vincent, 1989; Pascal *et al.*, 1988). It has also been found useful for larger species such as the Japanese hare, *Lepus brachyurus brachyurus* (Ando *et al.*, 1992).

Olsen *et al.* (2014) assessed the use of telomere length measured by quantitative PCR from skin samples to age North Atlantic humpback whales, *Megaptera novaeangliae*.

While a significant correlation between telomere length and age was found, telomere length was highly variable between individuals of similar age, indicating that telomere length is an imprecise determinant of age in humpback whales.

12.4   Time-specific life-tables and survival rates

This method of building a demographic picture of a population may be used if it is justified to assume that over the period of life of the individuals in the table, the recruitment rate has been constant and the mortality rate within each age class has been steady (although, of course, it may change with age – the table will reveal this). Time-specific life-tables are also termed static, stationary or vertical life-tables.

In contrast to age-specific life-tables (p. 443), where successive observations of a cohort are need, the time-specific life-table is based solely on the age-grouping of the individuals collected at a single instant of time. Thus, a relative method (see Chapter 7) may be used for sampling the population, provided that it samples at random with respect to the different age groups. As we are studying relative abundance, the youngest age group may be equated with a convenient number (e.g. 1; 100; 1000), and the other values corrected accordingly to give the l_x column. The remaining columns of the table can then be calculated as described in Chapter 11 (p. 443). An example of this type of life-table and its calculation is that given by Paris & Pitelka (1962) for the woodlouse, *Armadillidium*.

A number of time-specific life-tables were constructed for the mosquito *Aedes aegypti* by Southwood *et al.* (1972); they were for limited parts of the yearly cycle when the assumption of steady recruitment appeared justified. Modified key factor analysis may be applied to these life-tables; it will indicate differences between populations under the different conditions (with *Aedes* in different breeding sites) at the period of steady recruitment, but it will not disclose anything about the periods of population change.

An extensive collection of age-grouping data is provided by the 'catch curves' obtained from commercial fishing, and their analysis has been described by Ricker (1944, 1948), Wohlschlag (1954), Beverton & Holt (1957) and Ebert (1973) and others. Another approach to the analysis is provided when log numbers are plotted against age, by the comparison of actual shape of the plot and the straight line that would result from a Slobodkin type III constant survival rate (Fig. 12.1). Temporal variations in the rate of recruitment and in both the age- and time-specific survival rates will affect the shape of the curve. Age-specific survival rates may be obtained from mark and recapture data (see Chapter 3), and a measure of recruitment can sometimes be obtained from another source, such as insects from emergence traps. By obtaining information from such other sources the departures of the time-specific survivorship curve (Fig. 12.1) from linearity can often be interpreted; if recruitment and time-specific survival rates are constant, then the fall-off with age in Fig. 12.1 implies increasing mortality with increasing age (Wohlschlag, 1954).

The hunting records of terrestrial vertebrates provide further extensive age-grouped data. Surveys of the proportions of various ages and sexes are made before and after hunting, and the numbers of the different groups killed are also known; from this

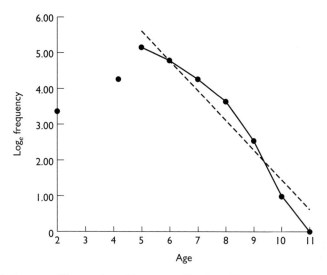

Figure 12.1   A time-specific survivorship curve. The logarithms of the frequencies plotted against age groups in a season's sample of the fish the least cisco, *Leucichthys sardinella* **(Data from Wohlschlag, 1954).**

information it is possible to calculate natality (referred to as production), survival rates and absolute population. The methods of computation have been reviewed by Hanson (1963) and Seber (1982), and some of these might be combined with the 'removal–sampling' approach and used with insects. Some of the methods do not require hunting returns; of these, the equation for the calculation of differential survival of age classes is of particular potential interest to insect ecologists (Hanson, 1963; Kelker & Hanson, 1964). It could be applied to a situation where generations overlap, but each was comparatively well synchronised; from biological knowledge we must select three times: t_1 shortly before generation II becomes adult; t_2 just after generation II has become adult; and t_3 a subsequent occasion. Assuming that the mortality of both sexes is equal, then the ratio of the survival of generation I to that of generation II during the period t_1 to t_2 is given by:

$$\frac{S_{II}}{S_I} = \frac{(\male t_2 - \female t_2)(\male t_3 \female t_1 - \male t_1 \female t_3)}{(\male t_3 - \female t_3)(\male t_2 \female t_1 - \male t_1 \female t_2)}$$

where the symbols represent the number of males and females in the samples on dates t_1, t_2, and t_3 as defined above (see also p. 93).

The use, in entomology, of age-grouping, time-specific life-tables and survivorship curve analysis of this type will probably prove most important in attempts to investigate the dynamics of the adult population. Detailed studies of this type will be necessary if the full meaning of k_0 (p. 449) is to be investigated.

Key factor analysis of age-specific life-tables or budgets (see Chapter 11) and survival analysis of time-specific life-tables provide a quantitative description of the population processes. From this, predictions may be made of future population trends or of the response of the population to some change: a control measure or harvesting. A simple management model could be constructed to utilise the information obtained in this way. However, both types of life-table are inappropriate tools to understand the population dynamics of the many populations in which there is continuous breeding and overlapping generations. As will be discussed below, it is possible to tackle the problem 'the other way round': that is, to predict population growth under 'Malthusian conditions' (i.e. 'unlimited' resources, when $r = dN/dt$ (see p. 446) and to compare the actual field results with those from the predictions. In most cases the same model may be used to predict the impact of change on the population processes.

12.4.1 *Physiological time*

These models predict the rate of population growth with time; population growth is dependent on generation time, which depends on development rate. Development rate is influenced by temperature and therefore, in strongly seasonal regions where the same stage may be exposed to significantly different temperatures, either a correction must be made for the different development rates or 'physiological time' may be used (Hughes, 1962, 1963; Gilbert *et al.*, 1976; Hardman, 1976; Atkinson, 1977). Theoretically, there is no reason why this should not always be used; however, it seems advisable to restrict it to models where it is really necessary. As most biologists can only easily visualise their populations against calendar time, the use of physiological time makes more difficult an intuitive check on the output of the model. Physiological time is commonly expressed as day degrees (D°) (Hughes, 1962; Hardman, 1976) or hour degrees (h°) (Atkinson, 1977), these being the cumulative product of total time × temperature (above the threshold). Allen (1976) describes a sine wave method for computing day degrees, and Scopes & Biggerstaff (1977) show how a temperature integrator may be usefully used in the field as an effective measure of accumulated physiological temperature. Jones *et al.* (1997) produced a life-table for koa seedworm based on degree–day demography, and point out that this considerably reduced the variability within the data set. For example, at different temperatures, the adult period varied from 72 to 32 days but, in contrast, on a physiological scale the observed range was only 2.7 to 6.2 days. Atkinson's (1977) studies suggest that the common practice of determining the physiological time for development at constant temperatures may be misleading when applied to fluctuating field conditions.

In most population models it is necessary to divide time, physiological or calendar, into intervals. The length of the interval is critical; if it is too large, the changes between intervals are too sudden and the heterogeneity within the time–interval age class too great. Too small an interval will add to the complexities and costs of the work, and the field data may be inadequate in quantity. Gilbert *et al.* (1976) tested a number of different time units and concluded that, for their aphid, a quarter instar period (a 'quip') was suitable; they make no claims for the generality of this unit, but it or a half-instar period seem to be of a reasonable order of magnitude to test initially. In their case, instar length was similar (in physiological time) for the first three instars; with other insects the instar-period might have to be based on the average or perhaps the shortest instar (excluding very short first instars) development time.

12.4.2   *Life-table parameters*

Predictive models require as input information on the magnitude and variation of the development times of the different stages: a complete example of this type of data is given by Gomez *et al.* (1977) for *Culex.* When key-factor or similar analysis is to be undertaken, it is debatable whether the mortalities of the different stages detected under apparently 'Malthusian conditions' (e.g. Hardman, 1976a) should be included in the prediction or whether, as seems safer, the prediction should assume 100% survival. Complete models, including the adult stage will require age-specific fecundity schedules.

12.4.3   *Recruitment in the field*

A reliable estimate of recruitment is essential: depending on the form of the model, this could be a measure of adult emergence or daily or annual egg production or some combination of the two (see also Chapters 10 and 11). These models generally assume that age distribution within a stage, at any one time, is effectively uniform.

12.4.4   *Empirical models*

The egg ratio method was developed by Edmondson (1968) for the estimation of birth rate in zooplankton communities. If F_e is the number of eggs observed per female and D is the mean development time of the eggs in days, then the number of offspring produced per female per day, B, is:

$$B = \frac{F_e}{D}$$

If the number of females alive on day 1 is designated 1, then the number on day 2 will be $1 + B$. The instantaneous birth rate, b, is then given by the equation:

$$b = \ln(1 + B).$$

Thus, the basic inputs for this model are an observed recruitment value (or in this case its equivalent) and an experimentally determined life-table parameter. The instantaneous rate of population change, r, is

$$r = b - m$$

where m is the instantaneous mortality rate and the size of the population given by:

$$N_t = N_0 e^{rt}$$

where N_0 and N_t are the size of the population on times 0 and t, respectively.

Population growth in the absence of death ($r = b$) can be predicted and the real growth rate, as expressed by the field population, can be compared with this (see Caswell, 1972 for details for the calculation of b if mortality exists).

The model of Southwood *et al.* (1972) allowed variable hatching and developmental times to be incorporated into the prediction. The size of the egg cohort of a particular time period is known from field observation. It is also necessary to know the developmental spectra of the various stages (e.g. the shape of the hatching curve for eggs of the same age). If physiological time is not used, but developmental periods differ at different seasons, this variation may also be incorporated. The model calculates the contribution of each egg cohort to the numbers of subsequent stages and then sums the contributions from each cohort to predict the total population of that stage in that time interval. These predictions may be compared with actual data from field sampling to estimate mortality levels: it is assumed that at a particular time the members of an age group suffer mortality at random, irrespective of their cohort origin. Vandermeer (1975) has queried the validity of this assumption which is also basic for the Lewis–Leslie matrix model if the age classes are too wide.

Shiyomi (1967) developed a model for the reproduction of aphids, recognising that as aphids are not distributed randomly in space, reproduction per parent will be variable and will conform to the negative binomial. It enables a prediction to be made of the probability that a certain number of offspring will arise on a plant from a particular density of females. This model is noteworthy as an early example of a model that takes into account a spatial factor: subsequent research has demonstrated the importance of spatial variation in ecological models.

12.4.5 *Intrinsic rate models and variable life-tables*

Hughes (1962, 1963) laid the foundations of this approach when he showed that stable age distributions often occurred in expanding aphid populations; under these conditions the potential rate of population increase could be predicted by a model based on the intrinsic rate of increase in Lotka's population growth equation. The models have been extensively developed in collaboration with N. Gilbert (Hughes and Gilbert, 1968; Gilbert and Hughes, 1971) and colleagues (Gilbert & Gutierrez, 1973; Gilbert *et al.*, 1976). They have been termed 'variable life-tables' by Gilbert *et al.* (1976), who give a detailed account of the basic philosophy and the construction of the models. The variable life-table method involves detailed quantitative study of all the factors causing population change in the target species with the aid of simulation models. The variable life-table also owes much in its methodology to C.S. Holling's component analysis, for the approach is to identify the major components of population change, insert these in the model, test the model against the field data, examine the places where it does not fit (as Gilbert *et al.* (1976) stress, poor fits to the model are the most interesting results) and then to modify that part of the model in a biologically meaningful way and re-run, comparing the prediction against field data. The initial components of a variable life-table might be expected to include the age fecundity schedule, the fecundity density relationship, the development schedule (normally on a physiological time-scale), the impact of density on development, and mortality schedules in relation to various natural enemies. These methodologies can simulate overlapping generations and changes in environmental variables such as temperature.

Entomologists, particularly those working on aphid population dynamics and control, have driven research in this field. The study by Gutierrez & Ponti (2013) on the relative contributions of biological and weather-driven factors in the control of the spotted alfalfa aphid, *Therioaphis maculate*, demonstrates present modelling capability and also the depth of knowledge required for successful analysis. The relative contributions of each control factor were estimated using a weather-driven physiologically based demographic system model. Variables within the model were consisting of alfalfa, the spotted alfalfa aphid, three introduced parasitoids, a native coccinellid beetle, a fungal pathogen and host plant resistance, together with daily weather data for 142 locations in Arizona and California. The potentially controlling factors were introduced to the model singly and in various combinations, and it was concluded that all the factors were needed to produce control of the pest throughout the study region. The model also allowed the relative magnitude of mortality factors to be ranked.

12.4.6 *Lewis–Leslie matrices and R packages*

During the 1940s, E.G. Lewis and P.H. Leslie both independently developed a deterministic model in matrix form for the age structure of a reproducing population. The general form of these models, in matrix notation, is:

$$a_t M = a_{t+1}$$

where M is a square matrix that describes the transition of the population over one time period and thus contains terms describing age-specific fecundities and survival rates, a_t is a column vector of the age structure of the population at time t (the start) and a_{t+1} is the column vector at time $t + 1$. Set out in full, it is:

$$\begin{bmatrix} f_0 & f_1 & f_3 & f_4 \\ P_0 & 0 & 0 & 0 \\ 0 & p_1 & 0 & 0 \\ 0 & 0 & p_3 & 0 \end{bmatrix} x \begin{bmatrix} a_0 \\ a_1 \\ a_2 \\ a_3 \end{bmatrix} = \begin{bmatrix} f_0 a_0 & f_1 a_1 & f_2 a_2 & f_3 a_3 \\ 0^{p_0 a_0} & 0 & 0 & 0 \\ 0 & p_1 a_1 & 0 & 0 \\ 0 & 0 & p_2 a_2 & 0 \end{bmatrix}$$

where the *f*-values represent the age-specific fecundity of each age class (0, 1, 2 and 3) as expressed in daughters born between t and $t + 1$ and $p_0 \ldots p_3$ represents the probability of a female (of the age indicated by the subscript) surviving into the next age class. It will be noted that these probabilities will always form a diagonal that commences at the left-hand end of the second row and ends (reaches the last row) in the penultimate column. In practice, f_0 is normally zero. The column vector $(a_0, \ldots a_3)$ represents the number of animals at each age group $i = 0$ to $t = 3$ at time t.

From any starting vector a_t the future population can be calculated by repeated multiplication so that after k time periods

$$a_{t+k} = M^k a_t$$

As k increases the age distribution gradually becomes constant. It can be shown that the long-term rate of growth of the population, λ (the per capita rate of increase, $r = \ln \lambda$, see p. 458) and the stationary (stable) age distribution are given by the dominant eigenvector and eigenvalue of matrix M. The population will increase in size when $\lambda > 1$. Eigenvalues and vectors are obtained by iteration using standard computer packages. Lucid accounts of the Lewis–Leslie matrix and its development are given by Williamson (1967, 1972) and Usher (1972).

Population growth from a starting vector considerably different from the stationary age distribution can show short-term oscillations as the population age structure changes. The magnitude of these oscillations is determined by the relative magnitudes of the eigenvalues of matrix M, which has led to the suggestion that they may be compared to assess the relative stability of different populations. In practice, because of natural variation in age-specific fecundities and survivals and, in particular, the introduction of non-linear features, such as density-dependence, at high population values, these approaches are probably of little value.

The male sex was introduced into the model by Williamson (1959) by doubling the number of rows and columns, whilst Pennycuick et al. (1968) produced a computer model incorporating density-dependence. They made the elements in the matrix M dependent on the size of the population (the sum of the appropriate vector column) by multiplying the values in the top row by a function F, and the p-values (in the line below the diagonal) by another function S. F and S are expressions that modify fecundity and survival to give sigmoid curves; they combine the present size of the population, the equilibrium population and the degree of density-dependence in a rather complex manner (May et al., 1974). The problem with such density-dependent models is that the eigenvalues and vectors can no longer be used to predict long-term population behaviour because the schedule of births and deaths is no longer constant.

A requirement of the Lewis–Leslie matrix is that the time intervals are exactly equal, yet insects are generally age-grouped by stages which are of unequal length. Lefkovitch (1965) considered this problem and describes a method of stage grouping taking account of development times and the influence of one stage on another. Many elements in the matrix will contain terms for more than one stage, and it is clear that unless a stable age distribution is reached the survival probabilities will change as the proportions of the stages in the elements changes. Vandermeer (1975) emphasises the need to make age categories as brief as possible, so that the assumption that all individuals within each category can be assumed to have the same ages (and survival probabilities) can be used without introducing a major source of error.

Longstaff (1977) has utilised these techniques to study the population dynamics of Collembola in culture. Besides deterministic models, he developed a stochastic model. The assumption was made that the parameter values were normally distributed around the mean and the actual value was selected by a random number generator. The accuracy of the deterministic and stochastic models varied, but the predictors were sufficiently close to the form of the actual populations for it to seem likely that the population parameters derived from the models (the density-dependent function and the equilibrium population) are meaningful.

The Lewis–Leslie Matrix approach has often been used in applied research on fish and mammals to gain insight into the response of a population to changes in age- (size-) selective mortality. Smith and Trout (1994) used this approach to model the effect of different rabbit control policies.

A concept which has proven useful, particularly in conservation studies, is the elasticity of the Leslie matrix. This measures the change in population growth, λ, for a small change in one of the matrix elements a_{ij}. The eigenvalues elasticity is given by:

$$e_{ij} = \frac{\partial log\,\lambda}{\partial log a_{ij}}.$$

Elasticity is used to identify critical phases in the life cycle that should be targeted by management, and to quantify how the population growth rate will respond to a perturbation to a particular part of the life cycle, such as juvenile survival or adult survival. For example, Richard *et al.* (2014) generated Leslie matrices for pairs of years in a study of feral horse dynamics on a Canadian island and found that changes in female adult survival had the greatest impact on population growth (greatest elasticity). A meta-analysis was undertaken by Van de Kerk *et al.* (2013), who studied 38 Leslie matrix models for Carnivora and examined the elasticity patterns amongst species and their correlations with life history characteristics such as size and life span as an aid to the development of management plans.

A number of useful computer packages are available in R for the analysis of matrix models. The R package popbio (Stubben & Milligan, 2007) offers basic functions for deterministic and stochastic matrix models to calculate long-term population growth rate, sensitivities and elasticities to changes in matrix elements. For the analysis of transient dynamics over the short term, Stott *et al.* (2012) developed the R package popdemo. If a discrete time step model is unsuitable, the R package IPMpack (Metcalf *et al.*, 2013) models demographic processes using continuous functions.

An example of a simple analysis of a population projection matrix, including the calculation of elasticity using popbio, is given below and the graphical output produced is shown in Fig. 12.2.

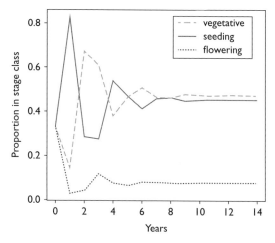

Figure 12.2 Output generated by the R package popbio using the function stage.vector.plot (p$stage.vectors, col=2:4). The code is given on p. 484.

```
library(popbio) #Open library
# Enter population projection matrix
# mean matrix from Freville et al., 2004
stages<-c("seedling", "vegetative", "flowering")
A<-matrix(c(
0, 0, 5.905,
0.368, 0.639, 0.025,
0.001, 0.152, 0.051
), nrow=3, byrow=TRUE,
dimnames=list(stages,stages)
)
n<-c(5,5,5)
#Calculate the population growth rate using 15 time steps
p<-pop.projection (A,n, 15)
#output results
p
#Calculate the damping ratio for the matrix
damping.ratio(A)
#calculates elasticity of the projection matrix
elast<-elasticity(A)
elast
#plots short-term dynamics
stage.vector.plot(p$stage.vectors, col=2:4)
```

References

Achionye Nzeh, C.G. (1994) Age and growth determination in *Sarotherodon galilaeus* by use of scales and opercular bones (Pisces: Cichlidae). *Revista de Biologia Tropical* **42**(1–2), 371–4.

Al Hassan, L.A.J. & Al Sayab, A.A. (1994) Eye lens diameter as an age indicator in the catfish, *Silurus triostegus*. *Pakistan J. Zool.* **26**(1), 81–2.

Allen, J.C. (1976) A modified sine wave method for calculating degree days. *Environ. Entomol.* **5**, 388–96.

Almeida, P.R., Moreira, F.M., Dimingos, I.M., Costa, J.L., Assis, C.A., & Costa, M.J. (1995) Age and growth of *Liza ramada* (Risso, 1826) in the River Tagus, Portugal. *Scientia Marina* **59**(2), 143–7.

Alvarado-Castillo, R.M. & Felix-Uraga, R. (1996) Determinación de la edad de *Istiophorus platypterus* (Pisces: Istiophoridae) al sur del Golfo de California, México. *Revista de Biologia Tropical* **44**(1), 233–9.

Ancrenaz, M. & Delhomme, A. (1997) Teeth eruption as a means of age determination in captive Arabian oryx, *Oryxleucoryx* (Bovidae, Hippotraginae). *Mammalia* **61**(1), 135–8.

Anderson, J.L. (1986) Age determination of the nyala *Tragelaphus angasi*. *South African J. Wildlife Res.* **16**(3), 82–90.

Anderson, J.R. (1964) Methods for distinguishing nulliparous from parous flies and for estimating the ages of *Fannia canicularis* and some other cyclorrophous Diptera. *Ann. Entomol.
Soc. Am.* **57**, 226–36.

Ando, A., Yamada, F., Taniguchi, A., & Shiraishi, S. (1992) Age determination by the eye lens
weight in the Japanese hare, *Lepus brachyurus brachyurus* and its application to two local
populations. *Sci. Bull. Faculty Agricult. Kyushu University* **46**(3–4), 169–75.

Antunes, C. & Tesch, F.W. (1997) A critical consideration of the metamorphosis zone when
identifying daily rings in otoliths of European eel, *Anguilla anguilla (L.). Ecol. Freshwater
Fish* **6**(2), 102–7.

Anwar, N.A., Richardson, C.A., & Seed, R. (1990) Age determination, growth rate and population structure of the horse mussel *Modiolus modiolus. J. Mar. Biol. Assoc. UK* **70**(2), 441–58.

Aprahamian, M.W. (1987) Use of the burning technique for age determination in eels
Anguilla anguilla (L.) derived from the stockings of elvers. *Fish. Res.* **6**(1), 93–6.

Atkinson, P.R. (1977) Preliminary analysis of a field population of citrus red scale, *Aonidiella
aurontii* (Maskell) and the measurement and expression of stage duration and reproduction for life tables. *Bull. Entomol. Res.* **67**, 65–87.

Baker, G.H. & Vogelzang, B.K. (1988) Life history, population dynamics and polymorphism
of *Theba pisana* (Mollusca: Helicidae) in Australia. *J. Appl. Ecol.* **25**, 867–87.

Barroso, C.M., Nunes, M., Richardson, C.A., & Moreira, M.H. (2005) The gastropod statolith:
a tool for determining the age of *Nassarius reticulatus. Mar. Biol.* **146**, 1139–44.

Beamish, F.W.H. & Medland, T.E. (1988) Age determination for lampreys. *Trans. Am. Fish.
Soc.* **117**(1), 63–71.

Berg, R. (1985) Age determination of eels, *Anguilla anguilla*: Comparison of field data with
otolith ring patterns. *J. Fish Biol.* **26**(5), 537–44.

Berman, M.S., McVey, A.L., & Ettershank, G. (1989) Age determination of Antarctic krill
using fluorescence and image analysis of size. *Polar Biol.* **9**(4), 267–72.

Beverton, R.J.H. & Holt, S.J. (1957) *On the dynamics of exploited fish populations.* Ministry of
Agriculture, Fisheries and Food, Gt Britain, HMSO, London.

Bianchini, J.J. & Delupi, L.H. (1990) Age determination of pampas deer (*Odocoileus bezoarti-
cus*) by comparative study of tooth development and wear. *Physis Seccion C los Continentes
y los Organismos Terrestres* **48**(114–115), 27–40.

Blouin, M.A. & Hall, G.R. (1990) Improved method for sectioning pectoral spines of catfish
for age determination. *J. Freshwater Ecol.* **5**(4), 489–90.

Brennan, J.S. & Cailliet, G.M. (1989) Comparative age-determination techniques for white
sturgeon in California (USA). *Trans. Am. Fish. Soc.* **118**(3), 296–310.

Britaev, T.A. & Belov, V.V. (1993) Age determination in polynoid polychaetes, using growth
rings on their jaws. *Zoologicheskii Zhurnal* **72**(11), 15–21.

Brown, H.A. (1990) Morphological variation and age-class determination in overwintering
tadpoles of the tailed frog, *Ascaphus truei. J. Zool.* **220**(2), 171–84.

Cailliet, G.M., Botsford, L.W., Brittnacher, J.G., Ford, G., Matsubayashi, M., King, A., Watters,
D.L., & Kope, R.G. (1996) Development of a computer-aided age determination system:
Evaluation based on otoliths of bank rockfish off California. *Trans. Am. Fish. Soc.* **125**(6),
874–88.

Camin, V., Baker, P., Carey, J., Valenzuela, J., & Arredondo, P.R. (1991) Biochemical age determination for adult Mediterranean fruit flies (Diptera: Tephritidae). *J. Econ. Entomol.* **84**(4),
1283–8.

Castanet, J. (1986) Skeletochronology of reptiles: III. Application. *Annales des Sciences
Naturelles Zoologie et Biologie Animale* **8**(3), 157–72.

Caswell, H. (1972) On instantaneous and finite birth rates. *Limnol. Oceanogr.* **17**, 787–91.

Cavallini, P. & Santini, S. (1995) Age determination in the red fox in a Mediterranean habitat. *Z. für Saeugetierkunde* **60**(3), 136–42.

Chae, H.Y. & Fujimaki, Y. (1996) Age determination of the hazel grouse based on skull pneumatization. *Jpn. J. Ornithol.* **45**(1), 17–22.

Chai, F.-C. & Dixon, S.E. (1971) A technique for ageing cocoons of the sawfly *Neodiprion serlifer* (Hymenoptera: diprionidae). *Can. Entomol.* **103**, 80–3.

Christophers, S.R. (1960) *Aedes aegypti (L.) The yellow fever mosquito.* Cambridge University Press, Cambridge.

Corbet, P.S. (1960) The recognition of nulliparous mosquitoes without dissection. *Nature* **187**, 525–6.

Corbet, P.S. (1961) The recognition of parous dragonflies (Odonata) by the presence of follicular relics. *Entomologist* **94**, 35–7.

Corbet, P.S. (1962a) Age-determination of adult dragonflies (Odonata). *Proc. XI Int. Congr. Entomol.* **3**, 287–9.

Corbet, P.S. (1962b) *A Biology of Dragonflies.* Witherby, London.

Corbet, P.S. (1963) Reliability of parasitic water mites (Hydracarina) as indicators of physiological age in mosquitoes (Diptera: Culicidae). *Entomol. Exp. Appl.* **6**, 215–33.

Corbet, P.S. (1970) The use of parasitic water-mites for age-grading female mosquitoes. *Mosquito News* **30**, 436–8.

Correia, J.P. & Figueiredo, I.M. (1997) A modified decalcification technique for enhancing growth bands in deep-coned vertebrae of elasmobranchs. *Environ. Biol. Fishes* **50**(2), 225–30.

Daly, H.V. (1961) Biological observations on *Hemihalictus lustrans*, with a description of the larva (Hymenoptera: Halictidae). *J. Kansas Entomol. Soc.* **34**, 134–41.

Davies, J.B., Corbet, P.S., Gillies, M.T., & McCrae, A.W.R. (1971) Parous rates in some Amazonian mosquitoes collected by three different methods. *Bull. Entomol. Res.* **61**, 125–32.

Dery, L.M. (1983) Use of laminated plastic to impress fish scales. *Progressive Fish Culturist* **45**, 88–9.

Detinova, T.S. (1962) Age-grouping methods in Diptera of medical importance with special reference to some vectors of malaria. *Monograph Series, World Health Organization* **47**, 13 pp.

Detinova, T.S. (1968) Age structure of insect populations of medical importance. *Annu. Rev. Entomol.* **13**, 427–50.

Drew, R.A.L. (1969) Morphology of the reproductive system of *Strumeta tryoni* (Froggatt) (Diptera: Trypetidae) with a method of distinguishing sexually mature males. *J. Aust. Entomol. Soc.* **8**, 21–32.

Dunn, E. (1951) Wing coloration as a means of determining the age of the Colorado beetle (*Leptinotarsa decemlineta* Say). *Ann. Appl. Biol.* **38**, 433–4.

Dyce, A.L. (1969) The recognition of multiparous and parous Culicoides without dissection. *J. Aust. Entomol. Soc.* **8**, 11–15.

Ebert, T.A. (1973) Estimating growth and mortality rates from size data. *Oecologia (Berl.)* **11**, 281–98.

Edmondson, W.T. (1968) A graphical model for evaluating the use of the egg ratio for measuring birth and death rates. *Oecologia* **1**, 1–37.

Emden, H.F.V., Jepson, W.F., & Southwood, T.R.E. (1961) The occurrence of a partial fourth generation of *Oscinel afrit* L. (Diptera: Chloropidae) in southern England. *Entomol. Exp. Appl.* **4**, 220–5.

Esteves, E., Simoes, P., Da Silva, H.M., & Andrade, J.P. (1995) Ageing of swordfish, *Xiphias gladius* Linnaeus, 1758, from the Azores, using sagittae, anal-fin spines and vertebrae. *Arquipelago Boletim da Universidade dos Acores Ciencias Biologicas e Marinhas* **13A**, 39–51.

Fandos, P., Orueta, J.F., & Aranda, Y. (1993) Tooth wear and its relation to kind of food: The repercussion on age criteria in *Capra pyrenaica*. *Acta Theriol.* **38**(1), 93–102.

Fedetov, D.M. (1947, 1955, 1960) The noxious little tortoise, *Eurygaster integriceps* Put. [In Russian.] 1 (272 pp.) and 2 (1947); 3 (278 pp.) (1955); 4 (239 pp.) (1960).

Fornasari, L., Pianezza, F., & Carabella, M. (1994) Criteria for the age determination of the Hawfinch *Coccothraustes coccothraustes*. *Ringing Migration* **15**(1), 50–5.

Francillon, H., Barbautl, R., Castanet, J., & De Ricqles, A. (1984) The biology of the desert toad *Bufo pentoni*: Age determination by the skeletochronological method, population structure and dynamics. *Revue D'Ecologie la Terre et la Vie* **39**(2), 209–24.

Francillon Viellot, H., Arntzen, J.W., & Geraudie, J. (1990) Age, growth and longevity of sympatric *Triturus cristatus*, *Triturus marmoratus* and their hybrids (Amphibia, Urodela): A skeletochronological comparison. *J. Herpetol.* **24**(1), 13–22.

Freville, H., Colas, B., Riba, M., Caswell, H., Mignot, A., Imbert, E., & Olivieri, I. (2004) Spatial and temporal demographic variability in the endemic plant species *Centaurea corymbosa* (Asteraceae). *Ecology* **85**, 694–703.

Galinou-Mitsoudi, S. & Sinis, A.I. (1995) Age and growth of *Lithophaga lithophaga* (Linnaeus, 1758) (Bivalvia: Mytilidae), based on annual growth lines in the shell. *J. Mollusc. Stud.* **61**(4), 435–53.

Garcia, F. (1993) Interpretation of shell marks for the estimation of growth of the European carpet clam *Ruditapes decussatus* L. of the Bay of Fos (Mediterranean Sea). *Oceanol. Acta* **16**(2), 199–203.

Gharzi, A. & Yari, A. (2013) Age determination in the Snake-eyed Lizard, *Ophisops elegans*, by means of skeletochronology (Reptilia: Lacertidae). *Zoology in the Middle East* **59**, 10–15.

Giglioli, M.E.C. (1963) Aids to ovarian dissection for age determination in mosquitoes. *Mosquito News* **23**, 156–9.

Gilbert, N. & Gutierrez, A.P. (1973) A plant–aphid–parasite relationship. *J. Anim. Ecol.* **42**, 323–40.

Gilbert, N., Gutierrez, A.P., Frazer, B.D., & Jones, R.E. (1976) *Ecological Relationships*. Addison-Wesley, Reading, San Francisco.

Gilbert, N. & Hughes, R.D. (1971) A model of an aphid population-three adventures. *J. Anim. Ecol.* **40**, 525–34.

Gillett, J.D. (1957) Age analysis of the biting cycle of the mosquito *Taeniorhynchus (Mansonioides) africana* Theobald, based on the presence of parasitic mites. *Ann. Trop. Med. Hyg.* **51**, 151–8.

Goddard, H.N. & Reynolds, J.C. (1993) Age determination in the red fox (*Vulpes vulpes* L.) from tooth cementum lines. *Gibier Faune Sauvage* **10**, 173–87.

Gomez, C., Rabinovich, J.E., & Machado-Allison, C.E. (1977) Population analysis of *Culex pipiensfatigans* Wielt. (Diptera: Culicidae) under laboratory conditions. *J. Med. Entomol.* **13**, 453–63.

Goodyear, C.P. (1997) Fish age determined from length: An evaluation of three methods using simulated red snapper data. *Fish. Bull.* **95**(1), 39–46.

Gordo, L.S. (1996) On the age and growth of bogue, *Boops boops* (L.), from the Portuguese coast. *Fish. Manag. Ecol.* **3**(2), 157–64.

Grana-Raffucci, F.A. & Appeldoorn, R.S. (1997) Age determination of larval strombid gastropods by means of growth increment counts in statoliths. *Fish. Bull.* **95**(4), 857–62.

Graves, G.R. (1997) Age determination of free-living male black-throated blue warblers during the breeding season. *J. Field Ornithol.* **68**(3), 443–9.

Gutierrez, A.P. & Ponti, L. (2013) Deconstructing the control of the spotted alfalfa aphid *Therioaphis maculata*. *Agric. Forest Entomol.* **15**, 272–84.

Hales, L.S.J. & Belk, M.C. (1992) Validation of otolith annuli of bluegills in a southeastern thermal reservoir. *Trans. Am. Fish. Soc.* **121**(6), 823–30.

Hamon, J., Grjedine, A., Adam, J.P., Chauvet, G., Coz, I., & Gruchet, H. (1961) Les méthodes devaluation de l'age physiologique des rnoustiques. *Bull. Soc. Entomol. Fr.* **669**, 137–61.

Hanson, W.R. (1963) Calculation of productivity, survival, and abundance of selected vertebrates from sex and age ratios. *Wildlife Monogr.* **9**, l–60.

Hardman, J.M. (1976) Life-table data for use in deterministic and stochastic simulation models predicting the growth of insect populations under Malthusian conditions. *Can. Entomol.* **108**, 897–906.

Hartman, G. (1992) Age determination of live beaver by dental X-ray. *Wildlife Soc. Bull.* **20**(2), 216–20.

Hartman, G.D. (1995) Age determination, age structure, and longevity in the mole, *Scalopus aquaticus* (Mammalia: Insectivora). *J. Zool.* **237**(1), 107–22.

Hassager, T.K. (1991) Comparison of three different otolith-based methods for age determination of turbot (*Scophthalmus maximus*). *Dana* **9**, 39–43.

Haydak, M.H. (1957) Changes with age in the appearance of some internal organs of the honeybee. *Bee World* **38**, 197–207.

Heise, S., Stubbe, M., Wieland, H., & Nessen, B. (1991) The analysis of the population structure of the common vole (*Microtus arvalis*, Pallas, 1779). *Z. für Angew. Zool.* **78**(1), 19–30.

Helldin, J.O. (1997) Age determination of Eurasian pine martens by radiographs of teeth in situ. *Wildlife Soc. Bull.* **25**(1), 83–8.

Henocque, Y. (1987) Age determination of the European lobster, *Homarus gammarus*, from its antennules. *Mer* **25**(1), 1–12.

Hesthagen, T. (1985) Validity of the age determination from scales of brown trout (*Salmo trutta* L.). *Inst. Freshwater Res. Drottingholm Rep.* **62**, 65–70.

Hoc, T.Q. & Charlwood, J.D. (1990) Age determination of *Aedes cantans* using the ovarian oil injection technique. *Med. Vet. Entomol.* **4**(2), 227–33.

Hoc, T.Q. & Wilkes, T.J. (1995) Age determination in the blackfly *Simulium woodi*, a vector onchocerciasis in Tanzania. *Med. Vet. Entomol.* **9**(1), 16–24.

Hohman, W.L. & Cypher, B.L. (1986) Age-class determination of ring-necked ducks (*Aythya collaris*). *J. Wildlife Manag.* **50**(3), 442–5.

Hughes, R.D. (1962) A method for estimating the effects of mortality on aphid populations. *J. Anim. Ecol.* **31**, 389–96.

Hughes, R.D. (1963) Population dynamics of the cabbage aphid, *Brevicoryne brassicae* (L.). *J. Anim. Ecol.* **32**, 393–424.

Hughes, R.D. & Gilbert, N. (1968) A model of an aphid population – a general statement. *J. Anim. Ecol.* **37**, 553–63.

Hutton, J.M. (1986) Age determination of living Nile crocodiles (*Crocodylus niloticus*) from the cortical stratification of bone. *Copeia* **2**, 332–41.

Iason, G.R. (1988) Age determination of mountain hares (*Lepus timidus*): A rapid method and when to use it. *J. Appl. Ecol.* **25**(2), 389–96.

Jean, Y., Bergeron, J.M., Bisson, S., & Larocque, B. (1985) Relative age determination of coyotes, *Canis latrans*, from southern Quebec (Canada). *Can. Field Nat.* **100**(4), 483–7.

Jones, M.G. (1971) Observations on changes in the female reproductive system of the wheat bulb fly, *Leptohylemyia coarctata* (Fall.). *Bull. Entomol. Res.* **61**, 55–68.

Jones, V.P., Tome, C.H.M., & Caprio, L.C. (1997) Life-tables for the koa seedworm (Lepidoptera: Tortricidae) based on degree-day demography. *Econom. Entomol.* **26**(6), 1291–8.

Kardos, E.H. & Bellamy, R.E. (1962) Distinguishing nulliparous from parous female *Culex larsalis* by examination of the ovarian tracheation. *Ann. Entomol. Soc Am.* **54**, 448–51.

Kay, R.F. & Cant, J.G.H. (1988) Age assessment using cementum annulus counts and tooth wear in a free-ranging population of *Macaca mulatta. Am. J. Primatol.* **15**(1), 1–16.

Kelker, G.H. & Hanson, W.R. (1964) Simplifying the calculation of differential survival of age-classes. *J. Wildlife Manag.* **28**, 411.

Kilada, R., Sainte-Marie, B., Rochette, R., Davis, N., Vanier, C., & Campana, S. (2012) Direct determination of age in shrimps, crabs, and lobsters. *Can. J. Fish. Aquat. Sci.* **69**, 1728–33.

Kishi, Y. (1971) Reconsideration of the method to measure the larval instars by use of the frequency distribution of head-capsule widths or lengths. *Can. Entomol.* **103**, 101–15.

Klomp, N.I. & Furness, R.W. (1992) A technique which may allow accurate determination of the age of adult birds. *Ibis* **134**(3), 245–9.

Kosminskii, R.B. (1960) The method of determining the age of the fleas *Leptopsylla segnis* Sch6nh. 1811 and *L. taschenbergi* Wagn. 1898 (Suctoria-Aphaniptera) and an experiment on the age analysis of a population of *L. Segnis*. [In Russian.]. *Med. Parazitol.* **29**, 590–4.

Kunitskaya, N.T. (1960) On the reproductive organs of female fleas and determination of their physiological age. [In Russian]. *Med. Parazitol.* **29**, 688–701.

Lagace, C.F. & Van Den Bosch, R. (1964) Progressive sclerotization and melanization of certain structures in males of a field population of *Hypera brunneipennis* (Coleoptera: Curculionidae). *Ann. Entomol. Soc. Am.* **57**, 247–52.

Lai, H.L., Gunderson, D.R., & Low, L.L. (1987) Age determination of Pacific cod, *Gadus macrocephalus*, using five ageing methods. *US Natl Mar. Fish. Serv. Fish. Bull.* **85**(4), 713–24.

Le Berre, R. (1966) *Contribution a l'étude biologique et écologique de* Simulium damnosium *(Diptera, Simuliidae)*. Paris, Mémoires ORSTOM.

Le Foll, D., Brichet, E., Reyss, J.L., Lalou, C., & Latrouite, D. (1989) Age determination of the spider crab *Maja squinado* and the European lobster *Homarus gammarus* by thorium-228 radium-228 chronology: Possible extension to other crustaceans. *Can. J. Fish. Aquat. Sci.* **46**(4), 720–4.

Lefkovitch, L.P. (1965) The study of population growth in organisms grouped by stages. *Biometrics* **21**, 1–18.

Lewis, D.J., Lainson, R., & Shaw, J.J. (1970) Determination of parous rates in phlebotomine sandflies with special reference to Amazonian species. *Bull. Entomol. Res.* **60**, 209–19.

Linley, J.R. (1965) The ovarian cycle and egg stage in *Leptoconops* (*Holoconops*) *becquaerti* (Kiefr.) (Diptera, Ceratopogaoidae). *Bull. Entomol. Res.* **56**, 37–56.

Longstaff, B.C. (1977) The dynamics of collembolan populations: a matrix model of single species population growth. *Can. J. Zool.* **55**, 314–24.

Loubens, G. & Panfili, J. (1992) Age determination of *Prochilodus nigricans* (Teleostei, Prochilodidae) in Beni (Bolivia): Setting of a procedure and application. *Aquatic Living Resources* **5**(1), 41–56.

Mann, R.H.K. & Steinmetz, B. (1985) On the accuracy of age-determination using scales from rudd, *Scardinius erythrophthalmus* of known age. *J. Fish Biol.* **26**(5), 621–8.

Marmontel, M., Nhea, T.J., Kochman, H.I., & Humphrey, S.R. (1996) Age determination in manatees using growth-layer-group counts in bone. *Mar. Mammal Sci.* **12**(1), 54–88.

Martin, H. (1996) Age determination in adults by analysis of dental cementum: Methods and limits. *Bull. Mem. Soc. d'Anthropologie de Paris* **8**(3–4), 433–40.

Mashkin, V.I. & Kolesnikov, V.V. (1990) Age determination of marmots (Marmota, Sciuridae) by the pattern of the wearing down of the chewing surface of teeth. *Zoologicheskii Zh.* **69**(6), 124–31.

Mason, D.R. (1984) Dentition and age determination of the warthog, *Phacochoerus aethiopicus*, in Zululand, South Africa. *Koedoe* **27**, 79–120.

May, R.M., Conway, G.R., Hassell, M.P., & Southwood, T.R.E. (1974) Time delays, density dependence and single-species oscillations. *J. Anim. Ecol.* **43**, 747–70.

Metcalf, C.J.E., McMahon, S.M., Salguero-Gómez, R., & Jongejans, E. (2013) IPMpack: an R package for integral projection models. *Methods Ecol. Evol.* **4**, 195–200.

Michener, C.D., Cross, E.A., Daly, H.V., Rettenmeyer, C.W., & Wille, A. (1955) Additional techniques for studying the behaviour of wild bees. *Insectes Sociaux* **2**, 237–46.

Michener, C.D. & Lange, R.B. (1959) Observations on the behaviour of Brazilian Halicted bees (Hymenoptera, Apoidea). IV. Augochloropsis, with notes on extralimital forms. *Am. Mus. Noritates* **1924**, 1–41.

Mochizuki, A., Shiga, M., & Imura, O. (1993) Pteridine accumulation for age determination in the melon fly, *Bactrocera* (*Zeugodacus*) *cucurbitae* (Coquillett) (Diptera: Tephritidae). *Appl. Entomol. Zool.* **28**(4), 584–6.

Moore, N.P., Cahill, J.P., Kelly, P.F., & Hayden, T.J. (1995) An assessment of five methods of age determination in an enclosed population of fallow deer (*Dama dama*). *Biol. Environ.* **95B**(1), 27–34.

Moreira, F., Costa, J.L., Almeida, P.R., Assis, C., & Costa, M.J. (1991) Age determination in *Pomatoschistus minutus* (Pallas) and *Pomatoschistus microps* (Kroyer) (Pisces: Gobiidae) from the upper Tagus estuary, Portugal. *J. Fish Biol.* **39**(3), 433–40.

Msangi, A. & Lehane, M.J. (1991) A method for determining the age of very young tsetse flies (Diptera: Glossinidae) and an investigation of the factors determining head fluorescent levels in newly emerged adults. *Bull. Entomol. Res.* **81**(2), 185–8.

Muirhead-Thomson, R.C. (1963) Practical entomology in malaria eradication. WHO (MHO/PA/62.63). Part I: 64–71.

Mullens, B.A. & Lehane, M.J. (1995) Fluorescence as a tool for age determination in *Culicoides variipennis sonorensis* (Diptera: Ceratopogonidae). *J. Med. Entomol.* **32**(4), 569–71.

Narimatsu, Y. & Munehara, H. (1997) Age determination and growth from otolith daily growth increments of *Hypoptychus dybowskii* (Gasterosteiformes). *Fish. Sci.* **63**(4), 503–8.

Nassar Montoya, F., Sainsbury, A.W., Kirkwood, J.K., & Du Boulay, G.H. (1992) Age determination of common marmosets (*Callithrix jacchus*) by radiographic examination of skeletal development. *J. Med. Primatol.* **21**(5), 259–64.

Neville, A.C. (1963) Daily growth layers in locust rubber-like cuticle, influenced by an external rhythm. *J. Insect Physiol.* **9**, 177–86.

Neville, A.C. (1970) Cuticle ultrastructure in relation to the whole insect. In: Neville, A.C. (ed.), *Insect Ultrastructure*. Symp. R. Entomol. Soc. Lond. **5**, 17–39.

Nishi, E. & Nishihira, M. (1996) Age-estimation of the Christmas tree worm *Spirobranchus giganteus* (Polychaeta, Serpulidae) living buried in the coral skeleton from the coral-growth band of the host coral. *Fish. Sci.* **62**(3), 400–3.

Officer, R.A., Gason, A.S., Walker, T.I., & Clement, J.G. (1996) Sources of variation in counts of growth increments in vertebrae from gummy shark, *Mustelus antarcticus*, and school shark, *Galeorhinus galeus*: Implications for age determination. *Can. J. Fish. Aquat. Sci.* **53**(8), 1765–77.

Ohtaishi, N., Kaji, K., Miura, S., & Wu, J. (1990) Age determination of the white-lipped deer *Cervus albirostris* by dental cementum and molar wear. *J. Mammal. Soc. Japan* **15**(1), 15–24.

Olsen, M.T., Robbins, J., Bérubé, M., Rew, M.B., & Palsbøll, P.J. (2014) Utility of telomere length measurements for age determination of humpback whales. *NAMMCO Scientific Publications*, **10**.

Oxenford, H.A., Hunte, W., Deane, R., & Campana, S.E. (1994) Otolith age validation and growth-rate variation in flyingfish (*Hirundichthys affinis*) from the eastern Caribbean. *Mar. Biol.* **118**(4), 585–92.

Paris, O.H. & Pitelka, A. (1962) Population characteristics of the terrestrial isopod *Armadillidium vulgare* in California grassland. *Ecology* **43**, 229–48.

Parsons, G.J., Robinson, S.M.C., Roff, J.C., & Dadswell, M.J. (1993) Daily growth rates as indicated by valve ridges in postlarval giant scallop (*Placopecten magellanicus*) (Bivalvia: Pectinidae). *Can. J. Fish. Aquat. Sci.* **50**(3), 456–64.

Parsons, G.R. (1993) Age determination and growth of the bonnethead shark *Sphyrna tiburo*: A comparison of two populations. *Mar. Biol.* **117**(1), 23–31.

Pascal, M., Damange, J.P., Douville, P., & Guedon, G. (1988) Age determination in pine vole *Pitymys duodecimcostatus* (De Selys-Longchamps, 1839). *Mammalia* **52**(1), 85–92.

Pennycuick, C.J., Compton, R.M., & Beckingham, L. (1968) A computer model for simulating the growth of a population, or of two interacting populations. *J. Theor. Biol.* **18**, 316–29.

Polat, N. & Gumus, A. (1995) Age determination of spiny dogfish (*Squalus acanthias* L. 1758) in Black Sea waters. *Israeli J. Aquacult. Bamidgeh* **47**(1), 17–24.

Pollock, J. (1969) Test for the mated status of male sheep blowflies. *Nature* **223**, 1287–8.

Quere, J.P. & Vincent, J.P. (1989) Age determination in wood mice (*Apodemus sylvaticus* L., 1758) by dry weight of the lens. *Mammalia* **53**(2), 287–94.

Raboud, C. (1986) Age determination of *Arianta arbustorum* (L.) (Pulmonata) based on growth breaks and inner layers. *J. Mollusc. Stud.* **52**(3), 243–7.

Raya, C.P., Fernandez Nunez, M., Balguerias, E., & Hernandez Gonzalez, C.L. (1994) Progress towards ageing cuttlefish *Sepia hierredda* from the northwestern African coast using statoliths. *Mar. Ecol. Prog. Ser.* **114**(1–2), 139–47.

Richard, E., Simpson, S.E., Medill, S.A., & McLoughlin, P.D. (2014) Interacting effects of age, density, and weather on survival and current reproduction for a large mammal. *Ecol. Evol.* **4**, 3851–60.

Richardson, C.A. (1987) Microgrowth patterns in the shell of the Malaysian cockle *Anadara granosa* (L.) and their use in age determination. *J. Exp. Mar. Biol. Ecol.* **111**(1), 77–98.

Richardson, C.A., Collis, S.A., Ekaratne, K., Dare, P., & Key, D. (1993) The age determination and growth rate of the European flat oyster, *Ostrea edulis*, in British waters determined from acetate peels of umbo growth lines. *Ices* **50**(4), 493–500.

Richardson, C.A., Seed, R., & Naylor, E. (1990) Use of internal growth bands for measuring individual and population growth rates in *Mytilus edulis* from offshore production platforms. *Mar. Ecol. Prog. Ser.* **66**(3), 259–65.

Richter, H. & McDermott, J.G. (1990) The staining of fish otoliths for age determination. *J. Fish Biol.* **36**(5), 773–80.

Ricker, W.E. (1944) Further notes on fishing mortality and effort. *Copeia* **1944**, 23–44.

Ricker, W.E. (1948) Methods of estimating vital statistics of fish populations. *Indiana Univ. Publ. Sci. Ser.* **15**, 101.

Rodhouse, P.G. & Hatfield, E.M.C. (1990) Age determination in squid using statolith growth increments. *Fish. Res.* **8**(4), 323–34.

Rosay, B. (1961) Anatomical indicators for assessing the age of mosquitoes: the teneral adult (Diptera: Culicidae). *Ann. Entomol. Soc. Am.* **54**, 526–9.

Rowe, F.P., Bradfield, A., Quy, R.J., & Swinney, J. (1985) The relationship between lens weight and age in the wild house mouse (*Mus musculus*). *J. Appl. Ecol.* **22**, 55–61.

Saunders, D.S. (1962) Age determination for female tsetse flies and the age composition of samples of *Glossina pallidipes* Aust., *G. palpalisfuscipes* Newst. and *G. brevipalpis*. Newst. *Bull. Entomol. Res.* **53**, 579–95.

Saunders, D.S. (1964) Age-changes in the ovaries of the sheep ked, *Melophagus ovinus* (L.) (Diptera: Hippoboscidae). *Proc. R. Entomol. Soc. Lond. A* **39**, 68–72.

Schiemenz, H. (1953) Zum Farbwechsel bei heimischen Heteropterenunter besonderer Berücksichtigung von Palomena Muls. and Rey. *Beitr. Entomol.* **3**, 359–71.

Schlein, I. & Gratz, N.G. (1972) Age determination of some flies and mosquitoes by daily growth layers of skeletal apodemes. *Bull. World Health Org.* **47**, 71–6.

Schlein, I. & Gratz, N.G. (1973) Determination of the age of some anopheline mosquitoes by daily growth layers of skeletal apodemes. *Bull. World Health Org.* **49**, 371–5.

Schneider, F. & Vogel, W. (1950) Neuere Erfahrungen in der Bekdmpfung der Kirschenfliege (*Rhagoletis cerasi*) Schweiz. *Z. Obst-weinbau* **59**, 37–47.

Scopes, N.E.A. & Biggerstaff, S.H. (1977) The use of a temperature integrator to predict the developmental period of the parasite *Aphidius matricariae* (Hal.). *J. Appl. Ecol.* **14**, 1–4.

Seber, G.A.F. (1982) *The Estimation of Animal Abundance and Related Parameters*. Griffin, London.

Sepulveda, A. (1994) Daily growth increments in the otoliths of European smelt *Osmerus eperlanus* larvae. *Mar. Ecol. Prog. Ser.* **108**(1–2), 33–42.

Sheehy, M.R.J., Greenwood, J.G., & Fielder, D.R. (1994) More accurate chronological age determination of crustaceans from field situations using the physiological age marker, lipofuscin. *Mar. Biol.* **121**(2), 237–45.

Sheehy, M.R.J., Shelton, P.M.J., Wickins, J.F., Belchier, M., & Gaten, E. (1996) Ageing the European lobster *Homarus gammarus* by the lipofuscin in its eyestalk ganglia. *Mar. Ecol. Prog. Ser.* **146**(1–3), 99–111.

Sheehy, M.R.J., Shelton, P.M.J., Wickins, J.F., Belchier, M., & Gaten, E. (1996) Correction of PREVIEWS 99518551. Ageing the European lobster *Homarus gammarus* by the lipofuscin in its eyestalk ganglia. *Mar. Ecol. Prog. Ser.* **143**(1–3), 99–111.

Shiyomi, M. (1967) A statistical model of the reproduction of aphids. *Res. Popul. Ecol.* **9**, 167–76.

Shurova, N.M. & Zolotarev, V.N. (1988) Seasonal growth layers in the shells of the mussel in the Black Sea (USSR). *Biologiya Morya* **1**, 18–22.

Singh, D. & Sharma, R.C. (1995) Age and growth of a Himalayan teleost *Schizothorax richardsonii* (Gray) from the Garhwal Hills (India). *Fish. Res.* **24**(4), 321–9.

Smirina, E.M. (1994) Age determination and longevity in amphibians. *Gerontology* **40**(2–4), 133–46.

Smith, G.C. & Trout, R.C. (1994) Using Leslie matrices to determine wild rabbit population growth and the potential for control. *J. Appl. Ecol.* **1994**, 223–30.

Southwood, T.R.E. & Leston. D. (1959) *Land and Water Bugs of the British Isles*. Warne and Sons, London.

Southwood, T.R.E., Murdie, G., Yasuno, M., Tonn, R.J., & Reader, T.M. (1972) Studies on the life budget of *Aedes aegypti* in Wat Samphaya, Bangkok, Thailand. *Bull. World Health Org.* **46**, 211–26.

Southwood, T.R.E. & Scudder, G.G.E. (1956) The bionomics and immature stages of the thistle lace bugs (*Tingis ampliata* H-S. and *T. cardui* L.; Hem., Tingidae). *Trans. Soc. Br. Entomol.* **12**, 93–112.

Squire, F.A. (1952) Observations on mating scars in *Glossina palpalis* (R.-D.). *Bull. Entomol. Res.* **42**, 601–4.

Stiefel, A. & Piechock, R. (1986) Circannual layers in the cementum of molar teeth in the beaver (*Castor fiber*) as an aid to exact age determination. *Zoologische Abhandlungen* **41**(14), 165–75.

Stott, I., Hodgson, D.J., & Townley, S. (2012) popdemo: an R package for population demography using projection matrix analysis. *Methods Ecol. Evol.* **3**, 797–802.

Stubben, C.J. & Milligan, B.G. (2007) Estimating and analyzing demographic models using the popbio package in R. *J. Statist. Software*, **22**, 11.

Taylor, R.D. (1988) Age determination of the African buffalo, *Syncerus caffer* (Sparrman) in Zimbabwe. *African J. Ecol.* **26**(3), 207–20.

Thomas, D.B. & Chen, A.C. (1989) Age determination in the adult screwworm (Diptera: Calliphoridae) by pteridine levels. *J. Econ. Entomol.* **82**(4), 1140–4.

Thomas, D.B. & Chen, A.C. (1990) Age distribution of adult female screwworms (Diptera: Calliphoridae) captured on sentinel animals in the coastal lowlands of Guatemala. *J. Econ. Entomol.* **83**(4), 1422–9.

Tian, Y. & Shimizu, M. (1997) Growth increment patterns in the shell of the cockle *Fulvia mutica* (Reeve) and their use in age determination. *Nippon Suisan Gakkaishi* **63**(4), 585–93.

Timmons, C., Hassell, A., Lauziere, I., & Bextine, B. (2011) Age determination of the glassywinged sharpshooter, *Homalodisca vitripennis*, using wing pigmentation. *J. Insect Sci.* **11**, 78.

Tomic-Carruthers, N., Robacker, D.C., & Mangan, R.L. (1996) Identification and age-dependence of pteridines in the head and adult Mexican fruit fly, *Anastrepha ludens*. *J. Insect Physiol.* **42**(4), 359–66.

Tucker, A.D. (1997) Validation of skeletochronology to determine age of freshwater crocodiles (*Crocodylus johnstoni*). *Mar. Freshwater Res.* **48**(4), 343–51.

Tucker, R. (1985) Age validation studies on the spines of the spurdog (*Squalus acanthias*) using tetracycline. *J. Mar. Biol. Assoc. UK.* **65**(3), 641–52.

Tyndale-Biscoe, M. & Hughes, D.R. (1969) Changes in the female reproductive system as age indicators in the bushfly *Musca vetustissima* Wlk. *Bull. Entomol. Res.* **59**, 129–41.

Tyndale-Biscoe, M. & Kitching, R.L. (1974) Cuticular bands as age criteria in the sheep blowfly, *Lucillia cuprina* (Wied.) (Diptera, Calliphoridae). *Bull. Entomol. Res.* **64**, 161–74.

Usher, M.B. (1972) Developments in the Leslie Matrix Model. In: Jeffers, J.N.R. (ed.), *Mathematical Models in Ecology.* Symp. Brit. Ecol. Soc. **12**, 29–60.

Usinger, R.L. (1966) *Monograph of Cimicidae.* The Thomas Say Foundation, Entomoligical Society of America.

Van de Kerk, M., de Kroon, H., Conde, D.A., & Jongejans, E. (2013) Carnivora population dynamics are as slow and as fast as those of other mammals: implications for their conservation. *PloS One* **8**(8), e70354.

Vandermeer, J.H. (1975) On the construction of the population projection matrix for a population grouped in unequal stages. *Biometrics* **31**, 239–42.

Van Dijk, T.S. (1972) The significance of the diversity in age composition of *Calathus melanocephalus* L. (Col. Carabidae) in space and time at Schiermannikoog. *Oecologia (Berl.)* **10**, 111–36.

Vidalis, K. & Tsimenidis, N. (1996) Age determination and growth of picarel (*Spicara smaris*) from the Cretan continental shelf (Greece). *Fish. Res.* **28**(4), 395–421.

Vlijm, L.S. & Van Dijk, T. (1967) Ecological studies on carabid beetles. II. General pattern of population structure in *Calathus melanocephalus* at Schiermonnikoog. *J. Morph. Okol. Tiere* **58**, 396–404.

Vogt, W.G. & Morton, R. (1991) Estimation of population size and survival of sheep blowfly, *Lucilia cuprina*, in the field from serial recoveries of marked flies affected by weather dispersal and age-dependent trappability. *Res. Pop. Biol.* **33**, 141–63.

Vogt, W.G., Woodburn, T.L., & Tyndale-Biscoe, M. (1974) A method of age determination in *Lucilia cuprina* (Wied) (Diptera, Calliphoridae) using cyclic changes in the female reproductive system. *Bull. Entomol. Res.* **64**, 365–70.

Waloff, N. (1958) Some methods of interpreting trends in field populations. *Proc. X Int. Congr. Entomol.* **2**, 675–6.

Williamson, M.H. (1959) Some extensions of the use of matrices in population theory. *Bull. Math. Biophys.* **21**, 13–17.

Williamson, M.H. (1967) Introducing Students to the Concepts of Population Dynamics. In: Lambert, J. (ed.), *The Teaching of Ecology. Symp. Br. Ecol. Soc.* **7**, 169–76.

Williamson, M.H. (1972) *The Analysis of Biological Populations*. Arnold, London.

Wohlschlag, D.E. (1954) Mortality rates of whitefish in an arctic lake. *Ecology* **35**, 388–96.

Worthington, D.G., Doherty, P.J., & Fowler, A.J. (1995) Variation in the relationship between otolith weight and age: Implications for the estimation of age of two tropical damselfish (*Pomacentrus moluccensis* and *P. wardi*). *Can. J. Fish. Aquat. Sci.* **52**, 233–42.

Xu, H., Ye, G.Y., Xu, Y., Hu, C., & Zhu, G.H. (2014) Age-dependent changes in cuticular hydrocarbons of larvae in *Aldrichina grahami* (Aldrich) (Diptera: Calliphoridae). *Forensic Sci. Int.* **242**, 236–41.

Young, E.C. (1965) Teneral development in British Corixidae. *Proc. R. Entomol. Soc. Lond. A.* **40**, 159–68.

Zuk, M. (1987) Age determination of adult field crickets: Methodology and field applications. *Can. J. Zool.* **65**, 1564–6.

13 Species Richness, Diversity and Packing

A great deal of time and expertise has been expended on the compilation of faunal lists for particular habitats, but the consequent increase in our understanding of the rules determining structure and function of animal communities or the impact on them of natural or mankind-induced change is still meagre. The body of quantitative theory on communities has continued to develop, initially following the lead given by the 'Hutchinson–MacArthur school' with an emphasis on niche and competition. The dominant view for much of the 20th century was that ecological communities were formed from species which coexisted because they differed in their niches. Inter-specific competition maintained niche differentiation and resource limitations and density-dependence constrained and stabilised population size. More recently, the field has been reinvigorated after Hubbell (2001, 2006) challenged the well-establish niche-based view of community structure by claiming that numerous macro-ecological patterns, including species abundance distributions (SADs) and species–area relationships (SARs), could be explained through his Unified Neutral Theory of Biodiversity and Biogeography. Hubbell's arguments were initially heavily influenced by observations on plants and the Barro Colorado Island tree data. He noted the high diversity of low-abundance tree species all seemingly inhabiting a similar niche. These species predominately interacted with their nearest neighbours, which almost always differed between individuals of the same species. Such inconsistency and diversity of interactions would not lead to competition driving species towards clear niche differentiation. The Unified Neutral Theory claims that species abundance changes by chance and is driven primarily by dispersal acting together with speciation and stochastic variation in birth and death rates. There are analogies between neutral theory and the theory of random genetic drift developed by evolutionary theorists in the 1960s.

Neutral and niche theories are still developing and, as often happens in science, some argue that they are both part of a more general unified theory (see Mathews and Whittaker, 2014, for a recent review). One example of a unifying model is Tilman's (2004) stochastic niche theory. Even when neutral theory does not give a good fit to the observations, which is frequently the case, it does offer a null hypothesis against which to test observations and has been influential in raising interest in the role of stochastic processes. Many animal communities can be partitioned into a core set of dominant species which are always present, adapted to specific niches, and under density-dependent control and a larger group of tourist species whose presence is driven by

Ecological Methods, Fourth Edition. P. A. Henderson and T. R. E. Southwood.
© 2016 John Wiley & Sons, Ltd. Published 2016 by John Wiley & Sons, Ltd.
Companion Website: www.wiley.com/go/henderson/ecologicalmethods

stochastic processes. An example of this partition is the Bristol Channel fish community reported by Magurran & Henderson (2003) and Henderson & Magurran (2014). As Chave (2004) notes in his review of neutral theory, the rare tourist species observed by Magurran & Henderson may fit a neutral model.

Theory, while directing the way we organise and analyse our data, is also posing hypotheses and suggesting new analytical approaches at a rate that almost defeats our ability to test; we still lack a sufficient body of long-term data sets of community abundance to test competing hypotheses.

Linked to increased interest in conservation, the term biological diversity or biodiversity (Wilson, 1988; Reaka-Kudla et al., 1997) – a measure of the total genetic and ecological diversity – has entered our vocabulary[1]. A task for ecologists is to give a quantitative measure to this rather spiritual concept. Only by creating and applying reliable measures of diversity can we measure how it varies both spatially and temporally and thus recognise the influences that create and destroy it. There are numerous measures of diversity or ways of estimating species richness, which method is to be preferred will depend on the quality of the data, the sampling effort, community properties such as species number, community stability and the objectives of the study.

In this chapter we will focus primarily on species diversity. It is axiomatic that the objectives and methods of analysis should be fully and carefully considered, before the field programme is undertaken. The methods of sampling are those outlined in Chapters 2 and 4–8, although the differential response of species to trapping is an indication for caution in respect of many of the methods in Chapter 7. The present chapter aims to describe the methods of handling and analysing data on animal communities (from guilds to continental faunas). Details of the theoretical studies and examples of fieldwork are given in MacArthur & Wilson (1967), MacArthur (1972), May (1973, 1976), McClure & Price (1976), Cody & Diamond (1975), Pianka (1976a,b) and Pimm (1991).

13.1 Diversity

Diversity, even restricted as it is (for the purposes of this chapter) to the variety of animal species, is one of those common-sense ideas that prove elusive and multifaceted when precise quantification is sought (Peet, 1974). A useful classification, due to Whittaker (1972), is:

α-diversity: the diversity of species within a community or habitat.

β-diversity: a measure of the rate and extent of change in species along a gradient, from one habitat to others.

γ-diversity: the richness in species of a range of habitats in a geographical area (e.g. island); it is a consequence of the alpha-diversity of the habitats together with the extent of the beta-diversity between them.

---

[1](The term biological diversity was probably first defined by Norse & McManus, 1980 and contracted to biodiversity by Rosen in 1985

α- and γ-diversity are thus qualities that simply have magnitude and could, theoretically, be described entirely by a single number (a scalar). β-diversity in contrast is analogous to a vector as it has magnitude and direction. Their descriptions therefore require different approaches.

The restriction of the measurement of diversity to the consideration of the numbers of species, and the failure to take account of the variety, or otherwise, of form represented by these species has been criticised by Van Valen (1965), Hendrickson & Ehrlich (1971) and Findley (1973). The latter has suggested that taxonomic distance between the components of a fauna should be used to give a measure of its 'phenetic packing'.

13.1.1 Description of α- and γ-diversity

When a fauna (or a flora) is sampled it is found that a few species are represented by a lot of individuals, and a large number of species by few individuals. These relative abundances may, in part, represent the basic pattern of niche utilisation in the community (or area) and also stochastic factors linked to migration between habitats. One approach to diversity has been to seek to expose the distributions that underlie the species patterns; others have been more pragmatic and sought the best description (e.g. Williams,1964). The result was an 'explosive speciation' of diversity indices (perhaps only rivalled in ecological methodology by 'new designs' for light traps), that initially brought confusion to the subject; to this, the iniquitousness of some relationships and the apparent constancy of certain numerical values have added a measure of mystique. The penetrating analysis by May (1975) of the underlying mathematical similarity of the assemblage of indices and the development of the diversity ordering approach (Tóthmérész, 1995) have now allowed the subject to develop more rationally. This process has also been aided by systematic reviews, such as Magurran (2013).

The estimate of total species richness S_{max} (see p. 498) is a straightforward measure of species diversity but does require high sampling effort as for small numbers of samples; the estimate is proportional to the number of samples collected. Such a measure only expresses one aspect of the data set. If the log number of individuals (or abundance) of each species is plotted against the rank, the plot will often approximate to a straight line for at least part of the plot (Fig. 13.1) (see also below and Fig. 13.4). Such plots emphasise that the species number (S): abundance of individuals (N) relationship has two features:

1. species richness;
2. equitability or evenness: the pattern of distribution of the individuals between the species [a faunal sample of 100 individuals representing 10 species could consist of ten individuals of each (extreme equitability) or, at the other extreme, 91 individuals of one species (the dominant) and one each of the other nine]. Both attributes of the data may be expressed using a model that describes the species-abundance relationship, if a suitable one can be identified. The popular alternative (which can easily be expressed to non-biologists such as planners) is to summarise both attributes with a single number – a diversity index.

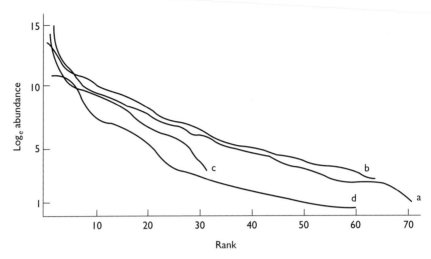

Figure 13.1   Rank order abundance plots for British inshore marine fish communities. (a) Sizewell; (b) Dungeness; (c) Kingsnorth; (d) West Thurrock (after Henderson, 1989).

13.1.2   Species richness

Species inventories for particular habitats or localities are frequently required for purposes such as conservation management. Because a complete census is rarely feasible the community must be sampled. An important problem that then arises is to estimate via sampling the total species (or other taxon) number, S_{max}, for the locality. This will give both a measure of the completeness of the inventory and also allow comparison with the species richness of other localities. An estimate of the maximum species number is also useful when assessing if the further information to be gained from continued sampling justifies the cost. In the following sections the different approaches that can be taken to estimate S_{max} are described. At present, no clear consensus as to the best approach is available, but it is clear that no method performs well when only a small proportion of the species present has been sampled. Walther & Morand (1998), in a study of parasite communities, found the Chao estimator (p. 503) used with presence–absence data to perform best. Toti *et al.* (2000) tested 11 species richness estimators on spider communities and concluded that the Michaelis–Menten means estimator performed best. King and Porter (2005) assessed a wide variety of methods to estimate the species richness of ants based on a data set comprising 1732 samples containing 94 species. They concluded that '… none of the estimators was stable, and their estimates should be viewed with skepticism.' A similar lack of success is reported from the analysis of museum collections. Petersen *et al.* (2003), using museum records data, concluded that… 'the first and second order Jackknife estimators yield the most accurate estimate of the number of collectable species in Denmark, while ACE, Bootstrap and Chao1 only provide light improvements over observed values. We find that all estimators underestimate the true diversity of Danish Asilidae and speculate that this performance is due to a discrepancy between the total and the collectable fauna in the region. Finally, we discuss the implications for species richness estimation and emphasise that for most terrestrial arthropod taxa these discrepancies are of such a magnitude that estimated species richness values may be dangerously low and of limited use in conservation decision making.'

In a review of species richness estimation methods, Reese *et al.* (2014) conclude that 'the most biased estimators were generally the most precise. This is a particularly dangerous combination because a precise and biased estimator can lead to a false sense of confidence when compared with an imprecise, but unbiased, estimator.' It can be concluded that species richness estimators should not be uncritically relied upon to give a useful estimate and when used some measure of their reliability should be presented. They tend to perform best when the community has been heavily sampled, and therefore they cannot be used to greatly reduce field expenditure. When sampling effort is sufficient for the species accumulation curve to be clearly decelerating towards an asymptote, fitting a Michaelis–Menten asymptotic curve is often the best option. In large communities open to migration the species accumulation curve may never form an asymptote, but eventually settle to an approximately constant, but low, rate of species accumulation. Henderson (2007), in a long-term study of a marine fish community, modelled species accumulation using a hyperbolic–linear model.

13.1.2.1 Extrapolating the species accumulation curve, rarefaction

The plot of the cumulative number of species, $S(n)$, collected against a measure of the sampling effort (n), is termed the 'species accumulation curve'. The sampling effort can be measured in many different ways; some examples are the number of quadrates taken, total number of animals handled, hours of observation or volume of water filtered. The species accumulation curve will only prove useful if applied to a defined habitat or area which is reasonably homogeneous.

Two illustrations of this curve (Fig. 13.2) show the cumulative species number of fish recorded from monthly samples collected between 1981 and 1995 at Hinkley Point in the lower Severn Estuary, and the cumulative number of species of Heteroptera against the number of individuals collected over several years from oak trees. The sampling method used for fish, described by Henderson & Holmes (1991), was to use the power station cooling water pumps as a sampler, and thus each month the volume of water filtered was a constant 3.24×10^5 m^3, giving a total volume sampled to the end of 1995 of about 5.18×10^7 m^3. As is normal when sufficient sampling effort is expended, the curve is asymptotic. Yet, even given this exceptionally high sampling effort, new species are still occasionally caught. The inshore fish fauna of the North Eastern Atlantic is well known, and Henderson (1989) argues that the total species number at this latitude in fully marine waters is about 80, although off-shore or deep water species may occasionally appear. Likewise, with the oak Heteroptera, when a limited number of trees in two nearby woods were sampled by knockdown (p. 149), yet more new species were found.

The order in which samples were taken alters the shape of a species accumulation curve, both because of random error and sample heterogeneity. In the case of the Hinkley Point data, which is a time series, more fish species were caught during the autumn than at other times of the year, so an accumulation curve commencing in September will rise faster than one starting in January. Further, examination of Fig. 13.2 shows no increase in species number between samples 44 and 76. These samples were collected during an exceptionally cold period when the inshore fish community was

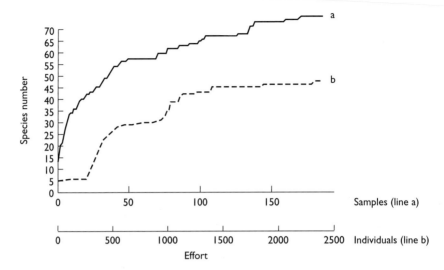

Figure 13.2 Cumulative species number of (a) fish recorded from monthly samples collected 1981 and 1995 collected at Hinkley Point in the lower Severn Estuary (after Henderson, 1989, with more recent data added); (b) Heteroptera from oak tree (after Southwood, 1996).

much reduced. To eliminate features caused by random or periodic temporal variation the sample order can be randomised. A useful procedure is to randomise the sample order r times and to calculate the mean and standard deviation of $S(n)$ over the r runs (Colwell & Coddington, 1994). This approach can also be applied to samples collected along transects, provided they are from a single community. The average species accumulation curve from 100 randomisations of the Hinkley Point data is shown in Fig. 13.3.

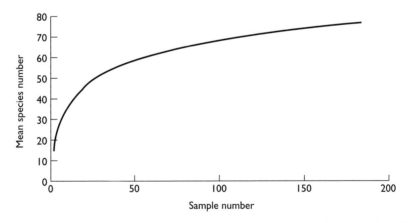

Figure 13.3 The average species accumulation curve from 20 randomisations of the Hinkley Point species accumulation curve shown in Fig. 13.2. The calculations were performed using species diversity and richness II software from Pisces Conservation Ltd., IRC House, Lymington SO41 8GN, England.

Extrapolation to estimate the total species complement (species richness) for the habitat, S_{max}, is only possible if the species accumulation curve is derived from a reasonably homogeneous (stable) community. So the first task is to look for heterogeneity. If species abundance, rather than presence–absence data are available, Colwell and Coddington (1994) suggest comparing the mean randomised species accumulation curve with the curve expected if all the individuals caught over all the samples were randomly assigned to the samples. If the expected curve rises more steeply from the origin, then heterogeneity is greater than could be explained by random sampling error alone. The expected curve can be computed in the following two ways. Both methods start by calculating the mean number of individuals per sample, n. The rarefaction approach generates the curve for x samples by selecting, without replacement, x random draws of n individuals from the pooled sum of the samples (Tipper, 1979; James & Rathbun, 1981). The expected number of species $E(S_n)$ in a sample of n individuals selected at random from a population of N individuals and S species is given by the hypergeometric distribution:

$$E(s_n) = S - \binom{N}{n}^{-1} \sum_{i=1}^{s} \binom{N - N_i}{n},$$ (13.1)

where N_i is the number of individuals of species i in the population.

The alternative, termed random placement, is easier to compute. With this method xn individuals are assumed to be randomly assigned to each of x collections (Coleman, 1981; Coleman et al., 1982). The increase in species number over a series of samples, $S(n)$, is calculated as:

$$S(n) = S_{TOT} - \sum_{i=1}^{i=S_{TOT}} (1 - \alpha)^{n_i},$$ (13.2)

where S_{TOT} is the total species number recorded, n_i the total number of individuals belonging to species i, n the sample number and α the fraction of the total sampling effort undertaken by sample n.

While non-asymptotic models for the species accumulation curve, based on either a log-linear or log-log relationship between $S(n)$ and n have been applied (Palmer, 1991), asymptotic models seem more appropriate. If it is assumed that the probability that the next individual captured will be a new species declines linearly with species number, then the species accumulation curve is the negative exponential function (Holdridge et al., 1971; Miller & Wiegert, 1989; Soberon & Llorente, 1993).

$$S(n) = S_{max}(1 - e^{-kn}),$$ (13.3)

where k is a fitted constant and n is the number of samples.

The asymptotic behaviour of the accumulation curve can also be modelled as the hyperbola:

$$S(n) = \frac{S_{max} n}{B + n},$$ (13.4)

where S_{max} and B are fitted constants. This is the Michaelis–Menten equation used in enzyme kinetics, and thus there is an extensive literature discussing the estimation of its parameters which unfortunately presents statistical difficulties. One approach favoured by Raaijmakers (1987) is to calculate S_{max} and B using their maximum likelihood estimators as follows:

If:

$$X_i = \frac{S(n)}{n}$$

and:

$$Y_i = S(n)$$

then:

$$\hat{B} = \frac{\bar{X}S_{yy} - \bar{Y}S_{xy}}{\bar{Y}S_{xx} - \bar{X}S_{xy}}$$

and:

$$\hat{S}_{max} = \dot{Y} + \hat{B}\bar{X} \qquad\qquad (13.5)$$

where S_{yy}, S_{xx} and S_{xy} are the sums of squares and cross-products of the deviations $Y_i - \bar{Y}$ and $X_i - \bar{X}$. While this method of estimating S_{max} has been criticised by Lamas et al. (1991), if it is decided to use the Michaelis–Menten equation, it seems the best approach available.

13.1.2.2   Using parametric models of relative abundance

Given quantitative data for species abundance models describing the relative abundance of species within the community such as the log-normal (see Section 13.1.3.3) can be used to estimate the total species complement, S_{max}. If the data fits a log-normal distribution then the number of unsampled species is given by the missing part of the distribution to the left of the veil-line (see Section 13.1.3.3 below). Alternative distributions which have been suggested include the Poisson-log-normal (Bulmer, 1974) and the log series (Williams, 1964) amongst others. It is unlikely that parametric methods yield reliable estimates for S_{max} and they should probably be avoided.

13.1.2.3   Non-parametric estimates

A variety of non-parametric estimators have been suggested. These were reviewed by Colwell and Coddington (1994) who, at that time, concluded that we had as yet insufficient experience to judge their relative merits. Subsequently, several new methods have been developed based on non-parametric maximum likelihood (NPML) estimation

(Norris and Pollock, 1998; Wang & Lindsay, 2005; Wang, 2010). These more recent methods are computationally complex and will not be described in detail here; however, software for their computation is described below. Studies using field data (as reviewed in Section 13.1.2 above) would indicate that non-parametric estimates are unreliable.

Using the observed number of species represented by one, a, or two, b, individuals in the sample, Chao (1984) derived the simple estimator:

$$\hat{S}_{max} = S_{obs} + (a^2/2b)$$ (13.6)

where S_{obs} is the actual number of species in the sample.

Chao (1987) gives the variance of this estimate as

$$var(\hat{S}_{max}) = b\left[\left(\frac{a/b}{4}\right)^4 + \left(\frac{a}{b}\right)^3 + \left(\frac{a/b}{2}\right)^2\right]$$ (13.7)

Note that when all the species have been observed more than twice $S_{max} = S_{obs}$, the census is considered complete. Equations (13.6) and (13.7) can be applied to presence–absence data when a is the number of species only found in one sample, and b is the number of species only found in two samples.

Two further estimators were suggested by Chao & Lee (1992). These are a modification of an early estimator calculated as Observed species number/coverage. Coverage is simply [1 – (number of species represented by only one individual/total number of individuals)].

Independently, Heltshe & Forrester (1983) and Burnham & Overton (1978) developed the first-order jack-knife estimator:

$$\hat{S}_{max} = S_{obs} + a(n - 1/n)$$ (13.8)

where n is the number of samples and a the number of species only found in one sample.

Heltshe & Forrester (1983) give the variance of this estimate as:

$$var(\hat{S}_{max}) = \frac{n-1}{n}\left(\sum_{0}^{S_{obs}} j^2 f_j - \frac{L^2}{n}\right)$$ (13.9)

where f_j is the number of samples holding j of the L species only found in one sample.

Further jack-knife estimators are discussed in Burnham & Overton (1979) and Smith & van Belle (1984).

A bootstrap estimate (see p. 60) of species richness can be calculated as follows (Smith & van Belle, 1984):

1. Randomly select with replacement n samples from the total available and calculate:

$$S_{max} = S_{OBS} + \sum_{i=1}^{S_{OBS}} (1 - p_i)^n$$ (13.10)

where p_i is the proportion of the n that has species i present.

2. Repeat Step 1 a large number of times – say 50 to 200 – and calculate the mean estimate of S_{max} and the variance as follows:

$$\text{var}(S_{max}) = \sum (1-p_i)^n \left[1 - (1-p_i)^n + \sum \sum \left\{ q_{i,j}^n - \left[(1-p_i)^n (1-p_j)^n \right] \right\} \right] \quad (13.11)$$

where q_{ij} is the proportion of the n bootstraps which hold both species i and j.

13.1.2.4 Software for calculating species richness and rarefaction

Widely utilised software packages to calculate non-parametric estimators are:

EstimateS (http://viceroy.eeb.uconn.edu/estimates/),
SPADE (http://chao.stat.nthu.edu.tw/software/SPADE/)
ws2m (http://eebweb.arizona.edu/diversity/) and
Species Diversity and Richness IV (http://www.pisces-conservation.com/indexsoft
    diversity.html).

The R-package SPECIES (Wang, 2011) offers functions to calculate a wide range of estimators, including the recently developed NPML methods. The following R code listing shows how simple SPECIES is to implement.

```
library("SPECIES")
#Use the supplied Malayan butterfly dataset of Fisher et al 1943
data("butterfly")
#Print out the data set
butterfly
#Note the data is in the form of a frequency table
#there are 118 species known from a single individual,
#74 species for which 2 individuals were caught etc.
# summing nˈj gives a total of 620 species
#First we will calculate Chao (1984) estimator plus 95% confidence intervals
chao1984(butterfly, conf = 0.95)
# Now we run the unconditional NPMLE of Norris Pollock (1998)
unpmle(butterfly, t = 15, C = 0, method = "W-L", b = 200, seed = NULL, conf = 0.95,
dis = 1)
```

The R package rich (Rossi, 2011) computes rarefaction curves and randomisation tests to compare the species richness of two communities. In addition, the R package vegan (Oksanen *et al.*, 2015) implements a large number of methods to calculate species accumulation curves and species richness. For many users this package will supply all the methods they require for community analysis. The data are entered in community matrix form, with the species as columns and the samples as rows. This matrix is best prepared in a spreadsheet such as Excel and then saved as a csv file to be loaded into R. The function specpool estimates total species richness based on species incidence over a series of samples or sites. The function estimateR generates estimates based on the pattern of abundance in a single sample.

The following R listing illustrates the application of vegan using the BCI data set supplied with the package.

```
library("vegan")                     #Get the vegan library
#Run the default species richness estimators; Chao; Jackknife; bootstrap
specpool(BCI)                        #We use the BCI data set in vegan
# Run estimators Chao and ACE for a single sample
k <-sample(nrow(BCI),1)              # randomly choose the sample (row) to run
estimateR(BCI[k,])                   # obtain richness estimates for sample k
## We can also study how the estimators change with sampling effort
pool <- poolaccum(BCI)               # shows how estimators change with sample (plot)
number
summary(pool, display = "chao")#summary data for the Chao estimator
#This produces a series of plots for Chao, Jackknife and Bootstrap estimators
plot(pool)
```

13.1.3 *Models for the* S:N *relationship*

The equitability of the species:abundance relationship will be a reflection of the underlying distributions. Four main groups can be recognised: the geometric series; the logarithmic series; the log-normal; and MacArthur's 'broken stick' model. The two former will give approximately straight lines when log abundance is plotted against rank (Fig. 13.1 and 13.4a), whilst the latter shows up as a straight plot when abundance is plotted against log rank (Fig. 13.4b) (Whittaker, 1972; May, 1975).

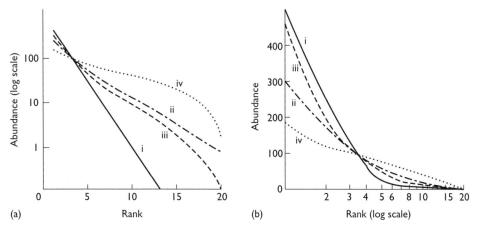

Figure 13.4 The rank abundance (or dominance–diversity) curves for different underlying distributions when plotted with the log scale on different axes: (a) log abundance (individuals): rank (species sequence); (b) abundance: log rank. i = geometric series, ii = log series, iii = lognormal, iv = MacArthur broken stick. (Adapted from Whittaker, 1972.) Note: the exact relative positions of the curves depend on the assumptions about the parameters. Whilst the broken stick and canonical log-normal are defined by one parameter (S_T, here taken as 20), geometric and log series and the general log-normal are defined by two parameters: S_T and J, where J = N_T/m, m being the number of individuals of the rarest species, often in practice one.

13.1.3.1 Geometric series

If a species occupies a fraction (k) of a habitat, another species the same fraction of the remainder, a third species k of what now remains and so on, the resulting rank-abundance distribution will, as originally shown by I. Motomura, be a geometric series (May, 1975). Termed the 'niche-pre-emption hypothesis', this distribution has often been used to describe floral diversity in temperate regions (Whittaker, 1970, 1972; McNaughton & Wolf, 1970). May (1975) shows that the Odum *et al.* (1960) formulation which is expressed in the terms of species and numbers (Fig. 13.1a), rather than rank:abundance (Fig. 13.2b) is the equivalent of this series.

13.1.3.2 Logarithmic (or log) series

Originally suggested as a suitable description of species abundance data by R.A. Fisher (Fisher *et al.*, 1943), this distribution has been criticised as lacking a theoretical justification at the level of species interaction. Kempton & Taylor (1974) and May (1975, 1976) have provided this. It was suggested as a model because the abundance of animals when ranged in rank-order is often approximated by the series:

$$ax, \frac{ax^2}{2}, \frac{ax^3}{3}... \tag{13.12}$$

The log series may be regarded as the convenient approximation to gamma model (of which the negative binomial is a well-known form; see p. 28) with maximal variance (Kempton and Taylor, 1974). Gamma models often arise as a result of a two-stage (compound) process (see p. 34); as May (1975) points out, if we envisage a geometric series type of niche-pre-emption, where the fractions of niche pre-empted have arisen from the arrival of successive species at uniform intervals of time, the arrival of the species at random time intervals would lead to the log series (see also Boswell & Patil, 1971). This distribution approximates to a straight line on a rank:log abundance plot (Fig. 13.4a), and the relationship between the number of species, S, and the number of individuals N, is given by:

$$S = \alpha log_e \left(1 + \frac{N}{\alpha}\right) \tag{13.13}$$

or in terms of a sampling parameter X, for the expected number of species in a total sample of N individuals, S_N is given by:

$$S_N = \frac{\alpha X^N}{N} \tag{13.14}$$

Williams (1964) gives a table and graph for estimating α, given species number and total number of individuals (see Fig. 13.9). In the R package vegan the function fisherfit fits a log series model: the following listing shows a simple application to the BCI data set.

```
library("vegan")          #Get the vegan library
k <-sample(nrow(BCI),1) # randomly choose the sample (row) to run
fisherfit(BCI[k,])        # Find Fisher's alpha
```

13.1.3.3   Log-normal distribution

Preston (1948) suggested that the log-normal distribution would give the best description of species-abundance patterns; the assumption being that individuals were distributed between species in accordance with the normal or Gaussian distribution and population growth is geometric (Williams, 1964). May (1975) enlarges on the biological reasons why the log-normal might be considered to apply to both opportunistic or equilibrium communities. Preston (1948, 1960, 1962a, b) considered the plot of the frequency of species against abundance classes on a log scale (Fig. 13.5). He used logarithms to the base 2 so that each class, or octave, involved a doubling of the size of the population. (The abundance classes could be on any logarithmic scale). The abundance of the species at the peak (mode) of the distribution is N. and the abundance octaves (R) are calculated:

$$R_i = \log_2 \left(\frac{N_i}{N_0} \right)$$ (13.15)

Preston suggested that sample size is usually too small to obtain the species in the lower (rarer) octaves; these species were, he said, hidden behind the 'veil line'. Field data will normally only represent those species to the right of the veil line. and therefore from Fig. 13.5 it will be seen that if the veil line is at N_0 or in any of the positive octaves the plot will approximate a straight line and be of the form shown in Fig. 13.4b

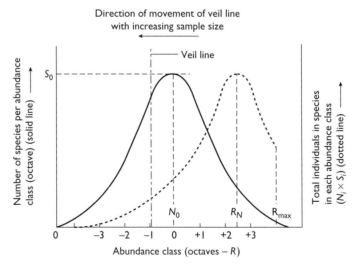

Figure 13.5   The log-normal representation of abundance and species relationships. (Adapted from May, 1975.)

(with axes reversed). However, following very intensive sampling of the Heteroptera in oak trees, Southwood (1996) found a bell-shaped curve; the over representation of the lowest abundance class (singletons) was due to vagrants. Preston recognised that the plot of individuals per class against abundance class (dotted line in Fig. 13.5) would not describe the complete 'bell', but be truncated at its crest (the 'canonical condition') so that $R_N = R_{max}$ (Fig. 13.5). This truncated log-normal, ending at the crest (unlike Fig. 13.5) is difficult to fit to data, although Bullock (1971a) and Gage & Tett (1973) did so. Kempton and Taylor (1974) and Bulmer (1974) showed that it is more tractable if regarded as a Poisson log-normal, described by the parameters S_{obs} and σ^2 (variance). As Preston (1948) noted, the dispersion constant

$$a = (2\sigma^2)^{\frac{-1}{2}}$$

is independent of sample size.

A useful ratio is

$$\gamma = \frac{R_N}{R_{max}} = \frac{log_2 2}{2a \left(log_e S_{obs}\right)^{1/2}}.$$

Preston (1962a,b) analysed a considerable body of field data and found that $\gamma \sim 1$ was general; likewise, Hutchinson (1953) that $a \sim 0.2$. A rational explanation for both has been given by May (1975); both arise from the mathematics inherent in the log-normal given a large value of S_{obs}.

In the R package vegan the function prestonfit groups species frequencies into doubling octave classes and fits Preston's log-normal model. Further, the function prestondistr fits a truncated lognormal without pooling into octaves. This example also shows the estimation of the species hidden by the veil and estimates total species richness using the function veiledspec. The following listing shows an application to the BCI data set.

```
library("vegan")                    #Get the vegan library
data(BCI)                           # Use the BCI data set in vegan
# prestonfit seems to need large samples
# lognormal model needs a large data set so species abundance over all samples is used
Model1_octaves <- prestonfit(colSums(BCI))     #fit lognormal with octaves
Model2_trunc <- prestondistr(colSums(BCI))     #fit truncated lognormal no pooling
Model1_octaves                      # print output
Model2_trunc                        #print output
plot(Model1_octaves)                # plot model1
lines(Model2_trunc, line.col="red") # show model 2 on plot
den <- density(log2(colSums(BCI)))  # Smoothed density
lines(den$x, ncol(BCI)*den$y, lwd=2) # Fairly similar to mod.oct
#Total species richness estimators
veiledspec(Model1_octaves)          # Species hidden by veil estimated as 10
veiledspec(Model2_trunc)            # Species hidden by veil estimated as about 6
```

13.1.3.4 MacArthur's 'broken stick' model

In a theoretical consideration of how niche hyper-volume might be divided, MacArthur (1957) postulated three models. One model – termed the 'broken stick' – assumes niche boundaries are drawn at random and the abundance of the ith most abundant species is given by:

$$N_i = \frac{N_T}{S_{obs}} \sum_{n=i}^{S_{obs}} \frac{1}{n} \qquad (13.16)$$

The mathematics of this model have been investigated by Webb (1974); it is more even than the log-normal and characterised simply by the parameter S_{obs}. A ratio commonly calculated in faunistic studies is J, defined as the total number of individuals (N_T) divided by the abundance of the rarest species. From the above equation it is clear that for the MacArthur model this will be N_T/N_{obs} and, as May (1975) points out, then $J = S_{obs}^2$.

13.1.4 Non-parametric indices of diversity

The description of the $S{:}N$ relationship in terms of the parameters of a model implies that the model is at least approximately applicable. It may be argued that non-parametric indices have the advantage that they make no assumptions of this type. There are also relative indices (e.g. Lloyd & Ghelardi, 1964) which compare the equitability of the distribution with that which would arise if the same sample was distributed according to a certain model, in this case the MacArthur model. Peet (1975) shows that such relative indices possess mathematically undesirable properties. Comparative reviews of relative merits of the indices described below are provided by Taylor (1978), Kempton (1979) and Magurran (2013).

13.1.4.1 Shannon–Wiener[2] function (H)

A function devised to determine the amount of information in a code, and defined:

$$H = - \sum_{i=1}^{S_{obs}} p_i \log_e p \qquad (13.17)$$

where p_i = the proportion of individuals in the ith species; or, in terms of species abundance:

$$H = \log_e N - \frac{1}{N} \sum_{i=1}^{\infty} (p_i \log_e p_i) n_i \qquad (13.18)$$

where n_i = the number of species with i individuals. The information measure is nits for base e and bits per individual for base 2 logarithms.

---

[2] Also referred to as the Shannon–Weaver function

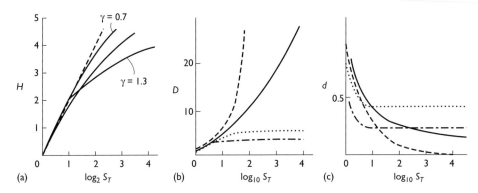

Figure 13.6 The performance of various diversity indices in relation to the total number of species (S_T) and the underlying model: log-normal (solid lines) ($\gamma = 1.0$ unless labelled otherwise), log series (. . . . .) $\alpha = 5$, geometric series (– – – –) $k = 0.4$, and MacArthur model (- - - - - -). (a) The Shannon–Wiener function. (b) The Simpson–Yule index. (c) The dominance index. (Adapted from May, 1975.)

Bulmer (1974) and May (1975) have shown that for a large sample with the log series model (Section 13.2.1.3.2):

$$H \cong 0.577 + \log_e \alpha,$$

where 0.577 is Euler's constant. Considering the behaviour of H with changing magnitude of S_{obs}, May (1975) concluded that it was an insensitive measure of the character of the $S{:}N$ relationship and is dominated by the abundant species (Fig. 13.6a).

13.1.4.2   Simpson–Yule index (D)

A diversity index proposed by Simpson (1949) to describe the probability that a second individual drawn from a population should be of the same species as the first.[3] The statistic, C (or Y) is given by:

$$C = \sum_i^{S_{obs}} p_i^2 \qquad (13.19)$$

where, strictly,

$$p_i^2 = \frac{N_i(N_i - 1)}{N_T(N_T - 1)},$$

but is usually approximated as:

$$p_i^2 = \left(\frac{N_i}{N_T}\right)^2$$

---

[3] A similar type of index had a few years earlier been proposed by G. Udney Yule to compare an author's characteristic vocabulary (frequency of different words in his writings).

where N_i is the number of individuals in the ith species and N_T the total individuals in the sample. The index is:

$$D = \frac{1}{C}$$

and the larger its value the greater the equitability (range $1 \to S_{OBS}$). The statistic $(1 - C)$ gives a measure of the probability of the next encounter (by the collector or any animal moving at random) being with another species (Hurlbert, 1971). May (1975) (Fig. 13.6b) shows that this index is strongly influenced for values of $S_{OBS} > 10$ by the underlying distribution.

13.1.4.3 Berger–Parker Dominance Index

This index is simple, both mathematically and conceptually, expressing the proportion of the total catch, N_T, that is due to the dominant species, N_{max}:

$$d = \frac{N_{max}}{N_T} \tag{13.20}$$

For reasonable values of S_{OBS} the index is not influenced by S_{OBS} and is relatively independent of the underlying model (Fig. 13.6c). May (1975) concludes that it seems 'to characterise the distribution as well as any (index), and better than most'.

13.1.4.4 Evenness (Equitability)

As diversity is at a maximum when all species within a community are equally abundant, a measure of evenness is the ratio of the observed diversity to the maximum possible for the observed species number. Hurlbert (1971) proposed a slightly more complex definition of evenness which considered both the maximum, D_{max} and minimum, D_{min} diversity:

$$E_H = \frac{D - D_{min}}{D_{max} - D_{min}} \tag{13.21}$$

Using the Shannon–Weiner index as an example, $D_{min} = 0$, $D_{max} = \log (S)$, where S is the total number of species:

$$E_H = \frac{H}{\log S}$$

13.1.4.5 McIntosh diversity measure

McIntosh (1967) suggested the dominance index:

$$D = \frac{N - U}{N - \sqrt{N}} \tag{13.22}$$

where N is the total number of individuals in the sample. U is calculated as:

$$U = \sqrt{\sum_i n_i^2}$$

and n_i is the number of individuals belonging to the ith species.

13.1.5 *Which model or index?*

The evaluation of the most appropriate model or index may be based on the consideration of their theoretical properties against our knowledge of ecology, or by testing them with field data for either their fit or their value in discrimination. Valuable studies in these respects are Bullock (1971a), Hurlbert (1971), Kempton & Taylor (1974), May (1975), Taylor *et al.* (1976), Routledge (1979) and Magurran (2013). There can be no universal 'best-buy', although there are rich opportunities for inappropriate usage! The objective of the analysis must be clearly conceived before the procedure is selected. It is certainly important to distinguish between data on organisms that are part of the trophic structure of the area (e.g. growing plants or birds seen foraging) and samples that may contain a number of animals that are outside their trivial ranges (e.g. suction trap catches at several metres above the vegetation). This is illustrated in studies by Kricher (1972) on the diversity of bird populations at different seasons, Southwood (1996) on insects from different species of oak tree, and Magurran and Henderson (2003) for fish in the Bristol Channel. It can be useful to separate the analysis of core permanent members of the community from the transient tourist species.

The use of models (e.g. MacArthur's) to gain some insight into resource apportionment may be justified in areal samples of 'trophically-related' organisms.

From theoretical studies, especially May (1975), it can be concluded that the models can be arranged in a series corresponding to maximal niche pre-emption or unevenness and moving through to a more uniform resource apportionment, thus:

(Uneven) Geometric series, log series, log-normal, MacArthur (Even)

This appears contrary to Kempton and Taylor's (1974) findings that, whereas most populations of macrolepidoptera were well described by the log series, those in unstable habitats with more 'rare' species were better described by the log-normal; it may be that the inclusion of 'additional species' has increased the evenness. May (1975) also concludes that the lognormal pattern may be expected with '…large or heterogeneous assemblies of species'.

Where comparative studies have included the log series, the conclusion has generally been that this provides the most 'comprehensible result' (e.g. Bullock, 1971a; Hurlbert, 1971; Kempton & Taylor, 1974; May, 1975; Taylor *et al.*, 1976; Magurran, 2013). Kempton and Taylor (1974) and Taylor *et al.* (1976) suggest that the 'basic environmental structure' of an area will determine the range of populations it can support, and on this assumption they are able to show that the log series is the best model. They point out that the rank: log abundance plot of the log series is almost linear over the mid region (see Fig. 13.4a), so α may be interpreted as the slope of this region '…where the moderately common species reflect most closely the nature of the environment' and fluctuate less violently, from year to year, than the more abundant species. However, there is no theoretical or empirical evidence to support this view, indeed the only generalisation that can be made is that the more abundant species are often those with the largest geographical ranges and thus possibly depend less on local conditions to determine abundance. They found that changes in α enable discrimination between diversities in areas where the environment was known to have changed. For the ecologist, the ability of a parameter or index to discriminate between changed conditions may be more relevant than the precision of the 'fit' of the underlying model. The richness and evenness parameters of the log-normal (S_{max} and σ^2) and the Shannon and Simpson indices were all found less useful as indications of environmental change and, even when the fit of the log series was not close, its robustness provided a meaningful α. Southwood *et al.* (2003) used Fisher's α to track changes in the community of Heteroptera over a 67-year period.

A novel concept of Taylor *et al.* (1976) is that, viewing diversity as a reflection of basic environmental structure, the two meaningful characteristics are not species richness and evenness, but the following.

1. Diversity as represented by the 'common α', the slope of the line as dominated by the moderately common species.
2. The fluctuations in numbers, from occasion to occasion (e.g. year to year) as shown by X of the log series [see Equation (13.14)]. This hypothesis has not, however, been further explored.

Perhaps because of small sample size and the low quality of the data, many published accounts do not take the parametric approach but simply present the Shannon–Wiener index, often to little purpose. Comparative studies have not given much support to the value of the non-parametric indices (Whittaker, 1972; May, 1975). The Shannon–Weiner index has often been found unsatisfactory (Bullock, 1971; Hurlbert, 1971; Taylor *et al.*, 1976; Routledge, 1979), and in spite of some apparently successful uses should in general be regarded as a distraction rather than an asset in ecological analysis. It is strongly influenced by species number and by the underlying model (Fig. 13.6a). The Simpson–Yule index is strongly influenced by the few dominant species, although bearing this limitation in mind it could be of value as an indicator of interspecific encounters. As is shown in Fig 13.7 for the Hinkley Point data, all of the non-parametric indices showed the same temporal pattern as species number and gave no additional insight. Thus, species number is often a straightforward measure for comparing diversity between samples collected in similar fashion. If the comparison is to be made between samples which differ in sampling effort, then estimates of total species richness, S_{MAX}, can be compared or rarefaction undertaken to produce a species richness for a standardised effort (see section 13.1.2.1).

Non-parametric diversity indices become useful when used for diversity-ordering (Tóthmérész, 1995). By generating a family of diversity indices it is possible to recognise non-comparable communities. If the objective of a study is to compare communities using a single non-parametric diversity measure, then diversity ordering must be undertaken.

13.1.6 *Comparing communities – diversity ordering*

Different diversity indices may differ in the ranking they give to communities (Hurlbert, 1971; Tóthmérész, 1995). An example from the latter report illustrates the

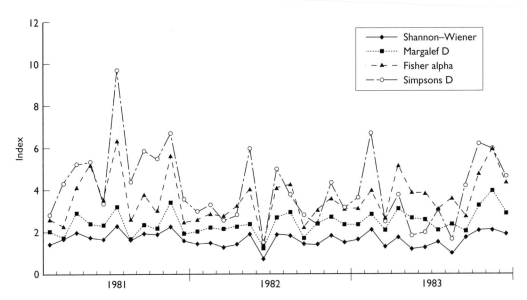

Figure 13.7   A comparison of a variety of diversity measures applied to monthly fish samples collected at Hinkley Point, England. In this case the data can be considered to be well ordered in that all the indices plotted show the same temporal pattern. Note that the Shannon–Wiener index is the least sensitive to change in the community.

point. Consider three imaginary communities with the following sets of species abundances for each of which diversity has been calculated using both Shannon–Wiener (H) and (2) Simpson's (D):

Community A: {33,29,28,5,5}, $H = 1.3808$, $D = 3.716$,
Community B: {42,30,10,8,5,5}, $H = 1.4574$, $D = 3.564$,
Community C: {32,21,16,12,9,6,4}, $H = 1.754$, $D = 5.22$,

Because $H(B) > H(A)$ it could be argued that B is the most diverse; however, as $D(A) > D(B)$ the opposite conclusion could also be entertained. Communities such as A and B which cannot be ordered are termed non-comparable. Such inconsistencies are an inevitable result of summarising both relative abundance and species number using a single number (Patil & Taillie, 1979). Diversity profiles offer a solution to this problem by identifying those communities that are consistent in their relative diversity. To identify these communities it is necessary to produce an expression which can generate the various indices by changing the value of preferably a single parameter.

Perhaps the most generally useful expression of this type is that due to Renyi (1961) which is based on the concept of entropy and defined as:

$$H_\beta = \left(\log \sum_{i=1}^{s} p_i^\beta \right) \Big/ (1 - \beta) \tag{13.23}$$

where β is the order ($\beta \geq 0$, $\beta \neq 0$), p_i the proportional abundance of the ith species and log the logarithm to a base of choice, often e.

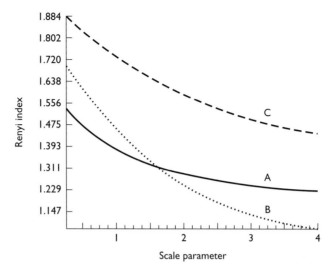

Figure 13.8 The diversity ordering of three artificial communities, which shows that C is the most diverse and that it is not possible to state which of A or B is the most diverse, because the relative magnitude of the index depends on the magnitude of the scale parameter.

Hill (1973) used an almost identical index N_a, which is related to H_β by the equality.

$$H_\alpha = \log(N_a)$$

Hill demonstrated that H_α for $a = 0, 1, 2$ gives the total species number, Shannon–Wiener's H and Simpson's D, respectively. Thus, by using different values of β, or 'a' a range of diversity measures can be generated from the Renyi or Hill equations. To test for non-comparability of communities, H_β is calculated for a range of β-values and the results presented graphically. Figure 13.8 shows the diversity ordering of the three artificial communities, A, B and C. This figure shows that C is always greater than A or B, and thus can be considered to be more diverse. As A and B cross over, one cannot be considered more diverse than the other by all reasonable measures. As described by Tóthmérész (1995), numerous other diversity index families have been suggested.

13.1.7 Procedure to determine α-diversity

In the light of present knowledge, the following may be taken as a guide to the analysis of diversity data for a single locality.

1. Plot graph(s) of log abundance on rank. It has always been important to examine the form of your data and this is easily done on a computer. Does the data form a straight line? Which are the species that depart most from it? Is there anything unusual in their biology (e.g. they could be vagrants)? The consideration of these graphs in the light of the earlier discussion in this chapter should indicate whether,

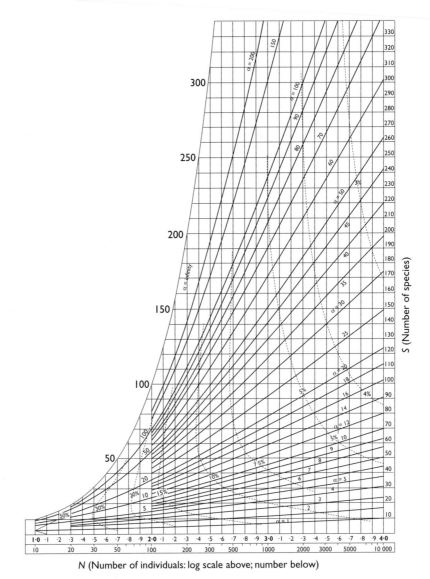

Figure 13.9   Nomograph for determining the index of diversity (α) for a number of species (*S*) and the number of individuals (*N*) in a random sample of a fauna. (Adapted from Williams, 1947.) Note that the standard errors contain only that component of variation due to sampling.

exceptionally (Taylor *et al.*, 1976), other models or indices should be used in addition to those outlined below. These graphs may be an excellent manner of presenting the data for publication (e.g. Sanders, 1969; Bazzaz, 1975; Taylor *et al.*, 1976; Southwood *et al.*, 1979; Henderson, 1989).

2. Plot the species accumulation curve(s) and calculate total species richness, S_{MAX}. The species accumulation curve gives insight into the sufficiency of the sampling

effort. As there is no best method to calculate S_{MAX} it is recommended that most of the methods given in Section 13.1.2 are explored. If the species accumulation curve is clearly de-accelerating, then fitting a hyperbolic (Michaelis–Menten) curve [Equation (13.4)] to estimate the asymptotic number is worth considering.

3. Determine α of the log series. This may be done rigorously by solving the maximum likelihood equation for α by numerical methods using a computer program (see section 13.1.3.2). For most purposes the value may be read off William's (1947) nomograph (Fig. 13.9). Taylor *et al.* (1976) point out that the true large sample variance is:

$$\text{var}(\hat{\alpha}) = \frac{\alpha}{-\log(1-X)}$$

The standard error contours on Fig. 13.9 exclude variation due to population fluctuations. Which variance is appropriate will clearly depend on the type of comparisons. If there are a series of samples it may be decided to:

(i) Calculate the 'common α' (Taylor *et al.*, 1976) by solving numerically:

$$\sum_{i=1}^{z} S_i = \sum_{i=1}^{z} \hat{a} \log\left[1 + \frac{N_1}{\hat{\alpha}}\right]$$

where z = total number of samples, and S and Nj are the *total* species and individuals in the ith sample.

(ii) Calculate the individual sampling factors (X) which indicate the magnitude of population fluctuation:

$$\hat{X}_i = 1 - e^{-\frac{S_t}{\hat{a}}}$$

4. Calculate the Berger–Parker dominance index [Equation (13.20)].
5. Calculate other parameters or statistics. This will be done where the graphs (1 above) reveal a special need: the appropriate equations are given earlier in the chapter.
6. If sites are to be compared, undertake diversity-ordering to ensure that they can be compared using a diversity index.

13.1.8 *Determining β-diversity*

β-diversity measures the increase in species diversity along transects, and is particularly applicable to the study of environmental gradients. It measures two attributes: the number of distinct habitats within a region; and the replacement of species by another between disjoint parts of the same habitat. Used together, α- and β-diversity can give an assessment of the diversity of an area (Routledge, 1977; Southwood *et al.*, 1979). Many of the methods were developed and extensively used in work on plant and bird ecology, and details will be found in Bullock (1971b), Whittaker (1972), Greig-Smith (1978),

Goodall (1973), Cody (1974, 1975), Mueller-Dombois & Ellenberg (1974), Cody & Diamond (1975), Pielou (1975) and Goldsmith & Harrison (1976). The general message from the review of α-diversity measurements was that the ecological insights gained were by no means proportional to the mathematical sophistication and complexity of the particular method; this message is also applicable to β-indices.

Wilson & Shmida (1984) considered the following six indices calculated using presence/absence data collected from a series of stations along a transect.

1. Whittaker's β_w

$$\beta_w = S/\alpha - 1 \tag{13.24}$$

where S = the total number of species and α the average species richness of the samples. All samples must have the same size (or sampling effort).

2. Cody's β_c

$$\beta_c = \frac{g(H) + l(H)}{2} \tag{13.25}$$

where $g(H)$ is the number of species gained and $l(H)$ is the number lost moving along the transect.

3. Routledge's β_R, β_I and β_E

$$\beta_R = \frac{S^2}{2r + S} - 1, \tag{13.26}$$

where S is the total species number for the transect and r the number of species pairs with overlapping distributions.

Assuming equal sample sizes,

$$\beta_I = \log(T) - \left[\left(\frac{1}{T}\right)\sum e_i \log(e_i)\right] - \left[\left(\frac{1}{T}\right)\sum \alpha_i \log(\alpha_i)\right] \tag{13.27}$$

where e_i is the number of samples along the transect in which species i is present, and α_i the species richness of sample i and T is Σe_i.

The third of Routledge's indices is simply

$$\beta_E = \exp(\beta_I) - 1 \tag{13.28}$$

4. Wilson & Schmida's β_T

$$\beta_T = \frac{[g(H) + l(H)]}{2\alpha} \tag{13.29}$$

where the parameters are defined as for β_c and β_w.

Based on an assessment of the essential properties of a useful index – the ability to detect change, additivity, independence of α and independence of sample size – Wilson & Schmida (1984) concluded that β_w was best. Harrison *et al.* (1992) applied this measure in modified form to the study of latitudinal gradients in a variety of aquatic and terrestrial groups.

A different approach to β-diversity is to measure diversity along a recognised gradient or at least a linear transect. The slope of the line will measure β-diversity, and sudden changes will reflect community or higher-order boundaries (Odum *et al.*, 1960). Pattern analysis, much used in plant ecology (Greig-Smith, 1978; Kershaw, 1973; Goldsmith & Harrison, 1976) adopts this approach on a small scale.

Whittaker (1972) summarised many studies on diversity gradients along 'coenoclines' by plotting the mean similarity coefficients (see p. 521) for each transect interval: the first part of the resulting curve is extended to zero distance, and this allows the 'threshold diversity' (C_O) to be estimated.

Wilson & Mohler (1983) discuss the calculation of β-diversity for quantitative data. A direct approach to β-diversity would be to compare the separate diversity indices, and the Shannon–Weaver index (or developments of it) has sometimes been used (e.g. Kikkawa, 1968; Hummon, 1974). However, in light of the shortcomings of this index as a measure of α-diversity (see above and Hurlbert, 1971; May, 1975; Taylor *et al.*, 1976), it seems inappropriate as a general technique. Mountford's (1962) index of similarity is based on the comparison of samples by the log series distribution but, as Bullock (1971a) points out, its assumption that the difference in distributions between different sites will be the converse of the similarities between samples from the same site (and some common α) may not be sound biologically. Bullock's (1971b) tests of the index on actual data tended to support these reservations. The most effective approach is based on the use of similarity indices, which are described below. Any of these similarity indices can be used to compare samples along a transect. Generally, the best index for presence/absence data is the Jaccard, and for quantitative data is the Morisita–Horn (Wolda, 1981, 1983; Magurran, 2013). Magurran (2013) gives a simple method for comparing the distribution of similarity coefficients between transects. Let

$$\beta = (a + b)(1 - J),$$

where a is the number of species in sample A, b the number in sample B and J the Jaccard similarity index (p. 522). This index can be calculated for sample pairs along each transect. The distribution of β for each transect can then be compared.

13.1.8.1 Partitioning β-diversity between species replacement and loss

The change in species composition along a gradient or transect can be partitioned into species replacement (termed turnover) and species loss (or gain) which is termed the 'nestedness'.

Baselga (2010) noted that the Sørensen dissimilarity index between two sites measures both turnover and nestedness and is defined as:

$$\beta_{Sor} = \frac{a + b}{2a + b + c}$$

where a is the number of species in common between the sites, b is the total species number at site 1 and c is the total species number at site 2.

In contrast, Simpson's dissimilarity index only measures turnover and is defined as:

$$\beta_{Sim} = \frac{min\,(b, c)}{a + min\,(b, c)}.$$

The nestedness between the sites is therefore simply

$$\beta_{nest} = \beta_{Sor} - \beta_{Sim}.$$

Similar arguments can be constructed based on Jaccard's dissimilarity.

It is therefore possible to decompose β-diversity gradients into turnover and nestedness components. The R package betapart (Baselga & Orme, 2012) undertakes the calculations for both spatial and temporal changes in assemblage composition.

The approach of Baselga is not the only one that can be taken to partition β-diversity.

Legendre & Cáceres (2013) view β-diversity as the total variance in the community matrix of species by samples. This is obtained by two routes, either by computing the sum-of squares of the species abundance data, or via a dissimilarity matrix. Total β-diversity can be partitioned into: (1) the species contributions which measure the degree of variation of species over the study area; and (2) local contributions which measure the uniqueness of the various sites.

13.2 Similarity and the comparison and classification of samples

When we compare the flora or fauna sampled at different localities we can approach the task by considering either the similarity or distinctness of their species assemblages. The conventional approach has been to measure similarity and numerous methods have been suggested, the most successful of which are described below. Legendre & Legendre (2012) give more complete accounts of similarity and distance measures. More recently, the term 'complementarity' has been introduced by Vane-Wright et al. (1991) to measure the difference in the biota between potential reserves. Conservationists often wish to maximise the amount of biodiversity protected for a given number and size of reserve; thus, they seek to maximise the complementarity. Colwell and Coddington (1994) have argued for a broadening of the concept to measure distinctness at all spatial scales.

Ecological data sets can often be organised as a two-dimensional matrix consisting of n samples (or stations) by S species. If the objective is to search for species interrelationships then order $S \times S$ matrix of species correlation's or similarities may be formed and what is termed an R analysis is undertaken. If the objective is to identify samples with similar communities, then the order $n \times n$ matrix of sample correlation's or similarities may be used in a Q analysis.

13.2.1   *Measures of complementarity*

Colwell and Coddington (1994) argue that the most appropriate measure is the Marczewski–Steinhaus distance:

$$c = \frac{\sum\limits_{i=1}^{S_{jk}} |X_{ij} - X_{ik}|}{\sum\limits_{i=1}^{S_{jk}} \max(X_{ij}....X_{ik})}, \tag{13.30}$$

where X_{ij} and X_{ik} are the presence–absence values (either 1 or 0) for species in species lists j and k. This is simply the complement of the familiar Jaccard index (see p. 522).

As a total measure for n lists, Colwell & Coddington (1994) quote Pielou as suggesting:

$$C_T = \frac{\sum U_{jk}}{n}, \tag{13.31}$$

where U_{jk} is the number of species only found in one of the two lists j and k and the summation is over all pairs of lists. C_T approaches $nS_T/4$ for large n, where S_T is the total species number in the combined species list.

13.2.2   *Similarity indices*

These are simple measures of either the extent to which two habitats have species in common (Q analysis) or species have habitats in common (R analysis). Binary similarity coefficients use presence–absence data, and following the introduction of computers more complex quantitative coefficients became practicable. Both groups of indices can be further divided into those which take account of the absence from both communities (double-zero methods) and those which do not. In most ecological applications it is unwise to use double-zero methods as they assign a high level of similarity to localities which jointly lack many species; double zeros are particularly prevalent in species-rich habitats such as the marine benthos or tropical arboreal arthropod assemblages. We therefore only describe binary indices that exclude double-zeros. A good account of similarity and distance measures is given in Legendre & Legendre (2012). Some methods use dissimilarity in their computations which is directly related to the similarity; for example Bray–Cutis dissimilarity is 1-Bray-curtis similarity.

13.2.2.1   Binary coefficients

These have been formulated in a number of slightly different ways and associated with even more originators. When comparing two sites, let a be the number of species held in common and b and c the numbers of species found at only one of the sites. When comparing two species the terms are similar; for example, a is the number of sites where they both were caught. The three simplest coefficients are:

Jaccard (1908) $C_j = a/(a + b + c)$

Sørensen (1948) $C_s = 2a/(2a + b + c)$

Mountford (1962) $C_M = \frac{2a}{2bc-(b+c)a}$

C_M was designed to be less sensitive to sample size than C_j or C_s; however, it assumes that species abundance fits a log-series model which may be inappropriate.

A more complex coefficient proposed by Fager (1957) for the study of zooplankton and used in an approach termed recurrent group analysis is:

$$C_F = \frac{a}{\sqrt{(a + b)(a + c)}} - \frac{1}{2\sqrt{a + c}}. \tag{13.32}$$

Following the evaluation of similarity measures using Rothamsted insect data, Smith (1986) concluded that no index based on presence/absence data was entirely satisfactory, but the Sørensen was the best of those considered.

13.2.2.2 Quantitative coefficients

Based purely on species number, binary coefficients give equal weight to all species and hence tend to place too much significance on the rare species whose capture will depend heavily on chance. Bray & Curtis (1957) brought abundance into consideration in a modified Sørensen coefficient and this approach, although criticised (Austin & Orloci, 1966), is widely used in plant ecology (Goldsmith & Harrison, 1976; Kent, 2012); the coefficient as modified essentially reflects the similarity in individuals between the habitats:

$$C_N = \frac{2jN}{aN + bN} \tag{13.33}$$

where aN = the total individuals sampled (N_T) in habitat a, bN = the same in habitat b and jN = the sum of the lesser values for the species common to both habitats (often termed W). The Bray–Curtis similarity index is frequently used for the analysis of benthic marine communities as the similarity measure of choice for ordination using multi-dimensional scaling.

However, for quantitative data, Wolda (1981, 1983) found that the only index not strongly influenced by sample size and species richness was the Moristita–Horn index:

$$C_{MH} = \frac{2 \sum (an_i . bn_i)}{(da + db)aN.bN} \tag{13.34}$$

where aN = the total number of individuals in site a, an_i = the number of individuals in the ith species in sample a, and

$$da = \frac{\sum an_i^2}{aN^2}.$$

Smith (1986) also concluded that for quantitative data the Moristita–Horn index was one of the most satisfactory.

The ready availability of computers now allows easy computation of Morisita's original index of similarity (Morisita, 1959):

$$C_m = \frac{2 \sum\limits^{n} X_{ij} X_{ik}}{(\lambda_1 + \lambda_2) N_j N_k} \tag{13.35}$$

where

$$N_j = \sum X_{ij}$$

$$N_k = \sum X_{ik}$$

$$\lambda_1 = \frac{\sum\limits^{n} [X_{ij}(X_{ij} - 1)]}{N_j(N_i - 1)}$$

and

$$\lambda_{12} = \frac{\sum\limits^{n} [X_{ik}(X_{ik} - 1)]}{N_k(N_k - 1)}$$

where X_{ij} and X_{ik} are the number of individuals of species i in habitats j and k, respectively. λ is Simpson's (1949) diversity index.

A considerably simpler, and frequently used, index is percent similarity (Whittaker, 1952) calculated using:

$$p = 100 - 0.5 \sum\limits_{i=1}^{S} |P_{a,i} - P_{b,i}| \tag{13.36}$$

where $P_{a,i}$ and $P_{b,i}$ are the percentage abundances of species i in samples a and b, respectively and S is total species number. This index takes little account of rare species.

13.2.2.3 Computation and display of indices

There are numerous similarity and distance measures available within R. The following R code listing calculates the Jaccard dissimilarity index and then presents the pattern of similarities between years using a heat map generated using the function R colddiss which is also listed below. The results of the output are shown in Plate 7.

```
library(gclus)        #package used to reorder the dissimilarity matrix
library(cluster)
library(vegan)        #package used for function vegdist
data <- read.table("C \\Hinkley annual data.csv", header = TRUE, sep = ",") #Load
Hinkley data
```

```
rnames <- data[,1]                          # assign labels in column 1 to "rnames"
mat_data <- data.matrix(data[,2:ncol(data)])   # transform column 2-n into a matrix
rownames(mat_data) <- rnames                 # assign row names
x <- as.matrix(mat_data)                     #put the data in a matrix
#calculate Jaccard's presence-absence dissimilarity
species.dis.jaccard <- vegdist(x, "jac", binary=TRUE)
#To produce a heat map to summarise similarities the function coldiss is used
# available at http://ichthyology.usm.edu/courses/multivariate/coldiss.R
#You will need to tell R where to find coldiss.R
source("coldiss.R")                          #Instruct R to accept input from the file
coldiss.R
coldiss(species.dis.jaccard, byrank=FALSE, diag=TRUE) #Call colddiss
#Second example using the transpose of the matrix x and Euclidean distance
col_distance = dist(t(x), method = "euclidian")
coldiss(col_distance, byrank=FALSE, diag=TRUE)
#Third example using the Bray-curtis distance measure on the transpose of x
col_distance.bray_curtis = vegdist(t(x))     #Bray-Curtis is the default for vegdist
coldiss(col_distance_bray_curtis, byrank=FALSE, diag=TRUE)


# coldiss()
# Color plots of a dissimilarity matrix, without and with ordering
# License: GPL-2
# Author: Francois Gillet, August 2009
"coldiss" <- function(D, nc = 4, byrank = TRUE, diag = FALSE)
{
  require(gclus)
  if (max(D)>1) D <- D/max(D)
  if (byrank) {
    spe.color = dmat.color(1-D, cm.colors(nc))
  }
  else {
    spe.color = dmat.color(1-D, byrank=FALSE, cm.colors(nc))
  }
  spe.o = order.single(1-D)
  speo.color = spe.color[spe.o,spe.o]
   op = par(mfrow=c(1,2), pty="s")
   if (diag) {
     plotcolors(spe.color, rlabels=attributes(D)$Labels,
        main="Dissimilarity Matrix",
        dlabels=attributes(D)$Labels)
     plotcolors(speo.color, rlabels=attributes(D)$Labels[spe.o],
        main="Ordered Dissimilarity Matrix",
        dlabels=attributes(D)$Labels[spe.o])
  }
  else {
```

```
plotcolors(spe.color, rlabels=attributes(D)$Labels,
    main="Dissimilarity Matrix")
plotcolors(speo.color, rlabels=attributes(D)$Labels[spe.o],
    main="Ordered Dissimilarity Matrix")
}
par(op)
}
```

13.2.3 *Multivariate analysis*

Multivariate analysis is used when the objective is to search for relationships between or classify objects (sites or species) which are defined by a number of attributes. Data sets can be large; for example, marine benthic or forest beetle faunal studies can easily require the analysis of a matrix of 100 samples (stations) by 350 species and thus multivariate analysis requires a computer. Curiously, most methods were developed before widespread availability of the computer and therefore did not find immediate general application. If the objective is to assign objects into a number of discrete groups, then cluster analysis should be considered. If there is no a priori reason to believe the objects will or could naturally fall into groups, then an ordination technique may be more suitable. Ordination assumes the objects form a continuum of variation, and the objective is often to generate hypotheses about the environmental factor(s) which mould community structure.

There are several motivations for using multivariate methods. In general, they are required because we need to search for the pattern of relationships between many variables simultaneously. Complex inter-relationships will not allow a useful analysis to be obtained by using each variable in isolation. The main motivations are:

- Classification: dividing variables or samples into groups with shared properties.
- Identifying gradients, trends or other patterns in multivariate data.
- Identifying which explanatory, independent or environmental variables are most influential in determining sample or community structure.
- Finally, and usually most importantly, we aim to distil the most important features from a set of data derived from an almost infinitely complex world, so that these can be presented clearly to others. This often entails displaying the main features in a two- or three-dimensional plot.

There is a considerable literature on multivariate techniques; useful texts for ecologists include Seal (1964), Cooley & Lohnes (1971), Digby & Kempton (1987), Henderson & Seaby (2008), Kent (2011), and Legendre & Legendre (2012).

13.2.3.1 Cluster analysis

When a number of sites or habitats are to be compared, the similarity measures of Section 13.2.2 can form the basis of cluster analysis which seeks to identify groups of sites or stations which are similar in their species composition. Alternatively, distance measures can be used to measure between object similarity (Digby & Kempton, 1987). Accounts

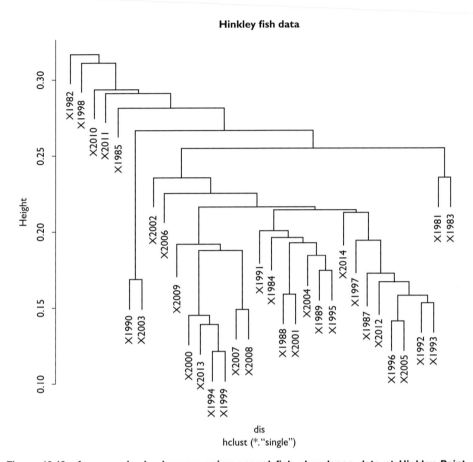

Figure 13.10 An example dendrogram using annual fish abundance data at Hinkley Point, Somerset. The dendrogram was generated using the R function hclust in the Vegan package to undertake hierarchical cluster analysis. The dissimilarity measure used was the Bray–Curtis. The R code used to generate this dendrogram is given in Section 13.2.3.1.

of the many methods that have been developed are given in Legendre & Legendre (2012) and Digby & Kempton (1987). These methods require a computer and many general statistical and specialised packages offer cluster analysis. For those unfamiliar with R, specialist packages such as Community Analysis Package (http://www.pisces-conservation.com/index.html?softcap.html$softwaremenu.html) can be used. The R function hclust can be used to undertake hierarchical cluster analysis. The following code lists a simple application using the Bray–Curtis dissimilarity measure, which is the default for the function vegdist, and single linkage agglomerative clustering. The result, shown in Fig. 13.10, indicates, for example, that 1981 and 1983 had similar species complements.

library(vegan)   #load the vegan package for Bray-Curtis
                                            dissimilarity
#Load Hinkley data: columns = species, rows=years
#The data give annual abundance of fish species between 1981 and 2014

```
data <- read.table("C:\\ Hinkley annual data.csv", header = TRUE, sep = ",") #read in
data from a csv file
rnames <- data[,1]                          # assign labels in column 1 to "rnames"
mat_data <- data.matrix(data[,2:ncol(data)]) # transform column 2-n into a matrix
rownames(mat_data) <- rnames                # assign row names
x <- as.matrix(mat_data)                    # makes a matrix of values
dis <- vegdist(t(x))                        # calculates Bray-Curtis dissimilarity on
                                              the transposed matrix
clus <-hclust(dis, method="single")         # Hierarchical clustering using single
                                              linkage
plot(clus, main = "Hinkley fish data")      # plot the dendrogram
```

There are numerous distance, similarity and dissimilarity measure available within R. For example, the dist function can be used to calculate the Euclidian distance use dist(x, method = "euclidian").

It is also possible using R to cluster both the samples and the species and produce a heat map which will order the species and samples so that all the species showing similar patterns of abundance cluster together. Plate 7 shows the output from the following code listing.

```
#load relevant packages
require(graphics); require(grDevices); require(RColorBrewer)
#set colour palette for heatmap
my_palette <- colorRampPalette(c("red", "yellow", "green"))(n = 299)
#Load Hinkley data
data <- read.table("C:\\Hinkley data.csv",header = TRUE, sep = ",")
rnames <- data[,1] # assign labels in column 1
mat_data <- data.matrix(data[,2:ncol(data)])
rownames(mat_data) <- rnames  # assign row names
mat_data=log10(mat_data+1)# log 10 transform
x  <- as.matrix(mat_data)
# calculate distances and produce clusters
row_distance = dist(x, method = "euclidian")
row_cluster = hclust(row_distance, method = "ward.D")
col_distance = dist(t(x), method = "euclidian")
col_cluster = hclust(col_distance, method = "ward.D")
rc <- rainbow(nrow(x), start = 0, end = .3)
cc <- rainbow(ncol(x), start = 0, end = .3)

hv <- heatmap(x, col=my_palette,  scale = "column",
          Rowv = as.dendrogram(row_cluster),
          Colv = as.dendrogram(col_cluster),
          xlab = "Years", #ylab =  "Fish Species",
          cexCol=0.9,
          cexRow=0.9,
          # main = "add title here if required"
          )
```

The heat map shown in Plate 7 classifies the species into differing abundance groups at the base there are the high abundance core specie which are always present. In contrast, the top of the map groups together the variable species which occasionally reach high abundance. The years (samples) are also arranged in terms of species similarity with the colder years of the 1980s forming a distinct cluster.

Classification methods comprise two principal types: hierarchical, where objects are assigned to groups which are themselves arranged into groups as in a dendrogram (see Fig. 13.10); and non-hierarchical, where the objects are simply assigned to groups. The methods are further classified as either agglomerative, where the analysis proceeds from the objects by sequentially uniting them, or divisive where all the objects start as members of a single group which is repeatedly divided. For computational and presentational reasons hierarchical–agglomerative methods are the most popular. However, divisive methods are frequently recommended because they initially divide the data into broad groups that can be used in classificatory schemes in the field.

The basic computational scheme used in cluster analysis can be illustrated using single linkage cluster analysis as an example. This is the simplest procedure and consists of the following steps.

1. Start with n groups each containing a single object (sites or species).
2. Calculate, using the similarity measure of choice, the array of between object similarities.
3. Find the two objects with the greatest similarity, and group them into a single object.
4. Assign similarities between this group and each of the other objects using the rule that the new similarity will be the greater of the two similarities prior to the join.
5. Continue steps 3 and 4 until one object is formed.

The results from a cluster analysis are usually presented in the form of a dendrogram (Fig. 13.10). The problem with all classification methods is that there can be no objective criteria of the best classification; indeed, even randomly generated data can produce a pleasing dendrogram. Always consider carefully if the groupings identified seem to make sense and reflect some feature of the natural world.

The Two-way indicator species analysis (TWINSPAN) procedure (Hill *et al.*, 1975) also produces dendrograms of the relationship between species and samples, but uses the reciprocal averaging ordination method to order the species and samples. This method is particularly attractive in studies where the objective is to classify communities so that field workers can quickly assign an area to a community type. This is because it identifies indicator species characteristic of each community. While once popular, the method has been subject to criticism in part because it is far from clear how exactly the result is generated and also because the pseudospecies concept is artificial. In vegetation science in particular, the method still has some popularity as the results can be useful. The Community Analysis Package (Pisces Conservation Ltd) offers an easy to use implementation of the procedure for Windows PCs.

13.2.3.2 Ordination

A number of ordination techniques are commonly used by ecologists, and it is not possible to give clear guidance as to the method that should be used. In community studies the aim of the analysis is to summarise the relationship between the samples (sites), and thus the best method is the one that gives the clearest and most interpretable picture. Hence, it may be appropriate to try a number of methods and compare the results. When there are clear differences between samples in terms of their species complements it is almost always the case that this will be shown by all of the commonly used ordination methods.

Typically, the sampling program has been designed to gain insight into what taxa are present at different localities, and to compare the results to see if they fall into some sort of pattern or classification. In many studies, the variables that can explain this pattern are not measured, or may not even be known. Possibly, after a pattern has been detected, an explanation might be inferred. For example, the distribution of benthic infauna observed might lead to an idea about the past human activity in the area. Methods which do not explicitly consider the explanatory variables include Principal Component Analysis (PCA), Correspondence Analysis (CA) and Non-metric Multidimensional Scaling (NMDS). Recently, because of its flexibility in choice of distance or dissimilarity measure, NMDS has become popular, particularly when used with the Bray–Curtis dissimilarity measure and quantitative species data.

If the objects or samples under study can be placed into groups, we often need to test if these groups are statistically significant. Analysis of Similarities (ANOSIM) is often the method of choice to test if members of a group are more similar to each other than they are to members of other groups. This randomisation test is particularly useful as it is generally applicable. With previously defined groups, Discriminant Analysis (DA) can be used to create a discriminant function to predict group membership. These predicted memberships can also be used to validate groups formed by Agglomerative Cluster Analysis (ACA).

There is also a group of methods for the analysis of situations where possible explanatory variables have been measured together with the descriptive variables. One of the most familiar is multiple regression, where a model is constructed in which a number of explanatory variables are used to predict the value of a dependent variable. Canonical Correspondence Analysis (CCA), a constrained ordination method based on correspondence and regression analysis. This method is presently by far the most popular constrained ordination method. CCA is termed a constrained method because the sample ordination scores are constrained to be a linear combination of the explanatory variables. Unlike unconstrained methods, CCA allows significance testing for the possible explanatory variables via randomisation tests.

A wide range of ordination techniques are available in R and the vegan package is frequently used. The following code listing undertakes the computation and graphical output for correspondence analysis and NMDS.

```
library(vegan) # We will use the vegan community analysis package
#Load data which is a species by sample array prepared in Excel and save as a csv file
test.csv <- read.table("C \\trans Hinkley fish.csv", header = TRUE, sep = ",")
```

```
mca <- cca (test.csv) # run a correspondence analysis
mca # Print results
plot(mca) #Plot results
# Below is an approach to a more attractive plot
plot(mca, type = "n")
points(mca, display = "sites", col = "blue", pch = 16)
text(mca, col = "red", dis = "sp")
#Now an example of Non-Metric Multidimensional Scaling
MDS_ord <- metaMDS(test.csv) #The default dissimilarity is the Bray-Curtis measure
MDS_ord
plot(MDS_ord, type="t")
```

13.3 Species packing

The coexistence of organisms, especially of apparent relatedness, has fascinated naturalists since the time of Alfred Russel Wallace, and the extent of interaction between them – the interspecific competition matrix – is the meeting ground of community and population ecology. Field observations on the distribution of organisms against a resource gradient will give a measure of the species packing in relation to this particular component of the species' niche, it is also relatively easy to determine the extent to which species overlap on a particular resource gradient. Thus, measures of 'niche breadth' (or width) and 'niche overlap' *on this resource gradient* can be obtained. It is tempting (and indeed desirable) to go beyond this, and having determined similar values on other resource gradients, to compound the overlaps in a hopefully appropriate manner to get a true measure of niche overlap. With this 'traveller's cheque' in hand the jump is made across the chiasm from community to population ecology; on arrival, the niche overlap measurements are converted to competition coefficients in classical Lotka–Volterra (or Gause) competition equations. (A safer route to these coefficients is by measuring the actual impact of one species on another; e.g. Vandermeer, 1969.) This analogy is not made to cast disparagement on those who have attempted the 'jump', for nothing is achieved (especially in ecology) by pessimistic contemplation of the difficulties, but rather to caution intending travellers in what is still an exciting area of ecology. Functionally, the analogy is intended to serve as an explanation as to why the explicit treatment in this section makes but a short part of the 'journey'. At present one can, as Green (1971) pointed out, easily demonstrate from field data that two species do not occupy the same niche, but not, without experiments, that they occupy the same niche.

In this section the methods of assessing the extent to which species occur together, the usage of a resource spectrum by one species, and the comparison of resource utilisation between species and niche overlap will be described.

13.3.1 *Measurement of interspecific association*

These measurements may be based either on presence–absence data or on abundance figures; as Hurlbert (1971) points out, presence–absence data is preferable if it is desired to measure the extent to which the two species' requirements are similar. Interspecific competition (and other factors) may lead to a 'misleading' lack of association if the measure is based on abundance. However, it would seem that whenever possible both types of analysis should be undertaken for a positive association on presence–absence data and a much weaker or negative one on abundance data would suggest (not prove) interspecific competition worthy of further analysis.

13.3.1.1   The departure of the distribution of presence or absence from independence

These methods measure the departure from independence of the distribution of the two species and *assume that the probability of occurrence of the species is constant for all samples.* Thus, if the distribution of two aphid predators were being compared it would only be legitimate to include samples that also contained the prey. A good example of the use of these indices is given by Evans & Freeman (1950), who measured the interspecific association of two species of flea on two different rodents (*Apodemus* and *Clethrionomys*); they found that there was a strong negative association on *Apodemus* and a moderate positive one on *Clethrionomys*, and suggested that the coarse and longer fur of *Clethrionomys* allowed the two fleas to avoid competition and exist together on that host. However, when applied to habitats whose uniformity is doubtful, these methods may give results of uncertain value, as was found by Macan (1954) in a study of the associations of various species of corixid bug in different ponds. In such a situation, some samples may be from habitats that are outside the environmental ranges of the species; this will have the effect of inflating the value for *d* in the table below and so lead to too many positive associations.

   Fager (1957) has demonstrated, by examples, that if two species are rare and therefore both are absent from most of the samples (this could, as just indicated, be due to unsuitable samples being included), a high level of association will be found. Conversely, if two species occur in most of the samples and so are nearly always found together, no association will be shown with these methods. (Although a biological association is obvious, it is equally correct to say that this does not depart significantly from an association that is due to chance and therefore does not necessarily imply any interspecific relationship.)

13.3.1.2   The two-dimensional Kolmogorov–Smirnov test

Garvey *et al.* (1998) proposed that a statistical technique used by astronomers could be used to detect the association between two species. The observations for each species are plotted as a scatter plot, each pair of coordinates (X, Y) is taken in turn and the number of points in each of the four surrounding quadrates (with (X, Y) as the origin) is counted. The counts for the four quadrates are expressed as proportions. These proportions are compared with those that would be expected if the two species were independent. The expected values are calculated by multiplying the counts of the proportions of observations $< X$ and $\geq X$ and $< Y$ and $\geq Y$. Within each quadrate the difference between the observed and expected proportion of points is determined and the maximum difference between the observed and expected over all the quadrates and points found. The original pairs of data points are then randomised 5000 times using a computer and the test calculations preformed on each randomisation. The statistical difference is then determined by comparing the observed maximum difference against the randomly created distribution of maximum differences.

13.3.1.3   The contingency table

The basis of these methods is the 2×2 contingency table.

		Species A		
		Present	Absent	Total
Species B	Present	a	b	a+b
	Absent	c	d	c+d
	Total	a+c	b+d	n = a+b+c+d

Such a table should always be drawn up so that A is more abundant than B, i.e. (a + b) < (a + c). A number of statistics are available for analysis of such a table, but the corrected chi-square (χ^2) makes fewest assumptions about the type of distribution, and the significance of the value obtained can be determined from tables available in standard statistical textbooks. It is calculated:

$$\chi^2 \frac{n\left[|ad - bc| - (n/2)\right]^2}{(a + c)(b + d)(a + b)(c + d)}$$
(13.37)

The test in this form is only valid if the expected numbers (if the distribution was random) are not less than 5; there is only one degree of freedom and so the 5% point is 3.84. Therefore, if a χ^2 of less than this is obtained, any apparent association could well be due to chance and further analysis should be abandoned. If the smallest expected number is less than 5, the exact test should be used.

13.3.1.4   Coefficients of association

If χ^2 is significant, then one of the coefficients of association may be used to give an actual quantitative value for comparison with other species. They are designed so that the coefficient has the same range as the correlation coefficient (r), i.e. $+1$ = complete positive association, -1 = complete negative association, 0 = no association. Cole (1949) reviews a number of coefficients and points out how, if the above interpretations of values from $+1$ to -1 are to hold for comparative purposes, the plot of the value of the coefficient against the possible number of joint occurrences should be linear; for several of the coefficients it is not, and for those in which it is, the plot does not pass through the zero. Some coefficients are given below.

1. *Coefficient of mean square contingency.* This coefficient makes no assumption about distribution, but it cannot give a value of $+1$ unless $a = d$ and b and $c = 0$. For less extreme forms of association it is useful and easily calculated:

$$C_{AB} = \sqrt{\frac{\chi^2}{n + \chi^2}} \qquad (13.38)$$

where C_{AB} = coefficient of association between A and B, n = total number of occurrences, and the χ^2 value is obtained as above.

This coefficient is recommended by Debauche (1962).

2. *Coefficient of interspecific association.* The coefficients originally designed by Cole (1949) have been shown by Hurlbert (1969) to be biased by the species frequencies. He showed this bias was considerably diminished if the coefficient is defined as:

$$C_{AB}^1 = \frac{ad - bc}{|ad - bc|} \left| \left(\frac{Obs\,\chi^2 - Min\,\chi^2}{Max\,\chi^2 - Min\,\chi^2} \right)^{\frac{1}{2}} \right| \qquad (13.39)$$

$Min\,\chi^2$ is the value of χ^2 when the observed a differs from its expected value ($\hat{a}$) by less than 1.0 (except when $a - \hat{a} = 0$ or $= 0.5$, the value of $Min\,\chi^2$ depends on whether ($ad - bc$) is positive or negative) and is given by:

$$Min\,\chi^2 = \frac{n^3(a - g[a])^2}{(a + b)(a + c)(c + d)(b + d)} \qquad (13.40)$$

where $g(\hat{a}) = \hat{a}$, rounded to the next lowest integer when $ad < bc$ or rounded to the next highest integer when $ad > bc$, if $\hat{a}$ is an integer then $g(\hat{a}) = \hat{a}$.

Max χ^2 is the value of χ^2 when a is as large or small as the marginal totals of the 2×2 table will permit formulated under specified conditions as follows:

$ad \geq bc,$

$ad < bc, a\ \chi^2 = \dfrac{(a + b)(b + d)n}{(a + c)(c + d)} \leq d\ \chi^2 = \dfrac{(a + b)(a + c)n}{(a + b)(c + d)}$

$ad < bc, a > d\ \chi^2 = \dfrac{(b + d)(c + d)n}{(a + b)(a + c)}$

$Obs\,\chi^2$ is calculated in the normal manner (see above). Pielou & Pielou (1968) describe a method for species of infrequent occurrence.

13.3.1.5 Proportion of individuals occurring together

Correlation may be utilised if the data can be normalised, otherwise some of the methods of comparing similarity between habitats may be used. For example, Sørensen's coefficient, as used by Whittaker & Fairbanks (1958), may be modified, to give the normal range of –1 (no association) to +1 (complete association):

$$I_{ai} = 2\left[\frac{J}{A+B} - 0.5\right]$$ (13.41)

where J = no. of individuals of A and B in samples where both species are present, and A and B = total of individuals of A and B in all samples.

13.3.2   Measurement of resource utilisation

Among the meaningful resources that may be partitioned, by apparently coexisting species are: food (defined by quality (type) and size); space (defined in Euclidian terms; e.g. height in vegetation); and physico-chemical characters (e.g. climate, substrate type) and time (seasonal and diel). This, by no means exhaustive, list gives an idea of the number of 'resource spectra' that contribute to a species' niche (Pianka, 1976b).

In considering the most appropriate way to measure resource utilisation it is useful to start from a theoretical basis (May & MacArthur, 1972). The extent of separation in the resource utilisation between potentially competing species is well expressed in terms of the separation measure (d) between the means of their curves and the widths (one standard deviation in the normal Gaussian curve) (w) (Fig. 13.11). In this theoretical situation (where resource utilisation functions are Gaussian and equal), then if the resource separation ratio (d/w) is less than 3 there will be some interaction between the species. Theoretically, the ratio will have a minimum value below which the competitive exclusion principle operates (the two species cannot coexist). May's (1973, 1974) studies suggest that one is the minimum value, but the conversion of this ratio to a competition coefficient is, as already indicated, fraught with biological difficulties (May, 1975; Abrams, 1975; Heck, 1976).

Resource utilisation curves are seldom normal; indeed, field data suggest that they vary greatly, being broad where resources are scarce, and narrow where they are more

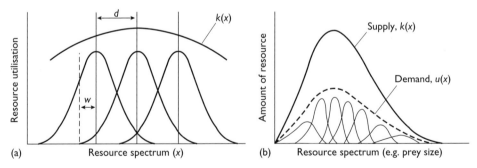

Figure 13.11   Theoretical resource-utilisation relationships. (a) The 'simple case' of three species with similar and normal) resource-utilisation curves; d, distance apart of means; w, standard deviation of utilisation, d/w, esource separation ratio = 2.5 (in these cases). (Adapted from May and MacArthur, 1972.) (b) The more typical case with varying resource-utilisation curves: broadest in the region of fewer resources and less interspecific competition. (Adapted from Pianka, 1976b.)

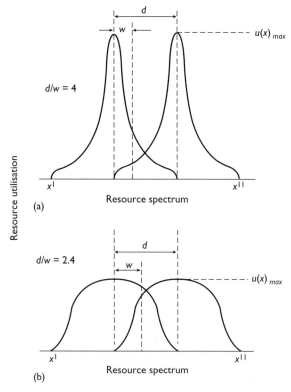

Figure 13.12   The effect of the shape of the utilisation curve on the separation measure (*d*) and hence species packing. Note that the two species with leptokurtic curves (a) and the two with platykurtic curves (b) both utilise the same quantity of resource (equal areas under curves) and the same range of resources (x^1–x^{11}), but the extent of utilisation, of which $u(x)_{max}$ is one measure, differs.

plentiful (Fig. 13.11b). The shape of the curve influences the extent of packing, as Roughgarden (1974) pointed out on mathematical grounds. This point is illustrated simply in Fig. 13.12, which represents two species pairs, the ones above having leptokurtic resource utilisation curves and the pair below platykurtic ones. It should be noted that each species takes an identical quantity of resource (the areas under the curves are equal) and the ranges ($x^1 \rightarrow x^{11}$) are equal, but when the lines for w (the distance from the mean of 68 % of the resource utilisation) are inserted their ratio (leptokurtic: platykurtic) is 3:5. The impact of this on the resource separation ratios (d/w) is shown. The difference implies that, if this was the only resource spectrum, four closely packed ($d/w \sim 1$) 'platykurtic species' could be replaced by seven closely packed 'leptokurtic species'. (These numerical values are, of course, precisely relevant only to Fig. 13.12.)

It will be noted that whereas the platykurtic species are not as closely packed as leptokurtic, the latter, for the same resource utilisation, have higher $u(x)_{MAX}$. If we imagine the supply of the resource $K(x)$ to be far above the utilisation (demand) level $u(x)$ in Fig. 13.12b, which could be due to predator pressure, then there will be no restrictions on the shapes of the curves that can replace the platykurtic ones, and tighter packing – as indicated above – is possible. There are two possible procedures for assessing species packing in respect of a resource dimension.

1. The d/w approach, in which the utilisation by each species is considered in turn, the mean and the width (w = 68% of the utilisation on one side of the mean) determined; the differences between the means giving d.
2. Proportional utilisation of the 'Simpson–Yule' index approach, in which the proportion of species utilising a particular segment of the resource gradient is determined (p_i), the breadth of utilisation ('niche breadth' of Levins, 1968) being equivalent to the Simpson–Yule index.

The former is conceptually simpler, as the investigator can visualise the 'areas' of resource occupied by the species, but it assumes a utilisation spectrum that approximates to the normal: thus, the resource dimension must be approximately ordered. It has seldom been used in field studies. The proportional utilisation approach makes no demands on the ordering on the resource dimension and has been widely used. Of course, p_i is simply the height of the utilisation curve at i, the curve whose width (in terms of one standard deviation) is given by w.

13.3.2.1   Species packing in terms of mean and width of resource utilisation spectrum ('*d/w* method')

This method can easily be used when the animal's use of the resource is for a discrete period and the resource spectrum is continuous: this applies especially to seasonal occurrence, diel occurrence and food items or other resources defined by size. Data on seasonal occurrence is widely available and often shows striking patterns of species packing (e.g. Greenslade, 1965; Shapiro, 1975; Southwood, 1978). The first steps are to find the mean and the total range of time of occurrence, which can usually be taken as twice the interval between the mean and the earliest occurrence. This method eliminates the extreme, end of season 'tail': these may be post-reproductives or late developers which will not survive winter (this may be checked) and will not affect the fitness of that species (they may however affect the species that occur with them, and then a new range will need to be found and used when assessing the impact of this putative competitor on those species). If the frequency of occurrence is approximately normally distributed (see p. 12) then the total range will be equivalent to $6w$ (three standard deviations on each side of the mean). If the frequency of occurrence shows kurtosis or other irregularity, w may be calculated by a method analogous to the graphical method for estimating the numbers in a stage (see Fig. 11.1). The mean is drawn in and the number of animal days (or other time period, e.g. 5 days) summed visually. Then, starting from the mean and working in vertical columns towards an end of the range, one sums the 'animal days' until 68% of the total that side of the mean have been accumulated. At this point a vertical line can be drawn and the distance of this to the mean approximates w. Experience with the graphical method for determining the numbers entering a stage (p. 432) has shown its robustness, and as a method of estimating resource utilisation it is simple and free of assumptions. Large sets of data are easily handled on a computer, but it is often necessary to make decisions about stragglers that should be based on biological knowledge about the specific and not on a general rule. This technique allows for different ws either side of the mean and for adjustments, as in the case of post-reproductives mentioned above.

When the values of w are different for the ith and jth species, a 'common w' $[w_{ij}]$ must be calculated:

$$w_{ij} = \left[w_i^2 + \frac{w_j^2}{2} \right]^{\frac{1}{2}}$$

(13.42)

The resource separation ratio is:

$$p_{ij} = \frac{d_{ij}}{w_{ij}}$$

where d_{ij} is the difference between the means of resource utilisation by the ith and jth species.

Species packing with respect to periodic variation at annual, tidal, or diurnal time scales must take account of the circularity of the resource and the ratio for a pair of species may be in the form $d/2w$, because both 'tails' can overlap. Greenslade (1963) treated data on the diurnal/nocturnal partitioning of Carabids in this manner.

Size may easily be arranged on a linear scale and it governs many aspects of resource utilisation. The size (and hence type) of food may show a reasonable relationship to body size (Hespenheide, 1975; Wilson, 1975; Davidson, 1977) or body part (e.g. bees' tongues, (Heinrich, 1976) or birds' bills (Karr & James, 1975); the trivial range is also approximately related to size when account is taken of feeding habits (predator versus herbivore) and habitat type (e.g. desert versus temperate grassland); in such an analysis for ambulatory animals, the span between the front and back legs is a better measure of size than body length.

Other resources, such as plant species or types for herbivores, could be arranged as the resource spectrum, but the ordering would require judgement analogous to Cody (1975) in the ordering of bird species (see above). Under such circumstances the more versatile proportional utilisation method is more objective.

13.3.2.2 Species packing in terms of proportional utilisation of different resource states (p_i method)

The breadth of utilisation ('niche breadth'; Levins, 1968) is given by the Simpson–Yule index applied to the distribution of the individuals between the resource states:

$$B = \frac{1}{\sum\limits_i^m p_i^2} = \frac{1}{\sum\limits_i^m \frac{N_i^2}{N_T^2}}$$

(13.43)

where p_i = proportion of individuals found in or using resource i, N_j = the number of individuals of the species in question in the ith resource state, N_T = the total number of individuals in all m resource states.

In this context the tendency of this index to 'undervalue' rare events is, probably, an advantage.

This method is versatile in that the 'resource states' do not have to be ordered along a continuum; they may, for example, be samples from m different categories of food type found by stomach content analysis or some other method. In comparative studies it may be useful to calculate the standardised niche breadth:

$$B_s = \frac{B-1}{m-1}. \qquad (13.44)$$

The calculation of B from a series of random samples from a heterogeneous environment is, however, open to criticism, for the degree of uniformity in distribution could well be influenced by the natural clumping (or otherwise) of the organism (see Chapter 2); one may claim that this is part of the resource, but this is not very illuminating. There is a considerable interpretative advantage if distinct resource states are sampled separately: for example, parts of the habitat that from other knowledge are known to be distinguished by ecologically significant factors. However, samples often reveal vagrants, transitory members of the fauna – insects that engage in upper air migration (e.g. aphids, many flies, coccinellids, etc.) regularly occur in places that are not part of *their* niche, as is also found with fish displaced by torrent, storm or current; Magurran & Henderson, 2003). Studies of colonies of Lepidoptera record the movement of individuals across terrain that is, again, not part of their niche. If these vagrants are relatively few, the Simpson–Yule index's character of emphasising the high probability occurrences will minimise their significance. A more general criticism (for the purposes of determining competition is that 'handfuls of animals' taken from their habitat may not represent usage of the same resource; one may live on the surface of the leaf blade, one on the stem; one may be active by day, another by night, and so on). Thus, although this approach is valid for the assessment of the *breadth of resources* used, its utilisation to calculate *'niche overlap'* must be handled with caution.

There is a further problem if the different resource states are present (in the samples) at very different frequencies. Colwell & Futuyma (1971) used as a measure of resource breadth the Shannon–Wiener index (see p. 509):

$$H = \sum pi \log pi$$

where p_i is the proportion of individuals using resource i.

They proposed weighting resource states so that each is equally represented. The weightings were defined, not in terms of physico-chemical or other measurements, but in terms of the rarity of the fauna. Sabath & Jones (1973) found that the weightings did not add to the information gained from unweighted matrices, and that the expansion of the matrix eventually prevented discrimination between species distributions. Another detailed study using the Colwell–Futuyma weightings was made by Hanski & Koskela (1977) on beetles in dung pats. This would seem to be an ideal 'test' of the concept for the habitat is discrete, its state may also be described by certain physical characters, and the beetles collected are a real assemblage which cannot necessarily be claimed for trap catches, a difficulty of which Sabath & Jones (1973) were well aware. Hanski & Koskela (1977) conclude that weighting by unusual species is unsatisfactory; however, they did weight 'resource states' in relation to the total number of individuals in each. Once again they found all the measurements of 'niche breadth' highly correlated, and concluded that the main results were fairly insensitive to weighting. It seems therefore that, in practice, the simple B has much to recommend it as a measure of the width of resource utilisation by a species.

Hurlbert (1977) argues that usage should be scaled by availability, so that Equation (13.48) is reformulated as:

$$B' = \frac{1}{\sum p_j^2 / a_j} \tag{13.45}$$

where a_j is the proportion of the total available resources consisting of resource j.

The variance of both B and B' is given by Smith (1982) as:

$$\text{var}(B) = \frac{4B^4 \left[\sum_{j=1}^{n} \frac{p_j^3}{a_j^3} - \left(\frac{1}{B} \right)^2 \right]}{\sum N_j}. \tag{13.46}$$

An alternative measure of niche breadth proposed by Smith (1982) is:

$$FT = \sum_{j=1}^{n} \sqrt{p_j a_j}, \tag{13.47}$$

which for large samples has estimated upper and lower 95% confidence:

$$\sin \left(x - \frac{1.96}{2\sqrt{\sum N_j}} \right)$$

and

$$\sin \left(x + \frac{1.96}{2\sqrt{\sum N_j}} \right)$$

respectively.

The angle is expressed in radians and $x = \arcsin(FT)$.

The proportional overlap in resource utilisation between two species (i and j) is (Schoener, 1968; Colwell & Futuyma, 1971):

$$\Theta_{ij} = 1 - 0.5 \sum_{h=1}^{m} p_{ih} - p_{jh} \qquad (13.48)$$

where p_{ih} = the proportion of species i in resource state h and p_{jh} = the proportion of species j in same resource states and m = the total number of resource states. (Note that Θ measures 'overlap', whilst p measures separation.) This method has been used with, for example, leaf-hoppers (McClure & Price, 1976) and wolf-spiders (Uetz, 1977).

Hurlbert (1978) defined niche overlap as 'the degree to which the frequency of encounter between two species is higher or lower than it would be if each species utilised each resource state in proportion to the abundance of that resource state' and so, L, niche overlap, is given by:

$$L = \sum \left(\frac{p_{ij} p_{ik}}{a_i} \right) \qquad (13.49)$$

where p_{ij} and p_{ik} are the proportion of resource i used by species j and k, respectively. Using computer simulation, Rickleffs & Lau (1980) showed niche overlap to be strongly influenced by sample size, while Smith & Zaret (1982) showed that the bias in its estimation is minimised when the numbers of the two species are equal. Niche overlap can also be measured using the Morisita–Horn similarity indices (p. 522).

13.3.3 Niche size and competition coefficients

All the methods discussed above provide information on species packing and their resource utilisation. These measures are themselves intrinsically interesting and may be used in studies of community ecology, island biogeography, and so on (e.g. Price, 1971; McClure & Price, 1976; Mühlenberg et al., 1977). The difficulty of equating resource utilisation measures with niche breadth has already been outlined. The utilisation indices (w and B) pose two problems: how may they be combined and, less easily answered, have all the components of the niche been assessed?

If species packing indices (p_{ij} and Θ_{ij}) are to be equated with niche overlap and hence competition attention also has to be paid to the renewability of resources, interactions at other trophic levels and the relative availability of other resources (May, 1973, 1974; Schoener, 1974; Levine, 1976; Pianka, 1976b).

Competition coefficients between the ith and jth species for one resource axis may be calculated (May, 1973, 1975):

1. from resource separation ratio:

$$a_{ij} = \exp \left[\frac{-d_{ij}^2}{4 \left| w_{ij}^2 \right|} \right]$$

2.   from proportional utilisation functions:

$$a_{ij} = \frac{\sum p_i p_j}{\sqrt{(\sum p_i)^2 (\sum p_j)^2}}$$

where each term is summed for the array of resource states along this dimension. This expression is due to Pianka (1973); May (1975) compares it with alternatives.

Weighted competition coefficients for food resources may be calculated (Schoener, 1974):

$$a_{ij} = \left(\frac{T_j}{T_i}\right)\left[\frac{\sum \left(\frac{p_{ik}}{f_k}\right)\left(\frac{p_{jk}}{f_k}\right) b_{ik}}{\sqrt{\sum \left(\frac{p_{ik}}{f_k}\right)^2 \left(\frac{p_{jk}}{f_k}\right)^2 b_{ik}}}\right]$$

where T_j/T_i = the ratio of the total number of food items consumed by an individual of the jth to that consumed by an individual of the ith species, measured over an interval of time that includes all regular fluctuations in consumption for both species; p_{ik} and p_{jk}, the frequencies of food type k in the diets of i and j, respectively, f_k, the frequency of food item k in the environment, b_{ik} = the net calories gained by an individual of i from one item of k, or (more approximately) the calories in an item of k, or (still more approximately) the biomass of an item of k (see Chapter 14). The summations are taken over all (m) of the food items eaten by one (i) of the two species. The equation is modified from Schoener's so as to put it in the same form as the basic Equation (13.59) (Pianka's expression). The above expression may also be used to obtain a coefficient in relation to foraging times in the same habitats; the T's are the ratios of total time, b_{ik} the calories obtained from an average item of food in habitat k, and the fs are omitted. As T_j/T_i will often approximate to unity, unless the average value of the food items in the different habitats are remarkably different, b_{ik} will not be significant, this will often reduce to the basic (Pianka's) expression. Thus, as indicated above (p. 538), weighting may not in practice always be necessary.

Competition coefficients between species may be expressed as a matrix – the diagonal values representing intraspecific interactions are unity (May, 1973, 1976).

Species generally interact in more than one resource dimension; as Schoener (1974) points out, the number of dimensions, their type, and the comparative extent of the interactions, are of intrinsic interest and already certain patterns are emerging. For example, predators partition a habitat through the diel cycle more often than other types.

The combination of competition coefficients from different resource axes is not straightforward except, as May (1975) has clearly shown, in the two limiting cases. When the two axes are totally independent, then the coefficients (a-values) should be multiplied; when the two axes are completely correlated, then the coefficients should be summed. A method for use with field data, allowing for various degrees of independence between resource axes, has not yet been devised.

Nevertheless, as there are so many difficulties over the assumptions of elegant theoretical models (May, 1973, 1974, 1975; Abrams, 1975; Heck, 1976; Armstrong, 1977), it seems important to determine from field observations which assumptions are justified and which refinements are, in practice, trivial. Field studies should be made over as many resource dimensions as possible, the d/w ratio used to determine the shape and separation of resource utilisation and where possible, the impact of perturbations studied (although here too there are difficulties in interpretation (Schoener, 1974).

References

Abrams, P. (1975) Limiting similarity and the form of the competition coefficient. *Theor. Popul. Biol.* **8**, 356–75.

Armstrong, R.A. (1977) Weighting factors and scale effects in the calculation of competition coefficients. *Am. Nat.* **111**, 810–12.

Austin, M.P. & Orloci, L. (1966) Geometric models in ecology. II. An evaluation of some ordination techniques. *J. Ecol.* **54**, 217–22.

Baselga, A. (2010) Partitioning the turnover and nestedness components of beta diversity. *Global Ecology and Biogeography* **19**, 134–43.

Baselga, A. & Orme, C.D.L. (2012) betapart: an R package for the study of beta diversity. *Methods Ecol. Evol.* **3**, 808–12.

Bazzaz, F.A. (1975) Plant species diversity in old field successional ecosystems in Southern Illinois. *Ecology* **56**, 485–8.

Boswell, M.T. & Patil, G.P. (1971) Chance mechanisms generating the logarithmetic series distribution used in the analysis of number of species and individuals. In: Patil, G.P., Pielou, E.C., & Wates, W.E. (eds), *Statistical Ecology*. Penn State University Press, Philadelphia.

Bray, J.R. & Curtis, C.T. (1957) An ordination of the upland forest communities of southern Wisconsin. *Ecol. Monogr.* **27**, 325–49.

Bullock, J.A. (1971) The investigations of samples containing many species. I. Sample description. *Biol. J. Linn. Soc.* **3**, 1–21.

Bullock, J.A. (1971) The investigations of samples containing many species. II. Sample comparison. *Biol. J. Linn. Soc.* **3**, 23–56.

Bulmer, M.G. (1974) On fitting the Poisson lognormal distribution to species abundance data. *Biometrics* **30**, 101–10.

Burnham, K.P. & Overton, W.S. (1978) Estimation of the size of a closed population when capture probabilities vary among animals. *Biometrika* **65**, 623–33.

Burnham, K.P. & Overton, W.S. (1979) Robust estimation of population size when capture probabilities vary among animals. *Ecology* **60**, 927–36.

Chao, A. (1984) Non-parametric estimation of the number of classes in a population. *Scand. J. Statist.* **11**, 265–70.

Chao, A. (1987) Estimating the population size for capture-recapture data with unequal matchability. *Biometrics* **43**, 783–91.

Chao, A. & Lee, S.-M. (1992) Estimating the number of classes via sample coverage. *J. Am. Statist. Assoc.* **87**, 210–17.

Chave, J. (2004) Neutral theory and community ecology. *Ecol. Lett.* **7**, 241–53.

Cody, M.L. (1974) *Competition and the Structure of Bird Communities*. Princeton University Press, Princeton, New Jersey.

Cody, M.L. (1975) Towards a theory of continental species diversity: bird distributions over Mediterranean habitat gradients. In: Cody, M.L. & Diamond, J.M. (eds), *Ecology and Evolution of Communities*. Harvard University Press, Cambridge, Mass.

Cody, M.L. & Diamond, J.M. (eds) (1975) *Ecology and Evolution of Communities*. Harvard University Press, Cambridge, Mass.

Cole, L.C. (1949) The measurement of interspecific association. *Ecology* **30**, 411–24.

Coleman, B.D. (1981) On random placement and species-area relations. *Mathematical Biosciences* **54**, 191–215.

Coleman, M.D., Mares, M.D., Willig, M.R., & Hsieh, Y.-H. (1982) Randomness, area, and species richness. *Ecology* **63**, 1121–33.

Colwell, R.K. & Coddington, J.A. (1994) Estimating terrestrial biodiversity through extrapolation. *Philos. Trans. Roy. Soc. B* **345**, 101–18.

Colwell, R.K. & Futuyma, D.J. (1971) On the measurement of niche breadth and overlap. *Ecology* **52**, 567–76.

Cooley, W.W. & Lohnes, P.R. (1971) *Multivariate Data Analysis*. John Wiley & Sons, New York, London.

Davidson, D.W. (1977) Species diversity and community organization in desert seed-eating ants. *Ecology* **58**, 711–24.

Debauche, H.R. (1962) The structural analysis of animal communities of the soil. In: Murphy, P.W. (ed.), *Progress in Soil Zoology*. Butterworths, London.

Digby, P.G.N. & Kempton, R.A. (1987) *Multivariate Analysis of Ecological Communities*. Chapman & Hall, London.

Evans, F.C. & Freeman, R.B. (1950) On the relationship of some mammal fleas to their hosts. *Ann. Entomol. Soc. Am.* **43**, 320–33.

Fager, E.W. (1957) Determination and analysis of recurrent groups. *Ecology* **38**, 586–95.

Findley, J.S. (1973) Phenetic packing as a measure of faunal diversity. *Am. Nat.* **107**, 580–4.

Fisher, R.A., Corbet, A.S., & Williams, C.B. (1943) The relation between the number of species and the number of individuals in a random sample of an animal population. *J. Anim. Ecol.* **12**, 42–58.

Gage, I. & Tett, P.B. (1973) The use of log-normal statistics to describe the benthos of Lochs Etive and Creran. *J. Anim. Ecol.* **42**, 373–82.

Garvey, J.E., Marschall, E.A., & Wright, R.A. (1998) From star charts to stoneflies: detecting relationships in continuous bivariate data. *Ecology* **79**, 442–7.

Goldsmith, F.B. & Harrison, C.M. (1976) Description and analysis of vegetation. In: Chapman, S.B. (ed.), *Methods in Plant Ecology*. Blackwell, Oxford, London.

Goodall, D.W. (1973) Sample similarity and species correlation. In: Whittaker, R.H. (ed.), *Handbook of Vegetation Science, Part V*. Junk, The Hague.

Green, R.H. (1971) A multivariate statistical approach to the Hutchinson niche: bivalve molluscs of central Canada. *Ecology* **52**, 543–56.

Greenslade, P.J.M. (1965) On the ecology of some British Carabid beetles, with special reference to life histories. *Trans. Soc. Br. Entomol.* **16**, 149–79.

Greig-Smith, P. (1978) *Quantitative Plant Ecology*. Butterworths, London.

Hanski, I. & Koskela, H. (1977) Niche relations among dung-inhabiting beetles. *Oecologia* **28**, 203–31.

Harrison, S., Ross, S.J., & Lawton, J.H. (1992) Beta-diversity on geographic gradients. *J. Anim. Ecol.* **61**, 151–8.

Heck, K.L. (1976) Some critical considerations of the theory of species packing. *Evol. Theory* **1**, 247–58.

Heinrich, B. (1976) Resource partitioning among some unsocial insects: bumblebees. *Ecology* **57**, 874–89.

Heltshe, J. & Forrester, N.E. (1983) Estimating species richness using the jackknife procedure. *Biometrics* **39**, 1–11.

Henderson, P.A. (1989) On the structure of the inshore fish community of England and Wales. *J. Mar. Biol. Assoc. UK* **69**, 145–63.

Henderson, P.A. (2007) Discrete and continuous change in the fish community of the Bristol Channel in response to climate change. *J. Mar. Biol. Assoc. UK* **87**, 589–98.

Henderson, P.A. & Holmes, R.H.A. (1991) On the population dynamics of dab, sole and flounder within Bridgwater Bay in the lower Severn Estuary, England. *Netherlands J. Sea Res.* **27**(34), 337–44.

Henderson, P.A. & Magurran, A.E. (2014) Direct evidence that density-dependent regulation underpins the temporal stability of abundant species in a diverse animal community. *Proc. Royal Soc. B: Biol. Sci.* **281**, 2014 1336.

Henderson, P.A. & Seaby, R.M. (2008) *A Practical Handbook for Multivariate Methods*. Pisces Conservation, Lymington, UK.

Hendrickson, J.A. & Ehrlich, P.R. (1971) An expanded concept of 'species diversity'. *Notulae Naturae Acad. Natur. Sci. Philadelphia* **439**, 1–6.

Hespenheide, H.A. (1975) Prey characteristics and predator niche width. In; Cody, M.L. & Diamond, J.M. (eds), *Ecology and Evolution of Communities*. Harvard University Press, Cambridge, Mass.

Hill, M.O. (1973) Diversity and evenness: a unifying notation and its consequences. *Ecology* **54**, 427–32.

Hill, M.O., Bunce, R.G.H., & Shaw, M.W. (1975) Indicator species analysis, a divisive polthetic method of classification and its application to a survey of native pinewoods in Scotland. *J. Ecol.* **63**, 597–613.

Holdridge, L.R., Grenke, W.C., Hatheway, W.H., Liang, T., & Tosi, J.A. (1971) *Forest Environments in Tropical Life Zones*. Pergamon Press, Oxford.

Hubbell, S.P. (2001) *The unified neutral theory of biodiversity and biogeography* (MPB-32) (Vol. 32). Princeton University Press.

Hubbell, S.P. (2006) Neutral theory and the evolution of ecological equivalence. *Ecology* **87**, 1387–98.

Hummon, W.D. (1974) SH: A similarity Index based on shared species diversity used to assess temporal and spatial relations among inter-tidal Marine Gastrotricha. *Oecologia* **17**, 203–20.

Hurlbert, S.H. (1969) A coefficient of interspecific association. *Ecology* **50**, 1–9.

Hurlbert, S.H. (1971) The non-concept of species diversity: a critique and alternative parameters. *Ecology* **52**(4), 577–86.

Hurlbert, S.H. (1978) The measurement of niche overlap and some derivatives. *Ecology* **59**, 67–77.

Jaccard, P. (1908) Nouvelles recherches sur la distribution florale. *Bull. Soc. vandoise Sci. nat.* **44**, 223–70.

James, F.C. & Rathbun, S. (1981) Rarefaction, relative abundance, and diversity of avian communities. *Auk* **98**, 785–800.

Karr, J.R. & James, F.C. (1975) Eco-morphological configurations and convergent evolution in species and communities. In: Cody, M.L. & Diamond, J.M. (eds), *Ecology and Evolution of Communities*. Harvard University Press, Cambridge, Mass.

Kempton, R.A. (1979) Structure of species abundance and measurement of diversity. *Biometrics* **35**, 307–22.

Kempton, R.A. & Taylor, L.R. (1974) Log-series and log-normal parameters as diversity discriminants for the *Lepidoptera. J. Anim. Ecol.* **43**, 381–99.

Kent, M. (2011) *Vegetation Description and Data Analysis: A Practical Approach.* John Wiley & Sons.

Kershaw, K.A. (1973) *Quantitative and Dynamic Plant Ecology.* Griffin, London.

Kikkawa, J. (1968) Ecological association of bird species and habitats in Eastern Australia; similarity analysis. *J. Anim. Ecol.* **37**, 143–65.

King, J.R. & Porter, S.D. (2005) Evaluation of sampling methods and species richness estimators for ants in upland ecosystems in Florida. *Environ. Entomol.* **34**, 1566–78.

Kricher, J.C. (1972) Bird species diversity: the effect of species richness and equitability on the diversity index. *Ecology* **53**, 278–82.

Lamas, G., Robbins, R.K., & Harvey, D.J. (1991) A preliminary survey of the butterfly fauna of Pakitza, Parque Nacional del Manu, Peru, with an estimate of its species richness. *Publ. Mus. Hist. nat. UNMSM* **40**, 1–19.

Legendre, P. & Cáceres, M. (2013) Beta diversity as the variance of community data: dissimilarity coefficients and partitioning. *Ecol. Lett.* **16**, 951–63.

Legendre, P. & Legendre, L.F. (2012) *Numerical Ecology*, Vol. 24. Elsevier.

Levine, S.H. (1976) Competitive interactions in ecosystems. *Am. Nat.* **110**, 903–10.

Levins, R. (1968) *Evolution in Changing Environments.* Princeton University Press, Princeton, New Jersey.

Lloyd, M. & Ghelardi, R.J. (1964) A table for calculating the 'equitability' component of species diversity. *J. Anim. Ecol.* **33**, 217–25.

Macan, T.T. (1954) A contribution to the study of the ecology of the Corixidae (Hemipt.). *J. Anim. Ecol.* **23**, 115–41.

MacArthur, R.H. (1957) On the relative abundance of bird species. *Proc. Natl Acad. Sci. USA* **43**, 293–5.

MacArthur, R.H. (1972) *Geographical Ecology. Patterns in the Distribution of Species.* Harper & Row, New York, London.

MacArthur, R.H. & Wilson, E.O. (1967) *The Theory of Island Biogeography.* Princeton University Press, Princeton, New Jersey.

Magurran, A.E. (2013) *Measuring Biological Diversity.* John Wiley & Sons.

Magurran, A.E. & Henderson, P.A. (2003) Explaining the excess of rare species in natural species abundance distributions. *Nature* **422**, 714–16.

Matthews, T.J. & Whittaker, R.J. (2014) Neutral theory and the species abundance distribution: recent developments and prospects for unifying niche and neutral perspectives. *Ecol. Evol.* **4**, 2263–77.

May, R.M. (1973) *Stability and Complexity in Model Ecosystems.* Princeton University Press, Princeton, New Jersey.

May, R.M. (1974) On the theory of niche overlap. *Theor. Popul. Biol.* **5**, 297–332.

May, R.M. (1975) Patterns of species abundance and diversity. In: Cody, M.L. & Diamond, J.M. (eds), *Ecology and Evolution of Communities.* Harvard University Press, Cambridge, Mass.

May, R.M. (1976) Models for two interacting populations. In: May, R.M. (ed.), *Theoretical Ecology.* Blackwell, Oxford.

May, R.M. & MacArthur, R.H. (1972) Niche overlap as a function of environmental variability. *Proc. Natl Acad. Sci. USA* **69**, 1109–13.

McClure, M.S. & Price, P.W. (1976) Ecotype characteristics of coexisting *Erythoneura* Leafhoppers (Homoptera: Cicadellidae) on sycamore. *Ecology* **57**, 928–40.

McIntosh, R.P. (1967) An index of diversity and the relation of certain concepts to diversity. *Ecology* **48**, 1115–26.

McNaughton, S.I. & Wolf, L.L. (1970) Dominance and the niche in ecological systems. *Science* **167**, 131–6.

Miller, R.J. & Wiegert, R.G. (1989) Documenting completeness, species-area relations, and the species-abundance distribution of a regional flora. *Ecology* **70**, 16–22.

Morisita, M. (1959) Measuring of interspecific association and similarity between communities. *Mem. Fac. Sci. Kyushu Univ. Series E* **3**, 65–80.

Mountford, M.D. (1962) An index of similarity and its application to classificatory problems. In: Murphy, P.W. (ed.), *Progress in Soil Zoology*. Butterworths, London.

Mueller-Dombois, D. & Ellenberg, H. (1974) *Aims and Methods of Vegetation Ecology*. John Wiley & Sons, New York, London, Sydney, Toronto.

Mühlenberg, M., Leipold, D., Mader, H.J., & Steinhauer, B. (1977) Island ecology of arthropods I & 11. *Oecologia* **29**, 117–44.

Norris J.L. & Pollock K.H. (1998) Non-parametric MLE for Poisson species abundance models allowing for heterogeneity between species. *Environ. Ecol. Statist.* **5** (4), 391–402.

Odum, H.T., Cantfon, J.E., & Kornicker, L.S. (1960) An organizational hierarchy postulate for the interpretation of species-individual distributions, species entropy, ecosystem evolution and the meaning of a species-variety index. *Ecology* **41**, 395–9.

Oksanen, J., Blanchet, F.G., Kindt, R., Legendre, P., Minchin, P.R., O'Hara, R.B., Simpson, G.L., Solymos, P., Stevens, H.H., and Wagner, H. (2015) vegan: Community Ecology Package. R package version 2.2-1. http://CRAN.R-project.org/package=vegan

Palmer, M.W. (1991) Estimating species richness: The second-order jackknife reconsidered. *Ecology* **72**, 1512–13.

Patil, G.P. & Taillie, C. (1979) An overview of diversity. In: Grassle, J.F., Patil, G.P., Smith, W., & Taillie, C. (eds), *Ecological Diversity in Theory and Practice*. International Cooperative Publishing House, Fairland, MD.

Peet, R.K. (1974) The measurement of species diversity. *Annu. Rev. Ecol. Syst.* **5**, 285–307.

Peet, R.K. (1975) Relative diversity indices. *Ecology* **56**, 496–8.

Petersen, F.T., Meier, R., & Larsen, M.N. (2003) Testing species richness estimation methods using museum label data on the Danish Asilidae. *Biodivers. Conserv.* **12**, 687–701.

Pianka, E.R. (1973) The structure of ligard communities. *Rev. Ecol. Syst.* **4**, 53–74.

Pianka, E.R. (1976a) Competition and Niche Theory. In: May, R.M. (ed.), *Theoretical Ecology*. Blackwell, Oxford.

Pianka, E.R. (1976b) *Evolutionary Ecology*. Harper and Row, New York.

Pielou, D.P. & Pielou, E.C. (1968) Association among species of infrequent occurrence: the insect and spider fauna of *Polyporus betulinus* (Builiard) flies. *J. Theoret. Biol.* **21**, 202–16.

Pielou, E.C. (1975) *Ecological Diversity*. John Wiley, New York.

Pimm, S.L. (1991) *The balance of nature?* The University of Chicago Press, Chicago.

Preston, F.W. (1948) The commonness and rarity of species. *Ecology* **29**, 254–83.

Preston, F.W. (1960) Time and space and the variation of species. *Ecology* **41**, 611–27.

Preston, F.W. (1962a) The canonical distribution of commonness and rarity. Part 1. *Ecology* **39**, 185–215.

Preston, F.W. (1962b) The canonical distribution of commonness and rarity. Part II. *Ecology* **39**, 410–32.

Price, P.W. (1971) Niche breadth and dominance of parasitic insects sharing the same host species. *Ecology* **52**, 587–96.

Raaijmakers, J.G.W. (1987) Statistical analysis of the Michaelis–Menten equation. *Biometrics* **43**, 793–803.

Reaka-Kudla, M.L., Wilson, D.E., & Wilson, E.O. (eds) (1997) *Biodiversity II.* Joseph Henry Press (US National Academy of Science), Washington DC.

Reese, G.C., Wilson, K.R., & Flather, C.H. (2014) Performance of species richness estimators across assemblage types and survey parameters. *Global Ecol. Biogeogr.* **23**, 585–94.

Renyi, A. (1961) On measures of entropy and information. In: Neyman, J. (ed.), *Proceedings of the 4th Berkeley Symposium on Mathematical Statistics and Probability.* University of California Press, Berkley, CA.

Rikleffs, R.E. & Lau, M. (1980) Bias and dispersion of overlap indexes – results of some Monte-Carlo simulations. *Ecology* **61**, 1019–24.

Rossi, J.-P. (2011) rich: An R Package to Analyse Species Richness. *Diversity* **3**, 112–20.

Roughgarden, J. (1974) Species packing and the competition function with illustrations from coral reef fish. *Theor. Pop. Biol.* **5**, 163–86.

Routledge, R.D. (1977) On Whittaker's components of diversity. *Ecology* **58**, 1120–7.

Routledge, R.D. (1979) Diversity indices: which ones are admissible. *J. Theor. Biol.* **76**, 503–15.

Sabath, M.D. & Jones, J.M. (1973) Measurement of niche breadth and overlap the Colwell–Futuyma method. *Ecology* **54**, 1143–7.

Sanders, B.L. (1969) Benthic Marine Diversity and the Stability-Time Hypothesis. In: *Diversity and Stability in Ecological Systems.* Brookhaven Symposium of Biology.

Schoener, T.W. (1968) The Anolis lizards of Bimini: resource partitioning in a complex fauna. *Ecology* **49**, 704–26.

Schoener, T.W. (1974) Resource partitioning in ecological communities. *Science* **185**, 27–39.

Seal, H. (1964) *Multivariate Statistical Analysis for Biologists.* Methuen and Co. Ltd, London.

Shapiro, A.M. (1975) The temporal component of butterfly species diversity. In: Cody, M.L. & Diamond, J.M. (eds), *Ecology and Evolution of Communities.* Harvard University Press, Cambridge, Mass.

Simpson, E.H. (1949) Measurement of diversity. *Nature* **163**, 688.

Smith, B. (1986) Evaluation of different similarity indices applied to data from the Rothamsted Insect Survey. University of York, York.

Smith, E.P. (1982) Niche breadth, resource availability and inference. *Ecology* **63**, 1675–81.

Smith, E.P. & van Belle, G. (1984) Non parametric estimation of species richness. *Biometrics* **40**, 119–29.

Smith, E.P. & Zaret, T.M. (1982) Bias in estimating niche overlap. *Ecology* **63**, 1248–53.

Soberon, M.J. & Llorente, B.J. (1993) The use of species accumulation functions for the prediction of species richness. *Conserv. Biol.* **7**, 480–8.

Sørensen, T. (1948) A method of establishing groups of equal amplitude in plant sociology based on similarity of species content and its application to analyses of the vegetation on Danish commons. *Biol. Skr. (K. danske vidensk. Selsk. N. S.)* **5**, 1–34.

Southwood, T.R.E. (1978) The components of diversity. In: *The Diversity of Insect Faunas.* Symposium, Royal Entomological Society of London.

Southwood, T.R.E. (1996) Natural communities: structure and dynamics. *Philos. Trans. R. Soc. Lond. B* **351**, 1113–29.

Southwood, T.R.E., Brown, V.K., & Reader, P.M. (1979) The relationship of insect and plant diversities in succession. *Biol. J. Linn. Soc.* **12**, 327–48.

Southwood, T.R.E., Henderson, P.A., & Woiwood, I.P. (2003) Stability and change over 67 years-the community of Heteroptera as caught in a light-trap at Rothamsted, UK. *Eur. J. Entomol.* **100**, 557–62.

Taylor, L.R. (1978) Bates, Williams, Hutchinson – a variety of diversities. In: Mound, L.A. & Warloff, N. (eds), *Diversity in Insect Faunas: 9th Symposium of the Royal Entomological Society.* Blackwell, Oxford.

Taylor, L.R., Kempton, R.A., & Woiwood, I.P. (1976) Diversity statistics and the log-series model. *J. Anim. Ecol.* **45**, 255–72.

Tilman, D. (2004) Niche tradeoffs, neutrality, and community structure: a stochastic theory of resource competition, invasion, and community assembly. *Proc. Natl Acad. Sci. USA,* **101**, 10854–61.

Tipper, J.C. (1979) Rarefaction and rarefiction – the use and abuse of a method in paleoecology. *Paleobiology* **5**, 423–34.

Toti, D.S., Coyle, F.A., & Miller, J.A. (2000) A structured inventory of Appalachian grass bald and heath bald spider assemblages and a test of species richness estimator performance. *J. Arachnol.* **28**, 329–45.

Tóthmérész, B. (1995) Comparison of different methods for diversity ordering. *J. Veg. Sci.* **6**, 283–90.

Uetz, G.W. (1977) Coexistence in a guild of wandering spiders. *J. Anim. Ecol.* **46**, 531–41.

Van Valen, L. (1965) Morphological variation and width of ecological niche. *Am. Nat.* **999**, 377–90.

Vandermeer, J.H. (1969) The community matrix and the number of species in a community. *Ecology* **50**, 362–71.

Vane-Wright, R.I., Humphries, C.J., & Williams, P.H. (1991) What to protect? Systematics and the agony of choice. *Biol. Conserv.* **55**, 235–54.

Walther, B.A. & Morand, S. (1998) Comparative performance of species richness estimation methods. *Parasitology* **116**, 395–405.

Wang, J.P. (2011) SPECIES: an R package for species richness estimation. *J. Statist. Software* **40**, 1–15.

Wang J.P. (2010) Estimating the species richness by a Poisson-compound gamma model. *Biometrika* **97** (3), 727–40.

Wang, J.P.Z. & Lindsay, B.G. (2005) A penalized nonparametric maximum likelihood approach to species richness estimation. *J. Am. Statist. Assoc.* **100**, 942–59.

Webb, D.J. (1974) The statistics of relative abundance and diversity. *J. Theor. Biol.* **43**, 277–92.

Whittaker, R.H. (1952) A study of summer foliage insect communities in Great Smoky Mountains. *Ecol. Monogr.* **22**, 1–44.

Whittaker, R.H. (1970) *Communities and Ecosystems.* Macmillan, London.

Whittaker, R.H. (1972) Evolution and measurement of species diversity. *Taxon* **21**, 213–51.

Whittaker, R.H. & Fairbanks, C.W. (1958) A study of plankton copepod communities in the Columbia basin, south eastern Washington. *Ecology* **39**, 46–65.

Williams, C.B. (1947) The logarithmic series and the comparison of island floras. *Proc. Linn. Soc. Lond.* **158**, 104–8.

Williams, C.B. (1964) *Patterns in the Balance of Nature and Related Problems in Quantitative Ecology.* Academic Press, London, New York.

Wilson, D.S. (1975) The adequacy of body size as a niche difference. *Am. Nat.* **109**, 769–84.

Wilson, E.O. (ed.) (1988) *Biodiversity*. National Academy Press, Washington, DC.

Wilson, M.V. & Mohler, C.L. (1983) Measuring compositional change along gradients. *Vegetatio* **54**, 129–41.

Wilson, M.V. & Shmida, A. (1984) Measuring beta diversity with presence–absence data. *J. Ecol.* **72**, 1055–64.

Wolda, H. (1981) Similarity indices, sample size and diversity. *Oecologia* **50**, 296–302.

Wolda, H. (1983) Diversity, diversity indices and tropical cockroaches. *Oecologia* **58**, 290–8.

14

The Estimation of Productivity and the Construction of Energy Budgets

The size of a population and the interactions between populations within an ecosystem may be expressed in terms of biomass (weight of living material) or energy content, as well as in numbers. The study of energy budgets generated great interest between the 1960s and 1980s, but has subsequently declined in popularity, probably because the great effort required to obtain the data did not generate sufficient novel insight to justify the additional costs incurred over the use of number or biomass as a currency. Biomass and energy are useful to ecologists in that they provide common units for the description of populations of animals and plants of different sizes. By the comparison of energy budgets, we may, for example, compare the strategies of warm-blooded mammals and cold-blooded reptiles or even whales and copepods. Such measures are particularly important in food-web studies, where the focus is on the quantification of the flux of energy or carbon between different trophic levels. A particularly good example of this type of study has been undertaken on the tiny Ythan estuary in Scotland (Baird & Milne, 1981; Raffaelli & Hall, 1996). Great effort is required to quantify each pathway within an ecosystem, a whole community study with the species coverage achieved in the Ythan cannot presently be produced for a large temperate estuary, and is almost inconceivable for a tropical floodplain or estuary. Comparing the energy flux of different pathways is one way in which we can assess the relative importance of the various interactions within a web (Paine, 1980, 1992). In studies of general predators, such as insectivorous birds, predation activity is often best expressed in units of biomass or energy. Conversely, if the energy requirements are known from metabolic measurements they may be used to predict food requirements in the field (Stiven, 1961). It must be remembered that the full richness of an organism's requirements and behaviour cannot be expressed in terms of energy. For example, food is more than simply an energy source, its quality in terms of specific amino acids, vitamins and other constituents will also be important (Boyd & Goodyear, 1971; Iversen, 1974; Schroeder, 1977; Onuf et al., 1977).

While energetic considerations are generally considered to place constraints on the length of food chains, and thus on the structure of food webs, energetic arguments have fulfilled their early promise only in respect of describing the trophic structure of ecosystems. They make only slight contributions to the development of plans for reserve management and the maintenance of biodiversity. Nor have they proved useful to our understanding of population dynamics and the prediction of population collapse or the explosive irruption of a pest or pathogen. It is probably because of these deficiencies that there appears to be less interest in framing population models in terms of energy now than there was during the 1960s and 1970s. The work undertaken during this period defined many present concepts and remains an important source of primary information. It is striking how little has been published on ecological energetics post 1990.

The concept of energy is fundamental in the consideration of the functioning of the ecosystem; Ivlev (1939, 1945) and Lindeman (1942) – now all classical papers – pointed the way to this approach, and the subject was reviewed by Gallucci (1973) and Wiegert (1976). Various trophic levels can be distinguished in the ecosystem. First, there are the primary producers (i.e. plants) that build up complex substances from simple inorganic substances utilising the energy from sunlight. The total energy retained by the plant as fixed carbon is usually referred to as the *gross production*, while the change in energy level (standing crop or stock) plus any losses due to grazing or mortality during the year, but less imports (e.g. migrants) is termed the *net production*. The amount of living material present at any given time is the *standing crop* or *biomass*. Unfortunately, the use of these terms has been confused, but MacFadyen (1963) has given a useful table of synonymy and Petrusewicz & MacFadyen (1970) have established standard definitions and notation. Information on primary productivity is given in Vollenweider (1969) and Chapman (1976).

The energy equations for an individual may be expressed:

Gross energy = Digestible energy = Metabolisable energy = Resting energy

or	+	+	+
Energy intake	Faecal waste↓	Urinary waste ↓	Activity
or			+
Ingestion			Growth
or			+
Consumption			Reproduction

(Note only the upper term partakes in the equation to the right.)

The energy budget of a population or a trophic level can be expressed in several equations (Petrusewicz & MacFadyen, 1970). These may be related to the equations above, noting that the energy used up in resting activity and the 'work' of growth is equated with respiration, so that:

$$C = D + F = A + U + F = R + P + (F + U) \tag{14.1}$$

where C = consumption (= ingestion), D = digestable energy, F = faecal waste, A = assimilation (metabolisable energy), U = urinary waste, R = respiration and P = productivity (= growth + reproduction).

This basic equation may be divided into a number of simpler equations for parts of the process, which are of value in the construction of energy budgets; for if only one term remains known then this may be found. Thus:

$$P = A–R \tag{14.2}$$

$$A = C–(F + U) \tag{14.3}$$

Productivity is the increase (growth) (Δ) in the biomass of the standing crop (B), plus production eliminated (E), through death, emigration, exuviae or other products (e.g. feathers of birds at moult, spiders' silk, byssus of mussels). So that:

$$P = \Delta B + E \tag{14.4}$$

One will also note from the composite equation that

$$P = Pg + Pr + E \tag{14.5}$$

where Pg = productivity as growth and Pr = productivity through the reproduction (the biomass of offspring).

These equations demonstrate the usefulness of energy, as opposed to simply biomass, in the description of this aspect of ecosystem or population dynamics. The early studies of Odum & Odum (1955) on a coral reef, of Odum (1957) on a hot spring, of Teal (1957) on a cold spring, and of Whittaker *et al.* (1974) on a deciduous forest exemplify the value of measurements of biomass and energy in describing and analysing the trophic structure of ecosystems. The quantification of the energy moving between ecological compartments or species, allowing the identification of pathways deemed important in terms of a greater energy flux was an objective in the 1970s International Biological Programme (IBP). This led to production of a series of handbooks (e.g. Ricker, 1968; Golley & Buechner, 1968; Petrusewicz & MacFadyen, 1970; Edmondson & Winberg, 1971; Grodzinski *et al.*, 1975) which may still be consulted for further information on the measurement of productivity and energy flow. Other useful sources are Paine's (1971) early review, Brafield & Llewellyn's (1982) introductory book, and the review of laboratory methods by Brafield (1985) which, while focused on fish, gives much general information.

14.1 Estimation of standing crop

If a detailed life-table or budget has been constructed (see Chapter 11) then the standing crop of the population at any given time can be determined by converting numbers to biomass (dry weight is usually the most appropriate measure). Given a known, calculated or experimentally obtained energy density for the organism, biomass can be further converted into energy (Brey *et al.*, 1988).

14.1.1 *Measurement of biomass*

Biomass can be expressed as wet weight, dry weight (DW), shell-free dry weight (SFDW), ash-free dry weight or as the amount of organic carbon present.

Wet weight for a terrestrial animal or plant is simply the living, freshly killed or harvested weight. For aquatic organisms, such as fish, it should be obtained after blotting to remove surface water. As wet weight can change with the state of dehydration of the organism and changes rapidly after death or following preservation, it is an unsuitable measure of biomass for comparative studies.

Biomass is generally expressed as dry weight obtained by drying a known wet weight of animals until a constant weight is achieved. Unfortunately, no standard protocol can be given. There is agreement that material should be dried at a low temperature to avoid the loss of volatile – especially lipoid – constituents, but there is no consensus as to what this temperature should be. For insects and other small organisms, freeze-drying or heating to a maximum of 60 °C is often recommended. Lovegrove (1966) suggested that plankton and other small organisms can be dried in a desiccator. Paine (1964) originally argued that for marine organisms with water of hydration in their skeletons cannot be dried below 100 °C; however, for organisms such as barnacles he concluded that drying could best be achieved at 80 °C (Paine, 1971). Atkinson & Wacasey (1983), following an extensive series of experiments, showed that a maximum energy density was estimated for benthic marine organisms when dried at 100 °C.

Shell-free-dry weight is used for molluscs and many marine organisms for which a high proportion of their weight comprises calcareous shell. Shells can be removed either pre or post drying by mechanical means (Atkinson & Wacasey, 1983) or by acid digestion in 1 M HCl. Acid decalcification probably results in the loss of some organic carbon. For all organisms, the ash-free dry weight will give a measure of the amount of organic material present and is determined by heating a sample of known dry weight in a muffle furnace at 500 °C for 4 h and subtracting the weight of the ash obtained from the original dry weight (Atkinson & Wacasey, 1983). However, at this temperature some inorganic material may also be lost; calcium carbonate for example is broken down at temperatures above 400 °C.

With small animals, especially plankton, it may be more convenient to work with the volume of the animals rather than the actual numbers (e.g. in Neess & Dugdale's method, see below). A number of devices for measuring the volume of small animals have been developed (e.g. Frolander, 1957; Stanford, 1973).

14.2 Determination of energy density

The energy content of a material may be determined directly by oxidation either by potassium dichromate in sulphuric acid (Teal, 1957), or by burning in oxygen and determining the amount of heat liberated. The latter method – bomb calorimetry – is most convenient and widely used in ecology; however, it involves drying the material and volatile substances can be lost. Hartman & Brandt (1995) gives a good account of the method as applied to fish.

Phillipson (1964) described a miniature non-adiabatic bomb (or ballistic) calorimeter that was simple and suitable for handling materials in the order of 5–100 mg dry weight. It is operated by placing the material in a platinum pan enclosed in the 'bomb'. The bomb is filled with oxygen until a pressure of 30 atmospheres is reached, cooled until the temperature (as indicated by a thermocouple) is steady, and then enclosed in a polystyrene insulating jacket. Passing an electric current through a thin wire that presses against it ignites the sample. The temperature rises rapidly, and this is recorded by the deflection of a potentiometer attached to the thermocouple. The maximum deflection in a firing is a measure of the caloric content of the sample. The instrument is calibrated by burning given weights of benzoic acid of known caloric value. This basic principle is still used by the computer-controlled adiabatic calorimeters manufactured by the Parr Instrument Company (see http://www.parrinst.com/products/oxygen-bomb-calorimeters/1341-plain-jacket-bomb-calorimeter/). To obtain an accurate estimate of the energy, density adjustments may be necessary to allow for the amount of fuse wire lost, acid production by sulphur and nitrogen, and the presence magnesium and calcium carbonates which decompose endothermically.

Naturally, not all the energy in an animal or plant is available to an animal that eats it, and quite frequently only a proportion of the potentially digestible energy, is assimilated. Verdun (1972) presents an extreme and iconoclastic view on availability. The assimilation rates for various types and quantities of food need to be determined. The total energy is, however, available to the ecosystem as micro-organisms can break down almost all organic compounds.

Values for the energy content of various plants have been published by Ovington & Heitkamp (1960), and of animals by Golley (1960), Slobodkin & Richman (1961) and Comita & Schindler (1963). There is, however, considerable variation in caloric value which, unrecognised, could be a major source of error in energy budgets. Both, plants (Boyd, 1969; Singh & Yadava, 1973; Caspers, 1977) and animals (Wiegert, 1965; Hinton, 1971; Wiegert & Coleman, 1970; Wissing & Hasler, 1971) vary from season to season. This variation is due to variations in the levels of fat, which accounts for much of the interspecific variation in energy density between animals (Slobodkin & Richman, 1961; Griffiths, 1977). Particularly large seasonal changes in stored fat levels are observed in birds; small passerines may double their body weight through fat accumulation prior to migration and even show marked diurnal variation (Gosler, 1996). Similarly large seasonal losses and gains are shown by fish; for example, Henderson *et al.* (1988) showed that sand smelt, *Atherina boyeri*, lived on stored fat over winter. It is also changed by mineral composition, which varies markedly in plants (Boyd & Blackburn, 1970; Caspers, 1977), insects (Wiegert, 1965) and crustaceans (Griffiths, 1977). Different stages of the same insect also vary markedly (Wiegert, 1965; McNeill, 1971; Hinton, 1971; Holter, 1974; Hagvar, 1975) (Table 14.1). In the froghoppers, *Philaenus* and *Neoplilaenus* the immature stages contain more ash than the adults, and the same appears to be true, to a lesser extent, of the willow leaf beetle, *Melosoma*, which shows a fall in caloric value throughout development. In contrast, in the dung beetle, *Aphodius*, it rose markedly during the last larval instar.

14.3 Estimation of energy flow

Energy flow in a population is usually estimated using one or more of the equations listed in the introduction to this chapter, especially Equations (14.2) and (14.3):

A(ssimilation) = R(espiration) + P(roduction)

A(ssimilation) = C(onsumption) − (F(aecal) + U(rinary waste)[1]

---

[1]Because faecal and urinary waste are mixed in insects, digestible energy (D) cannot be obtained (Reichle, 1969).

Table 14.1 Caloric values (Kcal/g dry wt or ash free dry wt*) of different stages of the same insect.

Insect	Data source	Egg	1	2	3	4	5	Pupa	Adult	♂	♀
Homoptera											
Neophilaenus*	Hinton, 1971		5.44	5.47	5.79	5.80	5.91			5.46–6.00	5.56–
Philaenus	Wiegert, 1965	6.31	4.58	5.27	5.45	5.53	5.71			5.49–5.95	5.16–6.08
Philaenus*	Wiegert, 1965	6.50	4.98	5.67	5.80	5.78	5.90			5.58–5.95	5.50–6.08
Heteroptera											
Leptopterna	McNeill, 1971	6.30		5.81			5.86				
Coleoptera											
Melosoma	Hagvar, 1975	6.16	5.70	5.28	5.24			5.71		6.05	6.06
Aphodius	Holter, 1975			4.67	4.66–5.80				5.68		
Orthoptera											
Encoptoloptius*	Bailey & Riegert, 1972		5.01	5.23	5.30	5.40	5.16			5.25	5.34
Chorthippus*	Gyllenberg, 1969			5.05	5.46	5.08				5.61	5.85

Frequently, two of the terms in the equations will be found directly, and the third by calculation. For example, for a particular animal it may be difficult to determine the respiratory rate in the field, but ingestion and consumption and waste excretion (egestion + excretion) per individual can be measured and these values could be converted to energy equivalents (kJ). Then, knowing the numbers of individuals in the field, the total energy assimilation (the energy income) of the population could be calculated. Assuming the numbers of individuals of different ages throughout the season and their energy densities are known, the (net) production can be calculated. The difference between assimilation and production will be a measure of respiration.

The construction of an energy budget requires the initial development of a numerical budget. Normally, it is necessary to calculate the standing crop and its energy equivalent (described above) and from this the production together with the independent measurement of either consumption and waste excretion rates or respiration. Seasonal variations will affect these values (e.g. Jonasson & Kristiansen, 1967); the energy budget will, perforce, contain a number of approximations and the more variables that can be determined the more reliance can be placed upon it; clearly, if all the terms in any of the equations are determined and their magnitudes prove compatible, when they are substituted in the equations, this provides valuable evidence as to their reliability. Thus, as with a numerical budget (p. 430) a detailed energy budget will expose its own errors. Typical examples of single-species energy budgets are Plaut *et al.* (1996) for the sea hare *Aplysia oculifera*, Chaabane *et al.* (1996) for the ground-beetle, *Abax ater*, Henen (1997) for the tortoise *Gopherus agassizii*, Richman (1958) for crustacea, Marchant & Nicholas (1974) for nematodes, Doohan (1973) for a rotifer, and Edmunds & Davies (1989) for the coral *Porites porites*. The studies of Wiegert (1964a,b), McNeill (1971), Lawton (1971), Bailey & Riegert (1972) and Holter (1974, 1975) may be taken as model examples of energy budget construction for insects. An exceptionally long-term study is that of Jenderedjian (1996), who uses data collected over a 60-year period to relate the energy budgets of oligocheata to primary production.

14.4 The measurement of production

Production consists of the total amount of material synthesised by the animal and is obtained by summing:

1. The increase in the standing crop during the season (or other time unit).
2. The biomass of all individuals that died or were eaten during the season.
3. The biomass of the total number of exuviae shed by all the individuals (those alive and those now dead) and of any other product, for example the spittle of cercopids (Wiegert, 1964a, the byssus of mussels (Kuenzler, 1961), the silk of many insects, mites and spiders.
4. The biomass of any reproductive products (sperms and seminal fluid, eggs), young individuals or adults that have left the area (i.e. emigration).

Strictly, the amount of nitrogenous waste excreted should also be measured. The biomass of any individuals that migrated into the population must be subtracted from the total obtained above.

The calculation of production from a budget, as was done for example by Wiegert (1964a), Golley & Gentry (1964), McNeill (1971) and Karo (1973), demands that the form of the mortality curve, the actual population at various stages and immigration and emigration are accurately known. The weights of the various products can be determined in the laboratory, but caution should be exercised in assuming that laboratory determinations hold under field conditions.

A somewhat different approach was developed by Allen (1951) and Neess & Dugdale (1959) for the study of the production of chironomid larvae in a lake. Instead of separating the numbers into different age categories (as in a budget) and then multiplying the age categories by their respective biomasses, Neess & Dugdale use the product of these variables as expressed in the field in terms of weight.

The net production of a developing larval population is the outcome of two opposing processes:

Increase in weight of the individual:

$$k_g Q_t = dQ_t/dt \tag{14.1}$$

and:

Fall in numbers:

$$- k_m N_t = dN_t/dt \tag{14.2}$$

where Q_t = weight of the individual at time t, N_t = number of individuals at time t, k_g = growth rate, and k_m = mortality rate. These equations (14.1 and 14.2) can be combined by the elimination of t and integrated to give:

$$\log_e Q_t^{1/k_g} = \log_e CN_t^{1/k_m}$$

or:

$$Q_t^{-k_m} = CN_t^{k_g} \tag{14.3}$$

The constant C may be substituted by an expression using Q_t or N_t by noting at time $t = 0$, when Q_0 = weight of a newly hatched larva and N_0 = the total hatch that:

$$Q_t = Q_0 \bigg/ \left(N_0/N_t \right)^{-k_g/k_m} \tag{14.4}$$

For given values of N_0 and Q_0 the shape of the curves describing the relationship between Q and N is determined solely by the ratio k_g/k_m; if the ratio remains constant, the growth–survivorship curve will trace out the same path. Neess & Dugdale noted that within the range of values of the ratio up to 2, large changes occur in the area under the curve for small variations in the actual ratio. Now, the area under the curve corresponds to the actual net production. Therefore, if it can be shown that k_g/k_m is a constant, a smooth curve may be drawn of numbers on mean weight and the area beneath it found. As Equation (14.9) can be rearranged to:

$$\log Q_t = \log Q_0 + \left(k_g/k_m\right) \log \left(N_0/N_t\right) \tag{14.5}$$

The constancy of the ratio may be tested by plotting $\log N$ against $\log Q$; a straight line will indicate that the ratio is constant.

The numbers per unit area can then be plotted against the mean weight, a smooth curve (an 'Allen curve') drawn and the area below it found (ACDE in Fig. 14.1). This is the actual net production in biomass per unit area (the units corresponding of course to those of Q_t and N_t).

A number of terms may be defined by reference to Fig. 14.1.

ACDE Actual net production.
BCED Actual net recruitment or increase in standing crop due to this cohort = Q_pN_p.
ABE   Directly recycled production.

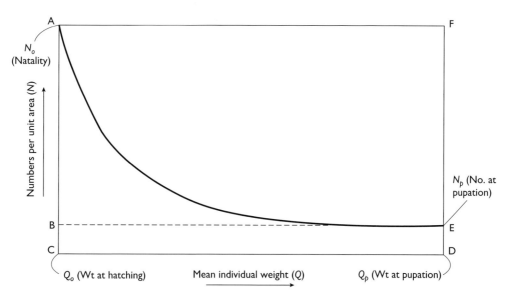

Figure 14.1 **The relationship between numbers and mean individual weight in an insect population during larval development; the area below the curve may be measured to give actual net production. The letters indicate areas representing other productivity terms (see text). (Adapted from Neess & Dugdale, 1959.)**

ACDF   Potential net production = $N_0 Q_P$.
AEF   Lost potential net production.
ABEF   Lost potential recruitment.

This method is most appropriate when the development of the individuals of a generation is closely synchronised, or where a particular cohort can be recognised and distinguished from others, as in dragonflies (Benke, 1976).

14.5 The measurement of feeding and assimilation

14.5.1 The quality of the food eaten

In some instances, such as with many arboreal lepidopterous larvae, the nature of an animal's food may be determined by observation in the field; observations in the laboratory may also be used, but with caution, for many animals will eat unnatural foods under artificial conditions. Another method frequently used by ornithologists and ichthyologists is to examine the crop or stomach contents (p. 385). Such methods may be used with chewing insects and crustaceans and the remains of different types of plant or animal identified under the microscope (e.g. Hanna, 1957; Mulkern & Anderson, 1959; Henderson & Walker, 1986). This approach can be quantified using a sedimentation technique (Brown, 1961). In insects which suck plants the actual tissue – phloem, xylem, mesophyll cells – from which the nutriment is obtained should be determined. The salivary sheath (Miles, 1972) or the position of the stylets (Wiegert, 1964b; Pollard, 1973) may sometimes be kept in place by a high-voltage shock (Ledbetter & Flemion, 1954) and are precise indications, but the examination of a freshly damaged leaf may sometimes be satisfactory (McNeill, 1971).

The identification of the meals of predators by DNA and other biochemical techniques is discussed in Chapter 10 (p. 388), where reference is also made to the use of radioisotope-tagged prey. Following a pioneer study by Pendleton & Grundmann (1954), radioisotopes have often been used to trace out food webs (e.g. Odum & Kuenzler, 1963; Paris & Sikora, 1965; Coleman, 1968; Van Hook, 1971; Shure, 1973; Walker *et al.*, 1991). The food plant or primary food source is tagged and subsequent arrival of radioactivity (assessed quantitatively by counting) in the animals demonstrates a trophic link. The rate of build-up of radioactivity in the animal has been taken as proportional to the trophic distance; for example, a peak in activity soon after tagging would indicate a herbivore, and a later and lower peak a predator or saphrophagous species. However, as Shure (1970) points out, such interpretations should be made with care because herbivores may change their host plant and those moving to the labelled plant during the experiment would show accumulation curves resembling those of a predator. Appropriate field observation can obviate such misinterpretations. Trophic links in soil invertebrates may be investigated using pesticides for the selective removal of animal groups (Edwards *et al.*, 1969), a method analogous to the 'insecticidal check-method' (p. 389). The forage-ratio and methods of assessing food preferences are discussed in Chapter 10 (p. 395).

14.6   Feeding and assimilation rates

There are four basic approaches to the measurement of these rates and related parameters:

- Radiotracer – measuring the passage of radioisotopes from the food to the animal and their subsequent loss.
- Gravimetric – by the direct weighing of the food, the faeces and the animal.
- Indicator – by marking the food with an inert non-absorbed indicator; the increase in its concentration as it passes through the alimentary canal measures the amount of matter absorbed.
- Faecal – by measuring faecal output and relating this to consumption.

14.6.1   Radiotracer techniques

Radiotracers may be used singly or as a double-marking method. When used singly there are two approaches. The amount of radioisotope used may be small relative to the equilibrium body burden of the animal, so that the rate of increase of assimilation of the isotope is linearly related to the rate of assimilation of the food. Examples are the studies of Engelmann (1961), Strong & Landes (1965), Hubbell *et al.* (1965), Malone & Nelson (1969) and Moore *et al.* (1974). Alternatively, the body burden may be increased so as to reach equilibrium when the following relationship holds:

$$r = \frac{KQ_e}{a} \tag{14.6}$$

where r = feeding rate (measured in µCi/day), a = proportion of ingested nuclide assimilated, K = the elimination constant[2] = [0.693 – biological half-life] (in days) and Q_e = the whole-body radioactivity (in µCi) in a steady-state equilibrium. Therefore, if the whole-body radioactivity is measured and the assimilation rate and the biological half-life are known, the feeding rate may be calculated. This approach has been developed by Crossley (1963b, 1966), who worked in an environment heavily contaminated with cesium-137. This isotope is almost completely assimilated, so that the term 'a' can be dropped from the above equation. Crossley (1963b) found that the biological half-life appeared to be linearly related to body weight, except in the pupal stage, of course, when there was no elimination. However, as Odum & Golley (1963) point out and Hubbell *et al.* (1965) confirm, the biological half-life of an isotope in a given animal is variable and will be influenced by temperature, activity, food and other factors, so that the half-life in the field is likely to be different from that determined in the laboratory.

---

[2]This formula only applies to isotopes with long half-lives, so that the effective half-life = the biological half-life. Where radioactive decay is significant then this loss rate needs to be added to the biological half-life.

Working on the isopod, *Armadillidium*, Hubbell *et al.* (1965) also show that the percentage of the isotope ingested that is actually assimilated will vary with the feeding rate; they used ^{85}Sr and, with this biologically significant element (it is utilised in the exoskeleton), its assimilation rate paralleled the actual assimilation of nutriments from the food. The biological half-life of an isotope may be different in the two sexes, as Williams & Reichle (1968) found in a chrysomelid beetle.

In conclusion, therefore, before radioisotopes can be used to estimate feeding and assimilation rates in field populations, it is necessary to determine the following.

1. The variability of the rate of assimilation of the isotope under field conditions.
2. The variability of the biological half-life of the isotope under field conditions.
3. Whether the level of isotope in the organism is low relative to the equilibrium level or whether it is high, a steady state having been reached. This is done from the equation already given, but some measure of *r* and *a* must be obtained experimentally so that the expected equilibrium body burden of radionuclide (Q_e) can be compared with that actually observed.
4. That if the isotope is likely to be generally present in the media, as in studies with aquatic predators and labelled prey, it is only taken-up by feeding. Dragonfly larvae appear to absorb ^{65}Zn on their surface (Kormondy, 1965).

For biologically indeterminate (i.e. with no specific metabolic function) trace substances, whether radioactive or not, Fagerström (1977) considers that the biological half-life will be proportional to body weight raised to $(1 - b)$, where *b* is the exponent relating body weight to metabolic rate, normally taken as 0.8.

Feeding (or ingestion) and assimilation can usually be separated, for when an animal is fed on radioisotope-tagged food and the fall-off of radioactivity with time is measured, it will be found that at first the curve is steep, reflecting the loss of the ingested but non-assimilated component. Later, the curve becomes shallow and this represents the elimination rate of the assimilated isotope (Odum & Golley, 1963) (Fig. 14.2). Furthermore, Odum & Golley point out that the assimilated isotope is eliminated in different ways depending on the element; for example, iodine mainly through the exuviae, zinc mainly by excretion, especially after moulting, or with the female, in the eggs.

Calow & Fletcher (1972) developed a double-tracer method in which the isotopes are used as a modification of the indicator technique (see below). They point out that the indicator and gravimetric methods assume that all material in the faeces is derived from the food, although strictly this is not true. By using two isotopes – one, carbon (^{14}C) that is biologically active, and the other chromium (^{51}Cr) that is not absorbed from the gut – they were able to allow for errors arising from this assumption. These two isotopes may be distinguished by their different radiations: β from ^{14}C (detected by liquid scintillation counting) and α from ^{51}Cr (detected by crystal scintillation counting). If their levels (counts/min; ct/min) are determined in the food and faeces, then:

$$\%A = 1 - \frac{\text{ct/min}\,^{51}\text{Cr (food)}}{\text{ct/min}\,^{14}\text{C (food)}} \times \frac{\text{ct/min}\,^{14}\text{C (faeces)}}{\text{ct/min}\,^{51}\text{Cr (faeces)}} \times 100 \qquad (14.7)$$

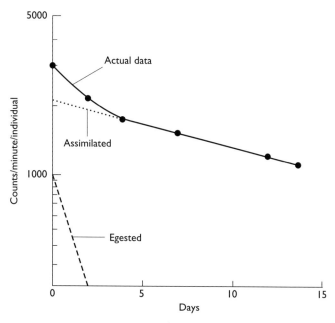

Figure 14.2 **The fall-off of radioactivity with time after ingestion showing the separation of the component that is assimilation from the component that is egested. (Data from Odum & Golley, 1963.)**

^{14}C could be replaced by another absorbed isotope (e.g. ^{3}H) and ^{51}Cr by a non-absorbed isotope.

The proper homogenisation of material for accurate scintillation counting is sometimes difficult; with ^{14}C -labelled material this may be overcome by combustion. Burnison & Perez (1974) describe a simple apparatus for this.

14.6.2 Gravimetric techniques

The simplest approach to the measurement of ingestion and one that is frequently applicable is to weigh the food before and after the animal has fed (e.g. Phillipson, 1960; Strong & Landes, 1965; Hubbell et al., 1965; Lawton, 1970). If all values are reduced to dry weight, errors due to the loss of moisture during feeding will be eliminated; otherwise this must be determined from controls.

Aquatic predators may 'lose' part of the body contents of their prey during feeding. This would be most significant when the prey is chewed, and large relative to the predator; Dagg (1974) found that an amphipod could lose nearly 40% of its 'apparent' meal in this way. The quantity of leaf litter destroyed by soil or aquatic arthropods may be determined by exposing a known quantity in mesh (e.g. nylon) bags (Crossley & Hoglund, 1962; Wiegert, 1974).

In the laboratory the faeces may usually be collected and their dry weight and energy density found. Those of the faeces, will, of course, differ from that of the food (Hubbell et al., 1965) and the difference in energy content between ingestion and egestion may be more striking than the actual weight differences. The collection in the field of the total amount of faeces produced by a population is impossible (p. 338), but a measure could be obtained for a small number of arboreal larvae under semi-experimental conditions.

Hubbell *et al.* (1965) have compared the results of gravimetric and radiotracer methods using [85]Sr for determining the feeding and assimilation rates of the woodlouse, *Armadillidium*, and have found close agreement. The only other weight that can be conveniently measured is the increase in weight of the animal (Fewkes, 1960; Evans, 1962); this is a measure of the amount of growth. When converted to dry weights the following two equations hold:

Assimilation = Production + Respiration
Production = Growth + Exuviae and other Products

Clearly, if respiration can be measured, the growth + weight of the products may be used to give an estimate of assimilation. If wet weights are used, allowance must also be made for the loss of weight due to the evaporation of water from the insect (Strong & Landes, 1965).

Just as the rate of assimilation of radioisotopes has been found to vary under different conditions (p. 562), so too does the nutritional assimilation rate and the relation between assimilation and production. For example, Sushchenya (1962) found with the brine shrimp (*Artemia salina*) that the greater the intake of food the smaller the percentage of it that is assimilated, but a slightly higher percentage of the assimilated energy was used for growth at the higher levels of food intake. Likewise, in the grasshopper, *Poecelocerus*, when food availability was lowered, a lower proportion of the assimilated energy was used for growth (Muthukrishnan & Delvi, 1974). Assimilation efficiencies are also influenced by temperature conditions (in poikilotherms) and even relatively small differences in food, such as slugs on lettuces and potatoes (Davidson, 1977a). Thus, caution needs to be exercised in the extrapolation of laboratory-determined values to field conditions.

14.6.3 Indicator methods

Assimilation can also be measured if the food can be marked with an indicator that is easily measured quantitatively, is non-toxic at the concentrations used, and is not absorbed by the gut. The percentage assimilation (%A) of the food is given by:

$$\%A = 1 - \left(\frac{conc\_indicator\_in.food/unit\_dry\_wt}{conc\_indicator\_in\_faeces/unit\_dry\_wt} \right) \times 100 \qquad (14.8)$$

Chromic oxide has been widely used as an indicator in studies on vertebrate nutrition. Corbett *et al.* (1960) and McGinnis & Kasting (1964) have shown how the chemical or, more conveniently, chromic oxide paper, may be used with insects as an indicator to measure assimilation from finely divided food. As the indicator must be mixed homogeneously throughout the food, this method will be inappropriate in many instances, especially with natural diets, but Holter (1973) found it very suitable for measuring food consumption in the larvae of the dung beetle, *Aphodius*. For salmon, digestibility has been assessed using yttrium oxide (Hatlen *et al.*, 2015).

14.6.4   *Measurement of faecal output*

Provided the assimilation rate is not too sensitive to variations in food quantity, for which there is some evidence (Muthukrishnan & Delvi, 1974), faecal output should be related to consumption and, once standardised for the particular food, could be used for this purpose. Mathavan & Pandian's (1974) studies with various lepidopterous larvae suggest it is reliable in this group (see also p. 339). Lawton (1971) used gut clearance times to give an independent estimate of consumption in the larva of a damsel fly, and Humphreys (1975) obtained a measure of the food consumed in the field by wolf spiders in the 14 days prior to capture from the guanine content of the excreta during the following seven days.

14.7   The measurement of the energy loss due to respiration and metabolic process

In the laboratory, metabolic rate can be measured directly by measuring the heat output, or indirectly by measuring oxygen consumption. Field measurement is more problematical and a number of approaches are summarised below. The measurement of the energy expenditure or respiration of animals while living naturally has proved particularly difficult. For terrestrial animals the doubly labelled water technique has been successful (Nagy, 1994) (p. 570). This technique cannot be applied to aquatic organisms such as fish, where one potential approach for the quantification of field metabolic activity is to monitor the physiology of a free animal using physiological telemetry (p. 572). The doubly labelled water method is favoured for field studies on birds (p. 570).

14.7.1   *Calorimetric*

The amount of energy lost through respiration may be determined directly by the measurement of the heat given out by the animal but, as some energy will have been used to vaporise water from the animal, this will be less than the total energy loss and a correction must be applied. It is difficult to work with small animals; few studies have been made with insects using this approach (Heinrich, 1972).

14.7.2   *The exchange of respiratory gases*

The energy equivalents of oxygen and carbon dioxide

During metabolic processes, in which energy is liberated, oxygen is utilised and carbon dioxide produced; the proportion of carbon dioxide evolved, to oxygen used, depends on the actual metabolic process and is referred to as the respiratory quotient (RQ) expressed as:

$$RQ = \frac{\text{Carbon dioxide produced}}{\text{Oxygen utilised}} \qquad (14.9)$$

Thus, if two of the three values are known the third may be found; the gases may be measured by a number of different techniques and the *RQ* by noting the change in volume at constant pressure and temperature when the animal respires in an enclosed space.

The amount of energy liberated in the process can be determined from the quantity of oxygen (or carbon dioxide) involved, if the *RQ* is known and certain assumptions are made as to the actual metabolic processes (long-accepted values are as follows: lipid, 19.8 kJ l^{-1} O_2 or 27.8 kJ l^{-1} CO_2: carbohydrate, 21.1 kJ l^{-1} for either gas; protein in uricotelic species, 18.7 kJ l^{-1} O_2 or 25.4 kJ l^{-1} CO_2; protein in ureotelic species, 19.2 kJ l^{-1} O_2 or 23.8 kJ l^{-1} CO_2). It has been more recently demonstrated that the assumptions underlying these relationships are often violated (Walsberg & Hoffman, 2005), and should therefore not be accepted without critical evaluation.

If carbohydrates alone are being utilised for energy *RQ* = 1 and 0.198 cm^3 (at S.T.P.) of oxygen will be consumed, and the same volume of carbon dioxide produced during the liberation of one calorie of energy (1 kcal = 4.18 kJ); with fats alone *RQ* = 0.707 and the consumption of 0.218 cm^3 of oxygen and production of 0.154 cm^3 of carbon dioxide will result from the liberation of one calorie. Tables are widely available of the energetic equivalents of unit volumes of oxygen and carbon dioxide for different *RQ*-values, assuming that carbohydrate and fat utilisation are the only metabolic processes involved (Table 14.2). Ecologists have frequently approximated by using the *RQ* obtained from experiments as equivalent to a 'nitrogen (N)-free RQ', disregarding the effect of protein and other metabolisms.

If greater accuracy is to be obtained, the quantity of nitrogen excreted must be measured (Shaw & Beadle, 1949), enabling the component of the gases due to protein catabolism to be established and the 'N-free RQ' determined. The respiratory exchange involved in the production of 1 g of nitrogen from the catabolism of a protein will depend on the proportions by weight of the various elements in its molecule; Kleiber (1961) explains the necessary calculations.

However, even this approach that accounts for fat, carbohydrate and protein catabolism does not include all the processes contributing to the *RQ*. Many others may theoretically influence the *RQ*, and as these have *RQ*-values outside the 0.7–1.0 range

Table 14.2   The caloric equivalents of oxygen and carbon dioxide for various values of *RQ*, due to utilisation of different proportions of carbohydrates and fats. (Data from Brody, 1945).

RQ	Oxygen kcal*/litre	Carbon dioxide kcal*/litre	% oxygen consumed by carbohydrate component
0.70	4.686	6.694	0.0
0.75	4.729	6.319	14.7
0.80	4.801	6.001	31.7
0.85	4.683	5.721	48.8
0.90	4.924	5.471	65.9
0.95	4.985	5.247	82.9
1.00	5.047	5.047	100.0

*1 kcal = 4.18 kJ.

small amounts may modify the composite *RQ* disproportionately; for example, alcohol breakdown has an *RQ* = 0.67, and the synthesis of fat from carbohydrate an *RQ* = 8.0 (Kleiber, 1961).

It is discouraging for the ecologist to note that Cahn (1956) and Kleiber (1961) report studies that show that the estimate of energy expended, obtained indirectly (from the *RQ*), may be found to deviate by as much as 25% from the true value calculated by direct calorimetry. As MacFadyen (1963) has pointed out, when doubt exists as to the exact catabolic process, but assuming that these involve only fats, carbohydrates and proteins, errors will be reduced if calculations are based on the oxygen uptake rather than on the carbon dioxide produced. This is because the minimum volume of oxygen that is required to produce 1 calorie is 90.8% of the maximum volume, but the minimum volume of carbon dioxide that results from the production of 1 calorie is as little as 77.7% of the maximum.

14.7.3   *The respiratory rate*

Starved warm-blooded animals exhibit a level of respiration referred to as the basic metabolism, but there is no comparable standard available for poikilothermic animals. For fish, which have been the subject of much energetics research, energy respired is divided between resting metabolism (energy consumed on tissue maintenance), activity metabolism (energy consumed in movement) and feeding metabolism (energy used digesting food). For cod, *Gadus morhua*, 10–17% of the ingested energy was used for digestion with the cost increasing with temperature (Soofiani & Hawkins, 1982). Fry (1947) defined the difference between basal and active metabolic rates as the 'scope for activity'. It is important to remember that although respiration rate and body weight show in general a linear relationship (Brody, 1945; Engelmann, 1961), with poikilotherms the rate may be varied by many factors such as temperature, age, season, oxygen concentration and even the type of respirometer used for the measurement (Allen, 1959; Golley & Gentry, 1964; Cossins & Bowler, 1987; Hack, 1997; Lyytikainen & Jobling, 1998). When respiration rate is being related to the weight of an invertebrate, it is necessary to eliminate individual variations in weight due to the contents of the gut (Allen, 1959).

14.7.3.1   Gas analysis

The precise analysis of the respiratory gas to determine the proportions of oxygen and carbon dioxide is described in detail in textbooks such as Kleiber (1961) and Kay (1964). The methods are extensively reviewed in Petrusewicz & MacFadyen (1970), Edmondson & Winberg (1971) and Grodzinski *et al.* (1975); only a general outline will be given here; normally it is necessary to modify the precise form of the instrument to suit the particular animal and investigation. The main types of respirometer, for both the gaseous phase and dissolved gaseous, are illustrated in Fig. 14.3; some comparisons of their performance are given in Table 14.3; detailed comparative data is given by Lawton & Richards (1970). Cech (1990) reviews the merits of different respirometers for studies on fish.

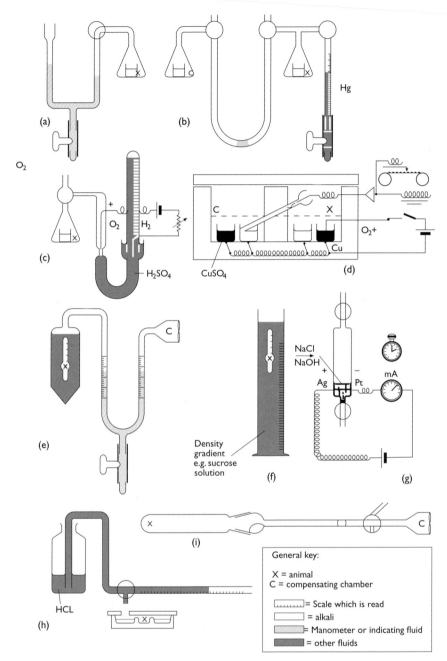

Figure 14.3 (*Opposite*.) **Summary sketches of the main types of respirometer (other than calorimeters). (Adapted from Petrusewicz & Macfadyen, 1970.) (a) Constant-volume Warburg open. (b) Compensating with mechanical restoration. (c) Simple electrolytic—open. (d) Compensating automatic. (e) Cartesian diver. (f) Gradient diver. (g) Kopf's method of CO2 determination. (h) Conway microdiffusion method. (i) Constant pressure.**

Table 14.3 Comparison of the performance of respirometers (Data from Petrusewicz & Macfadyen, 1970).

(1) Respirometer	(2) Measures O_2, CO_2, cals	(3) Container volume (ml)	(4) Normal rates (μl/h)	(5) Normal accuracy	(6) Automation	(7) Sensitivity: smallest detectable change (μl)
WARBURG Normal	O_2 and CO_2	25	10–500	2 μl/hr	N	0.5
Small vessels	O_2 and CO_2	5	2–50	1 μl/hr	N	0.5
GILSON Normal	O_2 and CO_2	16	5–500	2 μl/hr	P	0.5
DIVER Holter	O_2 and CO_2	0.005–0.05	0.01–1.0	5%	P	0.0001
Gregg and Lints	O_2 and CO_2	0.005–0.05	0.01–1.0	5%	Y	0.005
Gradient	O_2 and CO_2	0.001–0.05	0.01–1.0	3%	Y	0.0001
PHILLIPSON automatic	O_2 and CO_2	50	50–300	8%	Y	10
MACFADYEN electrolytic	O_2 and CO_2	7	1–100	1 μl/h	Y	0.01
Capillary const. vol.	O_2	1.0	2.0 ca	5%	P	0.05
Const. pressure	O_2	1–10	0.1–5	2%	N	0.05
CONWAY microdiffusion	CO_2	20	50–1000 +	20 μl/h	N	20
JENSEN CO_2 diffusion	CO_2	10	1.0 ca	0.05 μl/h	Y	—
CALVET microcalorimeter	heat	15–100	20–1000 +	1%	Y	0.2
WINKLER (dissolved oxygen)	O_2	10	100 ca	2%	N	30

Notes: All quantities are converted to equivalents of oxygen volumes, regardless of the parameter actually measured.

(5) Accuracy is expressed as the expected standard error of a series of readings over the normal range or, where accuracy declines at low ranges, as the probable minimum detectable rate.

(6) The column 'Automation' indicates whether application of automatic recording is possible (Y), inherently impossible (N), or potentially possible (P).

14.7.3.2   Isotopes and double-labelled water

McClintock & Lifson (1958) showed that the total carbon dioxide production of an animal could be determined by using double-labelled water. The heavy isotope of hydrogen is eliminated only as water, but that of oxygen (^{18}O) is eliminated both as carbon dioxide and as water; the difference between the two elimination rates will give a measure of the carbon dioxide evolved. This method is of great value to the ecologist for the determination of respiratory rates under natural conditions; the animals can be fed with double-labelled water, released in the field, recaptured at some later time and the levels of the isotopes determined by mass spectrometry (neutron activation, p. 119) and scintillation counting. This technique has been used, for example, for birds (Le Febvre, 1964; Utter & Le Febvre, 1973) and lizards (Nagy, 1975; Bennett & Nagy, 1977). The experimental approach is illustrated by a study on three desert rodents, *Acomys russatus*, *Acomys cahirinus* and *Sekeetamys calurus* undertaken by Degan *et al.* (1986). The study animals were injected with water containing 97 atoms % ^{18}O and 50 μCi tritium per ml. They were retained in captivity for 2 hours to allow this water to mix with their body fluids, a blood sample was taken, and they were returned to the wild. Three days later they were recaptured and a second blood sample collected. The water was extracted from the blood samples by micro-distillation, after which the tritium and ^{18}O levels were determined by liquid scintillation and Gamma-Romatic counting, respectively. The total water volume of each animal was calculated by the dilution of the ^{18}O, water flux by the decline in tritium specific activity and CO_2 production by the declines through time of the specific activities of tritium and ^{18}O (Lifson & McClintock, 1966).

Kam & Degen (1997) present a modelling approach which allows the estimation of all energy budget components in free-living animals using data supplied from a doubly-labelled water experiment.

The doubly-labelled water method (DLWM) is particularly useful for flying and diving birds (Shaffer, 2011). In a study using rhinoceros auklets, *Cerorhinca monocerata*, Shirai *et al.*, (2012) concluded that the absolute percentage error between the DLWM and those obtained from a respirometric chamber was around 8%.

14.7.3.3   Analysis in the gaseous phase for air-breathing animals

The animal is enclosed in a chamber and the changes caused by its respiration in the composition of the air in the chamber determined in some way. Several techniques are based on the absorption of the evolved carbon dioxide by an alkali solution (e.g. sodium hydroxide) so that the pressure or volume of the air will change proportionally to the consumed oxygen. In constant-volume respirometers (e.g. the Warburg; Dixon, 1951), the reduction in pressure is measured with a manometer.

Alternatively, the pressure can be kept constant and the volume allowed to change. This change can be measured, for example, by the movement of an oil droplet in a capillary tube; Smith & Douglas (1949) developed an apparatus of this type that was subsequently modified by Engelmann (1961), Wiegert (1964a) and Davies (1966). The change in volume can also be utilised in ultramicrorespirometers that operate on the principle of the Cartesian diver (Zeuthen, 1950; Kay, 1964; Wood & Lawton, 1973).

The amount of carbon dioxide evolved may be determined by alkali titration (Itô, 1964). Arlian (1973) describes methods of construction. An experiment using these types of respirometer cannot be continued for long, as the respiration may become abnormal as an increasing proportion of the available oxygen is utilised.[3]

If the respiratory rate is to be recorded over a period of time the oxygen utilised by the animal must be replaced. A number of electrolytic respirometers have been developed in which the change in volume resulting from the consumption of oxygen switches on an electrolysis apparatus that generates oxygen; when the previous volume has been restored the current is switched off. The activity of the compensating oxygen generator is recorded and after suitable calibration this may be converted to give the volume of oxygen utilised (MacFadyen, 1961; Phillipson, 1962; Annis & Nicol, 1975; Arnold & Keith, 1976). In all these volumetric or manometric methods a constant temperature must be maintained throughout the experiment, usually by using a water bath. Under certain circumstances the production of ozone is a hazard.

The quantities of carbon dioxide and oxygen in the air may also be assessed by methods utilising differences in the thermal conductivity, viscosity, magnetic susceptibility and other properties of the gases (Kleiber, 1961; Kay, 1964).

14.7.3.4   Analysis of dissolved gases for completely aquatic species

The animal must be enclosed in a limited volume of water out of contact with the atmospheric air. Studies can be undertaken using closed systems in which the study organism is sealed into a chamber in which the gradual decline in oxygen concentration is measured (e.g. Paul *et al.*, 1988). Care must be taken with this approach to ensure that the oxygen level does not drop to levels that suppress respiration. The alternative approach is to use a continuous-flow respirometer in which the oxygen concentration of the water flowing into and out of the chamber holding the animal is measured (e.g. Yammamoto *et al.*, 1998). Figure 14.4 gives diagrams of both types of experimental apparatus.

The gases dissolved in the water can be determined by a number of methods. Chemical methods, especially the 'Winkler' method, have been widely used (e.g. Teal, 1957; Richman, 1958; Beyers & Smith, 1971; Yammamoto *et al.*, 1998); the procedures of the Winkler and another titration technique, the phenosafranine, are described by Dowdeswell (1959). The accuracy of these can, however, be impaired by the presence of ferrous iron, nitrites and other 'impurities' in the water (Allee & Oesting, 1934).

Polarometry or polarography in which the concentration of dissolved oxygen is measured electrolytically has also been used in many ecological studies, and is described by Kay (1964). Oxygen electrodes that are particularly suitable for ecological work were developed by Carey & Teal (1965) and are now widely available from commercial manufacturers. They may also be used to measure the oxygen concentration in aquatic environments.

------

[3]Wightman (1977) *N.Z. J. Zool.* **4**, 453–69 considers that closed-vessel respiration rates need correction by ×2.5.

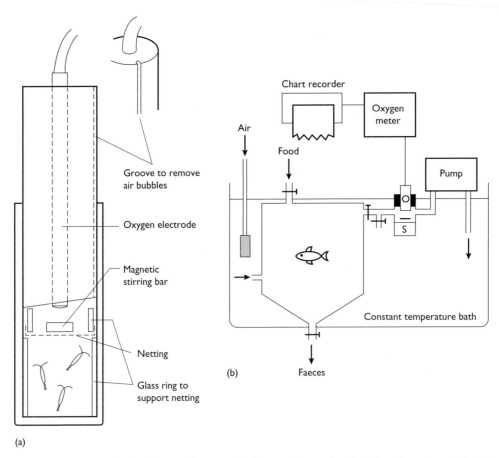

Figure 14.4 **Analysis of dissolved gases for completely aquatic species. (a) Closed system. (Adapted from Omori & Ikeda, 1984.) (b) Continuous-flow respirometer. (Adapted from Brafield & Llewellyn, 1982.)**

A tonometric method has been developed by Jones (1959) for use with very small volumes of fluid, and is thus more applicable to samples of fluid from within animals than from without! A small bubble of air is enclosed with the fluid, and the carbon dioxide, oxygen and nitrogen in it are allowed to reach diffusion equilibrium with these gases in the fluid. The composition of the bubble can then be analysed.

Gasometric methods involve the expulsion of the gases from the fluid and their subsequent analysis. Slyke & Van Neill (1924) extracted the gases under vacuum and the quantities of both oxygen and carbon dioxide can be determined (Kay, 1964). Only the oxygen and nitrogen can be measured by the method of Scholander *et al.* (1955), in which the gases are expelled from the sample by the addition of acid and a carbonate.

14.7.3.5 The use of telemetry to measure activity

Data on the activity of an animal can be used to infer energy expenditure. These techniques are particularly useful for fish. Information about the physiological activity of the fish is sent using either radio or ultrasonic signals. Metabolic activity has been estimated using heart-rate (Lucas *et al.*, 1993), tail-beat (Ross *et al.*, 1981), opercular muscle electromyograms (Rogers & Weatherley, 1983) and axial muscle electromyograms (Briggs & Post, 1997). All of these measurements give information on the activity of the animal which can be converted to energy expenditure using relationships obtained via laboratory experiments. However, trauma of capture and attachment of transmitting packages to animals may disturb their natural metabolism.

More recently, data loggers that incorporate tri-axial accelerometers have been used to record the acceleration of fish. For example, Wright *et al.* (2014) implanted surgically in the peritoneum of bass, *Dicentrarchus labrax*, G6a electronic data loggers (Cefas Technology; dimensions: 40 mm × 28 mm × 16.3 mm, 18.5 g in air, 6.7 g in seawater). From laboratory observations they concluded that dynamic body acceleration obtained from the data loggers scaled linearly with oxygen consumption, and can be used to measure the energetic costs of swimming. As always with such methods, the trauma associated with handling and surgery must be carefully considered. Wright *et al.* (2014) were not able to establish that the fish were unaffected by tagging.

14.8   The energy budget, efficiencies and transfer coefficients

The final energy budget may be presented in various forms: emphasis may be placed on the population studied (e.g. Lawton, 1970, 1971; McNeill, 1971), trophic links (e.g. Gyllenberg, 1970; Shure, 1973) or on biomass turnover and features of the ecosystem (e.g. Teal, 1962; Jonasson & Kristiansen, 1967; Burky, 1971). Budgets are generally expressed in joules or kilocalories (kcal; 1 kcal = 4186 joules) or carbon (1 g C ~ 0.5357 kcal).

14.8.1   *The energy budget of a population (or trophic level)*

A graphic representation of the energy budget of a caddis fly population is given by Otto (1975) (Fig. 14.5). This budget is noteworthy as it emphasises that a system is seldom without 'imports' (e.g. imagines from downstream and other streams) and 'exports' (e.g. predation of the adults by terrestrial animals that move away from the area). Furthermore, trophic links may cross habitat barriers (e.g. the predation of the larvae by crows).

Energy budgets can usefully be summarised and compared if the efficiencies of various processes are calculated. There have been many terms applied to these (Kozlovsky, 1968), but the following – which largely follow Wiegert (1964a) – are now recognised as standard (Petrusewicz & MacFadyen, 1970) [the notation follows Equation (14.1) at the start of this chapter]:

Assimilation/consumption	*A/C*
Production/assimilation or 'net' production efficiency	*P/A*
Production/consumption efficiency	*P/ C*
Production/respiration ratio	*P/ R*

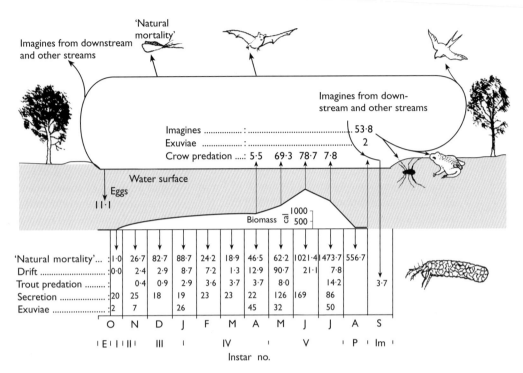

Figure 14.5 **The energy budget of a population of the caddis-fly, *Potamophylax cingulatus*, calculated on a monthly basis in terms of cal#m². (Adapted from Otto, 1975.)**

As Wiegert (1964a) emphasised, the terms (numerator and denominator) in these ratios should be precisely defined and the ratios described by the symbols (e.g. P/R ratio) or by a name that indicates these. The efficiencies, which represent energy transfers within organisms, are usually expressed as percentages.

Wiegert & Evans (1966) published an interesting table of P/C efficiencies; some values for other invertebrates are given in Table 14.4. The year-to-year variations (Gyllenberg, 1970; McNeill, 1971; Lawton, 1971) in the same population, which give an indication of the 'condition' of the population, its food supply and environment, show that too much stress must not be placed on small differences. However, not surprisingly, herbivores have lower P/C values than predators and homeotherms are lower than poikilotherms. The distinction between the two latter groups is clearly shown when known P/R ratios are plotted; there are two linear relationships, the homeotherms having a higher respiratory expenditure for the same production (McNeill & Lawton, 1970; Wiegert, 1976).

14.8.2 *Energy transfer across trophic links*

The pathways of energy transfer in an ecosystem are logically considered against a flow model of the type proposed by Wiegert & Owen (1971) (Fig. 14.6). The energy flow within the 'boxes' is described by the efficiencies or ratios listed above.

Table 14.4 Some energy efficiencies for populations of different invertebrates showing inter- and intraspecific variations.

Organism		Efficiencies			Source
		A/C%	P/A%	P/C%	
Predators	Year 1	90	52	47	Lawton, 1971
Pyrrhosoma (Odonata)	Year 2	88	48	42	Lawton, 1971
Phonoctonus (Het.)				c.45	Evans, 1962
Herbivores	Year 1	29	58	17	McNeill, 1971
Leptopterna (Het.)	Year 2	36	51	18	McNeill, 1971
	Year 3	36	52	19	McNeill, 1971
	Year 4	34	59	20	McNeill, 1971
	Year 5	28	55	16	McNeill, 1971
Rhynchaenus (Col.)	Larva I	24		20	Grimm, 1973
	Larva II	24		17	Grimm, 1973
	Larva III	20		14	Grimm, 1973
Melasoma (Col.) la.		c.72	35	24	Hågvar, 1975
Danaus (Lep.)		19	45	21	Schroeder, 1976
Euchaetias (Lep.)			49	20	Schroeder, 1977
Potamophylax (Trich.)			45	8	Otto, 1975
Saprophytic/microbial feeders					
Aphodius (Col.)		8	51	4	Holter, 1975
Pelodera (Nematoda)		60	38	22	Marchant & Nicholas, 1974
Brachionus (Rotifer) adult		19	57	11	Doohan, 1973

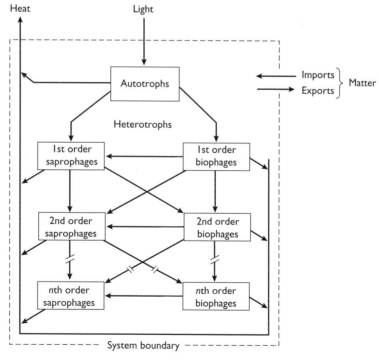

Figure 14.6 Energy transfer in an ecosystem: dioristic model. (Adapted from Wiegert & Owen, 1971.)

The rate at which energy is moves between 'boxes' is best described as:

Production turnover rate $P_t/B = \Theta_p$
Elimination turnover rate $E_t/B = \Theta_E$

where B = the average biomass over time unit i, and P and E are the total production and elimination values over time, respectively. These two rates [which are, of course, related; see Equation (14.4)] express the rate at which the production of the population (or trophic level) is being passed-on to the rest of the ecosystem. The transfer of energy across particular links between boxes [e.g. between the autotroph and the first-order biophage (or herbivore)] may be expressed as:

Harvest/Production transfer coefficient = E''_t/P'_t and
Yield/Production transfer coefficient = C''/P'_t

where E''_t = the total biomass of the 'donor' (prey) eliminated by the next level biophage during time t, P'_t = total production of the donor during this time period and C''_t = total consumption by the biophage or saprophage of the donor during period t. It will be noted that for the links to the biophages (the boxes on the right of Fig. 14.6) there are two transfer coefficients. The 'harvest transfer coefficient' represents the removal of production to the biophage, i.e. the kill or cull by a predator. However, as Dagg (1974) found, a predator – especially if its prey is large relative to it – may only consume about half the energy it harvests. Its actual consumption, the yield to it, the flow across one of the links between the boxes in Fig. 14.6, is given by the 'yield transfer coefficient'.

It is important to make these distinctions between the *efficiencies within* organisms, the *rates* at which production is passed on through trophic links, the impact (in energy terms) of one trophic level on that below it (*Harvest transfer coefficient*) and the actual energy flow along a particular link (*Yield transfer coefficient*). Ratios, other than those given here, may be useful in a particular context, but it should always be made clear whether the energy flow described is within a trophic level or between levels, or spans both levels and links.

14.9   Identification of ecological pathways using stable isotopes

Both, animals and plants can show selective uptake for the different isotopes of a number of elements, and the characteristic isotope ratios produced can be transferred to their consumers. While unable to give precise information on the food sources used by an individual, isotope ratios can be used to infer the major sources of energy for an ecological group. An example of the application of these methods is the study of Forsberg *et al.* (1993) on the carbon sources of Amazonian fish. Floodplain fish may enter the forest during the high-water season, so that it was unclear how much of their carbon was obtained from forest sources and how much from the extensive floating grasslands. These grasses (principally *Echinochloa polystachya* and *Paspalum repens*) use the C_4 photosynthetic pathway and contribute an estimated 52% of total primary production in the floodplain. The forest trees, periphyton and phytoplankton, which together are responsible for the remaining 48% of production, use the C_3 pathway. Plants using the

C_3 and C_4 pathways differ considerably in their carbon isotope ratio because C_3 plants select the lighter ^{12}C isotope. Using the $^{13}C/^{12}C$ ratio of 35 fish species, as measured using mass spectrometry, it was found that C_4 macrophytes accounted for only 2.5–17.6% of the carbon in fish. The major source of fish carbon was phytoplankton, which only contributed 8% to the total floodplain primary production but surprisingly 32% of the carbon in fish. This counterintuitive finding would have been almost impossible to obtain by other means because of the difficulty of identifying detritus from fish guts, and also because of the logistical problems posed by the spatial extent, instability and biodiversity of the habitat.

Stable isotopes are frequently used by bio-archeologists and other scientists seeking to reconstruct past events. For example, Henderson *et al.* (2014) used $\delta^{13}C$ and $\delta^{15}N$ stable isotope ratios in sections of tooth dentine from skeletons to study the nutrition and health of children.

Stable isotope studies have become particularly important for the quantification of food–web interactions and the flows of carbon and nitrogen through ecosystems following the availability of continuous flow isotope ratio mass spectrometers (CF-IRMS) during the 1990s. The application of the technique to the study of entire food webs from bacteria to top predators is reviewed by Middelburg (2014). Stable isotope ratios of nitrogen ($\delta^{15}N$) and carbon ($\delta^{13}C$) are frequently used to study food chain length and the trophic structure of ecosystems. Perkins *et al.* (2014) studied both laboratory-raised food chains and wild systems to test if stable isotopes gave reliable estimates of trophic position in systems comprising four trophic levels. They concluded that they gave reliable estimates of both trophic position of the organism and the trophic breadth of the foods consumed.

14.10 Assessment of energy and time costs of strategies

Ecologists and evolutionary biologists have become increasingly interested in the quantitative comparison of different strategies used by animals for activities such as foraging or reproduction. Energy must be used in order to acquire food, and the efficiency of this investment is the subject of optimal foraging theory (e.g. Townsend & Winfield, 1985). Strategies can be compared between species, and given polymorphism, also within a species. Stam *et al.* (1996) for example, compared the strategies of different clones of the parthenogenetic collembolan *Folsomia candida*. The ultimate criterion of strategic success, of course, is reproductive success, but valuable indications of the relative costs and benefits can be obtained from the consideration of time and energy budgets (Emlen, 1966; Schoener, 1971; Van Valen, 1976; Norberg, 1977; Wells & Clarke, 1996). The methods utilised will often be similar to those described above, for example the measurement of respiratory rate, energy value of 'harvest', but these will have to be related to particular activities. In essence the following are usually required:

1. A time budget of the various activities.
2. The rate of energy expenditure characteristic of each type of activity
3. The rate of energy gain food harvested or usually yielded (food ingested) for each type of activity.

The effects of protein and energy content on trophic strategies of tilapia was studied by Bowen *et al.* (1995), who concluded that herbivores access plentiful energy but are constrained by lack of protein, while omnivores follow a compromise which gives access to plentiful protein and energy.

Studies of this type have been made on territoriality and resources (e.g. McNab, 1963; Smith, 1968; Gill & Wolf, 1975a,b) on speed and pattern of movement in relation to resource availability (e.g. Calow, 1974) and foraging strategies (e.g. Mukerji & Le Roux, 1969; Heinrich, 1972; Charnov, 1976; Davidson, 1977b).

References

Allee, W.C. & Oesting, R. (1934) A critical examination of Winkler's method for determining dissolved oxygen in respiration studies with aquatic animals. *Physiol. Zool.* **7**, 509–41.

Allen, K.R. (1951) The Horokiwi Stream, a study of a trout population. *N. Z. Mar. Dep. Fish. Rep. DIV Bull.* **10**, 1–238.

Allen, M.D. (1959) Respiration rates of worker honeybees of different ages and at different temperatures. *J. Exp. Biol.* **36**, 92–101.

Annis, P.C. & Nicol, C.R. (1975) Respirometry system for small biological samples. *J. Appl. Ecol.* **12**, 137–41.

Arlian, L.G. (1973) Methods for making a cartesian diver for use with small arthropods. *Ann. Entomol. Soc. Am.* **66**, 694–5.

Arnold, D.I. & Keith, D.E. (1976) A simple continuous-flow respirometer for comparative respirometry changes in medium-sized aquatic organisms. *Water Res.* **10**, 261–4.

Atkinson, E.G. & Wacasey, J.W. (1983) Caloric equivalents for benthic marine organisms from the Canadian Arctic. *Can. Tech. Rep. Fish. Aquat. Sci.* **1216**, 32.

Bailey, C.G. & Riegert, P.W. (1972) Energy dynamics of *Encoptolophus sordidus cortalis* (Scudder) (Orthoptera: Acrididae) in a grassland ecosystem. *Can. J. Zool.* **51**, 91–100.

Baird, D. & Milne, H. (1981) Energy flow in the Ythan Estuary, Aberdeenshire, Scotland. *Estuarine Coastal Shelf Sci.* **13**, 455–72.

Benke, A.C. (1976) Dragonfly production and prey turnover. *Ecology* **57**, 915–27.

Bennett, A.F. & Nagy, K.A. (1977) Energy expenditure in free-ranging lizards. *Ecology* **58**, 697–700.

Beyers, R.I. & Smith, M.H. (1971) A calorimetric method for determining oxygen concentration in terrestrial situations. *Ecology* **52**, 374–5.

Bowen, S.H., Lutz, E.V., & Ahlgren, M.O. (1995) Dietary protein and energy as determinants of food quality: trophic strategies compared. *Ecology,* **76**, 899–907.

Boyd, C.E. (1969) The nutritive value of three species of water weeds. *Economic Botany* **239**, 123–7.

Boyd, C.E. & Blackburn, R.D. (1970) Seasonal changes in the proximate composition of some common aquatic weeds. *Hyacinth Control J.* **8**, 42–144.

Boyd, C.E. & Goodyear, C.P. (1971) Nutritive quality of food in ecological systems. *Arch. Hydrobiol.* **69**, 256–70.

Brafield, A.E. (1985) Laboratory studies of energy budgets. In: Tytler, P. & Calow, P. (eds), *Fish Energetics: New Perspectives.* John Hopkins University Press, Baltimore.

Brafield, A.E. & Llewellyn, M.J. (1982) *Animal Energetics.* Chapman & Hall, London.

Brody, S. (1945) *Bioenergetics and Growth.* Hafner, New York, 1023 pp.

Brey, T., Rumohr, H., & Ankar, S. (1988) Energy content of macroinvertebrates: general conversion factors from weight to energy. *J. Exp. Mar. Biol. Ecol.* **117**, 217–78.

Briggs, C.T. & Post, J.R. (1997) Field metabolic rates of rainbow trout estimated using electromyogram telemetry. *J. Fish Biol.* **51**, 807–23.

Brody, S. (1945) *Bioenergetics and Growth*. Hafner, New York, 1023 pp.

Brown, D.S. (1961) The food of the larvae of *Chloeon dipierum* L. and *Baetis rhodani* (Pictet) (Insecta, Ephemeroptera). *J. Anim. Ecol.* **30**, 55–75.

Burky, A.J. (1971) Biomass turnover, respiration and interpopulation variation in the stream limpet *Ferrissia rivularis* (Say). *Ecol. Monogr.* **41**, 235–51.

Burnison, B.K. & Perez, K.T. (1974) A simple method for the dry combustion of 14C-labelled materials. *Ecology* **55**, 899–902.

Cahn, R. (1956) *La regulation des processus métaboloques dans l'organism*. Hermann et Cie, Paris.

Calow, P. (1974) Some observations on locomotory strategies and their metabolic effects in two species of freshwater gastropods, *Anylus fluuiatilis* and *Planorbis contortus* Linn. *Oecologia* **16**, 149–61.

Calow, P. & Fletcher, C.R. (1972) A new radio tracer technique involving ^{14}C and ^{51}Cr for estimating the assimilation efficiencies of aquatic primary consumers. *Oecologia* **9**, 155–70.

Carey, F.G. & Teal, J.M. (1965) Responses of oxygen electrodes to variables in construction, assembly and use. *J. Appl. Physiol.* **20**, 1074–7.

Caspers, N. (1977) Seasonal variations of caloric values in herbaceous plants. *Oecologia* **269**, 379–83.

Cech, J.J. (1990) Respirometry. In: Schreck, C.B. & Moyle, P.B. (eds), *Methods of Fish Biology*. American Fisheries Society, Bethesda, MD, pp. 335–62.

Chaabane, K., Loreau, M., & Josens, G. (1996) Individual and population energy budgets of *Abax ater* (Coleoptera, Carabidae). *Ann. Zool. Fennici* **33**(1), 97–108.

Chapman, S.B. (1976) Production ecology and nutrient budgets. In: Chapman, S.B. (ed.), *Methods in Plant Ecology*. Blackwell Scientific Publications, Oxford.

Charnov, E.L. (1976) Optimal foraging: attack strategy of a mantid. *Am. Nat.* **110**, 141–51.

Coleman, D.C. (1968) *Food webs of small arthropods of a broom sedge field studied with radio-isotope-labelled fungi*. Proc. IBP. Tech. Meeting on Methods of Study in Soil Zoology, Paris, UNESCO.

Comita, G.W. & Schindler, D.W. (1963) Calorific values of microcrustacea. *Science* **1409**, 1394–6.

Corbett, J.L., Greenhalgh, F.D., McDonald, I., & Florence, E. (1960) Excretion of chromium sesquioxide administered as a component of paper to sheep. *Br. J. Nutr.* **14**, 289–9.

Cossins, A.R. & Bowler, K. (1987) *Temperature Biology of Animals*. Chapman & Hall, London.

Crossley, D.A. (1963a) Consumption of vegetation by insects. In: Schultz, V. & Klement, A.W. (eds), *Radioecology*. Rheinhold, New York.

Crossley, D.A. (1963b) Movement and accumulation of radiostrontium and radiocesium in insects. In: Schultz, V. & Klement, A.W. (eds), *Radioecology*. Rheinhold, New York.

Crossley, D.A. (1966) Radio-isotope measurement of food consumption by a leaf beetle species, *Chrysomela knabi* Brown. *Ecology* **47**, 1–8.

Crossley, D.A. & Hoglund, M.P. (1962) A litter-bag method for the study of micro-arthropods inhabiting leaf litter. *Ecology* **43**, 571–3.

Dagg, M.J. (1974) Loss of prey body contents during feeding by an aquatic predator. *Ecology* **55**, 903–6.

Davidson, D.H. (1977a) Assimilation efficiencies of slugs on different food materials. *Oecologia* **26**, 267–73.

Davidson, D.W. (1977b) Foraging ecology and community organization in desert-seed eating ants. *Ecology* **58**, 725–37.

Davies, P. (1966) A constant pressure respirometer for medium-sized animals. *Oikos* **179**, 108–12.

Degan, A.A., Kam, M., Hazan, A., & Nagy, K.A. (1986) Energy expenditure and water flux in three sympatric desert rodents. *J. Anim. Ecol.* **55**, 421–9.

Dixon, M. (1951) *Manometric Methods as Applied to the Measurement of Cell Respiration and Other Processes.* Cambridge University Press, Cambridge.

Doohan, M. (1973) An energy budget for adult *Brachionus plicatilis* Muller (Rotatoria). *Oecologia (Berl.)* **13**, 351–62.

Dowdeswell, W.H. (1959) *Practical Animal Ecology.* Methuen, London.

Edmondson, W.T. & Winberg, G.G. (eds) (1971) *A Manual on Methods for the Assessment of Secondary Productivity in Fresh Waters.* I.B.P. Handbook, Blackwell, Oxford.

Edmunds, P.J. & Davies, P.S. (1989) An energy budget for *Porites porites* (Scleractinia), growing in a stressed environment. *Coral Reefs* **8**(1), 37–44.

Edwards, C.A., Reichle, D.E.C., & Rossley, D.A.J. (1969) Experimental manipulation of soil invertebrate populations for trophic studies. *Ecology* **50**, 495–8.

Emlen, J.M. (1966) The role of time and energy in food preferences. *Am. Nat.* **100**, 611–17.

Engelmann, M.D. (1961) The role of soil arthropods in the energetics of an old field community. *Ecol. Monogr.* **31**, 221–38.

Evans, D.E. (1962) The food requirements of *Phonoctonus nigrofasciatus* Stal (Hemiptera, Reduviidae). *Entomol. Exp. Appl.* **5**, 33–9.

Fagerström, T. (1977) Bodyweight, metabolic rate and trace substance turnover in animals. *Oecologia* **29**, 99–116.

Fewkes, D.W. (1960) The food requirements by weight of some British Nabidae (Heteroptera). *Entomol. Exp. Appl.* **3**, 231–7.

Forsberg, B.R., Araujo-Lima C.A.R.M., Martinelli, L.A., Victoria, R.L., & Bonsassi, J.A. (1993) Autotrophic carbon sources for fish of the central Amazon. *Ecology* **74**, 643–52.

Frolander, H.F. (1957) A plankton volume indicator. *J. Cons. Perm. Int. Explor. Mer.* **229**, 278–83.

Fry, F.E.J. (1947) Effects of environment on animal activity. *University of Toronto Studies in Biology Series* **55**, 1–62.

Gallucci, V.F. (1973) On the principles of thermodynamics and ecology. *Annu. Rev. Ecol. Syst.* **4**, 329–57.

Gill, F.B. & Wolf, L.L. (1975a) Economics of feeding territoriality in the Golden Winged Sunbird. *Ecology* **56**, 333–45.

Gill, F.B. & Wolf, L.L. (1975b) Foraging strategies and energetics of cast African sunbirds at mistletoe flowers. *Am. Nat.* **109**, 491–510.

Golley, F.B. (1960) Energy dynamics of a food chain of an old-field community. *Ecol. Monogr.* **30**, 187–206.

Golley, F.B. & Buechner, H.K. (eds) (1968) *A Practical Guide to the Study of the Productivity of Large Herbivores.* I.B.P. Handbook. Blackwell, Oxford.

Golley, F.B. & Gentry, J.B. (1964) Bioenergetics of the southern Harvester ant, *Pogonomyrmex badius. Ecology* **45**, 217–25.

Gosler, A.G. (1996) Environmental and social determinants of winter fat storage in the great tit *Parus major. J. Anim. Ecol.* **65**, 1–17.

Griffiths, D. (1977) Caloric variation in Crustacea and other animals. *J. Anim. Ecol.* **465**, 593–605.

Grimm, R. (1973) Zum energieumsutz phytophager Insekten im Buchenwald I. *Oecologia* **11**, 187–262.

Grodzinski, W., Klekowski, R.Z., & Duncan, A. (eds) (1975) *Methods for Ecological Bioenergetics. I.B.P. Handbook*. Blackwell, Oxford.

Gyllenberg, G. (1969) The energy flow through a *Chorippus parallelus* (Zett.) (Orthoptera) population in a meadow in Tyürmine, Finland. *Acta Zool. Fennici* **123**, 1–74.

Gyllenberg, G. (1970) Energy flow through a simple food chain of a meadow ecosystem in four years. *Ann. Zool. Fennici* **7**, 283–9.

Hack, M.A. (1997) The effects of mass and age on standard metabolic rate in house crickets. *Physiol. Entomol.* **22**(4), 325–31.

Hagvar, S. (1975) Energy budget and growth during the development of *Melasoma collaris* (Coleoptera). *Oikos* **26**, 140–6.

Hanna, H.M. (1957) A study of the growth and feeding habits of the larvae of four species of caddis flies. *Proc. R. Entomol. Soc. Lond. A* **329**, 139–46.

Hartman, K.J. & Brandt, S.B. (1995) Estimating energy density of fish. *Trans. Am. Fish. Soc.* **124**, 347–55.

Hatlen, B., Nordgreen, A.H., Romarheim, O.H., Aas, T.S., & Åsgård, T. (2015) Addition of yttrium oxide as digestibility marker by vacuum coating on finished pellets – A method for assessing digestibility in commercial salmon feeds? *Aquaculture* **435**, 301–5.

Heinrich, B. (1972) Energetics of temperature regulation and foraging in a bumblebee, *Bombus terricola* Kirby. *J. Comp. Physiol.* **77**, 48–64.

Henderson, P.A., Holmes, R.H.A., & Bamber, R.N. (1988) Size-selective overwintering mortality in the sand smelt, *Atherina boyeri* Risso, and its role in population regulation. *J. Fish Biol.* **33**, 221–33.

Henderson, P.A. & Walker, I. (1986) On the leaf-litter community of the Amazonian blackwater stream Tarumazinho. *J. Trop. Ecol.* **2**, 1–17.

Henderson, R.C., Lee-Thorp, J., & Loe, L. (2014) Early life histories of the London poor using δ13C and δ15N stable isotope incremental dentine sampling. *Am. J. Phys. Anthropol.* **154**, 585–93.

Henen, B.T. (1997) Seasonal and annual energy budgets of female desert tortoises (*Gopherus agassizii*). *Ecology* **78**(1), 283–96.

Hinton, J.M. (1971) Energy flow in a natural population of *Neophilaenus litlealus* (Homoptera). *Oikos* **22**, 155–71.

Holter, P. (1973) A chromic oxide method for measuring consumption in dung-eating *Aphodius* larvae. *Oikos* **24**, 117–22.

Holter, P. (1974) Food utilization of dung-eating *Aphodius* larvae (Scarabaeidae). *Oikos* **25**, 71–9.

Holter, P. (1975) Energy budget of a natural population of *Aphodius rupes* larvae (Scarabaeidae). *Oikos* **26**, 177–86.

Hubbell, S.P., Sikora, A., & Paris, O.H. (1965) Radiotracer, gravimetric and calorimetric studies of ingestion and assimilation rates of an isopod. *Health Physics* **11**, 1485–501.

Humphreys, W.F. (1975) The food consumption of a Wolf Spider, *Geolylcosa godeffroyi* (Aracridae: Lycoridae), in the Australian Capital Territory. *Oecologia (Berl.)* **18**, 343–58.

Itô, Y. (1964) Preliminary studies on the respiratory energy loss of a spider, *Lycosa pseudoannulata*. *Res. Pop. Ecol.* **6**, 13–21.

Iversen, T.M. (1974) Ingestion and growth in *Sericostoma personatum* (Trichoptera) in relation to the nitrogen content of the ingested leaves. *Oikos* **25**, 278–82.

Ivlev, V.S. (1939) Transformation of' energy by aquatic animals. *Int. Rev. Ges. Hydrobiol. Hydrogr.* **38**, 449–558.

Ivlev, V.S. (1945) The biological productivity of waters. [In Russian.]. *Usp. sovrem. Biol.* **19**, 98–120.

Jenderedjian, K. (1996) Energy budget of Oligochaeta and its relationship with the primary production of Lake Sevan, Armenia. *Hydrobiologia* **334**(1–3), 133–40.

Jonasson, P.M. & Kritriansen, J. (1967) Primary and secondary production in Lake Esrom. Growth of *Chironomus anthracitius* in relation to seasonal cycles of phytoplankton and dissolved oxygen. *Int. Rev. Ges. Hydrobiol.* **52**, 163–217.

Jones, J.D. (1959) A new tonometric method for the determination of dissolved oxygen and carbon dioxide in small samples. *J. Exp. Biol.* **36**, 177–90.

Kam, M. & Degen, A.A. (1997) Energy budget in free-living animals: A novel approach based on the doubly labeled water method. *Am. J. Physiol.* **272**, 1336–43.

Karo, J. (1973) An attempt to estimate the energy flow through the population of Colorado Beetle (*Leptinotarsa decemlineato* Say). *Ekologia Poiska* **21**, 239–50.

Kay, R.H. (1964) *Experimental Biology. Measurement and Analysis.* Chapman & Hall, London.

Kleiber, M. (1961) *The Fire of Life. An Introduction to Animal Energetics.* John Wiley, New York.

Kormondy, E.J. (1965) Uptake and loss of zinc-65 in the dragonfly *Plathemis lydia. Limnol. Oceanogr.* **10**, 427–33.

Kozlovsky, D.G. (1968) A critical evaluation of the trophic level concept. 1. Ecological efficiencies. *Ecology* **49**, 48–60.

Kuenzler, E.J. (1961) Structure and energy flow of a mussel population in a Georgia salt marsh. *Limnol. Oceanogr.* **6**, 191–204.

Lawton, J.H. (1970) Feeding and food energy assimilation in larvae of the damselfly *Pyrrhosoma nymphula* (Sulz.) (Odonata: Zygoptera). *J. Anim. Ecol.* **39**, 669–89.

Lawton, J.H. (1971) Ecological energetics studies on larvae of the damselfly *Pyrrhosoma nymphula* (Sulz.) (Odonata: Zygoptera). *J. Anim. Ecol.* **40**, 385–419.

Lawton, J.H. & Richards, J. (1970) Comparability of Cartesian diver, Gilson, Warburg and Winkler methods of measuring the respiratory rates of aquatic invertebrates in ecological studies. *Oecologia (Berl.)* **4**, 319–24.

Le Febvre, E.A. (1964) The use of D_2O^{18} for measuring energy metabolism in *Columba livia* at rest and in flight. *Auk* **81**, 403–16.

Ledbetter, M.C. & Flemion, F. (1954) A method for obtaining piercing-sucking mouth parts in host tissues from the tarnished plant bug by high voltage shock. *Contrib. Boyce Thompson Inst.* **17**, 343–6.

Lifson, N. & McClintock, R. (1966) Theory of the use of the turnover rates of body water for measuring energy and material balance. *J. Theor. Biol.* **180**, 803–11.

Lindeman, R.L. (1942) The trophic-dynamic aspect of ecology. *Ecology* **23**, 399–418.

Lovegrove, T. (1966) The determination of dry weight of plankton and the effect of various factors on the values obtained. In: Barnes, H. (ed.), *Some Contemporary Studies in Marine Science*. Hafner, New York, pp. 429–67.

Lucas, M.C., Johnstone, A.D.F., & Priede, I.G. (1993) Use of physiological telemetry as a method of estimating metabolism of fish in the natural environment. *Trans. Am. Fish. Soc.* **122**, 822–33.

Lyytikainen, T. & Jobling, M. (1998) The effect of temperature fluctuations on oxygen consumption and ammonia excretion of underyearling Lake Inari arctic char. *J. Fish Biol.* **52**, 1186–98.

MacFadyen, A. (1961) A new system for continuous respirometry of small air-breathing invertebrates under near-natural conditions. *J. Exp. Biol.* **38**, 323–43.

MacFadyen, A. (1963) *Animal Ecology. Aims and Methods*. Pitman, London, New York.

Malone, C.R. & Nelson, D.J. (1969) Feeding rates of freshwater snails (*Goniobasis clavaejormis*) determined with cobalt 60. *Ecology* **50**, 728–30.

Marchant, R. & Nicholas, A.L. (1974) An energy budget for the free-living Nematode Pelodera (Rhabditidae). *Oecologia (Berl.)* **16**, 237–52.

Mathavan, S. & Pandian, T.I. (1974) Use of faecal weight as an indicator of food consumption in some lepidopterans. *Oecologia (Berl.)* **15**, 177–85.

McClintock, R. & Lifson, N. (1958) Determination of the total carbon dioxide output of rats by the D_2O^{18} method. *Am. J. Physiol.* **192**, 76–8.

McGinnis, A.J. & Kasting, R. (1964) Chromic oxide indicator method for measuring food utilization in a plant-feeding insect. *Science* **144**, 1464–5.

McNab, B.K. (1963) Bioenergetics and the determination of home-range size. *Am. Nat.* **97**, 133–40.

McNeill, S. (1971) The energetics of a population of *Leptopterna doiabrata* (Heteroptera: Miridae). *J. Anim. Ecol.* **40**, 127–40.

McNeill, S. & Lawton, J.H. (1970) Annual production and respiration in animal populations. *Nature* **225**, 472–4.

Middelburg, J.J. (2014) Stable isotopes dissect aquatic food webs from the top to the bottom. *Biogeosciences* **11**, 2357–71.

Miles, P.W. (1972) The saliva of hemiptera. *Adv. Insect. Physiol.* **9**, 183–225.

Moore, S.T., Schuster, M.F., & Harris, F.A. (1974) Radioisotope technique for estimating lady beetle consumption of tobacco budworm eggs and larvae. *J. Econ. Entomol.* **67**, 703–5.

Mukerji, M.K. & Le Roux, E.J. (1969) A study of energetics of *Podisus maculiventris* (Hemiptera: Pentatomidae). *Can. Entomol.* **101**, 449–60.

Mulkern, G.B. & Anderson, J.F. (1959) A technique for studying the food habits and preferences of grasshoppers. *J. Econ. Entomol.* **52**, 342.

Muthukrishnan, I. & Delvi, M.R. (1974) Effect of ration levels on food utilisation in the grasshopper *Poecilacerus pictus*. *Oecologia (Berl.)* **16**, 227–36.

Nagy, K.A. (1975) Nitrogen requirement and its relation to dietary water and potassium content in the lizard *Sauromalus obesus*. *J. Comp. Physiol.* **104**, 49–58.

Nagy, K.A. (1994) Field bioenergetics of mammals: what determines field metabolic rates? *Aust. J. Zool.* **42**, 43–53.

Neess, I. & Dugdale, C. (1959) Computation of production for populations of aquatic midge larvae. *Ecology* **40**, 425–30.

Norberg, R.A. (1977) An ecological theory on foraging time and energetics and choice of optimal food-searching method. *J. Anim. Ecol.* **46**, 511–29.

Odum, E.P. & Golley, F.B. (1963) Radioactive tracers as an aid to the measurement of energy flow at the population level in nature. In: Schultz, V. & Klement, A.W. (eds), *Radioecology*. Rheinhold, New York.

Odum, E.P. & Kuenzler, E.J. (1963) Experimental isolation of food chains in an old field ecosystem with the use of phosphorus-32. In: Schultz, V. & Klement, A. (eds), *Radioecology*. Rheinhold, New York.

Odum, H.T. (1957) Trophic structure and productivity of Silver springs, Florida. *Ecol. Monogr.* **27**, 55–112.

Odum, H.T. & Odum, E.P. (1955) Trophic structure and productivity of a Windward coral reef community on Eniwetok Atoll. *Ecol. Monogr.* **25**, 291–320.

Omori, M. & Ikeda, T. (1984) *Methods in Marine Zooplankton Ecology*. J. Wiley & Sons, New York, 332 pp.

Onuf, C.P., Teal, L.M., & Valiela, I. (1977) Interactions of nutrients, plant growth and herbivores in a mangrove ecosystem. *Ecology* **58**, 514–26.

Otto, C. (1975) Energetic relationships of the larval population of *Potamophylax cingulatus* (Trichoptera) in a South Swedish stream. *Oikos* **26**, 159–69.

Ovington, J.D. & Heitkamp, D. (1960) The accumulation of energy in forest plantations in Britain. *J. Ecol.* **48**, 639–46.

Paine, R.T. (1964) Ash and calorie determination of sponge and opisthobranch tissues. *Ecology* **45**, 384–7.

Paine, R.T. (1971) The measurement and application of the calorie to ecological problems. *Annu. Rev. Ecol. Syst.* **2**, 145–64.

Paine, R.T. (1980) Food webs: Linkage interaction strength and community infrastructure. *J. Anim. Ecol.* **49**, 667–85.

Paine, R.T. (1992) Food-web analysis through the analysis of per capita interaction strength. *Nature* **355**, 73–5.

Paris, O.H. & Sikora, A. (1965) Radiotracer demonstration of Isopod herbivory. *Ecology* **46**, 729–34.

Paul, A.J., Paul, J.M., & Smith, R.L. (1988) Respiratory energy requirements of the cod *Gadus macrocephalus* Tilesius relative to body size, food intake, and temperature. *J. Exp. Mar. Biol. Ecol.* **122**, 83–9.

Pendleton, R.C. & Grundmann, A.W. (1954) Use of phosphorus-32 in tracing some insect-plant relationships of the thistle, *Cirsium undulatum. Ecology* **35**, 187–91.

Perkins, M.J., McDonald, R.A., van Veen, F.F., Kelly, S.D., Rees, G., & Bearhop, S. (2014) Application of nitrogen and carbon stable isotopes (δ15N and δ13C) to quantify food chain length and trophic structure. *PloS One* **9**, e93281.

Petrusewicz, K.M.A. & MacFadyen, A. (1970) *Productivity of Terrestrial Animals. Principles and Methods.* Blackwell, Oxford.

Phillipson, J. (1960) The food consumption of different instars of *Mitopus morio* (F.) (Phalangiida) under natural conditions. *J. Anim. Ecol.* **29**, 299–307.

Phillipson, J. (1962) Respirometry and the study of energy turnover in natural systems with particular reference to harvest spiders (Phalangiida). *Oikos* **13**, 311–22.

Phillipson, J. (1964) A miniature bomb calorimeter for small biological samples. *Oikos* **159**, 130–9.

Plaut, I., Borut, A., & Spira, M.E. (1996) Lifetime energy budget in the sea hare *Aplysia oculifera. Comp. Biochem. Physiol. A* **113**(2), 205–12.

Pollard, D.G. (1973) Plant penetration by feeding aphids (Hemiptera: Aphoidea): a review. *Bull. Entomol. Res.* **62**, 631–714.

Raffaelli, D.G. & Hall, S.J. (1996) Assessing the relative importance of trophic links in food webs. In: Polis, G.A. & Winemiller, K.O. (eds), *Food Webs: Integration of Patterns and Dynamics.* Chapman & Hall, London.

Reichle, D.E. (1969) Measurement of elemental assimilation by animals from radioactive retention patterns. *Ecology* **50**, 1102–4.

Richman, S. (1958) The transformation of energy by *Daphnia pulex. Ecol. Monogr.* **28**, 273–91.

Ricker, W.E. (1968) *Methods for Assessment of Fish Production in Fresh Waters.* Blackwell, Oxford.

Rogers, S.C. & Weatherley, A.H. (1983) The use of opercular muscle electromyograms as an indicator of fish activity in rainbow trout, *Salmo gardneri* Richardson, as determined by radiotelemetry. *J. Fish Biol.* **23**, 535–47.

Ross, L.G., Watts, W., & Young, A.H. (1981) An ultrasonic biotelemetry system for continuous monitoring of tail-beat rate from free-swimming fish. *J. Fish Biol.* **18**, 479–90.

Schoener, T.W. (1971) Theory of feeding strategies. *Annu. Rev. Ecol. Syst.* **2**, 369–404.

Scholander, P.F., Vandam, L., Claff, C.L., & Kanwisher, J.W. (1955) Microgasometric determination of dissolved oxygen and nitrogen. *Biol. Bull. Woods Hole* **109**, 328-34.

Schroeder, L.A. (1976) Energy, matter and nitrogen utilization by larvae of the monarch butterfly *Danaus plexippus* (Danaidae: Lepidoptera). *Oikos* **27**, 27-31.

Schroeder, L.A. (1977) Energy, matter and nitrogen utilization by larvae of the milkweed tiger moth *Euchretias egle*. *Oikos* **28**, 27–31.

Shaffer, S.A. (2011) A review of seabird energetics using the doubly labeled water method. *Comp. Biochem. Physiol. A: Molec. Integr. Physiol.* **158**, 315–22.

Shaw, I. & Beadle, L.C. (1949) A simplified ultra-micro Kjeldahl method for the estimation of protein and total nitrogen in fluid samples of less than 1.0 m. *J. Exp. Biol.* **26**, 15–23.

Shirai, M., Ito, M., Yoda, K., & Niizuma, Y. (2012) Applicability of the doubly labelled water method to the rhinoceros auklet, *Cerorhinca monocerata*. *Biology Open* **1**, 1141–5.

Shure, D.J. (1970) Limitations in radio-tracer determination of consumer trophic positions. *Ecology* **51**, 899–901.

Shure, D.J. (1973) Radionuclide tracer analysis of trophic relationships in an old-field ecosystem. *Ecol. Monogr.* **43**, 1–19.

Singh, J.S. & Yadava, P.S. (1973) Caloric values of plant and insect species of a tropical grassland. *Oikos* **24**, 186–94.

Slobodkin, L.B. & Richman, S. (1961) Calories/gm in species of animals. *Nature* **191**, 299.

Slyke, D.D. & Van Neill, J.M. (1924) The determination of gases in blood and other solutions by extraction and manometric measurement. *J. Biol. Chem.* **61**, 523–73.

Smith, A.H. & Douglas, J.R. (1949) An insect respirometer. *Ann. Entomol. Soc. Am.* **42**, 14–18.

Smith, C.C. (1968) The adaptive nature of social organization in the genus of three squirrels *Tamiasciurus*. *Ecol. Monogr.* **38**, 31–63.

Soofiani, N.M. & Hawkins, A.D. (1982) Energetic costs at different levels of feeding in juvenile cod, *Gadus morhua* L. *J. Fish Biol.* **21**, 577–92.

Stam, E.M., Van de Leemkule, M.A., & Ernsting, G. (1996) Trade-offs in the life history and energy budget of the parthenogenetic collembolan *Folsomia candida* (Willem). *Oecologia* **107**(3), 283–92.

Stanford, J.A. (1973) A centrifuge method for determining live weights of aquatic insect larvae, with a note on weight loss in preservative. *Ecology* **54**, 449–51.

Stiven, A.E. (1961) Food energy available for and required by the blue grouse chick. *Ecology* **42**, 547–53.

Strong, F.E. & Landes, D.A. (1965) Feeding and nutrition of *Lygus hesperus* (Hemiptera: Miridae). II. An estimation of normal feeding rates. *Ann. Entomol. Soc. Am.* **58**, 309–14.

Sushchenya, L.M. (1962) [Quantitative data on nutrition and energy balance in *Artemia salina* (L.)] [In Russian]. *Doklady Akad. Nauk. S.S.S.R.* **143**, 1205–7.

Teal, J.M. (1957) Community metabolism in a temperate cold spring. *Ecol. Monogr.* **27**, 283–302.

Teal, J.M. (1962) Energy flow in the saltmarsh ecosystems of Georgia. *Ecology* **43**, 614–24.

Townsend, C.R. & Winfield, I.J. (1985) The application of optimal foraging theory to feeding behaviour in fish. In: Tyler, P. & Calow, P. (eds), *Fish Energetics: New Perspectives*. Croom Helm, London.

Utter, J.M. & Le Febvre, E.A. (1973) Daily energy expenditure of purple martins (*Progne subis*) during the breeding season: estimates using D_2O^{18} and time budget methods. *Ecology* **54**, 597–604.

Van Hook, R.I. (1971) Energy and nutrient dynamics of spiders and orthopteran populations in a grassland ecosystem. *Ecol. Monogr.* **41**, 1–26.

Van Valen, L. (1976) Energy and evolution. *Evolut. Theory* **1**, 179–229.

Verdun, J. (1972) Caloric content and available energy in plant matter. *Ecology* **53**, 982.

Vollenweider, R.A. (1969) *A Manual on Methods of Measuring Primary Production in Aquatic Environments*. Blackwell, Oxford.

Walker, I., Henderson, P.A., & Sterry, P.S. (1991) On the patterns of biomass transfer in the benthic fauna of an Amazonian black-water river, as evidenced by 32P label experiment. *Hydrobiologia* **215**, 153–62.

Walsberg, G.E. & Hoffman, T.C. (2005) Direct calorimetry reveals large errors in respirometric estimates of energy expenditure. *J. Exp. Biol.* **208**, 1035–43.

Wells, M.J. & Clarke, A. (1996) Energetics: The costs of living and reproducing for an individual cephalopod. *Philos. Trans. Roy. Soc. Lond. B* **351**, 1083–104.

Whittaker, R.H., Bormann, F.H., Likens, G.E., & Siccama, T.G. (1974) The Hubbard Brook ecosystem study: Forest biomass and production. *Ecol. Monogr.* **44**, 233–52.

Wiegert, R.G. (1964a) The ingestion of xylem sap by meadow spittle bugs, *Philaenus spumarius* (L.). *Am. Midl. Nat.* **71**, 422–8.

Wiegert, R.G. (1964b) Population energetics of meadow spittle bugs (*Philaenus spumarius* L.) as affected by migration and habitat. *Ecol. Monogr.* **34**, 225–41.

Wiegert, R.G. (1965) Intraspecific variation in calories/g of meadow spittle bugs (*Philaenus spumarius* (L.). *BioScience* **15**, 543–5.

Wiegert, R.G. (1974) Litterbag studies of microarthropod populations in three South Carolina old fields. *Ecology* **55**, 94–102.

Wiegert, R.G. (ed.) (1976) *Ecological Energetics. Benchmark Papers in Ecology*. Dowden, Hutchinson & Ross, Pennsylvania.

Wiegert, R.G. & Coleman, D.C. (1970) Ecological significance of low oxygen consumption and high fat accumulation by *Nasutitermes costalis* (Isoptera: Termitidae). *BioScience* **20**, 663–5.

Wiegert, R.G. & Evans, F.C. (1967) Investigations of secondary productivity in grasslands. In: Petrusewicz, K. (ed.), *Secondary Productivity of Terrestrial Ecosystems*. Polish Academy of Science, Warsaw, Cracow, pp. 499–578.

Wiegert, R.G. & Owen, D.F. (1971) Trophic structure, available resources and population density in terrestrial vs. aquatic ecosystems. *J. Theor. Biol.* **30**, 69–81.

Williams, E.C. & Reichle, D.F. (1968) Radioactive tracers in the study of energy turnover by a grazing insect (*Chrysochus amatus* Fab.: Coleoptera Chrysomelidae). *Oikos* **19**, 10–18.

Wissing, T.E. & Hasler, A.D. (1971) Intraseasonal change in caloric content of some freshwater invertebrates. *Ecology* **52**, 371–3.

Wood, T.G. & Lawton, J.H. (1973) Experimental studies on the respiratory rates of mites (Acari) from beech-woodland leaf litter. *Oecologia (Berl.)* **12**, 169–91.

Wright, S., Metcalfe, J., Hetherington, S., & Wilson, R. (2014) Estimating activity-specific energy expenditure in a teleost fish, using accelerometer loggers. *Mar. Ecol. Progr. Ser,* **496**, 19–32.

Yammamoto, T., Ueda, H., & Higashi, S. (1998) Correlation among dominance status, metabolic rate and otolith size in masu salmon. *J. Fish Biol.* **52**, 281–90.

Zeuthen, E. (1950) Cartesian diver respirometer. *Biol. Bull. Mar. Lab. Woods Hole* **98**, 139–43.

Zuntz, N. & Schumberg, H. (1901) *Studien zur einer Physiologie des* Marsches, Berlin, p. 361.

15 Studies at Large Spatial, Temporal and Numerical Scales and the Classification of Habitats

Technological advances over the past 40 years have greatly increased the opportunities for ecologists to study population and community processes over large spatial and temporal distances and to follow changes in large populations. A number of important developments have made this possible. First, a variety of sensors on board satellites have produced a massive amount of remote-sensed data. This has allowed the analysis of spatial and temporal variation in both physical variables such as temperature, rainfall or the area of open water, and biological variables such as the distribution of closed forest, open grassland or phytoplankton blooms. Second, the rapid reduction in the cost and size of powerful computers has made widely available to ecologists the computational power to analyse large data sets and digital images on personal computers. Third, linked to improved computational power, the internet has allowed researchers to utilise large numbers of enthusiastic, often amateur, investigators. The growth of citizen science projects over the last decade has been dramatic; a list of some of these projects is available at http://en.wikipedia.org/wiki/List_of_citizen_science_projects. The final process aiding large-scale studies is the gradual accumulation of long-term data. Ecology is a recent science with origins in the 20th century. Ecologists now have time-series for a far greater number of populations or communities than were available to earlier generations of researchers (Leigh & Johnston, 1994). In addition, the time length of established long-term studies is gradually increasing to the point where time series analysis techniques used by the physical sciences are applicable. For example, the statistical and numerical techniques developed by astronomers are now becoming applicable. All of these developments offer new opportunities to ecologists seeking to understand the way in which populations are organised and change through time. They also offer great advantages to conservation biologists seeking to identify, map and protect vulnerable species or ecosystems. In this chapter, we do not aim to give a full account of the wide range of techniques used for the organisation and analysis of large-scale spatial and temporal data sets. Rather, we aim to introduce the methods available and direct attention to some of the numerous books and articles which cover these topics in detail. Our purpose with respect to spatial data is to introduce via some examples the ways in which geographical information systems and remote sensing data can be used by ecologists and conservationists. For temporal data, we introduce some

Ecological Methods, Fourth Edition. P. A. Henderson and T. R. E. Southwood.
© 2016 John Wiley & Sons, Ltd. Published 2016 by John Wiley & Sons, Ltd.
Companion Website: www.wiley.com/go/henderson/ecologicalmethods

Table 15.1   The frequency at which the water level declined below selected water levels during the annual low-water season between 1902 and 1984 in the Reservá Mamirauá floodplain between the Rios Japurá and Solimões in the Amazon Basin, Brazil (Henderson et al., 1998). The values are given for an arbitrary 0 m datum, which is the depth of the shallow lakes and channels found in the region. The frequencies were calculated using PORTOBRAS data.

Water depth (m)	Frequency of occurrence
−2	Once every 25 years
−1	Once every 10 years
0	Every 4–5 years
1	Every 2–3 years
2	Every 2 years
3	8 out of 10 years
4	9 out of 10 years

elementary techniques for the analysis of the variability in population size and community structure.

The definition of large scale requires reference to the generation time and range of the organism. A long-term (large temporal scale) population study of an annually reproducing insect would be expected to include annual population estimates for at least 10 years. An equivalent study of an amoeba, that can reproduce daily, might be completed in a few weeks. However, if the focus of a long-term study is the role of seasonal variation in determining population number, then it is likely that a study will need at least 25 years of data, irrespective of the size of the organism. Climatic variation and many biological processes almost invariably produce autocorrelated time series. As the primary requisite for the detection of any pattern are independent observations, any autocorrelation increases the length of time series that must be studied. Gaston & McArdle (1994) give as a rule of thumb that, if a time series has n observations and an autocorrelation coefficient of r, then the effective sample number is $(1 - r^2)n$. As an illustration of the typical levels of variability experienced by animals Table 15.1 gives the frequency at which different low-water levels have been recorded in the Amazonian floodplain last century (Henderson et al., 1998). Levels less than −2 metres will have a profound effect on the aquatic fauna, as most of the lakes and channels in the floodplain would be dry. Such an event occurs, on average, once every 25 years. Thus, the fauna must have evolved to deal with conditions that may require a very long time series to actually observe. Even for the most temperate climate, anything shorter than 15 years is unlikely to include a representative selection of the possible seasonal patterns and events that the population could experience. Large spatial scales can be defined either in terms of the range of the population, or by the extent of a particular ecosystem such as a floodplain. Large-scale studies of single species usually aim to analyse distribution or density over the entire population or metapopulation range (Hanski, 1981, 1982, 1987; Hanski & Gilpin, 1991; Hassell et al., 1991; Gruttke & Engels, 1998).

15.1   Remote sensing data from satellites

Remote sensing is the measurement of reflected, emitted or back-scattered electromagnetic radiation from the Earth's surface using instruments placed at a distance, most often on a satellite, although aircraft are frequently used. An introduction to remote sensing resources is available at: http://rsd.gsfc.nasa.gov/rsd/RemoteSensing.html. There is a huge literature on remote sensing, much of which is accessible online; recent books include Barrett (2013), Konecny (2014), and for oceans Martin (2014). A list of remote sensing satellites is available at: http://claudelafleur.qc.ca/Scfam-remotesensing.html. It is normal to collect the information digitally as a stream of data representing the picture elements (the pixels) of the final image. The area of the Earth's surface represented by a single pixel corresponds to the maximum resolution for the detector. The spatial resolutions of the receiving instruments differ widely, and the value assigned to each pixel is the average for the corresponding land area. Pixels correspond to areas ranging from square kilometres for meteorological satellites to a few square centimetres for aircraft-borne sensors. For satellite-borne sensors, the temporal resolution, the time between repeated measurements at one locality, depends on the orbit. Polar orbiting satellites pass over different parts of the Earth's surface on each orbit, so that there may be a number of days between repeat measurements of the same locality. Geostationary satellites orbit at the equator at the same speed as the Earth so that they remain above the same locality, allowing relatively small temporal resolutions.

Digital data derived from sensors must be processed to obtain useful images that display to best effect the features of interest. A frequent problem is cloud cover that may, if images for an area are available over a relatively long period, frequently be removable by combining the images. The wavelengths chosen for study will depend on the variables of interest. The radiation at different wavelengths received from a pixel may be used to calculate an index that detects an important attribute. Early instruments such as those on the NOAA-AVHRR satellite data offered information on reflectance in the visible (red) and near-infrared regions of the spectrum, and these were used to give an index of vegetation cover which could be used to forecast potential famine. The normalised-difference vegetation index (NDVI) is:

$$NDVI = \frac{(NIR - VIS)}{(NIR + VIS)}$$

where *VIS* and *NIR* stand for the spectral reflectance measurements in the visible and near-infrared regions, respectively (see: http://earthobservatory.nasa.gov/Features/MeasuringVegetation/measuring_vegetation_2.php).

Remote sensing can be a powerful tool for studying vegetation particularly in remote areas, such as tropical forest habitats in the Amazon basin, which are still poorly understood with respect to the distribution, patchiness and area of different habitat types. These habitats present a considerable challenge for traditional survey methods because of the high species richness, the large area to be studied, and the general inaccessibility of the habitats. From the mid-1990s considerable data was becoming available. For example, Tuomisto *et al.* (1994) analysed Landsat-Multi Spectral Scanner (MSS) images from the Peruvian Amazon to detect and delimit vegetation types and geological formations. The authors used Landsat MSS images rather than those from the higher-resolution Landsat TM because of the lower cost per unit area and reduced computational requirements. A variety of methods were tried to give the best resolution of vegetation types using ERDAS and ILWAS software, and the satisfactory results were obtained using colour composites and histogram equalisation. Pixel-by-pixel classification techniques were also applied, but these proved unsatisfactory because of the patchiness of the habitat, which often results in quite different electromagnetic spectra from adjacent pixels. The resulting images showed all the major features of Amazonian floodplain systems, including scrole-swale topography caused by lateral channel migration, back swamps, inundation and *terra firme* forest. Using these images, it was possible to calculate the amount of the different types of major habitat available and to appreciate its patchiness. The resolution was, however, insufficient to measure the amount of floodplain water habitat much of which, during the low-water season, is in the form of narrow channels or small pools. Tuomisto *et al.* (1994) noted the need for local knowledge and ground truth surveys in order to check the different vegetation types were being distinguished. They concluded that some habitats could not be distinguished by their electromagnetic spectra alone. It is still the case that ground truth surveys are important.

Animal distribution is often closely linked to that of vegetation. Arseneault *et al.* (1997) used Landsat Thematic Mapper (TM) imagery to study the influence of caribou grazing on lichen biomass and percentage ground cover, while buzzard, *Buteo buteo*, nesting areas were predicted with the aid of vegetation cover indices derived from satellite imagery (Austin *et al.*, 1996). Remote sensing and vegetation indices are particularly useful for predicting habitat suitability for animals in arid regions (e.g. Verlinden & Masogo, 1997), the measurement of deforestation (e.g. Blackman, 2013; Buchanan *et al.*, 2013), erosion (e.g. Dwivedi *et al.*, 1997; Reiche *et al.*, 2012), and submerged seaweed and seagrass distribution (e.g. Ferguson & Korfmacher, 1997; Purkis & Roelfsema, 2015). Erosion and other landscape features can also be monitored using unmanned vehicles and small drones (e.g. D'Oleire-Oltmanns *et al.*, 2012).

Climatic variables may also be the basis of a study. The abundance and distribution of tsetse flies (*Glossina* spp.) in West Africa, as has been found for other insect vectors, are sensitive to climatic variables such as temperature, humidity and rainfall (Bursell, 1957; Rogers & Randolph, 1991; Hay *et al.*, 1996; Rogers *et al.*, 1996). Unfortunately, there are insufficient meteorological data available from African ground stations to allow the spatial distribution of tsetse flies to be predicted. Using images collected by the NOAA and Metosat meteorological satellites, it was possible to create variables that were correlated with temperature, rainfall and saturation deficit, as would be measured at ground-based meteorological stations. Because vegetation both responds to and changes atmospheric moisture, the NDVI (see above) was used to predict saturation deficit. Temperature close to the Earth's surface was estimated from infrared channels 4 and 5 of the NOAA/AVHRR satellite, using the relationship of Price (1984). The estimation of land temperature using NOAA/AVHRR imagery has recently been discussed by Andersen (1997). Rainfall is obviously related to cloud cover, and this can be estimated using a variable called cold-cloud duration (CCD), which measures the amount of convective cloud present. A Fourier series was used to summarise the seasonal climate for each pixel, and the mean, annual, biannual and triannual amplitudes and phases used in a discriminant analysis to predict tsetse density. Rogers *et al.* (1996) found that the presence–absence of tsetse fly was predicted with accuracies ranging from 67 to 100%, and that the thermal data was the best predictor of tsetse fly distribution. Similar success in the prediction of tsetse flies was found by Kitron *et al.* (1996) for Kenya, where Landsat TM band 7, which is associated with the moisture content of soil and vegetation, emerged as being consistently highly correlated with fly density. Other examples of the use of satellite remote sensing to study important insects include: forecasting locust swarms (Hielkema, 1990); using rainfall estimated by remote sensing to predict Senegalese grasshopper, *Oedaleus senegalensis*, outbreaks (Burt *et al.*, 1995); monitoring the abundance of the mosquito, *Aedes albifasciatus* (Gleiser *et al.*, 1997); and the identification of urban breeding sites of the mosquito *Culex annulirostris* (Dale & Morris, 1996). Washino & Wood (1994) review the application of remote-sensed imagery to the study of arthropod disease vectors.

15.2 Remote sensing using piloted and unmanned aircraft

Remote sensing can also be undertaken from aircraft; using photography, video or direct observation. Recently, small drones have become readily available for photographic and video surveys. Wint (1998) discusses the situations where aircraft have an advantage over satellite remote sensing. High-altitude surveys (1700–3500 m above ground level) are useful for mapping, particularly vegetation; with high-resolution photography (conventional and digital) the condition of the plants may be recognised, damage levels evaluated, and indirect assessment made of pest populations (Frenz & Karafiat, 1958; Aldrich *et al.*, 1959; Klein, 1973; Harris *et al.*, 1976; Wallen *et al.*, 1976; Everitt *et al.*, 1997). Airborne sensors can also be used for measuring lake primary production (see Chapter 14) (George, 1997) and hence may be useful for predicting fish biomass or eutrophication.

Numerical information on populations of large animals is obtained from low-level surveys (150–700 m above ground level). As Wint (1998) points out, visual observations are more reliable (and often cheaper) than automatic methods which may fail to detect animals, for example those sheltering under trees. He recommends a combination of direct visual counting with direct photography of any group of more than 10 individuals. The sampling strategy may be either stratified or systematic. Wint (1998) indicates the conditions that determine the optimal choice for a given situation and describes methods for calculating the standard errors of the estimates.

Besides gathering primary information, aircraft surveys may be used to check the interpretation of satellite remote sensing.

It is clear that drones have great potential applicability for wildlife surveys. For example, Vermeulen *et al.* (2013) assessed an Unmanned Aircraft System (UAS) to survey large mammals in the Nazinga Game Ranch in the south of Burkina Faso. The Gatewing ×100™ equipped with a Ricoh GR III camera was used to test both animal reaction as the UAS passed overhead, and also animal visibility on the images. No reaction was recorded as the UAS passed at a height of 100 m. Observations, made on a set of more than 7000 images, revealed that only elephants, *Loxodonta Africana*, were easily visible, small mammals were not visible. The main drawback found was the 45-minute flight endurance of the UAS. Increased endurance is required before piloted aircraft could be replaced.

15.3 Long-term studies

Most long-term studies have been of pests, food animals or species that give valuable products, such as fur. Perhaps the most famous examples are the Hudson Bay Companies records of mammals from northern Canada (Elton, 1927; Elton & Nicholson, 1942). Time series of fish catches have been widely used to understand the exploitation of stocks, but these have, in a particularly acute form, a problem that affects most long-term data sets, namely the efficiency of sampling and/or its methods changing with time. Fishing boats have tended to become more powerful and fishing more efficient as modern sonar and navigation aids became available. A standardisation of sampling effort throughout the length of the time series is desirable, but may be impossible to achieve. Numerous studies of long-term changes in marine ecosystems are described in Bachelet & Castel (1997).

Starting with the classical work of Schwerdfeger (1941), there are many long-term studies on forest pests or other insects on long-lived vegetation. Klimetzek (1990) describes a 180-year series for certain defoliators in pine forests of southern Germany, whilst the Rothamsted Insect Survey (Woiwod & Harrington, 1994; http://www.rothamsted.ac.uk/insect-survey) provides faunal information on a national scale. Bird surveys such as counts of migratory birds and ringing studies have also produced many time series. Examples of the application of such data are the analysis of variability in North American migratory bird counts (Keitt & Stanley, 1998) and the study on bird ringing data from the Wash, England (Rehfisch *et al.*, 1996). A good example of a large-scale volunteer manned project is the BTO/JNCC/RSPB Breeding Bird Survey (BBS), which monitors the populations of the UK's common breeding birds (http://www.bto.org/volunteer-surveys/bbs). The North American Breed Bird Survey commenced in the 1966 (http://www.mbr-pwrc.usgs.gov/bbs/genintro.html).

Through international cooperation, a Global Population Dynamics Database has been established (http://www3.imperial.ac.uk/cpb/databases/gpdd, which comprises approximately 5000 data sets for 1400 different species. The vast majority was collected in the Northern hemisphere and it is biased in favour of insects, mammals, fish and birds. Curiously, there are few long-term studies of higher plants. These time series may be used by ecologists to compare against their own time series. They can also be used to gain insight into the range of dynamical behaviour shown by animal populations and to identify if different taxa display characteristic dynamical properties.

Time series may also be used for forecasting, but great caution is required. The simplest approach is to extrapolate a trend, but while this may be successful in the short term, it will almost certainly result in ridiculous long-term forecasts, because real populations can neither decline nor rise indefinitely. A more sophisticated approach is to fit an autoregressive integrated moving average model (ARIMA). This is often referred to as the Box–Jenkins method as it was advocated in Box & Jenkins (1976). While such models may give an impressive fit to the data, they are based solely on the existing time series. If conditions alter so that different biological processes become dominant, the forecasts can be misleading. For example, Henderson & Corps (1997) showed that bass, *Dicentrarchus labrax*, juveniles of 2 and 3 years of age will cannibalise fish in their first year of life. During the 1980s, bass numbers were small because of low water temperatures so that the suppression of the new recruits by the older year classes was undetectable. After the exceptional recruitment following the warm summer of 1989 the following two year classes were suppressed by cannibalism. An ARIMA model based on the 1980s time series would have greatly overestimated recruitment in the early 1990s. Population forecasts over time scales greater than that at which reproduction and death will have replaced the entire population should only be made where there is considerable biological understanding.

Time series can also be reconstructed using fossils. Examples include the reconstruction of fish abundance time series using scales preserved in the sediments and studies of changes in vegetation using paleontological data (e.g. Larsen & MacDonald, 1998).

15.3.1 *Planning spatial and temporal sampling*

The guidance given in Chapter 2 on experimental design and planning should be consulted. In a typical study, sample sites are arranged over a spatial grid which is sampled at regular intervals in time. Generally, it is required to compare the abundance of animals over the grid at different points in time. A plot of abundance over the grid can be represented on a three-dimensional graph as a surface (e.g. Elkinton & Liebhold &, 1990). Statistical comparison of such surfaces requires that sampling has the following features (Legendre & McArdle, 1997):

1. The same locations used on each sampling occasion.
2. A regular grid is used. Stratified and random sampling methods are inferior.
3. The samples are replicated at each sampling station on each sampling occasion.
4. If possible, these replicates should be from around the fixed station at positions chosen at random. If the replicates are too close together, they will not be independent.
5. A balance is needed between spatial resolution (more sampling stations) and the power to detect changes (more replication at each station) (see p. 16).

15.3.2 *The classification of time series*

The shape of abundance time series is likely to reflect the demographic strategy of the animal. For example, an opportunistic species able to exploit vacant space and be capable of rapid reproduction will tend to produce sudden bursts in abundance when conditions are favourable. Ibanez & Fromentin (1995) describe a methodology for allocating abundance time series to one of the following six general forms.

1. Erratic series – shown by species which occasionally invade.
2. Series with periods of low abundance – exhibited by opportunistic species and those experiencing environmental stress.
3. Trending series – shown by species showing little seasonal recruitment and responding either to a favourable (positive slope) or unfavourable (negative slope) set of conditions.
4. Seasonal series – shown by species with clear seasonal recruitment but regulated to maintain an approximately constant long-term abundance.
5. Seasonal and trending series – species with seasonal recruitment and experiencing a change in conditions.
6. Spiky series – shown by permanently present species, which possibly display chaotic or limit cycle dynamics plus stochastic variability.

Examples of these six types of dynamics are shown in Fig. 15.1. While it may be possible to allocate time series to one of these six classes by visual examination, a rather less subjective method would be desirable. Given time series for a large suite of species, Ibanez & Fromentin (1995) used four synoptic attributes derived from the original

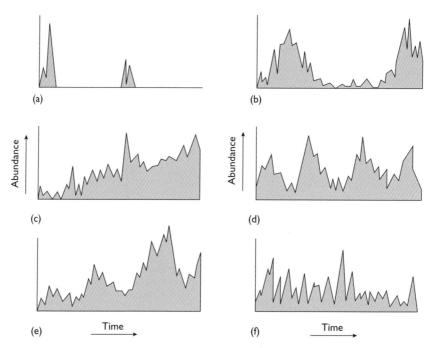

Figure 15.1   Examples of the types of dynamics displayed by populations. (Adapted from Ibanez & Fromentin, 1995.) (a) Erratic series; (b) Series with periods of low abundance; (c) Trending series; (d) Seasonal series; (e) Seasonal trending series; (f) Spiky series.

time series in a flexible clustering method to divide the species into six groups. These attributes, which measured the trend, variability, and seasonality of the original series, were obtained using a technique called Eigenvector Filtering (Colebrook, 1978). This is a method for time series decomposition into filtered and residual series using principal components analysis (PCA) which allows the main features of the series to be visualised. It is available as a function within the pastecs R package (Package for Analysis of Space-Time Ecological Series). The code listing below shows an example application to a monthly time series of mysid abundance stored as a simple csv file of monthly abundances. Because the samples were collected at regular intervals the date is added using the ts function.

```
library(pastecs) #Library holding Eigenvector filtering function
#Open data set a list of abundances
SS <- scan("C:\\Users\\Peter\\Documents\\R working\\Schistomysis spiritus
hink no missing.csv")
#Make a time series - add dates for regular monthly sampling
SStimeseries <- ts(SS, frequency=12, start=c(1981,1))
# Look at Autocorrelation to choose lag
acf(SStimeseries)
#Undertake time series decomposition using EVF
#Autocorrelation not significant after lag 3
melo.evf <- decevf(SStimeseries, lag=3, axes=1:2)
#Plot orginal, filtered and residual series
plot(melo.evf, col=c(1))
# Generate superimposed plot of original and filtered data
plot(melo.evf, col=c(1, 4), xlab="Months", stack=FALSE, resid=FALSE,lpos=c(0,
60000))
```

15.3.3 Time series analysis

This section gives a brief introduction to time series analysis from an ecological perspective; more detail about the mathematics and application of techniques introduced can be found in books by Chatfield (1994), Diggle (1990), Tong (1990) and Wei (1990). An ecological time series, X_t, normally comprises observations, $X_1, X_2 \ldots\ldots X_n$ of some population attribute such as population size or density, obtained from data collected at discrete periods in time. Generally, the interval between observations, Δ, should, whenever possible, be held constant. However, when this is not possible, a constant step time series can often be generated from the original set. Sometimes, this is done by removing some observations, though one of the commonest reasons for variable time steps is a missed observation which then has to be estimated by interpolation. If it is possible to collect independent replicate time series, then this will be statistically advantageous but it can rarely be achieved. While cost may be a factor it is often physically impossible; for example, there may only be one population that can be studied. In the following account, we will assume that only one time series is available for each attribute, and that the time steps are constant.

At the outset, it is important to note that series analysis requires that the number of observations, n, in a series should be large if the conclusions are to be statistically valid. Probably, $n = 100$ is a minimum requirement.

The analytical approach here assumes that the time series can be decomposed into number of separate components, each of which contributes to the total variability of the observed series. These components can be classified as trends, periodic or cyclic patterns and noise or random variation. Quite often the objective of an ecological study is to identify trends and any periodic changes and to assess their magnitude relative to the general noise in the time series. It is the teasing apart of these components that is the objective of time series analysis. An example of a time series broken down into these three components is shown in Fig. 15.2. This example clearly shows a powerful

Decomposition of additive time series

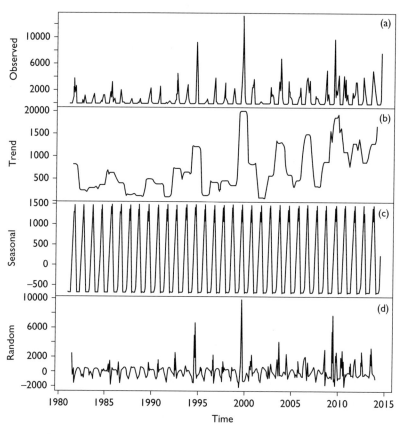

Figure 15.2 The decomposition of the components frequently observed in population time series. (a) Original time series of monthly abundance of the prawn, *Paciphaea sivado*; (b) Trend, note the increase in number after about 2005; (c) Cyclic component, in this case a 12-month seasonal cycle; (d) Noise or random component. The figure was generated in R using the function decompose (see p. 597 for the R code).

seasonal signal and a rising trend from about 2005. This output was generated by the following R code.

```
#Enter data from a csv file data as a single column with header removed
SS <- scan("C:\\ R working\\Paciphaea sivado no missing.csv")
```

```
#Make a time series file and structure data as monthly data
#This statement defines a time series of monthly data and sets the start date as
January 1982
#The original csv data set did not include dates
SStimeseries <- ts(SS, frequency=12, start=c(1981,1))
```

```
#Now we will decompose into trend season and random components
SStimeseriescomponents <- decompose(SStimeseries)
#Plot decomposition components
plot(SStimeseriescomponents)
```

The data, which are monthly counts for the prawn, *Paciphaea sivado*, in the Bristol Channel, are held as a single column in a csv file with no date. The ts function forms the time series data structure used by R. This is possible because the data were collected every month commencing in January 1981. R offers other functions to decompose a time series for example, plot(stl(SStimeseries,s.window="periodic")) uses the function stl to undertake a seasonal trend decomposition using loess smoothing.

It is common practice that the first stage of an analysis is almost always an investigation of the series for trends. This is because later stages of analysis frequently require series to be stationary. In a stationary series the mean and variance are constant with time. Thus, even if the trend is of no interest it needs to be identified and removed before further analysis can proceed. Trends are often defined as long-term changes in the mean level. In population studies we often look for trends of increasing or decreasing number. However, it is clear that there is no way to distinguish between long-term cycles, for which only a small part of its cycle is covered by the period of study, and a trend. Indeed, Granger (1966) defines trend in the mean as all cyclic components whose wavelength exceeds the length of the observed series. There are two simple statistical procedures to test for the existence of a trend (Kendall & Stuart, 1966). The number of positive differences between successive observations in the series can be counted and a difference-sign test applied. Alternatively, the observations can be re-ordered from largest to smallest and this new series compared for resemblance against the original series using Kendall's rank-order correlation coefficient.

Once a trend has been identified, it needs to be removed from the series. Quite likely it is also of ecological interest, in which case an appropriate model may be found to describe it. The simplest model is a linear equation that can be fitted by regression. If the trend is exponential, as may occur with a growing population, then a linear model can be applied after initial logarithmic transformation. Trends need not be linear or exponential, and more generally a polynomial model may be fitted although this may be difficult to interpret biologically. Once a model has been fitted, the residuals are obtained by subtracting the model from the original series to create a detrended series.

Two other approaches frequently used to remove trend are the calculation of moving averages and differencing. Differencing provides a simple method to remove a trend. The first difference of a series x_t is defined as:

$$Dx_t = x_t - x_{t-1}$$

First, differencing will remove a linear trend, as can be shown by calculating the first difference of the simple series showing a linear trend 1,2,3,4,5 for which the first differences are 1,1,1,1. Higher-order differences can be calculated by repeated application of a first-order difference. Second-order differencing will remove the trend from a series which is accelerating at a constant rate. Generally, repeated differencing is undertaken until the series appears stationary, although it is rare to use higher than second-order differencing. In R the diff function calculates differences.

Differencing has a similar effect to subtracting a moving average from the original series. Moving averages are particularly useful for removing periodic patterns from data such as seasonality. As shown in Fig. 15.3, if data are collected monthly, then a 12-month moving average will remove a strong seasonal signal which may be obscuring other features within the series. The R code to undertake a 12-month moving average on the regularly collected monthly data shown in Fig. 15.3 is listed below.

```
#Enter data from a csv file data as a single column with header removed
SS <- scan("C:\\Users\\Peter\\Documents\\R working\\Paciphaea sivado no miss-
ing.csv")
#Make a time series file and structure data as monthly data
SStimeseries <- ts(SS, frequency=12, start=c(1981,1))
#Calculate 12 month moving average
12month_ma= filter(SStimeseries, sides=2, c(.5, rep(1,11), .5)/12)
#The following longer code could be used and which makes the averaging explicit
#ma= filter(SStimeseries, sides=2,
filter=c(1/24,1/12,1/12,1/12,1/12,1/12,1/12,
1/12,1/12,1/12,1/12,1/12,1/24))
#Plot result
par(mfrow=c(2,1)) # A 2 by 1 panel of plots
ts.plot(SStimeseries, 12month_ma, lty=2:1, col=1:2, lwd=1:2, main="P. Sivado with
moving average superimposed")
residuals = SStimeseries- 12month_ma
plot(residuals, main = "residuals from moving average")
```

Not all series can be made stationary by differencing. In some cases a series may have a constant mean (stationary in the mean) but variable variance. If the variance of a series is proportional to its level, then a square-root transformation will give a constant variance. The method of Box and Cox (see Chapter 2) to select a suitable transformation for stabilising the variance can be applied.

The detrended and hopefully stationary series can then be analysed for periodic components. A series is defined as periodic with periodicity, P, if $x_t = x_{t+P}$ for all t. Typical examples of periodic pattern in biological time series are those caused by seasonality. Figure 15.3 shows a strong seasonality in the abundance of the prawn, *Paciphaea*

P. Sivado with moving average superimposed

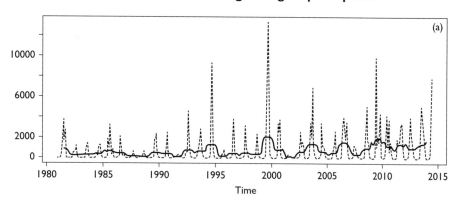

Residuals from moving average

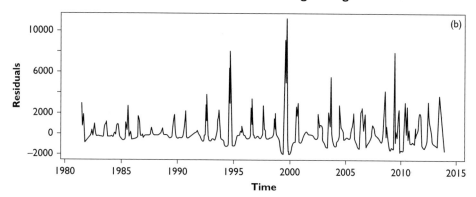

Figure 15.3   The effect of performing a 12-month moving average on a time series of monthly collected population abundance data. (a) The original time series of abundance of the prawn *Paciphaea sivado* with overlaid moving average; (b) The residual difference between the original series and the moving average. Figure generated using R (see p. 598 for the R code).

sivado, in the Severn Estuary, England. Cycles can also be caused by biological interactions such as delayed-density dependence or even cannibalism between age classes
(Henderson & Corps, 1997). Classical examples of periodicity in animal abundance are
those of northern temperate mammals (Elton & Nicholson, 1942; Finerty, 1980). Weak
periodic signals may be difficult to detect if a series contains a lot of noise.

The autocorrelation coefficients of a series measure the correlation between observations at different times apart. It is calculated in a manner analogous to a simple correlation coefficient, except that the pairs of values both derive from the same series. The
time or distance between the pairs of observations is normally termed the 'lag'. The
simplest approach to search for periodic components is to examine the correlogram;
this is a plot of the autocorrelation of the series over a range of lags. The seasonality
in the *Paciphaea sivado* time series is shown in Fig. 15.4 by the clear 12-month periodicity in the correlogram. This correlogram also shows the normal time series feature that
observations collected close together (small lag) tend to be more similar (higher positive

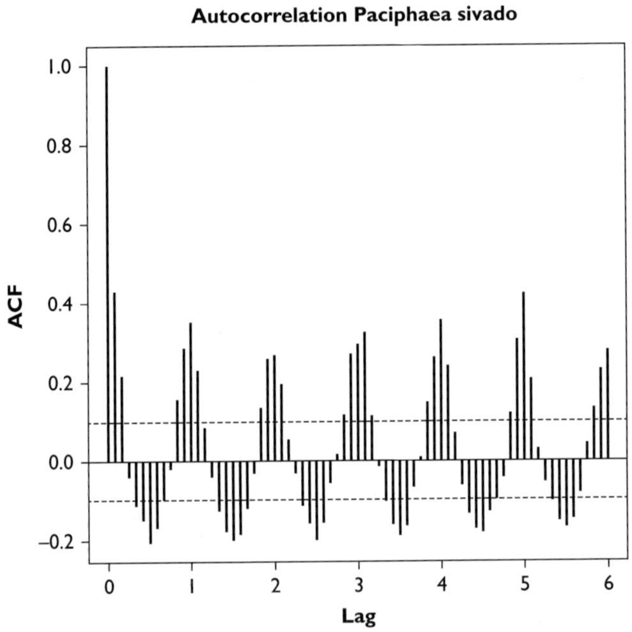

Autocorrelation Paciphaea sivado

Figure 15.4 The correlogram of the time series of abundance of the prawn *Paciphaea sivado* (see Figs 15.2 and 15.3). The x-axis is in years, so that there is a significant correlation after a lag of one year, indicating a significant 12-month cyclical seasonal variation in abundance. Figure generated using R (see p. 600 for the R code).

autocorrelation) than those further apart. Of course, with a lag of 0, the autocorrelation is +1 because each observation is being compared with itself. The autocorrelation at any particular lag is significantly different from zero at the 5% level if it lies outside the range $\pm 2/\sqrt{n}$. Chatfield (1994) recommends that this test be applied to the correlogram of the residuals of a time series model to determine if the residuals are random and thus the model a good fit. An autocorrelation plot in R is obtained using the acf function as shown in the code listing, below, which was used to produce Fig. 15.4.

```
#Enter data from a csv file data as a single column with header removed
SS <- scan("C:\\Users\\Peter\\Documents\\R working\\Paciphaea sivado no miss-
ing.csv")
#Make a time series file and structure data as monthly data
SStimeseries <- ts(SS, frequency=12, start=c(1981,1))
#Calculate and plot the autocorrelation
acf(SStimeseries, lag.max=72, main="Autocorrelation Paciphaea sivado")
```

Some of the most powerful tools for investigating the properties of time series are based on the Fourier representation of a series:

$$x_t = \sum_{k=0}^{k=n/2} (a_k \cos \omega_k t + b_k \sin \omega_k t) \tag{15.1}$$

where $\omega_k = 2\pi k/n$, $k = 0,1,\ldots\ldots, n/2$ are Fourier frequencies.

The terms a_k and b_k are the Fourier coefficients and are given by the expressions:

$$a_k = \frac{1}{n} \sum_{t=1}^{n} x_t \cos \omega_k t$$

for $k = 0$ and $n/2$ if n is even, and:

$$a_k = \frac{2}{n} \sum_{t=1}^{n} x_t \cos \omega_k t$$

for other values of k and:

$$b_k = \frac{2}{n} \sum_{t=1}^{n} x_t \sin \omega_k t \tag{15.2}$$

for $k = 1,2,3\ldots\ldots(n-1)/2$.

Thus, in the Fourier representation, the original series is described as the summation of a series of cosine and sine terms. Using this approach, a time series may be analysed in terms of the sinusoidal behaviour at a range of frequencies. While analysis based on the correlogram is in the time domain, studies based on the Fourier transform are termed 'frequency domain analyses'.

Using Perseval's relationship, we can partition the total variance shown by the series between the different frequencies because

$$\sum_{t=1}^{n} x_t^2 = na_0^2 + \frac{n}{2} \sum_{t=1}^{(n-1)/2} \left(a_k^2 + b_k^2 \right)$$

if n is odd and

$$\sum_{t=1}^{n} x_t^2 = na_0^2 + \frac{n}{2} \sum_{t=1}^{(n-1)/2} \left(a_k^2 + b_k^2 \right) + na_{n/2}^2 \tag{15.3}$$

if n is even.

This allows a time series to be analysed in a manner analogous to the more familiar analysis of variance. In tabulated form, the analysis of variance for a time series is as shown in Table 15.2.

The periodogram is formed by plotting these sums of squares (Table 15.2) against frequency. The periodogram is a direct estimate of the power spectrum and can be examined for information about the time series. However, it is an inconsistent estimate of

Table 15.2   Analysis of variance for a time series.

Source	Degrees of freedom	Sum of squares
Frequency ω_0	1	na_0^2
Frequency ω_1	2	$n/2(a_1^2 + b_1^2)$
Frequency ω_2	2	$n/2(a_1^2 + b_1^2)$
Frequency $\omega_{(n-1)/2}$	2	$n/2(a_{(n-1)/2}^2 + b_{(n-1)/2}^2)$
Frequency $\omega_{n/2}$	1	$na_{n/2}^2$
Total	n	Σx_t^2

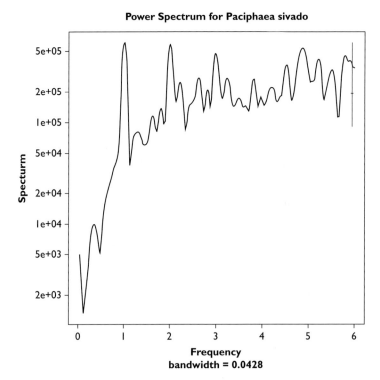

Power Spectrum for Paciphaea sivado

Figure 15.5 A typical example of a power spectrum for a population time series. The original time series is the monthly abundance of the prawn *Paciphaea sivado* in the Severn Estuary, 1981–2014. Figure generated using R (see p. 602 for the R code).

the spectrum because it uses n observations to estimate $n/2$ quantities and thus shows considerable noise. If, as is normal for ecological time series, we can assume that the underlying spectrum is a continuous function, then the periodogram can be smoothed in some way to produce an estimate of the spectrum. A number of different smoothing procedures are used, including a simple moving average of the periodogram.

In practice, the availability of time series analysis software that usually employs the fast Fourier transform algorithm has made the calculation and plotting of the spectrum easy and rapid. A typical example of the power spectrum for a population time series is that for the prawn, *Paciphaea sivado*, is shown in Fig. 15.5. This was generated using the R function spectrum using the following code listing.

```
#Enter data from a csv file data as a single column with header removed
SS <- scan("C:\\Users\\R working\\Paciphaea sivado no missing.csv")
#Make a time series file and structure data as monthly data
SStimeseries <- ts(SS, frequency=12, start=c(1981,1))
#First detrend the series by taking the first difference
SStimeseriesdiff1 <- diff(SStimeseries, differences=1)
#Undertake spectral analysis
spectrum(SStimeseriesdiff1, taper = 0.1,spans = c(3,5))
```

A visual examination of the spectrum shows the presence of a number of peaks; the first is a small peak at a frequency of about 0.5 cycle per year; the second, which is the main seasonal cycle with a frequency of 1; and then a series of higher frequency cycles of which the 6-month (two cycles per year) is marginally dominant. In this example, the data were collected monthly so that the highest frequency, termed the Nyquist frequency, that can be studied is $\frac{1}{2} \times 1 = 0.5$ cycles per month. In general, the Nyquist frequency is calculated as $1/2\Delta$, where Δ is the time between samples. The lowest frequency that can be studied is $2/n\Delta$. Tests to determine if the observed peaks in the power spectrum are significant were developed initially by Fisher, and useful worked examples of such tests are given in Chapter 12 of Wei (1990).

15.3.4 Detecting synchrony

While rarely detected, a number of important studies have investigated synchrony in the population dynamics between species. Well-known studies include microtine rodents and their predators (Henttonen *et al.*, 1987), some British butterfly populations (Pollard, 1991) and capercaillie, *Tetrao urogallus*, and grouse, *Lagpopus*, populations in Finland (Ranta *et al.*, 1995). The general analytical approach taken is first to standardise all the time series to zero mean and unit variance so that site- or species-specific differences in density are removed from the series, and then to calculate the cross-correlation (Diggle, 1990; Chatfield, 1994). Studies have also been made of synchrony between different populations of the same species (Barbour, 1990).

Methods have been developed to fit models to populations following a logistic growth pattern within a spatial framework, within which the metapopulations may become synchronised (Dennis *et al.*, 1998).

15.3.5 Measuring temporal variability

The abundance of every animal changes through time and, while some species have remarkably stable populations, others change by orders of magnitude in a few generations. The study of the degree of variation shown by populations is of considerable theoretical and applied interest. In general, it would be anticipated that the more variable a population the more likely it is to become extinct because it will occasionally reach low numbers where stochastic processes dominate (May, 1973). Historically, biologists have used a surprising number of measures of variability, many of which are clearly inappropriate.

The variance of population time series usually increases with the length of the time series. Thus, care should be taken when comparing the variability of time series of different lengths. By far the safest approach is to only compare values obtained from contemporaneous time series of the same length. However, if the comparison is between species with greatly differing generation times, then each series should be adjusted for the same number of generations. The increase in variance with time can be caused by three main agents (Gaston & McArdle, 1994). First, there tends to be autocorrelation in the environment so that, on average, the population abundance in temporally close sampling periods is more similar than those far apart. Second, autocorrelation is also generated within the population; small populations beget small populations and vice versa. Third, in the short time series common in biological studies there is often a trend caused by changes in the habitat. As yet, it is unclear whether, given time series of the length likely to be obtained, the variance will stabilise.

Most ecologists have considered that variability should measure proportional change, so that a change in density from 0.01 to 0.1 is equivalent to a change from 10 000 to 100 000 (Gaston & McArdle, 1994). If this is accepted, then variability is a measure of the deviation from the mean value expressed as a proportion of the mean. Gaston & McArdle (1994) consider that the two most appropriate measures of variation are the standard deviation of the log abundance values, $SD[\log(N)]$, or the coefficient of variation of the abundance values, $CV(N)$. If the time series contains no zeros, then $SD[\log(N)]$ is the more natural measure (Williamson, 1984). However, many abundance time series include zeros, and the commonly used expedient of using $SD[\log(N+1)]$ should be avoided and for these cases $CV(N)$ should be used.

When the population shows a linear trend it may be useful to measure the variation about the trend for which the most suitable measure is $SD[\log(R)]$ where $R = N_{t+1}/N_t$ (Williamson, 1984). When such a series contains zeros, then no suitable measure exists.

Zeros arise either because the species is absent (termed 'structural zeros') or because it was not included in the samples ('sampling zeros'). It is normally impossible to distinguish between these possibilities. In principle, zeros caused by the absence of the species would not be included in the calculation of the variability as they would tend to exaggerate the variability shown by the population when present. However, zeros arising from sampling variation should be included, otherwise the population variation will be underestimated. Unless there are strong biological reasons to believe that the species was not locally present, then zeros should be included and the coefficient of variation used as a measure of variation. As Gaston & McArdle (1994) argue, the sampling unit size for a study should generally be chosen to be large enough to eliminate zeros. Such an approach is unavailable in long-term studies where a single sampling method such as grab sampling is applied to a whole community. The pattern of relative species abundance is often such that millions of the commoner species would be caught in samples that were large enough to ensure that the rarer 50% of species were always to be present (see Fig. 13.1 for a typical aquatic species abundance pattern).

15.3.6 Detecting break-points

A frequent requirement in large spatial or temporal scale studies is to identify regions or points in time where abrupt change occurs. R offers a range of packages for the detection of structural change in both spatial and temporal data, including *breakpoint* and *strucchange*. The R package bfast (Breaks For Additive Season and Trend) is particularly useful for ecological time series as it integrates the decomposition of time series into trend, season, and random components, with methods for detecting and characterising break-points within time series. Figure 15.6 shows an example of break-point detection in a monthly data set of abundance of the mysid *Schistomysis spiritus* collected in the Bristol Channel from January 1982 to March 2015. The break-point analysis detected a change in the trend around 2008 when the gently rising trend in abundance

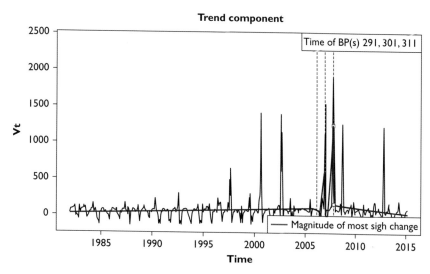

Figure 15.6   An example of the detection of a break point in trend for a population time series. The original time series is the monthly abundance of the mysid *Schistomysis spiritus* in the Bristol Channel, 1982–2015. Figure generated using the Package bfast (see p. 605 for the R code).

observed since 1982 was reversed. Figure 15.6 was generated using the following R code:

```
#Enter data from a csv file data as a single column with header removed
SS <- scan("C:\\Users\\ R working\\Schistomysis spiritus hink no missing.csv")
#The original csv data set did not include dates
SStimeseries <- ts(SS, frequency=12, start=c(1982,1)) #Print data set
#Plot dataset
plot.ts(SStimeseries)
#Now we will use BFAST
library (bfast)
# ratio of distance between breaks (time steps) and length of the time series
(rdist <- 10/length(SStimeseries))
fit <- bfast(SStimeseries,h=rdist, season="harmonic", max.iter=1)
plot(fit)
plot(fit,type="trend",largest=TRUE)
plot(fit,type="all")
citation("bfast")
```

15.3.7   *Determining if a species has become extinct*

It is difficult to determine with confidence that a species has become extinct because of the need to mount an extensive search for a small number of individuals over the entire geographical range. To aid in the assessment of the likelihood of extinction, a number of statistical methods based on the sequence of sightings through time rather than abundance data have been proposed. One of the most frequently applied mathematical models is optimal linear estimation (OLE), also called the Weibull or Weibull extreme value model (Roberts & Solow, 2003; Solow, 2005). OLE has been proposed as a tool to assess for inclusion in the IUCN Critically endangered – possibly extinct (Collen *et al.*, 2010). Unlike the stationary Poisson model, the OLE model does not assume that detection probability is constant. Clements *et al.* (2013) used experimental microsm data to test the reliability of the OLE method to estimate extinction date. While noting that the method generally gave accurate and precise estimates, they cautioned that this result depended on the observer behaviour, particularly changes in search effort and the quality of species identification.

The r package sExtinct offers a range of methods to estimate extinction date. The data is entered as a data.frame with two columns, the first holding the year of sighting, the second holding the number of sightings in each year. Years not entered are assumed to have zero sightings. The example data set is:

	Years	Sightings
1	1907	1
2	1910	1
3	1915	3
4	1916	4
5	1920	3
6	1925	1
7	1930	2
8	1931	1

The package is easy to use, as shown by the following code which undertakes all the methods available.

```
library(sExtinct) #Open library
data(example.data) #Use example data
example.data # Print example data
run.all(example.data,0.05,2012,FALSE,FALSE) #Run all methods
```

This produces the following output:

	Test	Estimate
1	OLE	1935
2	Strauss	1944
3	Solow1993; Eq. 2	1936
4	Solow 2005; Eq. 7	1941
5	Robson	1950
6	Burgman	1942

These methods should be applied with care as they can lead to errors. A recent example of an erroneously claimed extinction was for the Aldabra banded snail, *Rhachistia aldabrae*, declared extinct in 2007 (Gerlach, 2007). Fortunately for biodiversity, but unfortunately for the author, it was re-discovered at the UNESCO World Heritage Site of Aldabra Atoll, Seychelles in 2014. This particular error was significant as the claimed extinction of the snail was attributed to climate change (For further information, see: http://www.the-scientist.com/?articles.view/articleNo/41223/title/Snail-Revival-Raises-Peer-Review-Debate/.) The reason for this failure was probably insufficient search effort.

15.4 Geographical information systems

A geographical information system (GIS) is a database system in which the data are referenced with respect to their geographical coordinates on the Earth's surface. While this definition does not mention computers, in practice a powerful computer is at the core of a GIS which usually will be expected to retrieve, display and print quickly large amounts of data. GISs are becoming widely used in the fields of agriculture, conservation biology, environmental monitoring, epidemiology and forestry (Burrough & McDonnell, 1998).

GISs are most useful for studying temporal variation in the spatial distribution of species. Typically, a GIS holds data derived from many different sources. For example, animal abundance at set locations may be estimated from field samples, while vegetation cover is estimated from satellite imagery. There is usually a temporal component to the database so that changes in a physical or biological variables can be mapped through time. For the better-studied parts of the world large amounts of spatially organised data are now becoming available; for example, temperature records for Great Britain have recently become available on disk (Lennon & Turner, 1995).

A typical example of a GIS application in biology was the system developed for Projeto Mamirauá (see: http://www.mamiraua.org.br/pt-br for present information about this sustainable reserve). The project objective was to create a management plan for a large reserve in the floodplain of the Rios Solimões and Japurá in the Upper Amazon Basin. The multidisciplinary research team, comprising more than 50 specialists, worked on the project for more than three years. The reserve covers an area of about 1 250 000 ha and holds more than 1000 water bodies. At the outset, a database was required to hold place names and map community boundaries. As data accumulated on the flora and fauna within the reserve, it was natural to record data with reference to its geographical locality. Further, the regular seasonal flooding caused by the rise in flow of the main rivers created the need for time series data. However, the creation of such a database is not easy and creates considerable organisational challenges. It is almost inevitable that the time needed to create, implement and maintain the GIS will be underestimated. A frequent result is that field workers are collecting data before the database is ready for use, and this creates a backlog of data that often is in an unsuitable format for rapid entry. Once completed, such a GIS offers the prospect that ecologists can rapidly gain access to huge amounts of information and justify the cost. In the final stage, with a reserve management team in place, they have hopefully used the GIS to aid their implementation of the management plan and monitor the reserve.

15.5 Detection of density dependence in time series

Where sufficient detail is available, key-factor analysis (see Chapter 11) is normally the most powerful method for detecting density dependence (Vickery & Nudds,1991). However, many data sets of the type discussed in this chapter consist solely of a sequential series of estimates of one particular age or of the population at regular intervals. These investigations are extensive in time, whilst those permitting key-factor analysis are intensive (see Chapter 1). A number of approaches have been developed for the detection of density dependence in long time series (Rothery,1998).

The underlying approach is to test the null hypothesis (H_0) that the data can be produced by a random walk, with or without drift. A statistic must be calculated that will be capable of rejecting the null hypothesis in favour of the alternative hypothesis (H_a) that there is density dependence.

15.5.1 Bulmer's (1975) test

This is based on a Gompertz model of population growth and is appropriate for time-series that display no obvious trend. The test statistic (R) is:

$$R = \frac{\sum_{i=1}^{k}\left(X_i - \bar{X}\right)^2}{\sum_{i=1}^{k-1}\left(X_{i+1} - \bar{X}_i\right)^2} \tag{15.4}$$

where Xi is the log abundance in year i and k is the number of observations. The null hypothesis that there is no density dependence is rejected at the 5% level if R is smaller than about $0.25 + (k - 2)0.0366$. If the population is either increasing or decreasing, this will increase the value of R and make density-dependence undetectable using this method, and either of the following methods should be used instead. Bulmer also produced a modified test statistic $R*$, for use when measurement error is appreciable. The following R code listing calculates both R and $R*$ for a simple data set comprising a column of observations.

```
#First we need to input the data into a vector
#This is a simple column of annual (regularly observed) abundance estimates
Abundance_data <- scan("C:\\Users\\ R working\\Bulmer\\Bulmer_data.csv")
#Check the data is OK
Abundance_data
#First log the abundances NB zeros are not allowed!!
Log_Abundance_data <- log(Abundance_data)
#Find the number of observations
Obs <- length(Abundance_data)
#Calculate mean
Abund_mean <- mean(Log_Abundance_data)
```

```
#Calculate square of N(t+1)-N(t)
U <- 0
for (n in 1:trunc(Obs-1))
{U <- U+(Log_Abundance_data[n+1]-Log_Abundance˙data[n])^2}
V <- 0
for (n in 1:trunc(Obs))
{V <- V+(Log_Abundance_data[n]-mean(Log_Abundance_data))^2}
# Calculated test statistic
test_stat <- V/U
cat("Bulmer's test statistic R =", test_stat)
#Significance
RI <- 0.25+(Obs-2)*0.0366
cat("For density dependence to be significant at 5% level R must be smaller than", RI)
#Calculation on Bulmer's R*
W <- 0
for (n in 1:trunc(Obs-2))
{W <- W+(Log_Abundance_data[n+2]-Log_Abundance_data[n+1])
*(Log_Abundance_data[n]-mean(Log_Abundance_data))}
R_star <- W/V
cat("Bulmer's test statistic R* (used when measurement error is appreciable) =", R_star)
#Significance
R_star_0.05 <- (-13.7/Obs)+(139/(Obs^2))-(613/(Obs^3))
cat("For density dependence to be significant at 5% level R* must be smaller than",
R_star_0.05)
```

15.5.2 *Pollard* et al.'s *(1987) randomisation test*

A population subject to annual census for k years will yield a series $N_1, N_2,..., N_k$ estimates of population size. If this series is transformed by taking the natural logarithm, $X_i = \log_e N_i$ ($i = 1$ to k) then the following three time series models for the population can be considered:

1. $X_{t+1} = X_t + e_i$
2. $X_{t+1} = r + X_t + e_i$
3. $X_{t+1} = r + b\, X_t + e_i$

In these equations, r and b are parameters that determine the rate of change of the Population, and e_i are random, independent, variables. Model 1 is a simple random walk, model 2 describes a population that is changing without density-dependence (random walk with drift), and model 3 with $b \neq 1$, is a density-dependent model that is also termed the Gompertz equation.

For a density-independent population, the change in the $\log_e$ population between years, $D_i = (X_{t+1} - X_t)$, are just random fluctuations and could have occurred in any order. The procedure is therefore to calculate a test statistic, T, for the observed series and determine by Monte Carlo simulation if this value is so extreme that the null hypothesis of density-independence can be rejected.

To compare the density-dependent model (3) against a random walk model (1), the test statistic $T(1,3)$ is

$$\sum_{i=1}^{k} (X_{i-1} - X_i)^2 \tag{15.5}$$

and to compare model 3 against a changing population model 2 the test statistic, $T(2,3)$ is

$$\frac{\sum_{i=1}^{k} (X_{i+1} - m_2)^2 - B \sum_{i=1}^{k-1} (X_i - m_1) (X_{i+1} - m_2)}{\sum_{i=1}^{k-1} (X_{i+1} - X_i)^2 - (X_k - X_1)^2 \big/ (k-1)} \tag{15.6}$$

where

$$B = \sum_{i=1}^{k-1} (X_i - m_1),$$

$$m_1 = \sum_{i=1}^{n-1} X_i \big/ (k-1),$$

and

$$m_2 = \sum_{i=1}^{n-1} X_{i+1} \big/ (k-1).$$

To test the null hypothesis of density independence a computer is used to undertake the following procedure:

1. Use the observed $X_1, X_2, \ldots, X_k$ to compute the test statistic (either $T(1,3)$ or $T(2,3)$).
2. Calculate the D_i-values and either calculate all possible sets of X_is that can be produced by a permutation of the sequence of D_i-values given a fixed starting value of X_1 (there will be $(k-1)$ sequences). If k is >7, then randomly permutate the D_i-values to produce a large number of sequences starting from X_1. 1000 such sequences is probably sufficient.
3. For each sequence of X_i-values calculate the test statistic $T(1,3)$ or $T(2,3)$.
4. If less than 5% of the computed T-values are less than or equal to the observed T-value calculated in step 1, then reject the null hypothesis that the population dynamics are density-independent.

15.5.3 Dennis and Taper's (1994) bootstrap approach

This can be viewed as a development of Pollard *et al.*'s (1987) randomisation test. It makes more efficient use of the data to produce a more powerful test providing a stochastic logistic equation

$$N_{t+1} = N_t exp\,(r + bN_t + \sigma Z_t) \tag{15.7}$$

is an appropriate model. N_t and N_{t+1} are the population sizes at times t and $t + 1$, respectively and r, b and σ are parameters to be estimated. The population growth rate in the absence of density dependence or random noise is given by r. For values of $b < 0$, b gives a measure of density dependence. The term $\sigma\,Z_t$ introduces random noise, as Z_t is a random variable with 0 mean and variance of 1. If density-dependent regulation is acting then a model with $a > 0$, $b < 0$ will give a superior fit to the data than either a random walk ($a = 0$, $b = 0$) or random walk with drift ($a > 0$, $b = 0$) model. The test proceeds by fitting the random walk (only σ to be estimated), the random walk with drift (r and σ to be estimated) and density-dependent (r, b and σ to be estimated) models to the data. The fit of these models is then compared using a likelihood ratio test for which the confidence limits are found using a bootstrap procedure. Dennis and Taper (1994) describe the computational steps required, which can be programmed on a personal computer. They also discuss extensions of this approach for situations where a logistic model is inappropriate.

   Various simulation studies have been undertaken to compare the relative performances of these tests, but no clear conclusion emerges (Rothery, 1998). The underlying model must be considered for its appropriateness for the data in question. It is clear that the ability to detect density-dependence increases with the length of the time series and in some situations times series spanning 100 years or more may be required. Density-dependence can be detected in short time series for populations for which the Gompertz or logistic equations offer a good description of the dynamics.

15.5.4 Using a battery of approaches to detect density dependence

Henderson & Magurran (2014) used a conservative approach to detection of density dependence based on a battery of five methodologies as follows:

1. A non-linear relationship between log population change and log population size, or the presence of a threshold when the relationship abruptly changes provides evidence of density dependence (Freckleton *et al.*, 2006). The presence of a threshold, if suspected from a visual inspection of the plot, can be tested using the Chow test for structural breaks using the R code listed below (p. 612). A simple linear negative relationship provides insufficient support for density dependence because it can be generated by census error.
2. Density dependence is consistent with a log population change – log population size relationship with a slope > -1. Note that a random walk with measurement error generates a gradient of between 0 and -1, and therefore a negative value

within this range is not necessarily indicative of density dependence. However, measurement error acts against the observation of a slope of > -1. Accordingly, a slope above -1 in the presence of measurement error is convincing support for density-dependent regulation.

3. The R and $R*$ tests of Bulmer (1975) to detect density dependence were applied to all time series with no zero annual abundances. When zero abundances occurred, these tests were applied to sections of the time series not holding zeros, providing these sections spanned more than 12 years. Bulmer's tests are conservative as they may not detect density dependence with measurement error. Bulmer notes that '...R always provides a more powerful test than $R*$ and is therefore to be preferred unless appreciable errors of measurement are suspected'.

4. For the most abundant species, growth and mortality of the age classes present could be followed through time, and analyses to detect negative changes in growth, recruitment or survival linked to increased population density could be undertaken.

5. Species that were regularly unrecorded and never found in large numbers or biomass were considered not to show evidence for density dependence if their time series could not be statistically distinguished from a random time series.

R code for Chow test for structural breaks.

```
#Read in the data from the file
whiting <- read.table("C:/Users//Whiting DD data.csv", header=TRUE, sep=",")
#Print out the data to check it is OK
summary
whiting
# Run three regressions (1 unrestricted, 2 restricted to above and below break)
r.reg = lm(logchange ~ logN, data = whiting)
ur.reg1 = lm(logchange ~ logN, data = whiting[whiting$logN > 3.2,])
ur.reg2 = lm(logchange ~ logN, data = whiting[whiting$logN < 3.2,])
# review the regression results
summary(r.reg)
summary(ur.reg1)
summary(ur.reg2)
# Calculate sum of squared residuals for each regression
SSR = NULL
SSR$r = r.reg$residuals^2
SSR$ur1 = ur.reg1$residuals^2
SSR$ur2 = ur.reg2$residuals^2
# K is the number of regressors in our model
K = r.reg$rank
# Computing the Chow test statistic (F-test)
numerator = (sum(SSR$r) - (sum(SSR$ur1) + sum(SSR$ur2))) / K
denominator = (sum(SSR$ur1) + sum(SSR$ur2)) / (nrow(whiting) - 2*K)
chow = numerator / denominator
chow
```

```
# Calculate P-value
1-pf(chow, K, (nrow(whiting) - 2*K))
# Plot the results
plot(whiting,main="")
# restricted model
abline(r.reg, col = "red",lwd = 2, lty = "dashed")
# restricted model 1
segments(0, ur.reg2$coefficients[1], 3.2, ur.reg2$coefficients[1]+3.2*ur.reg2$
coefficients[2], col= 'blue')
# restricted model 2
segments(3.25, ur.reg1$coefficients[1]+3.25*ur.reg1$coefficients[2],
3.5, ur.reg1$coefficients[1]+3.5*ur.reg1$coefficients[2], col= 'blue')
```

15.6 Citizen science projects

Automated data acquisition and digital recording have resulted in the formation of massive data sets that can overwhelm the analytical resources of researchers. The development of the Web has allowed projects to harness the aid of huge numbers of volunteer analysts. The largest and most successful citizen science organisations is Zooniverse (https://www.zooniverse.org/), a citizen science web portal owned and operated by the Citizen Science Alliance. It is home to the internet's largest and most popular citizen science projects. The organisation started with the Galaxy Zoo project and now hosts numerous projects in the fields of astronomy, ecology, cell biology, humanities, and climate science. Zooniverse projects require human volunteers to complete research tasks. By 2014, the Zooniverse community consisted of more than one million registered volunteers and has generated more than 70 publications. There is a daily news website called 'The Daily Zooniverse' which gives information on the different projects.

A good example of the type of ecological project undertaken is Penguin Watch (http://www.penguinwatch.org/#/), run by Tom Hart and colleagues. Since 2009, the Penguin Lifelines project at the University of Oxford (www.penguinlifelines.org) has developed a camera-monitoring programme comprising of 50 cameras throughout the Southern Ocean and along the Antarctic Peninsula, overlooking colonies of Gentoo, Chinstrap, Adélie, and King penguins. The cameras take images of the penguins year-round. Volunteers annotate hundreds of thousands of images individually, marking adult penguins, chicks and eggs in the images by clicking on the centre of each object. By March 2015, a total of 1 546 372 images had been classified by 14 602 volunteers. Computer analysis software has yet to reach the level of recognition or accuracy of the human eye.

15.7 Ecosystem services

From the late 1990s, discussion and research into what are termed 'ecosystem services' has grown greatly. Ecosystem services are the benefits derived from ecosystems by mankind. The importance of wildlife conservation to ensure our well being is discussed by Hambler & Canney (2013). The state and future prospects for ecosystem services are reviewed by the millennium ecosystem assessment (MEA, 2005) (http://www.millenniumassessment.org/en/index.html). According to The Economics of Ecosystems and Biodiversity (TEEB) (http://www.teebweb.org/), ecosystem services can be categorised into four main types:

1. Provisioning services: products we obtain from ecosystems such as food, fresh water, wood and medicines.
2. Regulating services: benefits obtained from the regulation of ecosystem processes such as climate regulation and water purification.
3. Habitat services: habitat provision to maintain the viability of populations.
4. Cultural services: non-material benefits to individuals such recreation and pleasure.

The available evidence indicates that mankind requires diverse and extensive ecosystems to reliably supply the ecosystem services that are required. For example, Kremen & Chaplin-Kramer (2007) concluded that the insect providers of pollination and pest control which are vital to agriculture are best provided by complex landscapes that include natural and agricultural components. It is clear that research into ecosystem services has to be on a large spatial and temporal scale. Research objectives include: (1) identification of ecosystem service providers, e.g. the species of pollinators; (2) quantification of functional roles and relationships between species providing services; (3) identification of environmental conditions that influence service providers; and (4) determining the spatial and temporal scales over which ecosystem service providers operate. All the ecological methods presented within this book can be used in pursuit of these objectives.

TESSA (The Toolkit for Ecosystem Service Site-based Assessment) offers resources to aid the evaluation of ecosystem services (Peh *et al.*, 2013), and is available at: http://www.birdlife.org/worldwide/science/assessing-ecosystem-services-tessa.

15.8 Habitat classification

15.8.1 *Qualitative*

Zoologists frequently delimit their communities by reference to plants or environmental factors. The most universal classification of habits is that of Elton & Miller (1954):

1. Terrestrial system
   (a) Formations:
       - *Open-ground type* – if any dominant plants, these not more than 15 cm high.
       - Field type – dominant life form coincides with field layer, usually not more than 2 m in height.
       - *Scrub type* – dominant life form does not exceed a shrub layer, height generally not over 7.6 m.
       - *Woodland type* – trees dominant life form.

(b)   Vertical layers:
  - Subsoil and rock.
  - Topsoil.
  - Ground zone, including low-growing vegetation, less than 15 cm.
  - Low canopy – up to about 7.6 m.
  - High canopy.
  - Air above vegetation.
2.   Aquatic system.
  (a)   Formation types: see Table 15.3.
  (b)   Vertical layers.
  - Bottom, light, dark zones, water mass, light and dark zones – free water not among vegetation.
  - Submerged vegetation.
  - Water surface – upper and under surface of film of floating leaves. Emergent vegetation-reed swamp and similar vegetation, the bases of which are in the water.
  - Air above vegetation.
3.   Aquatic–terrestrial transition system – defined further by body type with which it occurs and by vegetational systems corresponding to the terrestrial system.
4.   Subterranean system – caves and underground waters.
5.   Domestic system.
6.   General system.
  - Dying and dead wood.
  - Macro-fungi.
  - Dung.
  - Carrion.
  - Animal artefacts – nests, etc.
  - Human artefacts – fence posts, straw stacks, etc.

Further division of the habitat into communities may be made on the type of plant. In soil (MacFadyen, 1952, 1954), marine benthic and freshwater studies (Whittaker &

Table 15.3   Formation types in an aquatic system.

	Very small	Small	Medium	Large	Very large
Still	Tree hole	Small pond <17 m^2	Pond <400 m^2	Large pool or tarn <40 ha	Lake or sea
Slow	Trickle gutter	Ditch; field dyke	Canal; river bank water		
Medium	Trickle	Lowland brook	Lowland river	Lowland large river	River estuary
Fast	Spring	Upland weir; small torrent; stream	Large torrent; stream		
Vertical or steep	Water drip; pipe outlet	Small weir; waterfall	Large weir; medium waterfall	Large waterfall	

Fairbanks, 1958), divisions are frequently based on the fauna itself using multivariate techniques see pp. 530–535.

15.8.2 Quantitative

The quantification of the characters of the habitat especially plants, in general terms was pioneered by MacArthur and MacArthur (1961), who assessed foliage height diversity in trees to relate to bird diversity. MacArthur (1972) suggested the best measure of foliage height diversity was D of the Simpson-Yule index (p. 510), where the various values of p_i are the proportion of the vegetation at various heights or at the different layers in the Eltonian system (see above). A quick method of assessing 15 inter-related variables in arboreal vegetation is described by James & Shugart (1970) and utilised by James (1971) to demonstrate the vegetational characteristics of bird habitats. Southwood *et al.* (1979) distinguished between structural diversity (the distribution of vegetation in three-dimensional space) and architectural diversity (the variety of plant parts – bark, leaves, flowers, etc.).

Numerical methods for classifying vegetation are reviewed by Kent (2011) and Jongman *et al.* (1995). In the past, one of the most popular vegetation classification methods was two-way indicator species analysis (TWINSPAN), originally developed by M.O. Hill. The analysis results in a hierarchical classification in which different communities are distinguished by the presence or relative abundance of indicator species. The use of indicator species is particularly attractive for field ecologists as it allows field sites to be quickly assigned to a predefined community. A problem with the method is the use of pseudospecies, which is an artificial concept in which a single species is divided into different abundance levels, and also the fact that it is difficult to understand how the method actually works. However, the results can be impressive. In marine benthic studies TWINSPAN, which was once quite widely used for the classification of animal communities, has been replaced by Multi-Dimensional Scaling. In the study of plant communities, reciprocal averaging and its variant 'detrended correspondence analysis' (DECORAMA) are still widely used. These techniques, although applicable to animal communities, are little used. The fundamental difference between terrestrial and marine ecology below the photic zone is the absence of large plants in the marine habitat. Thus, the most important factor determining the presence and abundance of terrestrial animals – the species of plants that are present – is no longer acting and the focus switches to sediment structure.

In a review of habitat characteristics, Southwood (1977, 1988) stressed the need to define spatial heterogeneity in relation to the trivial and migratory ranges of the animal, and temporal heterogeneity in relation to generation time. Stability in space, the length of time that the same location remains suitable for the species (durational stability), and stability in time, the seasonal fluctuations in 'favourableness', should be distinguished; such temporal changes, together with the general level of unfavourableness constitute the 'adversity axes' of Whittaker. It was proposed that most habitats could be characterised against the two axes of adversity and durational stability and that suites of bionomic adaptations were related to the position of the organism's habitat in this chart.

References

Aldrich, R.C., Bailey, W.F., & Heller, R.C. (1959) Large scale 70mm color photography techniques and equipment and their application to a forest sampling problem. *Photogrammetric Eng.* **25**, 747–54.

Andersen, H.S. (1997) Land surface temperature estimation based on NOAA-AVHRR data during the HAPEX-Sahel experiment. *J. Hydrol.* **189**(1–4), 788–814.

Arseneault, D., Villeneuve, N., Boismenu, C., LeBlanc, Y., & Deshaye, J. (1997) Estimating lichen biomass and caribou grazing on the wintering grounds of northern Quebec: An application of fire history and Landsat data. *J. Appl. Ecol.* **34**(1), 65–78.

Austin, G.E., Thomas, C.J., Houston, D.C., & Thompson, D.B.A. (1996) Predicting the spatial distribution of buzzard *Buteo buteo* nesting areas using a geographical information system and remote sensing. *J. Appl. Ecol.* **33**(6), 1541–50.

Barbour, D.A. (1990) Synchronous fluctuations in spatially separated populations of cyclic forest insects. In: Watt, A.D., Leather, S.R., Hunter, M.D., & Kidd, N.A. (eds), *Population Dynamics of Forest Insects*. Intercept, Andover, UK, pp. 339–46.

Bachelet, G. & Castel, J. (eds) (1997) *Long-Term Changes in Marine Ecosystems. Oceanologica Acta.* **20**, 329 pp.

Barrett, E.C. (2013) *Introduction to Environmental Remote Sensing*. Routledge.

Blackman, A. (2013) Evaluating forest conservation policies in developing countries using remote sensing data: An introduction and practical guide. *Forest Policy Economics*, **34**, 1–16.

Box, G.E.P. & Jenkins, G.M. (1976) *Time Series Analysis: Forecasting and Control*. Holden-Day, San Francisco.

Buchanan, G.M., Fishpool, L.D., Evans, M.I., & Butchart, S.H. (2013) Comparing field-based monitoring and remote-sensing, using deforestation from logging at Important Bird Areas as a case study. *Biol. Conserv.* **167**, 334–8.

Bulmer, M.G. (1975) The statistical analysis of density dependence. *Biometrics* **3**, 901–11.

Burrough, P.A. & McDonnell, R.A. (1998) *Principles of Geographical Information Systems*. Oxford University Press, Oxford.

Bursell, E. (1957) The effect of humidity on the activity of tsetse flies. *J. Exp. Biol.* **34**, 42–51.

Burt, P.J.A., Colvin, J., & Smith, S.M. (1995) Remote sensing of rainfall by satellite as an aid to *Oedaleus senegalensis* (Orthoptera: Acrididae) control in the Sahel. *Bull. Entomol. Res.* **85**(4), 455–62.

Chatfield, C. (1994) *The Analysis of Time Series: An Introduction*. Chapman & Hall, London.

Clements, C.F., Worsfold, N., Warren, P., Collen, B., Blackburn, T., Clark, N., & Petchey, O.L. (2013) Experimentally testing an extinction estimator: Solows Optimal Linear Estimation model. *J. Anim. Ecol.* **82**, 345–54.

Colebrook, J.M. (1978) Continuous plankton records: overwintering and annual fluctuations in abundance of zooplankton. *Mar. Biol.* **84**, 261–5.

Collen, B., Purvis, A., & Mace, G.M. (2010) When is a species really extinct? Testing extinction inference from a sighting record to inform conservation assessment. *Diversity Distributions* **16**, 755–64.

Dale, P.E.R. & Morris, C.D. (1996) *Culex annulirostris* breeding sites in urban areas: Using remote sensing and digital image analysis to develop a rapid predictor of potential breeding areas. *J. Am. Mosquito Control Assoc.* **12**, 316–20.

Dennis, B., Kemp, W.P., & Taper, M.L. (1998) Joint density dependence. *Ecology* **79**, 426–41.

Diggle, P.J. (1990) *Time Series: A Biostatistical Introduction*. Clarendon Press, Oxford.

D'Oleire-Oltmanns, S., Marzolff, I., Peter, K.D., & Ries, J.B. (2012). Unmanned Aerial Vehicle (UAV) for monitoring soil erosion in Morocco. *Remote Sensing* **4**, 3390–416.

Dwivedi, R.S., Sankar, T.R., Venkataratnam, L., Karale, R.L., Gawande, S.P., Rao, K.V.S., Senchaudhary, S., Bhaumik, K.R., & Mukharjee, K.K. (1997) The inventory and monitoring of eroded lands using remote sensing data. *Int. J. Remote Sensing* **18**(1), 107–19.

Elkinton, J.S., & Liebhold, A.M. (1990) Population dynamics of gypsy moth in North America. *Annu. Rev. Entomol.* **35**, 571–96.

Elton, C.S. (1924) Periodic fluctuations in the numbers of animals: their causes and effects. *Br. J. Exp. Biol.* **2**, 119–63.

Elton, C.S. & Miller, R.S. (1954) The ecological survey of animal communities with a practical system of classifying habitats by structural characters. *J. Ecol.* **42**, 460–96.

Elton, C. & Nicholson, M. (1942. The ten-year cycle in numbers of the lynx in Canada. *J. Anim. Ecol.* **11**, 215–44.

Everitt, J.H., Richerson, J.V., Karges, J.P., & Davis, M.R. (1997) Using remote sensing to detect and monitor a western pine beetle infestation in west Texas. *Southwest. Entomol.* **22**(3), 285–92.

Ferguson, R.L. & Korfmacher, K. (1997) Remote sensing and GIS analysis of seagrass meadows in North Carolina, USA. *Aquat. Bot.* **58**(3-4), 241–58.

Freckleton, R.P., Watkinson, A.R., Green, R.E., & Sutherland, W.J. (2006) Census error and the detection of density dependence. *J. Anim. Ecol.* **75**, 837–51.

Franz, J. & Karafiat, H. (1958) Eigen sich Kartierung und Scrienphotographic von Tannenläsen für Massenwechsel-studien? *Z. Agnew. Entomol.* **43**, 100–12.

Gaston, K.J. & McArdle, B.H. (1994) The temporal variability of animal abundances: measures, methods and patterns. *Philos. Trans. Roy. Soc. Lond. B* **345**, 335–58.

George, D.G. (1997) The airborne remote sensing of phytoplankton chlorophyll in the lakes and tarns of the English Lake District. *Int. J. Remote Sensing* **18**(9), 1961–75.

Gerlach, J. (2007) Short-term climate change and the extinction of the snail *Rhachistia aldabrae* (Gastropoda: Pulmonata). *Biol. Lett.* **3**, 581–5.

Gleiser, R.M., Gorla, D.E., & Almeida, F.F.L. (1997) Monitoring the abundance of *Aedes* (*Ochlerotatus*) *albifasciatus* (Macquart 1838) (Diptera: Culicidae) to the south of Mar Chiquita Lake, central Argentina, with the aid of remote sensing. *Ann. Trop. Med. Parasitol.* **91**(8), 917–26.

Granger, C.W.J. (1966) The typical shape of an economic variable. *Econometrica* **34**, 150–61.

Gruttke, H. & Engels, H. (1998) Metapolulation structure of *Carabus problematicus* in a fragmented landscape, significance of simulation results for conservation. In: Baumgärtner, J., Brandmayr, P., & Manly, B.F.J. (eds), *Population and Community Ecology for Insect Management and Conservation*. Balkema, Rotterdam.

Hambler, C. & Canney, S.M. (2013) *Conservation*. Cambridge University Press.

Hanski, I. (1981) Coexistence of competitors in patchy environment with and without predation. *Oikos* **37**, 306–12.

Hanski, I. (1982) Dynamics of regional distribution: the core and satellite species hypothesis. *Oikos* **38**, 210–21.

Hanski, I. (ed.) (1987) *Ecological significance of spatial and temporal variability. Annales Zoologici Fennici* **19**, 21–7.

Hanski, I. & Gilpin, M. (1991) Metapopulation dynamics: brief history and conceptual domain. *Biol. J. Linn. Soc.* **42**, 3–16.

Harris, M.K., Hart, W.G., Davis, M.R., Ingle, S.J., & Van Cleave, H.W. (1976) Aerial photography shows caterpillar infestation. *Pecan Quarterly* **10**, 12–18.

Hassell, M.P., Comins, H.N., & May, R.M. (1991) Spatial structure and chaos in insect population dynamics. *Nature* **353**, 255–8.

Hay, S.I., Tucker, C.J., Rogers, D.J., & Packer, M.J. (1996) Remotely sensed surrogates of meteorological data for the study of distribution and abundance of arthropod vectors of disease. *Ann. Trop. Med. Parasitol.* **90**, 1–19.

Henderson, P.A., Hamilton, W.D., & Crampton, W.G.R. (1998) Evolution and diversity in Amazonian floodplain communities. In: Newbury, D.M., Prins, H.H.T., & Brown, N.D. (eds), *Dynamics of Tropical Communities.* Blackwell Science, Oxford.

Henderson, P.A. & Corps, M. (1997) The role of temperature and cannibalism in inter-annual recruitment variation of bass, *Dicentrarchus labrax* (L) in British waters. *J. Fish Biol.* **50**, 280–95.

Henderson, P.A. & Magurran, A.E. (2014) Direct evidence that density-dependent regulation underpins the temporal stability of abundant species in a diverse animal community. *Proc. R. Soc. B: Biol. Sci.* **281**(1791), 20141336.

Henttonen, H., Oksanen, T., Jortikka, A., & Haukisalmi, V. (1987) How much do weasels shape microtine cycles in the northern Finnoscandian taiga? *Oikos* **50**, 353–65.

Hielkema, J.U. (1990) Satellite environmental monitoring for migrant pest forecasting by FAO: the ARTIMIS system. *Philos. Trans. Roy. Soc. B* **328**, 705–17.

Ibanez, F. & Fromentin, J.-M. (1995) Un typologie a partir de la forme des series chronologiques (TFS). *Oceanol. Acta* **20**, 11–25.

Jongman, R.H.G., Ter Braak, C.J.F., & Van Tongeren, O.F.R. (eds) (1995) *Data Analysis in Community and Landscape Ecology.* Cambridge University Press, Cambridge.

Keitt, T.H. & Stanley, H.E. (1998) Dynamics of North American breeding bird populations. *Nature* **393**, 257–60.

Kendall, M.G., Stuart, A., & Ord, J.K. (1983) *The Advanced Theory of Statistics*, Vol. 3, 4th edn, Griffin, London.

Kent, M. (2011) *Vegetation Description and Data Analysis: A Practical Approach.* John Wiley & Sons.

Finerty, J.P. (1980) *The Population Ecology of Cycles in Small Mammals.* Yale University Press, New Haven, 234 pp.

Kitron, U., Otieno, L.H., Hungerford, L.L., Odulaja, A., Brigham, W.U., Okello, O.O., Joselyn, M., Mohamed Ahmed, M.M., & Cook, E. (1996) Spatial analysis of the distribution of tsetse flies in the Lambwe Valley, Kenya, using Landsat TM satellite imagery and GIS. *J. Anim. Ecol.* **65**(3), 371–80.

James, F.C. (1971) Ordination of habitat relationships among breeding birds. *Wilson Bull.* **83**, 215–36.

James, F.C. & Shugart, H.H. (1970) A quantitative method of habitat description. *Audubon Fld. Notes* **24**(6), 727–36.

Klein, W.H. (1973) Beetle killed pine estimates. *Photogrammetric Eng.* **39**, 385–8.

Klimetzek, D. (1990) Population dynamics of pine-feeding insects: a historical survey. In: Watt, A.D., Leather, S.R., Hunter, M.D., & Kidd, N.A.C. (eds), *Population Dynamics of Forest Insects.* Intercept, Andover, UK, pp. 3–10.

Konecny, G. (2014) *Geoinformation: Remote Sensing, Photogrammetry and Geographic Information Systems.* CRC Press.

Kontkanen, P. (1957, January) On the delimitation of communities in research on animal biocoenotics. *Cold Spring Harbor Symp. Quant. Biol.* **22**, 373–8.

Kremen, C. & Chaplin-Kramer, R. (2007) Insects as providers of ecosystem services: crop pollination and pest control. In: *Insect Conservation Biology: Proceedings, 23rd Symposium Royal Entomological Society*, pp. 349–82.

Larsen, C.P.S. & MacDonald, G.M. (1998) An 840-year record of fire and vegetation in a boreal white spruce forest. *Ecology* **79**, 106–18.

Leigh, R.A., & Johnston, A.E. (eds) (1994) *Long-term Experiments in Agricultural and Ecological Sciences*. Proceedings of a Conference to celebrate the 150th Anniversary of Rothamsted Experimental Station, held at Rothamsted, 14–17 July 1993. CAB International, Wallingford.

Legendre, P. & McArdle, B.H. (1997) Comparison of surfaces. *Oceanoligica Acta* **20**, 27–41

Lennon, J.J. & Turner, J.R.G. (1995) Predicting the spatial distribution of climate: temperature in Great Britain. *J. Anim. Ecol.* **64**, 370–92.

MacArthur, R.H. (1972) *Geographical Ecology: Patterns in the Distribution of Species*. Harper & Row, New York, London, 269 pp.

MacArthur, R.H. & MacArthur, J.W. (1961) On bird species diversity. *Ecology* **42**, 594–8.

MacFadyen, A. (1952) The small arthropods of a Molinia fen at Cothill. *J. Anim. Ecol.* **21**, 87–167.

MacFadyen, A. (1954) The invertebrate fauna of Jan Mayen Island (East Greenland). *J. Anim. Ecol.* **23**, 261–97.

Martin, S. (2014) *An Introduction to Ocean Remote Sensing*. Cambridge University Press, Cambridge.

May, R.M. (1973) *Stability and Complexity in Model Ecosystems*. Princeton University Press, Princeton, New Jersey, 235 pp.

Peh, K.S.H., Balmford, A., Bradbury, R.B., Brown, C., Butchart, S.H., Hughes, F.M., & Birch, J.C. (2013) TESSA: a toolkit for rapid assessment of ecosystem services at sites of biodiversity conservation importance. *Ecosystem Serv.* **5**, 51–7.

Purkis, S. & Roelfsema, C. (2015) 11 Remote Sensing of Submerged Aquatic Vegetation and Coral Reefs. *Remote Sensing of Wetlands: Appl. Adv.* 223 pp.

Pollard, E. (1991) Synchrony of population fluctuations: the dominant influence of widespread factors on local butterfly populations. *Oikos* **60**, 7–10.

Price, J.C. (1984) Land surface temperature measurement for the split window channels of the NOAA 7 advanced very high resolution radiometer. *J. Geophys. Res.* **89**, 7231–7.

Ranta, E., Lindstrom, J., & Linden, H. (1995) Synchrony in tetraonid population dynamics. *J. Anim. Ecol.* **64**, 767–76.

Reiche, M., Funk, R., Zhang, Z., Hoffmann, C., Reiche, J., Wehrhan, M., & Sommer, M. (2012) Application of satellite remote sensing for mapping wind erosion risk and dust emission-deposition in Inner Mongolia grassland, China. *Grassland Sci.* **58**, 8–19.

Rehfisch, M.M., Clark, N.A., Langston, R.H.W., & Greenwood, J.D. (1996) A guide to the provision of refuges for waders: an analysis of 30 years of ringing data from the Wash, England. *J. Appl. Ecol.* **33**, 673–87.

Roberts, D.L. & Solow, A.R. (2003) Flightless birds: when did the dodo become extinct? *Nature*, **426**, 245.

Rogers, D.J. & Randolph, S.E. (1991) Mortality rates and population density of tsetse flies correlated with satellite imagery. *Nature* **351**, 739–41.

Rogers, D.J., Hay, S.I., & Packer, M.J. (1996) Predicting the distribution of tsetse flies in West Africa using temporal Fourier processed meteorological satellite data. *Ann. Trop. Med. Parasitol.* **90**, 225–41.

Rothery, P. (1998) The problems associated with the identification of density dependence in population data. In: Dempster, J.P. & McLean, I.F.G. (eds), *Insect Populations*. Kluwer Academic Publishers, Dordrecht, pp. 97–133.

Schwerdtfeger, F. (1941) Ober die Ursachen des massenwechsels der Insekten. *Zanger. Ent.* **28**, 254–303.

Solow, A.R. (2005) Inferring extinction from a sighting record. *Math. Biosci.* **195**, 47–55.

Southwood, T.R.E. (1977) Habitat, the template for ecological strategies? *J. Anim. Ecol.* **46**, 337–65.

Southwood, T.R.E. (1988) Tactics, strategies and templates. *Oikos* **52**, 3–18.

Southwood, T.R.E., Brown, V.K., & Reader, P.M. (1979) The relationships of plant and insect diversities in succession. *Biol. J. Linn. Soc.* **12**, 237–348.

Tong, H. (1990) *Non-Linear Time Series: A Dynamical Approach.* Oxford University Press, Oxford.

Tuomisto, H., Linna, A., & Kalliola, R. (1994) Use of digitally processed satellite images in studies of tropical rain forest vegetation. *Int. J. Remote Sensing* **15**, 1595–610.

Vickery, W.L. & Nudds, T.D. (1991) Testing for density-dependent effects in sequential censuses. *Oecologia* **85**, 419–23.

Verlinden, A. & Masogo, R. (1997) Satellite remote sensing of habitat suitability for ungulates and ostrich in the Kalahari of Botswana. *J. Arid Environ.* **35**(3), 563–74.

Vermeulen, C., Lejeune, P., Lisein, J., Sawadogo, P., & Bouché, P. (2013) Unmanned aerial survey of elephants. *PloS One*, **8**, e54700.

Wallen, V.R., Jackson, H.R., & MacDurmid, S.W. (1976) Remote sensing of corn aphid infestation, 1974 (Hemiptera: Aphididae). *Can. Entomol.* **108**, 751–4.

Washino, R.K. & Wood, B.L. (1994) Application of remote-sensing to arthropod vector surveillance and control. *Am. J. Trop. Med. Hyg.* **50** (Suppl.), 134–44.

Wei, W.W.S. (1990) *Time Series Analysis: Univariate and Mulivariate Methods.* Addison-Wesley, Redwood City, California.

Williamson, M. (1984) The measurement of population variability. *Ecol. Entomol.* **9**, 239–41.

Wint, G.R.W. (1998) Rapid resource assessment and environmental monitoring using low level aerial surveys. In: *Drylands: Sustainable Use of Rangelands into the Twenty-First Century.* Proceedings, International Workshop on the Sustainable Use of Rangelands and Desertification Control. International Fund for Agricultural Development, Rome; Ministry of Agriculture and Water, Jeddah, pp. 277–302.

Whittaker, R.H. & Fairbanks, C.W. (1958) A study of plankton copepod communities in the Columbia Basin, southeastern Washington. *Ecology*, **39**, 46–65.

Woiwod, I.D. & Harrington, R. (1994) Flying in the face of change. In: Leigh, R.A. & Johnston, A.E. (eds), *Long-term Experiments in Agricultural and Ecological Sciences.* CABI, Wallingford, UK, pp. 321–42.

Index